Indian Statistical Institute Series

Editors-in-Chief

Antar Bandyopadhyay, Indian Statistical Institute, New Delhi, Delhi, India

Pradipta Bandyopadhyay, Indian Statistical Institute, Kolkata, West Bengal, India

Bhramar Mukherjee, University of Michigan–Ann Arbor, Ann Arbor, MI, USA

B. Sury, Indian Statistical Institute, Bengaluru, Karnataka, India

Associate Editors

Atanu Biswas, Indian Statistical Institute, Kolkata, India

B. S. Daya Sagar, Indian Statistical Institute, Bengaluru, India

Mohan Delampady, Indian Statistical Institute, Bengaluru, India

Ashish Ghosh, Indian Statistical Institute, Kolkata, India

S. K. Neogy, Indian Statistical Institute, New Delhi, India

Rituparna Sen, Indian Statistical Institute, Bengaluru, India

C. R. E. Raja, Indian Statistical Institute, Bengaluru, India

The **Indian Statistical Institute Series**, a Scopus-indexed series, publishes high-quality content in the domain of mathematical sciences, bio-mathematics, financial mathematics, pure and applied mathematics, operations research, applied statistics and computer science and applications with primary focus on mathematics and statistics.

Launched at the 125th birth Anniversary of P.C. Mahalanobis, the series will publish high-quality content in the form of textbooks, monographs, lecture notes, and contributed volumes. Literature in this series are peer-reviewed by global experts in their respective fields, and will appeal to a wide audience of students, researchers, educators, and professionals across mathematics, statistics and computer science disciplines.

Alok Goswami · B. V. Rao

Measure Theory for Analysis and Probability

Springer

Alok Goswami
School of Mathematical and Computational
Science
Indian Association for the Cultivation
of Science
Jadavpur, Kolkata, West Bengal, India

B. V. Rao
Department of Mathematics
Chennai Mathematical Institute
Kelambakkam, Tamil Nadu, India

ISSN 2523-3114 ISSN 2523-3122 (electronic)
Indian Statistical Institute Series
ISBN 978-981-97-7928-4 ISBN 978-981-97-7929-1 (eBook)
https://doi.org/10.1007/978-981-97-7929-1

Mathematics Subject Classification: 28A05, 28A25, 28A35, 60A10, 60B10, 60E10, 60F05, 60F07, 60G15, 60G17

The original submitted manuscript has been translated into English. The translation was done using artificial intelligence. A subsequent revision was performed by the author(s) to further refine the work and to ensure that the translation is appropriate concerning content and scientific correctness. It may, however, read stylistically different from a conventional translation.

© The Editor(s) (if applicable) and The Author(s), under exclusive license to Springer Nature Singapore Pte Ltd. 2025

This work is subject to copyright. All rights are solely and exclusively licensed by the Publisher, whether the whole or part of the material is concerned, specifically the rights of translation, reprinting, reuse of illustrations, recitation, broadcasting, reproduction on microfilms or in any other physical way, and transmission or information storage and retrieval, electronic adaptation, computer software, or by similar or dissimilar methodology now known or hereafter developed.
The use of general descriptive names, registered names, trademarks, service marks, etc. in this publication does not imply, even in the absence of a specific statement, that such names are exempt from the relevant protective laws and regulations and therefore free for general use.
The publisher, the authors and the editors are safe to assume that the advice and information in this book are believed to be true and accurate at the date of publication. Neither the publisher nor the authors or the editors give a warranty, expressed or implied, with respect to the material contained herein or for any errors or omissions that may have been made. The publisher remains neutral with regard to jurisdictional claims in published maps and institutional affiliations.

This Springer imprint is published by the registered company Springer Nature Singapore Pte Ltd.
The registered company address is: 152 Beach Road, #21-01/04 Gateway East, Singapore 189721, Singapore

If disposing of this product, please recycle the paper.

Table of Contents

To the Reader 5

Chapter 1 7
Measure Theory: Why and What

Chapter 2 17
Measures: Construction and Properties

2.1 Measures: from Semifields to Fields 17
2.2 Caratheodory Extension Theorem 27
2.3 Monotone Class Theorem 43
2.4 Completeness and Completion 48
2.5 Approximation 50
2.6 Extension for Arbitrary Measures 53
2.7 Borel-Cantelli Lemma 55
2.8 Measures on Real line 56
2.9 Measures in Higher Dimensions 67
2.10 Regularity of Radon Measures 75
2.11 Probability Measures 77
2.12 Independence 80
2.13 Additional Exercises 85

Chapter 3 91
Measurable Functions and Integration

3.1 Measurable Functions 92
3.2 Integration 98
3.3 Limits and Integral 109
3.4 Integrals: Some Important Inequalities 116
3.5 Measurable Maps 122
3.6 Additional Exercises 129

Chapter 4 134
Random Variables and Random Vectors

4.1 Random Variables 134
4.2 Random Vectors 139
4.3 Independence for Random Variables and Vectors 142
4.4 Expectation 145

4.5 Moments, Variance and Covariance — 153
4.6 Additional Exercises — 157

Chapter 5 — 161
Product Spaces

5.1 Products of Measure Spaces (2-fold) — 162
5.2 Finite Products of Measure Spaces — 168
5.3 Measure Kernels — 170
5.4 Additional Exercises — 174

Chapter 6 — 179
Radon-Nikodym Theorem and L_p spaces

6.1 Radon-Nikodym Theorem, Lebesgue Decomposition — 179
6.2 L_p-Spaces — 193
6.3 Approximation of L_p Functions. — 200
6.4 Complex Measurable Functions and L_p-Spaces — 203
6.5 Additional Exercises — 211

Chapter 7 — 218
Convergence and Laws of Large Numbers

7.1 Convergence Concepts — 218
7.2 Convergence Concepts for Random Variables — 220
7.3 Laws of Large Numbers — 229
    7.3.1 Weak Law of Large Numbers — 231
    7.3.2 Strong Law of Large Numbers — 233
7.4 Three Series Theorem and Zero One Law — 241
7.5 Convergence Concepts for Random Vectors — 246
7.6 Additional Exercises — 247

Chapter 8 — 251
Weak Convergence and Central Limit Theorem

8.1 Weak Convergence — 251
8.2 Characteristic Functions — 260
8.3 Inversion Formula — 266
8.4 Moments and Characteristic Functions — 270
8.5 Characteristic Functions and Weak Convergence — 272
8.6 Characteristic Functions: Some Characterizations — 277

Table of Contents

8.7 Classical Central Limit Theorem	281
8.8 Lindeberg-Feller-Lévy Central Limit Theorem	283
8.9 Weak Convergence in Higher Dimensions	287
8.10 Additional Exercises	293

Chapter 9 296
Conditioning: The Right Approach

9.1 Conditional Expectation	297
9.2 Conditional Probabilities and Regular Conditional Probabilities	307
9.3 Regular Conditional Distributions	314
9.4 Additional Exercises	319

Chapter 10 322
Infinite Products

10.1 Product of a Sequence of Measurable Spaces	323
10.2 Constructing Probabilities on Countably Infinite Product Spaces	324
10.3 Constructing Probabilities on Uncountable Product Spaces	333
10.4 Kolmogorov-Chentsov Continuity Criterion	337
10.5 Additional Exercises	341

Chapter 11 344
Brownian Motion: A Brief Journey

11.1 Path Properties of SBM	347
11.2 Markov and Strong Markov Properties	357
11.3 Additional Exercises	370
11.4 Supplementary Section: RKHS	374

Index 381

To the Reader

This is a book on measure theory and its use in probability. But we believe that anyone seeking a strong foundation in measure theory for pursuing study in modern analysis will also gain significantly from the book. You would wonder if we need one more book on the subject. Our answer is simple: each musician renders a *raga* in her own style and the listener would always find something new – a *meend* here or an *andolan* there. It is rather difficult to give notations for these *gamakams*. Having said that, let us have a brief overview of the contents. A somewhat more detailed account is given in Chapter 1.

The book starts by addressing head-on the most fundamental question a reader is likely to ask, that is, 'Why measure theory?' Chapter 1 not only presents the motivation, but also establishes the inevitability of measure theory for gaining a good and proper understanding of even some basic concepts of probability theory. In the subsequent two chapters, we go at a *vilambit* tempo. First we construct measures, frequently pausing to explain why we are carrying a brick here or some cement there. We also discuss distribution functions in dimensions more than one and there you may find some facts refreshingly noteworthy. Then we get to learn measurable functions, the idea of integration and study the basic theorems. This is a traditional fare. In Chapter 4, we set up a dictionary between this story and probability theory. We discuss probabilistic concepts like mean, variance, moment generating function and conclude with a special discourse on Gaussian distribution. The next chapter introduces finite product spaces and ties it up with the concept of independence from probability theory.

In Chapter 6, we take up more advanced material like Radon-Nikodym theorem, L_p spaces and their duals. We hope you will enjoy the walk through the proof of the Radon-Nikodym Theorem and feel 'I could have done it'. A natural instinct guides its proof and we shall not ride on signed measures. Instead, we use this result later in understanding signed measures, Jordan and Hahn decompositions. We conclude with approximation theorems. Even if you are not interested in the later parts of the book, the material discussed so far should make you feel comfortable to get into several areas of modern analysis.

The next two chapters take you to two (out of the many) basic melodies of probability theory. Chapter 7 takes you through uniform integrability, various notions of convergence, Three Series Theorem and Laws of Large Numbers. Chapters 8 discusses Central Limit Theorem after two intermezzi. First is on weak convergence including Pólya's theorem and local limit theorem. Second is characteristic functions which includes inversion formula, uniqueness and Lévy Continuity theorems. In Chapter 9, we discuss conditional probabilities and conditional distributions. We hope you will appreciate the delicate issues involved here and the pitfalls. This is a basic and serious topic, which plays the theme music in many situations, but usually gets only cursory attention.

In Chapter 10, we discuss probabilities on infinite product spaces, which is essential for constructing i.i.d. random variables, Markov chains, continuous time processes or even to do Ergodic Theory. The last chapter takes up Brownian Motion, where we are content with only a small *alaap*. You will see some basic path properties and the vital strong Markov property as well as reflection principle.

This will equip you with basic essentials of probability theory and leave you at an important junction. Your sublime levels should dictate the path you would like to take. For your *Riyaz*, exercises are sprinkled throughout the text and also at the end of each chapter.

This composition is based on courses given by us at the Indian Statistical Institute for more than a decade. Over the years, we have consulted and learnt from several books and we thank all those authors. We had the opportunity to learn from the great master, the late Ashok Maitra, and we hope his inimitable style and indomitable spirit percolate through these pages.

We made a conscious decision to not include a bibliography. For most topics, the list of good relevant literature is surely long, but, at this day and age, one can get it on one's screen in a jiffy. That task is left for enthusiastic readers.

Finally, when it comes to acknowledgements, the name that has to come at the top of the list is that of the Indian Statistical Institute, to which we both are hugely indebted for our academic and teaching careers. Both of us got the opportunity to teach this topic an umpteen number of times to some of the best classes in that institute and that is where the seeds of this book were sown. We would also like to acknowledge the support from the respective institutions of our current affiliations. We thank Chennai Mathematical Institute and the Indian Association for the Cultivation of Science for helping us in continuing and completing the work. We also thank the National Institute for Science Education and Research, Bhubaneswar, for extending joint invitation to us. It was during our joint stay there on two successive occasions that some major portions of the manuscript were written. It has been a rather long journey from the time that the idea of the book was conceived till a completed manuscript was ready to be submitted. At the end of this journey, we cannot but convey our sincere thanks to our teachers who inspired us, to our friends and colleagues, young and old, who lent their unwavering support and to the outstanding students who forced us to think deeper and deeper. We dedicate the book to all of them.

Chapter 1

Measure Theory: Why and What

To understand the 'why' and 'what' of measure theory, a good place to start is to take a relook at the idea of Riemann integration that one learns in a basic calculus course. Suppose we are given a real-valued function f on a closed bounded interval $[a, b]$. For the sake of simplicity, let us assume that $f \geq 0$. If you draw the graph of this function, then it will be a curve lying above the x-axis. The issue, on which integration theory in calculus is built, is to find the area of the region enclosed under the curve between $x = a$ and $x = b$. Of course, the answer to the problem is easy when f is a constant function (in which case the enclosed region would be a rectangle) or, more generally, when the graph of f is a straight line (in which case the region would be a trapezium). In both cases, school geometry provides simple formulae for these areas.

What does one do in other cases? As one learns in a basic calculus course, Riemann's theory proposed the following solution. The main idea is to divide the interval $[a, b]$ into a large number of small subintervals and to pretend that, within each subinterval, the function f is constant. This constant value of f on each subinterval can be chosen to be any one from the different values that f actually takes in that subinterval. This would essentially mean that, on each of these subintervals, the actual area of the region under the graph is approximated by a rectangle. The area of each of these rectangles can be obtained using the the classical formula and then adding the areas of all these rectangles would give us an approximate value of the total area of the enclosed region. This seems to be a very simple and natural idea, although it gives us only an approximation and not the exact value of the area under question. So, a legitimate question to ask would be: how good is this approximation?

To answer this question, note that the error in the approximation really creeps in through the assumption that, within each subinterval, the function takes a constant value, while in reality, the given f need not actually be a constant function on these subintervals. Thus, the error in this approximation will depend on how much the actual values of the function at various points within a subinterval differ from the one single value that has been chosen as the "constant value" of f on that subinterval. What this means is that this method would yield a reasonably good approximation for the actual area as long as the values of the function at different points within a subinterval are not too far apart from any one of these values. To put it differently, what is being asked for is that if the subintervals are taken to be small enough, then the values taken by f within any subinterval are sufficiently close to one another. In calculus, this property of the given function is precisely what is called 'continuity' property.

Theory of Riemann integration is essentially a formalization of this idea. The approximate areas described above, that is, the sums of areas of approximating rectangles, are called Riemann sums. The theory proves that, if the function f is continuous on $[a, b]$, then by letting the partitions get finer and finer, one can make the Riemann sums converge to a limit and moreover, the limit does not depend on the specific choice of the "constant values" for the function on the subintervals. One would, therfore, be justified in defining this limit as the exact value of the area that we were looking for. This is called the *Riemann integral* of f over $[a, b]$ and is denoted by $\int_a^b f(x)dx$. The obvious question that naturally presents itself now is, what to do if f does not have the property described above, that is, if f is not continuous.

About fifty years later, another mathematician named Henri Lebesgue came up with a striking, yet very natural, idea. To describe this idea, let us recall that the reason why Riemann's method may fail in case f is not continuous is that, in that case, approximating the different values of the function on a subinterval by any one particular value may result in large errors. Here is a possible alternative. Assume, for the sake of simplicity, that f takes values inside a closed bounded interval S. Incidentally, Riemann's theory assumes, at the outset, that the function f in question is a bounded function. Thus, the assumption we are making here is nothing different, though we will see later that Lebesgue's theory does not ultimately need it. In any case, under this assumption, suppose we divide this interval S, called the *codomain*, into a large number of "small" disjoint subintervals, say, $S_1, \cdots, S_n$. The point to emphasize here is that we are partitioning the codomain of the function f, rather than its domain. This is a simple but the most crucial departure from Riemann's idea. It is clear that this would induce a partition of the domain $[a, b]$ into disjoint subsets $A_1 \cup \cdots \cup A_n$, where A_i is defined to be precisely the set of points in $[a, b]$ which are mapped inside S_i. However, it is very important to note that this induced partition of $[a, b]$ may not be a partition into sub-intervals.

From the construction itself, it is clear that, as long as the subintervals $S_1, \ldots, S_n$ are small enough, the different values taken by f at points of the set A_i, for any A_i, are automatically close to one another. This is because, for each i, all the points of A_i are mapped by f into the set S_i, which is a "small" subinterval of S. Therefore, the error would be automatically small if, for each i, the different values of f on A_i are approximated by a single constant value, which could be any number h_i, chosen from S_i. We may then approximate the area under the graph of f on each A_i by that of a "rectangle" of height h_i and "base"

Chapter 1. Measure Theory: Why and What

A_i and then the sum of the areas of these rectangles would give us an approximation of the required area, which is guaranteed to be a good approximation. Thus, the strategy is the same as in Riemann's theory, except that the sets A_i here play the role of the subintervals of $[a, b]$ in Riemann's theory. The important plus point now is that the construction itself ensures that the approximation has to be close to the actual area, with any desired degree of accuracy. Further, the finer the partition of S, the better would be the approximation and one may expect that in the end, we would get a limit which can be taken to represent as the actual area.

In a nutshell, this idea, due to Lebesgue, essentially mimics Riemann's idea with the only (but crucial) exception that, instead of starting with a partition of the domain of the function, one starts with a partition of its codomain and then considers the "induced" partition on the domain. This avoids requiring continuity of f as a necessity for ensuring closeness of the approximation. We will see later that the property of continuity is indeed an 'almost' indisposable necessity for Riemann's theory.

All of this sounds very simple. But, there is a little, but very crucial, catch here! Recall that, as already pointed out, the sets $A_1, \cdots, A_n$, defined above, do not necessarily partition the interval $[a, b]$ into subintervals. So, when we talk about approximating the area under the graph of f on each A_i by that of a "rectangle" on "base" A_i, how do we define the area of this approximating rectangles? Since the "base" A_i need not be an interval, it is not at all clear what is meant by "length" of the base, which is essential for computing area of a rectangle. This brings us straight to the core issue, which is that, in order to implement this new idea of calculating areas, we need to first be able to define a notion of length for general subsets of the real line, rather than just intervals. And, of course, the notion of length defined for general subsets cannot be arbitrary. They have to be, first of all, consistent with the existing notion of length for intervals and, secondly, they have to satisfy reasonable properties that a notion of 'length' is expected to satisfy.

Lebesgue confronted this problem head on and showed that such an extension of the notion of length from intervals to other subsets, satisfying desirable properties, is indeed possible. Further, this extension does cover a very large class of subsets of the real line, though not all subsets. All of these resulted in the development of a new theory of integration, known as Lebesgue integration, which goes beyond Riemann integration theory. Learning this integration theory is one of the core thrusts of measure theory. However, it turns out that the scope of the theory does not need to be limited to integrating functions on the real line or to just finding areas under the graph. To get the motivation for a somewhat more abstract set-up, we now turn our attention to a completely different context.

Probability theory is a branch of mathematics that aims to build a structured mathematical theory for what are called "random experiments". A random experiment is, roughly speaking, one where there are multiple possible outcomes and an underlying chance mechanism makes the experiment result in one of those. The simplest example is that of tossing a coin that could result either in Head or in Tail, each with a certain chance. The mathematical set-up starts by identifying the set of all posible outcomes of a random experiment, called the *sample space* and denoted by Ω. Subsets of Ω are called *events* and to each event $A \subset \Omega$, one attaches a number $P(A)$, taking value in $[0,1]$, which is called the *probability* of A. Of course, the mathematical theory requires that these assignment of probabilities to various events satisfy certain desirable properties, usually called "axioms of probability".

In a first course in probability, one starts by learning the most elementary, though quite natural, algorithm for assigning probabilities to various events. The idea is very simple. One first assigns, to each point $\omega \in \Omega$, a number p_ω, to be called the probability of ω. While doing this, all that one needs to ensure is that the numbers p_ω lie in $[0,1]$ and that they add up to one. This is dictated by the axioms of probability, that requires probabilities to be non-negative numbers and the 'total probability' to be 1. Now, for *any* $A \subset \Omega$, one defines $P(A)$ by just adding up the numbers p_ω over all ω belonging to A. This simple idea works fine for a large class of random experiments, or to be more precise, for all random experiments where the sample space Ω is countable. That all the "axioms of probability" hold good is guranteed by the way $P(A)$, $A \subset \Omega$, are defined here. However, this simple idea doesn't always work! Here is an example.

Let us consider the random experiment where a "fair" coin is tossed repeatedly an infinite number of times. The assumption of "fairness" simply means that every single toss of the coin results in a Head or a Tail, with probability $\frac{1}{2}$ each. Abbreviating Heads and Tails by H and T respectively, the sample space for our experiment of infinite number of tosses of the coin will clearly consist of all infinite sequences of Hs and Ts. The question now is : what should be the probability p_ω to be assigned to a point ω in Ω? Writing any such point ω as $\omega = (\omega_1, \omega_2, \cdots, \omega_n, \cdots)$, where each ω_n is either a H or a T, it should be clear that the probability of the outcome ω should be no more than the probability of the event that "the first toss results in ω_1" which, by the assumed fairness of the coin equals $1/2$. But then, by the same token, p_ω should also be no more than the probability of the event that "the first three tosses result in ω_1, ω_2 and ω_3 respectively" which has to be $1/8$. This is because, the fairness hypothesis implies that the 8 different outcomes of the first three tosses must all have the same probability. Repeating this argument, one can easily conclude that p_ω has to be bounded above by 2^{-n}, for every n. But, since p_ω has to be non-negative, the above condition entails that the only possible value of p_ω is $p_\omega = 0$.

So where have we landed? We are confronted with a situation where for every point ω, the only logical choice for the probability of ω, subject to natural properties of probability, is $p_\omega = 0$. But then it becomes pointless to try to define probabilities of events by adding probabilities of points in it. Thus, the simple algorithm, that we mentioned at the start, fails! Do we give up here and say that a proper mathematical theory of probability cannot be built for experiments like infinite tosses of a coin? Instead of doing that, let us try to think of some alternative ideas to resolve the impasse.

Let us start by trying to define probabilities of events directly, whenever possible, in a meaningful and logical way, without having to go through probabilities of points in those events. In a way, we have done that already! Let us, for example, consider the event A that "the first two tosses both result in Heads". This is clearly the set of points ω for which both ω_1 and ω_2 are H. Can we directly assign a probability to this event? It should be clear that, for this particular event, we may as well restrict ourselves to the random experiment of two tosses of a fair coin. This is because the results of the rest of the tosses would have no bearing on this event. But then the probability of the event in question would simply be the probability that in two tosses of a fair coin, both result in Head and that probability is just $\frac{1}{4}$. Thus, in the original experiment of infinite tosses, it would be logical to assign $\frac{1}{4}$ as probability of the event A. In general, if for any n, we consider an event A that completely specifies the outcomes of *only* the first n tosses, a natural thing to do would be to restrict to the experiment of n tosses of a fair coin and view A as just one of the 2^n outcomes of this restricted experiment. This would result in 2^{-n} as a logical assignment of probability of the event A in the original experiment of infinite tosses. What all these mean is that in the random experiment of infinite tosses of a fair coin, we have identified some special events, to which one can assign probabilities directly in a logical manner. The big question now is how to extend the assignment of probabilities from this special class of events to more general subsets of Ω. Of course, the extension has to be done in such a way that the final assignment of probabilities to all events satisfies the natural and desirable "axioms of probability".

Having described two problems in two seemingly very different contexts, we are now going to establish that the two problems are essentially of the same character. In both cases, we have a set X and a special class $\mathcal{S}$ of subsets of X on which we have a natural notion of some 'measure', a term loosely used here for assgnment of non-negative numbers to sets. In case of Lebesgue's integration theory, the set X is the real line, the special class $\mathcal{S}$ is the class of all intervals and the 'measure' in question is length. In the example from probability, X is the sample space Ω consisting of all infinite sequences of Hs and Ts, the special class $\mathcal{S}$ is the collection of those events where only a finite number of initial coordinates are completely specified and the 'measure' here is, of course, probability.

In both cases, our aim is to extend the measure in question from $\mathcal{S}$ to as large a class of subsets as possible in a way so that this extended measure on the larger class of sets satisfies some desirable and natural conditions. Tackling this issue is the first step in the subject of measure theory. The first major and fundamental result of measure theory, that describes how and under what conditions such an extension can be achieved, is the well-known Caratheodory Extension Theorem. It must be mentioned that, before Caratheodory, Lebesgue had already achieved this extension in the special context of the 'length' measure on the real line. Caratheodory's fundamental contribution was to identify the problem in the abstract set-up, give it a proper formulation and provide a complete solution. The solution shows not only how to extend a measure, but also describes the class of subsets to which this extension is possible, and, under additional assumptions, also guarantees uniqueness of such an extension. The reader may rest assured that the method of extension as outlined by Caratheodory is so natural that if one understands it properly, one would realize that there is no other way! The details of this is one of the main topics of Chapter 2.

It is worth pointing out at this stage what all can be achieved by the above extension. For one, we turn to Lebesgue's scheme of finding areas under the graph of a real function or, in other words, defining integrals of real functions on an interval. As discussed earlier, Lebesgue's idea was to partition the domain of the function in a way which will ensure that approximating the given function by a constant function within each set in the partition, results in small error. The upshot would be that the actual area under the graph of the function can be approximated, upto a desired level of accuracy, by a sum of areas of "rectangles", provided, of course, a measure of 'length' can be assigned to the sets in the partition that form the 'base' sets of these rectangles. In other words, Lebesgue's scheme will work fine for functions f for which the associated sets A_i, as described earlier, belong to the class of subsets of the real line to which length measure can be extended. While this still puts some restriction on f, the good news is that the restriction turns out to be much weaker than requiring continuity. Thus Caratheodory Extension enables Lebesgue's theory to define integrals of many more functions than the Riemann theory permits.

But the impact of Caratheodory Extension goes far beyond Lebesgue integration theory. Firstly, on real line itself, one may consider measures other than the length measure and define integrals with respect to such measures. As special cases of these, one would be able to capture the theory of finite sums, infinite series and Riemann-Stieltje's integrals, all under one umbrella of a general theory of integration. Secondly, the above method will also allow us to consider measures on subsets of not just the real line, but of any abstract set. We may then be able to discuss integration of real functions defined on abstract spaces with respect to measures on them. This has far reaching implications. Chapter 3 describes this general theory of integration of real-valued functions on an abstract set with respect to an arbitrary measure on the set.

One of the most significant impacts of this theory has been on the mathematical theory of probability. As pointed out earlier, the theory of probability starts with an assignment of probabilities to subsets of a sample space Ω, subject to certain axioms being satisfied. Clearly, this amounts to nothing but construction of a measure on subsets of the sample space, that will satisfy the axioms of probability. The earlier example of a random experiment involving infinite tosses of a fair coin illustrated that there is no simple-minded shortcut for such a construction, that always works. This is where Caratheodory theory comes into play and once we have a 'probability measure' on Ω, the general theory of integration as mentioned above will naturally apply for real functions on Ω. In probability theory, such real functions on the sample space Ω are called 'real random variables' and the integrals of random variabels with respect to probability measures are what are called 'expectations' or 'expected values'. What is the advantage of identifying expectations as integrals with respect to probability measures? In a first coures in probability, expected value is usually defined only for two special kinds of random variables and there again by two different formulas in the two cases. For a "discrete" random variable, expected value is defined as a sum, while for a random variable with a "continuous density", it is defined as a Riemann integral. In addition to possibly creating the false impression that all random variables are either discrete or have continuous densities, this causes some serious and unnecessary impediments for the mathematics of expected value. At the very least, one should realize that, for proving properties of expectations, it is both unnecessary and incorrect to bank on the dichotomy of "discrete" random variables and random variables with "continuous density". The measure theoretic formulation of expected value as an integral with respect to a probability measure, frees one from this limitation by defining expectations for all random variables at one go and thereby allowing the results of general integration theory to apply automatically. What must also be pointed out is that the formula for expectation as a sum in the discrete case and as a Riemann integral in case of continuous density, can be derived as special cases of the abstract integral. Chapter 4 discusses this measure theoretic formulation of probability.

Another fundamental contribution of the measure theoretic framework in probability sometimes goes unnoticed. It is commonplace in undergraduate probability to come across statements like "let X be a Poisson random variable" or "consider a random variable with the standard Normal distribution", etc. Such statements are often accepted overlooking a basic issue. Basic probability defines a random variable as a real function on a sample space and its distribution arises from the probability assignment of events in the sample space. So to have a random variable, we must first have a sample space, equipped with a probability assignment on events. So if a student is suddenly asked to "consider" a random variable with certain specific distribution, it would be natural for her to wonder where the sample space is. In fact, it would be perfect for her to ask: does there exist a sample space and a random variable defined on it, which has the specified distribution? Once again, it is the Caratheodory Extension Theorem that completely resolves the issue. The theorem gurantees that, given any probability distribution on the real line, there always is a sample space along with a probability assignment and a random variable defined on it, with that specified distribution.

If all these do not seem to be compelling enough, here is another very important issue that makes measure theoretic setting absolutely essential — almost unavoidable — for probability theory. Any student of probability would agree that conditional distribution is a fundamental concept in probability. For two random variables X and Y, one sometimes needs to talk about the "conditional distribution" of X, given that Y takes the value y. Appealing to the classical definition of conditional probability for this would require the conditioning event $\{Y = y\}$ to have positive probability. Things, therefore, would be alright if Y is a discrete random variable. But it is not at all clear how to define it, when Y is not discrete. One very special case that is touched upon in a first course in probability, is when the pair (X, Y) has a joint density and, in that case, one learns a formula for what is called the "conditional density" of X, given $Y = y$. However this formula, by itself, is completely ad hoc. It is not derived from the classical, and somewhat natural, notion of conditional probability nor has it got any immediate and visible connection with that. The measure theoretic framework of probability resolves this serious issue completely, by providing a unified definition of conditional distribution that does not call for any restriction on the conditioning random variable Y. This is taken up in Chapter 9.

Just as length can be viewed as a measure on the real line, one can view areas, volumes etc., as measures in higher dimensions. In school geometry, we learn that for a rectangle, area is just the product of two lengths and, for a rectangular parallelepiped, volume is a product of three lengths. Is this connection of area or volume with length restricted only to special sets like rectangles or rectangular parallelepipeds? As we will see, the connection extends far beyond these special sets. For example, the area measure on the plane can also be derived as a product of the length measures on the two axes and the volume measure as a product of three length measures. Of course, the idea of constructing 'product measures' can be carried out in any abstract setting as well and gives us a useful way of constructing new measures out of known ones. This is the topic of Chapter 5.

An interesting and useful consequence of the theory of product measures is that, under a simple criterion as stated in what we will learn as Fubini's theorem, integration with respect to product measures can be carried out by iterated single variable integrals. This is a measure theoretic extension of how one computes multivariate Riemann integrals in calculus. Further, the theory of integration with respect to product measures bring many seemingly different concepts under one umbrella and presents a unified theory. For example, the double (or triple) summation and the double (or triple) integrals and also joint sum-integral operations are but different manifestations of the same concept, namely, integrals with respect to product measures. As a result, the one single criterion, stated in Fubini's theorem, determines when we can interchange orders of summation or orders of integration or even interchange a sum and an integral.

Having developed abstract integration theory in Chapter 3 and having investigated properties of integral somewhat extensively, it seems only appropriate to dig deeper and discuss some very important results that are of huge significance in analysis as well as in probability. Radon-Nikodym theorem is one such result to which a passing reference is made in Chapter 3. We return to it in Chapter 6 and prove the theorem. The notion of singularity and the Lebesgue Decomposition Theorem then come as companions. We then take these further by proving these results for what are called "finite signed measures". The beautiful Hahn-Jordan decomposition appears along the way. The next big thing taken up in this chapter is introducing L_p-spaces and identifying their "dual spaces" for $1 \leq p < \infty$. The somewhat different picture in the case $p = \infty$ is discussed in a series of exercises at the end of the chapter and it may interest the inquisitive reader. To complete the circle of ideas, we also discuss complex measurable functions and the associated complex L_p-spaces.

In Chapters 7 and 8, we take up what can be described as some of the fundamental "limit theorems" in probability. While it is an undisputed fact that these limit theorems form the backbone of much of classical statistical theory and methodology, it is worth mentioning that they have important bearings in mathematics as well. A simple proof of Weierstrass Approximation offers a small illustration. All the different concepts of convergence are discussed in Chapter 7, except weak convergence, which is set aside exclusively for Chapter 8. The important notion of uniform integrability comes up as a bridge between convergence in probability and L_p convergence. Proving the various "laws of large numbers" is, of course, the main thrust. The celebrated Maximal Inequality, Three-Series Theorem and Zero-One Law, all due to Kolmogorov, make their appearance in this connection. Borel's Normal Number Theorem, Glivenko-Cantelli Lemma and Jessen-Wintner's Theorem are included as interesting applications.

As mentioned above, Chapter 8 deals exclusively with convergence in distribution or what is called 'weak convergence'. After discussing several useful characterizations of weak convergence, we introduce the concept of 'Characteristic functions', that enjoy a special place in the context of weak convergence due to Lévy Continuity Theorem. A separate section is devoted to Pólya criterion and Bochner's characterization. Of course, the final focus is on Central Limit Theorem (CLT). We first prove the classical CLT for i.i.d. random variables, using the conventional tools of characteristic functions and Lévy Continuity Theorem. Then the Lindeberg-Feller-Lévy CLT is proved using Lindeberg's excellent idea of an "independent coupling" of the given triangular array with an i.i.d. sequence of $N(0,1)$ random variables. The reader may find this non-classical idea interesting. A brief discussion on weak convergence for random vectors ends the chapter.

The extremely useful Cramer-Wold device naturally comes up in this context.

As mentioned earlier, resolving the subtle issues concerning conditional probability and conditional distribution through a necessary and appropriate use of the measure theoretic setup is dealt with in Chapter 9.

In Chapter 10, we extend the concept of product measure spaces discussed in Chapter 5, to infinite products as well, under the stipulation, of course, that the underlying component measures are all probability measures. It must be pointed out here that such extensions to infinite products are not merely an artifice. In fact, this theory allows one to construct, in a canonical way, different kinds of random processes, including sequences of independent random variables with specified distributions and also processes with given finite dimensional distributions. The passage from countable products to uncountable products is luckily very easy and paves way for constructing stochastic processes on uncountable index sets, with prescribed finite dimensional distributions. To illustrate this, we make a brief introduction to one of the most exciting topics in probability theeory, namely, the Standard Brownian Motion. We prove Kolmogorov's Consistency Theorem and Kolmogorov-Chentsov Continuity Criteria and use these to present one of the classical constructions of Brownian motion.

Chapter 11, which came as an afterthought, was mainly propelled by not being able to resist the temptation to undertake a quick journey through the ever-fascinating nuances of Brownian motion. Having defined and constructed Standard Brownian Motion in Chapter 10, it seemed unjust to deprive an inquisitive reader from having at least a quick peep into the mind-boggling erratic nature of Brownian trajectories and also one of the most fundamental probabilistic property of Brownian motion, namely, the Strong Markov Property.

We end this introductory chapter by saying that our intention in this introduction was to simply convince the reader that, firstly, in a large number of situations, bringing in measure theory is inevitable, and secondly, the subject and the underlying ideas are not so incomprehensible as is sometimes perceived. In fact, if anything, the theory is a very natural one. The discussion in this chapter, if carefully and seriously studied, should convince the reader that measure theory should be read and that it is readable. With this hope, the reader is extended a very warm welcome to the vast and beautiful, simple yet intriguing mathematics of measure theory.

Chapter 2

Measures: Construction and Properties

In this chapter we tackle the first major step in measure theory. To recapitulate, we noted in the last chapter that to pursue with Lebesgue's idea of integration, a fundamental requirement is to be able to assign a notion of length to arbitrary subsets of the real line.m Similarly, in the infinite coin tossing experiment, we have the sample space Ω, consisting of all infinite sequences of H and T, and our aim is to assign probabilities to arbitrary subsets of Ω, starting from the naturally assigned probabilities to some special subsets.

2.1 Measures: from Semifields to Fields

The commonality of the above two seemingly different situations was pointed out in the previous chapter. In both cases, we have an underlying non-empty set Ω and a special class $\mathcal{S}$ of subsets of Ω, on which we already have an existing notion of the relevant measure defined for these subsets. The goal is to extend this measure to a wider class of subsets in such a way that it satisfies some natural and desirable properties. This is the problem that we are now going to address.

Before we start, let us first observe that in both the contexts described above, the special class $\mathcal{S}$ has a few common features. Firstly, in both cases, the whole set Ω belongs to the class $\mathcal{S}$. Secondly, for any two sets in $\mathcal{S}$, their intersection is in $\mathcal{S}$. It should be kept in mind that on $\mathbb{R}$, both the empty set and all singleton sets are regarded as intervals and length is defined for both. Another common feature, which is not so obvious is that, in both contexts, the complement of any set in $\mathcal{S}$ can be written as a finite disjoint union of sets from $\mathcal{S}$. In case of real line, it is clear that complement of an interval is either an interval itself or a disjoint union of two intervals. To see this in coin tossing experiment, let us consider, for example, the event A that the first and third tosses are Heads, while the second toss is a Tail. It is then easy to see that then A^c equals the disjoint union $B \cup C \cup D$; where B is the event that first toss is a Tail, C is the event that first two tosses are Heads and D is the event that the first toss is a Head, but the next two tosses are Tails. The reader should now be able to convince herself that all events belonging to $\mathcal{S}$ have this property that their complements are finite disjoint unions of events in $\mathcal{S}$.

The above properties of the special class $\mathcal{S}$ will play a crucial role in the extension program that we are going to undertake. That makes the following definition natural and important.

DEFINITION **2.1.1.** A collection $\mathcal{S}$ of subsets of a non-empty set Ω is called a *semifield* on Ω if (i) $\Omega \in \mathcal{S}$, (ii) $S_1, S_2 \in \mathcal{S}$ implies that $S_1 \cap S_2 \in \mathcal{S}$ and (iii) $S \in \mathcal{S}$ implies that $S^c = S_1 \cup S_2 \cup \cdots \cup S_n$ for some $n \geq 1$ and disjoint sets $S_1, \cdots S_n$ in $\mathcal{S}$.

By repeated use of property (ii), it should be clear that $S_1, \cdots S_n \in \mathcal{S}$ for any $n \geq 1$ would imply $\bigcap_{i=1}^{n} S_i \in \mathcal{S}$. We express this by saying that $\mathcal{S}$ is closed under finite intersections. Another important observation is that properties (i) and (iii) together imply that the empty set $\emptyset$ belongs to $\mathcal{S}$. At this point, let us introduce a terminology which will be convenient and will be frequently used in the sequel.

DEFINITION **2.1.2.** For any non-empty class $\mathcal{C}$ of subsets of a non-empty set Ω, we will say that a set $A \subset \Omega$ is a *$\mathcal{C}$-set* to mean that $A \in \mathcal{C}$.

EXERCISE **2.1.3.** *If $\mathcal{S}$ is a semifield, any finite union of $\mathcal{S}$-sets can be written as a finite disjoint union of $\mathcal{S}$-sets.*

Having talked about measures in a loose sense so far, we shall now formalize the notion. What is clear is that a measure should be an assignment of non-negative numbers to sets. Following standard terminology, such an assignment will be referred to as a *non-negative set function*. Since "length measures" of some intervals may be infinite, we should allow measures of sets to take the value $+\infty$. Of course, the emptyset, in any context, should have zero measure. Finally, measures of disjoint sets should add up and give the measure of their union. All these lead to the following.

DEFINITION **2.1.4.** Let $\mathcal{S}$ be a semifield on a non-empty set Ω. By a *measure* on $\mathcal{S}$ is meant a set function $\mu : \mathcal{S} \to [0, \infty]$ satisfying the properties that (i) $\mu(\emptyset) = 0$ and (ii) if $S_1, S_2, \cdots$ are disjoint $\mathcal{S}$-sets with $\bigcup_i S_i \in \mathcal{S}$, then $\mu(\bigcup_i S_i) = \sum_i \mu(S_i)$.

Property (ii) above is usually referred to as additivity or, more specifically, 'countable additivity'. One easily sees that (i) and (ii) imply 'finite additivity' as well. Let us put a note of caution here. Here and in the sequel, statements or claims accompanied by phrases like *"one easily sees that"* should not be accepted in good faith and should actually be verified right away by the reader.

It may be noted here that in the above additivity property, we are considering sums (finite or infinite) whose terms are elements of $[0, \infty]$. To handle such sums, we adopt the following convention. In case at least one of the summands is ∞, the sum (finite or infinite) is defined to be ∞. The sum is also defined to be ∞, if it is an infinite series of finite summands that does not converge. The reader would surely find this convention reasonable.

2.1. Measures: from Semifields to Fields

DEFINITION 2.1.5. A measure μ on a semifield $\mathcal{S}$ is called a *finite measure* if $\mu(\Omega) < \infty$ and is called a *probability measure* if $\mu(\Omega) = 1$.

Clearly, any finite measure μ on a semifield $\mathcal{S}$ can be written as $\mu = \alpha \cdot \nu$, where ν is a probability measure and α is a nonnegative real number. Here and later, we use the equality $\mu = \alpha \cdot \nu$ to mean that $\mu(S) = \alpha\nu(S)$ for all $S \in \mathcal{S}$.

Any measure μ on a semifield $\mathcal{S}$ has "monotonicity property" (see Exercise 2.13.1), from which it follows that if μ is a finite measure (resp., a probability measure), then $\mu(S) \leq \mu(\Omega) < \infty$ (resp, $\mu(S) \leq 1$), for all $S \in \mathcal{S}$.

While it is convenient to work with finite measures, many important measures will not be finite measures. The "length measure" on the class of intervals is one such. However, it has the important property that $\Omega = \mathbb{R}$ can be written as a countable union of intervals, each of which has finite length. This property is expressed by saying that the length measure on the class of intervals is "σ-finite".

DEFINITION 2.1.6. A measure μ on a semifield $\mathcal{S}$ is said to be σ-*finite on* $\mathcal{S}$, if there exist sets $S_n \in \mathcal{S}$, $n \geq 1$, with $\mu(S_n) < \infty$ for each n, such that $\bigcup_n S_n = \Omega$.

As we will see later, an important feature of σ-finite measures is that results proved for finite measures can often (though not always) be easily extended to σ-finite measures. Accordingly, σ-finite measures have a special place in the theory.

We now proceed to confront our basic problem that we started out to settle, whether it is for implementing Lebesgue's new idea of integration or for assigning probabilities of various events in our example of infinite tosses of a fair coin. Here is a general formulation of the problem. Suppose we have a non-empty set Ω, a semifield $\mathcal{S}$ on Ω and a measure μ on $\mathcal{S}$. Can we extend this μ to a measure on a larger class of subsets and, if so, how and to what class? The starting point of measure theory is to settle this problem.

We start by exploring a natural and gradual path towards getting such an extension. An obvious first step would be to extend μ to sets obtained by taking finite disjoint unions of $\mathcal{S}$-sets. The natural definition of the μ-value of such a set would be to set it equal to the sum of μ-values of the constituent $\mathcal{S}$-sets. Since a semifield is, in general, not closed under finite disjoint unions, this would extend μ to a possibly larger class. But we should always keep our vigilance against possible loopholes. And therefore, the first thing we should ask is : is the above prescription unambiguous? What if, for two different finite collections of disjoint $\mathcal{S}$-sets, say, $S_1, S_2, \cdots, S_m$ and $T_1, T_2, \cdots, T_n$, the unions $\bigcup_i S_i$ and $\bigcup_j T_j$ result in the same set? In such a case, either of $\sum_i \mu(S_i)$ and $\sum_j \mu(T_j)$ is a potential candidate for μ-value of the union. So, in order to remove any ambiguity from our proposed first step of extension, we need to ensure that, in such a case, both sums give the same value. Once we get through this, the next important question would be whether the larger class obtained this way is a semifield and whether μ so defined is a measure on this larger class.

To properly address these issues, let us denote $\mathfrak{F}$ to be the class of subsets of Ω obtained by taking all possible finite disjoint unions of $\mathcal{S}$-sets. What we have proposed is an extension of μ from $\mathcal{S}$ to $\mathfrak{F}$. We first observe some properties of this new class $\mathfrak{F}$. Clearly, $\mathcal{S} \subset \mathfrak{F}$ and hence $\emptyset, \Omega \in \mathfrak{F}$. Next, let A and B be any two $\mathfrak{F}$-sets. This means that $A = S_1 \cup S_2 \cup \cdots \cup S_m$ and $B = T_1 \cup T_2 \cup \cdots \cup T_n$, where $S_1, S_2, \cdots, S_m$ are disjoint $\mathcal{S}$-sets and so also are $T_1, \cdots, T_n$. Clearly, $A \cap B = \bigcup_i \bigcup_j (S_i \cap T_j)$. Note that, $\mathcal{S}$ being a semifield, $S_i \cap T_j$ is an $\mathcal{S}$-set for each pair (i, j). Further, for any two distinct pairs (i, j) and (i', j'), the sets $S_i \cap T_j$ and $S_{i'} \cap T_{j'}$ are clearly disjoint. This means that $A \cap B \in \mathfrak{F}$, implying that, just like $\mathcal{S}$, the class $\mathfrak{F}$ is also closed under finite intersections.

However, $\mathfrak{F}$ has an important new property, that was not available for $\mathcal{S}$, namely, that $\mathfrak{F}$ is closed under complementation. To see this, take any $A \in \mathfrak{F}$, which means $A = S_1 \cup \cdots \cup S_n$, for some disjoint $\mathcal{S}$-sets $S_1, S_2, \cdots, S_n$. We have to show that $A^c \in \mathfrak{F}$. By property (iii) of a semifield, S_i^c, for each i, is a finite disjoint union of $\mathcal{S}$-sets and, therefore, $S_i^c \in \mathfrak{F}$, for each i. Since $\mathfrak{F}$ has already been shown to be closed under finite intersections, it follows that $A^c = \bigcap_{i=1}^{n} S_i^c \in \mathfrak{F}$.

To summarize, what we just proved is that the class $\mathfrak{F}$, consisting of all possible finite disjoint unions of $\mathcal{S}$-sets, contains the semifield $\mathcal{S}$ and is closed under finite intersections and complementations. The two closure properties mentioned above make $\mathfrak{F}$ a field on Ω, as per the following definition.

DEFINITION 2.1.7. A class $\mathfrak{F}$ of subsets of a non-empty set Ω is said to be a *field* if: (i) $\Omega \in \mathfrak{F}$, (ii) $A_1 \in \mathfrak{F}, A_2 \in \mathfrak{F}$ implies $A_1 \cap A_2 \in \mathfrak{F}$ and (iii) $A \in \mathfrak{F}$ implies $A^c \in \mathfrak{F}$.

Clearly a field is a semifield. An easy application of De Morgan's laws shows that a field is closed under finite unions as well. Further, De Morgan's laws also imply that, if in the definition above, condition (ii), requiring closure under finite intersections, is replaced by a condition requiring closure under finite unions, that would give us an equivalent definition of a field.

Finally, at the risk of making a loose statement, we claim that a field is closed under any finite number of set operations, that is, if you take a finite number of sets from a field and construct a new set by applying any finite number of set operations on these sets, then the resulting set is also in the field. One should think about it and see why this is true.

Here are a couple of easy exercises based on simple properties of fields, semi-fields and measures.

EXERCISE 2.1.8. *If $\{A_n : n \geq 1\}$ is a sequence of sets in a field $\mathcal{F}$, then show that there is a sequence $\{B_n; n \geq 1\}$ of disjoint $\mathfrak{F}$-sets such that $\bigcup_n A_n = \bigcup_n B_n$. This is called* disjointification.

EXERCISE 2.1.9. *If μ is a σ-finite measure on a field $\mathfrak{F}$, then show that there are disjoint sets $F_n \in \mathfrak{F}$, $n \geq 1$ with $\mu(F_n) < \infty$ for each n, such that $\bigcup_n F_n = \Omega$. Show that the same result holds if the field $\mathfrak{F}$ above is replaced by a semifield $\mathcal{S}$.*

2.1. Measures: from Semifields to Fields

What we proved before Definition 2.1.7, is the first assertion of the next proposition. We leave the proof of the second assertion as an easy exercise.

PROPOSITION **2.1.10.** If $\mathcal{S}$ is a semifield on a non-empty set Ω, then the class $\mathfrak{F}$ consisting of all finite disjoint unions of $\mathcal{S}$-sets is a field which includes the semifield $\mathcal{S}$. Further, if $\widetilde{\mathfrak{F}}$ is any field on Ω that includes $\mathcal{S}$, then $\widetilde{\mathfrak{F}}$ includes the field $\mathfrak{F}$ as well.

Here and in what follows, when we say that a class $\mathcal{H}$ '*includes*' or '*contains*' another class $\mathcal{C}$, what we mean is that every set that belongs to the class $\mathcal{C}$ also belongs to $\mathcal{H}$ and we often write $\mathcal{H} \supset \mathcal{C}$ for this. The same is also sometimes expressed by saying that $\mathcal{C}$ is '*contained in*' $\mathcal{H}$ and we write $\mathcal{C} \subset \mathcal{H}$. This is consistent with the set theoretic notion of inclusion because a class of subsets of Ω may itself be viewed as a set whose elements are certain subsets of Ω.

Before proceeding further, let us see some important examples of fields.

EXAMPLE **2.1.11.** We already know one example of a field on $\Omega = \mathbb{R}$, namely, the class $\mathfrak{F}$ consisting of all finite disjoint unions of intervals. This is because the class of all intervals is known to be a semifield on $\mathbb{R}$. As always, the empty set here is regarded as an interval.

EXAMPLE **2.1.12.** Using the same reason as in the previous example, the class $\mathfrak{F}$ consisting of all finite disjoint unions of intervals of the form $(a, b]$ and (a, ∞), with $-\infty \leq a < b < \infty$, can be seen to constitute a field on $\mathbb{R}$.

EXAMPLE **2.1.13.** On any infinite set Ω, the collection $\mathfrak{F}$ of all $A \subset \Omega$ such that either A or A^c is finite, is a field. This is referred to as the finite-cofinite field.

EXAMPLE **2.1.14.** On any uncountable set Ω, consider the collection $\mathfrak{F}$ of all $A \subset \Omega$ such that either A or A^c is countable. It turns out that this class is not just a field, but actually a "σ-field", a concept to be introduced later. Accordingly, it is usually referred to as the countable-cocountable σ-field.

EXAMPLE **2.1.15.** On the set Ω of all infinite sequences of zeros and ones, declare a subset $A \subset \Omega$ as "finite-dimensional" if there exists $m \geq 1$ such that for every choice $\epsilon_1, \cdots, \epsilon_m$ of zeros and ones, the set $\{\epsilon_1\} \times \cdots \{\epsilon_m\} \times \{0,1\} \times \{0,1\} \times \cdots$ is either a subset of A or disjoint from A. The class $\mathfrak{F}$ consisting of all finite dimensional subsets of Ω, can be seen to be field.

Returing to Proposition 2.1.10, what it essentially asserts is that the class $\mathfrak{F}$ consisting of all possible finite disjoint unions of $\mathcal{S}$-sets, is the 'smallest' field on Ω containing the semifield $\mathcal{S}$. For this reason, it is often called the field '*generated*' by $\mathcal{S}$. Let us point out here that the class $\mathfrak{F}$ also equals the collection of all finite unions (not necessarily disjoint) of sets in $\mathcal{S}$ (see Exercise 2.1.3).

DEFINITION **2.1.16.** Given a nonempty collection $\mathcal{C}$ of subsets of Ω, the smallest class of sets which contains the class $\mathcal{C}$ and is a field, is called the *field generated by* $\mathcal{C}$ and is denoted as $\mathfrak{F}(\mathcal{C})$.

A natural question is, given a non-empty collection $\mathcal{C}$ of subsets of a non-empty set Ω, whether there always exists a smallest field on Ω containing $\mathcal{C}$. The answer is 'yes', as the next two propositions are going to establish. The proofs of these propositions are fairly easy and left as exercises.

PROPOSITION 2.1.17. Given any non-empty family **C** of fields on Ω, their 'intersection', that is, the class consisting of precisely those sets that belong to each of the fields in the family **C**, is itself a field. (This is called the 'intersection' field.)

PROPOSITION 2.1.18. Given a nonempty class $\mathcal{C}$ of subsets of Ω, consider the family **C** of all fields containing $\mathcal{C}$. Then the intersection of all fields in **C** is the smallest field containg $\mathcal{C}$, that is, equals $\mathfrak{F}(\mathcal{C})$.

Note that the family **C** in Proposition 2.1.18 is non-empty, because the 'power set' $\mathcal{P}(\Omega)$, namely, the class of all subsets of Ω, is always a field containg any given class $\mathcal{C}$ of subsets of Ω.

We like to point out here that an assertion like the one in Proposition 2.1.17 does not hold good for families of semifields. For example, on $\Omega = \{1, 2, 3\}$,

$$\mathcal{S}_1 = \{\emptyset, \{3\}, \{1, 2\}, \Omega\} \quad \text{and} \quad \mathcal{S}_2 = \{\emptyset, \{1\}, \{2\}, \{3\}, \Omega\}$$

are two semifields, but $\mathcal{S}_1 \cap \mathcal{S}_2 = \{\emptyset, \{3\}, \Omega\}$ is not a semifield.

In general, given a class $\mathcal{C}$, one may not always be able to have an explicit description of all the sets in $\mathfrak{F}(\mathcal{C})$. We give below examples of some special cases, where such an explicit description can be given. The reader is advised to verify the claims made.

EXAMPLE 2.1.19. As already seen, the field generated by a semifield consists precisely of all possible finite disjoint unions (equivalently, all finite unions) of sets from the semifield.

In particular, the field on $\mathbb{R}$, generated by the class of all intervals (resp., all intervals with rational end points), consists precisely of all finite disjoint unions (equivalently, all finite unions) of intervals (resp., intervals with rational end points).

EXAMPLE 2.1.20. If $\mathcal{C} = \{A\}$, where A is a non-empty, proper subset of Ω, then $\mathfrak{F}(\mathcal{C}) = \{\emptyset, A, A^c, \Omega\}$.

EXAMPLE 2.1.21. If $\mathcal{C} = \{A_1, A_2, \cdots, A_k\}$ be a finite partition of Ω by non-empty sets, then $\mathfrak{F}(\mathcal{C})$ consists of all possible unions (including the empty union) of the sets in $\mathcal{C}$. Thus, $\mathfrak{F}(\mathcal{C})$ consists of exactly 2^k many sets.

REMARK 2.1.22. Interestingly enough, it turns out that (See Exercise 2.13.2), the field generated by *any finte class* of subsets of a non-empty set Ω (hence in particular, any finite field on Ω) is necessarily generated by a finite partition of Ω by non-empty sets. As a consequence, the number of sets in any finite field (equivalently, in any field generated by a finte class) *has to be* 2^k, for some $k \geq 1$.

2.1. Measures: from Semifields to Fields

EXAMPLE 2.1.23. If $\mathcal{C} = \{A_\alpha, \alpha \in \Lambda\}$ is an infinite partition of Ω by non-empty sets, then one can easily see that $\mathfrak{F}(\mathcal{C})$ consists precisely of all finite unions of sets in $\mathcal{C}$ and the complements of all such unions, that is,

$$\mathfrak{F}(\mathcal{C}) = \{A \subset \Omega : \text{ either } A \text{ or } A^c \text{ is a finite union of sets in } \mathcal{C}\}$$

A special case of this is when Ω is an infinite set and $\mathcal{C}$ consists of all its singleton subsets, that is, $\mathcal{C} = \{\{\omega\} : \omega \in \Omega\}$. In this case, $\mathfrak{F}(\mathcal{C})$ is just the finite-cofinite field described in Example 2.1.13.

REMARK 2.1.24. From the description of $\mathfrak{F}(\mathcal{C})$ in Example 2.1.23, it is easy to see that the number of sets in $\mathfrak{F}(\mathcal{C})$ would be countably infinite, if $\mathcal{C}$ is countably infinite, while $\mathfrak{F}(\mathcal{C})$ would be an uncountable class, if $\mathcal{C}$ is so.

Let us now return to the issue of extension of a measure μ given on a semifield $\mathcal{S}$. As was already discussed, a natural first step towards this should be to extend μ to the class $\mathfrak{F}$ consisting of all possible finite disjoint unions of $\mathcal{S}$-sets, which happens to be the field generated by $\mathcal{S}$.

Since any $A \in \mathfrak{F}$ is a disjoint union of finitely many $\mathcal{S}$-sets, say, $A_1, \ldots, A_n$, the required additivity property of measures dictates then that we define $\mu(A)$ as $\mu(A) = \sum_{i=1}^{n} \mu(A_i)$. But, as explained earlier, the first hurdle now is to ensure that this is well defined. More specifically, suppose a set A has two different representations as finite disjoint unions of $\mathcal{S}$-sets, say, $A = \bigcup_{i=1}^{n} A_i$ and $A = \bigcup_{j=1}^{m} B_j$. We have to then verify whether $\sum_{i=1}^{n} \mu(A_i) = \sum_{j=1}^{m} \mu(B_j)$. Observe that for each i, we have $A_i = A_i \cap A = \bigcup_{j=1}^{m} (A_i \cap B_j)$, where $A_i \cap B_1, \ldots, A_i \cap B_m$ are disjoint $\mathcal{S}$-sets. Additivity of μ on $\mathcal{S}$ implies $\mu(A_i) = \sum_{j=1}^{m} \mu(A_i \cap B_j)$, for each i. But, by the same argument, we would also have $\mu(B_j) = \sum_{i=1}^{n} \mu(A_i \cap B_j)$ for each j. It follows that $\sum_{i=1}^{n} \mu(A_i) = \sum_{j=1}^{m} \mu(B_j)$, since both sums equal $\sum_{i=1}^{n} \sum_{j=1}^{m} \mu(A_i \cap B_j)$. Assuming that the reader can convince herself that the interchange of order of summation here is justified, this shows that the proposed definition of μ on $\mathfrak{F}$ is unambiguous. Thus, we now have a $[0, \infty]$-valued set function on $\mathfrak{F}$. Of course, μ-values of $\mathcal{S}$-sets remain unchanged, which means that the μ on $\mathfrak{F}$ is an extension of the given measure μ on $\mathcal{S}$.

We now show that μ defines a measure on $\mathfrak{F}$. Since $\emptyset \in \mathcal{S}$ and μ is a measure on $\mathcal{S}$, we have $\mu(\emptyset) = 0$. Next, let $A_1, A_2, \cdots$ be a sequence of disjoint $\mathfrak{F}$-sets with $A = \bigcup_i A_i \in \mathfrak{F}$. We are to show that $\mu(A) = \sum_i \mu(A_i)$. For each i, since $A_i \in \mathfrak{F}$, we must have $A_i = \bigcup_{k=1}^{n_i} B_{ik}$, where B_{ik}, $1 \le k \le n_i$, are disjoint $\mathcal{S}$-sets. At the same time, $A \in \mathfrak{F}$ means that $A = \bigcup_{j=1}^{n} B_j$, for disjoint $\mathcal{S}$-sets $B_1, \ldots, B_n$. Now, for each j, we have $B_j = B_j \cap A = \bigcup_{i \ge 1} \bigcup_{k=1}^{n_i} (B_j \cap B_{ik})$, where $\{B_j \cap B_{ik}, i \ge 1, 1 \le k \le n_i\}$

is a countable collection of disjoint $\mathcal{S}$-sets. Therefore, by the additivity property of μ on $\mathcal{S}$, we have
$$\mu(B_j) = \sum_{i=1}^{\infty} \sum_{k=1}^{n_i} \mu(B_j \cap B_{ik}), \quad \text{for each } j.$$
Adding over j, we get
$$\sum_{j=1}^{n} \mu(B_j) = \sum_{j=1}^{n} \sum_{i=1}^{\infty} \sum_{k=1}^{n_i} \mu(B_j \cap B_{ik}) = \sum_{i=1}^{\infty} \sum_{k=1}^{n_i} \sum_{j=1}^{n} \mu(B_j \cap B_{ik}),$$
where the interchange of order of summation is valid, because the terms are all non-negative.

Now note that, for each i, the sets $B_j \cap B_{ik}, 1 \leq j \leq n, \ 1 \leq k \leq n_i$ form a finite number of disjoint $\mathcal{S}$-sets, whose union equals $A \cap A_i = A_i$. Therefore, by the definition of μ on $\mathfrak{F}$, we have, for each i,
$$\mu(A_i) = \sum_{k=1}^{n_i} \sum_{j=1}^{n} \mu(B_j \cap B_{ik}).$$
Using this in the previous equality, we finally get
$$\sum_{j=1}^{n} \mu(B_j) = \sum_{i=1}^{\infty} \mu(A_i).$$
But, by the definition of μ on $\mathfrak{F}$, the left hand side in the above equality is $\mu(A)$. We thus get $\mu(A) = \sum_{i=1}^{\infty} \mu(A_i)$, proving the required additivity of μ on $\mathfrak{F}$.

Now that we have been able to successfully extend the measure μ given on the semifield $\mathcal{S}$ to a measure on the field $\mathfrak{F}$, it should be clear that there could not be any other way of extending μ to a measure on $\mathfrak{F}$. It should also be clear that, if the given measure μ on $\mathcal{S}$ is a finite measure (resp. a probability measure, resp. a σ-finite measure) then so will be its extension to $\mathfrak{F}$. Note that, we have denoted the extension of the given measure μ on $\mathcal{S}$ to the larger class $\mathfrak{F}$ also by the same notation μ and we will continue to do so. The next proposition captures what we have just achieved.

PROPOSITION **2.1.25.** Given a measure μ on a semifield $\mathcal{S}$, there is a unique extension of μ to a measure on the field $\mathfrak{F}$ generated by $\mathcal{S}$.

Having extended the given measure μ on a semifield $\mathcal{S}$ to a measure μ on the field $\mathfrak{F} = \mathfrak{F}(\mathcal{S})$, our next aim is to try and see if we can extend it further to an even larger class of sets. But before that, it will be useful to record some properties of measures on a field.

PROPOSITION **2.1.26.** Let μ be a measure on a field $\mathfrak{F}$ on a non-empty set Ω.
(a) (monotonicity) If A, B are $\mathfrak{F}$-sets with $A \subset B$, then $\mu(A) \leq \mu(B)$.
(b) (strong additivity) If A, B are $\mathfrak{F}$-sets, then $\mu(A \cup B) + \mu(A \cap B) = \mu(A) + \mu(B)$.
(c) (continuity from below) If $A_n, n \geq 1$ are $\mathfrak{F}$-sets with $A_n \subset A_{n+1}$, for each n, and if $A = \bigcup_n A_n \in \mathfrak{F}$, then $\mu(A_n) \uparrow \mu(A)$.
(d) (continuity from above) If $A_n, n \geq 1$, with $A_n \supset A_{n+1}$, for each n, are $\mathfrak{F}$-sets, such that, $A = \bigcap_n A_n \in \mathfrak{F}$ and if $\mu(A_{n_0}) < \infty$, for some $n_0 \geq 1$, then $\mu(A_n) \downarrow \mu(A)$.

2.1. Measures: from Semifields to Fields

It is important to note here that, in part (d) of the above proposition, the hypothesis "$\mu(A_{n_0}) < \infty$ for some n_0" cannot, in general, be dispensed with (see Exercise 2.1.30 (b)).

Before proceeding to prove the proposition, we need to clarify some conventions on inequalities to be followed. The inequality $a \leq b$ will have its usual meaning when a and b are real numbers. Besides, the inequalities $-\infty \leq a$ and $a \leq \infty$ would always hold for any $a \in [-\infty, \infty]$. The usual interplay between order and addition, that one is familiar with, would continue to remain valid. In particular, for a, b, c, d in $[0, \infty]$, $a \leq b$ and $c \leq d$ imply that $a + c \leq b + d$. One should be careful though about strict inequalities. For example, $a < b$ would not imply $a + c < b + c$, unless c is finite.

Also, for two sets A and B, we use the standard notation $A \setminus B$ to denote $A \cap B^c$. Clearly, if A and B are any two sets belonging to a field, then $A \setminus B$ also belongs to that field.

Proof of Proposition 2.1.26. For (a), we first use additivity of μ to see that, for $\mathfrak{F}$-sets A, B with $A \subset B$, we have $\mu(B) = \mu(A) + \mu(B \setminus A)$. Now, non-negativity of μ gives $\mu(A) \leq \mu(B)$.

For (b), we observe that $A \cup B = A \cup (B \setminus A)$ and $B = (B \setminus A) \cup (A \cap B)$. Therefore, using additivity of μ, we get

$$\mu(A \cup B) + \mu(A \cap B) = \mu(A) + \mu(B \setminus A) + \mu(A \cap B) = \mu(A) + \mu(B).$$

To prove (c), we define $B_1 = A_1$ and $B_n = A_n \setminus A_{n-1}$, for $n \geq 2$. Clearly, $\{B_n\}$ is a sequence of disjoint $\mathfrak{F}$-sets, such that, $\bigcup_{k=1}^{n} B_k = A_n$, for each $n \geq 1$, and $\bigcup_n B_n = \bigcup_n A_n = A \in \mathfrak{F}$. Hence

$$\mu(A) = \sum_n \mu(B_n) = \lim_n \sum_{k=1}^{n} \mu(B_k) = \lim_n \mu(A_n),$$

which, together with (a), gives $\mu(A_n) \uparrow \mu(A)$.

Finally, for proving (d), we define $B_n = A_n \setminus A_{n+1}$, for each $n \geq n_0$, and observe that A_n, for each $n \geq n_0$, equals the disjoint union $A \cup B_n \cup B_{n+1} \cup \cdots$ of $\mathfrak{F}$-sets, implying that

$$\mu(A_n) = \mu(A) + \sum_{k=n}^{\infty} \mu(B_k), \quad \text{for each } n \geq n_0.$$

In particular, $\sum_{k=n_0}^{\infty} \mu(B_k) \leq \mu(A_{n_0}) < \infty$, so that, $\sum_{k=n}^{\infty} \mu(B_k) \downarrow 0$ as $n \to \infty$. It follows that $\mu(A_n) \downarrow \mu(A)$. $\square$

For a sequence $\{A_n\}$ of subsets of Ω, we say that A is the "increasing limit" of the sequence of sets $\{A_n\}$ and write $A_n \uparrow A$ to mean that $A_n \subset A_{n+1}$, for each n, and $A = \bigcup_n A_n$. Similarly, we say that A is the "decreasing limit" of the sequence of sets $\{A_n\}$ and write $A_n \downarrow A$, to mean that $A_n \supset A_{n+1}$, for each n, and $A = \bigcap_n A_n$. One can use this notation for the statements in parts (c) and (d) of Proposition 2.1.26. For example, the statement in part (c) can be rephrased as : $A_n \in \mathfrak{F}$, for $n \geq 1$, and $A_n \uparrow A \in \mathfrak{F} \Longrightarrow \mu(A_n) \uparrow \mu(A)$.

EXERCISE 2.1.27. Let μ be a measure on a field $\mathfrak{F}$.
(a) (Sub-additivity) Part (b) of Proposition 2.1.26 implies $\mu(A \cup B) \leq \mu(A) + \mu(B)$, for any two $\mathfrak{F}$-sets A, B. Using this, show that $\mu\big(\bigcup_n A_n\big) \leq \sum_n \mu(A_n)$, for any finite or infinite sequence $\{A_n\}$ of $\mathfrak{F}$-sets. Assume $\bigcup_n A_n \in \mathfrak{F}$, in case $\{A_n\}$ is an infinite sequence.
(b) Show that, for any three $\mathfrak{F}$-sets A, B and C,
$$\mu(A \cup B \cup C) + \mu(A \cap B) + \mu(B \cap C) + \mu(C \cap A) = \mu(A) + \mu(B) + \mu(C) + \mu(A \cap B \cap C).$$
In particular, if μ is a finite measure, then
$$\mu(A \cup B \cup C) = \mu(A) + \mu(B) + \mu(C) - \mu(A \cap B) - \mu(B \cap C) - \mu(C \cap A) + \mu(A \cap B \cap C).$$
(c) Extend (b) to the case of any n sets $A_1, A_2, \ldots, A_n$ in $\mathfrak{F}$, where $n > 3$.

The next proposition is an important result which often comes in very handy in verifying whether a given non-negative and finitely additive set function on a field is a measure.

PROPOSITION 2.1.28. Let μ be a $[0, \infty]$-valued set function on a field $\mathfrak{F}$ with $\mu(\emptyset) = 0$. Suppose that μ is finitely additive on $\mathfrak{F}$, that is, $\mu\big(\bigcup_{i=1}^n A_i\big) = \sum_{i=1}^n \mu(A_i)$, for any finite collection $A_1, \ldots, A_n$ of disjoint $\mathfrak{F}$-sets. Then, any one of the following conditions is sufficient for μ to be a measure on $\mathfrak{F}$.
(a) (continuity from below) $A_n \in \mathfrak{F}$, $n \geq 1$ and $A_n \uparrow A \in \mathfrak{F}$ imply that $\mu(A_n) \uparrow \mu(A)$.
(b) (continuity from above) $A_n \in \mathfrak{F}$, $n \geq 1$ and $A_n \downarrow A \in \mathfrak{F}$ imply that $\mu(A_n) \downarrow \mu(A)$.
(c) (continuity from above at emptyset) $A_n \in \mathfrak{F}$, $n \geq 1$, $A_n \downarrow \emptyset$ imply that $\mu(A_n) \downarrow 0$.

Proof. Since $\mu(\emptyset) = 0$ is given, we only need to verify countable additivity of μ on $\mathfrak{F}$ to show that μ is a measure.
We first show that (a) is sufficient. So, we assume that (a) holds and take a sequence $\{B_n\}$ of disjoint $\mathfrak{F}$-sets with $B = \bigcup_n B_n \in \mathfrak{F}$. Then, using condition (a) and the given hypothesis of finite additivity of μ, we get
$$\mu(B) = \lim_n \mu\Big(\bigcup_{k=1}^n B_k\Big) = \lim_n \sum_{k=1}^n \mu(B_k) = \sum_{n=1}^\infty \mu(B_n),$$
establishing countable additivity of μ on $\mathfrak{F}$.
Next, since condition (b) implies (c), we complete the proof by just showing that condition (c) is sufficient for μ to be countably additive. Let $\{B_n\}$ be a sequence of disjoint $\mathfrak{F}$-sets with $B = \bigcup_n B_n \in \mathfrak{F}$. Clearly, $C_n = \bigcup_{k>n} B_k = B \setminus \big(\bigcup_{k=1}^n B_k\big) \in \mathfrak{F}$, for each n. Using the hypothesis of finite additivity, we then get
$$\mu(B) = \mu\big(\bigcup_{k=1}^n B_k\big) + \mu(C_n) = \sum_{1}^n \mu(B_k) + \mu(C_n).$$
But $C_n \downarrow \emptyset$ implying, by (c), that $\mu(C_n) \downarrow 0$. The required result follows from this by letting $n \to \infty$ in the above. □

From part (c) of Proposition 2.1.26 and Proposition 2.1.28, it follows that, for a non-negative set function μ on a field, satisfying $\mu(\emptyset) = 0$, finite additivity plus continuity from below is equivalent to μ being a measure on the field. However, the same cannot be said about continuity from above or continuity from above at the emptyset. They are sufficient but not necessary (See Exercise 2.1.30 (b)).

Before continuing further, we pause to see some simple examples of measures on a field. More non-trivial and useful examples of measures will have to wait until we complete our extension programme.

EXAMPLE 2.1.29. Let $\mathfrak{F}$ be the finite-cofinite field on $\Omega = \mathbb{N}$, the set of natural numbers. Then, one can easily see that each of the following set functions μ_1, μ_2, μ_3 and μ_4 defines a measure on $\mathfrak{F}$.
(a) $\mu_1(A) = |A|$, $A \in \mathfrak{F}$. $|A|$ denotes the cardinality A.
(b) $\mu_2(A)$ equals 1 or 0 according as $n_0 \in A$ or $n_0 \notin A$, where n_0 is a fixed natural number.
(c) $\mu_3(A) = |A(p)|$, where $A(p)$ denotes the set of prime numbers belonging to A.
(d) $\mu_4(A) = \sum_n a_n \mathbf{1}_A(n)$, where $\{a_n\}$ is a sequence of non-negative reals and $\mathbf{1}_A(n)$, for each n, equals 1 or 0 according as $n \in A$ or $n \notin A$.

EXERCISE 2.1.30. (a) Show that, in Example 2.1.29, the measures μ_1, μ_2 and μ_3 are all special cases of μ_4.
(b) Consider the measure μ_1 in Example 2.1.29. Observe that for each $n \geq 1$, the set $A_n = \{n, n+1, \cdots\}$ belongs to $\mathfrak{F}$ and that $A_n \downarrow \emptyset$, while $\mu_1(A_n) = \infty$ for all $n \geq 1$. Does it contradict Proposition 2.1.26 (d)?
(c) On $\mathfrak{F}$ as in Example 2.1.29, consider the set function μ defined as $\mu(A) = 0$ or 1 according as A is finite or A^c is finite. Show that μ is not a measure, although it is finitely additive.

2.2 Caratheodory Extension Theorem

Let us now get back to the extension programme. Given what we have done so far, we can now start with a measure μ on a field $\mathfrak{F}$ and try to extend it to a larger class of sets. A natural class of sets for the first step of extension of μ beyond $\mathfrak{F}$ would be the class of countable disjoint unions of $\mathfrak{F}$-sets. Of course, for a field $\mathfrak{F}$, the class of all possible countable disjoint union of $\mathfrak{F}$-sets is the same as the class of all possible countable unions of $\mathfrak{F}$-sets (See Exercise 2.1.8). With this in mind, let us denote $\mathfrak{F}_\sigma$ to be the class of all possible countable unions of $\mathfrak{F}$-sets. Here and in the sequel, a "countable union" will mean either a finite or a countably infinite union. While finite unions of $\mathfrak{F}$-sets are already in $\mathfrak{F}$, countably infinite unions need not be. Thus, in general, $\mathfrak{F} \subsetneq \mathfrak{F}_\sigma$. Clearly, $\mathfrak{F}_\sigma$ is closed under countable unions.

To extend μ to $\mathfrak{F}_\sigma$, the natural way, as dictated by the required additivity property of measures, would be to represent an $\mathfrak{F}_\sigma$-set as a countable disjoint union of $\mathfrak{F}$-sets and then define the μ-value of the $\mathfrak{F}_\sigma$-set as the sum of the μ-values of these constituent $\mathfrak{F}$-sets. However, it is convenient to define μ on $\mathfrak{F}_\sigma$ in a slightly different way, though the end result would be the same. We first observe that $\mathfrak{F}_\sigma$ also equals the class of increasing limits of sequences of $\mathfrak{F}$-sets.

Indeed, if $A = \bigcup_{n=1}^{\infty} B_n$, where $B_n \in \mathfrak{F}$ for each n, then $A_n = \bigcup_{k=1}^{n} B_k \in \mathfrak{F}$, for each n, and $A_n \uparrow A$. Thus

$$\mathfrak{F}_\sigma = \{A \subset \Omega : A_n \uparrow A \text{ for some sequence } \{A_n\} \text{ of } \mathfrak{F}\text{-sets }\}. \quad (2.2.1)$$

Proposition 2.1.26(c) now offers the following natural alternative for defining μ on $\mathfrak{F}_\sigma$. Given $A \in \mathfrak{F}_\sigma$, we get a sequence $\{A_n\}$ of $\mathfrak{F}$-sets with $A_n \uparrow A$ and define

$$\mu(A) = \lim_n \mu(A_n). \quad (2.2.2)$$

Since $\mu(A_n)$ is nondecreasing in n (by Proposition 2.1.26(a)), the limit in (2.2.2) exists. Before showing that this gives us the same end result as what we would have obtained by viewing $\mathfrak{F}_\sigma$-sets as countable disjoint unions of $\mathfrak{F}$-sets and using additivity for assigning μ-values, our first task is to ensure that our definition of $\mu(A)$ as in (2.2.2) is unambiguous. Here is an useful observation, which, among other things, will settle this issue.

PROPOSITION **2.2.1.** Suppose $\{A_n\}$ and $\{B_m\}$ are two sequences of $\mathfrak{F}$-sets with $A_n \uparrow A$ and $B_m \uparrow B$. If $A \subset B$, then $\lim_n \mu(A_n) \leq \lim_m \mu(B_m)$.

Proof. Fixing any n, it is easy to see that $\{A_n \cap B_m : m \geq 1\}$ is a sequence of $\mathfrak{F}$-sets increasing to the set $A_n \cap B = A_n \in \mathfrak{F}$. So, by Proposition 2.1.26(c), $\lim_m \mu(A_n \cap B_m) = \mu(A_n)$. Since $\mu(B_m) \geq \mu(A_n \cap B_m)$, for each m, one gets $\lim_m \mu(B_m) \geq \mu(A_n)$. This being true for each n, the required result now follows by letting $n \to \infty$. □

An immediate consequence of the above proposition is that it settles the issue of well-definedness of the extension of μ to $\mathfrak{F}_\sigma$ as proposed in (2.2.2). If $\{A_n\}$ and $\{B_m\}$ are two sequences of $\mathfrak{F}$-sets, both increasing to the same set, then Proposition 2.2.1 guarantees that $\lim_n \mu(A_n) = \lim_m \mu(B_m)$. In other words, for $A \in \mathfrak{F}_\sigma$, our proposed definition of $\mu(A)$ will not depend on the specific choice of a sequence $\{A_n\}$ of $\mathfrak{F}$-sets increasing to A.

In particular, if we represent a set $A \in \mathfrak{F}_\sigma$ as a countable disjoint union of $\mathfrak{F}$-sets, say, $A = \bigcup_n C_n$, then $A_n = \bigcup_{k=1}^{n} C_k \in \mathfrak{F}$, for each n, and $A_n \uparrow A$, so that according to our definition (2.2.2), $\mu(A) = \lim_n \mu(A_n)$. But, for each n, $\mu(A_n) = \sum_{k=1}^{n} \mu(C_k)$, and so $\mu(A) = \lim_n \sum_{k=1}^{n} \mu(C_k) = \sum_{n=1}^{\infty} \mu(C_n)$. This shows that we get the same extension of μ to $\mathfrak{F}_\sigma$ irrespective of whether we view $\mathfrak{F}_\sigma$-sets as increasing limits of of $\mathfrak{F}$-sets and use continuity from below to define μ or view them as countable disjoint unions of $\mathfrak{F}$-sets and use additivity.

Finally, note also that $\mathfrak{F} \subset \mathfrak{F}_\sigma$ and further, Proposition 2.1.26(c) ensures that μ remains unchanged on $\mathfrak{F}$, thus making μ on $\mathfrak{F}_\sigma$ really an extension of the given measure μ on $\mathfrak{F}$.

Let us now observe some properties of the set function μ on $\mathfrak{F}_\sigma$. First, Proposition 2.2.1 establishes monotonicity of μ on $\mathfrak{F}_\sigma$, namely, that $A, B \in \mathfrak{F}_\sigma$ and $A \subset B$ imply $\mu(A) \leq \mu(B)$. Next, suppose that A and B are $\mathfrak{F}_\sigma$-sets. If $\{A_n\}$

2.2. Caratheodory Extension Theorem

and $\{B_n\}$ are sequences of $\mathfrak{F}$-sets increasing to A and B respectively, then clearly $\{A_n \cup B_n\}$ as well as $\{A_n \cap B_n\}$ are sequences of $\mathfrak{F}$-sets with $A_n \cup B_n \uparrow A \cup B$ and $A_n \cap B_n \uparrow A \cap B$. This shows, first of all, that $A \cup B$ and $A \cap B$ are both $\mathfrak{F}_\sigma$-sets. Further, by proposition 2.1.26(b),

$$\mu(A_n \cup B_n) + \mu(A_n \cap B_n) = \mu(A_n) + \mu(B_n), \quad \text{for every } n \geq 1,$$

from which it follows that

$$\mu(A \cup B) + \mu(A \cap B) = \mu(A) + \mu(B). \tag{2.2.3}$$

Finally, let $\{A_n\}$ be a sequence of $\mathfrak{F}_\sigma$-sets with $A_n \uparrow A$. Note that $\mathfrak{F}_\sigma$, being the class of all countable unions of $\mathfrak{F}$-sets, is closed under countable unions, and therefore, $A \in \mathfrak{F}_\sigma$. We propose to show that $\mu(A_n) \uparrow \mu(A)$. Towards this, let $\{A_{nm}, m \geq 1\}$, for each $n \geq 1$, be a sequence of $\mathfrak{F}$-sets increasing to A_n. Setting $B_m = \bigcup_{n \leq m} A_{nm}$, $m \geq 1$, it is easy to see that $\{B_m\}$ is an increasing sequence of $\mathfrak{F}$-sets with $B_m \subset A_m$, for each m. Denote $B = \bigcup_m B_m$, so that $B_m \uparrow B$. Now, for every n, one has the inclusion

$$A_{nm} \subset B_m \subset A_m \quad \text{for all } m \geq n. \tag{2.2.4}$$

Letting $m \to \infty$ in (2.2.4) and using $A_{nm} \uparrow A_n$, $B_m \uparrow B$ and $A_m \uparrow A$, we get

$$A_n \subset B \subset A \quad \text{for every } n$$

Now letting $n \to \infty$, we get $B = A$, which shows that $B_m \uparrow A$ and so $\mu(A) = \lim_m \mu(B_m)$, by definition of μ on $\mathfrak{F}_\sigma$. Incidentally, we already knew that $A \in \mathfrak{F}_\sigma$, but $B_m \uparrow A$ proves it again. Next, by monotonicity of μ on $\mathfrak{F}$, relation (2.2.4) gives us that, for every $n \geq 1$,

$$\mu(A_{nm}) \leq \mu(B_m) \leq \mu(A_m) \quad \text{for all } m \geq n.$$

Letting $m \to \infty$ and using $\mu(A) = \lim_m \mu(B_m)$, we first get

$$\mu(A_n) \leq \mu(A) \leq \lim_m \mu(A_m),$$

and then letting $n \to \infty$, we finally get $\mu(A) = \lim_n \mu(A_n)$, proving $\mu(A_n) \uparrow \mu(A)$. We sum up our findings on the extension of μ to $\mathfrak{F}_\sigma$ in the following proposition.

PROPOSITION **2.2.2.** The $[0, \infty]$-valued set function μ on $\mathfrak{F}_\sigma$ given by (2.2.1) is well defined and is an extension of the given measure μ on $\mathfrak{F}$. Further, this set function μ on $\mathfrak{F}_\sigma$ has the following properties.
(a) (monotonicity) If $A, B \in \mathfrak{F}_\sigma$ are such that $A \subset B$, then $\mu(A) \leq \mu(B)$.
(b) (strong additivity) $A, B \in \mathfrak{F}_\sigma$ implies that $A \cup B, A \cap B \in \mathfrak{F}_\sigma$ and,

$$\mu(A \cup B) + \mu(A \cap B) = \mu(A) + \mu(B).$$

(c) (continuity from below) If $\{A_n\}$ is an increasing sequence of $\mathfrak{F}_\sigma$-sets with $A_n \uparrow A$, then $A \in \mathfrak{F}_\sigma$ and $\mu(A_n) \uparrow \mu(A)$.

Having thus extended μ from $\mathfrak{F}$ to $\mathfrak{F}_\sigma$, we may next think of continuing in the same manner to extend μ further. Since taking countable unions of $\mathfrak{F}_\sigma$-sets do not yield any new class, we may now possibly consider the class of all possible countable intersections of sets in $\mathfrak{F}_\sigma$. This class of sets, usually denoted by $\mathfrak{F}_{\sigma\delta}$, will, in general, be larger than the class $\mathfrak{F}_\sigma$. Since $\mathfrak{F}_\sigma$ is closed under finite intersections, one can identify $\mathfrak{F}_{\sigma\delta}$ also as the class of all decreasing limits of sequences of $\mathfrak{F}_\sigma$-sets. This may prompt us to define the μ-value of a $\mathfrak{F}_{\sigma\delta}$-set as the decreasing limit of μ-values of the $\mathfrak{F}_\sigma$-sets in the sequence. However, there is a fundamental theoretical issue here. While continuity from below is always satisfied by a measure, continuity from above is not. So, an extention of μ to $\mathfrak{F}_{\sigma\delta}$, based on requiring continuity from above, may not be the correct thing to do. This problem can, of course, be skirted by restricting to the case when the given μ on $\mathfrak{F}$ is a finite measure. However, we would still have to verify that the proposed extension to $\mathfrak{F}_{\sigma\delta}$ is well defined and then establish appropriate properties of this extension. But, even after all this, our journey will be far from over. We can then venture into a larger class $\mathfrak{F}_{\sigma\delta\sigma}$ and then to an even larger class $\mathfrak{F}_{\sigma\delta\sigma\delta}$, and so on. No matter how natural this looks, a clear disadvantage is that the end of the journey is never in sight. This is so, because, in general, at every stage of this process, one would always be left with the possibility of extending μ, with all the right properties, to a strictly larger class. The other discouraging feature of this process is that at every stage, one has to verify several things before proceeding to the next step.

So a natural question is whether there is a 'better' alternative. At the cost of sacrificing naturality, let us be ambitious and see if we can assign a μ-value to *any arbitrary* subset $A \subset \Omega$, using only the set function μ extended to $\mathfrak{F}_\sigma$. This may seem over-ambitious at first sight, but let us exercise patience and see what we can achieve (and what we cannot). The reader is assured that she would not be disappointed at the end.

Consider an arbitrary subset $A \subset \Omega$. If one wants to assign a μ-value to the set A by using only the set function μ extended to $\mathfrak{F}_\sigma$-sets, there are two natural simple-minded options. We may try to approximate the set A either 'from outside', that is, by $\mathfrak{F}_\sigma$-sets that contain A or 'from inside', that is, by $\mathfrak{F}_\sigma$-sets that are contained in A. We select the first alternative. That this is the correct option will be established later.

If $B \supset A$ is an $\mathfrak{F}_\sigma$-set, then desired monotonicity property mandates that any possible candidate for μ-value of A must not exceed $\mu(B)$. In other words, $\{\mu(B) : B \in \mathfrak{F}_\sigma, B \supset A\}$ would be a set of upper bounds for any possible μ-value of A. Let us take the "least" of these upper bounds and call it $\mu^*(A)$, that is, let

$$\mu^*(A) = \inf\{\mu(B) : B \in \mathfrak{F}_\sigma, B \supset A\}. \qquad (2.2.5)$$

By Proposition 2.2.2(a), $\mu^*(A) = \mu(A)$, for $A \in \mathfrak{F}_\sigma$. Does it mean that we have thus been able to extend μ, in such a simple way, as a measure from $\mathfrak{F}_\sigma$ to one on all subsets of Ω? That would, of course, be too much to expect! Indeed, the set function μ^*, so defined on all subsets, will, in general, fail to be a measure — it is additivity that will fail. In fact, when the smoke clears, one would see that μ^* is not even finitely additive in general. This is precisely why we used the notation μ^* for this set function instead of μ.

2.2. Caratheodory Extension Theorem

So what do we do now? As already pointed out, μ^* defined on the class of all subsets of Ω may lack additivity, in fact, even finite additivity. One way to salvage the situation will be to look for as large a class of subsets as possible on which μ^* satisfies at least finite additivity. It will, indeed, turn out that this class of sets, in general, is an enormously large class. Further, this class will contain (and, in general, will be strictly larger than) each of the classes $\mathfrak{F}_\sigma$, $\mathfrak{F}_{\sigma\delta}$, $\mathfrak{F}_{\sigma\delta\sigma}$, $\cdots$. What is even more striking is that, on this class, μ^* will actually turn out to be countably additive also. This means that we will end up with an extension of the given μ on $\mathfrak{F}$ to a measure on a very large class of subsets of Ω.

But to achieve all of this, we need to first observe some properties of the set function μ^* defined on all subsets of Ω. Let us mention at this point that the set function μ^* defined on all subsets of Ω by the formula (2.2.5), is usually called an *outer measure* — 'outer' because, the underlying set A is being approximated here by $\mathfrak{F}_\sigma$-sets '*from outside*'.

It is clear from definition that. $\mu^*(A_1) \leq \mu^*(A_2)$ for any two subsets $A_1 \subset A_2$, that is, the set function μ^* on the class of all subsets satisfies the monotonicity property.

Next, let A_1 and A_2 be two subsets of Ω. If B_1 and B_2 are any $\mathfrak{F}_\sigma$-sets with $B_1 \supset A_1$ and $B_2 \supset A_2$, then $B_1 \cup B_2$ and $B_1 \cap B_2$ are $\mathfrak{F}_\sigma$-sets, by Proposition 2.2.2(b), with $B_1 \cup B_2 \supset A_1 \cup A_2$ and $B_1 \cap B_2 \supset A_1 \cap A_2$. Further, the same Proposition 2.2.2(b) also gives $\mu(B_1 \cup B_2) + \mu(B_1 \cap B_2) = \mu(B_1) + \mu(B_2)$. Using all of these and the definition of μ^*, we get

$$\mu^*(A_1 \cup A_2) + \mu^*(A_1 \cap A_2) \leq \mu(B_1) + \mu(B_2).$$

Taking infimum now over all $\mathfrak{F}_\sigma$-sets B_1 and B_2 with $B_1 \supset A_1$, $B_2 \supset A_2$, one has

$$\mu^*(A_1 \cup A_2) + \mu^*(A_1 \cap A_2) \leq \mu^*(A_1) + \mu^*(A_2),$$

an inequality, usually referred to as the 'strong subadditivity' property of μ^*.

Next, we show that if $\{A_n\}$ is any sequence of subsets of Ω with $A_n \uparrow A$, then $\mu^*(A_n) \uparrow \mu^*(A)$. By the monotonicity property of μ^*, the sequence $\{\mu^*(A_n)\}$ is monotonically increasing and also, $\mu^*(A_n) \leq \mu^*(A)$, for each n. This will imply that $\lim_n \mu^*(A_n)$ exists and satisfies $\lim_n \mu^*(A_n) \leq \mu^*(A)$. If $\lim_n \mu^*(A_n) = \infty$, then the inequality trivially becomes an equality. Let us, therefore, prove the required result now assuming that $\lim_n \mu^*(A_n) < \infty$, which, of course, implies that $\mu^*(A_n) < \infty$, for all n. Fixing an $\epsilon > 0$, we get, for each n, a set $B_n \in \mathfrak{F}_\sigma$ with $B_n \supset A_n$, such that, $\mu(B_n) \leq \mu^*(A_n) + 2^{-n}\epsilon$. Note that $B = \bigcup_n B_n \in \mathfrak{F}_\sigma$ and $B \supset \bigcup_n A_n = A$, implying that $\mu^*(A) \leq \mu(B)$. Taking $C_n = \bigcup_{k=1}^n B_k$, we clearly have $C_n \in \mathfrak{F}_\sigma$, for each n, and $C_n \uparrow B$. Therefore, $\mu(B) = \lim_n \mu(C_n)$, by Proposition 2.2.2(c). We now claim that, for every $n \geq 1$,

$$\mu(C_n) \leq \mu^*(A_n) + \sum_{k=1}^n \frac{\epsilon}{2^k}. \tag{2.2.6}$$

Assuming this claim to be true, it will follow, by letting $n \to \infty$, that

$$\mu(B) = \lim_n \mu(C_n) \leq \lim_n \mu^*(A_n) + \epsilon$$

Letting $\epsilon \downarrow 0$ and using $\mu^*(A) \leq \mu(B)$, we get $\mu^*(A) \leq \lim_n \mu^*(A_n)$. This, coupled with the earlier inequality $\lim_n \mu^*(A_n) \leq \mu^*(A)$, finally gives $\lim_n \mu^*(A_n) = \mu^*(A)$. The claim (2.2.6) is proved by induction. It clearly holds for $n = 1$. Let us assume that it holds for some n. Using the strong additivity of μ on $\mathfrak{F}_\sigma$ and the definition of C_{n+1}, we get

$$\mu(C_{n+1}) + \mu(C_n \cap B_{n+1}) = \mu(C_n) + \mu(B_{n+1}).$$

Induction hypothesis and the choice of B_{n+1} gives

$$\mu(C_{n+1}) + \mu(C_n \cap B_{n+1}) \leq \mu^*(A_n) + \sum_{k=1}^{n} \frac{\epsilon}{2^k} + \mu^*(A_{n+1}) + \frac{\epsilon}{2^{n+1}}.$$

Note now that $C_n \cap B_{n+1} \supset B_n \cap B_{n+1} \supset A_n \cap A_{n+1} = A_n$ and $C_n \cap B_{n+1} \in \mathfrak{F}_\sigma$, so that $\mu^*(A_n) \leq \mu(C_n \cap B_{n+1})$. Using this in the above inequality one gets

$$\mu(C_{n+1}) + \mu^*(A_n) \leq \mu^*(A_n) + \mu^*(A_{n+1}) + \sum_{k=1}^{n+1} \frac{\epsilon}{2^k}.$$

Since $\mu^*(A_n) < \infty$, we can cancel it from both sides to get (2.2.6) for $n+1$.

All the properties of the set function μ^* defined on all subsets of Ω, that have been proved above, are captured in the next proposition.

PROPOSITION **2.2.3**. *The function μ^* defined on the class of all subsets of Ω by (2.2.5) satisfies the following properties.*
(a) *(monotonicity) If $A \subset B$, then $\mu^*(A) \leq \mu^*(B)$.*
(b) *(strong subadditivity) $\mu^*(A \cup B) + \mu^*(A \cap B) \leq \mu^*(A) + \mu^*(B)$, for any two subsets A and B.*
(c) *(continuity from below) If $A_n \uparrow A$, then $\mu^*(A_n) \uparrow \mu^*(A)$.*

It is part (b) of Proposition 2.2.3 that underlines the main reason why the set function μ^* fails to be a measure on the class of all subsets. It asserts only subadditivity, and not additivity, for μ^*. To be specific, given two disjoint subsets A and B of Ω, all we can say at this point is that $\mu^*(A \cup B) \leq \mu^*(A) + \mu^*(B)$. Indeed, it can be shown that equality does not hold in general. Exercise 2.2.4 below provides an example. The reader will surely have many more examples, once she gets a grasp of what exactly is preventing equality.

EXERCISE **2.2.4**. *Consider the finite-cofinite field $\mathfrak{F}$ on $\mathbb{R}$ (or, in general, on any uncountable set). Define μ on $\mathfrak{F}$ by $\mu(F) = 0$ or 1, according as F is finite or cofinite.*
(a) *Show that μ is a measure (in fact, a probability measure) on $\mathfrak{F}$.*
(b) *With μ^* defined on all subsets as in (2.2.5), show that $\mu^*(A) = 1$, for any uncountable set A. Hence conclude that $\mu^*(A \cup B) < \mu^*(A) + \mu^*(B)$, for any two disjoint uncountable subsets A and B.*

In view of the continuity property of μ^* asserted in Proposition 2.2.3(c), it must be clear that the real problem here that stops μ^* from being a measure is only the lack of finite additivity. As it turns out, the source of the problem here is not so much the set function μ^* itself, but the fact that we are considering it on the class of 'all' subsets. It is quite conceivable that the same set function

2.2. Caratheodory Extension Theorem

μ^*, when restricted to an appropriate class of subsets, may indeed turn out to be finitely additive and therefore, in view of Proposition 2.2.3(c), may give us a measure on that class. This is what we now try to pursue. Indeed, we will try to identify the largest class of subsets of Ω, on which μ^* is finitely additive. Clearly, $\mathfrak{F}_\sigma$, and hence $\mathfrak{F}$, will automatically be included in that class.

However, to keep the execution of this plan somewhat simple, we are going to impose an additional restriction. It may seem to be a serious restriction at first sight, but we will see at a later stage that the restriction can be relaxed to a significant extent.

The restriction that we are going to now impose is that the measure μ on the field $\mathfrak{F}$, that we started with, is a probability measure, that is,

$$\boxed{\textbf{Assumption:}\ \mu(\Omega) = 1.} \qquad (2.2.7)$$

It would not make any difference if instead, we had assumed μ to be only a finite measure. This is because any finite measure, as already pointed out, is a non-negative multiple of a probability measure.

Assuming that μ is a probability measure on $\mathfrak{F}$, one can see, by taking $B = A^c$ in Proposition 2.2.3(b), that the corresponding μ^* satisfies $\mu^*(A) + \mu^*(A^c) \geq 1$ for every $A \subset \Omega$. Clearly, if A and A^c both came from some class of subsets on which μ^* is additive, then we would have had equality above. Let us therefore consider the class

$$\mathcal{A} = \{A \subset \Omega : \mu^*(A) + \mu^*(A^c) = 1\}. \qquad (2.2.8)$$

In view of what was observed above, it is clear that the class $\mathcal{A}$ may equivalently be defined as

$$\mathcal{A} = \{A \subset \Omega : \mu^*(A) + \mu^*(A^c) \leq 1\}. \qquad (2.2.9)$$

Since μ^* agrees with μ on $\mathfrak{F}$ and μ is a probability measure on $\mathfrak{F}$, the class $\mathcal{A}$ clearly includes the field $\mathfrak{F}$. We are eventually going to show that $\mathcal{A}$ is indeed the largest class of subsets of Ω, on which the set function μ^* is additive. In fact, we shall prove a number of interesting facts about the class $\mathcal{A}$ and about properties of the set function μ^* on $\mathcal{A}$.

Let us first show that $\mathcal{A}$ is itself a field. Firstly, Since $\mathfrak{F} \subset \mathcal{A}$, we have $\Omega \in \mathcal{A}$. Also from the definition of $\mathcal{A}$, it is immediate that $\mathcal{A}$ is closed under complementation. Next, let us take sets A and B belonging to $\mathcal{A}$, that is, $\mu^*(A) + \mu^*(A^c) = 1$, $\mu^*(B) + \mu^*(B^c) = 1$, which will imply $\mu^*(A) + \mu^*(B) + \mu^*(A^c) + \mu^*(B^c) = 2$. From the last equality, using strong subadditivity (Proposition 2.2.3(b)), we get

$$\mu^*(A \cup B) + \mu^*(A \cap B) + \mu^*(A^c \cup B^c) + \mu^*(A^c \cap B^c) \leq 2,$$

or equivalently,

$$\mu^*(A \cup B) + \mu^*\big((A \cup B)^c\big) + \mu^*(A \cap B) + \mu^*\big((A \cap B)^c\big) \leq 2. \qquad (2.2.10)$$

But, by strong subadditivity again, we also have

$$\mu^*(A \cup B) + \mu^*((A \cup B)^c) \geq 1 \text{ and } \mu^*(A \cap B) + \mu^*((A \cap B)^c) \geq 1. \qquad (2.2.11)$$

If strict inequality were to hold *even in one* of the two inequalities in (2.2.11), we would have obtained

$$\mu^*(A \cup B) + \mu^*((A \cup B)^c) + \mu^*(A \cap B) + \mu^*((A \cap B)^c) > 2, \qquad (2.2.12)$$

and that would contradict (2.2.10). This shows that both inequalities in (2.2.11) must actually be equalities, which implies that both $A \cup B$ as well as $A \cap B$ belong to $\mathcal{A}$. This completes the proof that the class $\mathcal{A}$ is a field on Ω.

We next show that the class $\mathcal{A}$ posseses an even stronger closure property, namely, that it is closed also under countably infinite unions. Since $\mathcal{A}$ is closed under complementation, this will imply that $\mathcal{A}$ is closed under countable intersections as well. To prove closure under countable unions, let $\{A_n, n \geq 1\}$ be a sequence of sets in $\mathcal{A}$. We are to show that $A = \bigcup_n A_n$ belongs to $\mathcal{A}$. From what has been proved already, we know $\bigcup_{k=1}^{n} A_k \in \mathcal{A}$. Since $A^c \subset \left(\bigcup_{k=1}^{n} A_k\right)^c$, we have,

$$\mu^*\left(\bigcup_{k=1}^{n} A_k\right) + \mu^*(A^c) \leq \mu^*\left(\bigcup_{k=1}^{n} A_k\right) + \mu^*\left(\left(\bigcup_{k=1}^{n} A_k\right)^c\right) = 1.$$

Letting $n \to \infty$ and using continuity of μ^* from below (Proposition 2.2.3(c)), we get $\mu^*(A) + \mu^*(A^c) \leq 1$, which by virtue of (2.2.9), proves that $A \in \mathcal{A}$.

The fact that the class $\mathcal{A}$ is not just a field, but is also closed under countable unions (finite as well as infinite), is going to be of utmost significance in the subsequent developments of the theory. We start with a defnition.

DEFINITION **2.2.5**. A class $\mathcal{A}$ of subsets of a non-empty set Ω is called a σ-field, if it contains the set Ω and is closed under complementations and countable unions.

Two of the simplest (and extreme) examples of σ-fields on a non-empty set Ω are $\mathcal{A} = \{\emptyset, \Omega\}$ and $\mathcal{A} = \mathcal{P}(\Omega) =$ the collection of all subsets of Ω. The first one is the smallest σ-field on Ω and is called the *trivial* σ-field, while the second one is the largest one and is sometimes referred to as the *discrete* σ-field.

Clearly a σ-field is always a field. On the other hand, it is easy to see from the defintions, that any finite field, that is, a field consisting of only finitely many sets, is a σ-field as well. It follows therefore, that the following finite fields seen earlier in Examples 2.1.20 and 2.1.21, are both examples of finite σ-fields :
(a) $\mathcal{A} = \{\emptyset, A, A^c, \Omega\}$ where A is any non-empty proper subset of Ω, and,
(b) $\mathcal{A} =$ all possible unions of sets in a finite partition of Ω by non-empty sets.

Since any finite σ-field is also a field, it follows from Remark 2.1.22, that the number of sets in *any finite σ-field* must equal 2^k for some $k \geq 1$.

From the above discussions, it follows that it is only an infinite field that may fail to be a σ-field. Exercise 2.2.6 gives an example of such a field (See also Exercise 2.13.2 for another example).

EXERCISE **2.2.6**. On $\Omega = \mathbb{N}$, consider the finite-cofinite field $\mathfrak{F}$, as defined in Example 2.1.13. Exhibit a sequence $\{A_n\}$ of $\mathfrak{F}$-sets, such that, $\bigcup_n A_n \notin \mathfrak{F}$ and hence conclude that $\mathfrak{F}$ is not a σ-field.

EXERCISE **2.2.7**. Recall the countable-cocountable class of subsets of an uncoutable set Ω, as described in Example 2.1.14 . Show that this class is a σ-field. We will refer to it in the sequel as the 'countable-cocountable' σ-field.

2.2. Caratheodory Extension Theorem

EXERCISE 2.2.8. *(a) Any field which is closed under countable unions (or closed under countable intersections) is a σ-field.*
(b) Any field which is closed under increasing limits (or closed under decreasing limits) is a σ-field.

Right after Definition 2.1.7 of a field, we made a remark that a field is closed under any finite number of set operations. A very similar statement can be made about σ-fields as well, except that now the phrase "finte nunber" can be strengthened to "countable number". In other words, if one takes a countable number of sets from a σ-field and constructs a new set by performing any countable number of set operations on these sets, then the resulting set will be in the σ-field. Once again, the reader should convince herself of the validity of this statement.

Before proceeding, let us have a quick recap of our journey so far in the problem of extension. Taking cognizance of assumption (2.2.7), we started with a probability measure μ given on a field $\mathfrak{F}$ and extended it first to $\mathfrak{F}_\sigma$. We then used that to define a set function μ^*, called the 'outer measure' or 'outer probability', on the class of all subsets. While μ^* is seen to have some nice properties, it fails, in general, to satisfy finite additivity on the class of all subsets. So, we turned our attention to searching for the largest class of sets on which μ^* would be finitely additive. This prompted us to define, in (2.2.8) (equivalently, in (2.2.9)), a certain class $\mathcal{A}$ of subsets of Ω and we showed that this class is not just a field, but a σ-field, containing the field $\mathfrak{F}$. We now show that the set function μ^*, when restricted to the σ-field $\mathcal{A}$, is finitely additive.

Towards this, let A_1 and A_2 be any two disjoint sets in $\mathcal{A}$. Disjointness implies $A_1^c \cup A_2^c = \Omega$ and therefore, $\mu^*(A_1^c \cup A_2^c) = 1$. From strong subadditivity and the fact that A_1, A_2 are $\mathcal{A}$-sets, we get

$$1 + \mu^*(A_1^c \cap A_2^c) \leq \mu^*(A_1^c) + \mu^*(A_2^c) = 2 - [\mu^*(A_1) + \mu^*(A_2)],$$

and hence,

$$\mu^*((A_1 \cup A_2)^c) = \mu^*(A_1^c \cap A_2^c) \leq 1 - [\mu^*(A_1) + \mu^*(A_2)].$$

Since $\mathcal{A}$ is a σ-field, we know $A_1 \cup A_2 \in \mathcal{A}$. This and the above inequality gives

$$\mu^*(A_1 \cup A_2) = 1 - \mu^*((A_1 \cup A_2)^c) \geq \mu^*(A_1) + \mu^*(A_2) \qquad (2.2.13)$$

On the other hand, from strong subadditivity and disjointness of A_1, A_2, we have

$$\mu^*(A_1 \cup A_2) \leq \mu^*(A_1) + \mu^*(A_2) \qquad (2.2.14)$$

From (2.2.13) and (2.2.14), we get $\mu^*(A_1 \cup A_2) = \mu^*(A_1) + \mu^*(A_2)$, proving finite addltivity of μ^* on $\mathcal{A}$.

This is a huge (almost the final) leap forward. Finite additivity of μ^* on the σ-field $\mathcal{A}$, as proved above, coupled with Propositions 2.1.28(a) and 2.2.3(c), will imply that μ^* is indeed countably additive, and, hence a probability, on $\mathcal{A}$.

We have thus proved that μ^* defines a probability measure on the σ-field $\mathcal{A}$. As noted earlier, the set function μ^* agrees with μ on $\mathfrak{F}_\sigma$ and, therefore, on $\mathfrak{F}$. Thus, μ^* gives us an extension of the probability measure μ given on the field $\mathfrak{F}$

to a probability measure on the σ-field $\mathcal{A}$. This gives us our first major result of the theory and is captured in the following theorem.

THEOREM **2.2.9**. Let $\mathfrak{F}$ be a field on a non-empty set Ω and μ be a probability measure on $\mathfrak{F}$. For any $A \subset \Omega$, define

$$\mu^*(A) = \inf \left\{ \lim_n \mu(B_n) : B_n \in \mathfrak{F},\ \{B_n\} \text{ increasing and } A \subset \bigcup_n B_n \right\} \quad (2.2.15)$$

and let $\mathcal{A} = \{A \subset \Omega : \mu^*(A) + \mu^*(A^c) = 1\}$. Then $\mathcal{A}$ is a σ-field containing $\mathfrak{F}$ and μ^* is a probability measure on $\mathcal{A}$ which agrees with μ on $\mathfrak{F}$.

It is easy to see that the definition of μ^* given by (2.2.15) above is no different from the earlier definition given by (2.2.5). It just helps us avoid bringing in the class $\mathfrak{F}_\sigma$ explicitly in the statement of the Theorem.

It is a good time now to take a quick stock of where we started and what we have achieved. Originally, we started with a measure μ on a semifield $\mathcal{S}$. We first extended μ as a measure to a larger class, namely, the field $\mathfrak{F}$ generated by $\mathcal{S}$. Recall that $\mathfrak{F}$ consists of all finite disjoint unions (equivalently, all finite unions) of $\mathcal{S}$-sets and μ was defined explicitly on $\mathfrak{F}$ by appealing to additivity.

The next job was to extend the measure obtained on $\mathfrak{F}$ to a larger class. At first, we extended μ to $\mathfrak{F}_\sigma$ by a natural explicit formula. However, continuing with that line of approach didn't seem to lead us to any tangible end. To pursue a different line of approach, we brought in the assumption that the given μ is a probability measure. The path to extension of μ that we pursued now was somewhat indirect. We first introduced a set function μ^*, called the outer probability, defined for all subsets. We then noted that this set function μ^*, which agrees with μ on $\mathfrak{F}$ may not, in general, be a measure on the class of all subsets. However, we were able to identify a class $\mathcal{A}$, which turns out to be a σ-field containing $\mathfrak{F}$, on which μ^* is a probability measure. Thus, we obtained an extension of the probability measure μ on $\mathfrak{F}$ to a probability measure μ^* on the σ-field $\mathcal{A}$.

At this point, it is perhaps worthwhile and important to examine our approach against a possible alternative approach, which is no less natural. Our approach, as described above, essentially consisted of first defining an 'outer approximation' μ^* for all subsets of Ω and then identifying the largest class of subsets on which μ^* is additive. A possible alternative to this could have been to also define an "inner approximation" μ_* for all subsets and then consider the collection of those sets for which the inner and outer approximations match. Of course, the issues that would then have to be settled are, firstly, whether this collection gives us a σ-field containing $\mathfrak{F}$, and, if so, then whether the common value of the two approximations, when restricted to this class, gives us a measure. Once these issues are settled in the affirmative, we could then ask whether this would give us anything different from the extension we have already obtained, as captured in Theorem 2.2.9.

Let us elaborate on these ideas, continuing to assume that, to start with, μ is a probability measure on $\mathfrak{F}$. The first important point to be made here is that,

2.2. Caratheodory Extension Theorem

unlike in the case of the outer measure μ^*, the class $\mathfrak{F}_\sigma$ with μ extended to it, will not be the appropriate class for defining the proposed inner approximation. As explained in the next paragraph (see also Exercise 2.2.10), the right class of sets to be used for inner approximations would be the class $\mathfrak{F}_\delta$, consisting of all possible countable intersections of $\mathfrak{F}$-sets. It is clear that $A \in \mathfrak{F}_\delta$ if and only if $A^c \in \mathfrak{F}_\sigma$. The class $\mathfrak{F}_\delta$ can also be equivalently described as the class of all decreasing limits of sequences of $\mathfrak{F}$-sets and this gives a natural extension of the given probability measure μ on $\mathfrak{F}$ to the class $\mathfrak{F}_\delta$, namely, $\mu(A) = \lim_n \mu(A_n)$ if $A_n \in \mathfrak{F}$ and $A_n \downarrow A$. One can establish the well-definedness of this extension, by proving an analogue of Proposition 2.2.1. Further, it is not difficult to formulate and prove an analogue of Proposition 2.2.2. The proposed inner approximation $\mu_*(A)$, for any subset $A \subset \Omega$, may now be defined as

$$\mu_*(A) = \sup\{\mu(C) : C \in \mathfrak{F}_\delta, C \subset A\} \qquad (2.2.16)$$

The set function μ_*, thus defined on the class of all susbsets, is called the 'inner measure' or 'inner probability'. One can then formulate and prove an analogue of Proposition 2.2.3. Alternatively, one could simply observe that $\mu_*(A) = 1 - \mu^*(A^c)$ and then apply Proposition 2.2.3 to get the appropriate properties of μ_*.

As proposed earlier, we may now denote $\mathcal{A}'$ to be the collection of all those sets $A \subset \Omega$, for which $\mu^*(A) = \mu_*(A)$. In view of $\mu_*(A) = 1 - \mu^*(A^c)$, it immediately follows *without any work* that the class $\mathcal{A}'$ is the same as $\mathcal{A}$ and this shows that, at the end, this approach leads to the same extension that we got before.

We next address the important question of why do we use $\mathfrak{F}_\sigma$ for outer approximations and $\mathfrak{F}_\delta$ for inner approximations? This is not difficult to explain. For example, when we are interested in getting an outer approximation for the measure of a set A, and we are allowed to go a little beyond $\mathfrak{F}$-sets and use sets in $\mathfrak{F}_\sigma$ and/or $\mathfrak{F}_\delta$ for our approximation, then using $\mathfrak{F}_\sigma$-sets rather than $\mathfrak{F}_\delta$-sets would be the right choice. This is because, for a general $A \subset \Omega$, the class $\mathfrak{F}_\sigma$ would offer us many more sets containing A, than the classes $\mathfrak{F}$ or $\mathfrak{F}_\delta$ would. In fact, as Exercise 2.2.10 below will show, using $\mathfrak{F}_\delta$-sets for outer approximation is as good as just using $\mathfrak{F}$-sets. One can similarly justify using $\mathfrak{F}_\delta$-sets instead of $\mathfrak{F}_\sigma$-sets for inner approximations (See Exercise 2.2.10 again).

EXERCISE **2.2.10.** *For any $A \subset \Omega$, show that*

$$\inf\{\mu(B) : B \in \mathfrak{F}_\delta, B \supset A\} = \inf\{\mu(B) : B \in \mathfrak{F}, B \supset A\}, \text{ and}$$
$$\sup\{\mu(B) : B \in \mathfrak{F}_\sigma, B \subset A\} = \sup\{\mu(B) : B \in \mathfrak{F}, B \subset A\}.$$

Of course, one could still ask: what is so wrong if for defining both inner and outer approximations, we just used sets in $\mathfrak{F}$ itself, rather than going for $\mathfrak{F}_\sigma$-sets or $\mathfrak{F}_\delta$-sets? Exercise 2.13.12 illustrates that this would lead to something terrible, or at least, very undesirable.

To complete this circle of thoughts, let us touch upon one more natural question that a curious reader may be inclined to ask and it is the following. Once we realize that, in order to define good outer or inner approximations for measures of general sets, we do need to use sets outside of $\mathfrak{F}$ along with a natural extension of μ to such sets, why should we stop at the classes $\mathfrak{F}_\sigma$ and $\mathfrak{F}_\delta$? It would seem quite plausible that we may do better if we use sets from even larger classes where μ has natural extensions. For example we could use sets from, say, the classes $\mathfrak{F}_{\sigma\delta}$ and $\mathfrak{F}_{\delta\sigma}$ or even $\mathfrak{F}_{\sigma\delta\sigma}$ and $\mathfrak{F}_{\delta\sigma\delta}$ and the natural extensions of μ to these classes to define inner or outer approximations for general sets. However, the fact of the matter is that, going these extra miles yields no pay-off at all. Indeed, as our results in Section 2.5 will reveal, nothing is gained by going beyond $\mathfrak{F}_\sigma$ for the outer approximation or beyond $\mathfrak{F}_\delta$ for the inner approximation.

Let us now return to our achieved extension μ^* of the given μ, obtained on the class $\mathcal{A}$. One disturbing feature of this extension, which may seem like a drawback, is that, unlike our first couple of steps of extension of μ from $\mathcal{S}$ to $\mathfrak{F}$ and then from $\mathfrak{F}$ to $\mathfrak{F}_\sigma$, our final extension μ^* is not given by a clearcut explicit formula. Given a set $A \in \mathcal{A}$, neither (2.2.5) nor (2.2.15) would be of much help in 'computing' the value of $\mu^*(A)$, except in some very special cases. But while accepting that this is how it is, a little reflection would convince the reader that this is not so unexpected. While sets in $\mathfrak{F}$ have an explicit description in terms of $\mathcal{S}$-sets and sets in $\mathfrak{F}_\sigma$ have an explicit description in terms of $\mathfrak{F}$-sets, no such description is available for all sets in $\mathcal{A}$ in terms of $\mathfrak{F}$-sets or $\mathfrak{F}_\sigma$-sets. Indeed, $\mathcal{A}$ would, in general, be a humongous class of sets compared to $\mathfrak{F}$. More explicitly, for an arbitrary $A \in \mathcal{A}$, it is not possible to pinpoint any specific sequence of set operations performed on some collection of $\mathfrak{F}$-sets that will give us the set A. It, therefore, stands to reason that no explicit formula for our extension μ^* to sets in $\mathcal{A}$, should be available. We will elaborate on this later. But at this point, a very important thing to note is that, no matter how one goes about extending the given probability μ on $\mathfrak{F}$ to a probability on $\mathcal{A}$, the extension would finally end up being same as μ^*. Thus, the extension to $\mathcal{A}$ is unique, irrespective of how the extension is obtained. This is left as an easy exercise for the reader in part (a) of Exercise 2.2.11

EXERCISE 2.2.11. *Let μ be a probability measure on a field $\mathfrak{F}$.*
(a) With μ^* and $\mathcal{A}$ as above, show that, if ν is any measure on $\mathcal{A}$ which agrees with μ on $\mathfrak{F}$, then ν must agree with μ^* on $\mathcal{A}$. (Hint: First show that $\nu = \mu$ on $\mathfrak{F}_\sigma$ and so, $\nu(A) \leq \mu^*(A)$, for all $A \in \mathcal{A}$).
(b) With $\mathcal{A}$ as above, suppose $\widetilde{\mathcal{A}}$ is a σ-field such that $\mathfrak{F} \subset \widetilde{\mathcal{A}} \subset \mathcal{A}$ and $\widetilde{\mu}$ is a measure on $\widetilde{\mathcal{A}}$ with $\widetilde{\mu} = \mu$ on $\mathfrak{F}$. Show that $\widetilde{\mu} = \mu$ on $\widetilde{\mathcal{A}}$.

We next address another important issue. In what was done above, the σ-field $\mathcal{A}$ on which μ^* gives us an extension of the given probability μ, was defined as $\mathcal{A} = \{A \subset \Omega : \mu^*(A) + \mu^*(A^c) = 1\}$. Thus, the class $\mathcal{A}$ is clearly seen to depend on the given probability μ. In other words, while we have shown that, given any probability measure on $\mathfrak{F}$, it can be extended to a σ-field containing $\mathfrak{F}$ and that the extension is unique, but that σ-field depends on the given probability. This may be a slightly disturbing issue. For reasons that will be clear as we go on, it would be desirable if, given a field $\mathfrak{F}$, we could identify *one single* σ-field

2.2. Caratheodory Extension Theorem

containing $\mathfrak{F}$, such that *every* probability measure on $\mathfrak{F}$ can be extended to that σ-field and further, the extension is unique. We proceed to show that this is indeed possible. The next definition is an anologue of Definition 2.1.16.

DEFINITION **2.2.12.** Given a non-empty collection $\mathcal{C}$ of subsets of Ω, the smallest class of subsets of Ω which contains $\mathcal{C}$ and is a σ-field is called the *σ-field generated by $\mathcal{C}$* and is denoted $\sigma(\mathcal{C})$. The class $\mathcal{C}$ is called a *generating class* for $\sigma(\mathcal{C})$.

That given any non-empty collection $\mathcal{C}$, there exists such a smallest σ-field containing $\mathcal{C}$ is established by following analogues of Propositions 2.1.17 and 2.1.18. Proofs of these are left as easy exercises.

PROPOSITION **2.2.13.** Given any family $\mathbf{C}$ of σ-fields on a non-empty set Ω, their 'intersection', that is, the class consisting of those sets that belong to every σ-field in the family $\mathbf{C}$, is itself a σ-field. This σ-field is called the *'intersection σ-field'*.

PROPOSITION **2.2.14.** Given a non-empty collection $\mathcal{C}$ of subsets of Ω, consider the family $\mathbf{C}$ of all σ-fields containing $\mathcal{C}$. (We know that $\mathcal{P}(\Omega)$ is always such a σ-field.) The intersection σ-field of the collection $\mathbf{C}$ is the smallest σ-field containing $\mathcal{C}$.

It is clear from Definition 2.2.12, that if $\mathcal{C}_1$ and $\mathcal{C}_2$ are two non-empty classes of subsets of Ω such that $\mathcal{C}_1 \subset \mathcal{C}_2$, then $\sigma(\mathcal{C}_1) \subset \sigma(\mathcal{C}_2)$. A strengthening of the above observation is given in part (a) of Exercise 2.2.15. It is a very useful result and is often used to examine if two different non-empty classes of subsets generate the same σ-field.

EXERCISE **2.2.15.** *(a) Given two non-empty classes $\mathcal{C}_1$ and $\mathcal{C}_2$ of subsets of Ω, show that $\sigma(\mathcal{C}_1) \subset \sigma(\mathcal{C}_2)$ if and only if $\mathcal{C}_1 \subset \sigma(\mathcal{C}_2)$.*
(b) Show that, for any non-empty collection $\mathcal{C}$ of subsets, $\sigma(\mathfrak{F}(\mathcal{C})) = \sigma(\mathcal{C})$.

It should be clear that, for any σ-field, except the trivial σ-field $\{\Omega, \emptyset\}$, there will always be many different generators. But the point of the next Exercise 2.2.16 is not just to illustrate that. Further, this will illustrate a nice application of part (a) of Exercise 2.2.15, but that is not the main point either. The primary reason why this is of significance is that the resulting σ-field on $\mathbb{R}$, discussed in Exercise 2.2.16 is going to be of utmost importance in the theory to come.

EXERCISE **2.2.16.** *Consider the following classes of subsets of $\mathbb{R}$: (i) the class of all intervals, (ii) the class of all bounded intervals, (iii) the class of all intervals with rational end points, (iv) the class of all open intervals, (v) the class of all intervals of the form $(a, b]$ with $-\infty < a < b < \infty$, (vi) the class of all intervals of the form $(-\infty, b]$ with $b \in \mathbb{R}$, (vii) the class of all intervals of the form (a, ∞) with $a \in \mathbb{R}$, (viii) the class of all open subsets of $\mathbb{R}$, (ix) the class of all intervals containing the point zero.*
Show that each of the above classes generate the same σ-field on $\mathbb{R}$. Show that all singleton sets and hence all countable sets are in this σ-field.

We had already noted that any finite field is a σ-field also. It, therefore, follows from what was seen in Examples 2.1.20 and 2.1.21, that

(a) if $\mathcal{C} = \{A\}$, where $\emptyset \subsetneq A \subsetneq \Omega$, then $\sigma(\mathcal{C}) = \{\emptyset, A, A^c, \Omega\}$,

(b) if $\mathcal{C}$ is a finite partition of Ω by non-empty sets, then $\sigma(\mathcal{C})$ consists of all possible unions (including the empty union) of sets in $\mathcal{C}$.

Further, it follows from Remark 2.1.22 that,

(c) if $\mathcal{C}$ is any non-empty finite class of subsets of Ω, then $\sigma(\mathcal{C}) = \mathfrak{F}(\mathcal{C})$ and the number of sets in $\sigma(\mathcal{C})$ is 2^k for some $k \geq 1$.

In Example 2.1.23, we considered the case when $\mathcal{C}$ is an infinite partition of Ω by non-empty sets and described $\mathfrak{F}(\mathcal{C})$. From the description, it followed that $\mathfrak{F}(\mathcal{C})$ is a countably infinite class or an uncountable class according as $\mathcal{C}$ is countably infinite or uncountable (See Remark 2.1.24). What can we say about $\sigma(\mathcal{C})$ is this case? The answer to this comes in Exercise 2.2.17.

EXERCISE **2.2.17.** *If* $\mathcal{C} = \{A_\alpha, \alpha \in \Lambda\}$ *is an infinite partition of* Ω *by non-empty sets, then show that*

$$\sigma(\mathcal{C}) = \{A \subset \Omega : \text{either } A \text{ or } A^c \text{ is a countable union of sets in } \mathcal{C}\}$$

and conclude that $\sigma(\mathcal{C})$ *is an uncountable class.*

A special case of Exercise 2.2.17 is when Ω is an infinite set and $\mathcal{C}$ consists of all singleton subsets of Ω. In that case, $\sigma(\mathcal{C})$ gives us the countable-cocountable σ-field on Ω, as described in Example 2.1.14. A further special case of this is when not only is $\mathcal{C}$ the class of all singleton subsets, but Ω is countable. It is clear that in this case $\sigma(\mathcal{C}) = \mathcal{P}(\Omega)$, the power set of Ω.

REMARK **2.2.18.** It has been noted earlier that if $\mathcal{C}$ is any non-empty finite class, then $\sigma(\mathcal{C})$ equals $\mathfrak{F}(\mathcal{C})$, both generated by a finite partition and contains 2^k many sets for some $k \geq 1$. In particular, any finite σ-field (equivalently, any finite field) contains 2^k many sets, for some $k \geq 1$.

In contrast, if $\mathcal{C}$ is an infinite partition of Ω, then $\mathfrak{F}(\mathcal{C})$ is countably infinite or uncountable according as $\mathcal{C}$ is countably infinite or uncountable (see Remark 2.1.24), whereas $\sigma(\mathcal{C})$ is always uncountable (see Exercise 2.2.17).

It turns out that, in general, any infinite σ-field (and not just the ones generated by an infinte partition) is *always uncountable* (see Exercise 2.13.4). An interesting upshot of this is that while it is possible for a field to be *either finite or countably infinite or uncountable*, a σ-field can only be *either finite or uncountable*.

Let us now return to the original problem of extending a measure μ given on a field $\mathfrak{F}$. Assuming that μ is probability measure, we had already obtained a probability measure μ^* on a σ-field $\mathcal{A}$ containing $\mathfrak{F}$, which is an extension of the given μ. Since the extension was also shown to be unique, we may and will, from now on, denote this extension also by μ. Note now that $\mathcal{A}$ being a σ-field containing $\mathfrak{F}$ implies that $\sigma(\mathfrak{F}) \subset \mathcal{A}$ and so μ restricted to $\sigma(\mathfrak{F})$ will be an extension of the probability measure μ from $\mathfrak{F}$ to $\sigma(\mathfrak{F})$. The important point

2.2. Caratheodory Extension Theorem

here is that while the σ-field $\mathcal{A}$ depends on μ, the σ-field $\sigma(\mathfrak{F})$ depends only on $\mathfrak{F}$. The upshot is that $\sigma(\mathfrak{F})$ is one σ-field containing $\mathfrak{F}$, to which *any* probability measure on $\mathfrak{F}$ can be extended and also, the extension is unique (see Exercise 2.2.11(b)).

Since any finite measure is a non-negative scalar multiple of a probability measure, the same will hold for any finite measure given on the field $\mathfrak{F}$. Recall next that, at the outset, we actually started with a semifield $\mathcal{S}$ and a measure μ given on $\mathcal{S}$. Our first step was an extension of μ from $\mathcal{S}$ to $\mathfrak{F} = \mathfrak{F}(\mathcal{S})$. The requirement of additivity makes this the **only possible**, and hence unique, extension. This, coupled with what was said above, makes μ on $\sigma(\mathfrak{F})$ actually a unique extension of the given finite measure μ on $\mathcal{S}$. Finally, since $\sigma(\mathfrak{F}) = \sigma(\mathcal{S})$ (see Exercise 2.2.15(b)), all of the above leads us to the next theorem.

THEOREM 2.2.19. Any finite measure μ given on a semifield $\mathcal{S}$ has a unique extension to the σ-field $\sigma(\mathcal{S})$ generated by $\mathcal{S}$.

With this, we have satisfactorily settled the problem of extension for finite measures given on a semifield. The natural next step would be to turn to infinite measures. While one may be tempted to follow the same route as for probability measures, but a closer scrutiny of what we did, would make one realise that μ being a probability measure was critical in our arguments. To handle general infinite measures, one would have to make significant changes in the argument and it becomes technically somewhat complicated. We will take it up in Section 2.6, but for now, we avoid the complication by assuming that the given measure on the semifield $\mathcal{S}$ is σ-finite on $\mathcal{S}$ and carry out the extension in that case. This will be the first illustration of what was said before, namely, that a result proved for finite measures can often be pushed through for σ-finite measures using a simple and standard algorithm.

Suppose we are given a σ-finite measure μ on a semifield $\mathcal{S}$. It is then clear that the extension of μ to the field $\mathfrak{F}$ generated by $\mathcal{S}$ is σ-finite on $\mathfrak{F}$. In fact, the converse of this is also true (see Exercise 2.13.13). The σ-finiteness of μ on $\mathfrak{F}$ implies that we can get disjoint sets $\Omega_n, n \geq 1$ belonging to $\mathfrak{F}$, with $\bigcup_n \Omega_n = \Omega$, such that, $\mu(\Omega_n) < \infty$, for each n (see Exercise 2.1.9). Now, for each n, the set function μ_n defined on $\mathfrak{F}$ by $\mu_n(A) = \mu(A \cap \Omega_n)$ is a finite measure on $\mathfrak{F}$ (see Exercise 2.13.10) and hence, by Theorem 2.2.19, has a unique extension to a finite measure, to be denoted also by μ_n, on $\sigma(\mathfrak{F})$. We now define a non-negative set function $\widetilde{\mu}$ on $\sigma(\mathfrak{F})$ by $\widetilde{\mu}(A) = \sum_n \mu_n(A)$. By the definition of the μ_n on $\mathfrak{F}$, it is clear that $\widetilde{\mu} = \mu$ on $\mathfrak{F}$. We now prove additivity of $\widetilde{\mu}$ on $\sigma(\mathfrak{F})$. For any sequence $\{A_k : k \geq 1\}$ of disjoint $\sigma(\mathfrak{F})$-sets, we have

$$\widetilde{\mu}(\bigcup_k A_k) = \sum_n \mu_n(\bigcup_k A_k) = \sum_n \sum_k \mu_n(A_k) = \sum_k \sum_n \mu_n(A_k) = \sum_k \widetilde{\mu}(A_k),$$

where interchange of order of summation is valid, since the terms are all non-negative. Thus, we have shown that $\widetilde{\mu}$ is a measure on $\sigma(\mathfrak{F})$ and is therefore an extension of the σ-finite measure μ on $\mathfrak{F}$.

We next show that the extension is unique. Let ν be any measure on $\sigma(\mathfrak{F})$ such that $\nu = \mu$ on $\mathfrak{F}$. In particular, for each n, we have $\nu(\Omega_n) = \mu(\Omega_n) < \infty$ and also $\nu(A \cap \Omega_n) = \mu(A \cap \Omega_n) = \mu_n(A)$ for all $A \in \mathfrak{F}$. Denoting by ν_n the set function on $\sigma(\mathfrak{F})$ defined by $\nu_n(A) = \nu(A \cap \Omega_n)$, it follows that ν_n is a finite measure on $\sigma(\mathfrak{F})$ which agrees with μ_n on $\mathfrak{F}$ and hence we must have $\nu_n = \widetilde{\mu}_n$ on $\sigma(\mathfrak{F})$. It now follows immediately that $\nu = \widetilde{\mu}$ on $\sigma(\mathfrak{F})$.

The argument above uses a specific partition $\{\Omega_n\}$ to define the extension. But the uniqueness of the extension shows that it does not matter which partition is used.

We have thus reached the first major landmark in measure theory known as the Caratheodory Extension Theorem.

THEOREM **2.2.20.** (Caratheodory Extension Theorem)
Let $\mathcal{S}$ be a semifield on a non-empty set Ω. Then, any σ-finite measure μ on $\mathcal{S}$ has a unique extension to a measure on $\sigma(\mathcal{S})$, necessarily σ-finite.

In the above, the role played by σ-finiteness of μ on $\mathcal{S}$ (equivalently on $\mathfrak{F}$) was essentially that it allowed us to express μ on $\mathfrak{F}$ as a sum of finite measures so that we could use the extension obtained earlier for finite measures. However, this is not the only reason for considering measures that are σ-finite on $\mathfrak{F}$, as illustrated by Example 2.2.21.

EXAMPLE **2.2.21.** Let $\Omega = (0,1]$ and let $\mathcal{S}$ be the semifield consisting of all sets of the form $(a,b]$, $0 \leq a \leq b \leq 1$. One can easily verify that every singleton subset of Ω is in $\sigma(\mathcal{S})$. As a consequence, both $D = (0,1] \cap \mathbb{Q}$ and D^c are $\sigma(\mathcal{S})$-sets. Consider set functions μ_1 and μ_2 on $\sigma(\mathcal{S})$ defined as $\mu_1(A) = |A \cap D|$ and $\mu_2(A) = |A \cap D^c|$, for $A \in \sigma(\mathcal{S})$. Clearly, μ_1 and μ_2 are both measures on $\sigma(\mathcal{S})$ with $\mu_1 \neq \mu_2$ on $\sigma(\mathcal{S})$. However, one easily sees that $\mu_1 = \mu_2$ on $\mathcal{S}$. Indeed, $\mu_1(S) = \mu_2(S) = \infty$, for every non-empty $S \in \mathcal{S}$. Thus we have two different extensions of the same measure on $\mathcal{S}$. Of course, the measure on $\mathcal{S}$ here is not σ-finite. It may be noted that, on $\sigma(\mathcal{S})$, the measure μ_1 is σ-finite, while μ_2 is not. To continue with this example, let us consider another measure μ_3 on $\sigma(\mathcal{S})$, defined as $\mu_3(A) = 2\mu_1(A)$. Once again, it can be easily seen that $\mu_3 = \mu_1$ on $\mathcal{S}$, while $\mu_1(\{r\}) \neq \mu_3(\{r\})$, for any $r \in (0,1] \cap \mathbb{Q}$. Note also that μ_3, just like μ_1, is σ-finite on $\sigma(\mathcal{S})$. Thus, μ_1 and μ_3 are two distinct σ-finite measures on $\sigma(\mathcal{S})$ that extend the same non-σ-finite measure on $\mathcal{S}$. This shows that the condition of σ-finiteness on $\mathcal{S}$ (equivalently, on $\mathfrak{F}$) in Caratheodory Extension Theorem is essential for the uniqueness of extension.

Getting back to our proof of Caratheodory Extension Theorem, one would recall that once we got the extension, its uniqueness was almost a trivial consequence of the manner in which the extension was obtained, namely, the formula for μ^* in terms of μ. However, we are now going to see a different way of establishing this uniqueness. This will involve first establishing and then using what happens to be an extremely powerful tool in much of measure theory.

2.3 Monotone Class Theorem

One of the important consequences of what we did in the last section is that a measure μ on a σ-field is uniquely determined by its restriction to any semifield generating the σ-field, provided, of course, μ is σ-finite on the semifield. The argument there was based on viewing μ on the σ-field as an extension of μ on the semifield generating it and then appealing to uniqueness of extension. However, since the measure here is already given on the σ-field, extension need not come to the fore and, therefore, it would seem more natural to try and argue this without bringing in the idea of extension. To make the point clearer, let us paraphrase the statement in question slightly differently. Let $\mathcal{A}$ be a σ-field and $\mathcal{S}$ a semifield generating $\mathcal{A}$. The statement in question is that, if μ is a measure on $\mathcal{A}$, which is σ-finite on $\mathcal{S}$, then any measure ν on $\mathcal{A}$, which agrees with μ on $\mathcal{S}$, *must* agree with μ on $\mathcal{A}$ also. This, in itself, happens to be a very important and useful fact in many contexts. And, we already know that the statement is correct. However, we got it as a consequence of the the whole theory of extension. It will be nice to be able to view this as a separate statement, in isolation from the extension machinery, and be able to get a direct proof.

We are going to see that this is indeed possible. But this will only be an upshot, and not the main objective of what we are now going to do. We are going to introduce and discuss an idea that will become one of the very important tools in the mathematics of measure theory.

Let us start by describing a very typical kind of problem that one has to often encounter in measure theory. We are given a σ-field $\mathcal{A}$ on a non-empty set Ω and we are required to show that every set in $\mathcal{A}$ has a certain property, say, $\mathfrak{p}$. This is, of course, only a generic description of the problem. Both the σ-field and the property $\mathfrak{p}$ will vary from context to context. Note that, except in some trivial situations, a direct verification of the assertion is impossible. The main difficulty is that, in most situations, one does not have any explicit description for all the sets that belong to the σ-field $\mathcal{A}$.

Confronted with this seemingly unsurmountable difficulty, the mathematics of measure theory comes up with interesting ways to tackle this problem. Here is the first idea. Suppose one is able to identify some special class $\mathcal{C}$ of subsets of Ω which generates the σ-field $\mathcal{A}$ and for which, property $\mathfrak{p}$ automatically holds or at least can be easily and directly verified to hold. One may then consider the class $\mathcal{G}$ of sets defined as $\mathcal{G} = \{A \in \mathcal{A} : A \text{ has property } \mathfrak{p}\}$. By what has been said above, we know $\mathcal{C} \subset \mathcal{G}$. If one is now able to show that $\mathcal{G}$ is a σ-field, then it will follow that $\mathcal{G} \supset \sigma(\mathcal{C}) = \mathcal{A}$, which will imply that $\mathcal{G} = \mathcal{A}$ and thus the required assertion would be proved.

The class $\mathcal{G}$ described above is often referred to as the class of 'good' sets and the technique is usually called the good sets technique or more commonly 'good sets principle'. This is an extremely important and frequently used technique in measure theory. Here is an example that illustrates an application of good sets principle.

EXAMPLE 2.3.1. On $\Omega = \mathbb{R}$, let $\mathcal{A}$ be the σ-field generated by the class of all bounded open intervals (a, b), $-\infty < a < b < \infty$. We want to show that, for every set $A \in \mathcal{A}$, the set $-A = \{-x : x \in A\}$ also belongs to $\mathcal{A}$. Of course, it is impossible to prove this assertion directly for each $A \in \mathcal{A}$, since we have no explicit description of all sets $A \in \mathcal{A}$. We apply the good sets principle. For $\mathcal{C}$, we take the given generating class of $\mathcal{A}$, that is, $\mathcal{C} = \{(a, b) : -\infty < a < b < \infty\}$. Since $-(a, b) = (-b, -a)$, sets in $\mathcal{C}$ clearly satisfy the required property. The class $\mathcal{G}$ here is $\mathcal{G} = \{A \in \mathcal{A} : -A \in \mathcal{A}\}$ and it is fairly easy to prove that $\mathcal{G}$ is a σ-field. Thus the assertion is proved.

Many more applications of good sets principle will appear throughout the book, including some in the exercises. It must be pointed out that in Example 2.3.1, a generating class $\mathcal{C}$ was already given and we could use that generating class as our $\mathcal{C}$ to apply the good sets technique. But, in many actual applications, we may only have the σ-field $\mathcal{A}$ and the property p to start with and it will be our responsibility to identify an appropriate generating class for $\mathcal{A}$, which can be used as $\mathcal{C}$ while applying this technique. Naturally, we need to choose $\mathcal{C}$ so that it is easy to verify property p for all $\mathcal{C}$-sets. Nevertheless, good sets principle turns out to be an extremely powerful tool in measure theory.

However, just like any other technique, good sets principle also has its limitations. To understand where problems may crop up, let us examine the two key steps of the technique. The first step is identifying some class $\mathcal{C}$ of subsets for which the property p in question is easily seen to hold. The only constraint on $\mathcal{C}$ is that it has to be a generating class for the σ-field $\mathcal{A}$. The next step is to consider the class $\mathcal{G}$ of 'good sets', as decsribed above, and show that it is a σ-field. It is this last step that may turn out to be difficult (even impossible) at times. The next example illustrates this.

EXAMPLE 2.3.2. Let μ and ν be probability measures on a σ-field $\mathcal{A}$. Suppose $\mathcal{S}$ is a semifield such that $\mathcal{A} = \sigma(\mathcal{S})$. We want to show that if $\mu = \nu$ on $\mathcal{S}$, then $\mu = \nu$ also on $\mathcal{A}$. If we were to apply good sets principle, then the semifield $\mathcal{S}$ would be a natural choice for the class $\mathcal{C}$, because $\mathcal{A} = \sigma(\mathcal{S})$ and also the property p : $\mu(A) = \nu(A)$ is given to hold for $A \in \mathcal{S}$. However, the reader can try and see that showing $\mathcal{G} = \{A \in \mathcal{A} : \mu(A) = \nu(A)\}$ to be a σ-field runs into difficulties. Indeed, showing $\mathcal{G}$ to be closed under countable (or even finite) unions is where we run into a dead end!

Of course, the assertion made in Example 2.3.2 is correct. We know it from the uniqueness part of Caratheodory Extension Theorem. But, in case one wants to prove this without bringing in the extension machinery, which is only very natural, good sets principle, as we saw, fails to work. So, do we have an alternative? Luckily, measure theory comes up with another powerful technique to handle situations where good sets principle may fail. We procced to describe it now. We start by defining a new and important type of classes of subsets of a set.

2.3. Monotone Class Theorem

DEFINITION **2.3.3.** A non-empty class $\mathcal{M}$ of subsets of a non empty set Ω is called a *monotone class* on Ω, if it is closed under monotone limits, that is,

$A_n \in \mathcal{M}$, for $n \geq 1$, and $A_n \uparrow A$ or $A_n \downarrow A$ imply that $A \in \mathcal{M}$.

A few important and useful facts follow as simple and straightforward consequences of the definition and hence are given here as exercises.

EXERCISE **2.3.4.** *(a) Show that a σ-field is a monotone class. Show that the converse is not true by observing that the class of all intervals is a monotone class on $\mathbb{R}$.*
(b) Show that a field, which is also a monotone class, is a σ-field.

EXERCISE **2.3.5.** *(a) Show that the intersection of any non-empty collection $\mathbf{M}$ of monotone classes on a non- empty set is itself a monotone class.*
(b) Hence deduce that given any non-empty class $\mathcal{C}$ of subsets of a non-empty set Ω, there is a smallest monotone class on Ω, containing $\mathcal{C}$, to be called the monotone class generated by $\mathcal{C}$ and to be denoted $\mathcal{M}(\mathcal{C})$.

We are now going to see one of the most useful results in measure theory. In handling problems of the kind described at the start of the section, this result provides a very powerful alternative in situations (like in Example 2.3.2), when good sets principle fails.

THEOREM **2.3.6.** (Monotone Class Theorem) *If $\mathcal{M}$ is a monotone class that contains a field $\mathfrak{F}$ on a non-empty set Ω, then $\mathcal{M}$ contains $\sigma(\mathfrak{F})$ also.*

Proof. Let $\mathcal{M}_0$ denote the smallest monotone class containing $\mathfrak{F}$ (which exists by Exercise 2.3.5(b)). The theorem will be proved, if we can show that $\mathcal{M}_0 \supset \sigma(\mathfrak{F})$. For this, it is enough to show that $\mathcal{M}_0$ is a σ-field, because this will make $\mathcal{M}_0$ a σ-field containing $\mathfrak{F}$. But, in turn, this will be proved if we only show that $\mathcal{M}_0$ is a field (see Exercise 2.3.4(b)).

We first show that $\mathcal{M}_0$ is closed under complementation. For this, we consider the class

$$\mathcal{M}_1 = \{A \in \mathcal{M}_0 : A^c \in \mathcal{M}_0\}$$

and show that $\mathcal{M}_1$ is a monotone class containing $\mathfrak{F}$. Since $\mathcal{M}_0$ is the smallest monotone class containing $\mathfrak{F}$, this would imply that $\mathcal{M}_1 \supset \mathcal{M}_0$, proving that $\mathcal{M}_0$ is closed under complementation. That $\mathfrak{F} \subset \mathcal{M}_1$ is clear because $\mathfrak{F} \subset \mathcal{M}_0$ and $\mathfrak{F}$ is a field and hence closed under complementation. Next, to prove that $\mathcal{M}_1$ is a monotone class, we take a sequence $A_n \in \mathcal{M}$, $n \geq 1$ with $A_n \uparrow A$ or $A_n \downarrow A$ and show that $A \in \mathcal{M}_1$. Now, $A_n \in \mathcal{M}_1$ implies $A_n \in \mathcal{M}_0$ as well as $A_n^c \in \mathcal{M}_0$, for each n. Further, $A_n \uparrow A$ (resp., $A_n \downarrow A$) implies that $A_n^c \downarrow A^c$ (resp., $A_n^c \uparrow A^c$). As $\mathcal{M}_0$ is a monotone class, it follows that both A and A^c belong to $\mathcal{M}_0$, implying that $A \in \mathcal{M}_1$, as was to be proved.

We next show that $\mathcal{M}_0$ is closed under finite intersections. As a first step, we show that if $A \in \mathfrak{F}$, then $A \cap B \in \mathcal{M}_0$, for all $B \in \mathcal{M}_0$. For this, we fix $A \in \mathfrak{F}$ and consider

$$\mathcal{M}_2 = \{B \in \mathcal{M}_0 : A \cap B \in \mathcal{M}_0\}.$$

Since $\mathfrak{F}$ is a field and $\mathfrak{F} \subset \mathcal{M}_0$, one easily gets $\mathfrak{F} \subset \mathcal{M}_2$. Next, using the facts that $B_n \uparrow B$ (resp., $B_n \downarrow B$) implies $A \cap B_n \uparrow A \cap B$ (resp., $A \cap B_n \downarrow A \cap B$) and $\mathcal{M}_0$ is a monotone class, one can easily see that $\mathcal{M}_2$ is a monotone class. It follows then that $\mathcal{M}_2 \supset \mathcal{M}_0$, proving that $A \cap B \in \mathcal{M}_0$ for all $B \in \mathcal{M}_0$. Since $A \in \mathfrak{F}$ was arbitrary, we have thus proved that $A \cap B \in \mathcal{M}_0$, for all $A \in \mathfrak{F}$ and all $B \in \mathcal{M}_0$. In our final step, we now fix $B \in \mathcal{M}_0$ and consider

$$\mathcal{M}_3 = \{A \in \mathcal{M}_0 : A \cap B \in \mathcal{M}_0\}.$$

From what has just been shown, we have $\mathfrak{F} \subset \mathcal{M}_3$. Same arguments as used above, can once again be used to show that $\mathcal{M}_3$ is a monotone class. This, as before, will imply that $\mathcal{M}_3 \supset \mathcal{M}_0$, proving finally that $A \cap B \in \mathcal{M}_0$, for all $A, B \in \mathcal{M}_0$, that is, $\mathcal{M}_0$ is closed under finite intersections. Thus, we have shown that $\mathcal{M}_0$ is a field and, as noted earlier, this completes the proof. □

Let us now see how Monotone Class Theorem helps us in handling problems where we are required to show that every set in a given σ-field $\mathcal{A}$ satisfies a certain property $\mathfrak{p}$. Suppose we are able to get hold of a field $\mathfrak{F}$ generating $\mathcal{A}$, such that property $\mathfrak{p}$ is known to hold (or can be easily shown to hold) for all sets in $\mathfrak{F}$. What the theorem says is that, in that case, it is enough to show that the class of sets satisfying property $\mathfrak{p}$ is a monotone class in order to conclude that property $\mathfrak{p}$ holds for every set in $\mathcal{A}$. The contrast with the 'good sets principle' is that here, in the first step, we have to have a field $\mathfrak{F}$ generating $\mathcal{A}$, and not just *any generating class* $\mathcal{C}$, whose sets satisfy property $\mathfrak{p}$. But at the next step, we only need to show now that the class of sets satisfying property $\mathfrak{p}$, is closed under monotone limits, rather than showing that it is a σ-field. Depending on the context, one or the other of the two techniques would be more appropriate.

Example 2.3.2 above was an illustration of a situation where using good sets principle did not prove to be of help. We are now going to show how monotone class theorem comes to our rescue and proves the assertion. In fact, we prove a stronger assertion where the measures are taken to be σ-finite on the semifield, and not just probability measures. Since this is going to be an important result, we state it as a theorem. The proof uses monotone class theorem and shows that uniqueness can be directly established without bringing in the extension theory.

THEOREM **2.3.7.** (Uniqueness Theorem) Let $\mathcal{S}$ be a semifield on a non-empty set Ω and let μ be a measure on $\sigma(\mathcal{S})$ which is σ-finite on $\mathcal{S}$. If ν is any measure on $\sigma(\mathcal{S})$ which agrees with μ on $\mathcal{S}$, then $\nu \equiv \mu$ on $\sigma(\mathcal{S})$.

Proof. Denoting $\mathfrak{F}$ to be the field generated by $\mathcal{S}$, we know that $\mathfrak{F}$ consists precisely of finite disjoint unions of $\mathcal{S}$-sets (see Proposition 2.1.10). Since μ and ν are given to agree on $\mathcal{S}$, it easily follows that $\nu \equiv \mu$ on $\mathfrak{F}$ as well. The problem, therefore, reduces to showing that, if μ is σ-finite on $\mathfrak{F}$, then $\nu \equiv \mu$ on $\mathfrak{F}$ implies that $\nu \equiv \mu$ on $\sigma(\mathfrak{F})$. (Recall that $\sigma(\mathcal{S}) = \sigma(\mathfrak{F})$, by Exercise 2.2.15(b).)

We first prove the result assuming that μ is a finite measure. In that case, ν is also a finite measure. In fact, $\nu(\Omega) = \mu(\Omega)$, since $\Omega \in \mathfrak{F}$. Using the fact

2.3. Monotone Class Theorem

that finite measures are continuous from both above and below (see Proposition 2.1.26), the class
$$\mathcal{M} = \{A \in \sigma(\mathfrak{F}) : \mu(A) = \nu(A)\}$$
can be easily shown to be a monotone class. But the hypothesis $\mu \equiv \nu$ on $\mathfrak{F}$ implies that $\mathfrak{F} \subset \mathcal{M}$. Monotone Class Theorem (Theorem 2.3.6) then asserts that $\mathcal{M} \supset \sigma(\mathfrak{F})$, which proves the result in case μ is a finite measure.

If μ is not finite, we use σ-finiteness of μ on $\mathcal{S}$ (and hence on $\mathfrak{F}$) to get disjoint sets $\Omega_n \in \mathfrak{F}$, $n \geq 1$, with $\Omega = \bigcup_n \Omega_n$, such that, $\mu(\Omega_n) < \infty$ for each n. We then have, for each n, $\nu(\Omega_n) = \mu(\Omega_n) < \infty$, and so $\mu_n(A) = \mu(A \cap \Omega_n)$ and $\nu_n(A) = \nu(A \cap \Omega_n)$, for $A \in \sigma(\mathcal{S})$, define finite measures on $\sigma(\mathcal{S})$, for each n. Further, the hypothesis $\nu \equiv \mu$ on $\mathfrak{F}$ implies that $\nu_n \equiv \mu_n$ on $\mathfrak{F}$, for each n. From what was proved above, we get $\mu_n \equiv \nu_n$ on $\sigma(\mathcal{S})$, for each n. Since $\mu = \sum_n \mu_n$ and $\nu = \sum_n \nu_n$ on $\sigma(\mathcal{S})$, the required result follows. □

A close scrutiny of the above proof reveals two things. Firstly, proving the result in case of finite measures is the main step, because then there is a routine extension to the σ-finite case. Next, for the finite case, continuity of finite measures from below and above guaranteed that the class $\mathcal{M}$ is a monotone class and so monotone class theorem became just the right tool. As we already saw earlier, good sets principle was not useful for this problem, because that would have required showing that the class $\mathcal{M}$ in the above proof is a σ-field. The real difficulty with that was in showing that the class $\mathcal{M}$ is closed under countable unions (or even finite unions). There would have been no difficulty at all if we were required to work with only countable disjoint unions. This leads us to describing another third alternative, which specifically addresses this and therefore provides one more useful tool for tackling problems of this kind. This is known as 'Dynkin's π-λ Theorem'. We need some definitions to start with.

DEFINITION 2.3.8. (a) A non-empty class $\mathcal{P}$ of subsets of a non-empty set Ω is called a π-*system* if it is closed under finite intersections.
(b) A class $\mathcal{L}$ of subsets of a non-empty set Ω is called a λ-*system* if $\emptyset \in \mathcal{L}$ and if $\mathcal{L}$ is closed under complementations and countable disjoint unions.

LEMMA 2.3.9. (a) Any λ-system $\mathcal{L}$ is closed under proper differences, that is, if A and B are sets in $\mathcal{L}$ such that $A \supset B$, then $A \setminus B \in \mathcal{L}$.
(b) If $\mathcal{L}$ is a λ-system and also a π-system, then $\mathcal{L}$ is a σ-field.

Proof. (a) $A, B \in \mathcal{L}$ and $A \supset B$ imply that A^c and B are disjoint $\mathcal{L}$-sets, so that $A^c \cup B \in \mathcal{L}$. Consequently, $A \setminus B = (A^c \cup B)^c \in \mathcal{L}$.
(b) We only have to show that $\mathcal{L}$ is closed under countable unions. First, observe that for any two sets $A, B \in \mathcal{L}$ we have $A \cap B \in \mathcal{L}$, since $\mathcal{L}$ is a π-system. But since $\mathcal{L}$ is also a λ-system and $A \supset A \cap B$, we will get, by part (a), that $A \cap B^c = A \setminus (A \cap B) \in \mathcal{L}$. It follows then that $A \cup B = (A \cap B^c) \cup B \in \mathcal{L}$. Thus, we have shown that $\mathcal{L}$ is closed under finite unions. Next, if $\{A_n\}$ is any sequence of $\mathcal{L}$-sets, then, by what has been proved, we have $A_1 \cup \cdots \cup A_{n-1} \in \mathcal{L}$, for every $n \geq 2$. By

the same argument as used above, we can get $B_n = A_n \cap (A_1 \cup \cdots \cup A_{n-1})^c \in \mathcal{L}$, for each $n \geq 1$. Taking $B_1 = A_1$, we get a sequence $\{B_n\}$ of disjoint sets in $\mathcal{L}$ and therefore $\bigcup_n B_n \in \mathcal{L}$. But $\bigcup_n A_n = \bigcup_n B_n$, thus showing that $\bigcup_n A_n \in \mathcal{L}$, and completing the proof. □

THEOREM **2.3.10.** (Dynkin's π-λ Theorem) If $\mathcal{L}$ is a λ-system containing a π-system $\mathcal{P}$, then $\mathcal{L}$ contains $\sigma(\mathcal{P})$.

Proof. Let $\mathcal{L}(\mathcal{P})$ denote the intersection of all λ-systems that contain $\mathcal{P}$. It is easy to check that $\mathcal{L}(\mathcal{P})$ is a λ-system and is the smallest λ-system containing $\mathcal{P}$. We show that $\mathcal{L}(\mathcal{P})$ is a π-system. By Lemma 2.3.9(b), this will imply that $\mathcal{L}(\mathcal{P})$ is a σ-field and hence $\mathcal{L}(\mathcal{P}) \supset \sigma(\mathcal{P})$. Since $\mathcal{L} \supset \mathcal{L}(\mathcal{P})$, this will complete the proof.

First, fix $A \in \mathcal{P}$ and let $\mathcal{L}_A = \{B \in \mathcal{L}(\mathcal{P}) : A \cap B \in \mathcal{L}(\mathcal{P})\}$. $\mathcal{L}_A$ clearly contains $\emptyset$ and is closed under countable disjoint unions. Next, if $B \in \mathcal{L}(\mathcal{P})$ is such that $A \cap B \in \mathcal{L}(\mathcal{P})$, then $B^c \in \mathcal{L}(\mathcal{P})$ and $A \cap B^c = A \setminus (A \cap B) \in \mathcal{L}(\mathcal{P})$. This shows that $\mathcal{L}_A$ is closed under complementation and is therefore a λ-system. Since $\mathcal{P} \subset \mathcal{L}_A$ clearly, we conclude $\mathcal{L}(\mathcal{P}) \subset \mathcal{L}_A$. This proves that $A \cap B \in \mathcal{L}(\mathcal{P})$, for any $A \in \mathcal{P}$ and any $B \in \mathcal{L}(\mathcal{P})$.

Finally, fix $B \in \mathcal{L}(\mathcal{P})$ and denote $\mathcal{L}_B = \{A \in \mathcal{L}(\mathcal{P}) : A \cap B \in \mathcal{L}(\mathcal{P})\}$. By what has been just proved, $\mathcal{P} \subset \mathcal{L}_B$. By similar arguments as above, one can now show that $\mathcal{L}_B$ is a λ-system and conclude that $\mathcal{L}(\mathcal{P}) \subset \mathcal{L}_B$. This shows that $A \cap B \in \mathcal{L}(\mathcal{P})$, for any $A, B \in \mathcal{L}(\mathcal{P})$, proving that $\mathcal{L}(\mathcal{P})$ is a π-system, as required. □

We will see several applications of Dynkin's π-λ Theorem in the sequel. Meanwhile, Here is an exercise that illustrates that Dynkin's π-λ Theorem can be used as an alternative to Monotone Class Theorem, in proving the Uniqueness Theorem (Theorem 2.3.7).

EXERCISE **2.3.11.** *(a) Use Dynkin's π-λ Theorem to prove that if two probability measures on a σ-field $\mathcal{A}$ agree on a π-system $\mathcal{P}$ that generates $\mathcal{A}$, then they must agree on $\mathcal{A}$.*
(b) Give an alternative proof of the Uniqueness Theorem (Theorem 2.3.7), using Dynkin's π-λ Theorem.

2.4 Completeness and Completion

In this section, we touch upon one issue that remains unanswered in what was done in Section 2.2. We started with a probability measure μ on a field $\mathfrak{F}$ and were able to extend it to a σ-field $\mathcal{A}$ containing $\mathfrak{F}$. But very soon, we decided to restrict μ to the smaller σ-field $\sigma(\mathfrak{F})$ and it was extension of μ to only $\sigma(\mathfrak{F})$ that was highlighted in Caratheodory Extension Theorem. The reason was, of course, that the σ-field $\mathcal{A}$ depends on μ, while it is desirable to identify one σ-field containing $\mathfrak{F}$, to which any given probability on $\mathfrak{F}$ can be extended. However,

2.4. Completeness and Completion

a natural question that arises is, whether this results in an irretrievable loss of information. To put it differently, once we restrict μ to $\sigma(\mathfrak{F})$, would it be possible for us now to retrieve the original extension of μ to $\mathcal{A}$? We are going to show now that this is indeed possible and therefore, there is no serious loss in restricting μ from $\mathcal{A}$ to $\sigma(\mathfrak{F})$.

DEFINITION 2.4.1. Let μ be a measure on a σ-field $\mathcal{A}$. Then $\mathcal{A}$ is said to be μ-*complete* if, $A \in \mathcal{A}$ and $\mu(A) = 0$ imply that $B \in \mathcal{A}$ for every $B \subset A$.

Let us introduce a terminology to be widely used in the sequel. For a σ-field $\mathcal{A}$ with a measure μ on it, sets $A \in \mathcal{A}$ with $\mu(A) = 0$ are referred to as μ-*null sets* (or simply *null sets*, when μ is clear from the context). Using this terminolgy, a μ-complete σ-field is precisely one in which every subset of any μ-null set belongs to the σ-field (and is, of course, a μ-null set itself).

If a σ-field is not μ-complete, then it is possible to enlarge it to a bigger σ-field and have a natural extension of μ, so that the larger σ-field, with the extended μ, becomes μ-complete. The step-by-step construction of this enlargement is given in the first three parts of Exercise 2.4.2 below.

EXERCISE 2.4.2. Let μ be a measure on a σ-field $\mathcal{A}$ on some non-empty set Ω. Denote
$$\mathcal{N} = \{B \subset \Omega : B \subset A \text{ for some } A \in \mathcal{A} \text{ with } \mu(A) = 0\}.$$
(a) Show that $\overline{\mathcal{A}} = \{A \cup N : A \in \mathcal{A}, N \in \mathcal{N}\}$ is a σ-field that contains $\mathcal{A}$..
(b) Show that if $A_1, A_2 \in \mathcal{A}$ and $N_1, N_2 \in \mathcal{N}$ are such that $A_1 \cup N_1 = A_2 \cup N_2$, then $\mu(A_1) = \mu(A_2)$.
(c) If $B = A \cup N$ for some $A \in \mathcal{A}$ and $N \in \mathcal{N}$, let us set $\overline{\mu}(B) = \mu(A)$. Show that $\overline{\mu}$ is well defined and defines a measure on $\overline{\mathcal{A}}$. Show also that $\overline{\mathcal{A}}$ is $\overline{\mu}$-complete.
(d) Suppose that $\mathcal{A}'$ is a σ-field containing $\mathcal{A}$ and μ' is a measure on $\mathcal{A}'$ which agrees with μ on $\mathcal{A}$. Show that, if $\mathcal{A}'$ is μ'-complete, then $\mathcal{A}' \supset \overline{\mathcal{A}}$ and μ' agrees with $\overline{\mu}$ on $\overline{\mathcal{A}}$.

The last part of Exercise 2.4.2 shows that the σ-field $\overline{\mathcal{A}}$ along with the measure $\overline{\mu}$, constructed in the first three parts, is the 'smallest enlargement' of the σ-field $\mathcal{A}$ and of the measure μ on it, that achieves completeness. The next definition makes this idea precise.

DEFINITION 2.4.3. Given a measure μ on a σ-field $\mathcal{A}$, the μ-*completion* of $\mathcal{A}$ is the smallest σ-field $\mathcal{A}_\mu$ containing $\mathcal{A}$, to which μ can be extended as a measure, such that, $\mathcal{A}_\mu$ is μ-complete.

As noted above, the existence of a unique μ-completion for any measure μ on a σ-field $\mathcal{A}$ is what Exercise 2.4.2 established. Indeed, the σ-field $\overline{\mathcal{A}}$ with the extension $\overline{\mu}$, given in that exercise is the μ-completion of $\mathcal{A}$. An ofen useful alternative description of the same μ-completion is given in the next exercise.

EXERCISE 2.4.4. Show that the $\overline{\mathcal{A}}$ described in the Exercise 2.4.2 is the same as the collection $\{A \Delta N : A \in \mathcal{A}, N \in \mathcal{N}\}$. Show also that if $A_1, A_2 \in \mathcal{A}$ and

$N_1, N_2 \in \mathcal{N}$ are such that $A_1 \Delta N_1 = A_2 \Delta N_2$, then $\mu(A_1) = \mu(A_2)$ and that this can be used to give an alternative definition of $\overline{\mu}$ on $\overline{\mathcal{A}}$.

Going back to Caratheodory Extension, we claim that the σ-field $\mathcal{A}$ to which the given probability measure μ on the field $\mathfrak{F}$ was originally extended, is nothing but the μ-completion of the σ-field $\sigma(\mathfrak{F})$.

To see this, take any subset $B \subset \Omega$. By the way μ^* was defined, one can easily see that there exist sets $C_n \in \mathfrak{F}_\sigma$, $n \geq 1$, with $C_n \supset B$, for each n, such that, $\mu(C_n) \downarrow \mu^*(B)$. Since $\mathfrak{F}_\sigma$ is closed under finite intersections (see Proposition 2.2.2(b)), we can and shall assume that $\{C_n\}$ is a monotonically decreasing sequence of of $\mathfrak{F}_\sigma$-sets. Denoting $C = \bigcap C_n$, continuity from above of the probability measure μ on $\sigma(\mathfrak{F})$, implies that $\mu(C_n) \downarrow \mu(C)$. Thus we have a set $C \in \sigma(\mathfrak{F})$, with $C \supset B$, such that $\mu^*(B) = \mu(C)$. Similarly, we can get a $\sigma(\mathfrak{F})$-set A with $A^c \supset B^c$ and $\mu^*(B^c) = \mu(A^c)$. Thus, for any $B \subset \Omega$, we can get sets C and A in $\sigma(\mathfrak{F})$, with $A \subset B \subset C$, such that $\mu^*(B) = \mu(C)$ and $\mu^*(B^c) = \mu(A^c)$. Now if $B \in \mathcal{A}$, then $\mu^*(B) = 1 - \mu^*(B^c)$, which implies $\mu(C) = 1 - \mu(A^c) = \mu(A)$. Since $A \subset C$ are both $\sigma(\mathfrak{F})$-sets, the set $C \setminus A$ is a μ-null set in $\sigma(\mathfrak{F})$. With $N = B \setminus A$, we have $B = A \cup N$, where $N \subset C \setminus A$ and so $N \in \mathcal{N}$. This shows that $\mathcal{A}$ is contained in the completion of $\sigma(\mathfrak{F})$.

For the converse part, it suffices to show that if M is any μ-null set in $\sigma(\mathfrak{F})$, then for any $N \subset M$, we have $N \in \mathcal{A}$, that is, $\mu^*(N) + \mu^*(N^c) \leq 1$ (see (2.2.9)). Clearly, $N \subset M$ implies $\mu^*(N) \leq \mu^*(M) = \mu(M) = 0$, and therefore, $\mu^*(N) + \mu^*(N^c) \leq \mu^*(N^c) \leq 1$, since $\mu^*(N^c) \leq \mu(\Omega) = 1$. This completes the proof of what we claimed and it is stated in the next theorem.

THEOREM **2.4.5.** Let μ be a probability measure on a field $\mathfrak{F}$ and $\mathcal{A}$ be the σ-field, as in Theorem 2.2.9, with μ extended to $\mathcal{A}$. Then $\mathcal{A}$ is the μ-completion of $\sigma(\mathfrak{F})$.

What this shows is that, if we know the probability measure μ on $\sigma(\mathfrak{F})$, we can always retrieve the σ-field $\mathcal{A}$ and also the probability μ on $\mathcal{A}$. In other words, there is no loss of information in restricting the probability μ to $\sigma(\mathfrak{F})$.

2.5 Approximation

In Section 2.2, we raised the question of why only sets from $\mathfrak{F}_\sigma$ were used to define 'outer' approximations for measures of arbitrary sets. Would we not get better appproximations if sets from larger classes like $\mathfrak{F}_{\sigma\delta}$ or $\mathfrak{F}_{\sigma\delta\sigma}$ etc., and the natural extension of the measure to these classes, were used? Similar questions were raised about using only sets from $\mathfrak{F}_\delta$ for 'inner' approximations. The results we are going to prove in this section will conclusively show that nothing would have been gained by going beyond $\mathfrak{F}_\sigma$-sets for outer approximations or beyond $\mathfrak{F}_\delta$-sets for inner approximations.

We deal with the case of probability measures first and then handle the case of σ-finite measures later. In both the theorems that follow, $\mathcal{A}$ will denote a σ-field on a non-empty set Ω and $\mathfrak{F}$ will denote any field that generates $\mathcal{A}$.

2.5. Approximation

THEOREM 2.5.1. Let μ be a probability measure on a σ-field $\mathcal{A}$. If $\mathfrak{F}$ is any field generating $\mathcal{A}$, then, for any $A \in \mathcal{A}$ and any $\epsilon > 0$,
(a) there exist sets $B \in \mathfrak{F}_\sigma$, $C \in \mathfrak{F}_\delta$, with $C \subset A \subset B$, such that, $\mu(B \setminus C) < \epsilon$, and
(b) there exists $F \in \mathfrak{F}$, such that, $\mu(A \Delta F) < \epsilon$.

Proof. Here is a one line proof of part (a). From what was argued earlier (see the paragraph following (2.2.16)), we have $\mu^*(A) = \mu_*(A) = \mu(A)$, for any $A \in \mathcal{A}$. The result now follows immediately from the definitions of μ^* and μ_*, given in (2.2.5) and (2.2.16).

However, we want to give a proof that avoids the use of μ_* and μ^* and is, therefore, more direct, though somewhat longer. This is a tailor-made situation for using the 'good sets principle'. Consider the class $\mathcal{G}$ of 'good sets', namely, sets in $\mathcal{A}$ that satisfy the stated property; in other words,

$$\mathcal{G} = \{A \in \mathcal{A} : \forall \epsilon > 0, \exists B \in \mathfrak{F}_\sigma, C \in \mathfrak{F}_\delta, \text{ with } C \subset A \subset B, \text{ such that}, \mu(B \setminus C) < \epsilon\}.$$

$\mathfrak{F} \subset \mathcal{G}$ clearly, so we have only to show that $\mathcal{G}$ is a σ-field. Since $\mathfrak{F} \subset \mathcal{G}$, we, of course, have $\Omega \in \mathcal{G}$. Next, using the fact that a set belongs to $\mathfrak{F}_\delta$ if and only if its complement belongs to $\mathfrak{F}_\sigma$, it can be easily seen that $\mathcal{G}$ is closed under complementation. Finally, to show that $\mathcal{G}$ is closed under countable unions, let $\{A_n\}$ be any sequence of sets in $\mathcal{G}$ and let $A = \bigcup_n A_n$. Since $A_n \in \mathcal{A}$ for each n, we have $A \in \mathcal{A}$. Now, let $\epsilon > 0$ be given. Since $A_n \in \mathcal{G}$ for each n, we can get, for each n, sets $B_n \in \mathfrak{F}_\sigma$, $C_n \in \mathfrak{F}_\delta$, with $C_n \subset A_n \subset B_n$, such that, $\mu(B_n \setminus C_n) < \epsilon/2^{n+1}$. Note that $\bigcup_{n \leq k} A_n \uparrow A$ as $k \to \infty$ and so we can get k such that $\mu(\bigcup_{n > k} A_n) = \mu(A) - \mu(\bigcup_{n \leq k} A_n) < \epsilon/2$. Let us take $B = \bigcup_n B_n$ and $C = \bigcup_{n \leq k} C_n$. Since $\mathfrak{F}_\sigma$ is closed under countable unions, we have $B \in \mathfrak{F}_\sigma$. On the other hand, $C_n^c \in \mathfrak{F}_\sigma$ for each n, and therefore $C^c \in \mathfrak{F}_\sigma$ (see Proposition 2.2.2(b)), which implies $C \in \mathfrak{F}_\delta$. Also, $C \subset A \subset B$ clearly. Finally, since $B \setminus C = (\bigcup_n B_n \setminus \bigcup_n C_n) \cup (\bigcup_{k > n} C_n) \subset (\bigcup_n (B_n \setminus C_n)) \cup (\bigcup_{k > n} A_n)$, we have $\mu(B \setminus C) < \sum_n 2^{-(n+1)} \epsilon + (\epsilon/2) = \epsilon$, showing that $A \in \mathcal{G}$. Thus, we have shown that $\mathcal{G}$ is a σ-field and that completes the proof of part (a).

For the proof of part (b), we once again use the 'good sets principle'. The class of good sets now is given by

$$\mathcal{G} = \{A \in \mathcal{A} : \forall \epsilon > 0, \exists F \in \mathfrak{F}, \text{ such that}, \mu(A \Delta F) < \epsilon\}.$$

That $\mathfrak{F} \subset \mathcal{G}$ is obvious and from this, it follows, in particular, that $\Omega \in \mathcal{G}$. Since $A \Delta F = A^c \Delta F^c$ for any two sets A, F, it is immediate that $\mathcal{G}$ is closed under complementation. Finally, take any sequence $\{A_n\}$ of sets in $\mathcal{G}$ and let $A = \bigcup_n A_n$. Given $\epsilon > 0$, first get k (as in the proof of part (a)), such that, $\mu(\bigcup_{n > k} A_n) < \epsilon/2$, and then, for each $n = 1, \cdots, k$, get $F_n \in \mathfrak{F}$ such that $\mu(A_n \Delta F_n) < \epsilon/2^{n+1}$. Taking $F = \bigcup_{n \leq k} F_n$, we have $F \in \mathfrak{F}$. It is also easy to see that $A \Delta F \subset \bigcup_{n \leq k}(A_n \Delta F_n) \cup (\bigcup_{n > k} A_n)$, and so, $\mu(A \Delta F) < \sum_{n \leq k}(\epsilon/2^{n+1}) + \epsilon/2 = \epsilon$, showing that $A \in \mathcal{G}$. We have thus proved that $\mathcal{G}$ is a σ-field and that completes the proof of part (b). □

Clearly, the assertions of the above theorem would remain valid for any finite measure μ as well. The theorem is enormously useful in the following way. When you are given a finite measure μ on a σ-field $\mathcal{A}$ and $\mathfrak{F}$ is any field generating $\mathcal{A}$, then what the above results say are the following. Any set $A \in \mathcal{A}$ can be sandwiched between two sets, one in $\mathfrak{F}_\delta$ and the other in $\mathfrak{F}_\sigma$, such that μ-measure of their difference is as small as we please. Similarly, given any set $A \in \mathcal{A}$, we can get a set $F \in \mathfrak{F}$ such that μ-measure of $A \triangle F$ is arbitrarily small. Incidentally, the first part also shows comprehensively, at least for finite measures on a field, that for getting 'outer' (respectively, 'inner') approximations for μ-measure of a set, one would not do any better by going beyond the class $\mathfrak{F}_\sigma$ (respectively, $\mathfrak{F}_\delta$).

We now extend the above results to σ-finite measures on a σ-field $\mathcal{A}$, provided, of course, the measure happens to be σ-finite on the field $\mathfrak{F}$ generating $\mathcal{A}$.

THEOREM **2.5.2.** Let μ be a measure on a σ-field $\mathcal{A}$. Suppose that $\mathfrak{F}$ is a field generating $\mathcal{A}$, such that, μ is σ-finite on $\mathfrak{F}$. Then, for any $A \in \mathcal{A}$ with $\mu(A) < \infty$ and for any $\epsilon > 0$,
(a) there exist sets $B \in \mathfrak{F}_\sigma$, $C \in \mathfrak{F}_\delta$, with $C \subset A \subset B$, such that, $\mu(B \setminus C) < \epsilon$, and
(b) there exists $F \in \mathfrak{F}$ such that $\mu(A \triangle F) < \epsilon$.

Proof. (a) Since μ is σ-finite on $\mathfrak{F}$, we can pick a sequence $\{\Omega_n\}$ of disjoint $\mathfrak{F}$-sets, with $\Omega = \bigcup_n \Omega_n$, such that $\mu(\Omega_n) < \infty$ for each n. If μ_n, for each n, is the finite measure defined on $\mathcal{A}$ by $\mu_n(A) = \mu(A \cap \Omega_n)$, then we have $\mu(A) = \sum_n \mu_n(A)$. Given any $A \in \mathcal{A}$ with $\mu(A) < \infty$ and any $\epsilon > 0$, we first get k, such that, $\sum_{n>k} \mu_n(A) < \epsilon/2$. Now for each $n \leq k$, we use Theorem 2.5.1(a) (and what was noted in the paragraph following its proof) to get $B_n \in \mathfrak{F}_\sigma$, $C_n \in \mathfrak{F}_\delta$, with $C_n \subset A \cap \Omega_n \subset B_n$, such that, $\mu_n(B_n \setminus C_n) < \epsilon/2^{n+1}$. The same theorem applied now, for each $n > k$, will give us $B_n \in \mathfrak{F}_\sigma$, with $A \cap \Omega_n \subset B_n$, such that, $\mu_n(B_n \setminus (A \cap \Omega_n)) < \epsilon/2^{n+1}$. Since $\Omega_n \in \mathfrak{F}$, we can safely assume that $B_n \subset \Omega_n$, for each n (replacing B_n by $B_n \cap \Omega_n$, if necessary). Taking $B = \bigcup_n B_n$ and $C = \bigcup_{n \leq k} C_n$, we have $C \subset A \subset B$ and also $B \in \mathfrak{F}_\sigma$, $C \in \mathfrak{F}_\delta$, as argued in the proof of Theorem 2.5.1(a). It is easy to see that $B \setminus C$ equals the disjoint union $\bigcup_{n \leq k}(B_n \setminus C_n) \bigcup \bigcup_{n > k} B_n$, so that

$$\mu(B \setminus C) = \sum_{n \leq k} \mu(B_n \setminus C_n) + \sum_{n > k} \mu(B_n)$$
$$= \sum_{n \leq k} \mu(B_n \setminus C_n) + \sum_{n > k} \mu(B_n \setminus (A \cap \Omega_n)) + \sum_{n > k} \mu(A \cap \Omega_n).$$

Note now that, $B_n \subset \Omega_n$, for each n, implies that, for each $n \leq k$, we have $\mu(B_n \setminus C_n) = \mu_n(B_n \setminus C_n) < \epsilon/2^{n+1}$, and, on the other hand, for each $n > k$, we have $\mu(B_n \setminus (A \cap \Omega_n)) = \mu_n(B_n \setminus (A \cap \Omega_n)) < \epsilon/2^{n+1}$. All these finally lead to $\mu(B \setminus C) < \sum_n \epsilon/2^{n+1} + \sum_{n > k} \mu_n(A) < \epsilon$, as was to be proved.

(b) One can, of course, employ the same technique as above, using the finite measures μ_n and applying Theorem 2.5.1(b). But, here is a simpler direct argument. Given any $A \in \mathcal{A}$ with $\mu(A) < \infty$ and any $\epsilon > 0$, we can use part (a), to get $B \in \mathfrak{F}_\sigma$, with $A \subset B$, such that, $\mu(B \setminus A) < \epsilon/2$. Since $\mathfrak{F}_\sigma$-sets are increasing limits of $\mathfrak{F}$-sets, using continuity from below and the fact that $\mu(B) < \mu(A) + \epsilon/2 < \infty$, we can get $F \in \mathfrak{F}$, with $F \subset B$, such that $\mu(B \setminus F) < \epsilon/2$. It is now easy to see that $A \Delta F \subset (B \setminus A) \cup (B \setminus F)$, whence it follows that $\mu(A \Delta F) < \epsilon$. □

REMARK 2.5.3. It is very important to note here that the hypothesis in Theorem 2.5.2 of μ being σ-finite on the field $\mathfrak{F}$, is necessary for the assertions of the theorem to hold. Exercise 2.13.16 illustrates that both parts (a) and (b) may fail, if μ is not σ-finite on $\mathfrak{F}$, even if it is σ-finite on $\mathcal{A}$.

Also, as illustrated by Exercise 2.13.17, the hypothesis $\mu(A) < \infty$ is necessary for both the assertions (a) and (b) in Theorem 2.5.2 to hold.

2.6 Extension for Arbitrary Measures

In the extension programme carried out in Sections 2.1 and 2.2, we originally started with an aribitrary measure μ on a semifield $\mathcal{S}$. This had a natural and explicit extension to a measure on the field $\mathfrak{F} = \mathfrak{F}(\mathcal{S})$, since $\mathfrak{F}$ consists precisely of all finite disjoint unions of $\mathcal{S}$-sets. We then extended μ to an even larger class $\mathfrak{F}_\sigma$ and we observed that μ, extended to $\mathfrak{F}_\sigma$, has many nice properties including strong additivity and continuity from below (see Proposition 2.2.2). Our next step was a drastic jump. Instead of pursuing with these incremental, but natural and explicit, extensions from one class to the next natural larger class of sets, we decided to get an 'extension' of μ at one go to all subsets. Using sets in $\mathfrak{F}_\sigma$ and their μ-values, we defined an 'outer approximation' μ^* for each subset. However, we saw that while the set function μ^* defined on all subsets has many nice properties, it lacks additivity property, in fact, lacks even finite additivity. Naturally, the idea then was to try and identify the largest class of subsets on which μ^* is finitely additive and then hope that this will be a large enough class and that μ^* will be a measure, and hence an extension of μ, on this class.

But, it is at this point that we brought in a seemingly restrictive assumption that the given μ is a probability measure (see (2.2.7)). The rest of the work in Section 2.2 was to show first that, for a probability measure μ on $\mathfrak{F}$ (equivalently, on $\mathcal{S}$), there is a unque extension to a probability measure on a σ-field $\mathcal{A}$ containing $\mathfrak{F}$. Fairly standard arguments then allowed us to conclude that the same result holds for measures μ that are σ-finite on $\mathfrak{F}$ (equivalently, on $\mathcal{S}$).

While it is true that working with a probability measure μ offered us some definite advantages, it is still natural to ask whether we could carry out the extension without bringing in that restriction. In this section, we address precisely that question and describe how the extension can be done for an arbitrary measure μ, not necessarily σ-finite, given on $\mathcal{S}$ (equivalently on $\mathfrak{F}$).

With μ^* defined as in (2.2.5) for all subsets, starting from an arbitrary measure μ on $\mathfrak{F}$, we modify our definition of the 'target' class $\mathcal{A}$ as

$$\mathcal{A} = \{A \subset \Omega : \mu^*(A \cap E) + \mu^*(A^c \cap E) \leq \mu(E) \text{ for all } E \in \mathfrak{F}\}. \quad (2.6.1)$$

It is not difficult to see (Exercise 2.6.1) that, in case μ is a probability measure on $\mathfrak{F}$, the class $\mathcal{A}$ defined above coincides with $\mathcal{A}$ defined in (2.2.9).

EXERCISE **2.6.1.** *Show that when μ is a probability measure on $\mathfrak{F}$, the class $\mathcal{A}$ defined in (2.6.1) coincides with the class $\{A \subset \Omega : \mu^*(A) + \mu^*(A^c) \leq 1\}$.*

We are now ready to state and prove an analogue of Theorem 2.2.9, which guarantees extension of an arbitrary measure on a field to a σ-field containing it.

THEOREM **2.6.2.** *Let μ be a measure on a field $\mathfrak{F}$ and let μ^* be as defined in (2.2.5). Then, the class $\mathcal{A}$ defined in (2.6.1) is a σ-field on Ω, containing $\mathfrak{F}$, and μ^* is a measure on $\mathcal{A}$.*

Proof. Since $\mu^* = \mu$ on $\mathfrak{F}$, one clearly has $\mathfrak{F} \subset \mathcal{A}$. Also, it is clear from the definition of $\mathcal{A}$ in (2.6.1), that it is closed under complementation. To proceed further, we first show that $\mathcal{A}$ is closed under finite unions. If $A, B \in \mathcal{A}$, then by the definition (2.6.1) of $\mathcal{A}$, we have, for every $E \in \mathfrak{F}$,

$$\mu^*(A \cap E) + \mu^*(A^c \cap E) \leq \mu(E) \text{ and } \mu^*(B \cap E) + \mu^*(B^c \cap E) \leq \mu(E). \quad (2.6.2)$$

Now, by the strong subadditivity of μ^*,

$$\mu^*((A \cup B) \cap E) + \mu^*((A \cap B) \cap E) \leq \mu^*(A \cap E) + \mu^*(B \cap E), \text{ and,}$$

$$\mu^*((A^c \cup B^c) \cap E) + \mu^*((A^c \cap B^c) \cap E) \leq \mu^*(A^c \cap E) + \mu^*(B^c \cap E).$$

Adding the above two inequalities and using (2.6.2), we get

$$\mu^*((A \cup B) \cap E) + \mu^*((A \cup B)^c \cap E) + \mu^*((A \cap B) \cap E)$$

$$+ \mu^*((A \cap B)^c \cap E) \leq 2\mu(E). \quad (2.6.3)$$

But by the strong subadditivity of μ^* once again, we also have,

$$\mu^*((A \cup B) \cap E) + \mu^*((A \cup B)^c \cap E) \geq \mu(E), \quad (2.6.4)$$

and, $\quad \mu^*((A \cap B) \cap E) + \mu^*((A \cap B)^c \cap E) \geq \mu(E). \quad (2.6.5)$

Now, if $\mu(E) < \infty$, then the above three inequalities force that all three of them must actually be equalities, while if $\mu(E) = \infty$, then equality holds in both (2.6.4) and (2.6.5) and hence in (2.6.3) also. Thus, we always have equalities in (2.6.4) and (2.6.5), which imply that both $A \cup B$ and $A \cap B$ are in $\mathcal{A}$. In particular, $\mathcal{A}$ is closed under finite unions. To prove that $\mathcal{A}$ is a σ-field, it is clearly enough to show now that $\mathcal{A}$ is closed under monotone increasing limits. For $A_n \in \mathcal{A}$ with $A_n \uparrow A$, we have, for all $E \in \mathfrak{F}$,

$$\mu^*(A_n \cap E) + \mu^*(A^c \cap E) \leq \mu^*(A_n \cap E) + \mu^*(A_n^c \cap E) \leq \mu(E),$$

where the first inequality uses $\mu^*(A^c \cap E) \leq \mu^*(A_n^c \cap E)$ (see Proposition 2.2.3(a)). By letting $n \to \infty$ and using Proposition 2.2.3(c), we get that, for all $E \in \mathfrak{F}$,

$$\mu^*(A \cap E) + \mu^*(A^c \cap E) \leq \mu(E),$$

which shows $A \in \mathcal{A}$, thus completing the proof that $\mathcal{A}$ is a σ-field.

To prove that μ^* is a measure on $\mathcal{A}$, we just show that μ^* is finitely additive on $\mathcal{A}$. The property of continuity from below (see Proposition 2.2.3(c)) will then imply that μ^* is countably additive on $\mathcal{A}$. So, for any two disjoint $\mathcal{A}$-sets A and B, we have to show that

$$\mu^*(A \cup B) = \mu^*(A) + \mu^*(B). \tag{2.6.6}$$

We first show that, for any $E \in \mathfrak{F}$ with $\mu(E) < \infty$,

$$\mu^*\big((A \cup B) \cap E\big) = \mu^*(A \cap E) + \mu^*(B \cap E) \tag{2.6.7}$$

Fix $E \in \mathfrak{F}$ with $\mu(E) < \infty$. By subadditivity of μ^*, we have

$$\mu^*\big((A \cup B) \cap E\big) \leq \mu^*(A \cap E) + \mu^*(B \cap E) \tag{2.6.8}$$

But since $(A^c \cup B^c) \cap E = E$, using subadditivity of μ^* again and (2.6.1), we get

$$\mu(E) + \mu^*\big((A^c \cap B^c) \cap E\big) \leq \mu^*(A^c \cap E) + \mu^*(B^c \cap E)$$

$$\leq 2\mu(E) - \big[\mu^*(A \cap E) + \mu^*(B \cap E)\big].$$

Now, if (2.6.8) were a strict inequality, then the above would lead to

$$\mu^*\big((A \cup B)^c \cap E\big) < \mu(E) - \mu^*\big((A \cup B) \cap E\big),$$

which contradicts what we have already proved, namely, that (2.6.4) is always an equality. This proves (2.6.7).

Coming back to (2.6.6), let us observe that it holds trivially if $\mu^*(A) = \infty$ or $\mu^*(B) = \infty$. Therefore it is enough to consider the case when both $\mu^*(A) < \infty$ and $\mu^*(B) < \infty$. In that case, by the definition of μ^*, there must exist $C, D \in \mathfrak{F}_\sigma$ with $A \subset C$, $B \subset D$, and $\mu(C) < \infty$, $\mu(D) < \infty$. Let $\{C_n\}$ and $\{D_n\}$ be sequences of $\mathfrak{F}$-sets such that $C_n \uparrow C$ and $D_n \uparrow D$. Put $E_n = C_n \cup D_n$ and $E = C \cup D$. It is clear that $E_n \in \mathfrak{F}$, $\mu(E_n) < \infty$, and $E_n \uparrow E$. By (2.6.7),

$$\mu^*\big((A \cup B) \cap E_n\big) = \mu^*(A \cap E_n) + \mu^*(B \cap E_n)$$

Letting $n \to \infty$ and using Proposition 2.2.4(c), one gets

$$\mu^*\big((A \cup B) \cap E\big) = \mu^*(A \cap E) + \mu^*(B \cap E).$$

Since $A \cup B \subset E$, (2.6.6) follows at once. And that completes the proof. □

Thus, we have obtained an extension of any arbitrary measure μ given on a field $\mathfrak{F}$ to a measure μ^* on a σ-field $\mathcal{A}$ containing $\mathfrak{F}$. Sets in the σ-field $\mathcal{A}$ are often called μ-measurable sets. As was proved earlier, this extension would be unique if μ is σ-finite on $\mathfrak{F}$. However, as shown by Example 2.2.21, uniqueness may not hold otherwise. Also, in case μ is σ-finite on the field $\mathfrak{F}$, the σ-field $\mathcal{A}$ turns out to be just the μ-completion of $\sigma(\mathfrak{F})$ (see Exercise 2.13.18).

2.7 Borel-Cantelli Lemma

For any sequence $\{A_n, n \geq 1\}$ of subsets of a non-empty set Ω, we define

$$\liminf A_n = \bigcup_{n=1}^{\infty} \bigcap_{k \geq n} A_k, \quad \limsup A_n = \bigcap_{n=1}^{\infty} \bigcup_{k \geq n} A_k.$$

From the above definition, one can easily see that $\liminf A_n$ consists of all points $\omega \in \Omega$, such that $\omega \in A_n$ for all but finitely many n, while $\limsup A_n$ consists of all points $\omega \in \Omega$, such that $\omega \in A_n$ for infinitely many n.

It is clear that, if $\mathcal{A}$ is a σ-field on Ω and $\{A_n\}$ is a sequence of $\mathcal{A}$-sets, then the sets $\liminf A_n$ and $\limsup A_n$ both belong to $\mathcal{A}$.

EXERCISE 2.7.1. *Show that $\liminf A_n \subset \limsup A_n$. Show that if either $A_n \downarrow A$ or $A_n \uparrow A$, then $\liminf A_n = \limsup A_n = A$.*

PROPOSITION 2.7.2. *Let μ be a measure on a σ-field $\mathcal{A}$. Then, for any sequence $\{A_n\}$ of sets in $\mathcal{A}$,*
$\mu(\liminf A_n) \leq \liminf \mu(A_n)$ *and, if μ is finite, $\mu(\limsup A_n) \geq \limsup \mu(A_n)$.*

Proof. Denoting $B_n = \bigcap_{k \geq n} A_k$, we have $B_n \uparrow \liminf A_n$ and so, by continuity of μ from below, $\mu(\liminf A_n) = \lim \mu(B_n)$. Since $\mu(B_n) \leq \mu(A_n)$ for each n, we have $\lim \mu(B_n) \leq \liminf \mu(A_n)$ and the required inequality follows. Proof of the other inequality, using now continuity from above of the finite measure μ, is similar and hence left as an exercise. $\square$

EXERCISE 2.7.3. *Show that, if μ is not a finite measure, then the inequality $\mu(\limsup A_n) \geq \limsup \mu(A_n)$ may not hold.*

The next theorem, commonly known as 'Borel-Cantelli Lemma', is the main result of this section. It is very simple, yet an extremely useful result. In Section 2.12, a *partial converse* of this in case of probability measures will be offered by Theorem 2.12.14 .

THEOREM 2.7.4. *(Borel-Cantelli Lemma) Let μ be a measure on a σ-field $\mathcal{A}$. If $\{A_n\}$ is a sequence of $\mathcal{A}$-sets such that $\sum_n \mu(A_n) < \infty$, then $\mu(\limsup A_n) = 0$.*

Proof. Denoting $B_n = \bigcup_{k \geq n} A_k$, we have $\mu(B_n) \leq \sum_{k \geq n} \mu(A_k)$ for each n. The hypothesis $\sum_n \mu(A_n) < \infty$ implies that, as $n \to \infty$, $\sum_{k \geq n} \mu(A_k) \to 0$, and so $\mu(B_n) \to 0$. Since $B_n \downarrow \limsup A_n$ and $\mu(B_1) \leq \sum_{n \geq 1} \mu(A_n) < \infty$, we apply Proposition 2.1.26 (d) to get $\mu(\limsup A_n) = \lim_n \mu(B_n) = 0$. $\square$

2.8 Measures on Real line

In this section, our aim is to construct measures on subsets of the real line by applying the general theory developed so far . A special case will, of course, be the length measure, which is more popularly known as the Lebesgue measure.

Let us recall that we have a notion of length for intervals and the collection of intervals forms a semifield on $\mathbb{R}$. If we can show that length is a measure on this semifield, then, noting that length is σ-finite on this semifield, Caratheodory Extension Theorem would guarantee a unique extension of length to a measure on the σ-field generated by the class of all intervals.

2.8. Measures on Real line

For convenience, instead of considering the class of *all* intervals, we consider a special class of intervals to be described now. We want to point out here that this special class of intervals will generate the same σ-field as the one generated by the class of all intervals. Let us denote

$$I_{a,b} = \{x \in \mathbb{R} : a < x \leq b\}, \text{ for } -\infty \leq a \leq b \leq \infty$$

Note that when both a and b are real with $a < b$, $I_{a,b}$ is just the interval $(a, b]$. The reader can easily see that $I_{-\infty,b} = (-\infty, b]$, when b is real, while $I_{a,\infty} = (a, \infty)$, for $a \geq -\infty$. Also, it is clear that $I_{a,b} = \emptyset$ when $a = b$. The reader should be able to verify that the class $\mathcal{S} = \{I_{a,b} : -\infty \leq a \leq b \leq \infty\}$ forms a semifield on $\mathbb{R}$. As seen before, the field $\mathfrak{F}$ generated by $\mathcal{S}$ consists of finite (disjoint) unions of sets in $\mathcal{S}$.

DEFINITION **2.8.1.** The σ-field generated by $\mathcal{S}$ (equivalently, the σ-field generated by $\mathfrak{F}(\mathcal{S})$) is called the *Borel σ-field* on $\mathbb{R}$, denoted by $\mathcal{B}(\mathbb{R})$ or simply by $\mathcal{B}$. Sets in $\mathcal{B}(\mathbb{R})$ are called *Borel subsets* of $\mathbb{R}$.

Let us, first of all, note that, by Exercise 2.2.16, $\mathcal{B}(\mathbb{R})$ is also the σ-field generated by the semifield consisting of all intervals. In fact, the same exercise presents several other generating classes for $\mathcal{B}(\mathbb{R})$.

How do the Borel subsets of $\mathbb{R}$ look like? Do they have any explicit identification mark? Unfortunately *no*, there is no *simple* characterization of Borel sets, except that they are sets that belong to the smallest σ-field on $\mathbb{R}$ containing all the $I_{a,b}$, $-\infty \leq a \leq b \leq \infty$. However, it is easy to see that all open subsets and, therefore, also all closed subsets of $\mathbb{R}$ are Borel sets. But then, so are all countable intersections of open sets and all countable unions of closed sets, and, so on.

Are there any subsets of $\mathbb{R}$ that are not Borel? Exercise 2.13.25 is aimed at showing that there are indeed non-Borel sets. However, that exercise also shows that one has to work very hard to exhibit a non-Borel set. What this means is that every set that can be described in a "reasonably" simple way will be a Borel set. In other words, even though Borel σ-field does not capture all subsets of $\mathbb{R}$, it is rich enough to contain all sets that we are likely to come across in practice.

EXERCISE **2.8.2.** Let D be any dense subset of $\mathbb{R}$. Then, show that,

$$\mathcal{B}(\mathbb{R}) = \sigma\{I_{a,b} : a < b, \ a, b \in D\}.$$

Use the above to conclude that there is a countable collection $\mathcal{C}$ of subsets of $\mathbb{R}$ such that $\sigma(\mathcal{C}) = \mathcal{B}(\mathbb{R})$, that is, $\mathcal{B}(\mathbb{R})$ is 'countably generated', according to the following Definition 2.8.3.

DEFINITION **2.8.3.** A σ-field $\mathcal{A}$ on Ω is said to be *countably generated* if it admits a countable generating class.

EXERCISE **2.8.4.** Let $f : \mathbb{R} \to \mathbb{R}$ be a continuous function and B a Borel set in $\mathbb{R}$. Show that the set $\{x \in \mathbb{R} : f(x) \in B\}$ is a Borel set. (Try using good sets principle). Show that the same is true also if $f : \mathbb{R} \to \mathbb{R}$ is a monotone function.

Having thus defined the Borel σ-field on $\mathbb{R}$, we are now ready to take up the next issue. Our aim is to now show that length indeed defines a measure on the semifield $\mathcal{S} = \{I_{a,b} : -\infty \leq a \leq b \leq \infty\}$. To be specific, we consider the set function λ defined on $\mathcal{S}$ by $\lambda(I_{a,b}) = b - a$ for $-\infty \leq a < b \leq \infty$. We want to show that λ defines a measure on $\mathcal{S}$. Clearly, $\lambda(\emptyset) = \lambda(I_{a,a}) = 0$. So, our only task is to show that λ is countably additive on $\mathcal{S}$. Since λ is clearly σ-finite on $\mathcal{S}$, this will allow us to have a unique extension of length on $\mathcal{S}$ to a measure λ on $\mathcal{B}(\mathbb{R})$ and even to its completion. This measure is what we will refer to as the *Lebesgue measure* on $\mathbb{R}$. The completion of the Borel σ-field with respect to λ will be called the *Lebesgue σ-field*, denoted by $\mathcal{L}$ and sets in $\mathcal{L}$ will be called *Lebesgue measurable sets*.

Coming now to the task of showing countable additivity of λ on $\mathcal{S}$, it turns out, however, that the same thing can be done, not just for the set function λ, but for a more general class of set functions on $\mathcal{S}$. Further, there is no special advantage in doing this only for λ. The basic argument remains the same. Therefore, that is what we are going to do. The advantage will clearly be that this will allow us, without any extra work, to construct many more measures on $\mathcal{B}(\mathbb{R})$, rather than just the Lebsgue measure λ. This immensely enriches the scope of measure theory and the associated integration theory to be developed later, by bringing together a wide variety of objects under one umbrella. A particularly significant application will be construction of probability measures on $\mathcal{B}(\mathbb{R})$.

To undestand the main idea, let us start with a measure μ on $\mathcal{B}(\mathbb{R})$. Let us further assume the measure μ to have the property that

$$\mu(I) < \infty \text{ for every bounded interval } I \subset \mathbb{R} \quad (2.8.1)$$

Consider the function G defined on $\mathbb{R}$ by

$$\begin{aligned} G(x) &= \mu(I_{0,x}) \quad \text{if} \quad x \geq 0 \\ &= -\mu(I_{x,0}) \quad \text{if} \quad x < 0. \end{aligned}$$

By the assumed property (2.8.1), the function G is real-valued. Let us now observe some interesting properties of the function G, which will be useful later for the construction of measures that was talked about above. First of all, it is fairly easy to see from the definition that G is non-decreasing. Let us next observe that G is continuous from the right everywhere, that is, $G(x+) = G(x)$ for all x. Indeed, let us fix $x \in \mathbb{R}$ and let $\{x_n\}$ be a sequence of real numbers with $x_n \downarrow x$. It is easy to see that, if $x \geq 0$, then $I_{0,x_n} \downarrow I_{0,x}$. On the other hand, if $x < 0$, then $x_n < 0$ for all large n and hence $I_{x_n,0} \uparrow I_{x,0}$. Right continuity of G, therefore, follows from Proposition 2.1.26. Since G is monotone, we can extend G to $[-\infty, \infty]$ by defining

$$G(-\infty) = \lim_{x \downarrow -\infty} G(x), \quad G(\infty) = \lim_{x \uparrow \infty} G(x). \quad (2.8.2)$$

Note that $G(\pm\infty)$ may take the values $\pm\infty$. But the important thing to observe now is that we have $\mu(I_{a,b}) = G(b) - G(a)$ for every $-\infty \leq a \leq b \leq \infty$. Thus, given any measure μ on $\mathcal{B}(\mathbb{R})$ satisfying property (2.8.1), we obtained a non-decreasing, right continuous function $G : \mathbb{R} \to \mathbb{R}$ such that

2.8. Measures on Real line

$$\mu(I_{a,b}) = G(b) - G(a), -\infty < a \leq b < \infty, \text{ and,}$$

$$\mu(I_{-\infty,b}) = G(b) - G(-\infty),\ b > -\infty,\quad \mu(I_{a,\infty}) = G(\infty) - G(a),\ a < \infty \quad (2.8.3)$$

where $G(\pm\infty)$ are as defined in (2.8.2). What is interesting and important, but by no means obvious, is that the converse is also true. That is, given any non-decreasing, right continuous function $G : \mathbb{R} \to \mathbb{R}$, there is a unique measure μ on $\mathcal{B}(\mathbb{R})$ such that (2.8.3) holds. Clearly, this measure μ will satisfy (2.8.1).

This is the crucial idea that will allow us to construct a very wide class of measures on $\mathcal{B}(\mathbb{R})$. Once we prove this, we can take the function $G(x) = x$ to get the Lebesgue measure on $\mathcal{B}(\mathbb{R})$. Of course, by taking other non-decreasing right continuous functions $G : \mathbb{R} \to \mathbb{R}$, we will be able to construct other measures on $\mathcal{B}(\mathbb{R})$. Indeed, every non-decreasing, right continuous G will give rise to a measure. We will consider some such specific functions later as illustrations. But right now, let us proceed towards proving this.

Let $G : \mathbb{R} \to \mathbb{R}$ be a non-decreasing right continuous function. Define $G(\pm\infty)$ by (2.8.2). We can now define a set function μ on $\mathcal{S}$ by the formula (2.8.3). We proceed to show that μ is a measure on $\mathcal{S}$. Non-negativity of the set function μ on $\mathcal{S}$ is immediate from the fact that G is non-decreasing. Also, $\mu(\emptyset) = \mu(I_{a,a}) = 0$. Thus the only challenge is to prove countable additivity of μ on $\mathcal{S}$. We do this by proving two lemmas, from which the required additivity of μ will follow.

LEMMA **2.8.5.** *Let $\{A_k\}$ be a countable family of disjoint sets in $\mathcal{S}$. If A is a set in $\mathcal{S}$ such that $\bigcup_k A_k \subset A$, then $\sum_k \mu(A_k) \leq \mu(A)$.*

Proof. Without loss of generality, we can and will assume throughout the proof that the sets A_k, given in the hypothesis, are all non-empty.

We first prove the assertion in case we have a finite family of disjoint $\mathcal{S}$-sets, say, $\{A_1, \cdots, A_n\}$ with $\bigcup_{k=1}^n A_k \subset A$, where $A \in \mathcal{S}$. Let $A_k = I_{a_k,b_k}$, $1 \leq k \leq n$ and $A = I_{a,b}$, where $-\infty \leq a_k < b_k \leq \infty$, $1 \leq k \leq n$ and $-\infty \leq a < b \leq \infty$.

The required result is trivially true for $n = 1$. Indeed, if $I_{a_1,b_1} \subset I_{a,b}$, then $a_1 \geq a$ and $b_1 \leq b$, so that $\mu(A_1) = G(b_1) - G(a_1) \leq G(b) - G(a) = \mu(A)$. We use induction to prove it for general n. By rearranging the sets, if necessary, we may and shall assume that $a_1 \leq a_2 \leq \cdots \leq a_n$. From the disjointness of $\{A_1, \cdots, A_n\}$ and from the fact that $\bigcup_{k=1}^n A_k \subset A$, one can easily see that we must have $a \leq a_1 < b_1 \leq a_2 < b_2 \leq \cdots \leq a_n < b_n \leq b$. But it is clear then that $\bigcup_{k=1}^{n-1} I_{a_k,b_k} \subset I_{a,a_n}$ and $I_{a_n,b_n} \subset I_{b_{n-1},b}$. The first inclusion, by the induction hypothesis, gives $\sum_{k=1}^{n-1} \mu(A_k) \leq \mu(I_{a,a_n}) = G(a_n) - G(a)$, while the second inclusion gives $\mu(A_n) = G(b_n) - G(a_n) \leq G(b) - G(a_n)$. Noting that $G(a_n)$ is finite (why?), we may add the above two inequalities to get $\sum_{k=1}^n \mu(A_k) \leq G(b) - G(a) = \mu(A)$, as required.

Finally, suppose we have a countably infinite family $\{A_1, A_2, \ldots\}$ of disjoint $\mathcal{S}$-sets with $\bigcup_k A_k \subset A$. Then, for each n, we have $\bigcup_{k=1}^n A_k \subset A$, and therefore, by what has been proved above, we have $\sum_{k=1}^{n} \mu(A_k) \leq \mu(A)$, for each n. Letting $n \to \infty$ now, we get $\sum_{k=1}^{\infty} \mu(A_k) \leq \mu(A)$. □

LEMMA 2.8.6. *Let $\{A_k\}$ be a countable family of sets in $\mathcal{S}$. If A is any set in $\mathcal{S}$ such that $A \subset \bigcup_k A_k$, then $\mu(A) \leq \sum_k \mu(A_k)$.*

Proof. Just as in the proof of the previous Lemma, we may and do assume that, for each k, $A_k \neq \emptyset$ and denote $A_k = I_{a_k, b_k}$, for $-\infty \leq a_k < b_k \leq \infty$. Also, there is nothing to prove if $A = \emptyset$ and so we assume $A = I_{a,b}$, where $-\infty \leq a < b \leq \infty$.

As in the previous lemma, we first prove it when $\{A_k\}$ is a finite family. Thus, given sets $A_k = I_{a_k,b_k}$, $1 \leq k \leq n$, and $A = I_{a,b}$ with $A \subset \bigcup_{k=1}^n A_k$, we are to show that $\sum_{k=1}^{n} \mu(A_k) \geq \mu(A)$. The assertion holds trivially for $n = 1$. We use induction to prove it for $n \geq 2$. Since $I_{a,b} \subset \bigcup_{k=1}^{n} I_{a_k, b_k}$, one can easily argue that we must have $b \leq b_k$ for some k. Without loss of generality, we can and do assume that $b \leq b_n$. Now, if $A \subset A_n$, then of course, $\mu(A) \leq \mu(A_n) \leq \sum_{k=1}^{n} \mu(A_k)$. On the other hand, if $A \not\subset A_n$, then that would mean $a < a_n$, and in that case, we must have $I_{a,a_n} \subset \bigcup_{k=1}^{n-1} A_k$ and hence $G(a_n) - G(a) = \mu(I_{a,a_n}) \leq \sum_{k=1}^{n-1} \mu(A_k)$, by the induction hypothesis. It then follows that $\mu(A) = G(b) - G(a) \leq G(b_n) - G(a) = G(a_n) - G(a) + G(b_n) - G(a_n) \leq \sum_{k=1}^{n} \mu(A_k)$.

Having proved the assertion in case of a finite family $\{A_k, 1 \leq k \leq n\}$, let us now consider an infinite sequence $\{A_1, A_2, \ldots\}$ of $\mathcal{S}$-sets, such that, $A \subset \bigcup_k A_k$, where A is a non-empty set in $\mathcal{S}$. Let us first assume that A is a bounded interval, that is, $A = I_{a,b}$ with $-\infty < a < b < \infty$. We fix an $\epsilon > 0$ and use right continuity of G to first get $a' \in (a, b)$, such that, $G(a') \leq G(a) + \epsilon/2$. Next, for each $k \geq 1$, we choose b'_k as follows. If $b_k = \infty$, we take $b'_k = b_k$. Otherwise, we use right continuity of G to get $b'_k > b_k$ such that $G(b'_k) \leq G(b_k) + \epsilon/2^{k+1}$. Note that this last inequality holds automatically even when $b'_k = b_k = \infty$. Now comes the most crucial step in our argument. By the way we have chosen a' and b'_k, $k \geq 1$, it follows that the closed bounded interval $[a', b]$ satisfies

$$[a', b] \subset I_{a,b} \subset \bigcup_k I_{a_k, b_k} \subset \bigcup_k (a_k, b'_k).$$

By the Heine-Borel theorem now, there exists an n such that $[a', b] \subset \bigcup_{k=1}^{n} (a_k, b'_k)$. But this would imply that $I_{a',b} \subset \bigcup_{k=1}^{n} I_{a_k, b'_k}$, and therefore, by the result already proved for a finite family, we get $\mu(I_{a',b}) \leq \sum_{k=1}^{n} \mu(I_{a_k, b'_k})$. Once again, from the way a' and b'_k, $k \geq 1$, were chosen, we have, on one hand, $\mu(A) \leq \mu(I_{a',b}) + \epsilon/2$ and, on the other, $\mu(I_{a_k, b'_k}) \leq \mu(A_k) + \epsilon/2^{k+1}$, for eack k. Using all of these, we finally get $\mu(A) \leq \sum_{k=1}^{n} [\mu(A_k) + \epsilon/2^{k+1}] + \epsilon/2 \leq \sum_{k=1}^{\infty} \mu(A_k) + \epsilon$. Since $\epsilon > 0$ was arbitrary, the result is now proved for the case when $A \in \mathcal{S}$ is a bounded interval. The proof for general $A \in \mathcal{S}$ can now be completed as follows. For any

2.8. Measures on Real line

$m \geq 1$, if we replace $A \in \mathcal{S}$ by the bounded interval $A \cap I_{-m,m} \in \mathcal{S}$, we have $A \cap I_{-m,m} \subset A \subset \bigcup_k A_k$, and therefore, by applying what we have just proved we get $\mu(A \cap I_{-m,m}) \leq \sum_{k=1}^{\infty} \mu(A_k)$. This being true for all $m \geq 1$, we let $m \to \infty$ to get $\lim_{m \uparrow \infty} \mu(A \cap I_{-m,m}) \leq \sum_{k=1}^{\infty} \mu(A_k)$. It is easy to see that $\lim_{m \uparrow \infty} \mu(A \cap I_{-m,m}) = \mu(A)$, from which the required result follows. □

As an immediate consequence of the Lemmas 2.8.5 and 2.8.6, it follows that μ defined on $\mathcal{S}$ by (2.8.3) is countably additive. Thus, we have proved that, given any non-decreasing, right continuous function $G : \mathbb{R} \to \mathbb{R}$, the set function μ defined on $\mathcal{S}$ by (2.8.3) is a measure on $\mathcal{S}$. Moreover, $\mu(I_{a,b}) = G(b) - G(a) < \infty$, whenever $-\infty < a \leq b < \infty$, and this clearly implies that the measure μ on $\mathcal{S}$ is σ-finite. Caratheodory Extension Theorem now guarantees that there is a unique extension of this μ on $\mathcal{S}$ to a measure μ on $\mathcal{B}(\mathbb{R})$. Further, this measure μ on $\mathcal{B}(\mathbb{R})$ will clearly satisfy the additional property that $\mu(A) < \infty$ for every bounded Borel set A. This is because any bounded subset of $\mathbb{R}$ is a subset of $I_{a,b}$ for some $-\infty < a < b < \infty$. A measure μ with this last property is called a 'Radon measure'.

DEFINITION 2.8.7. A measure μ on $\mathcal{B}(\mathbb{R})$ is called a *Radon measure* if it has the property that $\mu(A) < \infty$ for every bounded Borel set A. Radon measures on $\mathcal{B}(\mathbb{R})$ are sometimes also referred to as *Lebesgue-Stieltje's measures*.

Clearly, every Radon measure on $\mathcal{B}(\mathbb{R})$ is necessarily σ-finite, in fact, σ-finite on $\mathcal{S}$. But not all σ-finite measures on $\mathcal{B}(\mathbb{R})$ are Radon measures. For example, the measure μ on $\mathcal{B}(\mathbb{R})$ defined as $\mu(A) = |A \cap \mathbb{Q}|$ is clearly σ-finite on $\mathcal{B}(\mathbb{R})$, but is not a Radon measure, because every non-degenerate interval, bounded or not, will have infinite μ-measure.

Given a non-decreasing right continuous function $G : \mathbb{R} \to \mathbb{R}$, consider the unique Radon measure μ on $\mathcal{B}(\mathbb{R})$, determined by G, as above. Now, the function $H : \mathbb{R} \to \mathbb{R}$ defined as $H(x) = G(x) + c$, where c is any fixed real number, is clearly non-decreasing and right continuous. Defining $H(\pm\infty)$ in the same way as was done for G, we see that $H(b) - H(a) = G(b) - G(a) = \mu(I_{a,b})$, for all $I_{a,b} \in \mathcal{S}$. In other words, the functions G and H both determine the same Radon measure. Thus, the same Radon measure on $\mathcal{B}(\mathbb{R})$ may arise from more than one non-decreasing right continuous functions. However, just like the functions G and H above, any two non-decreasing, right continuous functions that give the same Radon measure, can differ by only an additive constant. Indeed, if H and G are two such functions giving rise to the same Radon measure μ then we must have $H(b) - H(a) = G(b) - G(a)$, for all pairs of real numbers $a < b$. In particular, this implies that $H(x) = G(x) + (H(0) - G(0))$, for all x.

DEFINITION 2.8.8. A function $G : \mathbb{R} \to \mathbb{R}$ is called a *distribution function* if it is non-decreasing and right continuous.

Using the terminogy introduced in the Definitions 2.8.7 and 2.8.8, what we have proved above is summarized in the following theorem.

THEOREM 2.8.9. *Given any Radon measure μ on $\mathcal{B}(\mathbb{R})$, there is a distribution function G on $\mathbb{R}$, such that $\mu(I_{a,b}) = G(b) - G(a)$, for every pair of real numbers $a < b$. Further, such a function G is unique up to an additive constant.*
Conversely, given any distribution function G on $\mathbb{R}$, there is a unique measure μ on $\mathcal{B}(\mathbb{R})$, such that $\mu(I_{a,b}) = G(b) - G(a)$ for all pairs of real numbers $a < b$. Further, this unique measure μ on $\mathcal{B}(\mathbb{R})$ is necessarily a Radon measure.

EXERCISE 2.8.10. *Show that each of the following functions on $\mathbb{R}$ is a distribution function: $G(x) = x$, $G(x) = 2x - 10$, $G(x) = [x] =$ the integer part of x, $G(x) = \int_0^x e^{-t} dt$ (with the convention that, for $x < 0$, $\int_0^x = -\int_x^0$)*

As seen above, any two distribution functions determining the same Radon measure may differ at most by an additive constant. Therefore, one can achieve uniqueness of the distribution function G by imposing an additional restriction like $G(0) = 0$. For finite measures however, there is a standard and natural way of achieving uniqueness of the associated distribution function. If μ is a finite measure on $\mathcal{B}(\mathbb{R})$, then $F(x) = \mu((-\infty, x])$ defines a real valued function on $\mathbb{R}$, which is a distribution function that determines μ. The function F has the additional property that $F(-\infty) = 0$. In fact, the function F defined above is the only distribution function associated to the measure μ, that satisfies $F(-\infty) = 0$. Of course, this F also has the additional property that $F(\infty) = \mu(\mathbb{R}) < \infty$.

Thus, there is a one to one correspondence between finite measures μ on $\mathcal{B}(\mathbb{R})$ and distribution functions $F : \mathbb{R} \to \mathbb{R}$, satisfying $F(-\infty) = 0$ and $F(\infty) < \infty$. In particular, there is a one to one correspondence between probability measures on $\mathcal{B}(\mathbb{R})$ and distribution functions F satisfying $F(-\infty) = 0$ and $F(\infty) = 1$. Such distribution functions are called *probability distribution functions on $\mathbb{R}$*. Since this is an important result, we state it below as a separate theorem, after formally defining probability distribution functions.

DEFINITION 2.8.11. *A non-decreasing, right continuous function $F : \mathbb{R} \to \mathbb{R}$ that also satisfies $\lim_{x \downarrow \infty} F(x) = 0$ and $\lim_{x \uparrow \infty} F(x) = 1$ is called a probability distribution function on $\mathbb{R}$.*

THEOREM 2.8.12. *There is a one to one correspondence between probability measures μ on $\mathcal{B}(\mathbb{R})$ and probability distribution functions F on $\mathbb{R}$, the correspondence being given by $\mu(I_{a,b}) = F(b) - F(a)$, for all pairs of real numbers $a < b$.*

DEFINITION 2.8.13. *For a probability measure μ on $\mathcal{B}(\mathbb{R})$, the unique probability distribution function F, associated to μ and given by $F(x) = \mu((-\infty, x])$, is called the probability distribution function of the probability μ.*

Let us briefly sum up what emerges from all that we have done above. Firstly, it should be amply clear to the reader that prescribing a measure μ on $\mathcal{B}(\mathbb{R})$ by

2.8. Measures on Real line

actually specifying the values of $\mu(B)$ for all Borel sets B is practically impossible, except for some very special measures μ. However, this difficulty can be circumvented for a large class of measures, namely, Radon measures, on $\mathcal{B}(\mathbb{R})$. What we just saw tells us that to specify a Radon measure on $\mathcal{B}(\mathbb{R})$, all we need to do is to specify a distribution function, that is, to specify a non-decreasing, right continuous function on $\mathbb{R}$ into $\mathbb{R}$, and we know that it determines a unique Radon measure. It should, however, be kept in mind that explicit evaluation of the Radon measure for arbitrary Borel sets starting from the distribution function remains an almost impossible task, except in some special cases. The point though is that specifying a distribution function is an immensely simpler job and we know that it does determine a unique Radon measure. The good news is that specifying the distribution function turns out to be enough for almost all the mathematics that we want to do with the corresponding Radon measure. Having said that, let us present a few examples of distribution functions (some given in Exercise 2.8.10) and try to understand the Radon measures they determine.

EXAMPLE 2.8.14. Let us consider $G(x) = x$. As mentioned earlier, the associated measure on $\mathcal{B}(\mathbb{R})$ is called the Lebesgue measure and is denoted by λ. The significance of the Lebsgue measure lies in the fact that, for any interval I, its measure $\lambda(I)$ equals the length of I. This is clearly true for intervals $I \in \mathcal{S}$. For intervals not in $\mathcal{S}$, it can be easily verified by properties of measures. For example, take a closed bounded interval $A = [a, b]$. If $a_n = a - n^{-1}$, then $I_{a_n,b} \downarrow A$ and $\lambda(I_{a_1,b}) < \infty$ so that $\lambda(A) = \lim_n [G(b) - G(a - n^{-1})] = b - a$. The special case $a = b$ gives us $\lambda(\{a\}) = 0$, for all real a. Thus λ on $\mathcal{B}(\mathbb{R})$ can be thought of as extending the notion of "length" from intervals to all Borel subsets of $\mathbb{R}$.

EXAMPLE 2.8.15. Consider the function $G(x) = 2x - 10$. The associated Radon measure μ can be easily seen to be nothing but the measure 2λ, that is, $\mu(B) = 2\lambda(B)$. Indeed, one can verify this directly for all sets in $\mathcal{S}$ and then appeal to the Uniqueness Theorem 2.3.7.

EXAMPLE 2.8.16. As a generalization of Example 2.8.15, we can take the distribution function $G(x) = \alpha x + \beta$, for any $\alpha > 0$ and $\beta \in \mathbb{R}$, and and see that the associated Radon measure is given by $\mu = \alpha \lambda$.

EXAMPLE 2.8.17. Consider the function $G(x) = [x]$. This is clearly a distribution function. But, unlike the last three examples, this G is not a continuous function. The integers constitute precisely all the discontinuity points of G. Denoting the associated Radon measure by μ, it is easy to see that $\mu(I_{a,b}) = G(b) - G(a) = |I_{a,b} \cap \mathbb{Z}|$. Now, by considering the σ-finite measure ν on $\mathcal{B}(\mathbb{R})$ given by $\nu(A) = |A \cap \mathbb{Z}|$, one can appeal to the Uniqueness Theorem 2.3.7 to get $\mu(A) = \nu(A) = |A \cap \mathbb{Z}|$, for all Borel sets A. This, therefore, is one of those special cases, where one gets an explicit formula for μ on all Borel sets.

EXERCISE 2.8.18. *Let G be a distribution function on $\mathbb{R}$ and denote μ to be the associated Radon measure on $\mathcal{B}(\mathbb{R})$. For any $x \in \mathbb{R}$, let $G(x-)$ denote the left*

limit of G at x, that is, $G(x-) = \lim_{y \to x-} G(y)$.
(a) For any bounded open interval (a, b), show that $\mu(a, b) = G(b-) - G(a)$.
(b) Show that $\mu(\{x\}) = G(x) - G(x-)$, for any $x \in \mathbb{R}$.
(c) Show that there can be at most countably many real numbers x, such that, $\mu(\{x\}) > 0$.
(d) Show that If $G(a) = G(b)$ for some pair of real numbers $a < b$, then $\mu(A) = 0$ for all Borel sets $A \subset (a, b]$.
(e) Show that the assertion (c) holds not just for Radon measures μ, but also for any σ-finite measure μ on $\mathbb{R}$.

The reader may look at Exercises 2.13.27 and 2.13.28 at the end of the chapter, which give some interesting illustrations of constructing new distribution functions from given ones and examining the relationship between the associated Radon measures.

EXAMPLE 2.8.19. $G(x) = \int_0^x g(t)dt$ where $g : \mathbb{R} \to \mathbb{R}$ is a non-negative function, which is 'locally' Riemann integrable, that is, Riemann integrable on every closed bounded interval. For $x < 0$, we follow the usual convention of interpreting $\int_0^x g(t)dt$ as $-\int_x^0 g(t)dt$. It is easy to see that G is a distribution function. Let μ denote the associated Radon measure. Of course, when $B \subset \mathbb{R}$ is any interval, $\mu(B)$ turns out to be just the Riemann integral of g over B and represents the area of the region over B under the graph of g. Thus, the measure μ on $\mathcal{B}(\mathbb{R})$ essentially extends the notion of integral of the locally Riemann integrable function g from integration over intervals to integration over general Borel sets and therefore, can be thought of as defining the concept of area, under the graph of g, over general Borel sets B. As an important consequence of the integration theory to be developed in Chapter 3, we will see that $\mu(B) = 0$ for any Borel set B with $\lambda(B) = 0$. In the terminology of Chapter 6, this means that μ is 'absolutely continuous' with respect to λ.

EXAMPLE 2.8.20. Let $\{p_n : n \geq 1\}$ be a sequence of positive real numbers and let $D = \{x_1, x_2, \cdots\}$ be a countable subset of $\mathbb{R}$. Let us assume that **at least one** of the following conditions hold: (a) the series $\sum_n p_n$ converges, (b) D is 'locally finite', that is, for every bounded interval I, the set $D \cap I$ is finite. Consider the function G on $\mathbb{R}$ defined as

$$G(x) = \sum_{\substack{n \\ 0 < x_n \leq x}} p_n, \ x \geq 0 \quad \text{and} \quad G(x) = -\sum_{\substack{n \\ x < x_n \leq 0}} p_n, \ x < 0. \quad (2.8.4)$$

Under the assumption that at least one of (a) and (b) hold, it is not difficult to see that G is real valued and that it is indeed a distribution function. Further, the set D is precisely the set of discontinuity points of G. From Exercise 2.8.18(b), it follows that the associated measure μ has mass p_n at the point x_n, that is, $\mu(\{x_n\}) = p_n$. Using the Uniqueness Theorem 2.3.7, as before, one can show that $\mu(B) = \mu(B \cap D) = \sum_B p_n$, for any Borel set B, where $\sum_B$ denotes sum over $\{n : x_n \in B \cap D\}$. In particular, $\mu(D^c) = 0$. Noting that $\lambda(D) = 0$, one sees that the character of this measure μ is diametrically opposite to the one in the previous

2.8. Measures on Real line

example. It puts its entire mass on the set D which has Lebesgue measure zero. This, in the terminology of Chapter 6 means that μ is 'singular' with respect to λ.

The measure μ is clearly finite if and only if (a) holds and, in that case, the function $F(x) = \sum_{n:x_n \leq x} p_n$ also serves as a distribution function for the measure μ. In the special case when $\sum_n p_n = 1$, the function F is a probability distribution function and the associated probability measure μ puts mass p_n at the point x_n.

We next discuss some important properties of Lebesgue measure. A very useful and not very surprising property of Lebesgue measure is its 'translation invariance', that is, if B is a Borel set and $\widetilde{B}$ is obtained by translating B through an additive constant, then $\widetilde{B}$ is again a Borel set and has the same Lebesgue measure as B. (see Exercise 2.13.19(a)). This agrees with our intuition that length of a set should remain unchanged by any shift of the set. Of course, any non-negative scalar multiple of λ will also have the same property. What is interesting is that, under a mild extra condition, Lebesgue measure can be shown (see Exercise 2.13.19(b)) to be the only (upto a scalar multiple) translation invariant measure on $\mathcal{B}(\mathbb{R})$.

We discuss another interesting property of Lebesgue measure. For each real number $x \in (0,1)$, consider the non-terminating decimal expansion of x. For example, for $x = .25$, we consider its nonterminating expansion $.24999999\cdots$. For each n, let $r_n(x)$ denote the proportion of times the digit 7 occurs in the first n places of the decimal expansion of x. Denoting $B = \{x \in (0,1) : \lim_n r_n(x) = \frac{1}{10}\}$, it should be clear that the set B does not have a simple description like a union of intervals or an intersection of unions of intervals. One may, of course, ask whether B is a Borel set, and if so, what is the value of $\lambda(B)$. It turns out that B is indeed a Borel set and, more interestingly, $\lambda(B) = 1$. The last fact is an interesting application of results in probability. It is, of course, not hard to realize that the digit 7 has no special role here. The result thus asserts that, for 'almost all' $x \in (0,1)$, each of the digits occurs with the same (asymptotic) frequency in the non-terminating decimal expansion of x. One expresses this by saying that almost all $x \in (0,1)$ are *normal numbers* and this result is known as Borel's Normal NumberTheorem. (see Exercise 7.6.12).

To discuss the next interesting fact on Lebesgue measure, we know that every singleton set has zero Lebesgue measure and therefore, the same is true of any countable subset of $\mathbb{R}$. It is natural to wonder whether countable sets are the only subsets of $\mathbb{R}$ with zero Lebesgue measure. In other words, are there uncountable sets with zero Lebesgue measure? The question becomes intriguing because it is not easy to think of any obvious example of such an uncountable set. Exercise 2.8.21 gives a very important example of a Borel set (in fact, a closed set), which is uncountable and has zero Lebesgue measure. The set C constructed there turns out to be an extremely important set in many different contexts and is well-known as the *Cantor set*.

EXERCISE **2.8.21**. *(Cantor set):* Let $I = [0,1]$ be the closed unit interval. Let

I_1 be the set obtained by removing, from I, its "open middle one-third", that is, the open interval $(1/3, 2/3)$. I_1 is thus a union of two disjoint closed intervals. Now, from each of these two closed intervals, the open middle one-third interval is removed, that is, the open intervals $(1/9, 2/9)$ and $(7/9, 8/9)$. This leaves us now with the set I_2, a disjoint union of four closed intervals. Next, the open middle one-third interval is removed from each of these four to get I_3, and so on.
(a) See that continuing in this manner produces a nested sequence $\{I_n\}$ of closed subsets of $[0, 1]$, where I_n, for each n, is a union of 2^n disjoint closed intervals, each of length $1/3^n$.
(b) Denoting $C = \bigcap I_n$, show that C is a closed set with $\lambda(C) = 0$.
(c) Show that $x \in C$ iff x has a unique ternary expansion $x = \sum_{k=1}^{\infty} \frac{a_k}{3^k}$ with $a_k \in \{0, 2\}$ for each k.
(d) Conclude from (c) that the set C is uncountable.

We end this section with an important observation which is very useful in some contexts. We start with a not too difficult exercise.

EXERCISE **2.8.22.** Let D be a countable dense subset of $\mathbb{R}$ and H a non-decreasing real-valued function on D. Show that

$$G(x) = \inf\{H(r) : r \in D, r > x\}, \ x \in \mathbb{R}$$

defines a distribution function G on $\mathbb{R}$, with $G(-\infty) = H(-\infty)$ and $G(\infty) = H(\infty)$. Show also that $G = H$ on D if and only if H is right continuous on D.

By taking $D = \mathbb{Q}$ in Exercise 2.8.22, we see that any non-decreasing right continuous real-valued function on $\mathbb{Q}$ is the restriction to $\mathbb{Q}$ of a distribution function on $\mathbb{R}$.

DEFINITION **2.8.23.** A non-decreasing, right continuous function $H : \mathbb{Q} \to \mathbb{R}$ is called a *distribution Function on* $\mathbb{Q}$.

In view of the comment made above, any distribution function on $\mathbb{Q}$ is the restriction to $\mathbb{Q}$ of a distribution function on $\mathbb{R}$ and hence determines a unique Radon measure on $\mathcal{B}(\mathbb{R})$. Thus we have the following theorem.

THEOREM **2.8.24.** Given any distribution function H on $\mathbb{Q}$ there is a unique Radon measure μ on $\mathcal{B}(\mathbb{R})$ such that $\mu(I_{s,r}) = H(r) - H(s)$ for every pair of rational numbers $s \leq r$.
Conversely, given a Radon measure μ on $\mathcal{B}(\mathbb{R})$ there is a distribution function H, unique upto additive constants, on $\mathbb{Q}$ such that the above relation holds.

Following are the analogues of Definition 2.8.23 and Theorem 2.8.24 for probability measures on $\mathcal{B}(\mathbb{R})$.

DEFINITION **2.8.25.** A function $H : \mathbb{Q} \to [0, 1]$ is called a *Probability distribution Function on* $\mathbb{Q}$, if it is non-decreasing, right continuous on $\mathbb{Q}$ and satisfies $\lim_{r \downarrow -\infty} H(r) = 0$, $\lim_{r \uparrow \infty} H(r) = 1$.

THEOREM 2.8.26. *There is a one-one correspondence between probability measures μ on $\mathcal{B}(\mathbb{R})$ and probability distribution functions H on $\mathbb{Q}$, satisfying the relation $\mu(I_{s,r}) = H(r) - H(s)$ for every pair of rational numbers $s \leq r$.*

2.9 Measures in Higher Dimensions

In this section, we are going to extend whatever was done on the real line, to higher dimensions, that is, to $\mathbb{R}^k$, $k > 1$. Let us first fix some notations. For $\mathbf{a} = (a_1, \ldots, a_k)$ and $\mathbf{b} = (b_1, \ldots, b_k)$ in $\mathbb{R}^k$, we write $\mathbf{a} \leq \mathbf{b}$ if $a_i \leq b_i$ for all i. We will use $\mathbf{a} < \mathbf{b}$ to mean $a_i < b_i$ for *all* i. It should be be noted here that $\mathbf{a} \leq \mathbf{b}$ and $\mathbf{a} \neq \mathbf{b}$ together do not imply that $\mathbf{a} < \mathbf{b}$, if $k > 1$. For example in $\mathbb{R}^2$, the two points $(1,1) \neq (1,2)$ satisfy $(1,1) \leq (1,2)$, but $(1,1) \not< (1,2)$.

With these notations set up, we are now ready to define analogues of the sets $I_{a,b}$ in $\mathbb{R}^k$. For $\mathbf{a}, \mathbf{b} \in \mathbb{R}^k$, denote
$$I_{\mathbf{a},\mathbf{b}} = \{\mathbf{x} \in \mathbb{R}^k : \mathbf{a} < \mathbf{x} \leq \mathbf{b}\}.$$
In case of $\mathbb{R}$, we also needed to include sets of the form $I_{-\infty,b}$, $I_{a,\infty}$, $I_{-\infty,\infty}$ in addition to $I_{a,b}$ for real $a \leq b$, in order to construct a semifield. The analogues of these in higher dimensions is a little more complicated. We need to allow points $\mathbf{a} = (a_1, \cdots, a_k)$, where some or all of the a_i may be $-\infty$ and the rest of the coordinates real, and also points $\mathbf{b} = (b_1, \cdots, b_k)$, where some or all the b_i may be ∞ and the rest of the coordinates real. In other words, the k-dimensional vectors we are allowing for the 'end-points' are $\mathbf{x} = (\mathbf{x_1}, \ldots, \mathbf{x_k})$, where $-\infty \leq x_i \leq \infty$, for all i. The inequalities $\mathbf{a} \leq \mathbf{b}$ and $\mathbf{a} < \mathbf{b}$ can still be defined in exactly the same manner and therefore sets $I_{\mathbf{a},\mathbf{b}} = \{\mathbf{x} \in \mathbb{R}^\mathbf{k} : \mathbf{a} < \mathbf{x} \leq \mathbf{b}\}$ will make sense. Note that, each such $I_{\mathbf{a},\mathbf{b}}$ is still a subset of $\mathbb{R}^k$. It is now fairly easy to see that the class $\mathcal{S}$ containing all such $I_{\mathbf{a},\mathbf{b}}$ forms a semifield on $\mathbb{R}^k$.

DEFINITION 2.9.1. *The σ-field on $\mathbb{R}^k$, generated by $\mathcal{S}$, is called the Borel σ-field on $\mathbb{R}^k$ and is denoted by $\mathcal{B}(\mathbb{R}^k)$. Sets in $\mathcal{B}(\mathbb{R}^k)$ are called Borel subsets of $\mathbb{R}^k$.*

A point $\mathbf{x} = (x_1, \ldots, x_k)$ is called a **rational point** in $\mathbb{R}^k$ if each x_i is rational. By a closed left half space in $\mathbb{R}^k$ we mean a set of the form $\{\mathbf{x} \in \mathbb{R}^k : x_i \leq c\}$, for some i and for some real c. One can similarly define a closed right half space.

EXERCISE 2.9.2. *Show that each of the following classes generate $\mathcal{B}(\mathbb{R}^k)$: (i) class of open sets, (ii) class of closed sets, (iii) class of open balls, (iv) class of closed balls, (v) class of open rectangles, (vi) class of all open balls centered at rational points and having rational radii, (vii) class of all closed left half spaces.*

EXERCISE 2.9.3. *Show that if $B_1, \ldots, B_k \in \mathcal{B}(\mathbb{R})$, then $B_1 \times \cdots \times B_k \in \mathcal{B}(\mathbb{R}^k)$.*

EXERCISE 2.9.4. *Show that if $B \in \mathcal{B}(\mathbb{R})$, then the set $\{(x,x) : x \in B\} \in \mathcal{B}(\mathbb{R}^2)$. Generalize this to higher dimensions.*

DEFINITION 2.9.5. *For $k \geq 1$, and $A \subset \mathbb{R}^k$, the Borel σ-field on A, denoted $\mathcal{B}(A)$, is defined by $\mathcal{B}(A) = \{B \cap A : B \in \mathcal{B}(\mathbb{R}^k)\}$. In the special case when $A \in \mathcal{B}(\mathbb{R}^k)$, one has $\mathcal{B}(A) = \{B \in \mathcal{B}(\mathbb{R}^k) : B \subset A\}$.*

Recall that in one dimension, every distribution function G on $\mathbb{R}$ determines a unique Radon measure μ on $\mathcal{B}(\mathbb{R})$ such that $\mu(I_{a,b}) = G(b) - G(a)$ for reals $a < b$. We seek the appropriate analogue of the concept of distribution function on $\mathbb{R}^k$ so that we can construct Radon measures on $\mathcal{B}(\mathbb{R}^k)$. The definition of Radon measures in higher dimensions is analogous to that on $\mathbb{R}$.

DEFINITION 2.9.6. *A measure μ on $\mathcal{B}(\mathbb{R}^k)$ is called a Radon measure on $\mathbb{R}^k$ if $\mu(A) < \infty$, for all bounded Borel subsets A of $\mathbb{R}^k$.*

Clearly, any Radon measure on $\mathbb{R}^k$ is σ-finite, in fact, σ-finite on the semifield $\mathcal{S}$. However, not all σ-finite measures on $\mathcal{B}(\mathbb{R}^k)$ are necessarily Radon measures. The reader may take a cue from the counter-example given on $\mathbb{R}$ to construct one on $\mathbb{R}^k$. Of course, any finite measure on $\mathcal{B}(\mathbb{R}^k)$ is a Radon measure.

We now come back to the construction of Radon measures from distribution functions. But, we first need to come up with the appropriate analogue of the notion of distribution functions in higher dimensions. For notational simplicity, we discuss the details for the case $k = 2$. The case of general $k > 2$ is conceptually the same, but just gets notationally complicated.

To begin with let us consider a finite measure μ on $\mathcal{B}(\mathbb{R}^2)$. Consider the function $G : \mathbb{R}^2 \to \mathbb{R}$ defined as

$$G(x_1, x_2) = \mu\{(y_1, y_2) : (y_1, y_2) \leq (x_1, x_2)\}.$$

It is easy to see that this function is right continuous, that is, for an $(x_1, x_2) \in \mathbb{R}^2$, if $x_1^n \downarrow x_1$ and $x_2^n \downarrow x_2$, then $G(x_1^n, x_2^n) \to G(x_1, x_2)$. Further, for $\mathbf{a}, \mathbf{b} \in R^2$ with $\mathbf{a} \leq \mathbf{b}$, one can use inclusion-exclusion formula to see that

$$\mu(I_{\mathbf{a},\mathbf{b}}) = \Delta_{\mathbf{a},\mathbf{b}} G, \qquad (2.9.1)$$

where $\Delta_{\mathbf{a},\mathbf{b}} G = G(b_1, b_2) - G(a_1, b_2) - G(b_1, a_2) + G(a_1, a_2)$. Since μ is a measure, we must have $\Delta_{\mathbf{a},\mathbf{b}} G \geq 0$.

One may recall that in $\mathbb{R}$, a distribution function G is related to the associated Radon measure μ through $\mu(I_{a,b}) = G(b) - G(a)$, for $a \leq b$. This implied that G must be non-decreasing. In $\mathbb{R}^2$, the property that $\Delta_{\mathbf{a},\mathbf{b}} G \geq 0$, for all $\mathbf{a}, \mathbf{b}$ in $\mathbb{R}^2$ with $\mathbf{a} \leq \mathbf{b}$, serves as the appropriate counterpart of the non-decreasing property of distribution functions on $\mathbb{R}$.

EXERCISE 2.9.7. *Consider the function $G : \mathbb{R}^2 \to \mathbb{R}$ defined as $G(x_1, x_2) = 1$, if $x_1 + x_2 \geq 0$ and $G(x_1, x_2) = 0$, if $x_1 + x_2 < 0$. Show that G is right continuous and coordinatewise non-decreasing everywhere. With $\mathbf{a} = (-1, -1)$ and $\mathbf{b} = (1, 1)$, show that $\Delta_{\mathbf{a},\mathbf{b}} G = -1 < 0$.*

DEFINITION 2.9.8. *A function $G : \mathbb{R}^2 \to \mathbb{R}$ is called a distribution function on $\mathbb{R}^2$ if (i) G is right continuous, that is, for any $(x_1, x_2) \in \mathbb{R}^2$, if $x_1^n \downarrow x_1$ and $x_2^n \downarrow x_2$, then $G(x_1^n, x_2^n) \to G(x_1, x_2)$, and (ii) $\Delta_{\mathbf{a},\mathbf{b}} G \geq 0$ for any $\mathbf{a}, \mathbf{b} \in \mathbb{R}^2$, $\mathbf{a} \leq \mathbf{b}$.*

As in $\mathbb{R}$, we first show that given any Radon measure on $\mathcal{B}(\mathbb{R}^2)$, there is a distribution function on $\mathbb{R}^2$, such that (2.9.1) holds for all $\mathbf{a}, \mathbf{b} \in \mathbb{R}^2$ with $\mathbf{a} \leq \mathbf{b}$.

2.9. Measures in Higher Dimensions

Let μ be a Radon measure on $\mathcal{B}(\mathbb{R}^2)$. Define $G: \mathbb{R}^2 \to \mathbb{R}$ by

$$G(x_1, x_2) = \begin{cases} \mu(I_{(0,0),(x_1,x_2)}) & \text{if } x_1 \geq 0, x_2 \geq 0 \\ -\mu(I_{(x_1,0),(0,x_2)}) & \text{if } x_1 \leq 0, x_2 \geq 0 \\ \mu(I_{(x_1,x_2),(0,0)}) & \text{if } x_1 \leq 0, x_2 \leq 0 \\ -\mu(I_{(0,x_2),(x_1,0)}) & \text{if } x_1 \geq 0, x_2 \leq 0 \end{cases} \qquad (2.9.2)$$

We leave it as an exercise for the reader to verify that G is right continuous and that (2.9.1) holds for all $\mathbf{a} \leq \mathbf{b}$. Thus, G is clearly a distribution function. Further, G determines μ in the sense that μ is the only measure on $\mathcal{B}(\mathbb{R}^2)$ satisfying (2.9.1) for all $\mathbf{a} \leq \mathbf{b}$. It needs to be highlighted here that the distribution function G given above may not, in general, be coordinatewise non-decreasing. This, in conjunction with Exercise 2.9.7, underlines an interesting fact that stands in sharp contrast to what happens on $\mathbb{R}$. For a function G on $\mathbb{R}^2$, the following two properties are not compatible (in the sense that neither of them imply the other): (i) G is non-decreasing coordinatewise, (ii) $\Delta_{\mathbf{a},\mathbf{b}}G \geq 0$ for all $\mathbf{a} \leq \mathbf{b}$. That (i) doesn't imply (ii) was illustrated by Exercise 2.9.7, while the reader should look at Exercise 2.9.19 and convince herself that (ii) doesn't imply (i). However, the lack of coordinatewise non-decreasing property for a distribution function in $\mathbb{R}^2$ is more than compensated by a useful property, stated in the next exercise.

EXERCISE **2.9.9.** *Let G be a distribution function on $\mathbb{R}^2$. Then, (i) for every fixed $\mathbf{a} \in \mathbb{R}^2$, the function $g_1(\mathbf{b}) = \Delta_{\mathbf{a},\mathbf{b}}G$, $\mathbf{b} \geq \mathbf{a}$, is coordinatewise non-decreasing, and (ii) for every fixed $\mathbf{b} \in \mathbb{R}^2$, the function $g_2(\mathbf{a}) = \Delta_{\mathbf{a},\mathbf{b}}G$, $\mathbf{a} \leq \mathbf{b}$, is coordinatewise non-increasing.*

Returning now to the distribution function G constructed in (2.9.2), it should be clear that, just as in the one-dimensional case, G is not the only distribution function determining the Radon measure μ. Any function that differs from G by an additive constant will also be a distribution function determining μ. However, unlike in $\mathbb{R}$, these are not the only other distribution functions determining the same μ (see Exercise 2.9.19). In fact, we will come back later with a few more interesting observations on distribution functions on $\mathbb{R}^2$.

We now come to the more important issue, namely, the converse of the above. The question is, given a distribution function $G : \mathbb{R}^2 \to \mathbb{R}$, whether there is a measure μ on $\mathcal{B}(\mathbb{R}^2)$ such that (2.9.1) holds for $\mathbf{a} \leq \mathbf{b}$. It is clear that if there is one, then it has to be unique, by the Uniqueness Theorem 2.3.7. Also, any such measure will be a Radon measure, because any bounded subset of $\mathbb{R}^2$ will be a subset of $I_{\mathbf{a},\mathbf{b}}$, for some $\mathbf{a}, \mathbf{b} \in \mathbb{R}^2$ with $\mathbf{a} < \mathbf{b}$ and $\mu(I_{\mathbf{a},\mathbf{b}}) < \infty$ by (2.9.1).

We now proceed to show that the answer to our above question is "yes". As in the one dimensional case, one may use (2.9.1) to define a non-negative set function on the class $\{I_{\mathbf{a},\mathbf{b}} : \mathbf{a}, \mathbf{b} \in \mathbb{R}^2, \mathbf{a} \leq \mathbf{b}\}$. Using now the properties of G as stated in Exercise 2.9.9, one can easily extend this μ to a set function on the semifield $\mathcal{S}$. For example, for reals a_1, a_2, b_2 with $a_2 < b_2$, we define
$$\mu\big(I_{(a_1,a_2),(\infty,b_2)}\big) = \lim_{b_1 \to \infty} \mu\big(I_{(a_1,a_2),(b_1,b_2)}\big) = \lim_{b_1 \to \infty} \Delta_{(a_1,a_2)(b_1,b_2)} G,$$
noting that the existence of the limit is guaranteed by the non-decreasing property stated in Exercise 2.9.9 (i).

REMARK 2.9.10. For any unbounded $\mathcal{S}$-set I, there is a sequence of bounded $\mathcal{S}$-sets I_n, such that $\mu(I_n) \uparrow \mu(I)$. One can take $I_n = I \cap R_n$ where $R_n = I_{(-n,-n),(n,n)}$.

Having thus defined the non-negative set function μ on $\mathcal{S}$, that satisfies (2.9.1), by definition, we have only to show now that μ is countably additive on $\mathcal{S}$. Since μ is clearly σ-finite on $\mathcal{S}$, we will then have a unique extension to a measure μ on $\mathcal{B}(\mathbb{R}^2)$. To show that μ is countably additive on $\mathcal{S}$, we go through several steps, stated in a series of lemmas.

LEMMA 2.9.11. Let $I = (a_1, a_2] \times (b_1, b_2]$ where $a_1 < a_2$ and $b_1 < b_2$ are real numbers. Consider partitions of $(a_1, a_2]$ and $(b_1, b_2]$ given by $a_1 = x_0 < x_1 < \cdots < x_m = a_2$ and $b_1 = y_0 < y_1 < \cdots < y_n = b_2$. If $I_{ij} = (x_i, x_{i+1}] \times (y_j, y_{j+1}]$ for $0 \le i < m, 0 \le j < n$, then $\mu(I) = \sum_{i,j} \mu(I_{ij})$.

In particular, if I and J are bounded $\mathcal{S}$-sets with $I \subset J$, then $\mu(I) \le \mu(J)$.

Proof. By the definition of μ, we have $\sum_{i,j} \mu(I_{ij}) = \sum_{ij} \Delta_{(x_i, y_j),(x_{i+1}, y_{j+1})} G$.

Now, for each fixed $0 \le i < m$, one has

$$\sum_{j=0}^{n-1} \Delta_{(x_i, y_j),(x_{i+1}, y_{j+1})} G = \sum_{j=0}^{n-1} [G(x_{i+1}, y_{j+1}) - G(x_i, y_{j+1}) - G(x_{i+1}, y_j) + G(x_i, y_j)]$$

$$= \sum_{j=0}^{n-1} [G(x_{i+1}, y_{j+1}) - G(x_{i+1}, y_j)] - \sum_{j=0}^{n-1} [G(x_i, y_{j+1}) - G(x_i, y_j)]$$

$$= [G(x_{i+1}, b_2) - G(x_{i+1}, b_1)] - [G(x_i, b_2) - G(x_i, b_1)],$$

the last equality following from the fact that both sums are telescopic sums. Summing this over i, $0 \le i < m$ gives

$$\sum_{i,j} \mu(I_{ij}) = \sum_{i=0}^{m-1} [G(x_{i+1}, b_2) - G(x_{i+1}, b_1)] - \sum_{i=0}^{m-1} [G(x_i, b_2) - G(x_i, b_1)]$$

$$= \sum_{i=0}^{m-1} [G(x_{i+1}, b_2) - G(x_i, b_2)] - \sum_{i=0}^{m-1} [G(x_{i+1}, b_1) - G(x_i, b_1)]$$

$$= G(a_2, b_2) - G(a_1, b_2) - G(a_2, b_1) + G(a_1, b_1),$$

once again using the fact that both sums are telescopic. Since the last expression equals $\Delta_{\mathbf{ab}} G = \mu(I)$, the proof of the first part is complete.

The second part follows by observing that if $I \subset J$, then there is a decomposition of J as in the earlier part, in which I is one of the sets $\{I_{ij}\}$. □

LEMMA 2.9.12. Let I be a bounded $\mathcal{S}$-set and $I_1, \ldots, I_p$ are disjoint non-empty $\mathcal{S}$-sets with $\bigcup_{k=1}^{p} I_k \subset I$. Then $\sum_{k=1}^{p} \mu(I_k) \le \mu(I)$.

Proof. Let $I = (a_1, a_2] \times (b_1, b_2]$ with real numbers $a_1 < a_2$ and $b_1 < b_2$. If $I_k = (a_{k1}, a_{k2}] \times (b_{k1}, b_{k2}]$, $1 \le k \le p$, then the hypothesis $\bigcup I_k \subset I$ implies that $(a_{k1}, a_{k2}] \subset (a_1, a_2]$ and $(b_{k1}, b_{k2}] \subset (b_1, b_2]$, for all k.

Let $a_1 = x_0 < x_1 < \cdots < x_m = a_2$ be the partition of $(a_1, a_2]$ induced by the points $\{a_{k1}, a_{k2}, 1 \le k \le p\}$ in $(a_1, a_2]$. Similarly, let $b_1 = y_0 < y_1 < \cdots < y_n = b_2$ be the partition of $(b_1, b_2]$ induced by the points $\{b_{k1}, b_{k2}, 1 \le k \le p\}$ in $(b_1, b_2]$.

2.9. Measures in Higher Dimensions

Let I_{ij}, $0 \leq i < m$ and $0 \leq j < n$ be as in Lemma 2.9.11. Fix k and let $\Lambda_k = \{(i,j) : 0 \leq i < m, 0 \leq j < n, I_{ij} \subset I_k\}$. Disjointness of the I_k implies that the Λ_k, for different k, are disjoint. Observe that $(i,j) \in \Lambda_k$ iff $a_{k1} \leq x_i \leq a_{k2}$ and $b_{k1} \leq y_j \leq b_{k2}$. Further, $\{I_{ij} : (i,j) \in \Lambda_k\}$ is precisely the decomposition of I_k, as in Lemma 2.9.11, by the partitions of $(a_{k1}, a_{k2}]$ and $(b_{k1}, b_{k2}]$, given by $\{x_i : a_{k1} \leq x_i \leq a_{k2}\}$ and $\{y_j : b_{k1} \leq y_j \leq b_{k2}\}$ respectively. This gives us $\mu(I_k) = \sum_{(i,j) \in \Lambda_k} \mu(I_{ij})$ by Lemma 2.9.11. Also Lemma 2.9.11 applied to I and $\{I_{ij} : 0 \leq i < m, 0 \leq j < n\}$ gives $\mu(I) = \sum_{i,j} \mu(I_{ij})$. Since μ is non-negative and Λ_k, $1 \leq k \leq p$, are disjoint, we get
$$\mu(I) = \sum_{i,j} \mu(I_{ij}) \geq \sum_{k=1}^p \sum_{(i,j) \in \Lambda_k} \mu(I_{ij}) = \sum_{k=1}^p \mu(I_k),$$
completing the proof. □

LEMMA 2.9.13. *Let I be a non-empty bounded $\mathcal{S}$-set. If $I_1, I_2, \cdots, I_n$ are disjoint $\mathcal{S}$-sets with $I \subset \bigcup_{k=1}^n I_k$, then $\mu(I) \leq \sum_{k=1}^n \mu(I_k)$.*

Proof. Because I is bounded, there is no loss of generality in assuming that each I_k, $1 \leq k \leq n$, is bounded (see Remark 2.9.10). We prove the result by induction. For $n = 1$, this is just a consequence of Lemma 2.9.11. Assume the result to be true for $n - 1$. Let $I = (a_1, a_2] \times (b_1, b_2]$. Since the point $(a_2, b_2) \in I$, it belongs to I_k for some k, and there is no loss of generality in assuming that $(a_2, b_2) \in I_n = (c_1, c_2] \times (d_1, d_2]$, say. Let $I' = (a_1, c_1] \times (b_1, b_2]$ and $I'' = (c_1, a_2] \times (b_1, d_1]$. Then I' and I'' are disjoint and $I' \cup I'' = I \cap I_n^c \subset \bigcup_{k=1}^{n-1} I_k$. Thus $I' \subset \bigcup_{k=1}^{n-1} [I_k \cap I']$ and $I'' \subset \bigcup_{k=1}^{n-1} [I_k \cap I'']$. Since $\mathcal{S}$ is closed under finite intersections, the induction hypothesis gives $\mu(I') \leq \sum_{k=1}^{n-1} \mu(I_k \cap I')$ and $\mu(I'') \leq \sum_{k=1}^{n-1} \mu(I_k \cap I'')$. But by Lemma 2.9.11, $\mu(I) = \mu(I') + \mu(I'') + \mu(I \cap I_n) \leq \sum_{k=1}^{n-1} \mu(I_k \cap I') + \sum_{k=1}^{n-1} \mu(I_k \cap I'') + \mu(I_n)$
$= \sum_{k=1}^{n-1} [\mu(I_k \cap I') + \mu(I_k \cap I'')] + \mu(I_n) \leq \sum_{k=1}^{n-1} \mu(I_k) + \mu(I_n)$ in view of Lemma 2.9.12, thus completing the proof. □

LEMMA 2.9.14. *Let I be a bounded $\mathcal{S}$-set. If $\{I_k, k \geq 1\}$ are $\mathcal{S}$-sets with $I \subset \bigcup_{k=1}^\infty I_k$, then $\mu(I) \leq \sum_{k=1}^\infty \mu(I_k)$.*

Proof. Since I is bounded, we can assume, without loss of generality, that each I_n is bounded. Let $I = (a_1, a_2] \times (b_1, b_2]$ and $I_k = (a_{k1}, a_{k2}] \times (b_{k1}, b_{k2}]$, for $k \geq 1$. Fixing $\epsilon > 0$, use right-continuity to get $\tilde{a}_1 \in (a_1, a_2)$ and $\tilde{b}_1 \in (b_1, b_2)$ such that $G(\tilde{a}_1, b_2) - G(a_1, b_2) + G(a_2, \tilde{b}_1) - G(a_2, b_1) < \epsilon/2$. Similarly, for each $k \geq 1$, first choose $\tilde{a}_{k2} > a_{k2}$ such that $G(\tilde{a}_{k2}, b_{k2}) - G(a_{k2}, b_{k2}) < \epsilon/2^{k+2}$ and then choose $\tilde{b}_{k2} > b_{k2}$ such that $G(\tilde{a}_{k2}, \tilde{b}_{k2}) - G(\tilde{a}_{k2}, b_{k2}) < \epsilon/2^{k+2}$. One can then easily check that
$$\mu(I) \leq \mu((\tilde{a}_1, a_2] \times (\tilde{b}_1, b_2]) + \frac{\epsilon}{2}, \quad \text{and}$$

$$\mu(I_k) \geq \mu((a_{k1}, \widetilde{a}_{k2}] \times (b_{k1}, \widetilde{b}_{k2}]) - \frac{\epsilon}{2^{k+1}}, \ k \geq 1 \qquad (2.9.3)$$

Since $[\widetilde{a}_1, a_2] \times [\widetilde{b}_1, b_2] \subset I \subset \bigcup_{k=1}^{\infty} (a_{k1}, a_{k2}] \times (b_{k1}, b_{k2}] \subset \bigcup_{k=1}^{\infty} (a_{k1}, \widetilde{a}_{k2}) \times (b_{k1}, \widetilde{b}_{k2})$, Heine-Borel Theorem implies that $[\widetilde{a}_1, a_2] \times [\widetilde{b}_1, b_2] \subset \bigcup_{k=1}^{n} (a_{k1}, \widetilde{a}_{k2}) \times (b_{k1}, \widetilde{b}_{k2})$, for some n, and hence, $(\widetilde{a}_1, a_2] \times (\widetilde{b}_1, b_2] \subset \bigcup_{k=1}^{n} (a_{k1}, \widetilde{a}_{k2}] \times (b_{k1}, \widetilde{b}_{k2}]$. Applying Lemma 2.9.13, we have $\mu((\widetilde{a}_1, a_2] \times (\widetilde{b}_1, b_2]) \leq \sum_{k=1}^{n} \mu((a_{k1}, \widetilde{a}_{k2}] \times (b_{k1}, \widetilde{b}_{k2}])$. Using (2.9.3) now, we get $\mu(I) \leq \sum_{k=1}^{n} \mu(I_k) + \sum_{k=1}^{n} \frac{\epsilon}{2^{k+1}} + \frac{\epsilon}{2} \leq \sum_{k=1}^{\infty} \mu(I_k) + \epsilon$. Since ϵ was arbitrary, the proof is complete. □

With Lemmas 2.9.12 and 2.9.14 now, we are ready to prove countable addititvity of μ on $\mathcal{S}$. Let $I \in \mathcal{S}$ and $\{I_k, k \geq 1\}$ be disjoint $\mathcal{S}$-sets such that $I = \bigcup_{k=1}^{\infty} I_k$. We first consider the case when I is bounded. Since for every n, $\bigcup_{k=1}^{n} I_k \subset I$, Lemma 2.9.12 gives $\sum_{k=1}^{n} \mu(I_k) \leq \mu(I)$. This being true for all n, we can let $n \to \infty$ to get $\sum_{k=1}^{\infty} \mu(I_k) \leq \mu(I)$. The other inequality comes from Lemma 2.9.14. Finally, if I is not bounded, the above argument would give $\mu(I \cap R_n) = \sum_{k} \mu(I_k \cap R_n)$, for all n, where R_n is as in Remark 2.9.10. By letting $n \to \infty$, we get $\mu(I) = \sum_{k} \mu(I_k)$, thus proving countable additivity of μ on $\mathcal{S}$.

EXERCISE **2.9.15.** Show that if $G : \mathbb{R}^2 \to \mathbb{R}$ is right continuous everywhere, meaning that, for any $\mathbf{a} \in \mathbb{R}^2$ and any sequence $\{\mathbf{a}^n\}$ in $\mathbb{R}^2$ decreasing coordinate-wise to $\mathbf{a} \in \mathbb{R}^2$, we have $\lim_n G(\mathbf{a}^n) = G(\mathbf{a})$, then for any $\mathbf{x} \in \mathbb{R}^2$ and any positive real number ϵ, there exists a positive real δ such that $\mathbf{y} \in I_{\mathbf{x}, \mathbf{x}+\delta \mathbf{e}}$ implies $|G(\mathbf{y}) - G(\mathbf{x})| < \epsilon$. Here $\mathbf{e}$ is the vector $(1, 1)$.

EXERCISE **2.9.16.** Show that for a non-decreasing $G : \mathbb{R}^2 \to \mathbb{R}$, right continuity of G is equivalent to right continuity in each coordinate.

We are now going to summarize the major results we get from what has been done. However, instead of stating the results only for $k = 2$, we state it for general k. As already noted, all the results are equally valid for any k and the proofs are essentially based on the same ideas, except that they get notationally messy. The interested reader may verify all the results for general k as exercises.

DEFINITION **2.9.17.** A function $G : \mathbb{R}^k \to \mathbb{R}$ is said to be a k-dimensional distribution function if (i) G is right continuous in each coordinate and (ii) $\Delta_{\mathbf{a},\mathbf{b}} G \geq 0$ for all $\mathbf{a}, \mathbf{b} \in \mathbb{R}^k$ with $\mathbf{a} < \mathbf{b}$, where

$$\Delta_{\mathbf{a},\mathbf{b}} G = \sum_{\epsilon \in \{0,1\}^k} (-1)^{k-\sum_i \epsilon_i} G(a_1 + \epsilon_1(b_1 - a_1), \ldots, a_k + \epsilon_k(b_k - a_k)).$$

2.9. Measures in Higher Dimensions

THEOREM 2.9.18. *Given a k-dimensional distribution function G, there is a unique Radon measure μ on $\mathcal{B}(\mathbb{R}^k)$, such that, $\mu(I_{\mathbf{a},\mathbf{b}}) = \Delta_{\mathbf{a},\mathbf{b}}G$, for all $\mathbf{a},\mathbf{b} \in \mathbb{R}^k$, $\mathbf{a} < \mathbf{b}$. Further, every Radon measure on $\mathcal{B}(\mathbb{R}^k)$ arises this way, although the associated distribution function is not unique.*

A special case is when G is the function $G(\mathbf{a}) = G(a_1, \cdots, a_k) = \Pi a_i$. One can easily verify that G is a distribution function. The corresponding Radon measure extends the notion of k-dimensional volume from k-dimensional rectangles to general Borel sets. This is known as the Lebesgue measure on $\mathcal{B}(\mathbb{R}^k)$, which will be denoted as λ^k. We will come back to this with a different way of looking at it when we discuss product measures in Chapter 5.

As noted in Theorem 2.9.18, given a Radon measures μ on $\mathcal{B}(\mathbb{R}^k)$, the distribution function associated to μ is not unique. However, unlike in the case of $\mathbb{R}$, two different distribution functions determining the same measure μ, need not differ just by an additive constant. Indeed, if G is a distribution function associated to μ, then any right continuous function H on $\mathbb{R}^k$ satisfying $\Delta_{\mathbf{a},\mathbf{b}}H = \Delta_{\mathbf{a},\mathbf{b}}G$ for all $\mathbf{a},\mathbf{b} \in \mathbb{R}^k$ with $\mathbf{a} \leq \mathbf{b}$, will also be a distributon function determining μ. On $\mathbb{R}^2$, the reader can easily verify that if G is a distribution function, then $H(x,y) = G(x,y) + H_1(x) + H_2(y)$, where H_1 and H_2 are any two real right continuous functions of one real variable, is another distribution function (and the only ones – see Exercise 2.9.19 (a)) determining the same measure as G.

EXERCISE 2.9.19. *(a) Show that if G and H are two distribution functions on $\mathbb{R}^2$, determining the same Radon measure, then $H(x,y) = G(x,y) + H_1(x) + H_2(y)$, where H_1 and H_2 are any two real right continuous functions of one real variable. (b) Show that if G and H are two distribution functions on $\mathbb{R}^k$, determining the same Radon measure, then $H(\mathbf{x}) = G(\mathbf{x}) + \sum_{i=1}^{k} H_i(\mathbf{x}^{(i)})$, where each H_i is a function of $(k-1)$ variables, which is right continuous in each coordinate, and $\mathbf{x}^{(i)}$ denotes the $(k-1)$-vector obtained by deleting the ith coordinate from $\mathbf{x}$.*

We now discuss the special case when μ is a probaility measure on $\mathbb{R}^k$. For one dimension, there is a standard way of making the distribution function unique when μ is a probability measure. Using the same idea one can define the distribution function of a probability measure on $\mathcal{B}(\mathbb{R}^k)$ so as to make the correspondence between the probability measure and its distribution function one-to-one.

DEFINITION 2.9.20. By a *Probability Distribution Function* on $\mathbb{R}^k$ is meant a function $F : \mathbb{R}^k \to [0,1]$ such that (i) F is right continuous, (ii) $\Delta_{\mathbf{a},\mathbf{b}}F \geq 0$ for every $\mathbf{a},\mathbf{b} \in \mathbb{R}^k$ with $\mathbf{a} \leq \mathbf{b}$, and, (iii) $F(\mathbf{x}) \to 0$ or 1, according as $\min_i x_i \to -\infty$ or $\min_i x_i \to \infty$.

EXERCISE 2.9.21. *Let $\{\mathbf{x}_n\}$ be a sequence in $\mathbb{R}^k$. Show that, if $\min_i x_{i,n} \to -\infty$, then for some j, $1 \leq j \leq k$, the sequence $\{x_{j,n}\}$ has a subsequence going to $-\infty$.*

It is fairly easy to see that if μ is a probability measure on $\mathcal{B}(\mathbb{R}^k)$, then the function $F(\mathbf{x}) = \mu((-\infty, x_1] \times \cdots \times (-\infty, x_k])$, $\mathbf{x} \in \mathbb{R}^k$, is a probability distribution function and is the unique probability distribution function on $\mathbb{R}^k$ that determines μ. As in the one-dimensional case, this F is called the *probability distribution function of the probability* μ. On the other hand, any probability distribution function on $\mathbb{R}^k$, by Theorem 2.9.18, determines a unique probability measure on $\mathcal{B}(\mathbb{R}^k)$. We thus have the following theorem.

THEOREM **2.9.22**. There is a one-one correspondence between probability measures μ on $\mathcal{B}(\mathbb{R}^k)$ and probability distribution functions F on $\mathbb{R}^k$, such that $\mu(I_{\mathbf{a},\mathbf{b}}) = \Delta_{\mathbf{a},\mathbf{b}} F$ for $\mathbf{a}, \mathbf{b} \in \mathbb{R}^k$ with $\mathbf{a} \leq \mathbf{b}$.

Just as in the case of one dimensions, one can restrict a probability distribution function on $\mathbb{R}^k$ to $\mathbb{Q}^k$ and get what is called a probability distribution function on $\mathbb{Q}^k$.

DEFINITION **2.9.23**. By a *Probability Distribution Function* on $\mathbb{Q}^k$ is meant a function $F : \mathbb{Q}^k \to [0,1]$ such that (i) F is right continuous, (ii) $\Delta_{\mathbf{s},\mathbf{r}} F \geq 0$ for every $\mathbf{s}, \mathbf{r} \in \mathbb{Q}^k$ with $\mathbf{s} \leq \mathbf{r}$ (iii) $F(\mathbf{r}) \to 0$ or 1, according as $\min_i r_i \to -\infty$ or $\min_i r_i \to \infty$.

The following is the k-dimensional analogue of Theorem 2.8.26

THEOREM **2.9.24**. There is a one-one correspondence between probability measures μ on $\mathcal{B}(\mathbb{R}^k)$ and probability distribution functions F on $\mathbb{Q}^k$, such that $\mu(I_{\mathbf{s},\mathbf{r}}) = \Delta_{\mathbf{s},\mathbf{r}} F$ for every pair $\mathbf{s} \leq \mathbf{r}$ in $\mathbb{Q}^k$.

We end this section with an observation of some interest. In whatever we have done so far, the only Radon measures on $\mathbb{R}$ (resp., on $\mathbb{R}^2$), for which we could construct a distribution function G that satisfies $\lim_{x \to -\infty} G(x) = 0$ (resp., $\lim_{x \wedge y \to -\infty} G(x,y) = 0$), are probability measures. The reader would, of course, see that the same can be done for finite measures as well. The question we wanted to ask is whether these are the only measures, for which this is possible. The next two results show that the same can be achieved even for some non-finite Radon measures, provide they sastisfy an additional condition.

THEOREM **2.9.25**. For a Radon measure μ on $\mathbb{R}$, the following are equivalent.
(a) $\mu\big((-\infty, a]\big) < \infty$, for some $a \in \mathbb{R}$.
(b) $\mu\big((-\infty, a]\big) < \infty$, for all $a \in \mathbb{R}$.
(c) μ admits a distribution function G satisfying $G(x) \to 0$, as $x \to -\infty$.

Proof. That bounded intervals have finite μ-measure establishes equivalence of (a) and (b). Further, if either of them hold, $G(x) = \mu\big((-\infty, x]\big)$ is real-valued and is clearly a distribution function for μ, satisfying the property in (c), by continuity from above. Finally, if (c) holds, then, for any $a \in \mathbb{R}$, continuity from below gives $\mu\big((-\infty, a]\big) = \lim_{n \to \infty} \big(G(a) - G(-n)\big) = G(a) < \infty$. □

THEOREM **2.9.26**. For a Radon measure μ on $\mathbb{R}^2$, the following are equivalent.
(a) $\mu\big((-\infty, a] \times (-\infty, b]\big) < \infty$, for all $(a,b) \in \mathbb{R}^2$.
(b) μ admits a distribution function G satisfying $G(x,y) \to 0$, as $x \to -\infty$ and $G(x,y) \to 0$, as $y \to -\infty$.
(c) μ admits a distribution function G satisfying $G(x,y) \to 0$, as $x \wedge y \to -\infty$.

Proof. If (a) holds, then the function $G(x,y) = \mu((-\infty, x] \times (-\infty, y])$, $(x,y) \in \mathbb{R}^2$, is real-valued and clearly gives a distribution function for μ that satisfies the conditions in both (b) and (c). That (c) implies (b) needs no proof. Finally, we just prove that (b) implies (a). Taking any $(a,b) \in \mathbb{R}^2$, we have $\mu((a-m,a] \times (b-n,b]) = G(a,b) - G(a-m,b) - G(a,b-n) + G(a-m,b-n)$, for all $m \geq 1$, $n \geq 1$. If we let $n \to \infty$, then using continuity from below on the left-hand-side and the hypothesis (b) on the right-hand-side, we get $\mu((a-m,a] \times (-\infty, b]) = G(a,b) - G(a-m,b)$. Letting $m \to \infty$ now, the same argument as above finally gives us $\mu((-\infty, a] \times (-\infty, b]) = G(a,b)$, which is finite. This completes the proof. □

REMARK 2.9.27. For any Radon measure μ on $\mathbb{R}$, it is clear that if $\mu((-\infty, a]) < \infty$ for some $a \in \mathbb{R}$, then the same holds for all $a \in \mathbb{R}$. However, a moment's reflection will convince the reader that the analogue of this does not hold on $\mathbb{R}^2$, namely, having $\mu((-\infty, a] \times (-\infty, b]) < \infty$ for some $(a,b) \in \mathbb{R}^2$ does not guarantee that the same holds for all $(a,b) \in \mathbb{R}^2$.

2.10 Regularity of Radon Measures

Let us recall a useful approximation property of σ-finite measures stated in Theorem 2.5.2. It says that if μ is a measure on a σ-field $\mathcal{A}$, which is σ-finite on a field $\mathfrak{F}$ generating $\mathcal{A}$, then any set in $\mathcal{A}$ can be approximated, in the sense of μ-measure, by $\mathfrak{F}_\delta$-sets from inside and by $\mathfrak{F}_\sigma$-sets from outside as closely as we want. For Radon measures on $\mathcal{B}(\mathbb{R}^k)$, for any $k \geq 1$, there is a special approximation property which uses the topology of $\mathbb{R}^k$. This is often referred to as 'regularity property' of Radon measures. What this property says is that the measure of any Borel set with finite measure can be approximated, as closely as we want, by compact sets from inside and open sets from outside.

THEOREM 2.10.1. *Let $k \geq 1$ and let μ be a Radon measure on $\mathcal{B}(\mathbb{R}^k)$. Then, for any $A \in \mathcal{B}(\mathbb{R}^k)$,*
$$\mu(A) = \sup\{\mu(K) : K \subset A, K \text{ compact}\} = \inf\{\mu(U) : U \supset A, U \text{ open}\}. \quad (2.10.1)$$
In particular, if $\mu(A) < \infty$, then for any $\epsilon > 0$, there is a compact set K_ϵ and an open set U_ϵ such that $K_\epsilon \subset A \subset U_\epsilon$ and $\mu(U_\epsilon \setminus K_\epsilon) < \epsilon$.

The following terminology is going to be used in the proof and also in the sequel. We will say that a Borel set A satisfies μ-*regularity*, if the equalities in (2.10.1) hold.

Proof. We first prove the result assuming that μ is a finite measure. We show that, for any Borel set A and any $\epsilon > 0$, there exist a compact set K_ϵ and an open U_ϵ, with $K_\epsilon \subset A \subset U_\epsilon$, such that, $\mu(U_\epsilon \setminus K_\epsilon) < \epsilon$.

We are going to use the 'good sets principle'. Denoting $\mathcal{G}$ to be the class of all Borel sets satisfying the above property for every $\epsilon > 0$, we are going to show that $\mathcal{G}$ contains $I_{\mathbf{a},\mathbf{b}}$ for every $\mathbf{a}, \mathbf{b} \in \mathbb{R}^k$ with $\mathbf{a} < \mathbf{b}$, and, that $\mathcal{G}$ is a σ-field.

Indeed, for any $\mathbf{a}, \mathbf{b} \in R^k$ with $\mathbf{a} < \mathbf{b}$ and for any $\epsilon > 0$, we can choose $\bar{\mathbf{a}} \in I_{\mathbf{a},\mathbf{b}}$ and $\bar{\mathbf{b}} > \mathbf{b}$ such that $\mu(I_{\bar{\mathbf{a}},\mathbf{b}}) > \mu(I_{\mathbf{a},\mathbf{b}}) - \epsilon/2$ and $\mu(I_{\mathbf{a},\bar{\mathbf{b}}}) < \mu(I_{\mathbf{a},\mathbf{b}}) + \epsilon/2$. This is a consequence of μ being a finite measure and hence continuous from both above and below. We now take $K_\epsilon = \{\mathbf{x} \in R^k : \bar{\mathbf{a}} \leq \mathbf{x} \leq \mathbf{b}\}$ and $U_\epsilon = \{\mathbf{x} \in R^k : \mathbf{a} < \mathbf{x} < \bar{\mathbf{b}}\}$. Then U_ϵ is clearly open and K_ϵ is closed and bounded, hence compact. Of course, by our choice of $\bar{\mathbf{a}}$ and $\bar{\mathbf{b}}$, we have $K_\epsilon \subset I_{\mathbf{a},\mathbf{b}} \subset U_\epsilon$ and $\mu(U_\epsilon \setminus K_\epsilon) < \epsilon$. This shows that $I_{\mathbf{a},\mathbf{b}} \in \mathcal{G}$.

We now show that $\mathcal{G}$ is a σ-field. Denoting $\mathbf{e}$ to be the vector $(1, \ldots, 1) \in R^k$, the sequence of compact sets $\bar{K}_n = \{\mathbf{x} \in R^k : -n\mathbf{e} \leq \mathbf{x} \leq n\mathbf{e}\}$ increases to R^k. To show that $R^k \in \mathcal{G}$, given any $\epsilon > 0$, take $U_\epsilon = R^k$ and $K_\epsilon = \bar{K}_n$, for n large enough, so that $\mu(\bar{K}_n^c) < \epsilon$. Once again, finiteness of μ is used here. To prove closure under complementation, suppose $A \in \mathcal{G}$ and let $\epsilon > 0$ be given. Choose a compact set K and an open set U such that $K \subset A \subset U$ and $\mu(U \setminus K) < \epsilon/2$. If we take $F_\epsilon = U^c$ and $V_\epsilon = K^c$, then $F_\epsilon \subset A^c \subset V_\epsilon$ and $\mu(V_\epsilon \setminus F_\epsilon) = \mu(U \setminus K) < \epsilon/2$. Clearly, V_ϵ is open, while F_ϵ is closed, but not necessarily compact. With the sets $\bar{K}_n$ as above, we take $K_\epsilon = F_\epsilon \cap \bar{K}_n$, for n sufficiently large, so that, $\mu(\bar{K}_n^c) < \epsilon/2$. Then we have a compact set $K_\epsilon \subset A^c$ and $\mu(V_\epsilon \setminus K_\epsilon) = \mu(V_\epsilon \setminus F_\epsilon) + \mu(F_\epsilon \setminus K_\epsilon) \leq \mu(V_\epsilon \setminus F_\epsilon) + \mu(\bar{K}_n^c) < \epsilon$. Finally, let $\{A_n, n \geq 1\}$ be a sequence of sets in $\mathcal{G}$ and let $A = \bigcup A_n$. Given $\epsilon > 0$, pick, for every n, a compact set K_n and an open set U_n, such that $K_n \subset A_n \subset U_n$ and $\mu(U_n \setminus K_n) < \epsilon/2^{n+1}$. Clearly $U_\epsilon = \bigcup_n U_n$ is an open set with $A \subset U_\epsilon$. While $F = \bigcup_n K_n \subset A$ and $\mu(U \setminus F) < \epsilon/2$, the set F need not be compact. However if we take $K_\epsilon = \bigcup_{n=1}^N K_n$, for an N sufficiently large, such that $\mu(F \setminus K_\epsilon) < \epsilon/2$, we will have a compact set K_ϵ with $K_\epsilon \subset A \subset U_\epsilon$ and $\mu(U_\epsilon \setminus K_\epsilon) < \epsilon$. Thus, we have shown that $\mathcal{G}$ is a σ-field, which completes the proof of the assertion in case μ is a finite measure.

Now let μ be any Radon measure on $\mathcal{B}(R^k)$. For each $n \geq 1$, if we denote B_n to be the open ball centered at $\mathbf{0} = (0, \ldots, 0)$ and of radius n, then $\mu_n(A) = \mu(A \cap B_n)$ for $A \in \mathcal{B}(R^k)$, defines a finite measure on $\mathcal{B}(R^k)$. Thus, by what has been proved above, every Borel set A satisfies μ_n-regularity, for each n. To show that every Borel set A satisfies μ-regularity, we first take any $\alpha < \mu(A)$ and get a compact set $K \subset A$, such that, $\mu(K) > \alpha$. Since $\alpha < \mu(A)$ is arbitrary, this will prove that $\mu(A) = \sup\{\mu(K) : K \subset A, K \text{ compact}\}$. Since $\mu_n(A) \uparrow \mu(A)$, therefore, for any $\alpha < \mu(A)$, we can get an N such that, $\alpha < \mu_N(A)$. Using μ_N-regularity of A, we can get a compact set $K \subset A$ such that $\mu_N(A \setminus K) < \mu_N(A) - \alpha$ and hence $\mu_N(K) > \alpha$. As a consequence, $\mu(K) \geq \mu_N(K) > \alpha$. To show the other part of μ-regularity, it suffices to show that, for any Borel set A, one has $\inf\{\mu(U) : U \supset A, U \text{ open}\} \leq \mu(A)$. For this, it is enough to show that, given $\epsilon > 0$, there is an open set $U \supset A$, such that, $\mu(U \setminus A) < \epsilon$. For each n, get open $U_n \supset A$ such that $\mu_n(U_n \setminus A) < \epsilon/2^n$. Clearly, $U_n \cap B_n$ is an open set, with $U_n \cap B_n \supset A \cap B_n$ and $\mu\big((U_n \cap B_n) \setminus (A \cap B_n)\big) = \mu_n(U_n \setminus A) < \epsilon/2^n$. Take $U = \bigcup_n (U_n \cap B_n)$. Then, U is clearly an open set with $U \supset \bigcup_n (A \cap B_n) = A$. Further, $\mu(U \setminus A) = \mu\big(\bigcup_n (U_n \cap B_n \setminus A \cap B_n)\big) \leq \sum_n \mu\big((U_n \cap B_n) \setminus (A \cap B_n)\big) < \epsilon$. This completes the proof. □

Sometimes, the first equality in (2.10.1) is called *'inner regularity'* for μ, while the second one is referred to as *'outer regularity'*. The following exercise shows that if μ is any σ-finite measure on $\mathcal{B}(\mathbb{R}^k)$, then every Borel set satisfies inner regularity for μ. It also shows, on the other hand, that outer regularity for μ may not, in general, be satisfied by all Borel sets.

EXERCISE **2.10.2.** *Let μ be any σ-finite measure on $\mathbb{R}^k$. Show that, for any Borel set A, $\mu(A) = \sup\{\mu(K) : K \subset A,\ K\ \text{compact}\}$.*
Give an example of a σ-finite measure μ on $\mathcal{B}(\mathbb{R})$ and a Borel set A for which $\mu(A) \neq \inf\{\mu(U) : A \subset U,\ U\ \text{open}\}$.

2.11 Probability Measures

A basic course in probability begins by representing outcomes of a random experiment as elements of a set, usually called the sample space and denoted by Ω. In elementary probability theory, Ω is finite most of the times and, occasionally, countably infinite. A probability $p(\omega)$ is assigned to each point $\omega \in \Omega$ and then, for every subset $A \subset \Omega$, probability of A is defined by $P(A) = \sum_{\omega \in A} p(\omega)$. The $p(\omega)$, for $\omega \in \Omega$, are non-negative numbers such that $\sum_{\omega \in \Omega} p(\omega) = 1$. This ensures that $P(\Omega) = 1$ and $0 \leq P(A) \leq 1$, for all $A \subset \Omega$. While probability thus defined, satisfies a number of nice properties and works well for some special situations, it was pointed out in Chapter 1, that this approach runs into serious limitations, whenever one goes beyond the very special types of random expreiments that are discussed in elementary probability. Indeed, we saw an example of a very natural experiment, where this method fails. The alternative turned out to be to directly assign probabilities to a special class of subsets and then hope to be able to extend this probability to other subsets, in a way so that certain desirable and natural properties of probability are maintained.

The developments in this chapter show that this is indeed possible, although we may not be able to extend to all subsets. In other words, we need to reconcile with the fact that, for the general theory of probability, we may not always have all subsets of the sample space as events, that is, sets to which probabilities can be assigned. We need to restrict ourselves to only a class of subsets, hopefully large enough, to which probabilities may be assigned. This may appear restrictive, but in reality, it is not so. This is because, in most situations, the class of subsets will be large enough for all practical purposes. Usually, all possible events of interest will be in the class. The case when sample space is countable and all subsets are events, does appear as a special case, in this set-up. This special case is often referred to as the **discrete case**.

Here is the mathematical set-up for probability in the measure theoretic framework. In this framework, a random experiment is mathematically modelled by a triplet $(\Omega, \mathcal{A}, P)$, where Ω represents the sample space, $\mathcal{A}$ represents a σ-field of events with a probability measure P defined on $\mathcal{A}$. The use of the notation P, instead of μ, for a probability measure is the standard convention in probability theory. The thing to be noted here is that the class of events is only a σ-field on Ω, usually a large one, but not the class of all subsets, in general. In what follows, such a triplet $(\Omega, \mathcal{A}, P)$ as above will be referred to as a Probability Space.

The special case when Ω is a countable set with $\mathcal{A} = \mathcal{P}(\Omega)$ and P on $\mathcal{A}$ defined as $P(A) = \sum_{\omega \in A} P(\{\omega\})$, is what we have earlier described as the discrete case. Here are some examples of probability spaces which are not discrete.

EXAMPLE 2.11.1. Let $\Omega = (0,1)$ and $\mathcal{A} =$ Borel σ-field on $(0,1)$. Then $P =$ Lebesgue measure on $\mathcal{A}$ gives a probability measure. This space $(\Omega, \mathcal{A}, P)$ is sometimes referred to as the Lebesgue space. This probability space is a natural model for the experiment of picking a random point from the interval $(0,1)$. However, we shall see later that this probability space has some kind of an universality, in the sense that, it is one probability space which can be used to capture many, if not all, random experiments.

EXAMPLE 2.11.2. Let Ω be the uncountable set of all infinite sequences of 0's and 1's. For every finite sequence $s = (\epsilon_1, \epsilon_2, \cdots, \epsilon_n)$ of 0's and 1's, let A_s denote the subset of Ω given by $A_s = \{\omega \in \Omega : \omega_i = \epsilon_i \ \forall \ 1 \leq i \leq n\}$. The class $\mathcal{S}$ consisting of all such sets A_s together with $\emptyset$ and Ω forms a semifield on Ω. P defined on $\mathcal{S}$ as

$$P(\emptyset) = 0; \quad P(\Omega) = 1; \text{ and } \quad P(A_s) = \frac{1}{2^{|s|}}, \text{ where } |s| = \text{length}(s)$$

can be shown to be a probability measure on $\mathcal{S}$ and hence extends uniquely to a probability measure on $\sigma(\mathcal{S})$. It should be noted that $\mathcal{A}$ here will not be the class of all subsets of Ω. This $(\Omega, \mathcal{A}, P)$ models an infinite number of tosses of a fair coin.

EXAMPLE 2.11.3. This is a generalization of Example 2.11.1. Take Ω to be any bounded Borel set in $\mathbb{R}^k$ with $\alpha = \lambda^k(\Omega) > 0$ and let $\mathcal{A} = \mathcal{B}(\Omega)$. Then, P defined on $\mathcal{A}$ by $P(A) = \lambda^k(A)/\alpha$, $A \in \mathcal{A}$ gives a probability measure on $\mathcal{A}$. This models picking a point from Ω at random.

Clearly, any probability measure P enjoys all the properties satisfied by finite measures. For the record, we list some of the useful properties here in a Proposition, which, of course, needs no proof.

PROPOSITION 2.11.4. Let P be a probability measure on a σ-field $\mathcal{A}$ on Ω. Then the following properties hold. All the sets that appear below are sets in $\mathcal{A}$.
(a) $A \subset B$ implies $P(A) \leq P(B)$. In particular, $0 \leq P(A) \leq 1$.
(b) $P(A^c) = 1 - P(A)$.
(c) $P(A \cup B) = P(A) + P(B) - P(A \cap B)$, and, more generally, for any $n \geq 2$,

$$P\left(\bigcup_1^n A_i\right) = S_1 - S_2 + \cdots + (-1)^{i-1} S_i + \cdots + (-1)^{n-1} S_n,$$

where S_k, for each $1 \leq k \leq n$, is sum of probabilities of all the k-fold intersections of sets from $\{A_1, \cdots, A_n\}$, that is, $S_k = \sum_{i_1 < \cdots < i_k} P\left(A_{i_1} \cap \cdots \cap A_{i_k}\right)$.

2.11. Probability Measures

(d) $P(\bigcup_n A_n) = \sum_n P(A_n)$, if $A_n \cap A_m = \emptyset$ for all $m \neq n$, or, more generally, if $P(A_m \cap A_n) = 0$, for all $m \neq n$.

(e) (Continuity) $A_n \downarrow A$ implies $P(A_n) \downarrow P(A)$ and $A_n \uparrow A$ implies $P(A_n) \uparrow P(A)$.

(f) (Borel-Cantelli Lemma) If $\sum_n P(A_n) < \infty$, then $P(\limsup A_n) = 0$.

As we will see later, an important role is played by probabilities on $(\mathbb{R}, \mathcal{B})$ and more generally on $(\mathbb{R}^k, \mathcal{B}^k)$. As seen in Theorem 2.9.22, there is a one-to-one correspondence between probabilities on $(\mathbb{R}^k, \mathcal{B}^k)$ and probability distribution functions on $\mathbb{R}^k$.

DEFINITION 2.11.5. A probability P on $\mathbb{R}^k$, or equivalently its distribution function F, is said to be *discrete* if there is a countable set $D \subset \mathbb{R}^k$ such that $P(D) = 1$.

By removing all those points $\mathbf{x}$ with $P(\{\mathbf{x}\}) = \mathbf{0}$, from the countable set as in Definition 2.11.5, we get a unique countable set $D \subset \mathbb{R}^k$ with $P(D) = 1$, such that, $P(\{\mathbf{x}\}) > 0$ for all $\mathbf{x} \in D$. This countable set D will be referred to as the *support* of P. Let $\{\mathbf{x_1}, \mathbf{x_2}, \cdots\}$ be an enumeration of the countable support set D and denote $p_i = P(\{\mathbf{x_i}\})$. Clearly, $p_i > 0$, for all i, and $\sum_i p_i = 1$. Further, the sequence $\{p_i\}$ completely determines P on $\mathcal{B}(\mathbb{R}^k)$ through the formula $P(A) = \sum_{i: \mathbf{x_i} \in A} p_i$, $A \in \mathcal{B}(\mathbb{R}^k)$. In other words, P is built up of the probabilities of its 'atoms', namely, the $\mathbf{x_i}$'s. It can be easily seen that the points $\mathbf{x_i}$ are points of discontinuity of the distribution function F. In case $k = 1$, these are the only points of discontinuity (see Example 2.8.20). In higher dimensions, the situation is slightly different. To clearly understand what is different in higher dimensions, let us consider $k = 2$. For any point $\mathbf{x} = (x_1, x_2) \in \mathbb{R}^2$, the left limit of the distribution function F at $\mathbf{x}$ is $F(\mathbf{x}-) = \lim_{\mathbf{y} < \mathbf{x}, \mathbf{y} \uparrow \mathbf{x}} F(\mathbf{y}) = P\{(-\infty, x_1) \times (-\infty, x_2)\}$, so that $F(\mathbf{x}) - F(\mathbf{x}-) = P(L_\mathbf{x})$, where

$$L_\mathbf{x} = \{\mathbf{y} : \mathbf{y} \leq \mathbf{x}, \mathbf{y} \not< \mathbf{x}\} = \{\{x_1\} \times (-\infty, x_2]\} \bigcup \{(-\infty, x_1] \times \{x_2\}\},$$

which is a union of half lines. Thus, F is discontinuous at $\mathbf{x}$ iff $P(L_\mathbf{x}) > 0$. From this, the reader may easily see that if F has a discontinuity at any one point $\mathbf{x}$, then there will be uncountably many discontinuity points of F. This is in stark contrast with what happens for $k = 1$ where a probability distribution function can have at most countably many discontinuities. Whatever we said above for $k = 2$ goes through for $k > 2$ as well.

DEFINITION 2.11.6. A probability P on $\mathcal{B}(\mathbb{R}^k)$ is called *continuous* if its distribution function is continuous everywhere. A probability P on $\mathcal{B}(\mathbb{R}^k)$ is called *nonatomic* if $P(\{\mathbf{x}\}) = 0$, for all $\mathbf{x} \in \mathbb{R}^k$.

It turns out that, for $k = 1$, a probability P is continuous if and only if it is nonatomic. This is because, if P is a probability on $\mathcal{B}(\mathbb{R})$ and F is its distribution function, then $P(\{x\}) = F(x) - F(x-)$, for every $x \in \mathbb{R}$. However, this does not hold for higher dimensions. Here is an example for $k = 2$.

EXAMPLE 2.11.7. Consider the probability P on $\mathcal{B}(\mathbb{R}^2)$ given by
$$P(A) = \tfrac{1}{2}\lambda(\{x : 0 \leq x \leq 1, (x,1) \in A\}) + \tfrac{1}{2}\lambda(\{y : 0 \leq y \leq 1, (1,y) \in A\}),$$
where λ is Lebesgue measure on $\mathbb{R}$. To make sure that the definition of P makes sense, it, of course, needs to be verified that the sets $\{x : 0 \leq x \leq 1, (x,1) \in A\}$ and $\{y : 0 \leq y \leq 1, (1,y) \in A\}$, for $A \in \mathcal{B}(\mathbb{R}^2)$, are Borel subsets of $[0,1]$, which is left as an exercise for the reader (think of 'good sets principle'). P is clearly non-atomic. However, one can easily see that all points of the form $(x,1)$, $x > 0$, and all points of the form $(1,y)$, $y > 0$, are discontinuity points of the corresponding distribution function F.

EXERCISE 2.11.8. Consider $F : \mathbb{R}^2 \to [0,1]$ defined as $F(x_1, x_2) = x_1^+ \wedge x_2^+ \wedge 1$, where x^+, for a real number x, denotes $\max\{x, 0\}$.
(a) Show that F is a continuous probability distribution function.
(b) Show also that the probability measure on $\mathcal{B}(\mathbb{R}^2)$, associated to F is given by $P(B) = \lambda(\{0 \leq x \leq 1 : (x,x) \in B\})$, for $B \in \mathcal{B}(\mathbb{R}^2)$. (It should be first verified that, for $B \in \mathcal{B}(\mathbb{R}^2)$, the set within braces is in $\mathcal{B}(\mathbb{R})$.)

EXERCISE 2.11.9. Let P be a continuous probability on $\mathbb{R}^k$.
(a) Show that, for every $\epsilon \in [0,1]$, there is a Borel set B such that $P(B) = \epsilon$.
(b) More generally, show that, given $p_1, \cdots, p_m$ with $p_i > 0$ for all i and $\sum p_i = 1$, there is a partition $A_1, A_2, \cdots, A_m$ of $\mathbb{R}^k$ by sets in $\mathcal{B}(\mathbb{R}^k)$, such that, $P(A_i) = p_i$. (Hint: Use continuity of the distribution function. For $k = 1$, it should be easy. For $k > 1$, one needs to work a bit harder.)

EXERCISE 2.11.10. Given a probability distribution function F on $\mathbb{R}$, show that either F is discrete or F is continuous or F can be uniquely written as a convex combination of a discrete probability distribution function and a continuous one. Does the above result hold for probability distribution functions on $\mathbb{R}^k$, $k > 1$?

2.12 Independence

The concept of independence is one of the most important ideas in probability theory. In fact, this is one of those concepts that are exclusive and typical to the domain of probability theory and do not have any meaning or significance in the arena of general measure theory.

In a first course in probability theory, one comes across the idea of independence of events. One first defines independence of two events and then (mutual) independence of any finite number of events. Let us recall the definitions and make some observations. Note that, in our present framework, we always have an underlying probability space $(\Omega, \mathcal{A}, P)$, wherein 'events' will always mean sets in the σ-field $\mathcal{A}$.

DEFINITION 2.12.1. Two events A_1 and A_2 are said to be *independent* if
$$P(A_1 \cap A_2) = P(A_1)P(A_2).$$
More generally, events $A_1, \cdots, A_n$ are said to be (*mutually*) *independent* if
$$P(A_{i_1} \cap \cdots \cap A_{i_k}) = P(A_{i_1}) \cdots P(A_{i_k}), \text{ for any } 1 \leq i_1 < \cdots < i_k \leq n. \quad (2.12.1)$$

2.12. Independence

Observe that the last condition involves more than one (actually, $2^n - n - 1$) equalities, and not just $P(A_1 \cap A_2 \cap \cdots \cap A_n) = P(A_1)P(A_2)\cdots P(A_n)$.

EXERCISE 2.12.2. *(a) Show that $A_1, \cdots, A_n$ are independent if and only if*

$$P(A_1^{\epsilon_1} \cap \cdots \cap A_n^{\epsilon_n}) = P(A_1^{\epsilon_1}) \cdots P(A_n^{\epsilon_n}), \text{ for all } \epsilon_1, \cdots, \epsilon_n \in \{0,1\}, \quad (2.12.2)$$

where, for a set A, the notations A^1 and A^0 are used for A and A^c respectively. (b) Show that if $A_1, \cdots, A_6$ are independent events, then (i) A_1, A_3^c and A_6 are independent events, (ii) $A_1 \cap A_3^c$, $A_2 \cup A_4 \cup A_6$ and A_5^c are independent events.

While in a first course in probability, the notion of independence remains limited to that of events, it turns out however that a more appropriate and mathematically more powerful way of thinking about and studying independence is through mutual independence of different classes of events, rather than independence of events. Indeed, that is going to be the approach that we are going to take here while discussing independence. Of course, the idea of independence of events is not going to be lost in this approach, as is illustrated by the following simple exercise and the observation following it.

EXERCISE 2.12.3. *Show that, for any two events A_1 and A_2, the following are equivalent: (i) A_1, A_2 are independent, (ii) A_1^c, A_2 are independent, (iii) A_1, A_2^c are independent, (iv) A_1^c, A_2^c are independent.*

To understand the point of Exercise 2.12.3, let us consider the two classes of events $\mathcal{A}_1 = \{\emptyset, A_1, A_1^c, \Omega\}$ and $\mathcal{A}_2 = \{\emptyset, A_2, A_2^c, \Omega\}$. What the exercise then says is that independence the two events A_1 and A_2 is equivalent to independence of the two classes $\mathcal{A}_1$ and $\mathcal{A}_2$, where the latter means independence of every pair of events C_1 and C_2, with $C_1 \in \mathcal{A}_1$ and $C_2 \in \mathcal{A}_2$. One may note here that both $\mathcal{A}_1$ and $\mathcal{A}_2$ are σ-fields. In fact, $\mathcal{A}_1 = \sigma(\{A_1\})$ and $\mathcal{A}_2 = \sigma(\{A_2\})$. Having thus established that independence of two events can be captured through (in fact, is equivalent to) independence of two σ-fields, we are now ready to formally introduce the notion of independence of classes of events.

DEFINITION 2.12.4. *Let $(\Omega, \mathcal{A}, P)$ be a probability space and let $\mathcal{C}_1, \cdots, \mathcal{C}_n$ be classes of subsets of Ω, with $\mathcal{C}_i \subset \mathcal{A}$, for each $i = 1,,\ldots,n$. The classes $\mathcal{C}_1, \cdots, \mathcal{C}_n$ are said to be independent, if for every choice of sets $C_1 \in \mathcal{C}_1, \cdots, C_n \in \mathcal{C}_n$, the events $C_1, \ldots, C_n$ are (mutually) independent.*

It is clear from the definition that if the classes $\mathcal{C}_1, \cdots, \mathcal{C}_n$ are independent, then any sub classes $\mathcal{C}_1' \subset \mathcal{C}_1, \cdots, \mathcal{C}_n' \subset \mathcal{C}_n$ will automatically be independent also. It should be noted that for every choice of $C_1, \cdots, C_n$, as in the above definition, independence of these events would require verification of either (2.12.1) or (2.12.2). Luckily, as the next exercise illustrates, the task gets siginificantly simpler, when Ω belongs to each of the classes.

EXERCISE 2.12.5. *Let $(\Omega, \mathcal{A}, P)$ be a probability space and let $\mathcal{C}_1, \cdots, \mathcal{C}_n$ be classes of $\mathcal{A}$-sets. Assume that $\Omega \in \mathcal{C}_i$, for each i. Show then that $\mathcal{C}_1, \cdots, \mathcal{C}_n$ are*

independent classes if and only if $P(C_1 \cap C_2 \cap \cdots \cap C_n) = P(C_1)P(C_2)\cdots P(C_n)$, for every choice of sets $C_1 \in \mathcal{C}_1, \ldots, C_n \in \mathcal{C}_n$.
Give a counter-example to show that the hypothesis that each class contains Ω, cannot be dispensed with.

The observation made in Exercise 2.12.5 clearly applies, in particular, when the classes of events are themselves semifields contained in $\mathcal{A}$ and that gives us the following nice result.

PROPOSITION 2.12.6. If $\mathcal{S}_1, \cdots, \mathcal{S}_n$ are semifields, each contained in $\mathcal{A}$, then $\mathcal{S}_1, \cdots, \mathcal{S}_n$ are independent if and only if
$P(A_1 \cap \cdots \cap A_n) = P(A_1)\cdots P(A_n)$, for any choice of sets $A_1 \in \mathcal{S}_1, \ldots, A_n \in \mathcal{S}_n$.

The importance of independence of a finite collection of semifields lies in the very important and extremely useful fact that it actually implies independence of the finite collection of σ-fields generated by those semifields. This is exactly what we are going to prove next. But, before that, let us quickly note the trivial fact that, if $\mathcal{C}$ is a class of sets contained in a σ-field $\mathcal{A}$, then $\sigma(\mathcal{C})$ is also contained in $\mathcal{A}$, which, in the terminolgy of Definition 2.12.7, is the same as saying that $\sigma(\mathcal{C})$ is a 'sub-σ-field' of $\mathcal{A}$.

DEFINITION 2.12.7. If $\mathcal{A}$ is a σ-field on Ω, then any σ-field $\mathcal{G}$ on Ω, such that $\mathcal{G} \subset \mathcal{A}$, is called a *sub-$\sigma$-field* of $\mathcal{A}$.

THEOREM 2.12.8. Let $(\Omega, \mathcal{A}, P)$ be a probability space and $\mathcal{S}_1, \cdots, \mathcal{S}_n$ be semifields on Ω, each contained in $\mathcal{A}$. Then the following are equivalent.
(a) $\mathcal{S}_1, \ldots, \mathcal{S}_n$ are independent.
(b) $P(A_1 \cap \cdots \cap A_n) = P(A_1)\cdots P(A_n)$, for any choice of $A_1 \in \mathcal{S}_1, \ldots, A_n \in \mathcal{S}_n$.
(c) $\sigma(\mathcal{S}_1), \ldots, \sigma(\mathcal{S}_n)$ are independent sub-σ-fields of $\mathcal{A}$.

Proof. Equivalence of (a) and (b) has been asserted in Proposition 2.12.6. Also, it is clear from the definition of independence that (c) implies (a) (see the observation made after Definition 2.12.4). Thus, we only need to verify that (b) implies (c).

Let us denote $\mathcal{A}_i = \sigma(\mathcal{S}_i)$, for $1 \leq i \leq n$. In view of Exercise 2.12.5 (or Proposition 2.12.6), all we need to show is that (b) implies

$$P(A_1 \cap \cdots \cap A_n) = P(A_1)\cdots P(A_n), \text{ for any } A_1 \in \mathcal{A}_1, \cdots, A_n \in \mathcal{A}_n \quad (2.12.3)$$

We prove (2.12.3) in steps. First, we fix $A_2 \in \mathcal{S}_2, \ldots, A_n \in \mathcal{S}_n$, and prove that (2.12.3) holds for all $A_1 \in \mathcal{A}_1$. For this, we may and do assume that $P(A_2 \cap \cdots \cap A_n) > 0$ (since otherwise, both sides of 2.12.3 equal 0, so nothing has to be proved). We define a set function Q_1 on $\mathcal{A}$ by

$$Q_1(A) = \frac{P(A \cap A_2 \cdots \cap A_n)}{P(A_2 \cap \cdots \cap A_n)}, \quad A \in \mathcal{A}.$$

It is easy to see that Q_1 is a probability on $\mathcal{A}$. Further, hypothesis (b) asserts that $Q_1 \equiv P$ on $\mathcal{S}_1$, from which it follows, by the uniqueness part of Caratheodory

2.12. Independence

Extension Theorem, that $Q_1 \equiv P$ on $\mathcal{A}_1$. But this proves that (2.12.3) holds for all $A_1 \in \mathcal{A}_1$ and $A_2 \in \mathcal{S}_2, \ldots, A_n \in \mathcal{S}_n$. As our next step, we fix $A_1 \in \mathcal{A}_1$ and $A_3 \in \mathcal{S}_3, \ldots, A_n \in \mathcal{S}_n$ and try to prove that (2.12.3) holds for all $A_2 \in \mathcal{A}_2$. Once again, we may assume that $P(A_1 \cap A_3 \cap \cdots A_n) > 0$ and use what was proved in the first step to see that the probability on $\mathcal{A}$ defined as

$$Q_2(A) = \frac{P(A_1 \cap A \cap A_3 \cdots \cap A_n)}{P(A_1 \cap A_3 \cap \cdots \cap A_n)}, \quad A \in \mathcal{A}$$

agrees wth P on $\mathcal{S}_2$. It follows that Q_2 must agree with P on $\mathcal{A}_2$. Thus, we have now proved that (2.12.3) holds for all $A_1 \in \mathcal{A}_1, A_2 \in \mathcal{A}_2$ and $A_3 \in \mathcal{S}_3, \ldots, A_n \in \mathcal{S}_n$. We can proceed step by step in the same way and complete the proof. $\square$

The proof of Theorem 2.12.8 rested crucially on using the uniqueness part of Caratheodory Extension theorem repeatedly. One can implement the same idea, but using this time the π-λ theorem (Theorem 2.3.10) in place of Caratheodory Extension theorem. This leads to the following analogue of Theorem 2.12.8, that replaces the semifields with π-systems. This turns out to be very handy sometimes. Since the proof, as pointed out, proceeds exactly along the same lines as that of Theorem 2.12.8, we omit the proof.

THEOREM **2.12.9**. Let $(\Omega, \mathcal{A}, P)$ be a probability space and suppose $\mathcal{P}_1, \cdots, \mathcal{P}_n$ are π-systems contained in $\mathcal{A}$. Then, the sub-σ-fields $\sigma(\mathcal{P}_1), \cdots, \sigma(\mathcal{P}_n)$ are independent if and only if the π-systems $\mathcal{P}_1, \cdots, \mathcal{P}_n$ are independent.

So far, we have discussed independence of only a finite number of classes of events. But, for some applications, it is necessary to talk about independence of an infinite number of classes of events. It is defined by simply requiring that every finite subcollection of these classes be independent.

DEFINITION **2.12.10**. Let $(\Omega, \mathcal{A}, P)$ be a probability space. A collection $\mathcal{C}_\alpha$, $\alpha \in \Delta$, of classes of $\mathcal{A}$-sets, are said to be (mutually) *independent* if for any $n \geq 2$ and any choice of $\alpha_1, \cdots, \alpha_n \in \Delta$, the classes $\mathcal{C}_{\alpha_1}, \cdots, \mathcal{C}_{\alpha_n}$ are independent.

The following two theorems are immediate consequences of Definition 2.12.10 and Theorems 2.12.8 and 2.12.9 and so the proofs are omitted.

THEOREM **2.12.11**. Let $(\Omega, \mathcal{A}, P)$ be a probability space and let $\mathcal{S}_\alpha$, $\alpha \in \Delta$, be a collection of semifields contained in $\mathcal{A}$. Then $\mathcal{S}_\alpha$, $\alpha \in \Delta$, are independent if and only if $\sigma(\mathcal{S}_\alpha)$, $\alpha \in \Delta$, are independent.

THEOREM **2.12.12**. Let $(\Omega, \mathcal{A}, P)$ be a probability space and let $\mathcal{P}_\alpha$, $\alpha \in \Delta$ be a collection of π-systems contained in $\mathcal{A}$. Then, $\mathcal{P}_\alpha$, $\alpha \in \Delta$ are independent if and only if $\sigma(\mathcal{P}_\alpha)$, $\alpha \in \Delta$ are independent.

We are next going to address a very important question that one is often confronted with in the context of independence. We start with a simple example. Suppose, we are given ten events A_i, $1 \leq i \leq 10$, that are known to be independent. Consider three events B_1, B_2 and B_3, where B_1 is constructed out of A_i, $1 \leq i \leq 5$ by some set operations, B_2 is constructed from A_i, $i = 6, 7$ and B_3 from A_i, $8 \leq i \leq 10$. The basic intuition of independence suggests then that B_1, B_2 and B_3 should be independent events and that intuition is indeed correct. But how does one prove this? Of course, if the exact descriptions of how the events B_1, B_2 and B_3 are constructed from the respective A_i's are available, then one can perhaps use the definition of independence of events and some standard properties of probability, to prove independence of B_1, B_2 and B_3. In fact, Exercise 2.12.2(b) was along these lines. But how do we prove a general result without requiring such specific descriptions. This is exactly where one of the many real advantages of introducing and working with the notion of independence of classes of events comes to the fore. From what we have seen so far, we know that independence of the events A_i, $1 \leq i \leq 10$ is equivalent to independence of σ-fields $\sigma(\{A_i\})$, $1 \leq i \leq 10$. In case one does not see it, one may use Theorem 2.12.9. Now comes the most important part. In fact, it will be an immediate consequence of our next theorem that independence of the σ-fields $\sigma(\{A_i\})$, $1 \leq i \leq 10$ implies independence of, say, the three σ-fields $\mathcal{G}_1 = \sigma(\{A_i, 1 \leq i \leq 5\}), \mathcal{G}_2 = \sigma(\{A_i, i = 6, 7\})$ and $\mathcal{G}_3 = \sigma(\{A_i, 7 \leq i \leq 10\})$. The events B_1, B_2 and B_3, by their construction, belong to $\mathcal{G}_1, \mathcal{G}_2$ and $\mathcal{G}_3$ respectively and so independence of B_1, B_2 and B_3 will follow.

THEOREM 2.12.13. *Let $(\Omega, \mathcal{A}, P)$ be a probability space and let $\mathcal{A}_\alpha$, $\alpha \in \Delta$, be a collection of sub-σ-fields of $\mathcal{A}$, which are independent. If $\{\Delta_\theta : \theta \in \Theta\}$ is any partition of Δ and if $\mathcal{G}_\theta = \sigma(\bigcup\{\mathcal{A}_\alpha : \alpha \in \Delta_\theta\})$, for each $\theta \in \Theta$, then the σ-fields $\mathcal{G}_\theta$, $\theta \in \Theta$, are mutually independent.*

Proof. It is easy to see that, for each $\theta \in \Theta$, the class

$$\mathcal{S}_\theta = \{A_{\alpha_1} \cap \cdots \cap A_{\alpha_n} : A_{\alpha_i} \in \mathcal{A}_{\alpha_i}, 1 \leq i \leq n, \alpha_1, \cdots, \alpha_n \in \Delta_\theta, n \geq 1\}$$

forms a semifield that generates the σ-field $\mathcal{G}_\theta$. Independence of the semifields $\mathcal{S}_\theta$, $\theta \in \Theta$, follows easily as an immediate consequence of the hypothesis of independence of the σ-fields $\mathcal{A}_\alpha$, $\alpha \in \Delta$. The required result now follows by appealing to Theorem 2.12.8. $\square$

We end with a very useful result which can be thought of as a partial converse of Borel-Cantelli Lemma (Proposition 2.11.4 (f)) and is often referred to as the 'Second Borel-Cantelli Lemma'.

THEOREM 2.12.14. (Second Borel-Cantelli Lemma)
If $\{A_n, n \geq 1\}$ is a sequence of mutually independent events in a probability space $(\Omega, \mathcal{A}, P)$ such that $\sum_n P(A_n) = +\infty$, then $P(\limsup A_n) = 1$.

Proof. : In view of the assumed independence of $\{A_n, n \geq 1\}$ and using the inequality $1 - x \leq e^{-x}$, one gets

$$P(\bigcap_{k=n}^{n+j} A_k^c) = \prod_{k=n}^{n+j} (1 - P(A_k)) \leq e^{-\sum_{k=n}^{n+j} P(A_k)}, \text{ for every } n \geq 1 \text{ and } j \geq 1.$$

Letting $j \to \infty$ and using continuity of probability and divergence of the series $\sum_n P(A_n)$, one gets $P(\bigcap_{k \geq n} A_k^c) = 0$. This being true for all n, we can conclude that $P(\bigcup_n \bigcap_{k \geq n} A_k^c) = 0$, from which $P(\limsup A_n) = 1$ follows, just by going to the complement. □

2.13 Additional Exercises

EXERCISE 2.13.1. *Show that, if $\mathcal{S}$ is a semifield on a non-empty set Ω, then any measure μ on $\mathcal{S}$ satisfies the monotonicity property, that is, $S_1, S_2 \in \mathcal{S}$ and $S_1 \subset S_2$ imply that $\mu(S_1) \leq \mu(S_2)$.*

EXERCISE 2.13.2. *Let $\mathcal{C} = \{C_1, C_2, \cdots, C_n\}$ be a finite collection of non-empty subsets of Ω. Consider sets of the form $D = \bigcap_{i=1}^{n} A_i$ where for every i, the set A_i is either C_i or C_i^c. Let $\mathcal{D}$ be the collection of all the non-empty sets obtained this way. Number of sets in $\mathcal{D}$ is at most 2^n. Show that $\mathcal{D}$ is a partition of Ω and $\sigma(\mathcal{C}) = \sigma(\mathcal{D}) = \mathfrak{F}(\mathcal{C}) = \mathfrak{F}(\mathcal{D})$. show that the number of sets in $\sigma(\mathcal{C})$ has to be 2^k for some k and hence conclude that the number of sets in any finite field (equivalently, finite σ-field) has to be some power of 2.*

EXERCISE 2.13.3. *Show that the collection of all finite unions of intervals constitutes a field on $\mathbb{R}$ but not a σ-field.*

EXERCISE 2.13.4. *Let $\mathcal{C} = \{C_1, C_2, \cdots\}$ be a countably infinite class of distinct subsets of Ω. Consider sets of the form $D = \bigcap_{i=1}^{\infty} A_i$, where for each i, the set A_i is either C_i or C_i^c. Let $\mathcal{D}$ be the collection of all the distinct non-empty sets D obtained this way.*
(a) Show that $\mathcal{D}$ gives a partition of Ω consisting of infinitely many sets and hence conclude that $\sigma(\mathcal{D})$ is uncountable.
(b) Show that $\sigma(\mathcal{D}) \subset \sigma(\mathcal{C})$ and hence conclude that $\sigma(\mathcal{C})$ is uncountable.
(c) Show that $\sigma(\mathcal{D}) = \sigma(\mathcal{C})$ if and only if $\mathcal{D}$ is countable.
(d) Conclude now that any infinite σ-field is uncountable.

EXERCISE 2.13.5. *As a special case of the above exercise, take $\Omega = \mathbb{R}$ and let $\mathcal{C}$ be the collection of all non-degenerate, non-empty intervals with rational end points. Identify $\mathcal{D}$ and $\sigma(\mathcal{D})$. Show that no set in $\mathcal{C}$ belongs to $\sigma(\mathcal{D})$.*

EXERCISE 2.13.6. *Let $\mathcal{C}$ be a class of subsets of Ω generating a σ-field $\mathcal{A}$. Let $\Omega' \subset \Omega$ be a non-empty subset. Show that $\mathcal{A}' = \{A \cap \Omega' : A \in \mathcal{A}\}$ is a σ-field on Ω' generated by $\mathcal{C}' = \{C \cap \Omega' : C \in \mathcal{C}\}$.*

EXERCISE 2.13.7. *Let $\mathcal{C}$ be a class of subsets of a non-empty set Ω and $\mathcal{A} = \sigma(\mathcal{C})$. Show that every $A \in \mathcal{A}$ belongs to $\sigma(\mathcal{C}_0)$ for some countable $\mathcal{C}_0 \subset \mathcal{C}$, possibly depending on A.*

EXERCISE 2.13.8. *(a) Let $\mathcal{S}_1$ and $\mathcal{S}_2$ be two semifields on a non-empty set Ω, generating the σ-fields $\mathcal{A}_1$ and $\mathcal{A}_2$ respectively. Denoting the smallest σ-field on Ω*

containing both $\mathcal{A}_1$ and $\mathcal{A}_2$ by $\mathcal{A}_1 \vee \mathcal{A}_2$, show that $\mathcal{S} = \{S_1 \cap S_2 : S_1 \in \mathcal{S}_1, S_2 \in \mathcal{S}_2\}$ is a semi-field generating $\mathcal{A}_1 \vee \mathcal{A}_2$.

More generally, show that if $\mathcal{S}_i$, $1 \leq i \leq n$, are semifields on Ω generating the σ-fields $\mathcal{A}_i$, $1 \leq i \leq n$, respectively, then $\mathcal{S} = \{S_1 \cap \cdots \cap S_n : S_i \in \mathcal{S}_i, 1 \leq i \leq n\}$ is a semifield and $\sigma(\mathcal{S}) = \mathcal{A}_1 \vee \cdots \vee \mathcal{A}_n$, the smallest σ-field on Ω, containing all the σ-fields $\mathcal{A}_i$, $1 \leq i \leq n$.

(b) Let $\{\mathcal{S}_\alpha, \alpha \in \Lambda\}$ be a family of semifields with $\mathcal{A}_\alpha = \sigma(\mathcal{S}_\alpha)$, $\alpha \in \Lambda$. Denoting $\bigvee_\alpha \mathcal{A}_\alpha$ to be the smallest σ-field on Ω, containing $\mathcal{A}_\alpha$ for all $\alpha \in \Lambda$, show that the class $\mathcal{S} = \{S_1 \cap \cdots \cap S_n : S_i \in \mathcal{S}_{\alpha_1}, 1 \leq i \leq n, \text{ for some } n \geq 1 \text{ and } \alpha_1, \ldots, \alpha_n \in \Lambda\}$ is a semifield generating $\bigvee_\alpha \mathcal{A}_\alpha$.

EXERCISE **2.13.9.** Recall that a σ-field is called 'countably generated' if there is a countable generating class.

(a) Show that, if $\mathcal{A}$ is a countably generated σ-field, then given any generating class $\mathcal{C}$ for $\mathcal{A}$, there is a countable subclass $\mathcal{C}' \subset \mathcal{C}$, that generates $\mathcal{A}$. [Hint: You may use Exercise 2.13.7.]

(b) Show that the countable cocountable σ-field on $\mathbb{R}$ is not countably generated. [This would show that a countably generated σ-field may contain a σ-field which is not countably generated. (How?)]

EXERCISE **2.13.10.** Let μ be a measure on a semifield $\mathcal{S}$ on Ω. Show that, for any set $S_0 \in \mathcal{S}$, the set function μ_0 defined on $\mathcal{S}$ by $\mu_0(S) = \mu(S \cap S_0)$ is a measure on $\mathcal{S}$ and that it is a finite measure if $\mu(S_0) < \infty$.

EXERCISE **2.13.11.** Let μ be a measure on a field $\mathfrak{F}$. Suppose $\{A_n\}$ is a sequence of $\mathfrak{F}$-sets such that $\mu(A_m \cap A_n) = 0$ for all $m \neq n$. Show that if $\bigcup_n A_n \in \mathfrak{F}$, then
$$\mu\left(\bigcup_n A_n\right) = \sum_n \mu(A_n).$$

EXERCISE **2.13.12.** Consider the real line and the length measure defined on the field $\mathfrak{F}$ consisting of all finite disjoint unions of intervals of the form $(a, b]$, $(-\infty, b]$, (a, ∞), $(-\infty, \infty)$, with $-\infty < a \leq b < \infty$. What would you get if you tried inner and outer approximation for length of $\mathbb{Q}$, the set of rationals, with sets from $\mathfrak{F}$.

EXERCISE **2.13.13.** Let μ be a measure on a field $\mathfrak{F}$ generated by a semifield $\mathcal{S}$. Show that μ is σ-finite on $\mathfrak{F}$ if and only if it is σ-finite on $\mathcal{S}$.

EXERCISE **2.13.14.** Let μ be a measure on a field $\mathfrak{F}$. Show that μ^* defined by (2.2.5) equals
$$\mu^*(A) = \inf\left\{\sum_n \mu(F_n) : \{F_n\} \text{ is a sequence of } \mathfrak{F}\text{-sets such that } \bigcup_n F_n \supset A\right\}.$$

EXERCISE **2.13.15.** Let P be a probability on a σ-field $\mathcal{A}$ on Ω and suppose $\Omega_0 \subset \Omega$ with $\Omega_0 \notin \mathcal{A}$ satisfies the property that $A \in \mathcal{A}$, $A \supset \Omega_0 \Longrightarrow \mu(A) = 1$.

(a) Show that $\mathcal{A}_0 = \{A \cap \Omega_0 : A \in \mathcal{A}\}$ is a σ-field on Ω_0. [The σ-field $\mathcal{A}_0$ is called the 'restriction' of $\mathcal{A}$ to Ω_0.]

(b) Define μ_0 on $\mathcal{A}_0$ by putting $\mu_0(A \cap \Omega_0) = \mu(A)$, for $A \in \mathcal{A}$. Show that μ_0 is

2.13. Additional Exercises

well-defined on $\mathcal{A}_0$ and that it defines a probability measure on $\mathcal{A}_0$.
[Sets Ω_0, as in this exercise, are called sets of "outer probability" one and the exercise shows that it is possible to restrict a given probability on a σ-field not only to sets in the σ-field with probability one, but also to sets not in the σ-field with outer probability one.]

EXERCISE **2.13.16.** On $\Omega = (0,1]$, consider the field $\mathfrak{F}$ consisting of all finite unions of intervals of the form $(a,b]$ with $0 \le a \le b \le 1$. Let $\mathcal{A} = \sigma(\mathfrak{F})$ and consider the measure μ on $\mathcal{A}$ defined as $\mu(A) = |A \cap \mathbb{Q}|$.
Show that μ is σ-finite on $\mathcal{A}$, but not σ-finite on $\mathfrak{F}$. Show also that, for every $A \in \mathcal{A}$ with $0 < \mu(A) < \infty$, both parts (a) and (b) of the Theorem 2.5.2 fail.

EXERCISE **2.13.17.** Consider the finite-cofinite field $\mathfrak{F}$ on $\mathbb{R}$. Show that $\mathcal{A} = \sigma(\mathfrak{F})$ equals the countable-cocountable σ-field.
Consider the measure μ on $\mathcal{A}$ defined as $\mu(A) = |A \cap \mathbb{N}|$. See that μ is σ-finite on $\mathfrak{F}$. Show however that both parts (a) and (b) of Theorem 2.5.2 fail for the set $A = \{2, 4, 6, \cdots\} \in \mathcal{A}$. Does this contradict Theorem 2.5.2?

EXERCISE **2.13.18.** Show that if μ is a σ-finite measure on a field $\mathfrak{F}$, then the σ-field $\mathcal{A}$ defined in (2.6.1) is the μ completion of $\sigma(\mathfrak{F})$.

EXERCISE **2.13.19.** (a) Show that for every Borel set B in $\mathbb{R}$ and every $c \in \mathbb{R}$, the sets $B + c = \{x + c : x \in B\}$ and $c \cdot B = \{c.x : x \in B\}$ are Borel sets. Further, $\lambda(B+c) = \lambda(B)$ and $\lambda(c \cdot B) = |c|\lambda(B)$.
(b) Let μ be a measure on $\mathbb{R}$ with $\mu(I_{0,1}) < \infty$. Show that if $\mu(B+c) = \mu(B)$, for all $B \in \mathcal{B}(R)$ and all $c \in R$, then there is a non-negative constant α such that $\mu(B) = \alpha\lambda(B)$ for all Borel sets B.

EXERCISE **2.13.20.** (a) Let $k > 1$. Show that, for every Borel set $B \in \mathcal{B}(\mathbb{R}^k)$, every $\mathbf{c} \in \mathbb{R}^k$ and every $c \in \mathbb{R}$, the sets $B + \mathbf{c} = \{\mathbf{x} + \mathbf{c} : \mathbf{x} \in B\}$ and $c \cdot B = \{c \cdot \mathbf{x} : \mathbf{x} \in B\}$ are Borel subsets of $\mathbb{R}^k$. Show also that $\lambda^k(B + \mathbf{c}) = \lambda^k(B)$ and $\lambda^k(c \cdot B) = |c|^k \lambda^k(B)$, where λ^k is the Lebesgue measure on $\mathcal{B}(\mathbb{R}^k)$.
(b) Let $k > 1$ and let μ be a measure on $\mathcal{B}(\mathbb{R}^k)$ with $\mu(I_{0,1}) < \infty$. Show that if μ is "translation-invariant", that is, $\mu(B+\mathbf{c}) = \mu(B)$, for all $B \in \mathcal{B}(\mathbb{R}^k)$ and all $\mathbf{c} \in \mathbb{R}^k$, than there is a non-negative constant $\alpha > 0$ such that $\mu(B) = c\lambda(B)$, for all Borel sets $B \subset \mathbb{R}^k$.
(c) Let $\mathbf{A}$ be a real orthogonal $k \times k$ matrix and let $T : \mathbb{R}^k \to \mathbb{R}^k$ denote the map $\mathbf{x} \mapsto \mathbf{Ax}$. Show that, for every set $B \in \mathcal{B}(\mathbb{R}^k)$, the set $T(B) = \{T(\mathbf{x}) : \mathbf{x} \in B\}$ is a Borel subset of $\mathbb{R}^k$ and that $\lambda^k(T(B)) = \lambda^k(B)$.

EXERCISE **2.13.21.** Let $\mathcal{L}$ denote the Lebesgue σ-field on $\mathbb{R}$, that is, $\mathcal{L}$ is the λ-completion of $\mathcal{B}(\mathbb{R})$, as defined in Section 2.8. Show that, for every $A \in \mathcal{L}$ and every $c \in R$, the sets $A + c$ and $c \cdot A$ belong to $\mathcal{L}$. Further, $\lambda(A+c) = \lambda(A)$ and $\lambda(c \cdot A) = |c|\lambda(A)$.

EXERCISE **2.13.22.** (a) Show that if $K \subset [0,1)$ is a compact set with $\lambda(K) > 0$, then the set $K - K = \{x - y : x, y \in K\}$ is a Borel set and contains a non-degenerate interval around 0.

(b) Show that if $A \in \mathcal{L}$ with $\lambda(A) > 0$, then the set $A - A = \{x - y : x, y \in A\}$ contains a non-degenerate interval around 0.

EXERCISE **2.13.23.** (a) Show that there exists a compact set $K \subset \mathbb{R}$ with positive Lebesgue measure which contains no non-degenerate interval.
(b) More generally, show that, if μ is any σ-finite measure on $\mathcal{B}(\mathbb{R})$ with $\mu(\{x\}) = 0$, for every $x \in \mathbb{R}$, then either $\mu \equiv 0$ or there exists a compact set $K \subset \mathbb{R}$ with $\mu(K) > 0$ such that K contains no non-degenerate interval.

EXERCISE **2.13.24.** (a) Show that, if P is a probability measure on $\mathcal{B}(\mathbb{R})$ such that, for each $B \in \mathcal{B}(\mathbb{R})$, $P(B)$ is either 0 or 1, then there must be an $x \in \mathbb{R}$, necessarily unique, with $P(\{x\}) = 1$.
(b) Show that the same property as in (a) holds if, in place of $\mathcal{B}(\mathbb{R})$, we have any countably generated σ-field $\mathcal{A}$ on a non-empty set Ω.
(c) Give an example of a probability P on the countable-cocountable σ-field $\mathcal{A}$ on $\mathbb{R}$ satisfying the hypothesis of (a), but $P(\{x\}) = 0$ for every $x \in \mathbb{R}$.

EXERCISE **2.13.25.** [In this exercise, you are going to show that there are subsets of $\mathbb{R}$ which are not Borel sets, in fact, not even Lebesgue measurable sets.]
Consider the equivalence relation $\sim$ on $\mathbb{R}$ defined as : $x \sim y$ if and only if $x - y$ is rational. From each of the equivalence classes resulting from the equivalence relation $\sim$, choose exactly one element $x \in [0, 1]$ and let $V \subset [0, 1]$ denote the set consisting of all the chosen points. Letting $\{r_1, r_2, \ldots\}$ to be an enumeration of all the rational numbers in $[-1, 1]$, denote $V_n = V + r_n$, $n \geq 1$.
(a) Show that the sets V_n, $n \geq 1$ are pairwise disjoint.
(b) Show that $[0, 1] \subset \bigcup_n V_n \subset [-1, 2]$.
(c) Using Exercise 2.13.21, conclude from (a) and (b) that $V \notin \mathcal{L}$.
(d) Show that neither $V - V$ nor $V^c - V^c$ contain a non-degenerate interval.
(e) Show that $V + r$ cannot contain a set of positive measure, for any rational r. Hence conclude that the outer Lebesgue measures of both V and V^c equal 1.
[The above construction of a non-Lebesgue measurable set is attributed to Vitali and the set V is usually referred to as a 'Vitali set'.]

EXERCISE **2.13.26.** Let P be a probability on a σ-field $\mathcal{A}$ on Ω and let $S \subset \Omega$ with $S \notin \mathcal{A}$.
(a) Show that $\widehat{\mathcal{A}} = \{(A_1 \cap S) \cup (A_2 \cap S^c) : A_1, A_2 \in \mathcal{A}\}$ is the smallest σ-field on Ω containing the σ-field $\mathcal{A}$ and also the set S.
In the next few steps, you will examine whether it is possible to extend P to a probability $\tilde{P}$ on $\widehat{\mathcal{A}}$ so as to have $\tilde{P}(S) = c$, where $c \in [0, 1]$ is any given number. Denoting $\alpha = \sup\{P(A) : A \in \mathcal{A}, A \subset S\}$ and $\beta = \inf\{P(A) : A \in \mathcal{A}, A \supset S\}$, it is clear that $\alpha \leq \beta$.
(b) Show that there are $\mathcal{A}$-sets $\underline{S}$ and $\overline{S}$ with $\underline{S} \subset S \subset \overline{S}$ such that $P(\underline{S}) = \alpha$ and $P(\overline{S}) = \beta$.
(c) Show that the sets $\underline{S}$ and $\overline{S}$ are essentially unique in the sense that if $\underline{S}_1$ and $\overline{S}_1$ are any other $\mathcal{A}$-sets satisfying the same properties as $\underline{S}$ and $\overline{S}$, then

2.13. Additional Exercises

$P(\underline{S}\Delta \underline{S}_1) = 0$ and $P(\overline{S}\Delta \overline{S}_1) = 0$.

(d) Show now that if either $c < \alpha$ or $c > \beta$, then there cannot be any extension $\tilde{P}$ of P to $\widehat{\mathcal{A}}$ with $\tilde{P}(S) = c$.

In what follows, assume that the given c satisfies $\alpha \leq c \leq \beta$.

(e) Show that (i) $A \subset \underline{S}$ and $A \in \widehat{\mathcal{A}}$ imply $A \in \mathcal{A}$, and, (ii) $A \subset \overline{S}^c$ and $A \in \widehat{\mathcal{A}}$ imply $A \in \mathcal{A}$.

(f) Show that any set $A \in \widehat{\mathcal{A}}$ can be written as a disjoint union of $A = A_1 \cup A_2 \cup A_3 \cup A_4$, where $A_1, \ldots, A_4$ are $\widehat{\mathcal{A}}$-sets satisfying $A_1 \subset \underline{S}$, $A_2 \subset S - \underline{S}$, $A_3 \subset \overline{S} - S$ and $A_4 \subset \overline{S}^c$.

(g) With the decomposition as in (f) above, show that there are $\mathcal{A}$-sets L_2 and L_3 with $L_2 \subset \overline{S}-\underline{S}$ and $L_3 \subset \overline{S}-\underline{S}$ such that $A_2 = L_2 \cap (\overline{S}-\underline{S})$ and $A_3 = L_3 \cap (\overline{S}-\underline{S})$.

(h) Show that the sets L_2 and L_3 in (g) are essentially unique in the sense that if M_2 and M_3 are any other $\mathcal{A}$-sets satisfying the same properties as L_2 and L_3 respectively, then $P(L_2 \Delta M_2) = 0$ and $P(L_3 \Delta M_3) = 0$.

Using the notations of (f) and (g) above, define a set function $\tilde{P}$ on $\widehat{\mathcal{A}}$ by setting
$$\tilde{P}(A) = P(A_1) + (c - \alpha)P(L_2) + (\beta - c)P(L_3) + P(A_4), \quad \text{for } A \in \widehat{\mathcal{A}}.$$

(i) Show that $\tilde{P}$ is well-defined and defines a probability on $\widehat{\mathcal{A}}$ such that $\tilde{P}(S) = c$ and $\tilde{P}(A) = P(A)$ for all $A \in \mathcal{A}$.

(j) Conclude that if $\overline{\mathcal{A}}$ is the P-completion of $\mathcal{A}$, then for any $S \subset \Omega$ with $S \notin \overline{\mathcal{A}}$, there are infinitely many extensions of P to a probability $\tilde{P}$ on the σ-field $\widehat{\mathcal{A}}$. In particular, if P denotes the Lebesgue measure restricted to the Borel σ-field $\mathcal{B}([0,1])$ on $[0,1]$, then show that, for any non-Lebesgue measurable set $A \subset [0,1]$, there are infinitely many extensions of P to the σ-field on $[0,1]$, generated by the class consising of all borel subsets and the set A.

EXERCISE **2.13.27**. Let G be a distribution function on $\mathbb{R}$.

(a) Show that $H(x) = \alpha G(x)$ is a distribution function for any $\alpha \geq 0$. How are measures associated to H and G related? What if we have taken $H(x) = G(\alpha x)$ for some $\alpha > 0$.

(b) If G' is another distribution function on $\mathbb{R}$, then show that $H = G + G'$ is a distribution function and examine how the measures associated to G and H related?

(c) Fix a and b with $-\infty \leq a < b \leq \infty$. Show that $H(x) = G((x \vee a) \wedge b)$ is a distribution function. Describe the measure associated with H in terms of that determined by G.

(d) Fix $-\infty < a < b < \infty$ and consider the function H defined by $H(x) = G(x \wedge a)$ for $x < b$; and $H(x) = G(x)$ for $x \geq b$. Show that H is a distribution function and describe the associated measure in terms of that determined by G.

EXERCISE **2.13.28**. If G_1 and G_2 are two distribution functions (respectively, probability distribution functions), then show that $G_1 \vee G_2$ and $G_1 \wedge G_2$ are both distribution functions (respectively, probability distribution functions).

EXERCISE **2.13.29**. Let μ_i, $i \geq 1$, be a sequence of finite measures on a measurable space $(\Omega, \mathcal{A})$, such that, for every $A \in \mathcal{A}$, the limit $\lim_{i \to \infty} \mu_i(A)$ exists and is finite. Denote $\mu(A) = \lim_{i \to \infty} \mu_i(A)$, for $A \in \mathcal{A}$.

(a) Show that μ defines a non-negative, finitely additive set function on $\mathcal{A}$ with $\mu(\emptyset) = 0$. Show also that there exists a finite constant $C > 0$, such that, $\sup_i \mu_i(A) \leq C$, for all $A \in \mathcal{A}$, and hence, $\mu(A) \leq C$, for all $A \in \mathcal{A}$.

Let $\{A_n\}$ be a sequence of $\mathcal{A}$-sets with $A_n \downarrow \emptyset$ and denote $a_n = \sup_i \mu_i(A_n)$, $n \geq 1$.

(b) Show that $\{a_n\}$ is a non-increasing sequence of non-negative real numbers and hence $\lim_n a_n$ exists.

Suppose, if possible, there exists $\epsilon > 0$, such that $a_n > \epsilon$, for all $n \geq 1$. $\quad (*)$

(c) Show that, for each $n \geq 1$ and every given $j \geq 1$, there exists $i > j$, such that $\mu_i(A_n) > \epsilon$. [Hint: Suppose, for some n, there exists j, such that, $\mu_i(A_n) \leq \epsilon$, for all $i > j$. Show then that, there exists $n' > n$ with $\mu_i(A_{n'}) \leq \epsilon$, for all i, contradicting $(*)$.]

(d) Show that there exist two sequences of integers $1 \leq n_1 < n_2 < n_3 < \cdots$ and $1 \leq i_1 < i_2 < i_3 < \cdots$, such that, $\mu_{i_k}(A_{n_k} \setminus A_{n_{k+1}}) > \epsilon/2$, for all $k \geq 1$.
[Hint: Taking $n_1 = 1$, get $i_1 \geq 1$ with $\mu_{i_1}(A_{n_1}) > \epsilon$ and then get $n_2 > 1 = n_1$, such that $\mu_{i_1}(A_{n_2}) < \epsilon/2$. Next, get $i_2 > i_1$, such that, $\mu_{i_2}(A_{n_2}) > \epsilon$ and then get $n_3 > n_2$ with $\mu_{i_2}(A_{n_3}) < \epsilon/2$. Repeat this process.]

(e) Denoting $B_k = A_{n_k} \setminus A_{n_{k+1}}$, $k \geq 1$, observe that B_k, $k \geq 1$ are disjoint $\mathcal{A}$-sets.

(f) For any fixed integer $m > 1$, let $S_j = \bigcup_{k \geq 1} B_{(k-1)m+j}$, for $j = 0, 1, \ldots, m-1$.
Show that $S_0, S_1, \ldots, S_{m-1}$ are disjoint $\mathcal{A}$-sets with the property that, for each j, there are infinitely many i, such that, $\mu_i(S_j) > \epsilon/2$.

(g) Choose and fix $m > 1$, such that, $C/m < \epsilon/2$, where $C > 0$ is as in (a). With $S_0, S_1, \ldots, S_{m-1}$ as in (f), show that $\mu(S_j) \geq \epsilon/2$, for each j and hence $\mu(\bigcup_j S_j) \geq m\epsilon/2$, which, by the choice of m, contradicts the observation in (a).

(h) From the contradiction, conclude that $(*)$ does not hold. Deduce that $a_n \to 0$ and hence $\mu(A_n) \to 0$. This being true for any sequence $\{A_n\}$ of $\mathcal{A}$-sets with $A_n \downarrow \emptyset$, conclude that μ is a measure on $\mathcal{A}$.

REMARK 2.13.30. What you have just proved in Exercise 2.13.29 is known as *Vitali-Hahn-Saks Theorem* and is very widely used in different branches of mathematics. Many standard proofs of this theorem use what is called Baire Category Theorem. You proved it using only basic properties of measures, not Baire Category Theorem.

Chapter 3

Measurable Functions and Integration

The primary aim of this chapter is to develop a general theory of integration, that is, a theory of integration for real valued functions defined on abstract spaces with respect to abstract measures. In particular, when applied with Lebesgue measure on the real line, this new integration theory will take us way beyond Riemann integration theory. It will enable us to define integrals for a much larger class of functions than the ones for which Riemann theory works, while, at the same time, the new integral will agree with the classical Riemann integral when the latter exists.

Before formally getting into the theory, let us recall Lebesgue's original idea of integration, as described in Chapter 1. Given a real valued function f on an interval $I \subset \mathbb{R}$ into an interval $J \subset \mathbb{R}$, the first step of Lebesgue's new idea is to partition the co-domain J into small subintervals $S_1, \cdots, S_n$. In fact, this is the most crucial departure from the Riemann theory, which starts by partitioning the domain I. Next, by considering $A_i = \{x \in I : f(x) \in S_i\}$, for each i, one gets an induced partition of I as $I = A_1 \cup \cdots \cup A_n$. This ensures that, for each i, if c_i is any arbitrarily chosen point in S_i, then the values $f(x)$, for all $x \in A_i$, will be close to c_i, because they all lie in the same small subinterval S_i. Therefore, the function which takes the constant value c_i on A_i, for each i, will be a fairly close approximation to the given function f, provided the subintervals S_i are sufficiently small. This is the crux of Lebesgue's idea, which proposes $\sum_{i=1}^{n} \left(c_i \times (\text{length of } A_i) \right)$ as an approximation for the integral of f over I. But, as noted in Chapter 1, the main problem now is that the sets A_i need not be intervals (or even finite disjoint unions of intervals) and so it is not clear what is meant by "length of A_i". This was the main issue that was handled in Chapter 2 and, as a consquence of what was done, we showed that the notion of length on $\mathbb{R}$ can be extended from intervals to a much larger class of sets, namely, the Borel sets, and even to the Lebesgue sets.

This means that Lebesgue's theory would work provided the sets A_i encountered above are Borel sets (or at least Lebesgue sets). From the way the sets A_i had been defined above, it is clear that requiring the sets A_i to be Borel sets (or Lebesgue sets) does boil down to putting a condition on the function f.

Thus, while Lebesgue's theory is a wonderful extension from Riemann's theory making the continuity condition on f unnecessary, some conditions still need to be imposed on the function f in order for the theory to go through. We will, of course, see that this condition is much weaker than continuity.

Before proceeding further, let us also observe that, in the above discussion, we could have used any other measure μ on $\mathcal{B}(\mathbb{R})$, instead of the length measure, to develop integration with respect to μ. In fact, we can push the theory even further by relaxing the requirement that the domain of the function be an interval $I \subset \mathbb{R}$. We can consider real-valued functions defined on any abstract non-empty set Ω. In order to carry out Lebesgue's essential idea, all we really need is a σ-field on Ω and a measure on the σ-field. The reader will see how it works.

Thus, even though we started with the simple motivation of trying to define an integration theory for real functions defined on intervals that allows many more functions than those covered by Riemann's theory, we will see that what we have achieved is much more. We will be able to develop a much more general and abstract theory of integration. This has far reaching consequences. In particular, the general theory of integration would include, under one umbrella, not only Riemann integrals but also finite or infinite sums, Riemann-Stieltjes integrals, expectations of random variables, among other things.

3.1 Measurable Functions

To proceed with the general theory of integration, let us first describe the set-up. The basic set-up would consist of a non-empty set, equipped with a σ-field and a measure on that σ-field. Such a triplet is called a measure space. Sometimes we may not want to fix a measure beforehand and start only with what is called a measurable space, namely, just a non-empty set equipped with a σ-field on it.

DEFINITION 3.1.1. By a *measurable space* is meant a pair $(\Omega, \mathcal{A})$, where Ω is a non-empty set and $\mathcal{A}$ is a σ-field on Ω. If μ is a measure on $\mathcal{A}$ then the triplet $(\Omega, \mathcal{A}, \mu)$ is called a *measure space*. If μ is a finite (resp. σ-finite) measure we call $(\Omega, \mathcal{A}, \mu)$ a *finite* (resp. σ-*finite*) *measure space*. A *probability space* is a measure space $(\Omega, \mathcal{A}, P)$, where the measure P on $\mathcal{A}$ is a probability measure.

Let $(\Omega, \mathcal{A}, \mu)$ be a measure space. The condition that a real function f on Ω needs to satsify in order that we may implement Lebesgue's idea to define integral of f, is captured in the following definition.

DEFINITION 3.1.2. Let $(\Omega, \mathcal{A})$ be a measurable space. A function $f : \Omega \to \mathbb{R}$ is called *measurable with respect to* $\mathcal{A}$ or simply $\mathcal{A}$-*measurable* if $\{\omega : f(\omega) \leq c\} \in \mathcal{A}$, for every $c \in \mathbb{R}$. A function $f : \mathbb{R} \to \mathbb{R}$ is said to be *Borel measurable* or simply *measurable* if it is $\mathcal{B}(\mathbb{R})$-measurable.

Note that the notion of $\mathcal{A}$-measurability of a real function f involves only the σ-field $\mathcal{A}$; it does not involve any measure on $\mathcal{A}$. If the σ-field $\mathcal{A}$ is fixed in the context, we will simply say that f is measurable to mean that f is $\mathcal{A}$-measurable.

3.1. Measurable Functions

It should be easy to see from Definition 3.1.2 that, if f is an $\mathcal{A}$-measurable function on Ω, then, for every interval $I \subset \mathbb{R}$, we have $\{\omega : f(\omega) \in I\} \in \mathcal{A}$. What the next result shows is that the seemingly simple condition in the Definition 3.1.2 of measurability actually implies much more.

PROPOSITION 3.1.3. Let $(\Omega, \mathcal{A})$ be a measurable space. Then a function $f : \Omega \to \mathbb{R}$ is $\mathcal{A}$-measurable if and only if $\{\omega : f(\omega) \in B\} \in \mathcal{A}$, for all Borel sets $B \subset \mathbb{R}$.

Proof. The 'if' part is trivial since, for all $c \in \mathbb{R}$, $(-\infty, c]$ is a Borel set in $\mathbb{R}$. Proof of the 'only if' part is a simple application of Good Sets principle. One can easily see that the class $\mathcal{G} = \{B \in \mathcal{B}(\mathbb{R}) : \{\omega : f(\omega) \in B\} \in \mathcal{A}\}$ is a σ-field. The result then follows, since measurability of f implies, by our definition, that $(-\infty, c] \in \mathcal{G}$ for all $c \in \mathbb{R}$. □

Proposition 3.1.3 shows that an equivalent criterion to define measurability, as is sometimes used, would be to require that $\{\omega : f(\omega) \in B\} \in \mathcal{A}$ for all Borel sets $B \subset \mathbb{R}$. Further, a similar application of the Good Sets principle as above, leads one easily to the following result.

PROPOSITION 3.1.4. If $\mathcal{C}$ is any class of subsets of $\mathbb{R}$ that generates $\mathcal{B}(\mathbb{R})$, then $f : \Omega \to \mathbb{R}$ is $\mathcal{A}$-measurable if and only if $\{\omega : f(\omega) \in C\} \in \mathcal{A}$, for every $C \in \mathcal{C}$.

This turns out to be an extremely useful result, since it provides one with several equivalent options when it comes to checking measurability of a given function. For example, one may use the various generating classes described in Exercise 2.2.16 to get different equivalent criteria for $\mathcal{A}$-measurability.

EXERCISE 3.1.5. *(a)* Show that every constant function is $\mathcal{A}$-measurable.
(b) Let $A \subset \Omega$ and $\alpha \neq \beta$ are real numbers. Show that the function f defined as $f(\omega) = \alpha$ for $\omega \in A$ and $f(\omega) = \beta$ for $\omega \notin A$, is an $\mathcal{A}$-measurable function if and only if $A \in \mathcal{A}$.

The simplest examples of measurable functions are what are known as indicator functions.

DEFINITION 3.1.6. For a subset $A \subset \Omega$ the *indicator function* of the set A, denoted I_A, is defined as
$$\begin{aligned} I_A(\omega) &= 1 \quad \text{if } \omega \in A \\ &= 0 \quad \text{otherwise} \end{aligned}$$

The following simple proposition is an easy consequence of Exercise 3.1.5 (b) and hence omitted.

PROPOSITION 3.1.7. I_A is $\mathcal{A}$-measurable if and only if $A \in \mathcal{A}$.

Since measurable functions will play an important role in the theory to come, let us list some useful properties showing that measurability is preserved under standard algebraic and lattice operations on functions.

PROPOSITION **3.1.8.** (a) If f is $\mathcal{A}$-measurable, then so is αf for any $\alpha \in \mathbb{R}$.
(b) If f, g are $\mathcal{A}$-measurable, then so are the functions $f + g$, fg, $\max\{f, g\}$ and $\min\{f, g\}$.
(c) If f is $\mathcal{A}$-measurable and $f(\omega) \neq 0$ for all ω, then the function $1/f$ is also $\mathcal{A}$-measurable.

Proof. (a) If $\alpha = 0$, then $\alpha f(\omega) = 0$ for all ω and is hence measurable. If $\alpha > 0$, then, for every $c \in \mathbb{R}$, $\{\omega : \alpha f(\omega) \leq c\} = \{\omega : f(\omega) \leq c/\alpha\} \in \mathcal{A}$, by definition. In case $\alpha < 0$, we have $\{\omega : \alpha f(\omega) \leq c\} = \{\omega : f(\omega) \geq c/\alpha\}$, which is in $\mathcal{A}$, by Proposition 3.1.3. Thus, in either case, αf is measurable.
(b) Denoting $\mathbb{Q}$ to be the set of rationals, it is easy to see that, for any $c \in \mathbb{R}$,

$$\{\omega : f(\omega) + g(\omega) > c\} = \bigcup_{r \in \mathbb{Q}} \{\{\omega : f(\omega) > r\} \cap \{\omega : g(\omega) > c - r\}\}.$$

Measurability of $f + g$ follows from this, by invoking Proposition 3.1.4 with the generating class $\mathcal{C} = \{(c, \infty) : c \in \mathbb{R}\}$.
In view of the above and Exercise 3.1.19, measurability of f and g will imply that f^2, g^2 and $(f + g)^2$ are all measurable. But then using (a) and the above, it follows that $fg = [(f + g)^2 - f^2 - g^2]/2$ is measurable.
Measurability of $\max\{f, g\}$ and $\min\{f, g\}$ follow from

$$\{\omega : \max\{f(\omega), g(\omega)\} \leq c\} = \{\omega : f(\omega) \leq c\} \cap \{\omega : g(\omega) \leq c\}$$

$$\{\omega : \min\{f(\omega), g(\omega)\} \leq c\} = \{\omega : f(\omega) \leq c\} \cup \{\omega : g(\omega) \leq c\}$$

(c) One has $\{\omega : 1/f(\omega) \leq c\} = \{\omega : f(\omega) < 0\} \cup \{\omega : f(\omega) \geq 1/c\} \in \mathcal{A}$, in case $c > 0$. One may similarly examine the set $\{\omega : 1/f(\omega) \leq c\}$, separately for the cases $c < 0$ and $c = 0$ to complete the proof. □

EXERCISE **3.1.9.** Let f, g be $\mathcal{A}$-measurable functions and $A \in \mathcal{A}$.
(a) Show that the function h defined as

$$\begin{aligned} h(\omega) &= f(\omega) \quad \text{if} \quad \omega \in A \\ &= g(\omega) \quad \text{if} \quad \omega \in A^c \end{aligned}$$

is measurable.
(b) Generalize the above with more than two functions and the same number of sets in $\mathcal{A}$.

An important consequence of Proposition 3.1.8 is that finite linear combinations of measurable functions are measurable. In particular, for real numbers $a_1, \cdots, a_n$ and sets $A_1, \cdots, A_n$ belonging to $\mathcal{A}$, the function $f = \sum_{i=1}^{n} a_i I_{A_i}$ is a measurable function. Such functions are called (real valued) simple functions.

DEFINITION **3.1.10.** By a (real valued) *simple $\mathcal{A}$-measurable function* is meant a function $f = \sum_{i=1}^{n} a_i I_{A_i}$, where $a_1, \cdots, a_n$ are real numbers and $A_1, \cdots, A_n$ are $\mathcal{A}$-sets.

3.1. Measurable Functions

Clearly, a simple function f takes only finitely many values. Conversely, any measurable function $f : \Omega \to \mathbb{R}$ taking finitely many values is a simple $\mathcal{A}$-measurable function. Indeed if $\{b_1, \cdots, b_k\}$ are the values taken by f, then $f = \sum b_j I_{B_j}$, where $B_j = \{\omega : f(\omega) = b_j\}$. Note that $\{B_1, \cdots, B_k\}$ gives a finite partition of Ω by $\mathcal{A}$-sets. We thus conclude that any simple $\mathcal{A}$-measurable function has a representation $\sum a_i I_{A_i}$ where $\{A_1, \cdots, A_n\}$ is a partition of Ω by $\mathcal{A}$-sets. Any representation of a simple function f as $\sum a_i I_{A_i}$ where $\{A_1, \cdots, A_n\}$ is a partition of Ω by $\mathcal{A}$-sets, is called a *canonical representation* of f.

EXERCISE 3.1.11. Show that the canonical representation $f = \sum a_i I_{A_i}$ is unique if the $\{a_i\}$ are required to be distinct.

The following proposition can be easily derived from the fact that simple $\mathcal{A}$-measurable functions are precisely those $\mathcal{A}$-measurable functions that take finitely many values.

PROPOSITION 3.1.12. If f, g are simple measurable functions and $\alpha \in \mathbb{R}$, then αf, $f + g$, $\max(f, g)$, $\min(f, g)$ are all simple measurable functions.

Our next concern will be the following. Suppose that we are given a sequence $\{f_n\}$ of real valued measurable functions. We would take the (pointwise) supremum or infimum of this sequence and ask whether we get a measurable function by doing so. Similar question would be asked about the (pointwise) limit of the sequence, if it exists. However we have to be careful. For example, taking supremum (or infimum) may not result in a real valued function (unless of course we assume that the sequence is bounded above (or below) pointwise). Same applies when we take limits because, for us, existence of limits allows infinite limits. Therefore we have to first decide what we mean by measurability of a function which may take values $\pm\infty$. As the next definition says, we do not change the criterion for measurability.

DEFINITION 3.1.13. Let $(\Omega, \mathcal{A})$ be a measurable space. A function f on Ω, taking values in $[-\infty, \infty]$, is called $\mathcal{A}$-measurable if $\{\omega : f(\omega) \leq c\} \in \mathcal{A}$ for all $c \in \mathbb{R}$.

EXERCISE 3.1.14. Show that for an $\mathcal{A}$-measurable function $f : \Omega \to [-\infty, \infty]$, the sets $\{\omega : f(\omega) = -\infty\}$ and $\{\omega : f(\omega) = \infty\}$ both belong to $\mathcal{A}$.

From now on, a measurable function will usually mean a $[-\infty, \infty]$-valued, also called *extended real valued*, measurable function. When we want to deal with a real valued measurable function, it would be explicitly stated.

PROPOSITION 3.1.15. For a sequence $\{f_n, n \geq 1\}$ of measurable functions on Ω,
(a) $\sup_n f_n$ and $\inf_n f_n$ are both measurable functions.
(b) $\limsup_n f_n$ and $\liminf_n f_n$ are both measurable functions.
(c) $f = \lim_n f_n$, if it exists pointwise, is a measurable function.

Proof. One can easily verify that, for $c \in \mathbb{R}$,

$$\{\omega : \sup_n f_n(\omega) \leq c\} = \bigcap_n \{f_n(\omega) \leq c\},$$
$$\{\inf_n f_n(\omega) < c\} = \bigcup_n \{f_n(\omega) < c\},$$
$$\{\limsup_n f_n(\omega) < c\} = \bigcup_n \{\sup_{m>n} f_m(\omega) < c\},$$
$$\{\liminf_n f_n(\omega) > c\} = \bigcup_n \{\inf_{m>n} f_m(\omega) > c\}.$$

These prove (a) and (b). Part (c) follows from (b). □

EXERCISE 3.1.16. *Show that, if $\{f_n, n \geq 1\}$ is any sequence of $\mathcal{A}$-measurable functions on Ω, then the sets*
$$\{\omega : \lim_n f_n(\omega) \text{ exists}\} \text{ and } \{\omega : \lim_n f_n(\omega) \text{ exists and is finite}\}$$
are both $\mathcal{A}$-sets.

EXERCISE 3.1.17. *Show that if $\{f_n, n \geq 1\}$ is any sequence of $\mathcal{A}$-measurable functions on Ω, then, for any $\mathcal{A}$-measurable function g on Ω, the function h defined on Ω as*
$$\begin{aligned} h(\omega) &= \lim_n f_n(\omega) &\text{if the limit exists} \\ &= g(\omega) &\text{otherwise} \end{aligned}.$$
is $\mathcal{A}$-measurable.

EXERCISE 3.1.18. *(a) Use Exercise 2.8.4 to see that any continuous (or monotone) function $f : \mathbb{R} \to \mathbb{R}$ is measurable.*
(b) Show that any right-continuous (or left-continuous) function $f : \mathbb{R} \to \mathbb{R}$ is measurable.

EXERCISE 3.1.19. *Suppose that $f : \Omega \to \mathbb{R}$ is $\mathcal{A}$-measurable and $g : \mathbb{R} \to \mathbb{R}$ is measurable. Then show that the function $h(\omega) = g(f(\omega))$ is an $\mathcal{A}$-measurable function on Ω. In particular, show that, for any $\mathcal{A}$-measurable function f on Ω, the functions $f + 13$, $[f] (= \text{integer part of } f)$, f^2, e^{-2f} and $\sin\left(\frac{3\pi}{4} f\right)$ are all $\mathcal{A}$-measurable.*

Having now allowed measurable functions to take values $\pm\infty$, a natural question is what happens to the properties listed in Proposition 3.1.8. The answer is that all those properties remain valid as long as the quantities are well-defined. For example, measurability of f and g would still imply measurability of $f + g$ provided the latter is well-defined everywhere. What do we mean by well-defined? If $f(\omega)$ and $g(\omega)$ are both real, then $f(\omega) + g(\omega)$ is well defined. Also there is no problem when at least one of them is real, since we define $\pm\infty + a = \pm\infty$ for any $a \in \mathbb{R}$. In case both $f(\omega)$ and $g(\omega)$ are infinite, $f(\omega) + g(\omega)$ still remains well-defined if both are ∞ or both are $-\infty$. We take $\infty + \infty = \infty$ and $(-\infty) + (-\infty) = -\infty$. So, the only case when $f(\omega) + g(\omega)$ remains undefined is if one of them is $+\infty$ and the other is $-\infty$. For products, the conventions are
$$a \cdot \infty = \infty, \ a \cdot (-\infty) = -\infty, \quad \text{for} \quad a \in (0, \infty];$$
$$a \cdot \infty = -\infty, \ a \cdot (-\infty) = \infty, \quad \text{for} \quad a \in [-\infty, 0);$$
$$\text{and} \quad 0 \cdot \infty = 0 \cdot (-\infty) = 0.$$

3.1. Measurable Functions

With these conventions, Proposition 3.1.8 remains valid even when f and g are $[-\infty, \infty]$-valued measurable functions. The details are left for the reader to verify in the next exercise. Of course, measurability of $f+g$ is to be asserted and proved only under the assumption that it is well-defined.

EXERCISE **3.1.20.** *Show that all the assertions of Proposition 3.1.8 remain valid for extended real valued measurable functions f and g as well. [Assume $f + g$ to be well-defined for the assertion on $f + g$.]*

EXERCISE **3.1.21.** *Let $(\Omega, \mathcal{A}, \mu)$ be a measure space and let $\overline{\mathcal{A}}$ be the μ-completion of $\mathcal{A}$. Show that an extended real valued function f on Ω is $\overline{\mathcal{A}}$-measurable iff there exists an $\mathcal{A}$-measurable function g such that $\overline{\mu}\{\omega : f(\omega) \neq g(\omega)\} = 0$, where $\overline{\mu}$ is the unique extension of μ to $\overline{\mathcal{A}}$.*

The above exercise when applied to the Borel σ-field on $\mathbb{R}$ and its λ-completion $\mathcal{L}$ (Lebesgue σ-field), says that a function $f : \mathbb{R} \to \mathbb{R}$ is *Lebesgue measurable* (meaning mesurable with respect to $\mathcal{L}$) iff there is a Borel measurable $g : \mathbb{R} \to \mathbb{R}$ such that $f = g$ outside a λ-null $\mathcal{L}$-set.

A reader who has carefully followed our brief introduction to Lebesgue's theory of integration in Chapter 1, would have realized that the idea is to approximate the functions to be integrated, by suitably chosen simple functions. One of the possible appropriate schemes of approximation is described below.

Suppose we are given an $\mathcal{A}$-measurable function f, possibly extended real valued, on Ω. Fix an integer $n \geq 1$ and consider the sets

$$A_{nk} = \{\omega : \frac{k}{2^n} \leq f(\omega) < \frac{k+1}{2^n}\}, \quad k = -n2^n, -n2^n + 1, \cdots, -1, 0, 1, \cdots, n2^n - 1,$$

and

$$A_{-n} = \{\omega : f(\omega) < -n\}; \quad A_n = \{\omega : f(\omega) \geq n\}.$$

By measurability of f, the above sets all belong to $\mathcal{A}$ and they form a finite partition of Ω. For each n, consider the real valued simple measurable function

$$f_n = \sum_{k=-n2^n}^{n2^n - 1} \frac{k}{2^n} I_{A_{nk}} + (-n) I_{A_{-n}} + n I_{A_n}.$$

We claim that $\lim_n f_n = f$. To see this, take any $\omega \in \Omega$. If $f(\omega) = -\infty$, then $f_n(\omega) = -n$ for all n, so that the result holds. The case $f(\omega) = \infty$ is similar. We next assume that $-\infty < f(\omega) < \infty$. But then, for all large n, $f(\omega) \in [-n, n)$ and therefore $\omega \in A_{nk}$ for some k. But this will imply, by definition of f_n on A_{nk}, that $0 \leq f(\omega) - f_n(\omega) < \frac{1}{2^n}$ for all large n. This completes proof of the claim.

Here are some easy exercises, which will be useful when we formally define integration in Section 3.2.

EXERCISE **3.1.22.** *If $f \geq 0$ then show that f_n are all non-negative and $f_n \uparrow f$.*

EXERCISE **3.1.23.** *If f is bounded, then show that f_n converges to f uniformly.*

EXERCISE 3.1.24. With f and the sets A_{nk}, A_n and A_{-n} as above, let

$$g_n = \sum_{k=0}^{n2^n-1} \frac{k}{2^n} I_{A_{nk}} + \sum_{k=-n2^n}^{-1} \frac{k+1}{2^n} I_{A_{nk}} + (-n)I_{A_{-n}} + nI_{A_n}.$$

Show that g_n, $n \geq 1$ is a sequence of simple functions with $|g_n| \leq |f|$ for all n, and $\lim_n g_n = f$.

For record-keeping, the next proposition summarizes the facts seen above, including those stated in the above exercises.

PROPOSITION 3.1.25. Given any $\mathcal{A}$-measurable $[-\infty, \infty]$-valued function f on Ω, there exists a sequence $\{f_n\}$ of real valued simple measurable functions such that $f_n \to f$ pointwise. Further, $\{f_n\}$ may be chosen so as to also satisfy $|f_n| \leq |f|$ for all n.

In case f is non-negative, the sequence $\{f_n\}$ may be chosen so that f_n, $n \geq 1$ are all non-negative and $f_n \uparrow f$.

If f is bounded, then the sequence $\{f_n\}$ may be chosen so as to converge to f uniformly.

3.2 Integration

Finally, we now proceed towards presenting the theory of integration. Functions to be integrated will be measurable functions on a measurable space $(\Omega, \mathcal{A})$. We also need a measure μ with respect to which the function will be integrated. So let us start with a measure space $(\Omega, \mathcal{A}, \mu)$. For most purposes, this will be a general measure space. Sometimes we may need the measure to be σ-finite or finite or even a probability measure, but we will make it explicit, when that is the case.

Our strategy for defining the integral would be roughly as follows. We will first define integrals of non-negative real valued simple functions. We will prove that the integral so defined satisfies desirable properties. We will then define integrals of general non-negative measurable functions through approximation by non-negative real valued simple functions and show that all the properties extend. At the final stage, a general measurable function will be written as the difference of two non-negative measurable functions and the integral will be defined as the difference of the integrals of these two functions, provided, of course, the difference is well-defined. We will again show that desirable properties of integrals are retained.

Let $(\Omega, \mathcal{A}, \mu)$ be a measure space. Let f be a non-negative real valued simple measurable function. If $f = \sum_{i=1}^{k} a_i I_{A_i}$ is a canonical representation, we define integral of f with respect to μ, to be denoted $\int f d\mu$, by

$$\int f d\mu = \sum_{i=1}^{k} a_i \mu(A_i).$$

3.2. Integration

Since the terms in the sum are all non-negative, the sum is always well defined. Note however that $\int f d\mu$ may take the value ∞. This will happen whenever $\mu(A_i) = \infty$ for some i with $a_i > 0$. An important point to note now is that the value of the integral $\int f d\mu$ does not depend on the particular canonical representation used to define it. To see this, let $f = \sum_{j=1}^{m} b_j I_{B_j}$ be another canonical representation for the same f. It is then easy to see that, for any pair (i,j) such that $A_i \cap B_j \neq \emptyset$, one must have $a_i = b_j$. As a consequence, for every pair (i,j) we have $a_i \mu(A_i \cap B_j) = b_j \mu(A_i \cap B_j)$. Using additivity of μ and the fact that $\{B_j, 1 \leq j \leq m\}$ and $\{A_i, 1 \leq i \leq k\}$ are both partitions of Ω, one obtains

$$\sum_{i=1}^{k} a_i \mu(A_i) = \sum_{i=1}^{k} \sum_{j=1}^{m} a_i \mu(A_i \cap B_j), \text{ while, } \sum_{j=1}^{m} b_j \mu(B_j) = \sum_{j=1}^{m} \sum_{i=1}^{k} b_j \mu(A_i \cap B_j),$$

from which equality of the two sums follows.

Having thus (unambiguously) defined the integral $\int f d\mu$ for non-negative real valued simple functions, let us observe its properties.

PROPOSITION 3.2.1. Let f and g be non-negative real valued simple measurable functions and α be any non-negative real number.

(a) (Linearity) Then, αf and $f + g$ are non-negative real valued simple functions and

$$\int (\alpha f) d\mu = \alpha \int f d\mu, \quad \int (f+g) d\mu = \int f d\mu + \int g d\mu.$$

(b) (Monotonicity) If $f \leq g$ pointwise, then $\int f d\mu \leq \int g d\mu$.

Proof. (a) Let $f = \sum_i a_i I_{A_i}$ and $g = \sum_j b_j I_{B_j}$ be canonical representations of f and g respectively. Then $\alpha f = \sum_i (\alpha a_i) I_{A_i}$ is a canonical representation of αf, so that

$$\int (\alpha f) d\mu = \sum_i \alpha a_i \mu(A_i) = \alpha \sum_i a_i \mu(A_i) = \alpha \int f d\mu.$$

Next, $f + g = \sum_i \sum_j (a_i + b_j) I_{A_i \cap B_j}$ is a canonical representation of $f + g$ and so

$$\int (f+g) d\mu = \sum_i \sum_j (a_i + b_j) \mu(A_i \cap B_j) = \sum_i \sum_j a_i \mu(A_i \cap B_j) + \sum_i \sum_j b_j \mu(A_i \cap B_j).$$

But, by the additivity of μ, the first sum in the right-hand-side above equals $\sum_i a_i \mu(A_i) = \int f d\mu$, while the second sum equals $\sum_j b_j \mu(B_j) = \int g d\mu$ and that proves the result.

(b) If $f = \sum_i a_i I_{A_i}$ and $g = \sum_j b_j I_{B_j}$ are canonical representations of f and g, then $f \leq g$ pointwise will imply that, for any pair (i,j) with $A_i \cap B_j \neq \emptyset$, we must have $a_i \leq b_j$. This implies that $a_i \mu(A_i \cap B_j) \leq b_j \mu(A_i \cap B_j)$ holds for every pair (i,j). The inequality $\int f d\mu \leq \int g d\mu$ now follows easily using ideas as above. □

COROLLARY 3.2.2. If $f = \sum_i a_i I_{A_i}$, with $a_i \geq 0$ for all i, is any representation of a non-negative real valued simple function f, then $\int f d\mu = \sum a_i \mu(A_i)$.

Proof. This is an immediate consequence of the linearity property proved in Proposition 3.2.1, if one simply observes that I_A, for any $A \in \mathcal{A}$, is a non-negative real valued simple function with a canonical representation $I_A = 1 \cdot I_A + 0 \cdot I_{A^c}$, so that $\int I_A \, d\mu = \mu(A)$. □

We now proceed to the next step which is to define $\int f d\mu$ for all non-negative (not necessarily real valued) $\mathcal{A}$-measurable functions. As mentioned earlier, the already defined notion of integrals of non-negative real valued simple functions are going to be used here. For any non-negative $\mathcal{A}$-measurable function f, we define

$$\int f d\mu = \sup \left\{ \int h d\mu : 0 \leq h \leq f, \ h \text{ real valued simple measurable} \right\}. \quad (3.2.1)$$

Here are some trivial observations. Firstly, the set on the right hand side is non-empty, because $h \equiv 0$ is a simple function satsifying $0 \leq h \leq f$. This also shows that $\int f d\mu \geq 0$, though it may take the value $+\infty$, since the set on the right may not be bounded above. Next, if f is a non-negative real valued simple function then the above gives the same value for $\int f d\mu$, as defined earlier. This follows from the monotonicity property proved in Proposition 3.2.1.

One seeming drawback of the above definition of $\int f d\mu$ is that it does not provide any straightforward formula and, therefore, is not often very convenient to work with when it comes to proving various properties of the integral. As a possible alternative definition, we may recall that, by Proposition 3.1.25, any non-negative measurable function f is an increasing limit of a sequence $\{f_n\}$ of non-negative real valued simple functions. This of course implies, by Proposition 3.2.1, that the sequence $\{\int f_n d\mu\}$ in a non-decreasing sequence in the set $[0, \infty]$ and hence has a limit in $[0, \infty]$. It would, therefore, be very natural to take the limit $\lim_n \int f_n d\mu$ as a possible definition of $\int f d\mu$. But, one immediately realizes that a possible problem with such a definition is that the value of $\int f d\mu$ would then seem to depend not just on f, but also on the sequence $\{f_n\}$ of simple functions that one uses to approximate f. On the other hand, the definition of $\int f d\mu$ given in (3.2.1) presents no such ambiguities. Luckily for us, the next result makes everything work out fine, by showing that, for any non-negative measurable f, *whatever be our choice* of a sequence $\{f_n\}$ of non-negative simple real valued functions with $f_n \uparrow f$, the limit $\lim_n \int f_n d\mu$ always equals $\int f d\mu$, as defined in (3.2.1).

PROPOSITION **3.2.3.** *Let f be a non-negative $\mathcal{A}$-measurable function and $\{f_n\}$ be any sequence of non-negative real valued simple functions increasing to f. Then*

$$\int f d\mu = \lim_n \int f_n \, d\mu.$$

Proof. By Propositin 3.2.1(b), $\{\int f_n d\mu\}$ is a non-decreasing sequence and so $\lim_n \int f_n d\mu$ exists. Further, since $0 \leq f_n \leq f$ for all n, the definition of $\int f d\mu$ implies $\int f_n \, d\mu \leq \int f d\mu$ for all n and therefore, we have $\lim_n \int f_n \, d\mu \leq \int f d\mu$. To show now that $\int f d\mu \leq \lim_n \int f_n \, d\mu$, it is enough to prove that if h is any

3.2. Integration

non-negative real valued simple function with $h \leq f$, then $\int h d\mu \leq \lim_n \int f_n d\mu$.
Let $h = \sum_{j=1}^{m} b_j I_{B_j}$ be a canonical representation. Fix $0 < \alpha < 1$ and define $\Omega_n = \{\omega : f_n(\omega) \geq \alpha h(\omega)\}$. Clearly, $\{\Omega_n, n \geq 1\}$ is an increasing sequence of $\mathcal{A}$-sets. We claim that $\Omega_n \uparrow \Omega$. If $h(\omega) = 0$, then $f_n(\omega) \geq 0 = \alpha h(\omega)$, so that $\omega \in \Omega_n$ for all n. On the other hand, if $h(\omega) > 0$, then $\alpha h(\omega) < h(\omega) \leq f(\omega)$ so that, for all large n, we must have $f_n(\omega) \geq \alpha h(\omega)$, that is, $\omega \in \Omega_n$. Thus, the claim is proved. Now, $f_n, f_n I_{\Omega_n}$ and $\alpha h I_{\Omega_n}$ are non-negative real valued simple functions satisfying $f_n \geq f_n I_{\Omega_n} \geq \alpha h I_{\Omega_n}$. Therefore, using Proposition 3.2.1 and Corollary 3.2.2, we get

$$\int f_n d\mu \geq \alpha \int h I_{\Omega_n} d\mu = \alpha \sum_{j=1}^{m} b_j \mu(B_j \cap \Omega_n), \text{ for each } n.$$

Letting $n \to \infty$ and using $\mu(B_j \cap \Omega_n) \uparrow \mu(B_j)$ for each j, we get

$$\lim_n \int f_n d\mu \geq \alpha \sum_{j=1}^{m} b_j \mu(B_j) = \alpha \int h d\mu.$$

Since the above inequality holds for any $\alpha < 1$, we now let $\alpha \uparrow 1$ to conclude that

$$\lim_n \int f_n d\mu \geq \int h d\mu,$$

as was to be shown. This completes the proof. □

The next exercise is an immediate application of the Proposition 3.2.3 and it shows that the integral of any non-negative simple function, even if it takes the value $+\infty$, is given by the same formula that was used for defining the integral of real valued non-negative simple functions.

EXERCISE **3.2.4.** Let $f = \sum_{i=1}^{k} a_i I_{A_i}$, with $A_i \in \mathcal{A}$ and $0 \leq a_i \leq \infty$, for each i. Show that $\int f d\mu = \sum_i a_i \mu(A_i)$.

We now proceed to prove linearity and monotonicity properties for integrals of non-negative measurable functions.

PROPOSITION **3.2.5.** Let f and g be non-negative measurable functions.
(a) (Linearity) $\int (\alpha f + \beta g) d\mu = \alpha \int f d\mu + \beta \int g d\mu$, for any $\alpha, \beta \in [0, \infty)$.
(b) (Monotonicity) If $f \leq g$ pointwise, then $\int f d\mu \leq \int g d\mu$.

Proof. Let $\{f_n\}$ and $\{g_n\}$ be any two sequences of non-negative real valued simple functions with $f_n \uparrow f$ and $g_n \uparrow g$. Then, by Proposition 3.2.1(a), $(\alpha f_n + \beta g_n)$ is a non-negative real valued simple function, for each n, and

$$\int (\alpha f_n + \beta g_n) d\mu = \alpha \int f_n d\mu + \beta \int g_n d\mu.$$

Since $(\alpha f_n + \beta g_n) \uparrow (\alpha f + \beta g)$, the required result follows from Proposition 3.2.3. Part (b) is an immediate consequence of the definition of $\int f d\mu$ and $\int g d\mu$, since, under the hypothesis, any non-negative real valued simple function h with $h \leq f$ will also satisfy $h \leq g$. □

We now come to the final step of defining integral for any $\mathcal{A}$-measurable function taking values in $[-\infty, \infty]$. For any extended real valued function f on Ω, the *positive part* of f and the *negative part* of f are defined as
$$f^+ = \max\{f, 0\}, \quad f^- = \max\{-f, 0\}.$$
By the definition, both f^+ and f^- are clearly non-negative functions. Also, if f is $\mathcal{A}$-measurable, then so are f^+ and f^- (see Proposition 3.1.8 (b)). The next easy exercise contains some useful facts about the positive and negative parts of a function.

EXERCISE **3.2.6.** *(a) Show that $f = f^+ - f^-$ and $|f| = f^+ + f^-$. Show also that the sets $\{\omega : f^+(\omega) \neq 0\}$ and $\{\omega : f^-(\omega) \neq 0\}$ are disjoint. Further, show that f is non-negative (resp. non-positive) if and only if $f^- = 0$ (resp. $f^+ = 0$).*
(b) Show that if $f = g - h$, where g, h are non-negative functions with
$$\{\omega : g(\omega) \neq 0\} \cap \{\omega : h(\omega) \neq 0\} = \emptyset,$$
then $g = f^+$ and $h = f^-$.

For any $\mathcal{A}$-measurable function f on Ω, its positive and negative parts f^+ and f^- are both non-negative measurable functions and therefore, their integrals $\int f^+ d\mu$ and $\int f^- d\mu$ are both defined. Since $f = f^+ - f^-$, it would be natural to define the integral of f by the formula $\int f d\mu = \int f^+ d\mu - \int f^- d\mu$. However, there is a slight problem with the righthand side. As integrals of non-negative measurable functions, each of $\int f^+ d\mu$ and $\int f^- d\mu$ can take the value $+\infty$. So, we have to exercise a little caution with $\int f^+ d\mu - \int f^- d\mu$. In case at least one of the two integrals is finite, the quantity $\int f^+ d\mu - \int f^- d\mu$ is well defined, though it may take the value $+\infty$ or $-\infty$. But when both the integrals equal $+\infty$, the difference is undefined and hence we cannot define $\int f d\mu$. This leads to the following.

DEFINITION **3.2.7.** For an $\mathcal{A}$-measurable function f on a measure space $(\Omega, \mathcal{A}, \mu)$, the *integral of f with respect to μ* is defined by the formula
$$\int f d\mu = \int f^+ d\mu - \int f^- d\mu,$$
provided at least one of the two integrals on the right side is finite. Otherwise, we say that $\int f d\mu$ does not exist (or is not defined). In case both the integrals $\int f^+ d\mu$ and $\int f^- d\mu$ are finite, the integral $\int f d\mu$ is also finite and we say that f is *μ-integrable* (or, simply, *integrable*).

It is worth noting here that f being μ-integrable is a stronger statement than saying $\int f d\mu$ exists. When $\int f d\mu$ exists, its value may still be $\pm\infty$. To say that f is μ-integrable means not only that $\int f d\mu$ exists but also that it is finite. Note for the record that if f is non-negative, then $f^- = 0$ and $f = f^+$, that is, $f \equiv f^+$. Thus, for a non-negative f, the above definition does not change $\int f d\mu$ from how it was defined earlier.

To avoid any possible misconception, it is important to point out here that it is not the fact that f is allowed to be extended real valued which is causing the integral $\int f d\mu$ to be not defined. This may happen for real valued f as well.

3.2. Integration

Having thus arrived at a definition of integral for all measurable functions, it is time to verify that this integral as defined in 3.2.7 satisfies all the properties that are naturally expected of integrals to satisfy.

THEOREM **3.2.8.** Let f, g be measurable functions on a measure space $(\Omega, \mathcal{A}, \mu)$.
(a) (Linearity) (i) If $\int f d\mu$ exists, then so does $\int (\alpha f) d\mu$, for any $\alpha \in R$, and it equals $\alpha \int f d\mu$.
(ii) Suppose $f(w) + g(w)$ is well defined for all w, the integrals $\int f d\mu$ and $\int g d\mu$ both exist and $\int f d\mu + \int g d\mu$ is well defined. Then the integral $\int (f+g) d\mu$ exists and equals $\int f d\mu + \int g d\mu$.
(b) (Monotonicity) If $f \leq g$ and both $\int f d\mu$ and $\int g d\mu$ exist, then $\int f d\mu \leq \int g d\mu$.

Proof. (a)(i) : If $\alpha \geq 0$, then $(\alpha f)^+ = \alpha f^+$, $(\alpha f)^- = \alpha f^-$ and the result follows from Proposition 3.2.5(a). If $\alpha < 0$, then $(\alpha f)^+ = (-\alpha) f^-$, $(\alpha f)^- = (-\alpha) f^+$ and, since $-\alpha > 0$, the required result follows again from Proposition 3.2.5(a).

(a)(ii) : We first claim that, under the hypothesis, one must have *either* both $\int f^+ d\mu < \infty$ and $\int g^+ d\mu < \infty$ or both $\int f^- d\mu < \infty$ and $\int g^- d\mu < \infty$. To see this, suppose first that one of $\int f^+ d\mu$ and $\int g^+ d\mu$, say, $\int f^+ d\mu$ equals $+\infty$. In this case, existence of $\int f d\mu$ forces that $\int f^- d\mu < \infty$, so that $\int f d\mu = \infty$. But then, the hypothesis that $\int f d\mu + \int g d\mu$ is well defined would have to imply that $\int g d\mu > -\infty$ and hence $\int g^- d\mu < \infty$. Thus, we have both $\int f^- d\mu < \infty$ and $\int g^- d\mu < \infty$. One can similarly show that if one of $\int f^- d\mu$ and $\int g^- d\mu$ equal $+\infty$, then $\int f^+ d\mu$ and $\int g^+ d\mu$ must both be finite. This proves the claim.

Existence of $\int (f+g) d\mu$ now follows easily. Indeed, one can first prove the inequalities $(f+g)^+ \leq f^+ + g^+$, $(f+g)^- \leq f^- + g^-$ easily and then use Proposition 3.2.5(b) and the claim proved above, to deduce that either $\int (f+g)^+ d\mu < \infty$ or $\int (f+g)^- d\mu < \infty$ and hence $\int (f+g) d\mu$ exists.

To finally prove $\int (f+g) d\mu = \int f d\mu + \int g d\mu$, we consider following equality that holds pointwise.

$$(f+g)^+ - (f+g)^- = f + g = f^+ - f^- + g^+ - g^- \qquad (3.2.2)$$

Change of sides gives

$$(f+g)^+ + f^- + g^- = (f+g)^- + f^+ + g^+ \qquad (3.2.3)$$

This change of sides does not need any justification if $(f+g)^+$ and $(f+g)^-$ are both finite, since in that case f and g must both be finite and so also are all the terms on the extreme right-hand side of Equation (3.2.2). However, a justification a required in other cases. Firstly, if the extreme left-hand side (hence also the extreme right-hand side) of Equation (3.2.2) equals $+\infty$, then one must have $(f+g)^- < \infty$ as well as $f^- < \infty$ and $g^- < \infty$. Thus, the change of sides would involve addition or subtraction of only real numbers. The case when the extreme left-hand side of Equation (3.2.2) equals $-\infty$, can be similarly handled.

Taking integrals of the non-negative measurable functions on the two sides of Equation (3.2.3) and applying Proposition 3.2.5(a), one gets

$$\int (f+g)^+ d\mu + \int f^- d\mu + \int g^- d\mu = \int (f+g)^- d\mu + \int f^+ d\mu + \int g^+ d\mu.$$

As proved earlier, we know that *either* both $\int f^+ d\mu < \infty$ and $\int g^+ d\mu < \infty$ (and hence $\int (f+g)^+ d\mu < \infty$ also) *or* both $\int f^- d\mu < \infty$ and $\int g^- d\mu < \infty$ (and hence $\int (f+g)^- d\mu < \infty$ also). Making the necessary changes of side for finite terms, one would be easily able rewrite the above equality as

$$\int (f+g)^+ d\mu - \int (f+g)^- d\mu = \int f^+ d\mu - \int f^- d\mu + \int g^+ d\mu - \int g^- d\mu,$$

whence the required result follows and that completes the proof of (a)(ii).

(b) : Since the required inequality will hold trivially if either $\int f d\mu = -\infty$ or $\int g d\mu = +\infty$, we are going to prove it assuming that $\int f d\mu > -\infty$ as well as $\int g d\mu < +\infty$. Observe that, $f \leq g$ implies that $f^+ \leq g^+$ and $f^- \geq g^-$. Now $\int f d\mu > -\infty$ implies $\int f^- d\mu < +\infty$ and so, $\int g^- d\mu < \infty$, by Proposition 3.2.5(b). In the same way, $\int g d\mu < +\infty$ will imply that $\int g^+ d\mu < \infty$ and so $\int f^+ d\mu < \infty$, by Proposition 3.2.5(b). Thus all the four integrals $\int f^+ d\mu$, $\int f^- d\mu$, $\int g^+ d\mu$ and $\int g^- d\mu$ are finite. Now, the inequality $\int f d\mu \leq \int g d\mu$ follows trivially. $\square$

PROPOSITION **3.2.9.** If $\int f d\mu$ exists, then $|\int f d\mu| \leq \int |f| d\mu$. Also, f is μ-integrable if and only if $\int |f| d\mu < \infty$.

Proof. The first statement is a consequence of the inequality $-|f| \leq f \leq |f|$ and Theorem 3.2.8 (b). Next, f is μ-integrable iff $\int f^+ d\mu$ and $\int f^- d\mu$ are both finite. However this is equivalent to $\int |f| d\mu = \int (f^+ + f^-) \, d\mu$ being finite, using linearity for one direction and monotonicity for the other direction. $\square$

As an immediate consequence of Proposition 3.2.9, we get the following result, which is very useful, particularly when we work on a probability space.

COROLLARY **3.2.10.** On a finite measure space $(\Omega, \mathcal{A}, \mu)$, any bounded measurable function is integrable.

Note that, in the above corollary, the assumption of finiteness of the measure μ is crucial. For example, on the measurable space $(\mathbb{R}, \mathcal{B}(\mathbb{R}))$, equipped with the Lebesgue measure λ, the constant function $f \equiv 1$ is not integrable (indeed, $\int f d\lambda = +\infty$), and, for the bounded measurable function $g = I_{(0,\infty)} - I_{(-\infty,0]}$, the integral $\int g \, d\lambda$ does not even exist.

What we are going to discuss next is analogous to something that is done in Riemann Integration Theory. Recall that if f is a Riemann-integrable function on closed bounded interval $[a, b]$, then it is also Riemann integrable on any closed bounded subinterval of $[a, b]$. Thus, we can talk about not only the integral $\int_a^b f(x) dx$, but also about the integrals $\int_c^d f(x) dx$ for any subinterval $[c, d] \subset [a, b]$. In the same vein, the integral $\int f d\mu$ that we have defined so far corresponds to integration of f over the whole space Ω. But we would also like to have an integral of f over appropriate subsets of Ω and that comes up in the next definition. Incidentally, from now on, we will sometimes use the phrase μ-*integral* to mean 'integral with respect to μ'.

3.2. Integration

DEFINITION **3.2.11.** Let $(\Omega, \mathcal{A}, \mu)$ be a measure space. For a measurable function f and a set $A \in \mathcal{A}$, the μ-integral of f over the set A is defined as

$$\int_A f d\mu = \int f I_A d\mu.$$

provided, of course, the integral on the right-hand side exists.

Note that, for any measurable function f and any $A \in \mathcal{A}$, the function $f I_A$ is a measurable function. According to the above definition, the integral $\int f d\mu$ that we had originally defined, is actually $\int_\Omega f d\mu$, but we will mostly continue writing it as $\int f d\mu$.

PROPOSITION **3.2.12.** For any non-negative measurable function f on a measure space $(\Omega, \mathcal{A}, \mu)$ and any set $A \in \mathcal{A}$,

$$\int_A f d\mu = \sup\left\{\int_A h d\mu : 0 \le h \le f,\ h\ \text{real valued simple}\right\}$$

Proof. By the definition,

$$\int_A f d\mu = \int f I_A\, d\mu = \sup\left\{\int g d\mu : 0 \le g \le f I_A,\ g\ \text{real valued simple}\right\}$$

Clearly, if h is a real valued simple function with $0 \le h \le f$, then $g = h I_A$ is real valued simple with $0 \le g \le f I_A$. Conversely, if g is real valued simple with $0 \le g \le f I_A$, then $0 \le g \le f$ and also $g = g I_A$. The result is immediate now. $\square$

PROPOSITION **3.2.13.** If $\int f d\mu$ exists, then so does $\int_A f d\mu$ for all $A \in \mathcal{A}$. Further, if f is integrable, then $\int_A f d\mu$ is finite for all $A \in \mathcal{A}$.

Proof. If $\int f^+ d\mu < \infty$, then so is $\int f^+ I_A d\mu = \int (f I_A)^+ d\mu$, by Proposition 3.2.5 (b) and therefore $\int_A f d\mu$ exists. The case when $\int f^- d\mu < \infty$ is similar. Also, the above two cases together prove the second part. $\square$

The next result will lead to a very important idea not just in integration thoery, but also in measure theory, in general.

PROPOSITION **3.2.14.** If $A \in \mathcal{A}$ is such that $\mu(A) = 0$, then for any measurable function f, the integral $\int_A f d\mu$ exists and equals zero.

Proof. For non-negative real valued simple $f = \sum a_i I_{A_i}$, we have $f I_A = \sum a_i I_{A_i \cap A}$ and so $\int_A f d\mu = \sum a_i \mu(A_i \cap A) = 0$. The result for any non-negative measurable f now follows from Proposition 3.2.12. Finally, if f is any measurable function, then $\int_A f^+ d\mu = \int_A f^- d\mu = 0$, by the previous case, and now $\int_A f d\mu = 0$ follows by simply noting that $(f I_A)^+ = f^+ I_A$ and $(f I_A)^- = f^- I_A$. $\square$

Prompted by the above proposition, we are now going to introduce an idea which is very crucial in integration theory. Suppose that $(\Omega, \mathcal{A}, \mu)$ is a measure space. For a real valued function f on Ω, we use the statement "$f \ge 0$ *everywhere*" to mean that $f(\omega) \ge 0$ for *all* $\omega \in \Omega$, or equivalently, that the set $\{\omega : f(\omega) \not\ge 0\}$ is *empty*. However, sometimes, we may have a function f

for which the set $\{\omega : f(\omega) \not\geq 0\}$ is not exactly empty, but 'negligible' in the sense that $\mu\{\omega : f(\omega) \not\geq 0\} = 0$. We will describe this situation by saying that $f \geq 0$ "*almost everywhere*", or to be more specific, "*μ-almost everywhere*". Of course, the concept of "almost everywhere" does not stop there. Given two functions f and g, we would say "$f = g$ *μ-almost everywhere*" to mean that $\mu\{\omega : f(\omega) \neq g(\omega)\} = 0$. Likewise, for a sequence $\{f_n\}$ of measurable functions, the statement "f_n *is increasing almost everywhere*" would mean that the set of $\omega \in \Omega$, for which $\{f_n(\omega)\}$ fails to be increasing has μ-measure zero. It is worthwhile to note that "$\{f_n(\omega)\}$ fails to be increasing" does not mean that "$\{f_n(\omega)\}$ is decreasing". It simply means that $f_{n+1}(\omega) \geq f_n(\omega)$ fails to hold for some n. Thus, what "f_n is increasing almost everywhere" exactly means is that "the set $\bigcup_n \{\omega : f_{n+1}(\omega) \not\geq f_n(\omega)\}$ has zero μ-measure". Also, for $\mathcal{A}$-sets A_1 and A_2, we would say "$A_1 \subset A_2$ μ-almost everywhere" to simply mean "$\mu(A_1 \cap A_2^c) = 0$".

DEFINITION 3.2.15. *A statement is said to hold μ-almost everywhere or simply almost everywhere* (to be abbreviated as $[\mu]$-*a.e.* or *a.e.*$[\mu]$ or simply *a.e.*) *if there is a set $A \in \mathcal{A}$ with $\mu(A) = 0$ such that the statement holds for all $\omega \notin A$.*

This concept will play an important role in much of what we will do. We are now going to see that, in the context of integration, a condition holding almost everywhere is often as good as it holding everywhere.

At this point, let us make an honest declaration about a certain liberty that we are going to take from now on. In the sequel, we would often make statements about functions without explicitly stating that we are talking about measurable functions on a measure space $(\Omega, \mathcal{A}, \mu)$ or make statements about sets without explicitly stating that they come from a σ-field $\mathcal{A}$. A reader, who is very particular about it, should fill in those blanks in the obvious way.

PROPOSITION 3.2.16. (a) *If $f = 0$ a.e.$[\mu]$, then $\int f d\mu = 0$.*
(b) *If $f \leq g$ a.e.$[\mu]$ and if $\int f d\mu$ and $\int g d\mu$ both exist, then $\int f d\mu \leq \int g d\mu$.*
(c) *If $f = g$ a.e.$[\mu]$ and if one of $\int f d\mu$ and $\int g d\mu$ exists, then so does the other and they are equal.*
(d) *If f is μ integrable, then f is finite a.e.$[\mu]$.*

Proof. (a) If $A = \{\omega : f(\omega) \neq 0\}$, then clearly $f = fI_A$. Since $\mu(A) = 0$ by the hypothesis, the result follows from Proposition 3.2.14.
(b) If $A = \{\omega : f(\omega) > g(\omega)\}$, then $\mu(A) = 0$ by hypothesis and so Proposition 3.2.14 gives us $\int fI_A d\mu = 0 = \int gI_A d\mu$, whence, using linearity of integrals, we get $\int f d\mu = \int fI_{A^c} d\mu$ and $\int g d\mu = \int gI_{A^c} d\mu$. But, by the definition of A, we have $fI_{A^c} \leq gI_{A^c}$ everywhere and so, by the monotonicity property of integrals, $\int fI_{A^c} d\mu \leq \int gI_{A^c} d\mu$ and that completes the proof.
(c) Let us assume that $\int f d\mu$ exists and show that $\int g d\mu$ exists and equals $\int f d\mu$. Setting $A = \{\omega : f(\omega) \neq g(\omega)\}$, we have $\mu(A) = 0$ and so $\int fI_A d\mu = 0 = \int gI_A d\mu$, by Proposition 3.2.14. Next, the assumption that $\int f d\mu$ exists, implies, by Proposition 3.2.13, that $\int fI_{A^c} d\mu$ exists. Since $fI_{A^c} = gI_{A^c}$ everywhere, $\int gI_{A^c} d\mu$ also

3.2. Integration

exists and equals $\int f I_{A^c} d\mu$. Theorem 3.2.8 (a)(ii) now implies that $\int g d\mu$ exists and equals $\int f I_{A^c} d\mu$, which, in turn, equals $\int f d\mu$.
(d) With $A = \{\omega : |f(\omega)| = \infty\}$, we have $|f| \geq |f| I_A$ and so $\int |f| d\mu \geq \int |f| I_A d\mu$. If $\mu(A) > 0$, then $\int |f| I_A d\mu = \infty \cdot \mu(A) = \infty$, contradicting integrability of f. □

REMARK 3.2.17. We make a very important remark here that will come into play many a times in the sequel in our statements on integrals. Proposition 3.2.16 provides us with a certain leverage that will often be conveniently exploited in the sequel. We will often make statements about an "integral" $\int f d\mu$, where the "function" f may not even be well-defined everywhere on Ω. However, in view of part (c) of Proposition 3.2.16, the integral $\int f d\mu$ will be still be meaningful and well-defined, whenever we can get hold of a set A with $\mu(A^c) = 0$, such that f is well-defined on A and the function $\tilde{f}$ given by $\tilde{f}(\omega) = \begin{cases} f(\omega) & \text{if } \omega \in A \\ 0 & \text{if } \omega \in A^c \end{cases}$, defines a measurable function on Ω. In that case, the notation $\int f d\mu$ is to be intrepreted as the integral $\int \tilde{f} d\mu$. Part (c) of Proposition 3.2.16 guarantees complete unambiguity in this interpretation of $\int f d\mu$. In particular, the reader would realize that it would make no difference if one decided to set $\tilde{f}$ to be any other constant, instead of 0, on A^c.

A well-known result in Riemann Integration theory says that if f is a non-negative continuous function on $[a,b]$ such that $\int_a^b f(x) dx = 0$, then $f \equiv 0$. Of course, if one is willing to sacrifice continuity, one may allow f to be non-zero at, for example, a finite set of points in $[a,b]$ and still have $\int_a^b f(x) dx = 0$. The next result is an analogue of that in the present context.

PROPOSITION 3.2.18. If $f \geq 0$ a.e.$[\mu]$ and $\int f d\mu = 0$, then $f = 0$ a.e.$[\mu]$.

Proof. Taking $A_n = \{\omega : f(\omega) > \frac{1}{n}\}$, $n \geq 1$, we have $\frac{1}{n} I_{A_n} \leq f I_{A_n} \leq f$ a.e.$[\mu]$, where the second inequality uses the hypothesis that $f \geq 0$ a.e.$[\mu]$. Using Proposition 3.2.16 (b), one gets $\frac{1}{n}\mu(A_n) \leq \int f d\mu = 0$, which implies $\mu(A_n) = 0$. Since this holds for all $n \geq 1$ and since $A_n \uparrow \{\omega : f(\omega) > 0\}$, as $n \uparrow \infty$, we use continuity of μ from below to get $\mu\{\omega : f(\omega) > 0\} = 0$. This, in conjunction with the hypothesis $f \geq 0$ a.e.$[\mu]$, proves $f = 0$ a.e.$[\mu]$. □

Going back to Riemann integrals again, one knows that $\int_a^b f(x) dx \geq 0$ may hold for a continuous f without $f(x)$ being non-negative at all $x \in [a,b]$. More generally, for continuous functions f and g, $\int_a^b f(x) dx \geq \int_a^b g(x) dx$ does not necessarily force $f(x) \geq g(x)$ to hold for all $x \in [a,b]$. However, if we have the stronger condition that $\int_c^d f(x) dx \geq \int_c^d g(x) dx$ holds for all $[c,d] \subset [a,b]$, then $f(x) \geq g(x)$ must necessarily hold for all x. Of course, if the continuity assumption is dropped, then violation of $f(x) \geq g(x)$ may be allowed at, for example, a finite set of points $x \in [a,b]$. The following is an analogue of this in the current context. However, for this, we need an additional assumption, namely, that the underlying measure is σ-finite.

PROPOSITION 3.2.19. Let $(\Omega, \mathcal{A}, \mu)$ be a σ-finite measure space. If f, g are two measurable functions satisfying $\int_A f d\mu \leq \int_A g d\mu$ for all $A \in \mathcal{A}$, then $f \leq g$ a.e.$[\mu]$.

Proof. Since μ is σ-finite, there exist $\mathcal{A}$-sets Ω_m, $m \geq 1$, with $\bigcup_m \Omega_m = \Omega$ such that $\mu(\Omega_m) < \infty$ for all m. For positive integers m, k and n, consider the set

$$A_{k,n}^m = \{\omega \in \Omega_m : g(\omega) < k, \ f(\omega) > -k, \ f(\omega) > g(\omega) + n^{-1}\}.$$

Using the hypothesis and the definition of $A_{k,n}^m$, we get

$$\int_{A_{k,n}^m} g \, d\mu \geq \int_{A_{k,n}^m} f \, d\mu \geq \int_{A_{k,n}^m} g \, d\mu + n^{-1} \mu(A_{k,n}^m) \qquad (3.2.4)$$

But the hypothesis and the definition of $A_{k,n}^m$ also gives us

$$-k\mu(\Omega_m) \leq -k\mu(A_{k,n}^m) \leq \int_{A_{k,n}^m} f \, d\mu \leq \int_{A_{k,n}^m} g \, d\mu \leq k\mu(A_{k,n}^m) \leq k\mu(\Omega_m) \qquad (3.2.5)$$

Since $k\mu(\Omega_m) < \infty$, (3.2.5) implies that $\int_{A_{k,n}^m} g \, d\mu$ is finite, so that from (3.2.4), we may conclude that $\mu(A_{k,n}^m) = 0$ and hence $\mu\left(\bigcup_{k,n,m} A_{k,n}^m\right) = 0$. Note now that, for any $\omega \notin \bigcup_{k,n,m} A_{k,n}^m$, we must have *either* $g(\omega) = \infty$ or $f(\omega) = -\infty$ or $f(\omega) \leq g(\omega)$. In other words, $f(\omega) \leq g(\omega)$ holds for all $\omega \notin \bigcup_{k,n,m} A_{k,n}^m$. □

REMARK 3.2.20. The proof shows that the hypothesis could be weakened to requiring that $\int_A f \, d\mu \leq \int_A g \, d\mu$ *only* for all $A \in \mathcal{A}$ with $\mu(A) < \infty$.

Proposition 3.2.19, along with the Remark 3.2.20, leads to two immediate corollaries.

COROLLARY 3.2.21. *If f, g are measurable functions on a σ-finite measure space $(\Omega, \mathcal{A}, \mu)$ such that $\int_A f \, d\mu = \int_A g \, d\mu$ for all $A \in \mathcal{A}$ with $\mu(A) < \infty$, then $f = g$ a.e.*

COROLLARY 3.2.22. *Let f be a measurable function on a σ-finite measure space $(\Omega, \mathcal{A}, \mu)$. If there exist real numbers $\alpha \leq \beta$, such that, for all $A \in \mathcal{A}$ with $\mu(A) < \infty$, one has $\alpha \mu(A) \leq \int_A f \, d\mu \leq \beta \mu(A)$, then $\alpha \leq f \leq \beta$ a.e.$[\mu]$.*

EXAMPLE 3.2.23. Consider any measurable space $(\Omega, \mathcal{A})$. Consider the measure μ on $\mathcal{A}$ given by $\mu(A) = \infty$ for all $A \neq \emptyset$. Considering the constant functions $f \equiv 2$ and $g \equiv 1$, one can see that σ-finiteness of μ in the above proposition can not be done away with.

REMARK 3.2.24. A basic and extremely important idea that emerges from the above discussion needs to be summarized and recorded here. Part (a) (ii) of Theorem 3.2.8 asserted that if $h(\omega) = f(\omega) + g(\omega)$ for all ω, then $\int h \, d\mu = \int f \, d\mu + \int g \, d\mu$, *whenever* the right side is defined. However, part (c) of Proposition 3.2.16 now allows us to assert that the same conclusion holds even under the weaker condition that $h(\omega) = f(\omega) + g(\omega)$ for a.e.$[\mu]$ ω. This is just one of many such instances where properties on integrals as stated earlier can be strengthened by replacing an "everywhere" condition with an appropriate a.e.$[\mu]$ condition. It is left as an exercise for the reader to identify all the results on integration where such a modification is possible and then to actually state and prove the modified results.

3.3 Limits and Integral

In the context of Riemann integral, a very important issue is that of interchange of limits and integrals. To be precise, for a sequence $\{f_n\}$ of Riemann integrable functions on $[a,b]$, for which $\lim_n f_n(x)$ exists and is Riemann integrable on $[a,b]$, when can one say that $\int_a^b \lim_n f_n(x)dx$ equals $\lim_n \int_a^b f_n(x)dx$? We are going to address precisely this question in the context of the integration that we are studying. While such interchanges are not always valid, we are going to get sufficient conditions under which they are valid. The two most fundamental results in this direction are known as 'Monotone Convergence Theorem' (MCT, for short) and 'Dominated Convergence Theorem' (DCT, for short).

THEOREM 3.3.1. (Monotone Convergence Theorem) If $\{f_n\}$ is a sequence of measurable functions on a measure space $(\Omega, \mathcal{A}, \mu)$ with $f_n \geq 0$, for all $n \geq 1$, and if $f_n \uparrow f$, then $\int f_n d\mu \uparrow \int f d\mu$.

Proof. By part (b) of Proposition 3.2.5, $\int f_n d\mu \leq \int f_{n+1} d\mu$ for all n, so that $\lim_n \int f_n d\mu$ exists. Also, $\int f_n d\mu \leq \int f d\mu$ for all n, and so $\lim_n \int f_n d\mu \leq \int f d\mu$. We proceed to prove that it is actually an equality.

For each n, let $\{f_{nm}, m \geq 1\}$ be a sequence of non-negative real valued simple functions such that $f_{nm} \uparrow f_n$, as $m \uparrow \infty$. Defining $g_m = \max_{1 \leq n \leq m} f_{nm}$, for each m, we get a sequence $\{g_m\}$ of non-negative real valued simple functions. It is easy to see that $g_m \leq g_{m+1}$, for all m, so that $g = \lim_m g_m$ exists and is a non-negative measurable function. By the definition of g_m, we have, for any fixed $n \geq 1$,

$$f_{nm}(\omega) \leq g_m(\omega) \leq f_m(\omega), \quad \text{for all } \omega \in \Omega \text{ and all } m \geq n. \quad (3.3.1)$$

Fixing $n \geq 1$ and taking limits as $m \to \infty$, we get

$$f_n(\omega) \leq g(\omega) \leq f(\omega), \quad \text{for all } \omega \in \Omega.$$

Now letting $n \to \infty$, we finally obtain $g(\omega) = f(\omega)$, for all $\omega \in \Omega$. We have thus proved that the sequence $\{g_m\}$ of non-negative real valued simple functions increase to f and therefore, by Proposition 3.2.3, $\int f d\mu = \lim_m \int g_m d\mu$. Next, integrating the functions in (3.3.1) and using monotonicity of integrals, we get

$$\int f_{nm} d\mu \leq \int g_m d\mu \leq \int f_m d\mu,$$

for all $n \geq 1$ and for all $m \geq n$. Taking limits as $m \to \infty$, gives us

$$\int f_n d\mu \leq \int f d\mu \leq \lim_m \int f_m d\mu.$$

The proof is now completed by taking limits, as $n \to \infty$, in the above. □

To prove the other fundamental result, namely DCT, we need the following result, known as Fatou's Lemma, which is an easy consequence of MCT. Let us point out though that while here we use it as an intermediate step to get DCT, the result, in itself, is very important and often found very useful.

LEMMA 3.3.2. (Fatou's Lemma) For any sequence $\{f_n\}$ of non-negative measurable functions on a measure space $(\Omega, \mathcal{A}, \mu)$,

$$\int \liminf_n f_n d\mu \leq \liminf_n \int f_n d\mu.$$

Proof. Define $g_n = \inf_{k \geq n} f_k$, for each n. Then, $\{g_n\}$ is a sequence of non-negative measurable functions with $g_n \uparrow \liminf_n f_n$, and therefore, by MCT,

$$\lim_n \int g_n d\mu = \int \liminf_n f_n d\mu.$$

But for each n, we have $g_n \leq f_n$ and so $\int g_n d\mu \leq \int f_n d\mu$. From this, one gets

$$\int \liminf_n f_n d\mu = \lim_n \int g_n d\mu \leq \liminf_n \int f_n d\mu. \qquad \square$$

THEOREM **3.3.3.** (Dominated Convergence Theorem) Let $\{f_n\}$ be a sequence of measurable functions on a measure space $(\Omega, \mathcal{A}, \mu)$, converging pointwise to f. If there is a μ-integrable function g such that $|f_n(\omega)| \leq g(\omega)$ for all n and all $\omega \in \Omega$, then $\int |f_n - f| \, d\mu \to 0$ and, in particular, $\int f_n d\mu \to \int f d\mu$, as $n \to \infty$.

Before proceeding with the proof of Theorem 3.3.3, it is important to point out that this is an instance where we are precisely in a situation that was elicited in Remark 3.2.17. The above theorem makes a statement about the "integral" $\int |f_n - f| \, d\mu$, but a careful reader would not overlook the fact that nothing in the hypotheses of the theorem guarantees that $f_n(\omega) - f(\omega)$ is well-defined for all ω. So, this is where what was explained in Remark 3.2.17 comes into play. The μ-integrability of g in the hypothesis of the theorem implies that the set $A = \{g = \infty\}$ has μ-measure 0. If we now define $\widetilde{f}_n = f_n$, $n \geq 1$ and $\widetilde{f} = f$ on A^c and set $\widetilde{f}_n = \widetilde{f} = 0$, $n \geq 1$, on A, then the hypotheses that $|f_n| \leq g$ pointwise for all n and $f_n \to f$ pointwise will imply that the functions $\widetilde{f}_n$, $n \geq 1$ and $\widetilde{f}$ are all well-defined real-valued measurable functions on Ω. Further, the set $\bigcup_n \{f_n \neq \widetilde{f}_n\} \bigcup \{f \neq \widetilde{f}\}$ has μ-measure 0. As explained in Remark 3.2.17, the reader should read the statements on f, f_n and $f_n - f$, $n \geq 1$, in the theorem (and also in its proof below), as really (the same) statements on $\widetilde{f}$, $\widetilde{f}_n$ and $\widetilde{f}_n - \widetilde{f}$, $n \geq 1$, respectively, though we will continue to use f_n, $n \geq 1$, and f.

Proof. The hypothesis that there is a μ-integrable function g such that $|f_n| \leq g$ pointwise for all n (and hence also $|f| \leq g$ pointwise) implies that each f_n as well as f are μ-integrable. For each n, define $h_n = 2g - |f_n - f|$. By the triangle inequality, each h_n is a non-negative measurable function and further, $h_n \to 2g$ pointwise. Now by Fatou's Lemma, we have

$$\int 2g \, d\mu \leq \liminf_n \int h_n d\mu = \int 2g \, d\mu - \limsup_n \int |f_n - f| d\mu.$$

Since $\int 2g \, d\mu$ is finite, the above inequality gives $\limsup_n \int |f_n - f| d\mu \leq 0$. But since $\int |f_n - f| d\mu \geq 0$ for each n, we have $\liminf_n \int |f_n - f| d\mu \geq 0$, which together with the above inequality on the lim sup yields $\int |f_n - f| d\mu \to 0$, as $n \to \infty$.
Finally, by Proposition 3.2.9, we have

$$\left| \int f_n d\mu - \int f d\mu \right| = \left| \int (f_n - f) d\mu \right| \leq \int |f_n - f| d\mu,$$

from which it follows that $\int f_n d\mu \to \int f d\mu$, as $n \to \infty$. $\qquad \square$

3.3. Limits and Integral

REMARK 3.3.4. In view of Remark 3.2.24, the hypotheses in MCT, Fatou's lemma, and DCT can be relaxed. For example, in MCT, it suffices to assume that $f_n \geq 0$, a.e.$[\mu]$, for each n and $f_n \uparrow f$, a.e.$[\mu]$ to conclude that $\int f_n d\mu \uparrow \int f d\mu$. Similarly, in DCT, we just need that $f_n \to f$ a.e.$[\mu]$ and that $|f_n| \leq g$ a.e.$[\mu]$, for each n, in order to assert the same conclusion. A similar relaxation can be done in the hypothesis of Fatou's lemma. The reader is encouraged to write down the precise statements of MCT, Fatou's lemma and DCT with the relaxed hypotheses and then derive them from Theorem 3.3.1, Lemma 3.3.2, Theorem 3.3.3 respectively, by appropriately using part (c) of Proposition 3.2.16. In the sequel, when we refer to MCT, Fatou's Lemma or DCT, we will often mean such "a.e. versions" with relaxed hypotheses.

On finite measure spaces, any real valued constant function is integrable and this leads to the following special case of DCT (a.e. version), known as 'Bounded Convergence Theorem' (BCT, for short). No proof is needed.

THEOREM 3.3.5. *(Bounded Convergence Theorem)* Let $\{f_n\}$ be a sequence of measurable functions on a finite measure space $(\Omega, \mathcal{A}, \mu)$, converging a.e.$[\mu]$ to a measurable function f. If there exists a real number M such that $|f_n| \leq M$ a.e.$[\mu]$, for each n, then $\int |f_n - f| d\mu \to 0$ and consequently $\int f_n d\mu \to \int f d\mu$, as $n \to \infty$.

From linearity we know that integrals and finite sums can be interchanged (provided, of course, they are defined). Applying MCT and DCT now, one can get sufficient conditions under which infinite sums and integrals can be interchanged.

THEOREM 3.3.6. Let $\{f_n, n \geq 1\}$ be a sequence of measurable functions on a measure space $(\Omega, \mathcal{A}, \mu)$. If *either* each f_n is non-negative a.e.$[\mu]$, *or* the non-negative measurable function $\sum_n |f_n|$ is μ-integrable, then

$$\int \Big(\sum_n f_n\Big) d\mu = \sum_n \int f_n d\mu.$$

REMARK 3.3.7. In both the statement and the proof of the above theorem, the ideas discussed in Remark 3.2.17 are once again called into action. Firstly, in case $f_n \geq 0$ a.e.$[\mu]$, for each n, if one denotes $A = \bigcap_n \{\omega : f_n \geq 0\}$, then $\mu(A^c) = 0$. Clearly, the infinite sum $\sum_n f_n(\omega)$ is well-defined and non-negative, for all $\omega \in A$ and that means that we can get a non-negative measurable function which equals $\sum_n f_n$, a.e. $[\mu]$. Following the policy laid down in Remark 3.2.17, we will write $\sum_n f_n$ to mean any such non-negative measurable function, that equals $\sum_n f_n$, a.e. $[\mu]$. In the second case, the hypothesis of integrability of the non-negative measurable function $\sum_n |f_n|$ will imply, by Proposition 3.2.9, that $\sum_n |f_n|$ is finite a.e. $[\mu]$ and hence the series $\sum_n f_n$ converges a.e.$[\mu]$. Once again, we will now write $\sum_n f_n$ to denote any measurable function defined on Ω, that equals the sum $\sum_n f_n$, wherever the latter converges. This measurable function will clearly be integrable, since $\big|\sum_n f_n\big| \leq \sum_n |f_n|$, wherever the series $\sum_n f_n$ converges.

Proof. When each f_n is non-negative a.e.$[\mu]$, then, for each n, the partial sum $g_n = \sum_{k=1}^{n} f_k$ is non-negative a.e.$[\mu]$. Since $g_n \uparrow \sum_k f_k$ a.e $[\mu]$, MCT (a.e. version) and linearity of integrals gives

$$\int \left(\sum_k f_k\right) d\mu = \lim_n \int g_n d\mu = \lim_n \sum_{k=1}^{n} \int f_k d\mu = \sum_k \int f_k d\mu.$$

In the other case, take $g = \sum_n |f_n|$. It follows that the partial sums g_n, as defined above, for $n \geq 1$, satisfy $|g_n| \leq g$ a.e $[\mu]$, for each n, and $g_n \to \sum_k f_k$ a.e.$[\mu]$. Since g is μ-integrable, DCT completes the proof. □

EXERCISE **3.3.8.** *Consider $\Omega = \mathbb{N}$ equipped with the σ-field $\mathcal{P}(\mathbb{N})$, the class of all subsets of $\mathbb{N}$, and let μ denote the counting measure, that is, $\mu(A)$, for any $A \subset \mathbb{N}$, equals the number of elements in A.*
(a) Show that real valued measurable functions f on $(\mathbb{N}, \mathcal{P}(\mathbb{N}))$ are nothing but all real sequences $\{f(n)\}$.
(b) Show that, for a non-negative f, the integral $\int f d\mu$ equals the sum $\sum_n f(n)$, which may take the value $+\infty$.
(c) Show that f is μ-integrable iff the infinite series $\sum f(n)$ converges absolutely and in that case $\int f d\mu = \sum_n f(n)$.
(b) Consider the function $f(n) = (-1)^{n+1}/n$, $n \in \mathbb{N}$. We know that the series $\sum f(n)$ converges. Does the integral $\int f d\mu$ exist?

By representing infinite series as an integral, the above exercise brings it under the umbrella of the general theory of integration and so one can apply results on general integrals, in particular, MCT, DCT and Fatou's Lemma, in the context of infinite series.

Our next aim is to relate Riemann integration on a closed bounded interval I to integration over I with respect to the Lebesgue measure λ. Let f be a bounded real valued function on a closed bounded interval $I = [a, b]$. Given any finite partition $\pi : a = t_0 < t_1 < \cdots < t_n = b$ of the interval I, we define the "upper" and "lower" functions h_π and g_π, relative to π, as

$$h_\pi(t) = f(a) I_{\{a\}}(t) + \sum_{j=1}^{n} M_j I_{(t_{j-1}, t_j]}(t), \quad g_\pi(t) = f(a) I_{\{a\}}(t) + \sum_{j=1}^{n} m_j I_{(t_{j-1}, t_j]}(t),$$

where $M_j = \sup\{f(t) : t_{j-1} < t \leq t_j\}$ and $m_j = \inf\{f(t) : t_{j-1} < t \leq t_j\}$. Clearly h_π and g_π are simple real valued measurable functions on the measure space $(I, \mathcal{B}(I), \lambda)$, where $\mathcal{B}(I)$ is the Borel σ-field on I (see Definition 2.9.5) and λ denotes "restriction" of Lebesgue measure to I. Denote $U_\pi = \int h_\pi d\lambda$ and $L_\pi = \int g_\pi d\lambda$. The reader must be able to quickly recognize that U_π and L_π are precisely what, in Riemann integration theory, go by the name of "upper" and "lower" Riemann sums of f, relative to the partition π.

Consider now a sequence $\{\pi_n\}$ of finite partitions of $[a, b]$, with $\pi_n \subset \pi_{n+1}$ and $\|\pi_n\| \downarrow 0$. We are using here the usual notation $\|\pi\|$ for the 'norm' of a partition π. Denote by $\{h_n\}$ and $\{g_n\}$ the corresponding sequences of upper and lower functions and let $\{U_n\}$ and $\{L_n\}$ be the associated sequences of upper and

3.3. Limits and Integral

lower Riemann sums. Clearly, $h_n \geq f \geq g_n$, for each n, by definition. Further, $\pi_n \subset \pi_{n+1}$ implies that the sequence $\{h_n\}$ is monotonically decreasing, while the sequence $\{g_n\}$ is monotonically increasing. Therefore, the pointwise limits $h = \lim_n h_n$ and $g = \lim_n g_n$ both exist and are measurable functions satisfying $h \geq f \geq g$. Since f is assumed to be bounded, h and g would both be bounded and so one can apply the Bounded Convergence Theorem on the finite measure space $(I, \mathcal{B}(I), \lambda)$ to get $U = \lim U_n = \int_I h d\lambda$ and $L = \lim L_n = \int_I g d\lambda$. Recall now that f is said to be *Riemann integrable* on $I = [a, b]$ if $U = L$ and in that case, the Riemann integral of f on I is defined as $\int_a^b f(x)dx = \alpha$ where $\alpha = U = L$. From what was seen above above, it follws that f is Riemann integrable on I iff $\int(h-g)d\lambda = 0$ or, equivalently, $h = f = g$ a.e.$[\lambda]$ (recall that $h \geq f \geq g$ holds everywhere).

We claim now that if $t \in [a, b]$ is not one of the countably many partition points in $\bigcup_n \pi_n$, then $h(t) = g(t)$ iff f is continuous at t (and of course $h(t) = f(t) = g(t)$ in that case). The verification of this claim is left to the reader as an exercise. Once this claim is verified, it will follow that Riemann integrability of f is equivalent to f being continuous a.e.$[\lambda]$. Note that, in that case, f equals the $\mathcal{B}(I)$-measurable function h a.e.$[\lambda]$ and is, therefore, measurable with respect to the λ-completion of $\mathcal{B}(I)$, called the "Lebesgue σ-field" on I (see Exs 3.1.21). Further, it is clear that, in that case, f is λ-integrable on I with $\int_I h d\lambda = \int_I f d\lambda = \int_I g d\lambda$. All these now lead to the following important result connecting Riemann integrals on bounded intervals with Lebesgue integrals.

THEOREM **3.3.9**. *A bounded real valued function f on a closed bounded interval $[a, b]$ is Riemann integrable iff f is continuous a.e.$[\lambda]$ on $[a, b]$, where λ is the Lebesgue measure on $[a, b]$. In that case, f is measurable with respect to the λ-completion of $\mathcal{B}(I)$ and λ-integrable on $[a, b]$ with the λ-integral being equal to the Riemann integral, that is, $\int_{[a,b]} f d\lambda = \int_a^b f(x)dx$.*

EXERCISE **3.3.10**. *Let G be a non-decreasing right continuous function on $\mathbb{R}$ and let μ be the unique Radon measure determined by G. Derive a result analogous to Theorem 3.3.9 connecting the Riemann-Stieltje's integral $\int_a^b f(x)dG(x)$ with the integral $\int_{(a,b]} f d\mu$, for a bounded real valued function f on $[a, b]$.*

REMARK **3.3.11**. **An important Notation**: For integrals of extended real valued measurable functions on $\mathbb{R}$ with respect to the Lebesgue measure, we will often use the notation $\int f(x)dx$ in place of $\int f d\lambda$ and, more generally, $\int_B f(x)dx$ in place of $\int_B f d\lambda$, where $B \in \mathcal{B}(\mathbb{R})$. In spite of notational similarity with Riemann integral, this should not, in general, be viewed as anything more than just an alternative notation for λ-integral. Of course, by virtue of Theorem 3.3.9 and Exercise 3.6.21, in case f is Riemann integrable on a closed bounded interval or f is non-negative and an improper Riemann integral of f on an unbounded interval exists, the λ-integral anyway agrees with the Riemann integral, thus justifying our notation. We will use similar notations in higher dimensions also. Thus, $\int f(\mathbf{x})d\mathbf{x}$ or, more generally,

$\int_B f(\mathbf{x})d\mathbf{x}$, for $B \in \mathcal{B}(\mathbb{R}^k)$, will be often used to just denote integrals with respect to the Lebesgue measure on $\mathcal{B}(\mathbb{R}^k)$.

The next result, which is a simple application of DCT, has many useful and important consequences.

THEOREM 3.3.12. (Scheffé's Lemma) Let $(\Omega, \mathcal{A}, \mu)$ be a measure space and let f and f_n, $n \geq 1$ be non-negative μ-integrable functions with $\int f_n d\mu = \int f d\mu$, for all n. If $f_n \to f$ a.e.$[\mu]$, then $\int |f_n - f| d\mu \to 0$.

Proof. The hypothesis $\int f d\mu = \int f_n d\mu$, equivalently, $\int (f - f_n)d\mu = 0$, for all n, implies $\int (f - f_n)^+ d\mu = \int (f_n - f)^- d\mu$, and hence $\int |f_n - f| d\mu = 2 \int (f - f_n)^+ d\mu$, for all n. Since $0 \leq (f - f_n)^+ \leq f$, for all n, and $(f - f_n)^+ \to 0$, as $n \to \infty$, DCT gives $\int (f - f_n)^+ d\mu \to 0$, from which the required result follows. □

The next theorem gives another important and useful application of DCT. Among other things, it gives a sufficient condition for 'differentiating under the integral sign'.

THEOREM 3.3.13. Let $I \subset \mathbb{R}$ be an open interval and $(\Omega, \mathcal{A}, \mu)$ be a measure space. Suppose that f is a real valued function on $I \times \Omega$, satisfying the following conditions:
(i) for each $\omega \in \Omega$, the function $f(\cdot, \omega) : I \to \mathbb{R}$ is continuous and
(ii) for each $t \in I$, the function $f(t, \cdot) : \Omega \to \mathbb{R}$ is $\mathcal{A}$-measurable.
If there exists a μ-integrable function g on Ω such that

$$|f(t,\omega)| \leq g(\omega), \quad \text{for all} \quad (t,\omega) \in I \times \Omega, \qquad (3.3.2)$$

then $\varphi(t) = \int f(t,\omega)d\mu$, $t \in I$, defines a real valued continuous function φ on I.
Further, if f satisfies the condition that
(iii) for each $\omega \in \Omega$, the function $f(\cdot, \omega) : I \to \mathbb{R}$ is differentiable everywhere with the derivative satisfying

$$|f'(t,\omega)| \leq h(\omega), \quad \text{for all} \quad (t,\omega) \in I \times \Omega, \qquad (3.3.3)$$

for some μ-integrable function h on Ω, then the function $f'(t, \cdot) : \Omega \to \mathbb{R}$, for each $t \in I$, is measurable and μ-integrable. Further, the function $\varphi : I \to \mathbb{R}$ defined above is differentiable with its derivative being given by

$$\varphi'(t) = \int f'(t,\omega)d\mu.$$

Proof. The condition (3.3.2) clearly implies that for each $t \in I$, the $\mathcal{A}$-measurable function $f(t, \cdot)$ is μ-integrable, so that $\varphi(t) = \int f(t,\omega)d\mu$ is real-valued. To see continuity of the function φ, let $\{t_n\}$ be a sequence in I with $t_n \to t \in I$. Using the hypothesis (i) and the condition (3.3.2), it follows from DCT that $\varphi(t_n) \to \varphi(t)$.

Turning now to the second part of the theorem, let us fix any $t \in I$ and a sequence $\{t_n\}$ in I with $0 < |t_n - t| \to 0$. Denoting

$$g_n(\omega) = \frac{f(t_n, \omega) - f(t, \omega)}{t_n - t}, \quad \omega \in \Omega,$$

3.3. Limits and Integral

g_n is clearly $\mathcal{A}$-measurable, by (ii). Since t and t_n are fixed here, we suppress dependence of g_n on them. The hypothesis of differentiability in (iii) implies that

$$g_n(\omega) \longrightarrow f'(t,\omega), \quad \text{for each } \omega, \qquad (3.3.4)$$

which gives $\mathcal{A}$-measurability of $f'(t,\cdot)$. Integrability of $f'(t,\cdot)$ follows from (3.3.3). Finally, for t and $\{t_n\}$ as above, we have, by the mean value theorem of calculus, $g_n(\omega) = f'(s_n,\omega)$, for every ω and n, where s_n is a point lying strictly between t_n and t. Of course, the point s_n, for each n, may possibly depend on ω. But in any case, by virtue of hypothesis (3.3.3), we would have $|g_n(\omega)| = |f'(s_n,\omega)| \leq h(\omega)$, for all n and ω. From this and (3.3.4), we get $\int g_n(\omega)d\mu \longrightarrow \int f'(t,\omega)d\mu$, using DCT. But, by linearity of integrals, we have

$$\int g_n(\omega) d\mu = \frac{\varphi(t_n) - \varphi(t)}{t_n - t}, \quad \text{for each } n,$$

and so we conclude that $\varphi'(t)$ exists and equals $\int f'(t,\omega)d\mu$. Since $t \in I$ was arbitrary, this completes the proof. □

REMARK 3.3.14. The proof actually shows that if one only wants continuity (resp. differentiability) of φ at a particular point $t_0 \in I$, then all one needs to assume is that (3.3.2) (resp. (3.3.3)) holds for t only in some open neighbourhood of t_0.

We end this section with another interesting application of MCT and DCT, leading to some very important concepts. These will be taken up later again, in greater detail, in Chapter 6.

THEOREM 3.3.15. Let f be a measurable function on $(\Omega, \mathcal{A}, \mu)$. Assume either that f is non-negative or that f is μ-integrable. Then the set function ν defined by

$$\nu(A) = \int_A f d\mu, \quad A \in \mathcal{A} \qquad (3.3.5)$$

is countably additive, in the sense that $\nu\left(\bigcup_n A_n\right) = \sum_n \nu(A_n)$, for any sequence $\{A_n\}$ of disjoint $\mathcal{A}$-sets.

REMARK 3.3.16. Note that, because f is assumed to be either non-negative or μ-integrable, $\nu(A) = \int_A f d\mu$ exists, for all $A \in \mathcal{A}$ (Proposition 3.2.13). Also, $\nu(A)$ is non-negative, if f is non-negative, while it is finite, if f is μ-integrable.

Proof. Let $\{A_n\}$ be a finite or infinite sequence of disjoint $\mathcal{A}$-sets and denote $A = \bigcup_n A_n$. Setting $f_n = fI_{A_n}$, we have $fI_A = \sum_n f_n$ and so $\nu(A) = \int \left(\sum_n f_n\right) d\mu$. Proving the required additivity therefore boils down to simply showing that $\int \left(\sum_n f_n\right) d\mu = \sum_n \int f_n d\mu$. In case $\{A_n\}$ is a finite sequence, this follows from linearity of integrals, while, in case $\{A_n\}$ is an infinite sequence, one needs to use Theorem 3.3.6. □

REMARK 3.3.17. If f is a non-negative measurable function on $(\Omega, \mathcal{A}, \mu)$, then the above theorem shows that the set function ν, defined by (3.3.5), is a measure on $\mathcal{A}$. Of course, if f is both non-negative and μ-integrable, then ν is a finite measure.

EXERCISE **3.3.18.** *Show that, if f is a non-negative real valued measurable function on a σ-finite measure space $(\Omega, \mathcal{A}, \mu)$, then ν defined by (3.3.5) is a σ-finite measure.*

A very important property of the measure ν, defined by (3.3.5) with a non-negative measurable f, is that for any $A \in \mathcal{A}$ with $\mu(A) = 0$, one has $\nu(A) = 0$ also. This is an immediate consequence of part (a) of Proposition 3.2.16. In other words, every μ-null set is also a ν-null set. This property is usually described by saying that ν is "absolutely continuous" with respect to μ (see also Chapter 6).

DEFINITION **3.3.19.** Let $(\Omega, \mathcal{A})$ be a measurable space and μ, ν be any two measures on $\mathcal{A}$. The measure ν is said to be *absolutely continuous* with respect to the measure μ, written $\nu \ll \mu$, if $\nu(A) = 0$ whenever $\mu(A) = 0$.

Thus, what we have just seen is that if f is a non-negative measurable function on a measure space $(\Omega, \mathcal{A}, \mu)$, then the set function ν defined by (3.3.5), is a measure on $\mathcal{A}$, such that, $\nu \ll \mu$. There is a very deep theorem, called the *Radon-Nikodym theorem* (to be proved in Chapter 6), which asserts that, in case μ is a σ-finite measure, then this is the only way that one can get measures ν on $\mathcal{A}$, that are absolutely continuous with respect to μ. More precisely, if $(\Omega, \mathcal{A}, \mu)$ is a σ-finite measure space and ν is a measure on $\mathcal{A}$ with $\nu \ll \mu$, then there is a non-negative measurable function f such that $\nu(A) = \int_A f d\mu$ for all $A \in \mathcal{A}$.

Going back to Theorem 3.3.15, we saw that if f is μ-integrable, then the set function ν, defined by (3.3.5), is a countably additive real valued set function. Of course, ν may not be a measure, since it may take negative values. However, the set functions ν^+ and ν^- defined by

$$\nu^+(A) = \int_A f^+ d\mu, \quad \nu^-(A) = \int_A f^- d\mu, \quad \text{for } A \in \mathcal{A}$$

are clearly finite measures and $\nu = \nu^+ - \nu^-$. Further, ν^+ and ν^- have disjoint supports; indeed, $\nu^+\{f \leq 0\} = 0$ and $\nu^-\{f > 0\} = 0$. There is a general result, called the *Jordan-Hahn Decomposition Theorem*, which says that any countably additive real valued set function ν on $\mathcal{A}$ always has such a decomposition as the difference of two finite measures that have disjoint supports.

As mentioned earlier, both these very important results, namely, Radon-Nikodym Theorem and Jordan-Hahn Decomposition Thorem will be discussed in detail and proved in Chapter 6.

3.4 Integrals : Some Important Inequalities

In this section, we are going to prove some important inequalities involving integrals, which turn out to be extremely useful in various contexts. We start with one which is perhaps the simplest and yet one of the most widely used inequalities.

3.4. Integrals : Some Important Inequalities

THEOREM **3.4.1.** (Chebycheff's inequality 1) Let $(\Omega, \mathcal{A}, \mu)$ be any measure space and f be any measurable function on Ω. Then, for all $a > 0$,

$$\mu(|f| \geq a) \leq \tfrac{1}{a} \int_{\{|f|>a\}} |f| d\mu \leq \tfrac{1}{a} \int |f| d\mu.$$

Proof. For all $a > 0$, one has $aI_{(|f|\geq a)} \leq |f|I_{(|f|\geq a)} \leq |f|$ pointwise. Note that all the three functions are non-negative measurable functions. The result now follows by integrating them and using the monotonicity property of integrals. □

The above inequality, which is more popularly known as *Chebycheff's inequality*, is also sometimes referred to as *Markov's inequality*.

We now introduce an important notation, which will not only make stating the subsequent inequalities a bit more convenient, but will also be used extensively in the latter chapters, in particular, in Chapter 6. Let $(\Omega, \mathcal{A}, \mu)$ be a measure space. For any measurable function f and for any $0 < p < \infty$, we write

$$\|f\|_p = (\int |f|^p d\mu)^{1/p}. \qquad (3.4.1)$$

Note that $|f|^p$ is a non-negative measurable function and $\|f\|_p \in [0, \infty]$, by definition. Propositions 3.2.16 and 3.2.18 imply that $\|f\|_p = 0$ iff $f = 0$ a.e.$[\mu]$.

Using this notation, we now state a more general version of Chebycheff's inequality and, in the sequel, we will refer to this as *Chebycheff's inequality*. Noting that $\{|f| > a\} = \{|f|^p > a^p\}$, for all $a > 0$ and all $0 < p < \infty$, this follows as an immediate consequence of Theorem 3.4.1.

THEOREM **3.4.2.** (Chebycheff's inequality) Let $(\Omega, \mathcal{A}, \mu)$ be any measure space and f be any measurable function on Ω. Then, for all $a > 0$, and all $p \in (0, \infty)$,

$$\mu(|f| \geq a) \leq \tfrac{1}{a^p} \int_{\{|f|>a\}} |f|^p d\mu \leq \tfrac{1}{a^p} \|f\|_p^p.$$

For a measurable function f, we defined $\|f\|_p$, only for $p \in (0, \infty)$. For $p = \infty$, we define it as follows

$$\|f\|_\infty = \inf\{a > 0 : \mu(|f| > a) = 0\} = \sup\{b > 0 : \mu(|f| > b) > 0\}. \qquad (3.4.2)$$

Just like $\|f\|_p$ for $p \in (0, \infty)$, it is clear from definition that $\|f\|_\infty \in [0, \infty]$ and also $\|f\|_\infty = 0$ iff $f = 0$ a.e.$[\mu]$. One can also easily see from the above definition that $|f(\omega)| \leq \|f\|_\infty$ for μ-almost every ω. Indeed, $\|f\|_\infty < \infty$ iff $|f| \leq a$ a.e.$[\mu]$ for some $a \in [0, \infty)$, and, in that case, $\|f\|_\infty$ is the *smallest* such a.

REMARK **3.4.3.** At this stage, we need to make an important remark. While we have now defined $\|f\|_p$ for all $p \in (0, \infty]$, our main focus in the sequel is going to be on $\|f\|_p$, for *only* $p \geq 1$. The primary reason for that, in a nutshell, is that it is *only* for $p \geq 1$ that the quantities $\|\cdot\|_p$ define what is called a "norm" (in fact, a "complete norm") on an appropriate vector space. This will be discussed and the corresponding vector spaces extensively studied in Chapter 6.

It is with the above-stated objective in mind that we now proceed to prove a couple of very important inequalities involving the quantities $\|f\|_p$, for $p \geq 1$.

To state the first inequality, we need to make one more definition. For any real number $p \in (1, \infty)$, a real number q such that $p^{-1} + q^{-1} = 1$ is called *conjugate index* of p. It is clear that $q \in (1, \infty)$ and is also unique, given by the formula $q = p/(p-1)$. The pair (p, q) is often referred to as a pair of *conjugate indices*. For $p = 1$ (resp., $p = \infty$), the conjugate index is taken to be $q = \infty$ (resp., $q = 1$).

THEOREM 3.4.4. (Hölder's inequality) Let (p, q) be a pair of conjugate indices in $[1, \infty]$. Then, for any two measurable functions f and g on a measure space $(\Omega, \mathcal{A}, \mu)$,

$$\int |fg| d\mu \leq \|f\|_p \|g\|_q. \tag{3.4.3}$$

Further, in case $p, q \in (1, \infty)$ and $\|f\|_p \|g\|_q < \infty$, equality holds in (3.4.3) if and only if $A|f|^p = B|g|^q$ a.e.$[\mu]$, for some real numbers A and B, where at least one of A and B is non-zero.

Proof. If the right hand side of (3.4.3) equals zero, then at least one of $\|f\|_p$ and $\|g\|_q$ equals zero, implying that at least one of f and g equals zero a.e $[\mu]$. But that means $fg = 0$ a.e.$[\mu]$ and so the left hand side of (3.4.3) will equal zero. Note also that, in this case, equality holds in (3.4.3). Further, since at least one of f and g equals zero a.e $[\mu]$, we will also have the equality $A|f|^p = B|g|^q$ a.e.$[\mu]$ satisfied, with at least one of A and B not equal to 0.

Assume next that both $\|f\|_p > 0$ and $\|g\|_q > 0$. If one of them equals ∞, then the right hand side equals ∞ and, in that case, (3.4.3) holds trivially.

Let us therefore assume now that $0 < \|f\|_p < \infty$, $0 < \|g\|_q < \infty$ and procced to prove (3.4.3). First, we consider the case $p = 1$, in which case $q = \infty$. Since $|g| \leq \|g\|_\infty$ a.e. $[\mu]$, we have $|fg| \leq |f| \cdot \|g\|_\infty$ a.e. $[\mu]$, from which we get (3.4.3), by integrating and using Theorem 3.2.8 (a)(i) and Proposition 3.2.16. The case $p = \infty, q = 1$ is similar.

Let us now consider the case $1 < p < \infty$, in which case $q = p/(p-1)$. Denoting $f^* = |f|/\|f\|_p$ and $g^* = |g|/\|g\|_q$, observe that $\|f^*\|_p = 1$ and $\|g^*\|_q = 1$. To prove (3.4.3), it is clearly enough to show that $\int f^* g^* d\mu \leq 1$.

We need to use here the simple fact that, for any two non-negative real numbers a and b, the inequality $ab \leq a^p/p + b^q/q$ holds. This is clearly true if either $a = 0$ or $b = 0$ and, in case $a > 0$ and $b > 0$, we have

$$ab = e^{\log ab} = e^{\frac{1}{p}\log a^p + \frac{1}{q}\log b^q} \leq e^{\log a^p}/p + e^{\log b^q}/q = a^p/p + b^q/q,$$

the inequality in the above coming from convexity of the function $x \mapsto e^x$.

Now the assumptions $0 < \|f\|_p < \infty$, $0 < \|g\|_q < \infty$ impliy that f^* and g^* are both non-negative real valued a.e.$[\mu]$ and so, using the above inequality, we get

$$f^* g^* \leq (f^*)^p/p + (g^*)^q/q \quad \text{a.e.}[\mu] \tag{3.4.4}$$

Integrating boh sides with respect to μ and using the fact that $1/p + 1/q = 1$, one easily gets $\int f^* g^* d\mu \leq 1$, as was required to be proved.

Finally, we are left to prove the last assertion only in this case. Note that, by Proposition 3.2.18, equality will hold in (3.4.3) iff equality holds a.e.$[\mu]$ in (3.4.4). Since exponential function is strictly convex, equality holds a.e.$[\mu]$ in (3.4.4) iff $\log(f^*)^p = \log(g^*)^q$ a.e.$[\mu]$, equivalently, $(f^*)^p = (g^*)^q$ a.e.$[\mu]$, that is, $|f|^p/\|f\|_p^p = |g|^q/\|g\|_q^q$ a.e.$[\mu]$, proving the assertion with $A = \|g\|_q^q$, $B = \|f\|_p^p$. □

3.4. Integrals : Some Important Inequalities

We state separately the following special case of Hölder's inequality, namely, for $p = q = 2$, because it is very well-known in a wide range of contexts and is popularly called the *Cauchy-Schwarz inequality*.

THEOREM 3.4.5. (Cauchy-Schwarz inequality) For any two measurable functions f and g on a measure space $(\Omega, \mathcal{A}, \mu)$,

$$\int |fg| d\mu \leq \|f\|_2 \|g\|_2, \qquad (3.4.5)$$

and, in case $\|f\|_2 \|g\|_2 < \infty$, equality holds in (3.4.5) if and only if $A|f| = B|g|$ a.e.$[\mu]$, for some real numbers A and B with at least one of them non-zero.

REMARK 3.4.6. The assertion on the necessary and sufficient condition for equality in (3.4.3) doesn't hold without the assumption $\|f\|_p \|g\|_q < \infty$. To see this, consider the functions $f(x) = x$ and $g(x) = 1/\sqrt{1+x^2}$ on the measure space $(\mathbb{R}, \mathcal{B}, \lambda)$. It is clear that $A|f| = B|g|$ a.e.$[\lambda]$ does not hold for any pair of constants A and B, while it is easy to see that $\int |fg| d\lambda = +\infty = \|f\|_2 \|g\|_2$.

The next result asserts that the quantities $\|\cdot\|_p$, for $1 \leq p \leq \infty$, satisfy a "triangle inequality", as is required for these to act as norms (see Remark 3.4.3). The proof uses Hölder's inequality and the usual triangle inequality of numbers.

THEOREM 3.4.7. (Minkowski inequality): Let $p \in [1, \infty]$. Then, for any two measurable functions f and g on a measure space $(\Omega, \mathcal{A}, \mu)$, for which the sum $f + g$ is well-defined a.e.$[\mu]$,

$$\|f + g\|_p \leq \|f\|_p + \|g\|_p \qquad (3.4.6)$$

Proof. For $p = 1$, (3.4.6) follows from the triangle inequality $|f + g| \leq |f| + |g|$ (wherever $f + g$ is well-defined) and properties of integrals. For $p = \infty$, we have $|f| \leq \|f\|_\infty$ a.e.$[\mu]$ and $|g| \leq \|g\|_\infty$ a.e.$[\mu]$, by the definition of $\|\cdot\|_\infty$. Using triangle ieqaulity once again, we have $|f + g| \leq |f| + |g|$ pointwise, provided $f + g$ is defined, and the right hand side is $[\mu]$-a.e. bounded above by $\|f\|_\infty + \|g\|_\infty$. The inequality $\|f + g\|_\infty \leq \|f\|_\infty + \|g\|_\infty$ now follows from the definition of $\|\cdot\|_\infty$. Let us now consider the case $1 < p < \infty$. The inequality (3.4.6) holds trivially if $\|f + g\|_p = 0$. On the other hand, by convexity of the map $x \mapsto |x|^p$, one has $|f + g|^p \leq 2^{p-1}(|f|^p + |g|^p)$, so that, $\|f + g\|_p = \infty$ would imply $\|f\|_p + \|g\|_p = \infty$ and hence the inequality (3.4.6) will hold trivially. Let us assume therefore that $0 < \|f + g\|_p < \infty$ and proceed to prove (3.4.6). Using the triangle inequality and the monotonicity and linearity properties of integrals, one gets

$$\int |f + g|^p d\mu \leq \int |f| |f + g|^{p-1} d\mu + \int |g| |f + g|^{p-1} d\mu$$

Hölder's inequality on the first term on the right hand side gives

$$\int |f| |f + g|^{p-1} d\mu \leq \|f\|_p \left[\int |f + g|^{(p-1)q} d\mu \right]^{1/q} = \|f\|_p (\|f + g\|_p)^{p/q}$$

Note that $(p-1)q = p$ for any pair of cojugate indices (p, q) in $(1, \infty)$. A similar argument on the second term would give

$$\int |g| |f + g|^{p-1} d\mu \leq \|g\|_p (\|f + g\|_p)^{p/q}$$

Combining the two terms, we get

$$\|f+g\|_p^p \le (\|f+g\|_p)^{p/q} \cdot (\|f\|_p + \|g\|_p),$$

which leads to (3.4.6), in view of $0 < \|f+g\|_p < \infty$ and the fact that $p - p/q = 1$. □

The aim of the following exercise is to illustrate that, for $p < 1$, the triangle inequality (3.4.6) fails to hold for $\|\cdot\|_p$, in general.

EXERCISE **3.4.8.** (a) Show that, for $0 < p < 1$, any two measurable functions f and g on a measure space satisfy the inequality $\|f+g\|_p \le 2^{1/p-1}(\|f\|_p + \|g\|_p)$.
(b) Let $\Omega = \{1, 2, \dots, n\}$, $\mathcal{A} = \mathcal{P}(\Omega)$ and μ the counting measure on $\mathcal{A}$. Use the functions $f = 2I_{\{1\}}$, $g = 2I_{\{2\}}$ to show that the constant $2^{1/p-1}$ in (a) is sharp.

Here is an easy but important consequence of Hölder's inequality, which says that for a finite measure space, finiteness of $\|f\|_s$ for some $s > 0$, implies finiteness of $\|f\|_r$, for all $0 < r < s$.

THEOREM **3.4.9.** Let $(\Omega, \mathcal{A}, \mu)$ be a finite measure space and f be a measurable function on Ω. If $\|f\|_s < \infty$ for some $s > 0$, then $\|f\|_r < \infty$ for all $0 < r < s$. Moreover, in case μ is a probability measure,

$$\|f\|_r \le \|f\|_s, \quad \text{for all } 0 < r < s. \tag{3.4.7}$$

Proof. Let $0 < r < s$ and take $p = s/r$. Then $p > 1$ and by applying Hölder's inequality with the two measurable functions $|f|^r$ and $g \equiv 1$, one gets

$$\int |f|^r d\mu \le \left(\int (|f|^r)^p d\mu\right)^{1/p} [\mu(\Omega)]^{1/q} = \left(\int |f|^s d\mu\right)^{r/s} [\mu(\Omega)]^{1/q},$$

or equivalently, $\|f\|_r \le \|f\|_s \left(\mu(\Omega)\right)^{1/(rq)}$. Since $\mu(\Omega) < \infty$, it follows that finiteness of $\|f\|_s$ implies that of $\|f\|_r$, for all $0 < r < s$.
Further, when μ is a probability measure, that is, $\mu(\Omega) = 1$, the inequality $\|f\|_r \le \|f\|_s$, for $0 < r < s$, is immediate. □

It is clear from the above proof that the hypothesis in Theorem 3.4.9 of μ being a finite measure, plays a crucial role. A natural question is : what, if anything, can we say in case of an infinite measure μ? We will answer this question later in Chapter 6 (see Remark 6.2.6). As we will see there, when the measure is not finite, there is no specific behaviour of a nature similar to the case of finite measures, that holds in general. Exercises 6.2.7 and 6.2.8 there will illustrate that, for infinite measures, one may encounter very divergent possible scenarios.

We end this section with one more important inequality that holds for probability spaces. To get a motivation for the inequality, consider an open interval $I \subset \mathbb{R}$ and recall that a convex function on I is a function $\varphi : I \to \mathbb{R}$ satisfying the inequality

$$\varphi(\alpha x_1 + (1-\alpha) x_2) \le \alpha \varphi(x_1) + (1-\alpha) \varphi(x_2), \quad \text{for all } x_1, x_2 \in I \text{ and } \alpha \in [0,1].$$

There is an interesting way to look at the above inequality. Consider the measurable space $(I, \mathcal{B}(I))$. The above inequality is simply saying that, for any $x_1, x_2 \in I$

3.4. Integrals : Some Important Inequalities

and any $\alpha \in [0,1]$, if we consider the probability measure μ on $(I, \mathcal{B}(I))$ given by $\mu(A) = \alpha I_A(x_1) + (1-\alpha)I_A(x_2)$, then, for the function $f(x) = x$ on $(I, \mathcal{B}(I), \mu)$,

$$\varphi(\textstyle\int f d\mu) \leq \int \varphi \circ f \, d\mu.$$

As will be seen in Lemma 3.4.11 below, any convex function φ on an open interval I is continuous and hence measurable. The next result, known as *Jensen's inequality*, generalizes the above inequality in a significant way.

THEOREM 3.4.10. (Jensen's inequality) *Let $(\Omega, \mathcal{A}, \mu)$ be a probability space and f be a μ-integrable function on Ω. Then for any convex function $\varphi : \mathbb{R} \to \mathbb{R}$,*

$$\varphi\left(\textstyle\int f d\mu\right) \leq \int \varphi \circ f \, d\mu, \qquad (3.4.8)$$

provided the integral $\int \varphi \circ f \, d\mu$ exists.

The following lemma gives us some important properties that any convex function on $\mathbb{R}$ has and these will be used for proving the above theorem.

LEMMA 3.4.11. *If $\varphi : \mathbb{R} \to \mathbb{R}$ is a convex function, then φ is continuous. Further, for every $x \in \mathbb{R}$, there are real numbers α and β, depending possibly on x, such that $\alpha + \beta x = \varphi(x)$ and $\alpha + \beta y \leq \varphi(y)$, for all $y \in \mathbb{R}$.*

Proof. From convexity of φ, one can easily see that, for any $a < b < c$,

$$(c-a)\varphi(b) \leq (c-b)\varphi(a) + (b-a)\varphi(c).$$

The following easy consequences of the above are left for the reader to verify :

(i) $\dfrac{\varphi(b) - \varphi(a)}{b-a} \leq \dfrac{\varphi(d) - \varphi(c)}{d-c}$, for any choice of real numbers $a < b \leq c < d$.

(ii) $\dfrac{\varphi(x) - \varphi(a)}{x-a} \leq \dfrac{\varphi(x) - \varphi(b)}{x-b}$, for any choice of real numbers $a < b < x$.

(iii) $\dfrac{\varphi(d) - \varphi(x)}{d-x} \geq \dfrac{\varphi(c) - \varphi(x)}{c-x}$, for any choice of real numbers $x < c < d$.

From these, one can easily deduce that, for every $x \in \mathbb{R}$, the limit $\lim_{y \to x-} \dfrac{\varphi(y) - \varphi(x)}{y-x}$ exists and further

$$\lim_{y \to x-} \frac{\varphi(y) - \varphi(x)}{y-x} = \sup_{y<x} \frac{\varphi(y) - \varphi(x)}{y-x} \leq \frac{\varphi(z) - \varphi(x)}{z-x}, \text{ for any } z > x \quad (3.4.9)$$

One can similarly show that, $\lim_{z \to x+} \dfrac{\varphi(z) - \varphi(x)}{z-x}$ exists for every $x \in \mathbb{R}$, and also

$$\lim_{z \to x+} \frac{\varphi(z) - \varphi(x)}{z-x} = \inf_{z>x} \frac{\varphi(z) - \varphi(x)}{z-x} \geq \frac{\varphi(y) - \varphi(x)}{y-x}, \text{ for any } y < x. \quad (3.4.10)$$

It follows from (3.4.9) and (3.4.10) that φ has, at every $x \in \mathbb{R}$, a finite left-hand-derivative, to be denoted $\varphi'(x-)$, as well as a finite right-hand-derivative, to be denoted $\varphi'(x+)$. Further, they satisfy $\varphi'(x-) \leq \varphi'(x+)$ at every $x \in \mathbb{R}$. Continuity of φ at every $x \in \mathbb{R}$, of course, is an immediate consequence of this. To prove the other assertion, let us fix $x \in \mathbb{R}$. If we take any β such that $\varphi'(x-) \leq \beta \leq \varphi'(x+)$ and set $\alpha = \varphi(x) - \beta x$, then clearly, $\varphi(x) = \alpha + \beta x$.

To show now that $\alpha + \beta y \leq \varphi(y)$ for all $y \neq x$, let us first take $y < x$. Since $\beta \geq \varphi'(x-)$, it follows from (3.4.9) that

$$\beta \geq \frac{\varphi(y) - \varphi(x)}{y - x} = \frac{\varphi(x) - \varphi(y)}{x - y} = \frac{\alpha + \beta x - \phi(y)}{x - y},$$

from which $\alpha + \beta y \leq \varphi(y)$ follows. For the case $y > x$, we can prove it similarly by noting that, by (3.4.10), $\beta \leq \varphi'(x+)$ implies that $\beta \leq \dfrac{\varphi(y) - \varphi(x)}{y - x}$. □

Proof of Jensen's Inequality. Since f is μ-integrable, we apply the second assertion of Lemma 3.4.11 with $x = \int f d\mu \in \mathbb{R}$ to get real numbers α and β such that $\varphi(\int f d\mu) = \alpha + \beta \cdot (\int f d\mu)$, while the inequality $\alpha + \beta f \leq \varphi \circ f$ holds pointwise. The inequality (3.4.8) now follows from monotonicity of integrals and the fact that, since μ is a probability measure, $\int (\alpha + \beta f) \, d\mu = \alpha + \beta \cdot (\int f d\mu) = \varphi(\int f d\mu)$. □

REMARK 3.4.12. (a) One can use Jensen's inequalilty and the fact that any finite measure is a scalar multiple of a probability measure to give an alternative proof of Theorem 3.4.9. [See Exercise 3.6.8.]
(b) While we have stated and proved Jensen's inequality only for convex functions defined on $\mathbb{R}$, one can prove an analogous result for convex functions defined on any open interval in $\mathbb{R}$. [See Exercise 3.6.9.]

3.5 Measurable Maps

In what we have done so far, we have talked about measurability only for extended real valued functions defined on a measurable space. This was the appropriate thing to do for developing integration theory. However, the concept of measurability need not be restricted only to extended real valued maps. One can and, in fact, needs to talk about measurability of maps on $(\Omega, \mathcal{A})$ taking values in any set E. What is required for this is that the set E also be equipped with a σ-field.

To define this concept, it will be convenient to first introduce the notion of, what are called, "inverse images of sets under functions". Let Ω and E be non-empty sets and let $f : \Omega \to E$ be any function. For any set $B \subset E$, the *inverse image of B under f*, written $f^{-1}(B)$ is defined as

$$f^{-1}(B) = \{\omega \in \Omega : f(\omega) \in B\}$$

Thus, for every $B \subset E$, the inverse image of B under f, namely $f^{-1}(B)$, gives us a subset of Ω. To avoid any possibility of notational confusion, let us make it clear that this concept has *nothing to do* with the usual concept of inverse function f^{-1}. Since f has not been assumed to be one-one and/or onto, the question of an inverse function f^{-1} doesn't even arise!

It is clear from definition that $f^{-1}(\emptyset) = \emptyset$ and $f^{-1}(E) = \Omega$. But an important property of inverse images is that the operation of taking inverse images commutes with usual set operations. The following are easy and left as exercises.

$$f^{-1}(B^c) = (f^{-1}(B))^c, \quad \text{for any } B \subset E, \text{ and,}$$

$$f^{-1}\left(\bigcup_\alpha B_\alpha\right) = \bigcup_\alpha f^{-1}(B_\alpha), \quad f^{-1}\left(\bigcap_\alpha B_\alpha\right) = \bigcap_\alpha f^{-1}(B_\alpha), \text{ for any family } B_\alpha \subset E.$$

3.5. Measurable Maps

Using the notation of inverse images, our definition 3.1.2 of measurability of a real valued fumction f on a measurable space $(\Omega, \mathcal{A})$ boils down to simply requiring that $f^{-1}((-\infty, c]) \in \mathcal{A}$, for all $c \in \mathbb{R}$. Of course, in Proposition 3.1.3, we saw that this turns out to be equivalent to requiring that $f^{-1}(B) \in \mathcal{A}$, for all $B \in \mathcal{B}(\mathbb{R})$. This leads us to the following general definition of measurability.

DEFINITION 3.5.1. Let $(\Omega, \mathcal{A})$ and $(E, \mathcal{E})$ be measurable spaces. A map $f : \Omega \to E$ is said to be *measurable*, or more specifically, $(\mathcal{A}, \mathcal{E})$-*measurable*, if

$$f^{-1}(B) \in \mathcal{A}, \quad \text{for all } B \in \mathcal{E}. \qquad (3.5.1)$$

It is clear that measurability of f is not just a condition on the function f, but also involves the σ-fields $\mathcal{A}$ and $\mathcal{E}$. We underline this dependence by often writing that $f : (\Omega, \mathcal{A}) \to (E, \mathcal{E})$ is a measurable map. It is obvious from the definition that if f is $(\mathcal{A}, \mathcal{E})$-measurable, then it remains measurable if we replace $\mathcal{A}$ by a larger σ-field on Ω and/or replace $\mathcal{E}$ by a smaller σ-field on E.

The next proposition is an analogue of Proposition 3.1.4 in the present context and, just like there, the proof of this is a simple application of the good sets principle and hence left as an exercise.

PROPOSITION 3.5.2. Let $\mathcal{C}$ be any family of subsets of E, generating the σ-field $\mathcal{E}$. Then, a map $f : (\Omega, \mathcal{A}) \to (E, \mathcal{E})$ is measurable if and only if

$$f^{-1}(B) \in \mathcal{A}, \quad \text{for all } B \in \mathcal{C}.$$

The following proposition states a simple yet useful general fact, of which, Exercise 3.1.19 was just a special case.

PROPOSITION 3.5.3. Let $f : (\Omega, \mathcal{A}) \to (E, \mathcal{E})$ and $g : (E, \mathcal{E}) \to (D, \mathcal{D})$ be measurable maps. Then $g \circ f : (\Omega, \mathcal{A}) \to (D, \mathcal{D})$ is a measurable map.

Proof. For $B \in \mathcal{D}$, one can easily see that $(g \circ f)^{-1}(B) = f^{-1}(g^{-1}(B))$. Measurability of g gives $g^{-1}(B) \in \mathcal{E}$ and then measurability of f gives $f^{-1}(g^{-1}(B)) \in \mathcal{A}$, completing the proof. □

EXERCISE 3.5.4. (a) *For $k, l \geq 1$, show that, if $f : \mathbb{R}^k \to \mathbb{R}^l$ is a continuous function, then $f : (\mathbb{R}^k, \mathcal{B}(\mathbb{R}^k)) \to (\mathbb{R}^l, \mathcal{B}(\mathbb{R}^l))$ is a measurable map.*
(b) *For each $i = 1, \cdots, k$, let $\pi_i : \mathbb{R}^k \to \mathbb{R}$ denote the i-th coordinate projection map defined as $\pi_i(x_1, \cdots, x_k) = x_i$, $(x_1, \cdots, x_k) \in \mathbb{R}^k$. Show that, for each i, the map $\pi_i : (\mathbb{R}^k, \mathcal{B}(\mathbb{R}^k)) \to (\mathbb{R}, \mathcal{B}(\mathbb{R}))$ is measurable.*
(c) *Show that a function $f : (\Omega, \mathcal{A}) \to (\mathbb{R}^k, \mathcal{B}(\mathbb{R}^k))$ is measurable iff, for each $i = 1, \cdots, k$, the composition $f \circ \pi_i : (\Omega, \mathcal{A}) \to (\mathbb{R}, \mathcal{B})$ is measurable.*

We noted earlier that a measurable map $f : (\Omega, \mathcal{A}) \to (E, \mathcal{E})$ remains measurable if we replace $\mathcal{A}$ by any larger σ-field on Ω. But if we replace $\mathcal{A}$ by a smaller σ-field, then f may or may not remain measurable. This leads us to the following interesting question. Let Ω, E be two non-empty sets and suppose we have a function $f : \Omega \to E$. Given now a σ-field $\mathcal{E}$ on E, we want to ask: what is the "smallest" σ-field on Ω, that will make the map f a measurable map? That there is such a smallest σ-field is clear. This is because the σ-field $\mathcal{P}(\Omega)$, consisting of all subsets of Ω, will obviously make f measurable and now the intersection of all σ-fields on Ω that make f measurable will be the smallest σ-field that we are looking for. However, what is interesting is that there is a very simple explicit description of this smallest σ-field. Indeed, using commutativity between inverse image formation and standard set operations, one can easily verify that the class $\mathcal{A}_0 = \{f^{-1}(B) : B \in \mathcal{E}\}$ forms a σ-field on Ω. On the other hand, from the definition of measurability, it is clear that, for any σ-field $\mathcal{A}$ on Ω, f is $(\mathcal{A}, \mathcal{E})$-measurable iff $\mathcal{A}_0 \subset \mathcal{A}$. Thus, $\mathcal{A}_0$ is the required smallest σ-field. It is called the σ-field on Ω *generated by f* and is denoted as $\sigma(f)$. It should not be forgotten that, throughout, the σ-field $\mathcal{E}$ on E is fixed.

The next proposition is often very handy in getting a more explicit identification of the sets belonging to $\sigma(f)$, as the following Exercise will illustrate. The proof of the proposition is an easy application of the good sets principle and is hence left for the reader to complete.

PROPOSITION 3.5.5. *If $\mathcal{C}$ is a class of subsets of E generating the σ-field $\mathcal{E}$, then the class $f^{-1}(\mathcal{C}) = \{f^{-1}(C) : C \in \mathcal{C}\}$ of subsets of Ω generates $\sigma(f)$. In short, $\sigma(f^{-1}(\mathcal{C})) = f^{-1}(\sigma(\mathcal{C}))$.*

EXERCISE 3.5.6. *Consider $\Omega = \mathbb{R}$ and $(E, \mathcal{E}) = (\mathbb{R}, \mathcal{B})$. For each of the following functions, give an explicit description of $\sigma(f)$ by clearly identifying which $\mathcal{B}(\mathbb{R})$-sets belong to $\sigma(f)$:*
(i) $f(x) = |x|$, (ii) $f(x) = x^2$, (iii) $f(x) = e^x$, (iv) $f(x) = \sin x$,
(v) $f(x) = [x] =$ integer part of x, (vi) $f(x) = x - [x] =$ fractional part of x.

REMARK 3.5.7. There is one little but important isssue that perhaps has not escaped attention of the careful reader and, therefore, needs to be addressed right away. The idea of measurability of real valued functions on a measurable space, as discussed at the start of Section 3.1, motivated us to introduce the more general concept of measurable maps $f : (\Omega, \mathcal{A}) \to (E, \mathcal{E})$. Real valued measurable functions then turn out to be just measurable maps in the special case when $(E, \mathcal{E}) = (\mathbb{R}, \mathcal{B})$. The question that was left untouched is: where do extended real valued measurable functions fit in? Clearly, we now have to take $E = [-\infty, \infty] = \mathbb{R} \cup \{\pm\infty\}$ and define an appropriate σ-field $\mathcal{E}$ on E, so that extended real-valued functions on a measurable space $(\Omega, \mathcal{A})$ turn out to be just $(\mathcal{A}, \mathcal{E})$-measurable maps on Ω into $E = [-\infty, \infty]$. As the next exercise shows, the solutiom is very simple, though it should be noted.

EXERCISE 3.5.8. *On $E = [-\infty, \infty]$, define $\mathcal{E}$ to be the smallest σ-field containing all Borel subsets of $\mathbb{R}$ and the singleton sets $\{-\infty\}$ and $\{\infty\}$.*
(a) Describe all the sets in $\mathcal{E}$ and also show that each of the following classes generate $\mathcal{E}$: (i) $\{[-\infty, c) : c \in \mathbb{R}\}$, (ii) $\{(c, \infty] : c \in \mathbb{R}\}$.
(b) Given an extended real valued function f on a non-empty set Ω, try to describe the sets in $\sigma(f)$.

3.5. Measurable Maps

Returning now to our main stream of discussion, given a measurable space $(E, \mathcal{E})$, we have identified the smallest σ-field $\mathcal{A}$ on a set Ω that makes a function $f : \Omega \to E$ to be $(\mathcal{A}, \mathcal{E})$-measurable. From here, natural next step would be to ask: what if we had considered two maps f and g on Ω into E? Given a σ-field $\mathcal{E}$ on E, it should be clear again that there is a smallest σ-field on Ω that makes both f and g measurable. As above, one only needs to note that $\mathcal{P}(\Omega)$ is one σ-field on Ω that makes every function on Ω measurable. However, the difference now is that, unlike in the case of a single function, we may not, in general, have a simple description of the sets that constitute the smallest σ-field. Matters do not change much if we take $f : \Omega \to E_1$ and $g : \Omega \to E_2$, where E_1 and E_2 are equipped with σ-fields $\mathcal{E}_1$ and $\mathcal{E}_2$ respectively. We now formalize and generalize this in the next definition.

DEFINITION 3.5.9. Let Ω be a non-empty set and $\{(E_\alpha, \mathcal{E}_\alpha), \alpha \in \Delta\}$ be a family of measurable spaces. Given a class of functions $\{f_\alpha : \Omega \to E_\alpha, \alpha \in \Delta\}$, the smallest σ-field on Ω which makes f_α measurable for all $\alpha \in \Delta$, is called *the σ-field generated by the class of functions* $\{f_\alpha, \alpha \in \Delta\}$ and is denoted by $\sigma(f_\alpha : \alpha \in \Delta)$.

Noting once again that the power set $\mathcal{P}(\Omega)$ is one σ-field on Ω that makes all the f_α measurable, it should be clear that there does exist a smallest σ-field on Ω that will make all $f_\alpha, \alpha \in \Lambda$, measurable. The next proposition is a generalization of Proposition 3.5.5, whose proof is again based on application of the good sets principle and hence left as an exercise. This proposition can sometimes help one to get a complete description of the σ-field generated by a class consisting of more than one function, as the exercise following it should illustrate.

PROPOSITION 3.5.10. For each α, let $\mathcal{C}_\alpha$ be a class of subsets of E_α that generates the σ-field $\mathcal{E}_\alpha$. Given a non-empty set Ω and functions $f_\alpha : \Omega \to E_\alpha$, for $\alpha \in \Delta$, the class $\{f_\alpha^{-1}(C_\alpha) : C_\alpha \in \mathcal{C}_\alpha, \alpha \in \Delta\}$ of subsets of Ω generates $\sigma(f_\alpha : \alpha \in \Delta)$.

EXERCISE 3.5.11. With $\pi_i, i = 1, \cdots, k$, as defined in Exercise 3.5.4 (b), show that $\mathcal{B}(\mathbb{R}^k) = \sigma(\pi_i : i = 1, \cdots, k)$.

EXERCISE 3.5.12. (a) Consider the real valued functions f_1 and f_2 on $\mathbb{R}^2$ given by $f_1(x_1, x_2) = x_1 \vee x_2$ and $f_2(x_1, x_2) = x_1 \wedge x_2$. Describe the σ-field $\sigma(f_1, f_2)$.
(b) Consider the following real valued functions on $\mathbb{R}^2$: $f_1(x, y) = [x] =$ integer part of x, $f_2(x, y) = [y] =$ integer part of y, $f_3(x, y) = \cos 2\pi x$ and $f_4(x, y) = \cos 2\pi y$. Describe the σ-fields $\sigma(f_1, f_2)$ and $\sigma(f_1, f_2, f_3, f_4)$.

If $(E, \mathcal{E})$ is a measurable space and $f : \Omega \to E$, then, as a special case of Proposition 3.5.3, we know that, for any extended real valued measurable function g on E, the extended real valued function $h = g \circ f$ on Ω is $\sigma(f)$-measurable. The next theorem asserts that this is how all $\sigma(f)$-measurable extended real valued functions on Ω arise.

THEOREM 3.5.13. Let $(E, \mathcal{E})$ be a measurable space and f be a function on some non-empty set Ω into E. Then an extended real valued function h on Ω is measurable with respect to $\sigma(f)$ if and only if $h = g \circ f$ for some extended real valued measurable function g on E. Further, if h is real valued, then g can be chosen to be a real-valued measurable function on E

Proof. In view of what was noted above, just the 'only if' part needs to be proved. We have to show that any extended real valued $\sigma(f)$-measurable function h on Ω must be of the form $h = g \circ f$, with g as stated. Firstly, if h is an indicator function, say, $h = I_A$, for some $A \in \sigma(f)$, then A must equal $f^{-1}(B)$ for some $B \in \mathcal{E}$. This gives us $h = I_{f^{-1}(B)} = I_B \circ f$. So, the required result holds with $g = I_B$. Having proved the required result for indicator functions h, one can easily extend it, by linearity, to real valued $\sigma(f)$-measurable simple functions h. Finally, if h is any extended real valued $\sigma(f)$-measurable function, then, by Proposition 3.1.25, we can get a sequence $\{h_n\}$ of real valued $\sigma(f)$-measurable simple functions, such that, $h_n \to h$ pointwise. But then, by what has been already proved, we have $h_n = g_n \circ f$, where g_n, for each n, is a real valued measurable function on E. Taking $g = \limsup g_n$, we get an extended real valued measurable function g on E and one can easily verify that $h = g \circ f$. In case the given h is real valued, we modify the definition of the g above to $g = \limsup g_n\, I_{\{-\infty < \limsup g_n < \infty\}}$ to get a real-valued measurable function on E, which will satisfy $h = g \circ f$. □

EXERCISE **3.5.14**. *In the proof of Theorem 3.5.13, we had $h_n \to h$ pointwise and $h_n = g_n \circ f$, for each n. Why was g defined as $\limsup g_n$ and not as $\lim g_n$?*

EXERCISE **3.5.15**. *Let $f_1, \cdots, f_k$ be real valued functions on a non-empty set E.
(a) Use Exercise 3.5.4 (c) to show that $\sigma(\{f_1, \cdots, f_k\})$ consists precisely of all sets of the form $\{x \in E : (f_1(x), \cdots, f_k(x)) \in B\}$ as B varies over $\mathcal{B}(\mathbb{R}^k)$.
(b) Show that an extended real valued function h on E is $\sigma(\{f_1, \cdots, f_k\})$-measurable iff $h = g \circ (f_1, \cdots, f_k)$, for some extended real valued measurable function g on $(\mathbb{R}^k, \mathcal{B}(\mathbb{R}^k))$. Further, if h is real valued, g can be chosen to be real valued. [Use Exercise 3.5.11, if necessary.]*

We next discuss how a measurable map may transport a measure from the σ-field on the domain space to that on the co-domain space and subsequently, how integration on the domain space can be transferred to integration on the co-domain space. This is called "Change of Variable" in this abstract integration theory. It is often a very useful tool that converts an integral of a real measurable function on an abstract set to an integral on the real line with respect to the transported/induced measure. We start with some preliminaries. Suppose we have a measurable map $f : (\Omega, \mathcal{A}) \to (E, \mathcal{E})$. Then, given any measure μ on $\mathcal{A}$, we can define a set function ν on $\mathcal{E}$ by $\nu(B) = \mu(f^{-1}(B))$, $B \in \mathcal{E}$. The next theorem asserts that ν defines a measure on $\mathcal{E}$, which is fairly easy to prove. It is called the *measure on $\mathcal{E}$ induced by f*, or simply, the *induced measure*. Sometimes, one denotes ν as μf^{-1}. It is easy to see that ν would be a finite measure (resp., a probability measure) on $\mathcal{E}$ according as μ is a finite measure (resp., a probability

3.5. Measurable Maps

measure) on $\mathcal{A}$. By taking μ to be any σ-finite, but not finite, measure on $\mathcal{A}$ and taking $(E, \mathcal{E})$ to be 'trivial' (that is, $\mathcal{E} = \{\emptyset, E\}$), one can see that σ-finite measures do not, in general, induce σ-finite measures.

THEOREM **3.5.16.** (Change of Variable Theorem) Let $(\Omega, \mathcal{A}, \mu)$ be a measure space and $f : (\Omega, \mathcal{A}) \to (E, \mathcal{E})$ be a measurable map. Then, the set function ν defined on $\mathcal{E}$ by

$$\nu(B) = \mu\bigl(f^{-1}(B)\bigr), \quad \text{for } B \in \mathcal{B},$$

is a measure on $\mathcal{E}$. [ν is called the *induced measure*.]
Further, for any extended real valued measurable function g on E,

$$\int_\Omega (g \circ f) d\mu = \int_E g d\nu. \tag{3.5.2}$$

REMARK **3.5.17.** As always, the equality of the two integrals in (3.5.2) in the theorem should be interpreted as asserting that if one of them exists, then so also does the other and they are equal.

Proof. From the facts that μ is a measure on $\mathcal{A}$ and that inverse images and set operations commute, one can easily verify that ν is a measure on $\mathcal{E}$. For the other part, the assertion (3.5.2) is immediate from the definition of ν, in case g is an indicator function, say, $g = I_B$ for some $B \in \mathcal{E}$. In case g is a non-negative real valued $\mathcal{E}$-measurable simple function, the equality in (3.5.2) then follows easily, using linearity. For general non-negative measurable g, one gets a sequence $\{g_n\}$ of non-negative real valued measurable simple functions with $g_n \uparrow g$ and gets (3.5.2) by appealing to MCT. Finally, for a general extended real valued measurable g, one uses the equality proved for the non-negative g^+ and g^- and argues that existence of any one of the two sides of (3.5.2) for g implies existence of the other side and the equality of the two. □

EXERCISE **3.5.18.** With f, ν and g as in Theorem 3.5.16 and any $B \in \mathcal{E}$, show that $\int_{f^{-1}(B)} (g \circ f) d\mu = \int_B g d\nu$ (with equality interpreted as in the above remark).

REMARK **3.5.19.** It was mentioned already that the "induced measure" ν is often denoted as μf^{-1}. Using this notation, the integral equality of Exercise 3.5.18 can be and is often written as $\int_{f^{-1}(B)} (g \circ f) d\mu = \int_B g \, d(\mu f^{-1})$.

REMARK **3.5.20.** As a special case of Theorem 3.5.16, if we take $(E, \mathcal{E}) = (R, \mathcal{B})$ and f a real valued measurable function on $(\Omega, \mathcal{A}, \mu)$, then the change of variable rule (3.5.2) asserts that $\int_\Omega (g \circ f) \, d\mu = \int_R g(x) \, d(\mu f^{-1})(x)$, for any measurable function $g : \mathbb{R} \to \mathbb{R}$. Thus integrating $g \circ f$ on Ω with respect to the measure μ is the same as integrating g on $\mathbb{R}$ with respect to the measure μf^{-1} induced by f. This happens to be an extremely powerful tool because it essentially reduces computing (or checking existernce of) the integral of *any* real measurable function on *any* measure space to doing the same for an integral of a real function on $\mathbb{R}$ with respect to an appropriate measure on $\mathcal{B}(\mathbb{R})$.

We end this section with a very useful result which goes by the name of *Monotone Class Theorem (MCT) for Functions*. This often provides a very handy tool when one tries to prove, in some context, that a specified property holds for all bounded measurable functions on some measurable space (or measure space). We start with a lemma.

LEMMA 3.5.21. *Let $\mathfrak{L}$ be a vector space of bounded real functions on a non-empty set Ω, that contains constant functions and is closed under limits of uniformly bounded and non-decreasing sequences of non-negative functions. Then $\mathfrak{L}$ is closed under uniform limits.*

Proof. For $g \in \mathfrak{L}$, let us denote $\|g\|_\infty = \sup\{|g(\omega)| : \omega \in \Omega\}$. Now, suppose $\{h_n\}$ is a sequence in $\mathfrak{L}$, converging uniformly to a function h, that is, $\|h_n - h\|_\infty \to 0$. This clearly implies that $\sup_n \|h_n\|_\infty < \infty$. Since $\mathfrak{L}$ is a vector space, we can assume that $\|h_n\|_\infty \leq 1$ for all n (by "normalizing" the h_n, if necessary). Further, by passing through a subsequence, if necessary, we may also assume that $\|h_{n+1} - h_n\|_\infty \leq 2^{-n}$ for each n. Put $g_n = h_n + 2 - 2^{-(n-1)}$. Clearly $g_n \in \mathfrak{L}$, $g_1 = h_1 + 1 \geq 0$, $g_{n+1} - g_n = h_{n+1} - h_n + 2^{-n} \geq 0$ and $\|g_n\|_\infty \leq \|h_n\|_\infty + 2$. By the hypothesis on $\mathfrak{L}$, we get $\lim g_n = h + 2 \in \mathfrak{L}$, which implies $h \in \mathfrak{L}$. □

THEOREM 3.5.22. *(MCT for functions) Let $\mathfrak{L}$ be a vector space of bounded real functions on a non-empty set Ω that contains constant functions and is closed under limits of uniformly bounded and non-decreasing sequences of non-negative functions. If $\mathfrak{C} \subset \mathfrak{L}$ is a class of functions closed under pointwise multiplication, then $\mathfrak{L}$ contains all bounded real functions that are measurable with respect to $\sigma(\mathfrak{C})$.*

Proof. Denoting $\mathfrak{D}$ to be the uniform closure of the algebra generated by $\mathfrak{C}$, it follows from Lemma 3.5.21 and properties of $\mathfrak{L}$, that $\mathfrak{D} \subset \mathfrak{L}$. We next want to observe that if $h \in \mathfrak{D}$, then $|h| \in \mathfrak{D}$ also. Indeed, if $\|h\|_\infty \leq M$, then using the fact that the continuous function $x \mapsto |x|$ on $[-M, M]$ is the uniform limit of a sequence of polynomials p_n, one obtains that $|h|$ is the uniform limit of $p_n(h)$. Since $p_n(h) \in \mathfrak{D}$ for each n, we get that $|h| \in \mathfrak{D}$. It now follows that if $h, g \in \mathfrak{D}$, then the functions $h \vee g = (|h - g| + h + g)/2$ and $h \wedge g = h + g - h \vee g$ are both in $\mathfrak{D}$. In particular, if $h \in \mathfrak{D}$, then for any $a \in \mathbb{R}$ and each $n \geq 1$, the function $g_n = \varphi_n(h, a) \in \mathfrak{D}$, where $\varphi_n(z, a) = [(a + \frac{1}{n} - z)^+] \wedge 1$. Clearly, $I_{(h \leq a)}$ is the monotone limit of $\{g_n\}$. We therefore conclude that $I_{(h \leq a)} \in \mathfrak{L}$, for any $h \in \mathfrak{D}$ and any $a \in \mathbb{R}$. More generally, if $h_1, \ldots, h_m \in \mathfrak{D}$, then, for real numbers $a_1, \ldots, a_m$, the function $I_{(\bigcap \{h_i \leq a_i\})}$ is the monotone limit of $\prod_{i=1}^{m} \varphi_n(h_i, a_i) \in \mathfrak{D}$ and therefore belongs to $\mathfrak{L}$. Denoting $\mathcal{P}$ to be the class of all finite intersections of sets of the form $\{h \leq a\}$, where $h \in \mathfrak{C}$ and $a \in \mathbb{R}$, it is clear that $\mathcal{P}$ is a π-system that generates $\sigma(\mathfrak{C})$ and what we have just proved is that $I_A \in \mathfrak{L}$, for every $A \in \mathcal{P}$. Using properties of $\mathfrak{L}$, it is easy to see that $\mathcal{L} = \{A \in \sigma(\mathfrak{C}) : I_A \in \mathfrak{L}\}$ is a λ-system. Using the π-λ Theorem, one concludes that $I_A \in \mathfrak{L}$ for all $A \in \sigma(\mathfrak{C})$. But then, $\mathfrak{L}$ being a vector space, it follows that every real valued simple $\sigma(\mathfrak{C})$-measurable function belongs to $\mathfrak{L}$. Using Proposition 3.1.25 (last part) and Lemma 3.5.21, one finally concludes that $\mathfrak{L}$ contains all bounded $\sigma(\mathfrak{C})$-measurable functions. □

Here is an interesting and useful corollary of the above theorem, the proof of which is left as an exercise.

COROLLARY **3.5.23.** *Let C denote the space of all real valued continuous functions on some interval $I \subset \mathbb{R}$ and let $\mathcal{C}$ be the σ-field on C generated by the class of all evaluation maps e_t, $t \in I$, on C, defined as $e_t(x) = x(t)$, $x \in C$. Let $\mathfrak{L}$ be a vector space of bounded $\mathcal{C}$-measurable real functions on C that contains constant functions and is closed under bounded pointwise limits. If $\mathfrak{L}$ contains all functions γ of the form $\gamma(x) = \prod_{i=1}^{k} g_i(x(t_i))$, where $\{t_1, \cdots, t_k\}$ vary over all non-empty finite subsets of I and $g_1, g_2, \cdots, g_k$ are bounded continuous real valued functions on $\mathbb{R}$, then $\mathfrak{L}$ contains all bounded $\mathcal{C}$-measurable functions.*

3.6 Additional Exercises

EXERCISE **3.6.1.** *Let f be a real measurable function on a measure space $(\Omega, \mathcal{A}, \mu)$. Show that the issue of existence of $\int f d\mu$ and the value of $\int f d\mu$ in case it exists, remains unchanged even if the underlying measure space is changed to $(\Omega, \mathcal{A}', \mu')$ where $\mathcal{A}' = \sigma(f)$ and μ' denotes the restriction of μ on $\mathcal{A}'$.*

EXERCISE **3.6.2.** *Suppose that f and g are measurable functions with $f \leq g$. Show that if either $\int f^- d\mu < \infty$ or $\int g^+ d\mu < \infty$, then both $\int f d\mu$ and $\int g d\mu$ exist and satisfy $\int f d\mu \leq \int g d\mu$.*

EXERCISE **3.6.3.** *If f is integrable, then show that $\{\omega : f(\omega) \neq 0\}$ is a countable union of sets of finite measure.*

EXERCISE **3.6.4.** *Let $(\Omega, \mathcal{A}, \mu)$ be a measure space and $\Omega_0 \in \mathcal{A}$. Consider the restriction of $\mathcal{A}$ to Ω_0, that is $\mathcal{A}_0 = \{A \in \mathcal{A} : A \subset \Omega_0\}$ and let μ_0 denote the restriction of μ to $\mathcal{A}_0$. Let f be an μ-integrable function and let f_0 be its restriction to Ω_0. Show then that $\int f_0 d\mu_0 = \int_{\Omega_0} f d\mu$.*
What if, we only assumed that the integral $\int f d\mu$ exists, instead of assuming that f is μ-integrable?

EXERCISE **3.6.5.** *Examine the condition for equality in (3.4.3) for the case $p = 1$.*

EXERCISE **3.6.6.** *Let f be a measurable function on $(\Omega, \mathcal{A}, \mu)$, which is not identically zero. Let $I = \{p \geq 0 : \int |f|^p d\mu < \infty\}$.*
(a) Show that if I is non-empty, then I is an interval (possibly degenerate). Show also that when I is non-empty, the function φ defined on I as $\varphi(p) = \log \int |f|^p d\mu$ is a convex function.
(b). Show that if μ is a finite measure then for any real valued measurable function f, the set $\{\theta \in \mathbb{R} : \int e^{\theta f} d\mu < \infty\}$ is always an interval, possibly degenerate, containing zero.

EXERCISE **3.6.7.** Let $(\Omega, \mathcal{A}, \mu)$ be a finite measure space and f a measurable function on Ω. Show that (i) the limit $\lim_{p\to\infty} \|f\|_p$ exists and equals $\|f\|_\infty$ and that (ii) $\|f\|_\infty < \infty$ if and only if $\sup\{\|f\|_p : 1 \le p < \infty\} < \infty$.

EXERCISE **3.6.8.** Use convexity of the function $x \mapsto |x|^p$, for $p > 1$, and Jensen's inequality to give an alternative proof of Theorem 3.4.9.

EXERCISE **3.6.9.** Let $I \subset \mathbb{R}$ be an open interval and $\varphi : I \to \mathbb{R}$ a convex function.
(a) Prove an anlogue of Lemma 3.4.11 for φ on I.
(b) Show that for any integrable function f on a probability space $(\Omega, \mathcal{A}, \mu)$, taking values in I, the integral $\int f d\mu$ takes values in I.
(c) Show that, if $\varphi \circ f$ is μ-integrable, then $\varphi(\int f dP) \le \int \varphi \circ f d\mu$.

EXERCISE **3.6.10.** Let $(\Omega, \mathcal{A}, \mu)$ be a measure space. Consider the measure ν on $\mathcal{A}$ given by $\nu(A) = \int_A f d\mu$, $A \in \mathcal{A}$, where f is a non-negative measurable function on $(\Omega, \mathcal{A}, \mu)$ (see Remark 3.3.17).
(a) Show that $\int h d\nu = \int h f d\mu$, for any non-negative measurable function h on $(\Omega, \mathcal{A})$.
(b) Show that if h is any extended real-valued measurable function on $(\Omega, \mathcal{A})$, then $\int h d\nu = \int h f d\mu$, in the sense that, if one of the integrals exists, then so does the other and, in that case, they are equal.
(c) Conclude, in particular, that h is ν-integrable if and only if hf is μ-integrable.

EXERCISE **3.6.11.** The quantities discussed in this exercise arise in Information Theory as well as in Statistics. Think of the 'log' function as an extended real valued measurable function defined on $[0, \infty]$ with $\log 0 = -\infty$ and $\log \infty = \infty$. Let $(\Omega, \mathcal{A}, P)$ be a probability space and let f, g be two non-negative measurable functions on $(\Omega, \mathcal{A}, P)$ with $\int f dP = \int g dP = 1$. Consider the probability measures μ and ν on $\mathcal{A}$ given by $\mu(A) = \int_A f dP$ and $\nu(A) = \int_A g dP$, for $A \in \mathcal{A}$ (see Remark 3.3.17).
(a) Show that the integral $\int \log(f/g) d\nu$ exists and takes values in $[-\infty, \infty)$.
(b) Show that the integral $\int \log(f/g) d\mu$ exists and takes values in $(-\infty, \infty]$.
(c) Show that $\int \log(f/g) d\nu = \int \log(f/g) d\mu$ if and only if $\mu \equiv \nu$ (equivalently, $f = g$ a.e.$[P]$).

EXERCISE **3.6.12.** Let $\{p_n, n \ge 1\}$ be a sequence of positive real numbers and $D = \{x_1, x_2, \ldots\}$ a countable subset of $\mathbb{R}$ as in Example 2.8.20. Assuming that at least one of the conditions stated there holds, let G be the distribution function defined in (2.8.4) and μ be the corresponding Radon measure on $\mathcal{B}(\mathbb{R})$. Let h be an extended real-valued measurable function on $(\mathbb{R}, \mathcal{B}(\mathbb{R}), \mu)$.
(a) Assuming h is non-negative, let $\{h_n\}$ be the sequence of non-negative real-valued simple functions defined as $h_n(x) = \sum_{i=1}^n (h(x_i) \wedge n) I_{\{x_i\}}(x)$, $x \in \mathbb{R}$. Show that $h_n \uparrow h$, a.e.$[\mu]$, and hence deduce that $\int h d\mu = \sum_i h(x_i) p_i$.
(b) For a general measurable h, show that $\int h d\mu$ exists if and only if at least one of the sums $S_1 = \sum_{i: h(x_i) \ge 0} h(x_i) p_i$ and $S_2 = \sum_{i: h(x_i) < 0} h(x_i) p_i$ is finite and, in that case,

$\int h \, d\mu = S_1 + S_2$.
(c) Show that h is μ-integrable if and only if the series $\sum_i h(x_i) p_i$ converges absolutely and, in that case, $\int h \, d\mu = \sum_i h(x_i) p_i$.

EXERCISE **3.6.13.** Let $f : \mathbb{R} \to \mathbb{R}$ be a non-negative real-valued measurable function such that $\int_B f \, d\lambda < \infty$, for every bounded borel set $B \subset \mathbb{R}$.
(a) Show that the function $G : \mathbb{R} \to \mathbb{R}$ defined as $G(x) = \int_{(0,x]} f \, d\lambda$, for $x \geq 0$ and $G(x) = -\int_{(x,0]} f \, d\lambda$, for $x < 0$, is a continuous distribution function on $\mathbb{R}$.
(b) Denoting μ to be the Radon measure on $\mathcal{B}(\mathbb{R})$ determined by G, show that $\mu(B) = \int_B f \, d\lambda$, for every $B \in \mathcal{B}(\mathbb{R})$.
(c) Use Exercise 3.6.10 to conclude that, if h is any extended real-valued measurable function on $(\mathbb{R}, \mathcal{B}(\mathbb{R}), \mu)$, then $\int h \, d\mu = \int h f \, d\lambda$, in the sense that if one of the integrals exist, then so does the other and, in that case, they are equal. In particular, h is μ-integrable if and only if hf is λ-integrable.

EXERCISE **3.6.14.** [A generalization of Exercise 3.6.12 to $\mathbb{R}^k$, for any $k \geq 1$.]
Let $\{p_n, n \geq 1\}$ be a sequence of positive real numbers and $D = \{\mathbf{x}_1, \mathbf{x}_2, \ldots\}$ a countable subset of $\mathbb{R}^k$ satisfying at least one of the conditions (i) $\sum_n p_n$ converges and (ii) $A \cap D$ is finite for all bounded $A \subset \mathbb{R}^k$.
(a) Show that $\mu(A) = \sum_n p_n I_A(\mathbf{x}_n)$, $A \in \mathcal{B}(\mathbb{R}^k)$ defines a Radon measure.
(b) Show that μ is a finite measure (resp., a probability measure) if and only if (i) holds (resp., $\sum_n p_n = 1$).
(c) Show that, for any extended real-valued measurable function h on $(\mathbb{R}^k, \mathcal{B}(\mathbb{R}^k))$, the integral $\int h \, d\mu$ exists if and only if at least one of $S_1 = \sum_n \max\{h(\mathbf{x}_n), 0\} p_n$ and $S_2 = \sum_n \min\{h(\mathbf{x}_n), 0\} p_n$ is finite and, in that case $\int h \, d\mu = S_1 + S_2$.
(d) Show also that h is μ-integrable if and only if the series $\sum_n h(\mathbf{x}_n) p_n$ converges absolutely and, in that case, $\int h \, d\mu = \sum_n h(\mathbf{x}_n) p_n$.

EXERCISE **3.6.15.** [A generalization of Exercise 3.6.13 to $\mathbb{R}^k$, for any $k \geq 1$.]
Let f be a non-negative real-valued borel measurable function on $\mathbb{R}^k$ satisfying the condition that $\int_A f \, d\lambda < \infty$, for any bounded borel set $A \subset \mathbb{R}^k$.
(a) Show that $\mu(A) = \int_A f \, d\lambda$, $A \in \mathcal{B}(\mathbb{R}^k)$ defines a Radon measure.
(b) Show that μ is a finite measure (resp., a probability measure) if and only if $\int f \, d\lambda < \infty$ (resp., $\int f \, d\lambda = 1$).
(c) Use Exercise 3.6.10 now to conclude that, if h is any extended real-valued measurable function on $(\mathbb{R}^k, \mathcal{B}(\mathbb{R}^k))$, then $\int h \, d\mu$ exists if and only if $\int h f \, d\lambda$ exists and, in that case, they are equal. In particular, h is μ-integrable if and only if hf is λ-integrable.

EXERCISE **3.6.16.** [This is a generalization of Exercise 3.5.12 (a)] Fix any $n \geq 2$ and consider the real valued functions $f_1, \ldots, f_n$ defined on $\mathbb{R}^n$ as follows. For $\mathbf{x} = (x_1, x_2, \cdots, x_n) \in \mathbb{R}^n$, define $f_1(\mathbf{x}) = \max\{x_i : 1 \leq i \leq n\}$ and i_1 to be any index such that $f_1(\mathbf{x}) = x_{i_1}$. Next, define $f_2(\mathbf{x}) = \max\{x_i : 1 \leq i \leq n, i \neq i_1\}$ and $i_2 \neq i_1$ to be any index such that $f_2(\mathbf{x}) = x_{i_2}$. In general, For any $1 \leq k < n$, having defined $f_1, \ldots, f_k$ and distinct indices $i_1, \ldots, i_k$, define

$f_{k+1}(\mathbf{x}) = \max\{x_i : 1 \leq i \leq n, i \neq i_1, \ldots, i_k\}$ and $i_{k+1} \neq i_1, \ldots, i_k$ to be any index such that $f_{k+1}(\mathbf{x}) = x_{i_{k+1}}$. Describe $\sigma(f_1, f_2, \ldots, f_n)$.

EXERCISE **3.6.17.** Show that the functions $g(t) = \int_0^\infty \cos(tx) e^{-x^2/2} dx$, $t \in \mathbb{R}$ and $h(t) = \int_0^\infty \sin(tx) e^{-x^2/2} dx$, $t \in \mathbb{R}$ are both continuously differentiable everywhere on $\mathbb{R}$ and find $g'(t)$ and $h'(t)$.

EXERCISE **3.6.18.** (a) Let $\mathcal{A}$ denote the countable-cocountable σ-field on an uncountable set Ω. Show that if f is any real-valued measurable function on $(\Omega, \mathcal{A})$, then there exists $c \in \mathbb{R}$, such that, $\{\omega : f(\omega) \leq c\}$ is cocountable and $\{\omega : f(\omega) < c\}$ is countable. Use this to conclude that there is a cocountable set $A \subset \Omega$ and a real number c_0, such that $f \equiv c_0$ on A.
(b) Take $\Omega = [a,b]$ for some $a < b$ and $\mathcal{A}$ as in (a). Consider the measure μ on $\mathcal{A}$, defined as $\mu(A) = |A|$, if $A \in \mathcal{A}$ is finite and $\mu(A) = \infty$, otherwise. Show that if f is any real-valued measurable function on $(\Omega, \mathcal{A}, \mu)$, which is integrable, then there must be a cocountable set $A \subset \Omega$, such that, $f \equiv 0$ on A.

EXERCISE **3.6.19.** For any function $f : \mathbb{R} \to \mathbb{R}$, its 'Oscillation function' is defined as follows. For $x \in \mathbb{R}$, denote $O_f(x, \delta) = \sup\{|f(y) - f(z)| : y, z \in (x - \delta, x + \delta)\}$, for each $\delta > 0$. Clearly, $O_f(x, \delta)$ decreases as $\delta \downarrow 0$, so that $O_f(x) = \lim_{\delta \downarrow 0} O_f(x, \delta)$ exists. The non-negative function O_f on $\mathbb{R}$ is called the Oscillation function of f. Show that, for any function f, its Oscillation function O_f is always measurable and hence deduce that set of continuity points of f is always a Borel set. [Consider the sets $\{x \in \mathbb{R} : O(x) < \alpha\}$, for $\alpha > 0$.]

EXERCISE **3.6.20.** Let $\mathcal{C}$ be a class of subsets of a non-empty set Ω. Show that $\mathcal{A} = \bigcup\{\sigma(\mathcal{C}_0) : \mathcal{C}_0 \subset \mathcal{C}, \mathcal{C}_0 \text{ is countable}\}$ is a σ-field and deduce that $\mathcal{A} = \sigma(\mathcal{C})$. Next, let $(E, \mathcal{E})$ be a measurable space and $\{f_\alpha, \alpha \in \Delta\}$ be a family of functions on Ω into E. Show that $\mathcal{A} = \bigcup\{\sigma(f_\alpha, \alpha \in \Delta_0) : \Delta_0 \subset \Delta, \Delta_0 \text{ is countable}\}$ is a σ-field and hence deduce that $\mathcal{A} = \sigma(f_\alpha, \alpha \in \Delta)$.

EXERCISE **3.6.21.** Let f be a real valued function defined on an unbounded interval I, for which the "improper Riemann integral" $\int_I f(x) dx$ exists and is finite.
(a) Show that f must be continuous a.e.$[\lambda]$ on I and hence measurable with respect to the Lebesgue σ-field on I (that is, the λ-completion of $\mathcal{B}(I)$).
(b) Show that, in case f is non-negative, then f is λ-integrable on I and the λ-integral agrees with the improper Riemann integral.
(c) Show that without the non-negativity hypothesis, the assertion in (b) may fail.

EXERCISE **3.6.22.** [A strengthening of MCT]
Let $\{f_n\}$ be a sequence of measurable functions on a measure space $(\Omega, \mathcal{A}, \mu)$, such that, $f_n \uparrow f$ a.e.$[\mu]$, for some measurable function f.
(a) Show that if there exists a μ-integrable function g, such that, $f_n \geq g$ a.e.$[\mu]$, for all n, then the integrals $\int f_n d\mu$, $n \geq 1$, and $\int f d\mu$ all exist and $\int f_n d\mu \uparrow \int f d\mu$.
(b) Show that, in part (a), if the assumption of μ-integrability of g is weakened to g being measurable with $\int g^- d\mu < \infty$, then also the same conclusion holds.

3.6. Additional Exercises

EXERCISE **3.6.23**. *Use Exercise 3.6.22 to state and prove a stronger version of Fatou's Lemma.*

EXERCISE **3.6.24**. *Let $(\Omega, \mathcal{A})$ be a measurable space and $\mathcal{S}$ a semifield generating the σ-field $\mathcal{A}$. Let $\mathcal{L}$ be a vector space of bounded real-valued functions on Ω, which is closed under bounded pointwise convergence, meaning that, if $\{f_n\}$ is any sequence of functions in $\mathcal{L}$ with $\sup_n \|f\|_\infty < \infty$, such that, $f_n \to f$, then f also belongs to $\mathcal{L}$. Show that if $I_S \in \mathcal{L}$, for all $S \in \mathcal{S}$, then $\mathcal{L}$ contains all bounded real-valued $\mathcal{A}$-measurable functions.*

EXERCISE **3.6.25**. *Consider a sequence of probability distribution functions $\{F_n\}$ on $\mathbb{R}$ with $F_n(x) = 0$ for $x \leq 0$ and $F_n(x) = 1$ for $x \geq 1$ for all n and defined as follows on $[0,1]$.*
$F_0(x) = x$, $F_1(x) = \frac{3x}{2} I_{[0,1/3]}(x) + \frac{1}{2} I_{(1/3,2/3)}(x) + \frac{3x-1}{2} I_{[2/3,1]}(x)$
F_2 *is the polygonal line with vertices at the points*

$$(0,0), (\tfrac{1}{9}, \tfrac{1}{4}), (\tfrac{2}{9}, \tfrac{1}{4}), (\tfrac{1}{3}, \tfrac{1}{2}), (\tfrac{2}{3}, \tfrac{1}{2}), (\tfrac{7}{9}, \tfrac{3}{4}), (\tfrac{8}{9}, \tfrac{3}{4}) \text{ and } (1,1),$$

and so on. There is an elegant recursive way of defining the sequence $\{F_n\}$ of functions on $[0,1]$. Start with $F_0(x) = x, 0 \leq x \leq 1$. For any $n \geq 0$, define F_{n+1} in terms of F_n by the following rule :

$$F_{n+1}(x) = \begin{cases} \frac{1}{2} F_n(3x), & \text{for } 0 \leq x \leq \frac{1}{3} \\ \frac{1}{2}, & \text{for } \frac{1}{3} \leq x \leq \frac{2}{3} \\ \frac{1}{2} + \frac{1}{2} F_n(3x-2), & \text{for } \frac{2}{3} \leq x \leq 1 \end{cases}$$

(Check that the above defintion is compatible at points $\frac{1}{3}$ and $\frac{2}{3}$.)
(a) Show that $\sup_{0 \leq x \leq 1} |F_{n+1}(x) - F_n(x)| \leq \frac{1}{2} \sup_{0 \leq x \leq 1} |F_n(x) - F_{n-1}(x)|$, for all $n \geq 1$.
(b) Show that the sequence $\{F_n\}$ converges uniformly to a probability distribution function F, which is continuous.
(c) Denoting P to be the unique probability on $\mathcal{B}(\mathbb{R})$ corresponding to the distributiom function F, show that $P(\{x\}) = 0$ for all x and $P(C) = 1$ where C is the Cantor set (see Exercise 2.8.21).

REMARK **3.6.26**. This distribution function F appearing in Exercise 3.6.25 is called the *Cantor distribution function* and the corresponding probability P on $\mathcal{B}(\mathbb{R})$ is called the *Cantor Distribution*. Part (c) of the above exercise reveals a ratder intriguing property of P, namrely, $P(\{x\}) = 0$ for all $x \in \mathbb{R}$ (that is, P has no point mass), while at same time, its total mass in concentrated on the Cantor set C, which has zero Lebesgue measure (See Exercise 2.8.21). In the language of Chapter 6, this last property essentially says that P is 'singular' with respect to λ (or simply, P is singular). The graph of the probability distribution function F on $[0,1]$ is often referred to as the *'Devil's staircase'* (try to guess why).

Chapter 4

Random Variables and Random Vectors

As was noted in Chapter 1, the framework of measure theory builds the proper mathematical foundation for probability theory, in a much more crucial way than perhaps it does for any other branch of mathematics. In Section 2.11, we elaborated the measure theoretic framework for probability theory. As mentioned earlier, Probability theory is the mathematics of random experiments and the mathematical set-up that is used to model a random experiment consists of a triplet $(\Omega, \mathcal{A}, P)$, called a *probability space*, where $(\Omega, \mathcal{A})$ is a measurable space and P a probability measure on $\mathcal{A}$. As was noted in Section 2.11, Ω represents the set of "possible outcomes" of a random experiment, $\mathcal{A}$ is the class of "events" and P the probability assignment to various events. It was emphasized that except in a very special case, namely, the "discrete case", the σ-field $\mathcal{A}$, consisting of "events', may not include all subsets of Ω. All of these and much more were discussed in Chapter 2. In this Chapter, we are going to see how the idea of measurable functions and the theory of integration discussed in the previous chapter, comes into play in probability theory. In course of this, we will be studying one of the central objects of interest in probability theory, namely, random variables and more generally, random vectors.

4.1 Random Variables

A reader who has had an exposure to probability theory at undergraduate level, may recall that a random variable is defined there simply as a real valued function on the sample space Ω, that is, simply as a function $X : \Omega \to \mathbb{R}$. It may be worth noting here that we have switched to the conventional notation used in probability theory, namely, denoting real valued functions on Ω by X, Y, etc., instead of f, g, etc, as was the practice in the previous chapters. We will follow this convention when we are in the arena of probability theory. Returning to the definition of a random variable X, it should be clear though that any meaningful definition should allow one to be able to talk about probabilities of subsets of Ω, that are described by X taking specific values. For example, if X is a random variable, then one should be able to talk about probabilities of the sets $\{\omega : X(\omega) = 10\}$, $\{\omega : X(\omega) > 13\}$, $\{\omega : -2 < X(\omega) \leq 5.2\}$, and so on. All these are, of course,

4.1. Random Variables

subsets of Ω. However, not all subsets of Ω, in general, would be events and have probabilities assigned to them. The only subsets Ω that are events and have probabilities assigned to them are those belonging to the σ-field $\mathcal{A}$. This leads us to the following question. What is the condition needed on a function $X : \Omega \to \mathbb{R}$ that would guarantee that the above sets and similar other sets will belong to $\mathcal{A}$? Any meaningful definition of a random variable must offer this guarantee. Definition 3.1.2 and Proposition 3.1.3 makes it clear that what is required is that the function X be $\mathcal{A}$-measurable function. This leads to the following (refined) definition of a random variable.

DEFINITION 4.1.1. A *random variable* on a probability space $(\Omega, \mathcal{A}, P)$ is an extended real valued $\mathcal{A}$-measurable function on Ω.

REMARK 4.1.2. Even though one is mostly interested in only real valued random variables, it becomes necessary sometimes to allow random variables to take $\pm\infty$ as possible values. As was done for general measurable functions, a random variable taking only real values will be explicitly referred to as a *real random variable*.

The reader couldn't have escaped noticing another change of notation. In the above definition, we wrote $(\Omega, \mathcal{A}, P)$ for the probability space, thereby denoting the probability measure by P. But, the reader must have encountered the same in Section 2.11 as well. In probability theory, it is customary to use P, Q, etc. to denote probability measures rather than μ, ν etc., as in measure theory.

EXERCISE 4.1.3. (a) If X is a real random variable on $(\Omega, \mathcal{A}, P)$, then so also is $Y = h(X)$, for any Borel measurable function $h : \mathbb{R} \to \mathbb{R}$.
(b) If $X_1, \cdots, X_n$ are real random variables on $(\Omega, \mathcal{A}, P)$ and $h : \mathbb{R}^n \to \mathbb{R}$ is $\mathcal{B}(\mathbb{R}^n)$-measurable, then $Y = h(X_1, \cdots, X_n)$ is a real random variable.

From Theorem 3.5.16, we know that any real random variable X on a probability space $(\Omega, \mathcal{A}, P)$ will induce a measure on $\mathcal{B}(\mathbb{R})$, namely $Q(B) = P(X^{-1}(B))$, for $B \in \mathcal{B}$. Further, as pointed out there, this induced measure Q will indeed be a probability measure on $\mathcal{B}(\mathbb{R})$. This induced probability PX^{-1} plays a significant role in the study of the random variable X.

DEFINITION 4.1.4. For a real random variable X on a probability space $(\Omega, \mathcal{A}, P)$, the induced probability $Q = PX^{-1}$ on $(\mathbb{R}, \mathcal{B})$ is called the *Probability Distribution* (or simply, the *Distribution*) of X and is also sometimes denoted by P_X. The associated probability distribution function (see Definition 2.8.13) is called the *Probability Distribution Function* (or simply, *Distribution Function*) of X, denoted by F_X.

Thus, the distribution P_X of a real random variable X captures complete information on probabilities of all events 'determined' by X, that is, all events of the form $\{\omega : X(\omega) \in B\}$, for $B \in \mathcal{B}(\mathbb{R})$. In the sequel, we will often use the abbreviation $\{X \in B\}$ to mean $\{\omega : X(\omega) \in B\}$. With this notation, we have $F_X(x) = P(X \leq x)$, for $x \in \mathbb{R}$.

Caratheodory Extension Theorem now gives us the following extremely important result, which almost forms the backbone of much of probability theory.

THEOREM 4.1.5. (Existence of Random Variable with given Distribution) Given any probability distribution function F on $\mathbb{R}$, there is a probability space $(\Omega, \mathcal{A}, P)$ and a real random variable X on it such that $F_X = F$.

Proof. In view of the Caratheodory Extension Theorem 2.2.20, we know that, given any probability distribution function F on $\mathbb{R}$, there is a unique probability P on $(\mathbb{R}, \mathcal{B})$ associated to F, that is, $P((-\infty, x]) = F(x)$, for all $x \in \mathbb{R}$. It is easy to see now that, on the probability space $(\mathbb{R}, \mathcal{B}, P)$, the randon variable X defined as $X(\omega) = \omega$ (that is, the identity map on $\mathbb{R}$) does the job. $\square$

While the above theorem does prove existence of random variables with any specified distribution functions, the probability in the above construction, and hence the probability space, depends on the specified distribution function, while the function defining the random variable remains the same, namely, the identity map, always. It would be nice if one could identify one canonical probability space $(\Omega, \mathcal{A}, P)$, such that, given any probability distribution function F, one can define an appropriate random variable X on this space such that $F_X = F$. The following interesting and very useful construction provides such a canonical probability space on which one can obtain a 'realization' of any specified probability distributiion function in the above sense.

THEOREM 4.1.6. Consider $(\Omega, \mathcal{A}, P)$ with $\Omega = (0, 1)$, $\mathcal{A} = \mathcal{B}((0, 1))$ and P, the Lebesgue measure on $\mathcal{A}$. Given any probability distribution function F on $\mathbb{R}$, there is a real random variable X on the probability space $(\Omega, \mathcal{A}, P)$, such that $F_X = F$.

Proof. Given a probability distribution function F, we consider the function G on $\Omega = (0, 1)$ defined as

$$G(u) = \sup\{x \in \mathbb{R} : F(x) < u\}, \quad \text{for } u \in (0, 1) \tag{4.1.1}$$

From the properties that $\lim_{x \to -\infty} F(x) = 0$ and $\lim_{x \to \infty} F(x) = 1$, one can easily see that, for every $u \in (0, 1)$, the set $\{x \in \mathbb{R} : F(x) < u\}$ is non-empty and bounded above, implying that G defines a real valued function on $(0, 1)$. Further, since F is non-decreasing, one has $\{x \in \mathbb{R} : F(x) < u\} \subseteq \{x \in \mathbb{R} : F(x) < u'\}$, for any $u, u' \in (0, 1)$ with $u < u'$, implying that G is non-decreasing. It follows that G is a real valued measurable function on $(\Omega, \mathcal{A})$ [See Exercise 3.1.18 (c)]. Next, using right continuity of F, one can show that $F(x) < u$ implies $x < G(u)$, equivalently, that $G(u) \le x$ implies $u \le F(x)$. On the other hand, $u \le F(x)$ implies $G(u) \le x$, by the definition of G. We have thus shown that, for any $x \in \mathbb{R}$ and any $u \in (0, 1)$, $G(u) \le x \iff u \le F(x)$. From all these, it now follows that $X = G$ defines a real random variable on the probability space $(\Omega, \mathcal{A}, P)$, such that, for any $x \in \mathbb{R}$,

$$P(X \le x) = P(\{u \in (0, 1) : G(u) \le x\}) = P(\{u \in (0, 1) : u \le F(x)\}) = F(x).$$

This shows $F_X = F$ and that completes the proof. $\square$

REMARK 4.1.7. The function G defined above is known as the *'quantile function'* associated to the probability distribution function F. More specifically, $G(u)$, for

4.1. Random Variables

$u \in (0,1)$, is called the '*u-th quantile*' of the probability given by the distribution function F. One can also think of G as some sort of an 'inverse function' of the function F. In fact, when the probability distribution function F is strictly increasing and continuous, making it a one-one, onto function on $\mathbb{R}$ to $(0,1)$, the function G is indeed the usual inverse function F^{-1}.

Given a real random variable X, its distribution P_X on $(\mathbb{R}, \mathcal{B})$ can be a discrete probability or a continuous probability on $\mathcal{B}$ [See Section 2.11 for the definition] or a mixture of the two types. A random variable X is called a *discrete random variable* or a *continuous random variable* according as P_X is a discrete probability or a continuous one.

X is a discrete random variable if and only if there is a countable set $D \subset \mathbb{R}$ such that $\sum_{x \in D} P(X = x) = 1$. In this case, the function $p_X : \mathbb{R} \to [0, 1]$ defined as

$$p_X(x) = P(X = x), \quad x \in \mathbb{R}$$

is called the *probability mass function (pmf)* of the random variable X. Clearly, $p_X(x) = 0$ for $x \notin D$ and $\sum_{x \in D} p_X(x) = 1$. It is clear that the distribution of a discrete random variable is completely captured by its pmf. Indeed, for any $B \in \mathcal{B}$, one has $P_X(B) = \sum_{x \in B} p_X(x) = \sum_{x \in B \cap D} p_X(x)$.

On the other hand, X is a continuous random variable iff F_X is continuous everywhere, or equivalently, $P(X = x) = 0$ for all $x \in \mathbb{R}$. A special case of this is when

$$F_X(x) = \int_{(-\infty, x]} f_X(u) du, \quad x \in \mathbb{R}, \qquad (4.1.2)$$

for some non-negative measurable function f_X on $\mathbb{R}$. In this case, the function f_X is called the *probability density function (pdf*, in short) of X. It is clear that $\int_\mathbb{R} f_X(x) dx = 1$. A small point needs to be made here. Instead of referring to f_X as 'the' probability density function of X, the correct statement would be to say that it is 'a' probability density function of X. The reason is that, from what we learnt in integration theory, any non-negative measurable function, which differs from f_X only on a Lebesgue null set, would be as good a candidate for a density function of X. However, we will continue to use the qualifier 'the', as long as the reader remains aware of this non-uniqueness. In fact, in most standard cases, density function would happen to be a continuous function on an interval and we accept that as 'the' density and work with it, even though we remain aware that, if we destroy the continuity at any finite number of points, the new function will work equally well as a density.

The special case when a continuous random variable X satisfies (4.1.2) is often described by saying that X is an *absolutely continuous* random variable, or more commonly, X is a *random variable with a density*. The reason for using the phrase 'absolutely continuous' will be somewhat clear in Chapter 6. However, a reader who is already familiar with the notion of absolute conitnuity of real functions may find it interesting to know that F_X being given by (4.1.2) is equivalent to absolute continuity of F_X as a real valued function on $\mathbb{R}$. Clearly, for a random variable X with a density, its pdf f_X completely determines the distribution, in the sense that, $P_X(B) = \int_B f_X(u)du$, for all $B \in \mathcal{B}$. It must be noted that while the pmf in the discrete case and the pdf in the density case behave in a similar manner in many ways, the pdf, unlike pmf, does not represent a probability. It should also be noted that not all continuous random variables are necessarily absolutely continuous. Indeed, a random variable with distribution function F, as given in Exercise 3.6.25, is continuous, but not absolutely continuous (see Remark 3.6.26). Of course, Theorem 4.1.6 guarantees the existence of such a random variable. However, a more explicit construction of such a random variable is given in Exercise 4.6.2.

EXERCISE **4.1.8.** (a) Show that, on a discrete probability space, any random variable is discrete. Show also that, it is possible to have discrete random variables on probability spaces which are not discrete.
(b) Show that, if X is a discrete random variable, then so is $h(X)$ for any measurable function $h : \mathbb{R} \to \mathbb{R}$. If X is continuous and $h : \mathbb{R} \to \mathbb{R}$ is measurable, is $h(X)$ necessarily continuous?

EXERCISE **4.1.9.** (a) If X and Y are discrete random variables, show that $X+Y$, $X-Y$, XY and X/Y, in case $Y \neq 0$, are all discrete random variables.
(b) Is the analogue of (a) true if X and Y are both continuous random variables, or, both X and Y have densities?

EXAMPLE **4.1.10.** We mention here some special examples of discrete random variables and absolutely continuous random variables, for possible future reference and also because the associaited distributions are presented in a basic course in probability as some of the "standard univariate distributions".
(a) Let n be a positive integer and $0 < \theta < 1$. Consider the function $p : \mathbb{R} \to [0,1]$ defined as
$$p(x) = \binom{n}{x}\theta^x(1-\theta)^{n-x}, \text{ for } x \in \{0, 1, \ldots, n\} \text{ and } p(x) = 0 \text{ otherwise.}$$
It is easy to see that p defines a pmf. A discrete random variable X with pmf p is said to have a Binomial Distribution with parameters (n, θ), and we write $X \sim \text{Bin}(n, \theta)$.
(b) Let $\lambda > 0$. A random variable X is said to have Poisson Distribution with parameter λ and we write $X \sim \text{Poi}(\lambda)$ if X has pmf given by
$$p(x) = e^{-\lambda}\frac{\lambda^x}{x!}, \text{ for } x = 0, 1, \ldots \text{ and } p(x) = 0 \text{ otherwise.}$$
(c) A random variable X is said to have Uniform Distribution on a bounded open interval (a, b) if it has pdf
$$f(x) = \frac{1}{b-a} I_{(a,b)}(x).$$
In this case we write $X \sim \text{U}(a,b)$. It is easy to verify that if $X \sim \text{U}(a,b)$, then for any $c \neq 0$ and $d \in \mathbb{R}$, the random variable $cX + d$ also has a uniform distribution.
(d) A random variable X with pdf
$$f(x) = \lambda e^{-\lambda x} I_{(0,\infty)}(x),$$

where $\lambda > 0$, is said to have an Exponential Distribution with parameter λ. One writes $X \sim \text{Exp}(\lambda)$. The reader can easily verify that $X \sim \text{Exp}(\lambda)$ implies $X/\alpha \sim \text{Exp}(\alpha\lambda)$ for any $\alpha > 0$.

(e) A random variable with pdf

$$f(x) = (\lambda/2)e^{-\lambda|x|}, \quad x \in \mathbb{R},$$

where $\lambda > 0$, is said to have Laplace Distribution with parameter λ and is written as $X \sim \text{Lap}(\lambda)$. It is again easy to check that $X \sim \text{Lap}(\lambda)$ implies $X/\alpha \sim \text{Lap}(\alpha\lambda)$ for any $\alpha > 0$.

(f) For any $\mu \in \mathbb{R}$ and $\sigma > 0$, it can be shown, with some work, that the function

$$f(x) = (\sqrt{2\pi}\sigma)^{-1} e^{-\frac{(x-\mu)^2}{2\sigma^2}}, \quad x \in \mathbb{R}$$

is a pdf. A random variable X with this pdf f is said to have a Normal Distribution with parameters (μ, σ^2). One writes $X \sim \text{N}(\mu, \sigma^2)$. The special case $\text{N}(0,1)$ is refered to as the 'Standard Normal Distribution'. One can verify quite easily that if $X \sim \text{N}(\mu, \sigma^2)$, then $aX + b \sim \text{N}(a\mu + b, a^2\sigma^2)$, for any $a \in \mathbb{R}$ and $b \neq 0$. In particular, $X \sim \text{N}(\mu, \sigma^2)$ iff $(X - \mu)/\sigma$ has the Standard Normal Distribution.

4.2 Random Vectors

DEFINITION 4.2.1. A measurable function $\mathbf{X} = (X_1, \cdots, X_k)$ on a probability space $(\Omega, \mathcal{A}, P)$ into $(\mathbb{R}^k, \mathcal{B}(\mathbb{R}^k))$ is called a k-dimensional Random Vector. The induced probability on $\mathcal{B}(\mathbb{R}^k)$, namely $P\mathbf{X}^{-1}$, also denoted by $P_{\mathbf{X}}$, is called the *Joint Distribution* of $\mathbf{X}$. The corresponding distribution function given by $F_{\mathbf{X}}(\mathbf{x}) = P_{\mathbf{X}}((-\infty, x_1] \times \cdots \times (-\infty, x_k])$, $\mathbf{x} = (x_1, \cdots, x_k) \in \mathbb{R}^k$, is a probability distribution function on $\mathbb{R}^k$, as defined in 2.9.20, and is called the *Distribution Function* of $\mathbf{X}$, sometimes more specifically, called *Joint Distribution Function* of $\mathbf{X}$.

An immediate consequence of Exercise 3.5.4(c) is that $\mathbf{X} = (X_1, \cdots, X_k)$ is a random vector on $(\Omega, \mathcal{A}, P)$ if and only if each X_i, $1 \leq i \leq k$, is a real random variable.

The following theorem is the k-dimensional analogue of Theorem 4.1.5 on the existence of real random variables with given probability distribution functions.

THEOREM 4.2.2. (Existence of random vectors with given joint distributions) Given any k-dimensional probability distribution function F, there exists a probability space $(\Omega, \mathcal{A}, P)$ and a random vector $\mathbf{X}$ on it, such that $F_{\mathbf{X}} = F$.

Proof As in the one dimensional case, one can take $(\Omega, \mathcal{A}) = (\mathbb{R}^k, \mathcal{B}(\mathbb{R}^k))$ and then take P to be the unique probability on $\mathcal{B}(\mathbb{R}^k)$ determined by the given F. It is clear then that the identity map $\mathbf{X}$ on $\mathbb{R}^k$ will do the job. □

A k-dimensional random vector $\mathbf{X}$ is said to be *discrete* if its distribution $P_{\mathbf{X}}$ is a discrete probability on $\mathcal{B}(\mathbb{R}^k)$, that is, $P_{\mathbf{X}}(D) = 1$, for some countable set $D \subset \mathbb{R}^k$. In this case, the function $p_{\mathbf{X}}(\mathbf{x}) = P_{\mathbf{X}}(\{\mathbf{x}\}) = P(\mathbf{X} = \mathbf{x})$, $\mathbf{x} \in \mathbb{R}^k$, is called the *joint probability mass function* (abbreviated as *joint pmf*) of $\mathbf{X}$.

Clearly, $p_\mathbf{X}(\mathbf{x}) = 0$ for all $\mathbf{x} \notin D$, and, the joint distribution $P_\mathbf{X}$ is completely determined by $p_\mathbf{X}$ through the formula $P_\mathbf{X}(B) = \sum_{\mathbf{x} \in B \cap D} p_\mathbf{X}(\mathbf{x})$, for $\mathbb{B} \in \mathcal{B}(\mathbb{R}^k)$.

A random vector $\mathbf{X}$ is said to be *continuous* if its joint distribution function is continuous everywhere. It should be noted that, unlike in case of a real random variable, this is not equivalent to $P_\mathbf{X}$ being non-atomic. In fact, as was illustrated in Example 2.11.7, the condition of non-atomicity, namely, $P(\mathbf{X} = \mathbf{x}) = 0$, for all $\mathbf{x} \in \mathbb{R}^k$, is strictly weaker than $F_\mathbf{X}$ being continuous.

As in case of real random variables, a special case of a random vector $\mathbf{X}$ being continuous is when there exists a non-negative measurable function $f_\mathbf{X}$ on $\mathbb{R}^k$, such that $\Delta_{\mathbf{ab}} F_\mathbf{X} = \int_{I_{\mathbf{ab}}} f_\mathbf{X}(\mathbf{u}) d\mathbf{u}$, for all $\mathbf{a} < \mathbf{b}$ in $\mathbb{R}^k$. The function $f_\mathbf{X}$, which is unique upto sets of zero λ^k-measure, is called the *joint probability density function* (abbreviated as *joint pdf*) of $\mathbf{X}$. In this case, we say that $\mathbf{X}$ is an *absolutely continuous random vector* or is a *random vector with a joint pdf*. Exactly as in the one-dimensional case, it is easy to see that $P_\mathbf{X}(B) = \int_B f_\mathbf{X}(\mathbf{u}) d\mathbf{u}$, for all $B \in \mathcal{B}(\mathbb{R}^k)$, and thus the joint pdf completely determines the distribution of $\mathbf{X}$, though, unlike joint pmf, it does not represent a probability.

EXERCISE 4.2.3. *(a) Let $\mathbf{X} = (X_1, X_2, \cdots, X_k)$ be a random vector. Show that $\mathbf{X}$ is discrete (resp. continuous) iff X_i is discrete (resp. continuous), for each i. (b) Show that if the random vector $\mathbf{X} = (X_1, \cdots, X_k)$ is absolutely continuous, then so is each X_i, $1 \leq i \leq k$. Give an example of a random vector $\mathbf{X} = (X_1, X_2)$ which is not absolutely continuous but both X_1 and X_2 are absolutely continuous.*

We end this section with a brief discussion on a very special k-dimensional probablity distribution that plays a central role in much of probability and statistics. This is a k-dimensional version of the Normal distribution on $\mathbb{R}$ mentioned in Example 4.1.10(f) of the previous section. We start with an exercise in calculus, on a k-variable integral.

EXERCISE 4.2.4. *(a) Let $\mathbf{Q}$ be a real symmetric $k \times k$ matrix and let $q(\cdot)$ denote the quadratic form given by $\mathbf{Q}$, that is, $q(\mathbf{x}) = \mathbf{x}'\mathbf{Q}\mathbf{x}$, $\mathbf{x} \in \mathbb{R}^k$. Show that the non-negative function $h(\mathbf{x}) = e^{-q(\mathbf{x})}$ is integrable on $\mathbb{R}^k$ if and only if $\mathbf{Q}$ is positive definite and that, in that case, the integral is $\sqrt{\pi^k / \det(\mathbf{Q})}$. [Use an orthogonal transformation $\mathbf{x} \mapsto \mathbf{A}\mathbf{x} = \mathbf{y} = (y_1, \cdots, y_k)'$, such that, $q(\mathbf{x}) = \alpha_1 y_1^2 + \cdots + \alpha_k y_k^2$, where $\alpha_1, \ldots, \alpha_k$ are the (real) eigenvalues of $\mathbf{Q}$ and then use Exercise 2.13.20 to see that $\int_{\mathbb{R}^k} e^{-q(\mathbf{x})} d\mathbf{x} = \int_{\mathbb{R}^k} \prod_j e^{-\alpha_j y_j^2} d\mathbf{y}$.]*
(b) Deduce that, if $\mathbf{\Sigma}$ is a real symmetric $k \times k$ matrix, then the function on $\mathbb{R}^k$ given by $f(\mathbf{x}) = (2\pi)^{-k/2} (\det(\mathbf{\Sigma}))^{-1/2} \exp\{-\frac{1}{2} \mathbf{x}' \mathbf{\Sigma}^{-1} \mathbf{x}\}$, $\mathbf{x} \in \mathbb{R}^k$, defines a probability density function on $\mathbb{R}^k$ if and only if $\mathbf{\Sigma}$ is positive definite.

DEFINITION 4.2.5. Let $\boldsymbol{\mu} \in \mathbb{R}^k$ be any k-dimensional (column) vector and $\mathbf{\Sigma}$ a real symmetric positive definite $k \times k$ matrix. A k-dimensional random vector $\mathbf{X}$ is said to have a k-dimensional *Gaussian/Normal distribution* with parameters $\boldsymbol{\mu}$ and $\mathbf{\Sigma}$ and we write $\mathbf{X} \sim \mathsf{N}_k(\boldsymbol{\mu}, \mathbf{\Sigma})$, if $\mathbf{X}$ is absolutely continuous with a joint probability density

4.2. Random Vectors

function given by

$$f(\mathbf{x}) = (2\pi)^{-k/2} \big(\det(\mathbf{\Sigma})\big)^{-1/2} \exp\{-\tfrac{1}{2}(\mathbf{x}-\boldsymbol{\mu})'\mathbf{\Sigma}^{-1}(\mathbf{x}-\boldsymbol{\mu})\}.$$

Note that in case $k = 1$, the above reduces to the $N(\mu, \sigma^2)$-distribution, for some $\mu \in \mathbb{R}$ and $\sigma > 0$ as in Example 4.1.10(f) of the previous section. The next result is very important and proves to be extremely useful in working with k-dimensional Gaussian distributions. Its proof is a straightforward application of the Jacobian Rule in multivariable integral calculus and hence omitted.

PROPOSITION 4.2.6. *Let $\mathbf{X} \sim N_k(\boldsymbol{\mu}, \mathbf{\Sigma})$. Then, for any k-dimensional column vector $\boldsymbol{\beta}$ and any real non-singular $k \times k$ matrix $\mathbf{A}$, the random vector $\mathbf{Y} = \mathbf{AX} + \boldsymbol{\beta}$ has a k-dimensional Gaussian distribution with parameters $\widetilde{\boldsymbol{\mu}} = \mathbf{A}\boldsymbol{\mu} + \boldsymbol{\beta}$ and $\widetilde{\mathbf{\Sigma}} = \mathbf{A}\mathbf{\Sigma}\mathbf{A}'$.*

EXERCISE 4.2.7. *Let $\pi(\cdot)$ denote a permutation of $\{1, \ldots, k\}$. Show that, if $\mathbf{X} \sim N_k(\boldsymbol{\mu}, \mathbf{\Sigma})$, then the random vector $\mathbf{Y} = (X_{\pi(1)}, \ldots, X_{\pi(n)})$ also has a k-dimensional Gaussian distribution and find the parameters.*

We now proceed to prove a series of important properties of the k-dimensional Gaussian distribution. The next proposition is a first step.

PROPOSITION 4.2.8. *Let $k > 1$ and let $\mathbf{X} = (X_1, \ldots, X_k)'$ have a $N_k(\boldsymbol{\mu}, \mathbf{\Sigma})$ distribution. Then, for any $1 \leq m < k$, the random vector $(X_1, \ldots, X_m)'$ has a m-dimensional Gaussian distribution.*

Proof. It is enough to prove the assertion only for $m = k - 1$, since the result for any $m < k$ can then be obtained by repeating this. Also, let us first assume that $\boldsymbol{\mu} = \mathbf{0}$. Thus, $\mathbf{X} \sim N_k(\mathbf{0}, \mathbf{\Sigma})$ and hence has density $f(\mathbf{x}) = K \exp\{-\tfrac{1}{2}\mathbf{x}'\mathbf{\Sigma}^{-1}\mathbf{x}\}$, where $K = (2\pi)^{-k/2}\big(\det(\mathbf{\Sigma})\big)^{-1/2}$. Denoting $\tfrac{1}{2}\mathbf{\Sigma}^{-1} = ((\beta_{ij}))$, we first claim that $\beta_{kk} \neq 0$. This is because otherwise, for any fixed $x_1, \ldots, x_{k-1}$, we will have real numbers a and b, depending *only* on $x_1, \ldots, x_{k-1}$, such that, $\exp\{-\tfrac{1}{2}\mathbf{x}'\mathbf{\Sigma}^{-1}\mathbf{x}\} = e^{-ax_k+b}$, which is clearly not integrable in x_k on $\mathbb{R}$ and that contradicts the fact that f is a probability density function and hence must be integrable on $\mathbb{R}^k$. Of course, a reader with some exposure to linear algebra could claim $\beta_{kk} \neq 0$ directly from positive definiteness of $\tfrac{1}{2}\mathbf{\Sigma}^{-1}$.

We now consider the non-singular linear transformation $\mathbf{x} \mapsto \mathbf{y} = \mathbf{Ax}$, given by $y_i = x_i$, for $1 \leq i \leq k-1$, and $y_k = \beta_{1k}x_1 + \cdots + \beta_{kk}x_k$. Easy algebra then shows that $\tfrac{1}{2}\mathbf{x}'\mathbf{\Sigma}^{-1}\mathbf{x} - \beta_{kk}^{-1}y_k^2 = \widetilde{q}(y_1, \ldots, y_{k-1})$, where $\widetilde{q}(\cdot)$ is a quadratic form in $(k-1)$ variables. But then, by Proposition 4.2.6, $\mathbf{Y} = \mathbf{AX}$ has a k-dimensional distribution with density given by $\widetilde{K} \exp\{\widetilde{q}(y_1, \ldots, y_{k-1}) + \beta_{kk}^{-1}y_k^2\}$, where $\widetilde{K} = (2\pi)^{-k/2}\big(\det(\mathbf{A}\mathbf{\Sigma}\mathbf{A}')\big)^{-1/2}$. The form of the density of the vector $\mathbf{Y}$ clearly shows that it factors into the product of a $(k-1)$-dimensional Gaussian density and a 1-dimensional Gaussian density. From this, one can easily conclude (how?) that the random vector $(Y_1, \ldots, Y_{k-1})'$ has a $(k-1)$-dimensional Gaussian distribution. But, by our construction, $(Y_1, \ldots, Y_{k-1})' = (X_1, \ldots, X_{k-1})'$ and thus the resullt is proved in case $\boldsymbol{\mu} = \mathbf{0}$. To prove the result for general $\boldsymbol{\mu}$, we use Proposition 4.2.6 to observe that $\mathbf{X} - \boldsymbol{\mu} \sim N_k(\mathbf{0}, \mathbf{\Sigma})$, so that, by what we have just proved, $(X_1-\mu_1,\ldots,X_{k-1}-\mu_{k-1})'$ will have a $(k-1)$-dimensional Gaussian distrbution. Applying Proposition 4.2.6 again, we conclude that $(X_1,\ldots,X_{k-1})'$ has a $(k-1)$-dimensional Gaussian distribution. This completes the proof. □

The next two results now follow as immediate consequences of Proposition 4.2.6, Exercise 4.2.7 and Proposition 4.2.8.

PROPOSITION **4.2.9.** If $\mathbf{X} \sim \mathsf{N}_k(\boldsymbol{\mu}, \boldsymbol{\Sigma})$, then for any $1 \leq m < k$ and any choice of $1 \leq k_1 < \cdots < k_m \leq k$, the random vector $(X_{k_1}, \ldots, X_{k_m})'$ has a m-dimensional Gaussian distribution.

PROPOSITION **4.2.10.** Let $\mathbf{X} \sim \mathsf{N}_k(\boldsymbol{\mu}, \boldsymbol{\Sigma})$. Then, for any non-zero (column) vector $\mathbf{a}$, the real random variable $\mathbf{a}'\mathbf{X}$ has a Normal distribution.

REMARK **4.2.11.** It is clear from Definition 4.2.5 that the parameters $\boldsymbol{\mu}$ and $\boldsymbol{\Sigma}$ determine the k-dimensional Gaussian distribution. In Section 4.5, we will see what these two parameters really represent. Once we see that, it will follow that the parameters of the m-dimensional Gaussian random vector $(X_{k_1},\ldots,X_{k_m})'$ of Proposition 4.2.8 are the m-dimensional vector $(\mu_{k_1},\cdots,\mu_{k_m})'$ and the $m\times m$ submatrix of $\boldsymbol{\Sigma}$, given by $((\sigma_{k_i k_j}))$. Similarly, it will follow that the Normal random variable $\mathbf{a}'\mathbf{X}$ of Proposition 4.2.10 has mean $\mathbf{a}'\boldsymbol{\mu}$ and variance $\mathbf{a}'\boldsymbol{\Sigma}\mathbf{a}$, with the latter being strictly positive by the positive definiteness of $\boldsymbol{\Sigma}$.

It turns out that the property proved in Proposition 4.2.10 actually characterizes the $\mathsf{N}_k(\boldsymbol{\mu}, \boldsymbol{\Sigma})$ distribution. This will be shown in Section 8.9. Further, we will also see that this property leads one to propose a slightly more general definition of k-dimensional Gaussian distributions, that allows the real symmetric matrix $\boldsymbol{\Sigma}$ to be just non-negative definite, rather than positive definite.

4.3 Independence for Random Variables and Random Vectors

We know that $\mathbf{X} = (X_1,\cdots,X_k)$ is a random vector on a probability space if and only if $X_1,\cdots,X_k$ are real random variables and the joint distribution $P_{\mathbf{X}}$ determines the distribution of each of the random variables X_i, $1 \leq i \leq k$. Indeed, for any $1 \leq i \leq k$ and any $B \in \mathcal{B}(\mathbb{R})$, one has $P_{X_i}(B) = P_{\mathbf{X}}(\widetilde{B})$, where $\widetilde{B} = \{(x_1,\cdots,x_k) \in \mathbb{R}^k : x_i \in B\}$. These probabilities P_{X_i}, for $1 \leq i \leq k$, are called '*marginal*' distributions in this context. A natural question is whether the marginal distributions, in turn, determine the joint distribution of the vector $\mathbf{X}$. The answer, in general, is 'no' and the following exercise illustrates that.

EXERCISE **4.3.1.** *Let* $\Omega = \{0,1\}\times\{0,1\}$, $\mathcal{A} = \mathcal{P}(\Omega)$ *and let P be the probability on* $\mathcal{A}$ *defined by* $P(\{\omega\}) = 1/4$, *for each* $\omega \in \Omega$. *Consider the real random variables U and V defined on* $(\Omega, \mathcal{A}, P)$ *as* $U(\omega_1, \omega_2) = \omega_1$ *and* $V(\omega_1, \omega_2) = \omega_2$. *Show then that the two random vectors* $\mathbf{X} = (X_1, X_2)$ *and* $\mathbf{Y} = (Y_1, Y_2)$, *where* $X_1 = U$, $X_2 = V$ *and* $Y_1 = Y_2 = U$, *have very different joint distributions, although they both have the same marginals, that is,* $P_{X_i} = P_{Y_i}$, $i = 1, 2$.

4.3. Independence for Random Variables and Random Vectors

Thus, the marginal distributions are always determined by, but do not in general determine, the joint distribution. However, there is one situation when the joint distribution of $\mathbf{X} = (X_1, \cdots, X_k)$ is indeed determined by the marginal distributions of $X_1, \cdots, X_k$ and that situation is when the random variables $X_1, \cdots, X_k$ are 'independent', a notion that we are now going to discuss.

In a first course of probability, independence of two real random variables X and Y is often defined by the condition that

$$F_{X,Y}(a_1, a_2) = F_X(a_1) F_Y(a_2), \quad \text{for all real } a_1, a_2. \tag{4.3.1}$$

Of course, when X and Y are both discrete, this is equivalent to

$$P(X = x, Y = y) = P(X = x) P(Y = y), \quad \text{for all } x, y \in \mathbb{R}.$$

One can easily see that, in general, (4.3.1) is equivalent to

$$P(X \in I_{ab}, Y \in I_{cd}) = P(X \in I_{ab}) P(Y \in I_{cd}),$$

for all choices of $-\infty \le a \le b \le \infty, -\infty \le c \le d \le \infty$.
By using the same idea as in the proof of Theorem 2.12.8, one can deduce that (4.3.1) is equivalent to:

$$P(X \in B_1, Y \in B_2) = P(X \in B_1) P(Y \in B_2) \text{ for all } B_1, B_2 \in \mathcal{B}(R) \tag{4.3.2}$$

We use (4.3.2) for our definition of independence for, not just two, but any finite number of real random variables.

DEFINITION **4.3.2**. Real random variables $X_1, \cdots, X_k$ defined on a probability space $(\Omega, \mathcal{A}, P)$ are said to be *mutually independent* if for any choice $B_1, \cdots, B_k$ of Borel subsets of $\mathbb{R}$,

$$P(X_i \in B_i, 1 \le i \le k) = \prod_{i=1}^{k} P(X_i \in B_i) \tag{4.3.3}$$

As noted above, the same idea as in the proof of Theorem 2.12.8 shows that (4.3.3) is equivalent to $F_{\mathbf{X}}(a_1, \cdots, a_k) = \prod_i F_{X_i}(a_i)$ for all $(a_1, \cdots, a_k) \in \mathbb{R}^k$. This last equality shows that when $X_1, \cdots, X_k$ are independent real trandom variables, the joint distribution function (and hence the joint distribution) of $\mathbf{X} = (X_1, \cdots, X_k)$ is completely determined by the marginal distribution functions.

In Section 2.12, we had discussed the notion of independence of classes of subsets [see Definition 2.12.4]. The notion of independence of real random variables can also be formulated in terms of independence of classes of sets, as is seen in the following easy exercise. Such a formulation is very useful in many ways.

EXERCISE **4.3.3**. *Show that real random variables $X_1, X_2, \cdots, X_k$ on $(\Omega, \mathcal{A}, P)$ are independent iff the σ-fields $\sigma(X_1), \sigma(X_2), \cdots, \sigma(X_k)$ are independent.*

We summarize all the above observations in the following proposition.

PROPOSITION **4.3.4**. Let $X_1, \cdots, X_k$ be real random variables on a probability space $(\Omega, \mathcal{A}, P)$. Then the following are equivalent.
(1) $X_1, \cdots, X_k$ are independent.
(2) $F_{\mathbf{X}}(a_1, \cdots, a_k) = \prod_1^k F_{X_i}(a_i)$ where $\mathbf{X} = (X_1, X_2, \cdots, X_k)$.
(3) $\sigma(X_1), \cdots, \sigma(X_k)$ are independent sub-σ-fields of $\mathcal{A}$.

An important and useful consequence of Proposition 4.3.4 is the following.

COROLLARY **4.3.5.** If $X_1, \ldots, X_k$ are independent real random variables, then for any choice of real valued measurable functions $h_1, \ldots, h_k$ on $\mathbb{R}$, the random variables $h_1(X_1), \ldots, h_k(X_k)$ are independent.

Proof. One needs to simply observe that $\sigma(h_i(X_i)) \subset \sigma(X_i)$, for each $1 \leq i \leq k$, and then appeal to Proposition 4.3.4. $\square$

REMARK **4.3.6.** In Definition 4.3.2 and what followed, we have only talked about independence of real random variables. However, we may sometimes need to discuss independence for extended real valued radndom variables as well. For that, one only needs to slightly modify Definition 4.3.2 by allowing the B_i, $i = 1, \ldots, k$, in (4.3.3), to also be $\{\infty\}$ or $\{-\infty\}$. Equivalence of (1) and (3) in Proposition 4.3.4 remains valid for the extended real valued case, the proof of which is left as an exercise.

Proposition 4.3.4 motivates the following definition for mutual independence of a finite number of random vectors.

DEFINITION **4.3.7.** Random vectors $\mathbf{X}_1, \ldots, \mathbf{X}_n$ are said to be *mutually independent* if the σ-fields $\sigma(\mathbf{X}_1), \ldots, \sigma(\mathbf{X}_n)$ are independent.

Here $\sigma(\mathbf{X}_i)$, for each i, stands for the σ-field generated by class of random variables constituting the random vector $\mathbf{X}_i$. It is not conceptually difficult to translate the above definition of independence of random vectors in terms of their joint distribution functions, provided one is ready to do it carefully. It should be noted here that the random vectors $\mathbf{X}_i$ are allowed to be of different dimensions.

As in the case of σ-fields, the notion of independence can be extended to cover any arbitrary collection of random variables rather than just a finite number. Also, in view of Remark 4.3.6, we do not need to restrict ourselves only to real random variables.

DEFINITION **4.3.8.** A collection $\{X_\alpha, \alpha \in \Delta\}$ of random variables on a probability space are said to be *mutually independent* if every finite subcollection is independent.

EXERCISE **4.3.9.** (a) Events $\{A_\alpha, \alpha \in \Delta\}$ are independent iff the random variables $\{I_{A_\alpha}, \alpha \in \Delta\}$ are independent.
(b) $\{X_\alpha, \alpha \in \Delta\}$ are mutually independent iff the σ-fields $\{\sigma(X_\alpha), \alpha \in \Delta\}$ are mutually independent.

We end with a very important and useful result of which Corollary 4.3.5 turns out to be just a special case. Roughly speaking, what the result says is that, given a collection of mutually independent random variables, if one constructs new random variables out of disjoint subcollections, then these new random variables must be mutually independent. This is not something unexpected. In fact, in basic probability theory, this is often accepted and used generously without a question. The reader should realize though that no matter how natural it may seem from an intuitive point of view, it does need a proper justification before

4.4. Expectation

one can use it. And once again, the measure theoretic framework turns out to be absolutely essential to provide a proper justification.

THEOREM 4.3.10. *Let $\{X_\alpha, \alpha \in \Delta\}$ be a collection of mutually independent random variables on a probability space. Let $\{\Delta_\beta, \beta \in \Theta\}$ be a partition of Δ and, for each $\beta \in \Theta$, let Y_β be a random variable which is measurable with respect to $\sigma(\{X_\alpha, \alpha \in \Delta_\beta\})$. Then the collection $\{Y_\beta, \beta \in \Theta\}$ are mutually independent.*

Proof. The assumed $\sigma(\{X_\alpha, \alpha \in \Delta_\beta\})$-measurability of Y_β simply says that $\sigma(Y_\beta) \subset \sigma(\{X_\alpha, \alpha \in \Delta_\beta\})$, for each $\beta \in \Theta$. The result now follows immediately from Theorem 2.12.13 and Exercise 4.3.9(b). □

EXERCISE 4.3.11. *Let $X_1, \cdots, X_k$ be independent real random variables.*
(a) *Show that, for any permutation $\pi = (\pi(1), \cdots, \pi(k))$ of $\{1, \cdots, k\}$, the random variables $X_{\pi(1)}, \cdots, X_{\pi(k)}$ are independent.*
(b) *Show that, for any choice of integers $0 = k_0 < k_1 < \cdots < k_l = k$ and measurable functions $h_j : \mathbb{R}^{k_j - k_{j-1}} \to \mathbb{R}$, for $j = 1, \cdots, l$, the random variables $Y_j = h_j(X_{k_{j-1}+1}, \cdots, X_{k_j})$, $1 \leq j \leq l$, are independent.*

EXERCISE 4.3.12. *Let $\mathbf{X} = (X_1, \ldots, X_k)$ be a random vector.*
(a) *Show that if $X_1, \ldots, X_k$ are discrete then $X_1, \ldots, X_k$ are independent iff the joint pmf of $\mathbf{X}$ equals the product of the marginal pmfs.*
(b) *Show that if $X_1, \ldots, X_k$ have densities $f_1, \ldots, f_k$ then $X_1, \ldots, X_k$ are independent iff $\mathbf{X}$ is absolutely continuous and has joint density function given by $f(x_1, \ldots, x_k) = \prod_i f_i(x_i)$.*

4.4 Expectation

Anyone familiar with probability theory would know that an important concept related to a random variable is that of its expected value, also called expectation or mean, denoted by $E(X)$. Let us quickly recall how this concept is introduced in an undergraduate course in probability. The expected value of a random variable is supposed to represent a weighted average of its set of values, with the weights being the probabilities of the different values. Indeed, for a discrete random variable, that is literally how its expected value is defined, namely, if X has pmf p_X, then its expected value is defined as

$$E(X) = \sum_x x p_X(x).$$

Since $p_X(x) = 0$ for x outside a countable set, the above sum reduces to either a finite sum or an infinite series. In the latter case, one puts absolute convergence of the infinite series as a pre-condition for $E(X)$ to be defined.

But for a random variable with density f_X, undergraduate probability proposes a completely different definition, namely,

$$E(X) = \int_\mathbb{R} x f_X(x) dx,$$

Of course the usual integrability conditions are imposed. However, it is by no means obvious from this formula how this formula captures the concept of weighted average of values. Not only is a sum replaced by an integral, but also pmf is replaced by pdf which does not represent probabilities. To make matters worse, one goes even further to define the expectation of a function $h(X)$ of such a random variable X by the formula

$$E(h(X)) = \int_{\mathbb{R}} h(x) f_X(x) dx.$$

The problem is essentially two fold. Firstly, the theoretical basis of such a formula for the expected value of a random variable X having a pdf, is not clear. Secondly, by offering two different formulas for expected value in two different cases, it automatically leaves open the question of how to define expectation for random variables which are neither of these two types. But at the same time, it goes on to prescribe a fomula for $E(h(X))$ as an integral, in case X has a pdf. Shouldn't it be logical to first ascertain what kind of a random variable $h(X)$ itself is and define $E(h(X))$ accordingly? It is worth noting here that no such logical problem arises when X is discrete (see Exercise 4.1.8(b)) and one can easily justify the formula $E(h(X)) = \sum_x h(x) p_X(x)$, provided the series converges absolutely.

All these should convince the reader that this issue requires a serious relook. Ideally, there should be *one and only one* definition of expected value for all random variables and the multiple definitions described above should be derived from the genreral definition as working formulas in special cases. In addition to setting a correct theoretical foundation, this will also have the definite advantage that there will be only one theory of expected value. A reader exposed to undergraduate probability would remember that while proving properties of expected value, how one needs to always split the proof into two cases, thereby also incorrectly ignoring other possibilities. The measure theoretic setup developed in the previous chapters rectifies this problem in the most perfect way. In fact, measure theoretic formulation of probability acts as a unifier in this context.

The idea is to define the expectation of any random variable as an integral of the kind introduced in Chapter 3. This would allow us to use the entire integration theory already developed there to easily derive properties of expectation. Interestingly enough (and luckily also), it will turn out that the different definitions of expected value prescribed in undergraduate probability can be derived from this one definition of $E(X)$, as just valid formulas for special cases,.

To present now the definition of expectation, let us recall that a random variable X is nothing but an extended real valued measurable function on a probability space $(\Omega, \mathcal{A}, P)$.

DEFINITION **4.4.1.** *Expected value* of a random variable X, denoted $E(X)$, is said to exist and is given by $E(X) = \int X dP$, provided the integral $\int X dP$ exists. In case the integral $\int X dP$ fails to exist, one says that $E(X)$ does not exist.

$E(X)$ is also called *Expectation* or *Mean* of X. It must be noted that we are allowing $E(X)$ to be defined, even if it is not finite. In particular, expectation

4.4. Expectation

$E(X)$ for a non-negative random variable X always exists, but may equal $+\infty$. If $\int X dP$ is finite, that is, X is P-integrable, we say that X has a finite expectation.

As a consequence of Definition 4.4.1, we get the following properties of expectation directly from the corresponding properties of integrals that were proved in Chapter 3. No further proofs are, therefore, needed for these.

THEOREM 4.4.2. (a) $X \leq Y$ a.s.$[P]$ implies $E(X) \leq E(Y)$, provided both expectations exist.
(b) If $E(X)$ exists, so does $E(\alpha X)$ for any $\alpha \in \mathbb{R}$ and $E(\alpha X) = \alpha E(X)$.
(c) If $X + Y$ is well defined, $E(X)$ and $E(Y)$ both exist and if $E(X) + E(Y)$ is well defined, then $E(X + Y)$ exists and equals $E(X) + E(Y)$.
(d) If $E(X)$ exists, then $|E(X)| \leq E|X|$; in particular, X has finite expectation if and only if $E|X| < \infty$.
(e) Any random variable X, which is bounded a.s.$[P]$, has a finite expectation.
(f) If $X \geq 0$ a.s.$[P]$ and $E(X) = 0$, then $X = 0$ a.s.$[P]$.
(g) If $E(X I_A) \leq E(Y I_A)$ for all $A \in \mathcal{A}$, then $X \leq Y$ a.s.$[P]$.
(h) If X has finite expectation, then X is real valued a.s.$[P]$.

THEOREM 4.4.3. (a) (MCT) $0 \leq X_n \uparrow X$ a.s.$[P]$ implies $E(X_n) \uparrow E(X)$.
(b) (Fatou's Lemma) $X_n \geq 0$ a.s.$[P]$ implies $E(\liminf X_n) \leq \liminf E(X_n)$.
(c) (DCT) If $X_n \to X$ a.s.$[P]$ and if there is a random variable Y with finite expectation, such that, $|X_n| \leq Y$ a.s.$[P]$, for all n, then $E|X_n - X| \to 0$ and, in particular, $E(X_n) \to E(X)$.

EXERCISE 4.4.4. For each $n \geq 1$, consider the function φ_n on $[0, \infty)$ defined by $\varphi_n(x) = \sum_{k=0}^{\infty} \frac{k}{2^n} I_{[\frac{k}{2^n}, \frac{k+1}{2^n})}(x)$. Show that if X is any non-negative real random variable, then $X_n = \varphi_n(X)$ is a sequence of non-negative discrete random variables with $X_n \uparrow X$ and $E(X) = \lim_n E(X_n)$.

The next theorem uses the following notations, that are just analogues of notations introduced in Chapter 3. For a random variable X, we denote

$$\|X\|_p = [E(|X|^p)]^{1/p}, \ 0 < p < \infty, \text{ and } \|X\|_\infty = \inf\{a \geq 0 : P(X > a) = 0\}$$

THEOREM 4.4.5. (a) (Chebycheff's inequality) For any random variable X and any $\lambda > 0$, $P(\{|X| > \lambda\}) \leq E(|X|^p)/\lambda^p = (\|X\|_p/\lambda)^p$, for every $p > 0$.
(b) For any random variable X, $\|X\|_r \leq \|X\|_s$, for $0 < r \leq s \leq +\infty$.
(c) (Hölder's inequality) For random variables X and Y, $E(|XY|) \leq \|X\|_p \|Y\|_q$, for any pair of conjugate indices $p, q \in [1, \infty]$. Further, in case $p, q \in (1, \infty)$ and $\|X\|_p \|Y\|_q < \infty$, equality holds if and only if $A|X|^p = B|Y|^q$ a.s.$[P]$, for some real A and B, with at least one of them non-zero.
(d) (CS inequality) For random variables X and Y, $E(|XY|) \leq \|X\|_2 \|Y\|_2$, and, in case $\|X\|_2 \|Y\|_2 < \infty$, equality holds if and only if $A|X| = B|Y|$ a.s.$[P]$, for some real A and B, with at least one of them non-zero.

(e) (Minkowski's inequality) For random variables X and Y and for any $p \in [1, \infty]$, $\|X+Y\|_p \leq \|X\|_p + \|Y\|_p$.

(f) (Jensen's inequality) If φ is a convex function on R, then for any random variable X, with both X and $\varphi(X)$ having finite expectations, $\varphi(E(X)) \leq E(\varphi(X))$.

REMARK **4.4.6.** Here is an interesting application:
Weierstrass Polynomial approximation theorem is a well-known result in Analysis. It says that any real continuous function on a closed bounded interval can be approximated uniformly by polynomials. Here is an elementary proof of this, based on probability. The proof is due to Bernstein and is based on Chebycheff's Inequality. It should be clear that it is enough to prove this for a continuous function on $[0,1]$. Given a continuous function f on $[0,1]$, we know that there is an $M > 0$ such that $|f(t)| \leq M$ for all $t \in [0,1]$. Let $\epsilon > 0$ be given. Since any real continuous function on $[0,1]$ is uniformly continuous, we can get $\delta > 0$ such that $|f(t) - f(s)| < \epsilon/2$, whenever $|t - s| < \delta$. Choose a positive integer n such that $n\epsilon\delta^2 > M$. Consider the n-th degree polynomial p_n defined by $p_n(t) = \sum_{k=0}^{n} \binom{n}{k} f(k/n) t^k (1-t)^{n-k}$. It is clear that, for $t \in [0,1]$, $p_n(t) = E(f(X/n))$, where $X \sim \text{Bin}(n,t)$. Thus, for any $t \in [0,1]$, we have $|f(t) - p_n(t)| \leq E|f(t) - f(X/n)| < (\epsilon/2) + 2MP(\{|X - nt| > n\delta\})$. By Chebycheff's Inequality, the second term in the last expression is bounded above by $2ME[(X - E(X))^2]/(n\delta)^2$. Easy algebra gives $E[(X - E(X))^2] = nt(1-t) \leq n/4$. Thus we get $\sup_t |f(t) - p_n(t)| < (\epsilon/2) + M/(2n\delta^2)$. The choice of n guarantees that the right hand side is $< \epsilon$. This proves the result. The family of n-th degree polynomials p_n, as defined above, are called *Bernstein Polynomials*.

The next result is just the Change of Variable theorem (Theorem 3.5.16), in the present context.

THEOREM **4.4.7.** (Change of Variable Theorem) For any real random variable X with distribution P_X and any extended real valued measurable function h on $\mathbb{R}$,

$$E(h(X)) = \int_{\mathbb{R}} h \, dP_X. \qquad (4.4.1)$$

More generally, for any k-dimensional random vector $\mathbf{X}$ with joint distribution $P_\mathbf{X}$ and any extended real valued measurable function h on $\mathbb{R}^k$,

$$E(h(\mathbf{X})) = \int_{\mathbb{R}^d} h \, dP_\mathbf{X}. \qquad (4.4.2)$$

In each of (4.4.1) and (4.4.2) above, equality is to be interpreted as saying "if any one side of the equality exists, then so does the other and they are equal".

An important takeaway of the above is that, for a real random variable X, the expectation of X, or of $h(X)$ for any extended real valued measurable h on $\mathbb{R}$, can be converted to an integral on $\mathbb{R}$ and it depends only on the distribution of X. Similarly, the expectation of any extended real valued measurable function of a random vector $\mathbf{X}$ depends only on the joint distribution of $\mathbf{X}$.

Converse of this, of course, holds trivially! Expectations of bounded measurable functions of a random variable X (resp., a random vector $\mathbf{X}$) – in fact, just

4.4. Expectation

$E(I_B(X))$, $B \in \mathcal{B}$, (resp., $E(I_B(\mathbf{X}))$, $B \in \mathcal{B}^k$,) – determine the distribution of X (resp., the joint distribution of $\mathbf{X}$). What is interesting and not completely trivial is that even expectations of bounded continuous functions of a random variable (or of a random vector) suffice. This is our next result.

THEOREM 4.4.8. If X and Y are real random variables (defined possibly on different probability spaces) such that $E(h(X)) = E(h(Y))$ for all bounded continuous functions $h : \mathbb{R} \to \mathbb{R}$, then $P_X = P_Y$.
Similarly, if $\mathbf{X}$ and $\mathbf{Y}$ are k-dimensional random vectors (defined possibly on different probability spaces) such that $E(h(\mathbf{X})) = E(h(\mathbf{Y}))$ for all bounded continuous functions $h : \mathbb{R}^k \to \mathbb{R}$, then $P_\mathbf{X} = P_\mathbf{Y}$.

Proof. We prove only the one-dimensional case. The proof for higher dimensions is similar and left as an exercise. Let X and Y be real random variables satisfying the hypothesis. To prove that $P_X = P_Y$, it is enough to show that $F_X = F_Y$. Fix $a \in \mathbb{R}$ and for each $n \geq 1$, consider the bounded continuous function $h_n : \mathbb{R} \to \mathbb{R}$ given by $h(x) = [n(a + \frac{1}{n} - x)^+] \wedge 1$. Indeed, h_n is simply the continuous function that takes the value 1 on $(-\infty, a]$, takes the value 0 on $[a + \frac{1}{n}, \infty)$ and is linear on $[a, a + \frac{1}{n}]$. It follows that $0 \leq h_n(x) \leq 1$ and that $h_n(x) \to h(x) = I_{(-\infty, a]}(x)$, for all $x \in \mathbb{R}$. By hypothesis, $E(h_n(X)) = E(h_n(Y))$ for each n. By applying DCT, one gets $E(h(X)) = E(h(Y))$, that is, $F_X(a) = F_Y(a)$. □

REMARK 4.4.9. The assertion of Theorem 4.4.8 can be strengthened significantly. What is indeed true is that the distribution of a random variable X (resp., a random vector $\mathbf{X}$) can be shown to be uniquely determined by the values of $E(h(X))$ (resp., $E(h(\mathbf{X}))$ for *only infinitely differentiable functions* h on $\mathbb{R}$ (resp., on $\mathbb{R}^k$) with compact support. The main idea is to first observe that one can replace 'bounded continuous functions' in Theorem 4.4.8 by 'bounded continuous functions with compact support' and then to show that bounded continuous functions with compact support can be 'uniformly approximated' by infinitely differentiable functions with compact support. This is illustrated for $\mathbb{R}$ in Exercises 4.6.3 and 4.6.4.

We are next going to validate the formulae for expectations in the discrete and absolutely continuous cases, that one learns in a first course in probability. Indeed, as a consequence of the Change of Variable Theorem (Theorem 4.4.7) and using the results of the Exercises 3.6.14 and 3.6.15 on integration with respect to discrete and absolutely continuous probability measure on $\mathbb{R}^k$, for any $k \geq 1$, one gets the following.

THEOREM 4.4.10. *Let $\mathbf{X}$ be a k-dimensional discrete random vector, for $k \geq 1$, and let h be any extended real-valued borel measurable function on $\mathbb{R}^k$.*
If $\mathbf{X}$ is discrete with joint pmf $p_\mathbf{X}$, then the random variable $h(\mathbf{X})$ has finite expectation iff the sum $\sum_\mathbf{x} h(\mathbf{x}) p_\mathbf{X}(\mathbf{x})$ is absolutely convergent, and, in that case,
$$E(h(\mathbf{X})) = \sum_\mathbf{x} h(\mathbf{x}) p_\mathbf{X}(\mathbf{x}).$$
If the random vector $\mathbf{X}$ has a joint density function $f_\mathbf{X}$, then the random variable $h(\mathbf{X})$ has finite expectation iff $\int_{\mathbb{R}^k} |h(\mathbf{x})| f_\mathbf{X}(\mathbf{x}) d\mathbf{x} < \infty$, and in that case,
$$E(h(\mathbf{X})) = \int_{\mathbb{R}^k} h(\mathbf{x}) f_\mathbf{X}(\mathbf{x}) d\mathbf{x}.$$

Note here, that a 1-dimensional random vector is simply a real random variable. Thus, the above shows that the definition of expectation for a function of a discrete or absolutely continuous random variable (or random vector), that one usually encounters in undergraduate probability, turn out to be just consequences of the general definition of expectation given here.

A very important and widely used result discussed in undergraduate probability is that, for two independent random variables, expectation of the product equals the product of expectations. However, due to the absence of measure theoretic formulation in undergraduate probability, proof of this result can only be given for the cases when either both random variables are discrete or they have a joint density. But, now that we have one single definition of expectation for any random variable as an integral, the result can be proved straightaway without any such restrictions. This once again illustrates the power of the measure theoretic framework.

THEOREM **4.4.11.** (a) Let $(\Omega, \mathcal{A}, P)$ be a probability space and let $\mathcal{A}_1$, $\mathcal{A}_2$ be two independent sub-σ-fields of $\mathcal{A}$. If W and Z are any two random variables, measurable with respect to $\mathcal{A}_1$ and $\mathcal{A}_2$ respectively, such that either both are non-negative or both have finite expectations, then $E(WZ)$ exists and equals $E(W)E(Z)$.
(b) Suppose $\mathbf{X}$ and $\mathbf{Y}$ are independent random vectors, of dimensions k and m respectively. Let h and g be extended real valued measurable functions on $\mathbb{R}^k$ and $\mathbb{R}^m$ respectively such that *either* h and g are both non-negative *or* both $h(\mathbf{X})$ and $g(\mathbf{Y})$ have finite expectations. Then $E\big(h(\mathbf{X})g(\mathbf{Y})\big)$ exists and equals $E(h(\mathbf{X}))E(g(\mathbf{Y}))$.

Proof. Since (b) follows easily from (a), by taking $\mathcal{A}_1 = \sigma(\mathbf{X}), \mathcal{A}_2 = \sigma(\mathbf{Y})$ and $W = h(\mathbf{X}), Z = g(\mathbf{Y})$, we need to just prove (a).
Towards proving (a), we first take $W = I_{A_1}, Z = I_{A_2}$, where $A_1 \in \mathcal{A}_1, A_2 \in \mathcal{A}_2$. From independence of $\mathcal{A}_1$ and $\mathcal{A}_2$, we easily get $E(WZ) = P(A_1 \cap A_2) = P(A_1)P(A_2) = E(W)E(Z)$. Using linearity property of expectation (parts (b) and (c) of Theorem 4.4.2), the result extends easily to the case when both W and Z are non-negative real-valued simple functions, measurable with respect to $\mathcal{A}_1$ and $\mathcal{A}_2$ respectively. Next, if both W and Z are non-negative, we can get sequences $\{W_n\}$ and $\{Z_n\}$ of non-negative real-valued simple random variables, with W_n and Z_n, for each n, measurable with respect to $\mathcal{A}_1$ and $\mathcal{A}_2$ respectively, such that $W_n \uparrow W$ and $Z_n \uparrow Z$. From what we have proved, we have $E(W_n Z_n) = E(W_n)E(Z_n)$, for each n. From the fact that $0 \leq W_n Z_n \uparrow WZ$ and using MCT(Theorem 4.4.3(a)), we easily get $E(WZ) = E(W)E(Z)$. Finally, we consider the case when $E(W)$ and $E(Z)$ both exist and are finite. In this case, the non-negative $\mathcal{A}_1$-measurable random variables W^+, W^- both have finite expectations and so do the non-negative $\mathcal{A}_2$-measurable random variables Z^+, Z^-. By what has been shown and using additivity (Theorem 4.4.2(b)), we get

$$E(W^+ Z^+ + W^- Z^-) = E(W^+)E(Z^+) + E(W^-)E(Z^-),$$

and also

$$E(W^- Z^+ + W^+ Z^-) = E(W^-)E(Z^+) + E(W^+)E(Z^-).$$

4.4. Expectation

In particular, the left side of both equalities above are finite. Noting now that
$$WZ = (W^+Z^+ + W^-Z^-) - (W^+Z^- + W^-Z^+),$$
the equality $E(WZ) = E(W)E(Z)$ follows easily from the linearity property of expectation ((b) and (c) of Theorem 4.4.2) and simple algebra. □

REMARK 4.4.12. An important point in the above theorem, which is sometimes overlooked, is that under the hypothesis of independence of **X** and **Z**, just finiteness of $E(h(\mathbf{X}))$ and $E(g(\mathbf{Y}))$ is enough to guarantee finiteness of $E(h(\mathbf{X})g(\mathbf{Y}))$. It should be noted that, in general, finiteness of expectations of two random variables does not guarantee that their product has finite expectation. One needs finiteness of appropriate higher powers of the two random variables. For example, finiteness of expectations of the squares of two random variables will, by the Cauchy-Schwarz inequality, guarantee that their product also has finite expectation.

EXERCISE 4.4.13. (a) Show that real random variables $X_1, \ldots, X_n$ are independent if and only if
$$E\Big(\prod_{j=1}^n h_j(X_j)\Big) = \prod_{j=1}^n E(h_j(X_j)), \tag{4.4.3}$$
for all bounded measurable functions $h_1, \cdots, h_n$ on $\mathbb{R}$.
(b) Show that if real random variables $X_1, \ldots, X_n$ satisfy (4.4.3) for all bounded continuous functions $h_1, \ldots, h_n$ (or even for all bounded continuous functions $h_1, \ldots, h_n$ with compact support) on $\mathbb{R}$, then $X_1, \cdots, X_n$ are independent.
(c) Generalize (a) and (b) for random vectors $\mathbf{X}_1, \ldots, \mathbf{X}_n$.

It is a standard result in Riemann integration theory that if f is a real valued continuous function on $\mathbb{R}$, then $\int_{-\infty}^{\infty} |f(x)|dx < \infty$ iff the tail integral goes to zero, that is, $\int_{(|x|>n)} |f(x)|dx \to 0$ as $n \to \infty$. The following result is an analogue of this for expectations of random variables. It says that a random variable has finite expectation iff the tail integral goes to zero.

THEOREM 4.4.14. A random variable X on $(\Omega, \mathcal{A}, P)$ has finite expectation if and only if $E(|X|I_{(|X|>n)}) \to 0$ as $n \to \infty$.

Proof. Recall that if X has finite expectation, then $|X| < \infty$ a.s. $[P]$, so that $|X|I_{(|X|>n)} \to 0$ a.s.$[P]$. Applying DCT, one gets the 'only if' part.
For the 'if' part, $E(|X|I_{(|X|>n)}) \to 0$ implies that $E(|X|I_{(|X|>n)}) < 1$ for some n. But then, $E(|X|) = E(|X|I_{(|X|\le n)}) + E(|X|I_{(|X|>n)}) \le n + 1 < \infty$. □

Chebycheff's inequality $P(|X| > n) \le E(|X|)/n$ asserts that, for X to have a finite expectation, the tail probabilities $P(|X| > n)$ have to go to zero. Theorem 4.4.14 asserts something much stronger. Since $nP(|X| > n) \le E(|X|I_{(|X|>n)})$, finiteness of $E(X)$, according to Theorem 4.4.14, makes it necessary that even $nP(|X| > n) \to 0$ as $n \to \infty$. Of course, just this condition does not happen to be sufficient to guarantee that $E(X)$ is finite. So a natural question is: can one get a necessary and sufficient condition on the tail probabilities (in the sense of how fast they should go to zero) for $E(X)$ to be finite? The next result gives a precise answer to this question. In fact, the result is slightly more general in the sense that it deals with finiteness of $E(|X|^p)$ for $p \geq 1$.

THEOREM **4.4.15**. For each $p \in [1, \infty)$, there are constants $0 < C_p < \infty$ and $0 < D_p < \infty$ such that, for any real random variable X,
$$p\,C_p \sum_{k=1}^{\infty} k^{p-1} P(|X| > k) \leq E(|X|^p) \leq 1 + p\,D_p \sum_{k=1}^{\infty} k^{p-1} P(|X| > k) \quad (4.4.4)$$

Proof. We are going to use the fact (see Exercise 4.4.17) that, for any $p \geq 1$, there are constants $0 < C_p < 1/p$ and $0 < D_p < \infty$, such that,
$$C_p k^{p-1} \leq (k+1)^{p-1} \leq D_p k^{p-1}, \text{ for all } k \geq 1. \quad (4.4.5)$$

Note that $E(|X|^p) = \sum_{k=0}^{\infty} E\big(|X|^p I_{(k < |X| \leq k+1)}\big)$, which gives us
$$\sum_{k=0}^{\infty} k^p P(k < |X| \leq k+1) \leq E(|X|^p) \leq \sum_{k=0}^{\infty} (k+1)^p P(k < |X| \leq k+1) \quad (4.4.6)$$

For $k \geq 0$, let $G(k) = P(|X| > k)$, so that $P(k < |X| \leq k+1) = G(k) - G(k+1)$. Using this in the second inequality in (4.4.6), one easily gets
$$E(|X|^p) \leq \sum_{k=0}^{\infty} (k+1)^p G(k) - \sum_{k=1}^{\infty} k^p G(k) \leq 1 + \sum_{k=1}^{\infty} [(k+1)^p - k^p] G(k).$$

Now, the mean value theorem applied to $\varphi(x) = x^p$ for $x > 0$ gives, for each $k \geq 1$, $(k+1)^p - k^p = p\theta_k^{p-1}$ with $\theta_k \in (k, k+1)$, implying $(k+1)^p - k^p \leq p(k+1)^{p-1}$. Using this and the second inequality in (4.4.5), one finally gets
$$E(|X|^p) \leq 1 + p\,D_p \sum_{k=1}^{\infty} k^{p-1} P(|X| > k).$$

For the other inequality, we use the first inequality in (4.4.6) to get
$$E(|X|^p) \geq \sum_{k=1}^{\infty} k^p G(k) - \sum_{k=2}^{\infty} (k-1)^p G(k) = \sum_{k=1}^{\infty} [k^p - (k-1)^p] G(k).$$

As before, we can use the mean value theorem and the first inequality in (4.4.5) to get $k^p - (k-1)^p \geq p(k-1)^{p-1} \geq pC_p k^{p-1}$, for each $k \geq 2$. For $k=1$, the inequality $k^p - (k-1)^p \geq pC_p k^{p-1}$ holds trivially because $Cp < 1/p$. Applying this in the above inequality for $E(|X|^p)$, we get
$$E(|X|^p) \geq p\,C_p \sum_{k=1}^{\infty} k^{p-1} P(|X| > k). \qquad \square$$

COROLLARY **4.4.16**. For any random variable X and any $p \geq 1$,
$$E(|X|^p) < \infty \text{ if and only if the series } \sum_{k=1}^{\infty} k^{p-1} P(|X| > k) \text{ converges.}$$

In particular, a random variable X has finite expectation if and only if the series $\sum_{k=1}^{\infty} P(|X| > k)$ converges.

EXERCISE **4.4.17**. Show that for any $p \geq 1$, there exist constants C_p and D_p with $0 < C_p < 1/p$ and $0 < D_p < \infty$, such that,
$$C_p k^{p-1} \leq (k+1)^{p-1} \leq D_p k^{p-1} \quad \text{for all } k \geq 1.$$
[*Hint:* The sequence $\{(1 + \frac{1}{k})^{p-1}, k \geq 1\}$ of positive real numbers converges.]

4.5 Moments, Variance and Covariance

We introduce a few definitions and notations that are very widely used in probability as well as in statistics. It will be convenient for us to use these notations in the sequel.

DEFINITION 4.5.1. For a random variable X and integer $k \geq 1$, we say that X has a finite k-th *moment* if $X \in L_k$ and in that case the real number $E(X^k)$ is called the *k-th moment* of X.

As we have already seen (Theorem 4.4.5(b)), if the k-th moment of a random variable X is finite, then so are the j-th moments of X, for all $j \leq k$. In particular, finiteness of second moment for a random variable implies finiteness of its mean also, which allows us to make the next definition.

DEFINITION 4.5.2. If X has a finite second moment, then the *variance* of X, written $V(X)$, is defined as $V(X) = E(X - E(X))^2$.

Clearly, $V(X) \geq 0$ by definition and equality holds iff X is degenerate (that is, $P(X = c) = 1$, for some $c \in \mathbb{R}$; indeed for $c = E(X)$). It is an easy exercise to see that $V(X) = E(X^2) - (E(X))^2$. Also, from the definition of variance, one can easily prove that $V(aX + b) = a^2 V(X)$.

DEFINITION 4.5.3. For two random variables X and Y with finite second moments, the *covariance* between X and Y, written $C(X, Y)$, is defined as

$$C(X,Y) = E[(X - E(X))(Y - E(Y))].$$

Note that, by Cauchy-Schwarz inequality, finiteness of second moments of X and Y implies that the covariance $C(X, Y)$ is finite.

Clearly, $C(X, Y) = C(Y, X)$. Also, from the definition of covariance, one can easily prove the following properties (assume that all the random variables below have finite second moments):
(i) $C(X, X) = V(X)$,
(ii) $C(X, Y) = E(XY) - E(X)E(Y)$,
(iii) $C(aX + b, a'Y + b') = aa' C(X, Y)$,
(iv) $C(aX + bY, Z) = aC(X, Z) + bC(Y, Z)$,
(v) $V(X + Y) = V(X) + V(Y) + 2C(X, Y)$,
(vi) If X, Y are independent, then $C(X, Y) = 0$ and $V(X + Y) = V(X) + V(Y)$.

EXERCISE 4.5.4. Show that the converse of (vi) above is not true, by exhibiting random variables X and Y on a probability space which are not independent, but have $C(X, Y) = 0$.

DEFINITION 4.5.5. If $\mathbf{X} = (X_1, \ldots, X_k)$ is a k-dimensional random vector, such that, $E(X_i)$ is finite for each i, the k-dimensional vector $(E(X_1), \ldots, E(X_k))$ is called the *mean vector* of $\mathbf{X}$ and is denoted by $M(\mathbf{X})$.
Further, if X_i, for each i, has finite second moment, then the $k \times k$ matrix $((\sigma_{ij}))$, where $\sigma_{ij} = C(X_i, X_j)$, for each $1 \leq i, j \leq k$, is called the *dispersion matrix* of $\mathbf{X}$ (also called the *variance-covariance matrix* of $\mathbf{X}$) and is denoted by $D(\mathbf{X})$.

In the sequel, whenever we mention $M(\mathbf{X})$ for a random vector $\mathbf{X}$, it has to be understood that finiteness of $E(X_i)$, for each i, is assumed. Similarly, when we mention $D(\mathbf{X})$, finiteness of second moment of X_i, for each i, is assumed, even if it is not explicitly stated.

Also, in the study of random vectors, it is often customary to think of vectors, random as well as non-random, as **column vectors** by default. In what follows (starting from the next exercise), we will follow that convention.

EXERCISE 4.5.6. (a) Show that for k-dimensional random vectors $\mathbf{X}$ and $\mathbf{Y}$ defined on a probability space, $M(\mathbf{X}+\mathbf{Y}) = M(\mathbf{X}) + M(\mathbf{Y})$; in particular, $M(\mathbf{X}+\mathbf{c}) = M(\mathbf{X}) + \mathbf{c}$, for any k-dimensional vector $\mathbf{c}$.
(b) Show that, if $\mathbf{X}$ is a k-dimensional random vector, then
(i) $M(a\mathbf{X}) = aM(\mathbf{X})$, $D(a\mathbf{X}) = a^2 D(\mathbf{X})$, for any real number a, and,
(ii) $E(\mathbf{a}'\mathbf{X}) = \mathbf{a}'M(\mathbf{X})$, $V(\mathbf{a}'\mathbf{X}) = \mathbf{a}'D(\mathbf{X})\mathbf{a}$, for any k-dimensional vetor $\mathbf{a}$.
(c) Show that for any k-dimensional random vector $\mathbf{X}$, its dispersion matrix $D(\mathbf{X})$ is a real symmetric non-negative definite matrix. Further, $D(\mathbf{X})$ is positive definite, unless there is a non-zero k-dimensional vector $\mathbf{a}$ such that $\mathbf{a}'\mathbf{X}$ is a degenerate real random variable.
(d) Show that $D(\mathbf{X}+\mathbf{c}) = D(\mathbf{X})$, for any k-dimensional random vector $\mathbf{X}$ and any k-dimensional vector $\mathbf{c}$.
(e) Show that $D(\mathbf{X}) = E\big((\mathbf{X}-M(\mathbf{X}))(\mathbf{X}-M(\mathbf{X}))'\big) = E(\mathbf{X}\mathbf{X}') - M(\mathbf{X})M(\mathbf{X})'$.
(f) Show that if $\mathbf{X}$ is a k-dimensional random vector, then for any real $m \times k$ matrix $\mathbf{A}$ and any m-dimensional vector $\mathbf{c}$, where $m \geq 1$, the mean vector and dispersion matrix of the m-dimensional random vector $\mathbf{Y} = \mathbf{A}\mathbf{X} + \mathbf{c}$ are given by $M(\mathbf{Y}) = \mathbf{A}M(\mathbf{X}) + \mathbf{c}$ and $D(\mathbf{Y}) = \mathbf{A}D(\mathbf{X})\mathbf{A}'$ respectively.

We now return to the k-dimensional Gaussian distribution $\mathsf{N}_k(\boldsymbol{\mu}, \boldsymbol{\Sigma})$, as defined in 4.2.5 and show that the parameters $\boldsymbol{\mu}$ and $\boldsymbol{\Sigma}$ actually represent the mean vector and the dispersion matrix respectively of the underlying random vector.

PROPOSITION 4.5.7. (a) If $\mathbf{X} = (X_1, \ldots, X_k) \sim \mathsf{N}_k(\mathbf{0}, \mathbf{I}_k)$, where $\mathbf{I}_k$ is the $k \times k$ identity matrix, then $X_1, \ldots, X_k$ are i.i.d. $\mathsf{N}(0,1)$ random variables.
(b) For any $\mathbf{X} \sim \mathsf{N}_k(\boldsymbol{\mu}, \boldsymbol{\Sigma})$, the mean vector $M(\mathbf{X})$ and the dispersion matrix $D(\mathbf{X})$ of $\mathbf{X}$ are given by $\boldsymbol{\mu}$ and $\boldsymbol{\Sigma}$ respectively.

Proof. (a) If $\mathbf{X} \sim \mathsf{N}_k(\mathbf{0}, \mathbf{I}_k)$, then the joint probability density function of $\mathbf{X}$, as given in Definition 4.2.5 reduces to
$$f(\mathbf{x}) = (2\pi)^{-k/2} \exp\{-(x_1^2 + \cdots + x_k^2)/2\},$$
which is just the product of k one-dimensional densities of the $\mathsf{N}(0,1)$ distributions. It follows, therefore (see Exercise 4.3.12 (b)), that $X_1, \ldots, X_k$ are independent random variables, with each having $\mathsf{N}(0,1)$ distribution.
(b) Since $\boldsymbol{\Sigma}$ is positive definite, we can get a real non-singular $k \times k$ matrix $\mathbf{B}$ such that $\mathbf{B}\boldsymbol{\Sigma}\mathbf{B}' = \mathbf{I}_k$. If we consider the random vector $\mathbf{Y} = \mathbf{B}(\mathbf{X} - \boldsymbol{\mu})$, then using Proposition 4.2.6, we will get $\mathbf{Y} \sim \mathsf{N}_k(\mathbf{0}, \mathbf{I}_k)$. Using part (a), we can easily see that $M(\mathbf{Y}) = \mathbf{0}$ and $D(\mathbf{Y}) = \mathbf{I}_k$. Denoting now $\mathbf{A} = \mathbf{B}^{-1}$, we have

$\mathbf{X} = \mathbf{AY} + \boldsymbol{\mu}$. Using Exercise 4.5.6 (f), we get $M(\mathbf{X}) = \mathbf{A}M(\mathbf{Y}) + \boldsymbol{\mu} = \boldsymbol{\mu}$ and $D(\mathbf{X}) = \mathbf{A}D(\mathbf{Y})\mathbf{A}' = \mathbf{A}\mathbf{A}' = \boldsymbol{\Sigma}$. □

REMARK 4.5.8. Now that we know that the parameters $\boldsymbol{\mu}$ and $\boldsymbol{\Sigma}$ of a k-dimensional Gaussian random vector $\mathbf{X}$ represent the mean vector $M(\mathbf{X})$ and the dispersion matrix $D(\mathbf{X})$ respectively, it follows that, for any $1 \leq k_1 < \cdots < k_m \leq k$, the parameters for the Gaussian random vector $(X_{k_1}, \ldots, X_{k_m})'$ must be $(E(X_{k_1}), \ldots, E(X_{k_m}))' = (\mu_{k_1}, \cdots, \mu_{k_m})'$ and $((C(X_{k_i}, X_{k_j})) = ((\sigma_{k_i k_j}))$ respectively (see Remark 4.2.11). Similarly, the mean and variance of the Normal random variable $\mathbf{a}'\mathbf{X}$ will be $\mathbf{a}'\boldsymbol{\mu}$ and $\mathbf{a}'\boldsymbol{\Sigma}\mathbf{a}$ respectively.

From Propositions 4.2.6, 4.2.9 and 4.5.7, one gets the following useful result, which essentially asserts that if $\mathbf{X} = (X_1, \ldots, X_k)$ has a k-dimensional Gaussian distribution, then any m-dimensional random vector (with $1 \leq m \leq k$), whose components are any m "*linearly independent*" linear combinations of $X_1, \ldots, X_k$, has m-dimensional Gaussian distribution with parameters that can be easily expressed in terms of those of $\mathbf{X}$.

PROPOSITION 4.5.9. Let $\mathbf{X} \sim \mathsf{N}_k(\boldsymbol{\mu}, \boldsymbol{\Sigma})$. Then for any $1 \leq m \leq k$ and any real $m \times k$ matrix $\mathbf{L}$ with $\operatorname{rank}(\mathbf{L}) = m$, the m-dimensional random vector $\mathbf{Y} = \mathbf{LX}$ has a $\mathsf{N}_m(\widetilde{\boldsymbol{\mu}}, \widetilde{\boldsymbol{\Sigma}})$-distribution, where $\widetilde{\boldsymbol{\mu}} = \mathbf{L}\boldsymbol{\mu}$ and $\widetilde{\boldsymbol{\Sigma}} = \mathbf{L}\boldsymbol{\Sigma}\mathbf{L}'$.

Proof. In case $m = k$, the matrix $\mathbf{L}$ is non-singular, so that the random vector $\mathbf{Y}$ is Gaussian with parameters $\mathbf{L}\boldsymbol{\mu}$ and $\mathbf{L}\boldsymbol{\Sigma}\mathbf{L}'$, by Proposition 4.2.6. In case $m < k$, one can get a non-singular $k \times k$ real matrix $\mathbf{A}$, whose first m rows are just the rows of $\mathbf{L}$. Propositions 4.2.6 and 4.2.9 imply that $\mathbf{Y}$ is Gaussian. That its parameters are $\widetilde{\boldsymbol{\mu}} = \mathbf{L}\boldsymbol{\mu}$ and $\widetilde{\boldsymbol{\Sigma}} = \mathbf{L}\boldsymbol{\Sigma}\mathbf{L}'$ follow from Proposition 4.5.7 and Exercise 4.5.6 (f). □

A consequence of Proposition 4.5.7 (b) is the extremely important and useful property of Gaussian distributions which asserts that if $\mathbf{X} = (X_1, \ldots, X_k)$ is a k-dimensional Gaussian random vector, then X_i and X_j are independent iff $C(X_i, X_j) = 0$. This can be seen easily by noting that $(X_i, X_j)'$ will be a 2-dimensional Gaussian random vector with a diagonal dispersion matrix, implying that the joiint density of (X_i, X_j) is a product of two one-dimensional normal densities. One should take note of this special property of Gaussian distribution in the backdrop of the fact that $C(X, Y) = 0$ does not, in general, imply independence of X and Y (Exercise 4.5.4). It must be kept in mind though that for the above to be true, we must have (X_i, X_j) to be jointly Gaussian.

We end this section with an important result which has a wide range of applications and gave huge impetus to the rich theory of what is nowadays known as "Large Deviations". We first state a general result and then apply it in a special case. We know that for any real random variable X, the set $\{\theta : E(e^{\theta X}) < \infty\}$ is an interval (possibly degenerate) containing $\theta = 0$ (see Exercise 3.6.6 (b)).

THEOREM 4.5.10. Let $X_1, \cdots, X_n$ be independent and identically distributed random variables on a probability space $(\Omega, \mathcal{A}, P)$. Denote $m(\theta) = E(e^{\theta X_1})$ and assume $I = \{\theta > 0 : m(\theta) < \infty\} \neq \emptyset$. Then for any $t > 0$,
$$P(X_1 + \cdots + X_n > t) \leq e^{-\theta t}[m(\theta)]^n, \text{ for all } \theta \in I,$$
and consequently,
$$P(X_1 + \cdots + X_n > t) \leq \inf_{\theta \in I} \{e^{-\theta t}[m(\theta)]^n\}. \tag{4.5.1}$$

Proof. For any $\theta \in I$, we have $P(X_1 + \cdots + X_n > t) = P(e^{\theta(X_1 + \cdots + X_n)} > e^{\theta t})$, which, by Chebycheff's inequality, is bounded above by $e^{-\theta t} E\left(\prod_{i=1}^n e^{\theta X_i}\right)$. Using the fact that $X_1, \cdots, X_n$ are independent and identically distributed, we have $E\left(\prod_{i=1}^n e^{\theta X_i}\right) = [m(\theta)]^n$. Thus we get $P(X_1 + \cdots + X_n > t) \leq e^{-\theta t}[m(\theta)]^n$, for all $\theta \in I$. Since this inequality holds for all $\theta \in I$, (4.5.1) follows. □

As it stands, the inequality (4.5.1) is not of much use, unless we can actually minimize $e^{-\theta t}[m(\theta)]^n$ over $\theta \in I$. It turns out that in many situations, it is possible to do so and in those cases, (4.5.1) provides a very powerful upper bound for the tail probabilities $P(X_1 + \cdots + X_n > t)$. One such special case is illustrated in the next result.

THEOREM 4.5.11. Let $X_1, \cdots, X_n$ be independent random variables, each taking the values ± 1 with probability $\frac{1}{2}$ each. Then for any $t > 0$,
$$P(|X_1 + \cdots + X_n| > t) \leq 2e^{-t^2/(2n)}.$$

Proof. The reader should convince herself that $P(|X_1 + \cdots + X_n| > t)$ equals $2P(X_1 + \cdots + X_n > t)$, for any $t > 0$, so that it suffices to prove the inequality $P(X_1 + \cdots + X_n > t) \leq e^{-t^2/(2n)}$, for $t > 0$. Now, $m(\theta) = E(e^{\theta X_1}) = \frac{e^\theta + e^{-\theta}}{2}$ and we leave it for the reader to verify that the last expression is bounded above by $e^{\theta^2/2}$, for all $\theta \in \mathbb{R}$. Simple calculus shows that $e^{-\theta t}(e^{\theta^2/2})^n$ attains a unique minimum value $e^{-t^2/(2n)}$, attained at $\theta = t/n$, and that completes the proof. □

Just to see how powerful this inequality is, let $\{X_n, n \geq 1\}$ be a sequence of mutually independent random variables, each taking values ± 1 with probability $\frac{1}{2}$ each. Denoting $S_n = X_1 + \cdots + X_n$, for each $n \geq 1$, one can use Theorem 4.5.11 to deduce that the series $\sum_n P(|S_n| > n\epsilon)$ converges, for any $\epsilon > 0$, however small. By the Borel-Cantelli Lemma (Theorem 2.11.4 (f)), we get that $P\big(\limsup_n \{\omega : |S_n(\omega)| > n\epsilon\}\big) = 0$, for every $\epsilon > 0$. In particular, denoting $A_j = \limsup_n \{\omega : |S_n(\omega)|/n > \frac{1}{j}\}$, $j \geq 1$, we have $P(A_j) = 0$, for every $j \geq 1$, and therefore $P(\bigcup_j A_j) = 0$. Recalling the definition of limsup of sets, the reader can easily see that $\omega \notin \bigcup_j A_j$ implies $S_n(\omega)/n \to 0$. Thus, what we have shown is that $(S_n/n) \to 0$ a.s.$[P]$. What we have just proved, for the special sequence X_n considered above, is what is generally known as the "Strong Law of Large Numbers". In a subsequent chapter, we will prove a much more general version of the Strong law of Large Numbers which will require more sophisticated tools.

4.6. Additional Exercises

The function $m(\theta)$ on I, introduced in Theorem 4.5.10, goes by the name of 'moment generating function' and plays an important role in probability and statistics.

DEFINITION 4.5.12. For a random variable X, let $I(X) = \{\theta \in \mathbb{R} : E(e^{\theta X}) < \infty\}$. The function $m_X : I(X) \to (0, \infty)$ defined as $m_X(\theta) = E(e^{\theta X})$ is called the *Moment Generating Function* (abbreviated as *mgf*) of X.

We already know (Exercise 3.6.6 (b)) that $I(X)$ is always an interval, possibly degenerate, containing zero and that $\log m_X$ is a convex function on $I(X)$. The moment generating function has interesting properties and contains useful information about the random variable. These are given as exercises below.

EXERCISE 4.5.13. (a) m_X is continuous on the set of all interior points of $I(X)$.
(b) If $Y = aX + b$, $a \neq 0$, then $I(Y) = \{\theta/a : \theta \in I(X)\}$ and $m_Y(\theta) = e^{b\theta} m_X(a\theta)$.
(c) If X is a bounded random variable, then $I(X) = \mathbb{R}$.
(d) If X is non-negative then $I(X) \supset (-\infty, 0]$.
(e) If X has a symmetric distribution, then $I(X)$ is symmetric around zero.
(f) If X and Y are independent, then $m_{X+Y} = m_X \cdot m_Y$ on $I(X) \cap I(Y)$.
(g) If for some $\epsilon > 0$, $(-\epsilon, \epsilon) \subset I(X)$, then m_X is infinitely differentiable on an open neighbourhood of 0; further, X has finite moments of all orders and $E(X^k) = m_X^{(k)}(0)$ for all $k \geq 1$. [This is the reason why m_X is called moment generating function.]
(h) Give examples to show that all the different possibilities for $I(X)$ as an interval containing zero are actually attained.

Throughout this chapter we have confined ourselves to real valued (or even extended real valued) random variables and k-dimensional random vectors. Occasionally we may have to consider measurable maps X on a probability space $(\Omega, \mathcal{A}, P)$ into an abstract space $(E, \mathcal{E})$ as random variables also. We will refer to such random variables as E-valued random variables. For such a random variable, the probability PX^{-1} on $(E, \mathcal{E})$ will be referred to as the distribution of X. The reader should convince herself that one can discuss notions like independence for such random variables also. However, since integration is defined only for extended real valued measurable functions, the concept of expectation is limited to extended real valued random variables only.

4.6 Additional Exercises

EXERCISE 4.6.1. (a) Show that if $\mathbf{X}$ is a k-dimensional random vector on a probability space $(\Omega, \mathcal{A}, P)$, then $\sigma(\mathbf{X})$ is a countably generated σ-field contained in $\mathcal{A}$. (Here, $k \geq 1$.)
(b) Conversely, if $\mathcal{G}$ is any countably generated sigma-field contained in $\mathcal{A}$, then show that there is a real random variable X on $(\Omega, \mathcal{A}, P)$, such that, $\mathcal{G} = \sigma(X)$.

EXERCISE **4.6.2.** Let $\{\xi_n\}$ be a sequence of independent random variables with $P(\xi_n = 0) = P(\xi_n = 1) = 1/2$.
(a) Show that, with probability one, the series $\sum_n \frac{2\xi_n}{3^n}$ converges and the random variable $X = \sum_n \frac{2\xi_n}{3^n}$ takes values in the Cantor set C (See Exercise 2.8.21).
(b) Show that the distribution function of X is the function F of Exercise 3.6.25 and hence deduce that, for any $x \in C$,
$$F(x) = \sum_n \tfrac{\delta_n}{2^n}, \text{ where } x = \sum_n \tfrac{2\delta_n}{3^n}.$$
(c) Using (a) and (b) above, conclude that X is a continuous random variable, which is not absolutely continuous. (In fact, in the language of Chapter 6, the distribution of X is 'singular' with respect to Lebesgue measure on $\mathbb{R}$.)

EXERCISE **4.6.3.** (a) Let $a < b$ be any two real numbers. Pick any two sequences $\{a_n\}$ and $\{b_n\}$ with $a < a_n < b < b_n$, for all n, such that, $a_n \downarrow a$ and $b_n \downarrow b$. For each n, construct a continuous function h_n taking values in $[0, 1]$, such that, $h_n(x) = 1$, for $x \in [a_n, b]$, and $h_n(x) = 0$, for $x \notin (a, b_n)$.
(b) Show that $h_n \to I_{(a,b]}$ pointwise, as $n \to \infty$.
(c) Show that if X and Y are two random variables (possibly on different probability spaces), such that $E\big(h(X)\big) = E\big(h(Y)\big)$, for all continuous functions $h : \mathbb{R} \to \mathbb{R}$ with compact support, then X and Y have the same distribution.

EXERCISE **4.6.4.** A real-valued function on $\mathbb{R}$ is said to be a C_∞ function, if it is infinitely differentiable everywhere.
(a) Consider the function $g(x) = ke^{-1/(1-x^2)}I_{(-1,1)}(x)$, $x \in \mathbb{R}$, where k is so chosen that $\int_\mathbb{R} g(x)dx = 1$. Show that g is a C_∞ function with compact support. For any real-valued λ-integrable function h on $\mathbb{R}$ and any $\epsilon > 0$, denote h_ϵ to be the function defined as
$$h_\epsilon(x) = \int_\mathbb{R} g_\epsilon(y)h(x-y)dy, \ x \in \mathbb{R},$$
where $g_\epsilon(x) = \frac{1}{\epsilon}g(x/\epsilon)$, $x \in \mathbb{R}$.
(b) Show that h_ϵ is a λ-integrable C_∞ function. [Use Exercise 5.4.12.]
(c) Show that, if h has compact support, than so also does h_ϵ, for every $\epsilon > 0$.
(d) Show that, $|h(x) - h_\epsilon(x)| \leq \sup\{|h(y) - h(x)| : y \in (x - \epsilon, x + \epsilon)\}$, for any $x \in \mathbb{R}$ and any $\epsilon > 0$.
(e) Let $h : \mathbb{R} \to \mathbb{R}$ be a continuous function with compact support. Choose a sequence $\{\epsilon_n\}$ of positive real numbers with $\epsilon_n \downarrow 0$ and writing h_{ϵ_n}, for each n, simply as h_n, show that $\{h_n\}$ is a sequence of C_∞-functions with compact support, that converges uniformly to h.
(f) Show now that X and Y are two random variables (possibly on different probability spaces), such that $E\big(h(X)\big) = E\big(h(Y)\big)$, for all C_∞-functions $h : \mathbb{R} \to \mathbb{R}$ with compact support, then X and Y have the same distribution. (You may use Exercise 4.6.3.)

4.6. Additional Exercises

EXERCISE **4.6.5**. Show that for any random variable X with finite expectation, there is a sequence $\{X_n\}$ of discrete random variables, each taking finitely many values, such that, $X_n \to X$ a.s.[P] and $E|X_n - X| \to 0$. [Use Exercise 4.4.4.]

EXERCISE **4.6.6**. [A simple generalization of what was seen in Remark 4.4.6.] Let $I \subset \mathbb{R}$ be a non-degenerate interval and $\{Z_{n,x}, n \geq 1\}$, for each $x \in I$, be a sequence of real random variabes with values in I, such that $E(Z_{n,x}) = x$, for all $n \geq 1$. Show that if $V(Z_{n,x}) \to 0$, as $n \to \infty$, for each $x \in I$, then $E(f(Z_{n,x})) \to f(x)$ pointwise for any bounded continuous function f on I. Further, show that the convergence is uniform on any subinterval of I, on which $V(Z_{n,x}) \to 0$ uniformly in x and f is uniformly continuous.

EXERCISE **4.6.7**. Let X be a real random variable on some probability space.
(a) Show that if G is any probability distribution function on $\mathbb{R}$, then $G(z - X)$, for each $z \in \mathbb{R}$, is a real radom variable taking values in $[0,1]$.
(b) Show that $F(z) = E(G(z - X))$, $z \in \mathbb{R}$ defines a probability distribution function on $\mathbb{R}$.
(c) Show that if G is continuous everywhere, then so is F.
(d) Assuming now that X is strictly positive, prove analogues of (a), (b) and (c) with $G(zX)$ and $F(z) = E(G(zX))$.

EXERCISE **4.6.8**. (a) Let h be a continuous, strictly increasing function on $[0,\infty)$ onto $[0,\infty)$ and let g be its inverse. Show that, if X is any non-negative real random variable, then $\sum_{k\geq 1} P(X > g(k)) \leq E(h(X)) \leq \sum_{k\geq 0} P(X > g(k))$.
(b) Conclude that a random variable X has finite jth moment if and only if $\sum_k P(|X| > k^{1/j})$ converges.

EXERCISE **4.6.9**. Let $I \subset \mathbb{R}$ be an open interval and let X be an absolutely continuous real random variable with density f that vanishes outside of I. Suppose g is a real-valued function on I, which is continuously differentiable everywhere on I with g' vanishing nowhere on I.
(a) Show that $J = g(I)$ is an open interval and $g: I \to J$ is strictly monotone.
(b) Show that the random variable $Y = g(X)$ has a density $\widetilde{f}$ given by
$$\widetilde{f}(y) = \begin{cases} f(g^{-1}y)/|g'(g^{-1}(y))|, & \text{if } y \in J \\ 0, & \text{if } y \notin J \end{cases}$$

EXERCISE **4.6.10**. (a) Let $U \subset \mathbb{R}^k$ be an open subset and let $\mathbf{X}$ be a k-dimensional random vector with joint density f that vanishes outside U. Suppose $\mathbf{g} = (g_1, \ldots, g_k): U \to \mathbb{R}^k$ be a continuously differentiable map, that is, for all $1 \leq i, j \leq k$, the partial derivatives $\frac{\partial g_i}{\partial x_j}$ exist and are continuous at every $\mathbf{x} = (x_1, \ldots, x_k) \in U$. Suppose further that $\mathbf{g}$ is one-to-one and that its Jacobian matrix $J(\mathbf{x}) = ((\frac{\partial g_i}{\partial x_j}))$ has a non-vanishing determinant $D(\mathbf{x})$ at all $\mathbf{x} \in \mathbf{U}$. By the inverse mapping theorem from analysis, $V = \mathbf{g}(U)$ is open. Show that the random vector $\mathbf{Y} = \mathbf{g}(\mathbf{X})$ has a joint density given by
$$\widetilde{f}(\mathbf{y}) = \begin{cases} f(\mathbf{g}^{-1}(\mathbf{y}))/|D(\mathbf{g}^{-1}(\mathbf{y}))|, & \text{if } \mathbf{y} \in V \\ 0, & \text{if } \mathbf{y} \notin V \end{cases}$$

EXERCISE 4.6.11. Let $\mathbf{X}$ be a k-dimensional random vector with $\mathbf{X} \sim N_k(\boldsymbol{\mu}, \boldsymbol{\Sigma})$, where $\boldsymbol{\mu} \in \mathbb{R}^k$ and $\boldsymbol{\Sigma} = ((\sigma_{ij}))$ is a symmetric positive definite matrix.
(a) Show that if there exists $1 \leq k_1 < k$, such that $\sigma_{ij} = 0$ for any $i \leq k_1$ and $j > k_1$, then $\mathbf{Y} = (X_1, \ldots, X_{k_1})'$ and $\mathbf{Z} = (X_{k_1+1}, \ldots, X_k)'$ are independent random vectors.
(b) Let $C_1 = \{i_1 < \cdots < i_m\}$ and $C_2 = \{j_1 < \cdots < j_l\}$ be two non-empty disjoint subsets of $\{1, \ldots, k\}$ such that $\sigma_{ij} = 0$ for all $i \in C_1$, $j \in C_2$. Show that $\mathbf{Y} = (X_{i_1}, \ldots, X_{i_m})'$ and $\mathbf{Z} = (X_{j_1}, \ldots, X_{j_l})'$ are independent random vectors.
(c) Generalize (a) to the case when $\boldsymbol{\Sigma}$ is a block diagonal matrix with m blocks.

EXERCISE 4.6.12. Let X be a discrete real random variable taking values in a countably infinite set $D = \{x_1, x_2, \ldots\} \subset \mathbb{R}$, with p.m.f. $p_X(x_n) = 2^{-n}$, $n \geq 1$. Show that, given any $\alpha \in [0, 1]$, there is a borel set A, such that, $P(X \in A) = \alpha$.

EXERCISE 4.6.13. (a) We saw that for any random variable X and any $\alpha > 0$, finiteness of $E(|X|^\alpha)$ implies $n^\alpha P(|X| > n) \to 0$ as $n \to \infty$. Give a counter-example to show that the converse does not hold in general.
(b) Show that if X is a random variable such that $n^\alpha P(|X| > n) \to 0$ as $n \to \infty$, for some $\alpha > 0$, then $E(|X|^\beta) < \infty$ for all $0 < \beta < \alpha$.
(c) Give a counter-example to show that the converse of (b) is not true.

EXERCISE 4.6.14. Consider $\Omega = [0,1] \times [0,1]$, equipped with its Borel σ-field $\mathcal{A} = \mathcal{B}([0,1] \times [0,1])$ and let P be the 2-dimensional Lebesgue measure λ^2 restricted to $\mathcal{A}$.
(a) Let $A \subset [0,1]$ be a non-Lebesgue measurable set with $\lambda^*(A) = 1$ and $\lambda_*(A) = 0$. Denote $\widehat{\mathcal{A}}$ to be the smallest σ-field on Ω, containing all the borel subsets and also the four sets $A \times A$, $A \times A^c$, $A^c \times A$ and $A^c \times A^c$. Use Exercise 2.13.26 to show that given any four non-negative real numbers adding to 1, there is an extension of P to a probability $\tilde{P}$ on $\widehat{\mathcal{A}}$ that assigns these prescribed values to the above four sets.
Recall (Definition 4.3.2) that random variables X and Y on a probability space are said to be independent if $P(X \in B_1, Y \in B_2) = P(X \in B_1)P(Y \in B_2)$, for any pair of borel subsets B_1, B_2 of $\mathbb{R}$.
H. Steinhauss proposed a slightly different definition, according to which two random variables X and Y on a probability space are to be called independent if $P(X \in B_1, Y \in B_2) = P(X \in B_1)P(Y \in B_2)$, for any two subsets B_1, B_2 of $\mathbb{R}$, not necessarily borel, as long as the sets $\{X \in B_1\}$ and $\{Y \in B_2\}$ belong to the underlying σ-field.
(b) Show that if two random variables are independent in the sense of Steinhauss, then they are independent according to our definition.
(c) Use the probability space $(\Omega, \widehat{\mathcal{A}}, \tilde{P})$ of part (a), to show that the converse of the statement in (b) does not hold.

Chapter 5

Product Spaces

In this chapter, we are going to focus on a concept for which the best motivation comes from the relationship between areas (or volumes) and length. From school geometry, we know that area of a rectangle (resp. volume of a rectangular parallelepiped) equals the product of lengths. Now that we have Lebesgue measure on $\mathcal{B}(\mathbb{R}^1)$ as an extension of length and, at the same time, Lebesgue measure on $\mathcal{B}(\mathbb{R}^k)$, for any k, as extension of k-dimensional volume, a natural question to ask would be: is there a connection?

Our construction of Lebesgue measure on $\mathcal{B}(\mathbb{R}^k)$, in Section 2.9, was self-contained and had no connection with one dimensional Lebesgue measure. We are now going to show that k-dimensional Lebesgue measure is indeed the "k-fold product", in a sense to be made precise, of Lebesgue measure on $\mathbb{R}$. Here is a quick preview of how it is done. For simplicity, let us consider the case $k = 2$. Sets of the form $B_1 \times B_2$, where B_1 and B_2 are one dimensional Borel sets, are called *measurable rectangles* or simply *rectangles*. While every such rectangle belongs to $\mathcal{B}(\mathbb{R}^2)$ (see Exercise 2.9.3), the class of rectangles does not form a σ-field and therefore does not give us all two dimensional Borel sets. Consider, for example, the set $D = \{(x_1, x_2) : x_1^2 + x_2^2 \leq 1\}$, which is a closed set and hence a Borel subset of $\mathbb{R}^2$, but D is clearly (?) not a rectangle. However, it turns out that the class $\mathcal{S}$ consisting of all measurable rectangles is a semifield on $\mathbb{R}^2$ and that $\sigma(\mathcal{S}) = \mathcal{B}(\mathbb{R}^2)$. Further, the set function μ defined on $\mathcal{S}$ by $\mu(B_1 \times B_2) = \lambda(B_1)\lambda(B_2)$, turns out to be a σ-finite measure on $\mathcal{S}$ and hence has a unique extension to $\mathcal{B}(\mathbb{R}^2)$. Finally, using an uniqueness argument we will be able to show that $\mu = \lambda^2$.

The scheme described above for $\mathcal{B}(\mathbb{R}^2)$, luckily, works in a completely abstract set-up and leads to a general notion of product σ-fields and product measures. In what follows we will only discuss two-fold products, but it can be easily seen that the same idea works for defining products of any finite number of σ-fields and measures.

5.1 Product of Measure Spaces (2-fold)

First, we describe the concept of 'product' of two measurable spaces. To be specific, given two measurable spaces $(\Omega_1, \mathcal{A}_1)$ and $(\Omega_2, \mathcal{A}_2)$, we would construct a new measurable space $(\Omega, \mathcal{A})$, which would represent, in an appropriate way, the 'product' of these two measurable spaces. The natural candidate for the set Ω is the cartesian product $\Omega = \Omega_1 \times \Omega_2$. To define the product σ-field on Ω, we start with the class $\mathcal{S}$ of subsets of Ω defined as

$$\mathcal{S} = \{A_1 \times A_2 : A_1 \in \mathcal{A}_1, A_2 \in \mathcal{A}_2\}.$$

Sets in $\mathcal{S}$ are called *measurable rectangles*. It is an easy exercise to verify that $\mathcal{S}$ is a semifield on Ω.

EXERCISE 5.1.1. *When is this class $\mathcal{S}$ a σ-field on Ω? When is $\mathcal{S}$ a field on Ω?*

DEFINITION 5.1.2. *The σ-field $\mathcal{A}$ on Ω, generated by the semifield $\mathcal{S}$, is called the product σ-field on Ω and is denoted by $\mathcal{A}_1 \otimes \mathcal{A}_2$.*
The measurable space $(\Omega_1 \times \Omega_2, \mathcal{A}_1 \otimes \mathcal{A}_2)$ is called the product measurable space.

In the special case when $(\Omega_1, \mathcal{A}_1) = (\Omega_2, \mathcal{A}_2) = (E, \mathcal{E})$, the product measurable space $(E \times E, \mathcal{E} \otimes \mathcal{E})$ will be denoted by $(E^2, \mathcal{E}^2)$.

EXERCISE 5.1.3. *Show that if $\mathcal{C}_i$, for $i = 1, 2$, is a class of subsets of Ω_i, that generates the σ-field $\mathcal{A}_i$, then the class $\mathcal{C} = \{A_1 \times A_2 : A_i \in \mathcal{C}_i, i = 1, 2\}$ is a generating class for $\mathcal{A}_1 \otimes \mathcal{A}_2$.*

EXERCISE 5.1.4. *Recalling the notation $\mathcal{B} = \mathcal{B}(\mathbb{R})$, show that $\mathcal{B} \otimes \mathcal{B} = \mathcal{B}(\mathbb{R}^2)$. Thus, in the notation introduced just after Definition 5.1.2, $\mathcal{B}^2 = \mathcal{B}(\mathbb{R}^2)$.*

EXERCISE 5.1.5. *(a) Let $(\Omega_1, \mathcal{A}_1), (\Omega_2, \mathcal{A}_2)$ be measurable spaces. Denoting π_1 and π_2 to be usual the coordinate projection maps on $\Omega_1 \times \Omega_2$ onto Ω_1 and Ω_2 respectively, show that, for each i, the map $\pi_i : (\Omega_1 \times \Omega_2, \mathcal{A}_1 \otimes \mathcal{A}_2) \to (\Omega_i, \mathcal{A}_i)$ is measurable. Show also that $\mathcal{A}_1 \otimes \mathcal{A}_2$ is the smallest σ-field on $\Omega_1 \times \Omega_2$ that makes both π_1 and π_2 measurable.*
(b) Show that if $(E, \mathcal{E})$ is any measurable space, then $f : E \to \Omega_1 \times \Omega_2$ is $(\mathcal{E}, \mathcal{A}_1 \otimes \mathcal{A}_2)$-measurable iff $f \circ \pi_i$ is $(\mathcal{E}, \mathcal{A}_i)$-measurable, for each $i = 1, 2$.

Having thus defined the product σ-field $\mathcal{A}_1 \otimes \mathcal{A}_2$ on the cartesian product $\Omega_1 \times \Omega_2$, we discuss an important property of sets in the product σ-field. This property will be particularly useful in our next step, that is, when we proceed to construct 'product measure'.

DEFINITION 5.1.6. *Fix any set $A \subset \Omega_1 \times \Omega_2$. For each $\omega_1 \in \Omega_1$, the section of A at ω_1 is defined to be the set $A(\omega_1) = \{\omega_2 \in \Omega_2 : (\omega_1, \omega_2) \in A\}$. This is also called the ω_1-section of A, or sometimes called the vertical section of A at ω_1.*
Similarly, for each $\omega_2 \in \Omega_2$, the section of A at ω_2 (also called the ω_2-section of A or, sometimes called the horizontal section of A at ω_2) is defined to be the set $A^*(\omega_2) = \{\omega_1 \in \Omega_1 : (\omega_1, \omega_2) \in A\}$.

5.1. Product of Measure Spaces (2-fold)

It is clear from the definition that for any $A \subset \Omega_1 \times \Omega_2$, $A(\omega_1) \subset \Omega_2$ for each $\omega_1 \in \Omega_1$ and $A^*(\omega_2) \subset \Omega_1$ for each $\omega_2 \in \Omega_2$. It is also quite easy to see that usual set operations commute with the operations of taking sections, vertical as well as horizontal. Indeed, for any $A \subset \Omega_1 \times \Omega_2$ and any $\omega_1 \in \Omega_1$, one can easily verify that $(A^c)(\omega_1) = (A(\omega_1))^c$. Also, if $\{A_\alpha\}$ is any family of subsets of $\Omega_1 \times \Omega_2$, then for any $\omega_1 \in \Omega_1$, one has $\left(\bigcup_\alpha A_\alpha\right)(\omega_1) = \bigcup_\alpha \left(A_\alpha(\omega_1)\right)$ and $\left(\bigcap_\alpha A_\alpha\right)(\omega_1) = \bigcap_\alpha \left(A_\alpha(\omega_1)\right)$. Similar results hold for horizontal sections as well.

PROPOSITION 5.1.7. If $A \in \mathcal{A}_1 \otimes \mathcal{A}_2$, then $A(\omega_1) \in \mathcal{A}_2$, for each $\omega_1 \in \Omega_1$ and $A^*(\omega_2) \in \mathcal{A}_1$, for each $\omega_2 \in \Omega_2$.

Proof. We just prove the assertion for the ω_1-sections. The proof for ω_2-sections requires the same arguments and is, therefore, left as an exercise.
If A is a measurable rectangle, say, $A = A_1 \times A_2$, where $A_i \in \mathcal{A}_i$, $i = 1, 2$, then it is easy to see from the definition of ω_1-sections, that $A(\omega_1)$ equals A_2 or $\emptyset$, according as $\omega_1 \in A_1$ or $\omega_1 \notin A_1$. Thus, $A(\omega_1) \in \mathcal{A}_2$ always. We complete the proof now using good sets principle. Let

$$\mathcal{G} = \{A \in \mathcal{A} : A(\omega_1) \in \mathcal{A}_2, \text{ for each } \omega_1 \in \Omega_1\}.$$

The above observation shows that $\mathcal{S} \subset \mathcal{G}$. Using the fact that set operations commute with the operation of taking ω_1-sections, it is easy to verify that $\mathcal{G}$ is a σ-field. Since the class $\mathcal{S}$ of measurable rectangles generate $\mathcal{A}_1 \otimes \mathcal{A}_2$, the proof is complete. □

We now proceed towards defining 'product measures' on $\mathcal{A}_1 \otimes \mathcal{A}_2$. Suppose we are given σ-finite measures μ_1 on $\mathcal{A}_1$ and μ_2 on $\mathcal{A}_2$. We want to define a σ-finite measure μ on the product σ-field $\mathcal{A}_1 \otimes \mathcal{A}_2$, which could legitimately be called the 'product' of the measures μ_1 and μ_2. A natural requirement for such a measure μ has to be that on measurable rectangles, it must actually be a product, that is, we must have

$$\mu(A_1 \times A_2) = \mu_1(A_1)\mu_2(A_2), \text{ for any measurable rectangle } A_1 \times A_2 \quad (5.1.1)$$

Since the class $\mathcal{S}$ of measurable rectangles is a semifield that generates the product σ-field, one possible option would be to show that the set fucntion μ defined on $\mathcal{S}$ by (5.1.1), is a σ-finite measure on $\mathcal{S}$. Caratheodory Extension Theorem would then give us a unique extension of μ to a measure on $\mathcal{A}_1 \otimes \mathcal{A}_2$.

However, we are going to take a different route. We will explicitly define a set function μ directly on the product σ-field and show that it is a σ-finite measure on $\mathcal{A}_1 \otimes \mathcal{A}_2$ and that it also satisfies (5.1.1). We do this in the following proposition. Let us first recall, from Proposition 5.1.7, that for any $A \in \mathcal{A}_1 \otimes \mathcal{A}_2$, the sections $A(\omega_1) \in \mathcal{A}_2$, for all $\omega_1 \in \Omega_1$, so that $\omega_1 \mapsto \mu_2(A(\omega_1))$ is a well-defined $[0, \infty]$-valued function on Ω_1.

PROPOSITION 5.1.8. Let $(\Omega_i, \mathcal{A}_i, \mu_i)$, $i = 1, 2$, be σ-finite measure spaces.
(a) For each $A \in \mathcal{A}_1 \otimes \mathcal{A}_2$, the map $\omega_1 \mapsto \mu_2(A(\omega_1))$ defines a non-negative extended real valued $\mathcal{A}_1$-measurable function on Ω_1.

(b) The set function μ defined on $\mathcal{A}_1 \otimes \mathcal{A}_2$ by

$$\mu(A) = \int \mu_2(A(\omega_1)) d\mu_1(\omega_1) \tag{5.1.2}$$

is a σ-finite measure on $\mathcal{A}_1 \otimes \mathcal{A}_2$. Further, the measure μ is the unique measure on $\mathcal{A}_1 \otimes \mathcal{A}_2$ satisfying (5.1.1).

Proof. (a) We first prove the result assuming that μ_2 is a finite measure. The proof of Proposition 5.1.7 shows that, if $A = A_1 \times A_2 \in \mathcal{S}$, then $\mu_2(A(\omega_1))$ equals $\mu_2(A_2)$ or 0, according as, $\omega_1 \in A_1$ or $\omega_1 \notin A_1$. Thus, for $A = A_1 \times A_2 \in \mathcal{S}$, we have $\mu_2(A(\omega_1)) = \mu_2(A_2) I_{A_1}(\omega_1)$, which is clearly an $\mathcal{A}_1$-measurable function of ω_1. If A is a finite disjoint union of $\mathcal{S}$-sets, say, $A = \bigcup_{i=1}^{n} (A_{1,i} \times A_{2,i})$, then using the commutativity between set operations and the operation of taking sections, one can easily see that $A(\omega_1)$ will be the disjoint union of the ω_1-sections of $A_{1,i} \times A_{2,i}$, $i = 1, \cdots, n$, for each $\omega_1 \in \Omega_1$. By virtue of additivity of μ_2, this will imply that $\mu_2(A(\omega_1)) = \sum_{i=1}^{n} \mu_2((A_{1,i} \times A_{2,i})(\omega_1)) = \sum_{i=1}^{n} \mu_2(A_{2,i}) I_{A_{1,i}}(\omega_1)$, which is a simple, non-negative real-valued $\mathcal{A}_1$-measurable function of ω_1. Thus, the required assertion is proved for all sets A belonging to the field consisting of finite disjoint unions of $\mathcal{S}$-sets. We now consider the class

$$\mathcal{M} = \{A \in \mathcal{A}_1 \otimes \mathcal{A}_2 : \omega_1 \mapsto \mu_2(A(\omega_1)) \text{ is } \mathcal{A}_1\text{-measurable}\}$$

and show that it is a monotone class. Once again, from commutativity between set operations and taking sections, it follows that, for any $A_n \subset \Omega_1 \times \Omega_2$, $n \geq 1$ with $A_n \uparrow A$ (resp., $A_n \downarrow A$), one has $A_n(\omega_1) \uparrow A(\omega_1)$ (resp., $A_n(\omega_1) \downarrow A(\omega_1)$), for each $\omega_1 \in \Omega_1$. Using continuity of μ_2 from below and from above and the fact that limits of measurable functions are measurable, one can easily see that $\mathcal{M}$ is a monotone class. An appeal to Monotone Class Theorem proves the assertion in case μ_2 is a finite measure. The reader may note that it is only in using continuity from above, where the assumption that μ_2 is a finite measure came into play.
To prove the result now for σ-finite μ_2, we use the standard technique of writing $\mu_2 = \sum_n \mu_{2,n}$, where $\mu_{2,n}$, for each n, is a finite measure on $\mathcal{A}_2$. Applying what was proved above, we have, for each n, the map $\omega_1 \mapsto \mu_{2,n}(A(\omega_1))$ is $\mathcal{A}_1$-measurable, for any $A \in \mathcal{A}_1 \otimes \mathcal{A}_2$. The proof of (a) is now complete by noting that $\mu_2(A(\omega_1)) = \sum_n \mu_{2,n}(A(\omega_1))$, for each $\omega_1 \in \Omega_1$ and any $A \in \mathcal{A}_1 \otimes \mathcal{A}_2$.
(b) Clearly, μ is a $[0, \infty]$-valued set function on $\mathcal{A}_1 \otimes \mathcal{A}_2$ with $\mu(\emptyset) = 0$. To prove countable additivity of μ, observe once again that, if $\{A_n\}$ is a sequence of disjoint sets in $\mathcal{A}_1 \otimes \mathcal{A}_2$, then $(\bigcup_n A_n)(\omega_1)$ equals the disjoint union $\bigcup_n A_n(\omega_1)$ and hence $\mu_2((\bigcup_n A_n)(\omega_1)) = \sum_n \mu_2(A_n(\omega_1))$, for each $\omega_1 \in \Omega_1$. That $\mu(\bigcup_n A_n) = \sum_n \mu(A_n)$ is now an immediate consequence of the definition of μ and Theorem 3.3.6. This proves that μ is a measure on $\mathcal{A}_1 \otimes \mathcal{A}_2$.
Next, if $A = A_1 \times A_2 \in \mathcal{S}$, then $\mu_2(A(\omega_1)) = \mu_2(A_2) \mathbf{I}_{A_1}(\omega_1)$, as seen in the proof of (a). It follows easily from the definition (5.1.2) that μ satisfies (5.1.1).
Finally, σ-finiteness of μ_1 and μ_2 implies that there exist sets $A_{1,m} \in \mathcal{A}_1$, $m \geq 1$ and $A_{2,n} \in \mathcal{A}_2$, $n \geq 1$ with $\bigcup_m A_{1,m} = \Omega_1$, $\bigcup_n A_{2,n} = \Omega_2$ such that $\mu_1(A_{1,m}) < \infty$, for all m, and $\mu_2(A_{2,n}) < \infty$, for all n. Clearly, $\{A_{1,m} \times A_{2,n}, m, n \geq 1\}$ is a countable

5.1. Product of Measure Spaces (2-fold)

familty of $\mathcal{S}$-sets with $\bigcup_{m,n}(A_{1,m} \times A_{2,n}) = \Omega_1 \times \Omega_2$ such that $\mu(A_{1,m} \times A_{2,n}) < \infty$ for each m,n, by (5.1.1). This shows that μ is a σ-finite measure – indeed σ-finite on the semifield $\mathcal{S}$, so that, by the uniqueness part of Caratheodory Extension Theorem, μ must be the unique measure on $\mathcal{A}_1 \otimes \mathcal{A}_2$ satisfying (5.1.1). The proof is now complete. □

DEFINITION 5.1.9. Given σ-finite measures μ_1 and μ_2 on $\mathcal{A}_1$ and $\mathcal{A}_2$ respectively, the σ-finite measure μ defined on the product σ-field $\mathcal{A}_1 \otimes \mathcal{A}_2$ by (5.1.2) is called the *product measure* and is denoted $\mu_1 \otimes \mu_2$.

With these notations, the following theorem summarizes what we have proved.

THEOREM 5.1.10. Given two σ-finite measure spaces $(\Omega_1, \mathcal{A}_1, \mu_1)$ and $(\Omega_2, \mathcal{A}_2, \mu_2)$, there is a unique (necessarily σ-finite) measure $\mu_1 \otimes \mu_2$ on $\mathcal{A}_1 \otimes \mathcal{A}_2$ such that

$$\mu_1 \otimes \mu_2(A_1 \times A_2) = \mu_1(A_1)\mu_2(A_2), \text{ for all } A_1 \in \mathcal{A}_1, A_2 \in \mathcal{A}_2$$

The σ-finite measure space $(\Omega_1 \times \Omega_2, \mathcal{A}_1 \otimes \mathcal{A}_2, \mu_1 \otimes \mu_2)$ is often referred to as the product of the two σ-finite measure spaces $(\Omega_1, \mathcal{A}_1, \mu_1)$ and $(\Omega_2, \mathcal{A}_2, \mu_2)$. One even writes

$$(\Omega_1 \times \Omega_2, \mathcal{A}_1 \otimes \mathcal{A}_2, \mu_1 \otimes \mu_2) = (\Omega_1, \mathcal{A}_1, \mu_1) \otimes (\Omega_2, \mathcal{A}_2, \mu_2)$$

If both μ_1 and μ_2 are finite (resp. probability) measures, then so is $\mu_1 \otimes \mu_2$.

In the definition of the product measure, as given in (5.1.2), we used vertical sections. However, from the fact that the product measure μ satisfies (5.1.1), it should be clear that it would not make any difference if we had used horizontal sections instead. Indeed, one would get the same product measure. For the sake of record, this is left as an exercise below.

EXERCISE 5.1.11. Show that for every $A \in \mathcal{A}_1 \otimes \mathcal{A}_2$, the map $\omega_2 \mapsto \mu_1(A^*(\omega_2))$ is an $\mathcal{A}_2$-measurable map on Ω_2 and that the product measure $\mu_1 \otimes \mu_2$, as defined in 5.1.9, satisfies $\mu_1 \otimes \mu_2(A) = \int \mu_1(A^*(\omega_2))d\mu_2(\omega_2)$.

A reader familiar with multivariate Riemann integrals would recall that most of the times, computing multivariate integrals boils down to computing iterated univariate Riemann integrals. The next result, which is an abstract analogue of this, is extremely useful. It shows that, under certain conditions, integration of a measurable function on a product space with respect to a product measure can be reduced to iterated single variable integrations.

THEOREM 5.1.12. (Fubini's Theorem) Let $(\Omega_1, \mathcal{A}_1, \mu_1)$ and $(\Omega_2, \mathcal{A}_2, \mu_2)$ be σ finite measure spaces and denote $(\Omega, \mathcal{A}, \mu) = (\Omega_1 \times \Omega_2, \mathcal{A}_1 \otimes \mathcal{A}_2, \mu_1 \otimes \mu_2)$.
(a) Let f be any $\mathcal{A}$-measurable function on Ω. Then, for each $\omega_1 \in \Omega_1$, the map $\omega_2 \mapsto f(\omega_1, \omega_2)$ on Ω_2 is $\mathcal{A}_2$-measurable and, for each $\omega_2 \in \Omega_2$, the map $\omega_1 \mapsto f(\omega_1, \omega_2)$ on Ω_1 is $\mathcal{A}_1$-measurable.
(b) If f is any non-negative $\mathcal{A}$-measurable function on Ω, then the maps $\omega_1 \mapsto \int_{\Omega_2} f(\omega_1, \omega_2)d\mu_2(\omega_2)$ and $\omega_2 \mapsto \int_{\Omega_1} f(\omega_1, \omega_2)d\mu_1(\omega_1)$ are non-negative measurable functions with respect to the σ-fields $\mathcal{A}_1$ and $\mathcal{A}_2$ respectively. Further,

$$\int_\Omega f d\mu = \int_{\Omega_1}\left(\int_{\Omega_2} f(\omega_1, \omega_2)d\mu_2(\omega_2)\right)d\mu_1(\omega_1) = \int_{\Omega_2}\left(\int_{\Omega_1} f(\omega_1, \omega_2)d\mu_1(\omega_1)\right)d\mu_2(\omega_2).$$

(c) If f is a μ-integrable function on Ω, then the integral $\int_{\Omega_2} f(\omega_1,\omega_2)d\mu_2(\omega_2)$ is well-defined and finite, for a.e ω_1 $[\mu_1]$, and the integral $\int_{\Omega_1} f(\omega_1,\omega_2)d\mu_1(\omega_1)$ is well-defined and finite, for a.e ω_2 $[\mu_2]$. Further, the maps $\omega_1 \mapsto \int_{\Omega_2} f(\omega_1,\omega_2)d\mu_2(\omega_2)$ and $\omega_2 \mapsto \int_{\Omega_1} f(\omega_1,\omega_2)d\mu_1(\omega_1)$ are measurable with respect to the σ-fields $\mathcal{A}_1$ and $\mathcal{A}_2$ respectively. Finally,

$$\int_\Omega f d\mu = \int_{\Omega_1}\left(\int_{\Omega_2} f(\omega_1,\omega_2)d\mu_2(\omega_2)\right)d\mu_1(\omega_1) = \int_{\Omega_2}\left(\int_{\Omega_1} f(\omega_1,\omega_2)d\mu_1(\omega_1)\right)d\mu_2(\omega_2).$$

REMARK 5.1.13. In part (c), we are talking about functions being measurable, when the functions are only known to be well defined outside some null sets. As noted in Remark 3.2.17, we come across such situations a number of times and we adapt the same convention as mentioned there. We define the function to be zero (or any constant of your choice) on the null set, where it is not defined, and mean measurability of this function, now defined everywhere, by measurability of the original a.e. well-defined function.

REMARK 5.1.14. In part (c), the hypothesis demands that f be μ-integrable. In case one is wondering how to check it before computing the integral, recall that μ-integrability of f simply means that the non-negative function $|f|$ has a finite integral and that can be checked by applying part (b). For the record, we note here that part (b) sometimes goes by the name of *Tonelli's Theorem*.

REMARK 5.1.15. Fubini's Theorem not only provides conditions under which an integral with respect to product measure can be computed through iterated single variable integrals, but also shows that, in such cases, the iteration can be done in any order. This is how the theorem is often put to use and referred to as 'change of order of integration'.

Proof. (a) We only prove $\mathcal{A}_2$-measurability of the map $\omega_2 \mapsto f(\omega_1,\omega_2)$, for each $\omega_1 \in \Omega_1$. Proof of $\mathcal{A}_1$-measurability of the map $\omega_1 \mapsto f(\omega_1,\omega_2)$, for each $\omega_2 \in \Omega_2$, is similar. We first argue that the result is true when $f = \mathbf{I}_A$, for $A \in \mathcal{A}$. If $A \in \mathcal{S}$, say, $A = A_1 \times A_2$, then the map $\omega_2 \mapsto f(\omega_1,\omega_2)$ is either the function $I_{A_2}(\omega_2)$ or the zero-function, depending on whether $\omega_1 \in A_1$ or $\omega_1 \notin A_1$. Thus, if $A \in \mathcal{S}$, then the map $\omega_2 \mapsto \mathbf{I}_A(\omega_1,\omega_2)$ is $\mathcal{A}_2$-measurable, for each ω_1. In case A is a finite disjoint union of $\mathcal{S}$-sets, say $A = \bigcup_i S_i$ then $\mathbf{I}_A = \sum_i \mathbf{I}_{S_i}$, so that, the map $\omega_2 \mapsto \mathbf{I}_A(\omega_1,\omega_2)$, for each $\omega_1 \in \Omega_1$, equals the (finite) sum of the $\mathcal{A}_2$-measurable maps $\omega_2 \mapsto \mathbf{I}_{S_i}(\omega_1,\omega_2)$ and is hence $\mathcal{A}_2$-measurable. Note that, the class of all finite disjoint unions of $\mathcal{S}$-sets form a field that generates $\mathcal{A}$. It is easy to see that the class of sets $A \in \mathcal{A}$, for which $\omega_2 \mapsto \mathbf{I}_A(\omega_1,\omega_2)$ is $\mathcal{A}_2$-measurable, for each $\omega_1 \in \Omega_1$, forms a monotone class. Using monotone class theorem, one concludes that, for all $A \in \mathcal{A}$, the map $\omega_2 \mapsto \mathbf{I}_A(\omega_1,\omega_2)$ is $\mathcal{A}_2$-measurable, for each $\omega_1 \in \Omega_1$. Having thus proved the result for $f = \mathbf{I}_A$, $A \in \mathcal{A}$, it is immediate that the result is true for any simple $\mathcal{A}$-measurable f. Since any $\mathcal{A}$-measurable function f on Ω is a pointwise limit of a sequence of simple $\mathcal{A}$-measurable functions, one can now

5.1. Product of Measure Spaces (2-fold)

deduce the required result for any $\mathcal{A}$-measurable f, using the fact that limits of $\mathcal{A}_2$-measurable maps have to be $\mathcal{A}_2$-measurable.

(b) If $f = I_A$ for any $A \in \mathcal{A}$, then the result follows from Proposition 5.1.1, Exercise 5.1.11 and the definition (5.1.2) of μ. Rest of the proof goes through by standard argument. From indicators, one can get it easily for non-negative real valued simple functions and then one uses MCT to deduce the result for any non-negative measurable function.

(c) Recalling the decomposition $f = f^+ - f^-$ and using the first assertion of part (b) on the non-negative measurable functions f^+ and f^-, one gets that the maps $\omega_1 \mapsto \int f^+(\omega_1, \omega_2) d\mu_2(\omega_2)$ and $\omega_1 \mapsto \int f^-(\omega_1, \omega_2) d\mu_2(\omega_2)$ are both non-negative $\mathcal{A}_1$-measurable functions on Ω_1. The hypothesis that f is μ-integrable implies that f^+ and f^- are both μ-integrable. Using now the second assertion of part (b), this would mean that the integrals $\int_{\Omega_1} \left(\int_{\Omega_2} f^+(\omega_1, \omega_2) d\mu_2(\omega_2) \right) d\mu_1(\omega_1)$ and $\int_{\Omega_1} \left(\int_{\Omega_2} f^-(\omega_1, \omega_2) d\mu_2(\omega_2) \right) d\mu_1(\omega_1)$ are both finite. Since finiteness of μ_1-integral of a $\mathcal{A}_1$-measurable function on Ω_1 implies that the function must be finite a.e.$[\mu_1]$, we may conclude from the above that both $\int f^+(\omega_1, \omega_2) d\mu_2(\omega_2)$ and $\int f^-(\omega_1, \omega_2) d\mu_2(\omega_2)$ are finite, for $[\mu_1]$-a.e. $\omega_1 \in \Omega_1$. It then follows that $\int f(\omega_1, \omega_2) d\mu_2(\omega_2)$ is well-defined (in fact, finite), for $[\mu_1]$-a.e. $\omega_1 \in \Omega_1$ and defines a measurable function on Ω_1 (see Remark 5.1.13). Similar argument can be applied to prove that the map $\omega_1 \mapsto \int f(\omega_1, \omega_2) d\mu_1(\omega_1)$ is well-defined (in fact, finite), for μ_2-a.e. $\omega_2 \in \Omega_2$, and defines a $\mathcal{A}_2$-measurable function on Ω_2. The final equality is a consequence of linearity of integrals. □

EXERCISE 5.1.16. *Suppose that f_1 and f_2 are measurable functions on $(\Omega_1, \mathcal{A}_1)$ and $(\Omega_2, \mathcal{A}_2)$ respectively. Show that the function f defined on $\Omega_1 \times \Omega_2$ by $f(\omega_1, \omega_2) = f_1(\omega_1) f_2(\omega_2)$ is measurable with respect to the product σ-field. Further, show that $\int f d\mu = \int f_1 d\mu_1 \int f_2 d\mu_2$, whenever all the three integrals exist, and, in particular, when f_1, f_2 are both non-negative or both integrable.*

EXERCISE 5.1.17. *(a) Let $(\Omega_1, \mathcal{A}_1)$ and $(\Omega_2, \mathcal{A}_2)$ be measurable spaces and let μ be a σ-finite measure on $\mathcal{A}_1 \otimes \mathcal{A}_2$. Suppose there exist measures μ_1 on $\mathcal{A}_1$ and μ_2 on $\mathcal{A}_2$, such that, for every pair of bounded non-negative measurable functions f_1 on Ω_1 and f_2 on Ω_2,*

$$\int_\Omega f_1(\omega_1) f_2(\omega_2) d\mu = \int_{\Omega_1} f_1(\omega_1) d\mu_1 \int_{\Omega_2} f_2(\omega_2) d\mu_2. \quad (5.1.3)$$

Show then that μ_1 and μ_2 are necessarily σ-finite and $\mu = \mu_1 \otimes \mu_2$.
(b) In the special case when $(\Omega_1, \mathcal{A}_1) = (\Omega_2, \mathcal{A}_2) = (\mathbb{R}, \mathcal{B})$, and μ is a Radon measure on $\mathcal{B}^2$, show that it is enough to have (5.1.3) hold for bounded continuous functions f_1 and f_2 on $\mathbb{R}$ or for just compactly supported continuous functions f_1 and f_2 on $\mathbb{R}$ in order for the conclusion as in part (a) to be true.

EXERCISE 5.1.18. *Let $(\Omega, \mathcal{A}, \mu)$ be a σ-finite measure space and let f be a non-negative measurable function on Ω. Denote $G = \{(\omega, t) \in \Omega \times [0, \infty) : t \leq f(\omega)\}$. Show that $G \in \mathcal{A} \otimes \mathcal{B}$ and $(\mu \otimes \lambda)(G) = \int f d\mu$. Identifying the left side of the* last equality as the 'area' of the region under the 'graph' of f, the result can be interpreted as saying that this area equals the integral $\int f d\mu$.

The next proposition is a very important and useful consequence of Fubini's Theorem. The main idea is essentially an extension of the previous Exercise. But because of the importance of the result, we put it here as a proposition.

PROPOSITION 5.1.19. *Let $(\Omega, \mathcal{A}, \mu)$ be a σ-finite measure space. Then, for any measurable function f on Ω and any $p \in [1, \infty)$,*

$$\int |f|^p d\mu = p \int_0^\infty t^{p-1} \mu(|f| > t) \, dt \qquad (5.1.4)$$

Proof. The reader can convince herself that unless f is finite a.e.$[\mu]$, both sides of (5.1.4) equals ∞. So, it is enough to prove it only for a real-valued measurable function f. Consider the product measure space $([0, \infty) \times \Omega, \mathcal{B}^+ \otimes \mathcal{A}, \lambda \otimes \mu)$, where $\mathcal{B}^+$ denotes the Borel σ-field on $[0, \infty)$ and λ the Lebesgue measure. As in Exercise 5.1.18, one can show that the set $G = \{(t, \omega) : 0 \le t < |f(\omega)|\}$ belongs to $\mathcal{B}^+ \otimes \mathcal{A}$ and so $g(t, \omega) = p\, t^{p-1} I_G(t, \omega)$ is a non-negative $\mathcal{B}^+ \otimes \mathcal{A}$-measurable function. Now, for any $\omega \in \Omega$, it is easy to see that $\int_0^\infty g(t, \omega) dt = |f(\omega)|^p$. That gives $\int |f|^p d\mu = \int_\Omega \int_0^\infty g(t, \omega) dt d\mu$. Using part (b) of Fubini's Thorem, one can interchange the order of integration on the right-hand-side and get (5.1.4). □

As a special case of the above proposition, we have the following result on random variables.

COROLLARY 5.1.20. *For any random variable X on a probability space $(\Omega, \mathcal{A}, P)$ and for any $p \in [1, \infty)$,*

$$E(|X|^p) = p \int_0^\infty t^{p-1} P(|X| > t) dt \qquad (5.1.5)$$

REMARK 5.1.21. Equation (5.1.4) above implies that, for a measurable function f on a σ-finite measure space $(\Omega, \mathcal{A}, \mu)$,

$$\int |f|^p d\mu < \infty \text{ if and only if } \int_0^\infty t^{p-1} \mu(|f| > t) \, dt < \infty \qquad (5.1.6)$$

In a way, this gives some sort of a precise quantitative estimate of how fast $\mu(|f| > t)$ has to decay as $t \to \infty$, so as to have $|f|^p$ integrable. Similarly, equation (5.1.5) gives a quantitative estimate of the rate at which the tail probability $P(|X| > t)$ has to decay as $t \to \infty$ in order for X to have a finite pth moment. The reader may compare this with the statement of Corollary 4.4.16.

EXERCISE 5.1.22. *Argue, without any further computations, that in (5.1.4) (resp., (5.1.5)), we can replace $\mu(|f| > t)$ by $\mu(|f| \ge t)$ (resp., $P(|X| > t)$ by $P(|X| \ge t)$) and the formula still remains valid.*

5.2 Finite Products of Measure Spaces

In the last section, the product of two σ-finite measure spaces was defined and studied in detail. In this section, we give a brief sketch of how one can analogously

5.2. Finite Products of Measure Spaces

define an n-fold product of σ-finite measure spaces, for any $n \geq 2$. All the details to be filled in are left for the reader as exercises.

For $i = 1, 2, \cdots, n$, let $(\Omega_i, \mathcal{A}_i, \mu_i)$, be σ-finite measure spaces. By a *measurable rectangle* is meant a subset A of the cartesian product $\overset{n}{\underset{1}{\times}} \Omega_i$ of the form $A = \overset{n}{\underset{1}{\times}} A_i$, where $A_i \in \mathcal{A}_i$, for $i = 1, \ldots, n$. Denoting $\mathcal{S}$ to be the class of all measurable rectangles, the σ-field on $\overset{n}{\underset{1}{\times}} \Omega_i$, generated by $\mathcal{S}$, is called the *finite product σ-field* and is denoted $\overset{n}{\underset{1}{\otimes}} \mathcal{A}_i$. The measurable space $\left(\overset{n}{\underset{1}{\times}} \Omega_i, \overset{n}{\underset{1}{\otimes}} \mathcal{A}_i \right)$, thus constructed, is called the n-fold *product measurable space*.

EXERCISE 5.2.1. (a) Show that the class $\mathcal{S}$ of all measurable rectangles is a semifield on $\overset{n}{\underset{1}{\times}} \Omega_i$.
(b) Show that if $\mathcal{C}_i$, for each $i = 1, \cdots, n$, is a class of subsets of Ω_i, that generates $\mathcal{A}_i$, then the class $\mathcal{C} = \{ \overset{n}{\underset{1}{\times}} C_i : C_i \in \mathcal{C}_i, i = 1, \ldots, n \}$ generates $\overset{n}{\underset{1}{\otimes}} \mathcal{A}_i$.

Finally, the product measure $\mu = \overset{n}{\underset{1}{\otimes}} \mu_i$ may be defined in an analogous way as in the case of two-fold products. One may either show that the set function μ defined on $\mathcal{S}$ by the formula $\mu\left(\overset{n}{\underset{1}{\times}} A_i \right) = \prod_{i=1}^{n} \mu_i(A_i)$ is a measure (necessarily σ-finite) on $\mathcal{S}$ and appeal to Caratheodory Extension Theorem to get a unique extension to the product σ-field $\overset{n}{\underset{1}{\otimes}} \mathcal{A}_i$. The other alternative, as we did in case of two-fold product, is to define μ directly on $\overset{n}{\underset{1}{\otimes}} \mathcal{A}_i$ by the formula

$$\mu(A) = \int_{\Omega_1} \cdots \int_{\Omega_n} I_A(\omega_1, \ldots, \omega_n) \, d\mu_n(\omega_n) \cdots d\mu_1(\omega_1), \text{ for } A \in \overset{n}{\underset{1}{\otimes}} \mathcal{A}_i. \quad (5.2.1)$$

EXERCISE 5.2.2. Show that μ given by the formula (5.2.1) is well-defined and defines the unique measure (necessarily σ-finite) on $\mathcal{A}$ such that

$$\mu\left(\overset{n}{\underset{1}{\times}} A_i \right) = \prod_{1}^{n} \mu_i(A_i), \quad \text{for } \overset{n}{\underset{1}{\times}} A_i \in \mathcal{S}.$$

The σ-finite measure μ, thus constructed, is called the n-fold *product measure*, denoted $\overset{n}{\underset{1}{\otimes}} \mu_i$ and the σ-finite measure space $\left(\overset{n}{\underset{1}{\times}} \Omega_i, \overset{n}{\underset{1}{\otimes}} \mathcal{A}_i, \overset{n}{\underset{1}{\otimes}} \mu_i \right)$ is called the n-fold *product measure space*.

EXERCISE 5.2.3. (a) State and prove an anologue of Fubini Theorem (5.1.12) for measurable functions on $\left(\overset{n}{\underset{1}{\times}} \Omega_i, \overset{n}{\underset{1}{\otimes}} \mathcal{A}_i, \overset{n}{\underset{1}{\otimes}} \mu_i \right)$.
(b) State analogues of Remarks 5.1.13 and 5.1.15 in the context of n-fold products.

EXERCISE 5.2.4. Denote the product space $\left(\overset{n}{\underset{1}{\times}} \Omega_i, \overset{n}{\underset{1}{\otimes}} \mathcal{A}_i \right)$ by $(\Omega, \mathcal{A})$.
(a) Show that, for each $i = 1, \ldots, n$, the coordinate projection map $\pi_i : \Omega \to \Omega_i$ defined by $\pi_i(\omega_1, \cdots, \omega_n) = \omega_i$ is an $\mathcal{A}_i$-measurable map on $(\Omega, \mathcal{A})$. Show also that $\mathcal{A} = \overset{n}{\underset{1}{\otimes}} \mathcal{A}_i$ is the smallest σ-field on Ω that makes all the coordinate projection maps π_i, $1 \leq i \leq n$, measurable.
(b) Show that, for any measurable space $(E, \mathcal{E})$, a map $f : (E, \mathcal{E}) \to (\Omega, \mathcal{A})$ is measurable iff $f \circ \pi_i : (E, \mathcal{E}) \to (\Omega_i, \mathcal{A}_i)$ is measurable, for each $i = 1, \ldots, n$.

(c) If μ_i, for each $i = 1, \cdots, n$, is a σ-finite measure on $\mathcal{A}_i$ and $\mu = \overset{n}{\underset{1}{\otimes}} \mu_i$ on $\mathcal{A}$, then show that $\mu_i = \mu\pi_i^{-1}$, for each $i = 1, \ldots, n$.

EXERCISE **5.2.5.** Let $(\Omega_i, \mathcal{A}_i, P_i)$, $1 \le i \le n$, be probability spaces and let $(\Omega, \mathcal{A}, P)$ be the product probability space. Denoting $\pi_i : \Omega \to \Omega_i$, $i = 1, \ldots, n$, to be the coordinate maps, show that the sub-σ-fields $\mathcal{C}_i = \sigma(\pi_i)$, $1 \le i \le n$, of $\mathcal{A}$ are mutually independent.

EXERCISE **5.2.6.** Let X_i, $i = 1, \cdots, n$ be real random variables on a probability space and let $\mathbf{X} = (X_1, \cdots, X_n)$. Show that independence of $\{X_i, i = 1, \cdots, n\}$ is equivalent to $P_\mathbf{X} = \overset{n}{\underset{i=1}{\otimes}} P_{X_i}$ on $\mathcal{B}^n$.

The last exercise shows that the notion of product probabilities is intimately connected with the notion of independence. To reinforce this connection, we now show that given probability spaces $(\Omega_i, \mathcal{A}_i, P_i)$, $i = 1, \cdots, n$, the *only* probability P on $\overset{n}{\underset{i=1}{\otimes}} \mathcal{A}_i$ with $P\pi_i^{-1} = P_i$, $i = 1, \cdots, n$, under which the sub-σ-fields $\mathcal{C}_i = \sigma(\pi_i)$, $i = 1, \cdots, n$, are independent, is the product probability $P = \overset{n}{\underset{i=1}{\otimes}} P_i$.

THEOREM **5.2.7.** Let $(\Omega_i, \mathcal{A}_i)$, $i = 1, \cdots, n$, be measurable spaces and let P be a probability on the product space $\left(\overset{n}{\underset{i=1}{\times}} \Omega_i, \overset{n}{\underset{i=1}{\otimes}} \mathcal{A}_i \right)$. Denote $\pi_i : \overset{n}{\underset{i=1}{\times}} \Omega_i \to \Omega_i$, for $i = 1, \cdots, n$, to be the coordinate projection maps and let $P_i = P\pi_i^{-1}$, $i = 1, \cdots, n$. Then, the following are equivalent:

(a) $P = \overset{n}{\underset{i=1}{\otimes}} P_i$.

(b) Under the probability P, the σ-fields $\mathcal{C}_i = \sigma(\pi_i)$, $i = 1, \cdots, n$, are independent.

Proof. Condition (a) is equivalent to
$$P(A_1 \times \cdots \times A_n) = \prod_{i=1}^n P_i(A_i), \text{ for all choices of } A_i \in \mathcal{A}_i, i = 1, \cdots, n.$$
But $P_i(A_i) = P(\pi_i^{-1}(A_i))$, $i = 1, \cdots, n$, and $A_1 \times \cdots \times A_n = \bigcap_{i=1}^n \pi_i^{-1}(A_i)$.
Thus (a) is equivalent to
$$P\left(\bigcap_{i=1}^n \pi_i^{-1}(A_i) \right) = \prod_{i=1}^n P(\pi_i^{-1}(A_i)), \text{ for all } A_i \in \mathcal{A}_i, i = 1, \cdots, n.$$
This completes the proof. □

EXERCISE **5.2.8.** (a) Given probability distribution functions F_i, $i = 1, \cdots, n$, on $\mathbb{R}$, show that there is a probability space $(\Omega, \mathcal{A}, P)$ and random variables $X_1, \cdots, X_n$ on $(\Omega, \mathcal{A}, P)$ with $F_{X_i} = F_i$, for each i, such that $X_1, \cdots, X_n$ are independent.
(b) Generalize the above so as to replace random variables $X_1, \cdots, X_n$ by random vectors $\mathbf{X}_1, \cdots, \mathbf{X}_n$ (possibly of different dimensions).

5.3   Measure Kernels

In the last two sections, we studied the concept of products of a finite number of σ-finite measure spaces. In this section, we are going to discuss a somewhat more

5.3. Measure Kernels

general concept. We will still be looking at finite products of measurable spaces, but the measure to be constructed on the product σ-field will not necessarily arise as the product of σ-finite measures given on the component σ-fields. Instead, we will consider a more general way of constructing σ-finite measures on the product σ-field. The idea behind such construction will be very natural and will turn out to be extremely useful in many contexts, in particular, in probability. As before, we will present the idea in complete detail in the case of 2-fold products and give a brief outline of how the same idea can be implemented for any finite product.

To understand the idea, let us start with two measurable spaces $(\Omega_1, \mathcal{A}_1)$ and $(\Omega_2, \mathcal{A}_2)$ and for a moment visualise the cartesian product $\Omega_1 \times \Omega_2$ as the plane with Ω_1 as the horizontal axis and Ω_2 as the vertical axis. Our definition of the product σ-field $\mathcal{A}_1 \otimes \mathcal{A}_2$ ensured that if A any subset of the plane that belongs to the product σ-field, then for every point ω_1 on the horizontal axis, the vertical section of A, namely, $A(\omega_1)$ belongs to $\mathcal{A}_2$. In section 5.1, we defined the "product" measure of A as follows. Given measures μ_1 and μ_2 on $\mathcal{A}_1$ and $\mathcal{A}_2$ respectively, we considered, for each point ω_1 on the horizontal axis, the vertical section $A(\omega_1)$ of A, then took the μ_2-measure of that section and finally integrated that measure over all ω_1 with respect to the measure μ_1. An important point to note here is that we used the same measure μ_2 for vertical sections of A at all points ω_1.

What if, given a subset A of the plane, we want to use different measures for the vertical sections of A at different points ω_1, instead of using the same measure μ_2 for $A(\omega_1)$ at all ω? What this would mean, in particular, is that even when the vertical sections of A at two different points ω_1 and ω_1' happen to be the same subsets of the vertical axis, we may not assign the same mass to $A(\omega_1)$ and $A(\omega_1')$. This, at first sight, may look like a somewhat artificial extension, but a moment's reflection would convince the reader that this will be very natural and the right thing to do in many situations. Consider, for example, a situation where the targeted measure on the plane represents mass of a set and the density varies from point to point on the horizontal axis.

If we agree to proceed with this idea of using different measures for the vertical sections $A(\omega_1)$ of a set A at different points ω_1, then we would end up with numbers $\mu_2(\omega_1, A(\omega_1))$ for various $\omega_1 \in \Omega_1$. Now we need to integrate these over ω_1 with respect to a measure μ_1 and that will requires measurability in ω_1. With all this motivation, we now proceed formally to make this idea precise.

DEFINITION 5.3.1. Let $(\Omega_1, \mathcal{A}_1)$, $(\Omega_2, \mathcal{A}_2)$ be two measurable spaces. By a *measure kernel* on $\Omega_1 \times \mathcal{A}_2$ is meant a function $\mu_2 : \Omega_1 \times \mathcal{A}_2 \to [0, \infty]$ such that,
(i) for each fixed $\omega_1 \in \Omega_1$, $\mu_2(\omega_1, \cdot)$ is a measure on $\mathcal{A}_2$, and,
(ii) for each fixed $A_2 \in \mathcal{A}_2$, $\mu_2(\cdot, A_2)$ is an $\mathcal{A}_1$-measurable function on Ω_1.

A measure kernel is called a *finite kernel* if for each $\omega_1 \in \Omega_1$, $\mu_2(\omega_1, \cdot)$ is a finite measure and is called a *σ-finite kernel* if there is a sequence $\{\Omega_{2,n}, n \geq 1\}$ of $\mathcal{A}_2$-sets satisfying $\Omega_2 = \bigcup_{n \geq 1} \Omega_{2,n}$ and $\mu_2(\omega_1, \Omega_{2,n}) < \infty$ for *every* $\omega_1 \in \Omega_1$.

It should be noted that σ-finiteness of a measure kernel μ_2 is stronger than just requiring that for every $\omega_1 \in \Omega_1$, the measure $\mu_2(\omega_1, \cdot)$ is a σ-finite measure.

We now proceed to construct a measure on $\mathcal{A}_1 \otimes \mathcal{A}_2$ starting from a σ-finite measure on $\mathcal{A}_1$ and a σ-finite measure kernel on $\Omega_1 \times \mathcal{A}_2$. The product measure discussed earlier will be a special case of this, because, given any σ-finite measure μ_2 on $\mathcal{A}_2$, one can define $\mu_2(\omega_1, \cdot) = \mu_2(\cdot)$, for all $\omega_1 \in \Omega_1$, and that will be a σ-finite measure kernel.

We start with an easy observation (Exercise 5.3.2) that says that any σ-finite measure kernel can be expressed as a sum of finite measure kernels. This will often be useful in deducing a result about σ-finite kernels by first proving it for finite measure kernels.

EXERCISE **5.3.2**. *Show that, given any σ-finite measure kernel μ_2 on $\Omega_1 \times \mathcal{A}_2$, there is a sequence $\mu_{2,n}$, $n \geq 1$, of finite measure kernels on $\Omega_1 \times \mathcal{A}_2$, such that $\mu_2(\omega_1, A_2) = \sum_n \mu_{2,n}(\omega_1, A_2)$, for each $\omega_1 \in \Omega_1$ and $A_2 \in \mathcal{A}_2$.*

PROPOSITION **5.3.3**. *If μ_2 is a σ-finite measure kernel on $\Omega_1 \times \mathcal{A}_2$, then, for each $A \in \mathcal{A}_1 \otimes \mathcal{A}_2$, the map $\omega_1 \mapsto \mu_2(\omega_1, A(\omega_1))$ on Ω_1 is $\mathcal{A}_1$-measurable.*

Proof. We first prove the result assuming μ_2 to be a finite measure kernel. Firstly, if $A = A_1 \times A_2 \in \mathcal{S}$, then by what we have seen repeatedly in Section 5.1, one easily gets $\mu_2(\omega_1, A(\omega_1)) = \mu_2(\omega_1, A_2) \cdot I_{A_1}(\omega_1)$ which is $\mathcal{A}_1$-measurable, since both $\mu_2(\cdot, A_2)$ and $I_{A_1}(\cdot)$ are $\mathcal{A}_1$-measurable. The same now can be easily seen to hold for sets A in the field $\mathcal{F}$ consisting of finite disjoint union of $\mathcal{S}$-sets. The proof can now be easily completed by showing that

$$\mathcal{M} = \{A \in \mathcal{A}_1 \otimes \mathcal{A}_2 : \omega_1 \mapsto \mu_2(\omega_1, A(\omega_1)) \text{ is } \mathcal{A}_1\text{-measurable}\}$$

is a monotone class. Finiteness of the measure kernel μ_2 will be needed here in showing that $\mathcal{M}$ is closed under decreasing limits.

Finally, one simply needs to use the result of Exercise 5.3.2 and what was proved above, to get the result in case μ_2 is a σ-finite measure kernel. □

THEOREM **5.3.4**. *(a) Let μ_1 be a σ-finite measure on $(\Omega_1, \mathcal{A}_1)$ and μ_2 a σ-finite measure kernel on $\Omega_1 \times \mathcal{A}_2$. Then*

$$\mu(A) = \int_{\Omega_1} \mu_2(\omega_1, A(\omega_1)) d\mu_1(\omega_1), \quad \text{for } A \in \mathcal{A}_1 \otimes \mathcal{A}_2, \tag{5.3.1}$$

defines a σ-finite measure on $\mathcal{A}_1 \otimes \mathcal{A}_2$.

(b) If f is any $\mathcal{A}_1 \otimes \mathcal{A}_2$-measurable function on $\Omega_1 \times \Omega_2$, which is either non-negative or μ-integrable then $\omega_1 \mapsto \int_{\Omega_2} f(\omega_1, \omega_2) \mu_2(\omega_1, d\omega_2)$ defines an $\mathcal{A}_1$-measurable function on Ω_1, which is well-defined and is non-negative, in case f is non-negative, while finite a.e.$[\mu_1]$, in case f is μ-integrable. Further,

$$\int_{\Omega_1 \times \Omega_2} f d\mu = \int_{\Omega_1} \Big(\int_{\Omega_2} f(\omega_1, \omega_2) \mu_2(\omega_1, d\omega_2) \Big) d\mu_1(\omega_1). \tag{5.3.2}$$

Proof. (a) For every $A \in \mathcal{A}_1 \otimes \mathcal{A}_2$, the map $\omega_1 \mapsto \mu_2(\omega_1, A(\omega_1))$ is, by Proposition 5.3.3, a non-negative $\mathcal{A}_1$-measurable function on Ω_1. Therefore, the integral

5.3. Measure Kernels

in (5.3.1) exists for all $A \in \mathcal{A}_1 \otimes \mathcal{A}_2$ and defines a non-negative set function on $\mathcal{A}_1 \otimes \mathcal{A}_2$, that satisfies $\mu(\emptyset) = 0$. One can now prove that μ is countably additive and hence a measure, by using arguments similar to those in the proof of Proposition 5.1.8(b) and using Theorem 3.3.6. Next, since μ_1 is σ-finite, we have a sequence $\{\Omega_{1,m}, m \geq 1\}$ of disjoint $\mathcal{A}_1$-sets with $\Omega_1 = \bigcup_{m \geq 1} \Omega_{1,m}$, such that $\mu_1(\Omega_{1,m}) < \infty$, for each $m \geq 1$. With $\Omega_{2,n} \in \mathcal{A}_2$, as in the definition 5.3.1 of σ-finite measure kernel μ_2, we define $E_{k,m,n} = \{\omega_1 \in \Omega_{1,m} : \mu_2(\omega_1, \Omega_{2,n}) \leq k\}$. It is easy to see that $\Omega_1 = \bigcup_{k,m,n} E_{k,m,n}$ and $E_{k,m,n} \in \mathcal{A}_1$ with $\mu(E_{k,m,n} \times \Omega_{2,n}) < \infty$, for each k, m, n. Clearly, $\bigcup_{k,m,n} E_{k,m,n} \times \Omega_{2,n} = \Omega_1 \times \Omega_2$ and that proves σ-finiteness of the measure μ (in fact, σ-finiteness on the semifield $\mathcal{S}$).

(b) For this, one uses the same standard argument that was used, for example, in the proof of Fubini's Theorem 5.1.12. To be specific, in case $f = I_A$ for some $A \in \mathcal{A}_1 \otimes \mathcal{A}_2$, the required result is essentially taken care of by the part (a) of the theorem. From this, the journey is through usual route, namely, first considering non-negative real-valued simple measurable functions f, then general non-negative measurable functions f, and finally any μ-integrable function f. $\square$

REMARK **5.3.5.** In the theorem above, we saw that if μ_1 is a σ-finite measure and μ_2 is a σ-finite measure kernel, then μ is a σ-finite measure. It may be instructive to know that even when μ_1 is a finite measure and μ_2 is a finite measure kernel, the measure μ is σ-finite, but need not be finite.

EXERCISE **5.3.6.** *Give an example of the phenomenon mentioned in Remark 5.3.5, that is, an example where μ_1 is a finite measure and μ_2 is a finite measure kernel, but yet the measure μ is not finite.*

DEFINITION **5.3.7.** A measure kernel on $\Omega_1 \times \mathcal{A}_2$ is called a *probability kernel* if, for every $\omega_1 \in \Omega_1$, the measure $\mu_2(\omega_1, \cdot)$ is a probability measure.

REMARK **5.3.8.** It is easy to see that if μ_1 is a probability measure on $\mathcal{A}_1$ and μ_2 is a probability kernel on $\Omega_1 \times \mathcal{A}_2$, then μ is a probability measure on $\mathcal{A}_1 \otimes \mathcal{A}_2$.

REMARK **5.3.9.** A question that one may ask is whether every σ-finite measure on $\mathcal{A}_1 \otimes \mathcal{A}_2$ arises as above through a σ-finite measure on $\mathcal{A}_1$ and a σ-finite measure kernel on $\Omega_1 \times \mathcal{A}_2$. While the answer, in general, is NO, we will see in a later chapter that this is indeed true in the special case when $\Omega_1 = \Omega_2 = \mathbb{R}$ and $\mathcal{A}_1 = \mathcal{A}_2 = \mathcal{B}$.

It should be clear to the reader that whatever we have done above with products of two spaces can be easily extended to finite products. We simply state this in the following theorem. The proof is omitted because it uses the same idea as in the case of 2-fold products, except that it just gets a bit messy notationally. The interested reader should try to write the proof herself.

THEOREM **5.3.10.** *Let $(\Omega_j, \mathcal{A}_j)$, $1 \leq j \leq n$, be measurable spaces and let $(\Omega, \mathcal{A})$ denote the product measurable space $\left(\underset{j=1}{\overset{n}{\times}} \Omega_j, \underset{j=1}{\overset{n}{\otimes}} \mathcal{A}_j \right)$.*

(a) If μ_1 is a σ-finite measure on $\mathcal{A}_1$ and if μ_j, for each $j = 2, \cdots, n$, is a σ-finite measure kernel on $\underset{i=1}{\overset{j-1}{\times}} \Omega_i \times \mathcal{A}_j$, then

$$\mu(A) = \int_\Omega I_A(\omega_1, \ldots, \omega_n) \mu_n(\omega_1, \ldots, \omega_{n-1}, d\omega_n) \cdots \mu_2(\omega_1, d\omega_2)\mu_1(d\omega_1), \ A \in \mathcal{A}$$

defines a σ-finite measure on $\mathcal{A}$.

In case μ_1 is a probability measure and each μ_j, $2 \leq j \leq n$, is a probability kernel, then μ is a probability measure.

(b) Further, for any measurable function f on $(\Omega, \mathcal{A})$ which is *either* non-negative or μ-integrable,

$$\int_\Omega f d\mu = \int_{\Omega_1} \int_{\Omega_2} \cdots \int_{\Omega_n} f(\omega_1, \ldots, \omega_n) \mu_n(\omega_1, \ldots, \omega_{n-1}, d\omega_n) \cdots \mu_2(\omega_1, d\omega_2) \mu_1(d\omega_1)$$

5.4 Additional Exercises

EXERCISE 5.4.1. Let $(\Omega, \mathcal{A})$ be a measurable space and $I \subset \mathbb{R}$ be a non-degenerate interval. Suppose f is a real-valued function on $I \times \Omega$, such that, (i) the map $f(t, \cdot) : \Omega \to \mathbb{R}$ is $\mathcal{A}$-measurable, for every $t \in I$ and (ii) the map $f(\cdot, \omega) : I \to \mathbb{R}$ is right-continuous everywhere on I, for each $\omega \in \Omega$. Show then that f is measurable with respect to the product σ-field $\mathcal{B}(I) \otimes \mathcal{A}$.

Show also that the same conclusion holds if in (ii) above, we replace "right-continuous" by "left-continuous".

EXERCISE 5.4.2. (a) Let $(\Omega_i, \mathcal{A}_i, \mu_i)$, $i = 1, 2$, be two σ-finite measure spaces. Show that, for any set $A \in \mathcal{A}_1 \otimes \mathcal{A}_2$, the following two statements are equivalent:
(i) $\mu_2(A(\omega_1)) = 0$, a.e. $[\mu_1]$.
(ii) $\mu_1(A^*(\omega_2)) = 0$, a.e. $[\mu_2]$.
(b) Generalize (a) to the case of any finite product of σ-finite measure spaces.

EXERCISE 5.4.3. Let $(\Omega_i, \mathcal{A}_i, \mu_i)$, $i = 1, \ldots, n$, be σ-finite measure spaces and let $\mathcal{A}$ denote the product σ-field $\underset{i}{\otimes} \mathcal{A}_i$ on the cartesian product $\Omega = \underset{i}{\times} \Omega_i$.
(a) Letting $\pi : (1, 2, \cdots, n) \mapsto (i_1, i_2, \cdots, i_n)$ denote a permutation of $\{1, 2, \ldots, n\}$, define a set function $\widetilde{\mu}$ on $\mathcal{A}$ by

$$\widetilde{\mu}(A) = \int_{\Omega_{i_1}} \cdots \int_{\Omega_{i_n}} I_A(\omega_1, \ldots, \omega_n)\, d\mu_{i_n}(\omega_{i_n}) \cdots d\mu_{i_1}(\omega_{i_1}), \text{ for } A \in \mathcal{A}.$$

Show that $\widetilde{\mu} \equiv \mu$, where μ is as defined in (5.2.1).
(b) Use this to state and prove an extension of Fubini's Theorem to any finite product space.

EXERCISE 5.4.4. Show that $\mathcal{B}(\mathbb{R}^k)$ is the k-fold product of $\mathcal{B}(R)$ and so we can write $\mathcal{B}^k$ for $\mathcal{B}(\mathbb{R}^k)$. Show also that λ^k is the k-fold product of the one-dimensional Lebesgue measure λ.

EXERCISE 5.4.5. (a) Let $(\Omega, \mathcal{A})$ be a measurable space and $D = \{(\omega, \omega) : \omega \in \Omega\}$. Show that $D \in \mathcal{A} \otimes \mathcal{A}$ if and only if there is a countably generated σ-field $\mathcal{A}_0 \subset \mathcal{A}$, such that $\{\omega\} \in \mathcal{A}_0$ for all $\omega \in \Omega$. [*Use Exercise 2.13.7.*]
(b) Let $\mathcal{A}$ be the countable-cocountable σ-field on $\mathbb{R}$. Examine whether the set $D = \{(x, x) : x \in \mathbb{R}\}$ belongs to $\mathcal{A} \otimes \mathcal{A}$ or not.

5.4. Additional Exercises

EXERCISE 5.4.6. Let $B = \{(x,y) : x^2 + y^2 \leq 1\}$. Show that $\lambda^2(B) = \pi$.

EXERCISE 5.4.7. Suppose that μ_i, $1 \leq i \leq k$, are Radon measures on $(\mathbb{R}, \mathcal{B})$ with associated distribution functions G_i, $1 \leq i \leq k$, respectively.
(a) Show that the function $G : \mathbb{R}^k \to \mathbb{R}$ defined by $G(x_1, \cdots, x_k) = \prod_1^k G_i(x_i)$ is a distribution function on $\mathbb{R}^k$.
(b) Denoting μ to be the unique Radon measure on $(\mathbb{R}^k, \mathcal{B}^k)$ determined by G, show that μ agrees with $\otimes \mu_i$ on the class $\mathcal{C} = \{I_{\mathbf{ab}} : \mathbf{a} \leq \mathbf{b}, \mathbf{a}, \mathbf{b} \in \mathbb{R}^k\}$ and hence conclude that $\mu = \otimes \mu_i$ on $(\mathbb{R}^k, \mathcal{B}^k)$.

EXERCISE 5.4.8. Use Fubini's theorem on an appropriate product measure space to show that if $\{a_{i,j}, i, j \geq 1\}$ are real numbers, then $\sum_{i,j} a_{i,j} = \sum_i (\sum_j a_{i,j}) = \sum_j (\sum_i a_{i,j})$, if either $a_{i,j} \geq 0$ for all (i,j) or $\sum_{i,j} |a_{i,j}| < \infty$.

In this chapter, product of two measure spaces was defined assuming that both are σ-finite measure spaces. The aim of the next exercise is to illustrate that assuming just one of the measures to be σ-finite is enough to define a "product measure".

EXERCISE 5.4.9. Consider two measure spaces $(\Omega_1, \mathcal{A}_1, \mu_1)$ and $(\Omega_2, \mathcal{A}_2, \mu_2)$ and let $(\Omega, \mathcal{A})$ denote $(\Omega_1 \times \Omega_2, \mathcal{A}_1 \otimes \mathcal{A}_2)$. Assume that μ_2 is sigma-finite. From what was proved in Section 5.1, the section $A(\omega_1) \in \mathcal{A}_2$, for every $A \in \mathcal{A}$ and $\omega_1 \in \Omega_1$ and the map $\omega_1 \mapsto \mu_2(A(\omega_1))$ is a non-negative $\mathcal{A}_1$-measurable map on Ω_1.
(a) Show that $\mu(A) = \int \mu_2(A(\omega_1)) d\mu_1$, $A \in \mathcal{A}$, is a well-defined set function and defines a measure on $\mathcal{A}$ satisfying $\mu(A_1 \times A_2) = \mu_1(A_1)\mu_2(A_2)$, for all $A_1 \in \mathcal{A}_1$ and $A_2 \in \mathcal{A}_2$. (We are not claiming that μ is the only measure satisfying this.)
(b) Show that the equality $\int f d\mu = \int \int f(\omega_1, \omega_2) d\mu_2(\omega_2) d\mu_1(\omega_1)$ holds for any $\mathcal{A}$-measurable function f on Ω, which is either non-negative or μ-integrable.
(c) Similarly assuming that μ_1 is σ-finite, define an appropriate measure on $\mathcal{A}$ and prove analogues of (a) and (b).
For the next two parts, let $(\Omega_1, \mathcal{A}_1) = (\Omega_2, \mathcal{A}_2) = ([0,1], \mathcal{B}([0,1]))$. Also, for a set $A \subset [0,1] \times [0,1]$ and $x, y \in [0,1]$, let A_x denote $A(x)$ and A_y^* denote $A^*(y)$.
(d) Denoting λ and ν to be respectively the Lebesgue measure and the counting measure on $([0,1], \mathcal{B}([0,1]))$, consider $\mu = \nu \otimes \lambda$ on $\mathcal{B}([0,1]) \otimes \mathcal{B}([0,1])$ (which, by Exercise 5.4.4, equals $\mathcal{B}([0,1] \times [0,1])$). Show that $\mu(A) < \infty$, if and only if, there is a countable set $D \subset [0,1]$ (depending on A), such that, $\lambda(A_x) = 0$ for all $x \notin D$ and $\sum_{x \in D} \lambda(A_x) < \infty$. In particular, show that if $A = \bigcup_n A_n$ with $\mu(A_n) < \infty$ for all n, then $\lambda(A_x) = 0$, for all but countably many $x \subset [0,1]$. State and prove analogous results for the product measure $\lambda \otimes \nu$.
(e) With μ as in (d) above, show that a measurable function f is μ-integrable if and only if there is a countable set D such that $f(x, \cdot) = 0$, a.e.$[\lambda]$, for all $x \notin D$ and $\sum_{x \in D} \int |f(x, \cdot)| d\lambda < \infty$. Prove analogous results for $\lambda \otimes \nu$-integrable functions.

EXERCISE 5.4.10. Let h be a strictly increasing continuously differentiable function on $[0, \infty)$ with $h(0) = 0$. Show that, if f is any measurable function on some σ-finite measure space $(\Omega, \mathcal{A}, \mu)$, then $\int h(|f|) d\mu = \int_0^\infty h'(t) \mu(\{|f| > t\}) dt$.

EXERCISE 5.4.11. [*This exercise may be seen as a contrast to Exercsie 5.4.4.*] Recall that the Lebesgue σ-field was defined earlier as the λ-completion of $\mathcal{B}(\mathbb{R})$. Let us denote it by $\mathcal{L}(\mathbb{R})$. Similarly, denote $\mathcal{L}(\mathbb{R}^2)$ to be the λ^2-completion of $\mathcal{B}(\mathbb{R}^2)$, called the **Lebesgue σ-field on $\mathbb{R}^2$**. Show that $\mathcal{L}(\mathbb{R}) \otimes \mathcal{L}(\mathbb{R}) \subsetneq \mathcal{L}(\mathbb{R}^2)$.

EXERCISE 5.4.12. Let $g : \mathbb{R} \to \mathbb{R}$ be function with compact support, which is continuously differentiable everywhere. For any integrable function f on $(\mathbb{R}, \mathcal{B}(\mathbb{R}), \lambda)$, define $f * g(x) = \int g(y) f(x-y) \, dy$, $x \in \mathbb{R}$. Show that
(a) $f * g : \mathbb{R} \to \mathbb{R}$ is an integrable function, and,
(b) $f * g$ is continuously differentiable with $(f * g)'(x) = \int g'(y) f(x-y) \, dy$.

EXERCISE 5.4.13. Let μ_1 and μ_2 be two σ-finite measures on $(\mathbb{R}, \mathcal{B})$ and let $\mu = \mu_1 \otimes \mu_2$ denote the product measure on $(\mathbb{R}^2, \mathcal{B}^2)$. Consider the measurable map $h : (\mathbb{R}^2, \mathcal{B}^2) \to (\mathbb{R}, \mathcal{B})$ given by $h(x, y) = x + y$, for $(x, y) \in \mathbb{R}^2$ and let $\nu = \mu h^{-1}$ denote the induced σ-finite measure on $(\mathbb{R}, \mathcal{B})$. This σ-finite measure ν on $(\mathbb{R}, \mathcal{B})$ is called the **convolution** of the σ-finite measures μ_1 and μ_2 and is denoted $\mu_1 * \mu_2$.
(a) Show that $\mu_1 * \mu_2(B) = \int_{\mathbb{R}} \mu_2(B-x) \mu_1(dx)$, for all $B \in \mathcal{B}$.
(b) Show that $(\mu_1, \mu_2) \mapsto \mu_1 * \mu_2$ defines a binary operation on the set of all σ-finite measures on $(\mathbb{R}, \mathcal{B})$, which is associative, commutative and admits an identity.
(c) Show however that, under the binary opeartion $*$, a sigma-finite measure μ does not admit an inverse, except when the measure μ is a non-zero (necessarily, finite!) point mass, that is, $0 < \mu(\mathbb{R}) = \mu(\{c\}) < \infty$, for some $c \in \mathbb{R}$.
(d) Show that $\mu_1 * \mu_2$ is a finite measure (resp., a probability measure) if μ_1 and μ_2 are both finite measures (resp., probability measures).
(e) Show that if at least one of μ_1 and μ_2 admits no point mass, then $\mu_1 * \mu_2$ also admits no point mass.
(f) Suppose that $\mu_1(B) = \int_B f_1 d\lambda$, for all $B \in \mathcal{B}$, where $f_1 : \mathbb{R} \to \mathbb{R}$ is a non-negative measurable function. Show that $f(z) = \int_{\mathbb{R}} f_1(z - x) d\mu_2(x)$, $z \in \mathbb{R}$, defines a measurable function on $\mathbb{R}$ and that $\mu_1 * \mu_2(B) = \int_B f d\lambda$, for all $B \in \mathcal{B}$. Prove an analogous result assuming that $\mu_2(B) = \int f_2 d\lambda$, $B \in \mathcal{B}$, for some non-negative measurable $f_2 : \mathbb{R} \to \mathbb{R}$.

REMARK 5.4.14. In the language of what will be discussed in Chapter 6, Exercise 5.4.13 asserts that if at least one of μ_1 and μ_2 is absolutely continuous with respect to λ, then $\mu_1 * \mu_2 \ll \lambda$ and $\frac{d(\mu_1 * \mu_2)}{d\lambda}(z) = \int_{\mathbb{R}} \frac{d\mu_1}{d\lambda}(z-x) \, d\mu_2(x)$, in case $\mu_1 \ll \lambda$ (resp., $= \int_{\mathbb{R}} \frac{d\mu_2}{d\lambda}(z-x) \, d\mu_1(x)$, in case $\mu_2 \ll \lambda$), for $z \in \mathbb{R}$.

EXERCISE 5.4.15. Let X, G and F be as in Exercise 4.6.7 and let P_X denote the distribution of X.
(a) Show that if P_G denotes the probability measure on $(\mathbb{R}, \mathcal{B})$ determined by G, then the probability measure P_F determined by F is given by $P_F = P_X * P_G$, where '$*$' is as defined in Exercise 5.4.13.
(b) Conclude that if G is a continuous distribution function, then so is F, while if

5.4. Additional Exercises

G has a density g then so has F with its density f given by $f(u) = E(g(u-X))$.
(c) Show that F is the distribution function of $X+Y$ where Y is a random variable with distribution function G and independent of X.
(d) Hence conclude that if X and Y are independent random variables and if any one of then has a continuous distribution function (resp., has a density), then $X+Y$ also has a continuous distribution (resp., has a density).

REMARK 5.4.16. If X and Y are independent with distribution functions G and H respectively, then the distribution function F of $X+Y$ as obtained in Exercise 5.4.15, is known as the 'convolution' of G and H and one writes $F = G*H = H*G$.

EXERCISE 5.4.17. Let X and Y be independent random variables on some probability space $(\Omega, \mathcal{A}, P)$. Assume $P(Y=0) = 0$.
(a) Let $Z = XY$. Show that if X has a continuous distribution (resp., has a density), then so does Z. Also, find a formula for the density of Z, in terms of the density of X.
(b) Let $h : \mathbb{R}^2 \to \mathbb{R}$ be a borel measurable function and let $Z = h(X,Y)$. Show that, $h(x,Y)$, for each $x \in \mathbb{R}$, is a real random variable and, if Z has finite expectation, then $h(x,Y)$ has finite expectation, for P_X-a.e. $x \in \mathbb{R}$, and the map $x \mapsto \varphi(x) = E(h(x,Y))$ defines a borel measurable function on $\mathbb{R}$ (define $\varphi(x) = 0$, if necessary, for x in a P_X-null set). Further, show that $\varphi(X)$ has finite expectation and $E(\varphi(X)) = E(Z)$.

EXERCISE 5.4.18. Show that, if X and Y are independent random variables, then $E(|X+Y|) < \infty$ implies that $E(|X|) < \infty$, $E(|Y|) < \infty$, and, more generally, $E(|X+Y|^p) < \infty$ implies that $E(|X|^p) < \infty$, $E(|Y|^p) < \infty$.

EXERCISE 5.4.19. Show that if $X_1, \ldots, X_n$ are independent random variables and if $X_1 + \cdots + X_n$ is degenerate, then $X_1, \ldots, X_n$ must all be degenerate.

EXERCISE 5.4.20. Denote by $(\Omega, \mathcal{A})$, the product space $(\Omega_1 \times \Omega_2, \mathcal{A}_1 \otimes \mathcal{A}_2)$. Consider the classes
$$\mathcal{V} = \{A_1 \times \Omega_2 : A_1 \in \mathcal{A}_1\}, \quad \mathcal{H} = \{\Omega_1 \times A_2 : A_2 \in \mathcal{A}_2\}. \quad (5.4.1)$$
(a) Show that both $\mathcal{H}$ and $\mathcal{V}$ are sub-σ-fields of $\mathcal{A}$ on Ω (called the **vertical** and **horizontal** sub-σ-fields respectively).
(b) Show that a probability measure μ on $\mathcal{A}$ satisfies
$$\mu(H \cap V) = \mu(H) \cdot \mu(V), \quad \text{for all } H \in \mathcal{H}, \ V \in \mathcal{V}, \quad (5.4.2)$$
if and only if $\mu = \mu_1 \otimes \mu_2$, for some probability measures μ_1 and μ_2 on the σ-fields $\mathcal{A}_1$ and $\mathcal{A}_2$ respectively.

EXERCISE 5.4.21. Consider the space $(\mathbb{R}^2, \mathcal{B}^2)$ and the set $D = \{(x,x) : x \in \mathbb{R}\}$. Let $\mathcal{B}(D)$ be the Borel σ-field on D, that is, $\mathcal{B}(D) = \{B \in \mathcal{B}^2 : B \subset D\} = \{B \cap D : B \in \mathcal{B}^2\}$. Show that the map $\varphi(x,x) = x$ is one-to one on D onto $\mathbb{R}$ such that both φ and φ^{-1} are measurable. Thus the set D sitting in $\mathbb{R}^2$ is a copy of $\mathbb{R}$ in all respects. (We say $(D, \mathcal{B}(D))$ is 'Borel isomorphic' to $(\mathbb{R}, \mathcal{B})$.)

EXERCISE **5.4.22.** Consider $(\mathbb{R}^2, \mathcal{B}^2)$ as in Exercise 5.4.21, but, instead of Ω described there, consider the set $\Omega^* = \{(4, y) : y \in \mathbb{R}\} \in \mathcal{B}^2$. State and prove analogues of the statements in Exercise 5.4.21.

EXERCISE **5.4.23.** If $f(x, y)$ is a measurable function on $(\mathbb{R}^2, \mathcal{B}^2)$ then we know that (i) the function $y \mapsto f(x, y)$ is measurable on $(\mathbb{R}, \mathcal{B})$ for each $x \in \mathbb{R}$ and (ii) the function $x \mapsto f(x, y)$ is measurable on $(\mathbb{R}, \mathcal{B})$ for each $y \in \mathbb{R}$. This is expressed by saying that f is separately measurable. Is the converse true? That is $f : \mathbb{R}^2 \to \mathbb{R}$ such that (i) and (ii) above hold, then is it measurable on $(\mathbb{R}^2, \mathcal{B}^2)$? Not necessarily! Here is an example to illustrate this.
Let $A \subset [0,1]$ be a non-Borel set (we know such sets exist). Consider the set $B = \{(x, x) : x \in A\}$. Show that I_B satisfies (i) and (ii) but is not measurable.

EXERCISE **5.4.24.** With a non-Borel $A \subset [0,1]$, let $C = B \cup \{(x,y) : x^2 + y^2 < 1\}$, where $B = \{(x, +\sqrt{1-x^2}) : x \in A\}$ or
$$B = \{(x, +\sqrt{1-x^2}) : x \in A\} \cup \{(x, -\sqrt{1-x^2}) : x \in A\}$$
Show that, in either case, $C \notin \mathcal{B}^2$, but each of its sections belongs to $\mathcal{B}^1 = \mathcal{B}$. (The interesting point here is that C is a convex set!)

EXERCISE **5.4.25.** Let μ be a finite measure on $(\mathbb{R}, \mathcal{B})$ with distribution function $F(x) = \mu\big((-\infty, x]\big)$. Show that $\int \big(F(x+a) - F(x)\big) dx = a\mu(\mathbb{R})$, for every $a \in \mathbb{R}$.

EXERCISE **5.4.26.** Consider the product space $(\mathbb{N} \times \mathbb{N}, \mathcal{P}(\mathbb{N}) \otimes \mathcal{P}(\mathbb{N}), \mu \otimes \mu)$, where μ is the counting measure on $\mathcal{P}(\mathbb{N})$. With $A_1 = \{(i,j) : j = i\}$ and $A_2 = \{(i,j) : j = i+1\}$, examine the 'iterated' integral and the 'double' integral for the function $f = I_{A_1} - I_{A_2}$ to see that Fubini's Theorem does not apply. What fails?

EXERCISE **5.4.27.** Consider the product space $(\mathbb{R} \times \mathbb{R}, \mathcal{B} \otimes \mathcal{B}, \lambda \otimes \lambda)$
(a) For the function $f(x,y) = yI_{(-1,1)}(y)$, examine the 'iterated' integral and the 'double' integral to see that Fubini's theorem does not apply. What fails?
(b) With $A_1 = \{(x,y) : 0 \leq x - y \leq 1\}$ and $A_2 = \{(x,y) : -1 \leq x - y < 0\}$, examine the 'iterated' integral and the 'double' integral for the function $f = I_{A_1} - I_{A_2}$ to see that Fubini's Theorem does not apply. What fails?
(c) Consider the real valued function f defined on $\mathbb{R}^2$ as
$$f(x,y) = \frac{\exp(x^2 y^2) \sin(xy)}{1 + x^4 y^4}, \quad \text{if} \ -1 \leq x, y \leq 1, \quad \text{and} \quad f(x,y) = 0, \ \text{otherwise.}$$
Examine the 'iterated' integral and the 'double' integral of f to see if Fubini's theorem works. If it does not, explain what fails.

Chapter 6

Radon-Nikodym Theorem and L_p spaces

6.1 Radon-Nikodym Theorem, Lebesgue Decomposition

In Chapter 3, we already introduced the notion of absolute continuity (see Definition 3.3.19) and then stated an important result, called the Radon-Nikodym Theorem. In this section, we will prove this theorem along with several other important results. Recall that if μ and ν are measures on a σ-field $\mathcal{A}$, then ν is said to be *absolutely continuous* with respect to μ, written $\nu \ll \mu$, if any μ-null set in $\mathcal{A}$ is also ν-null. We start with a simple result that gives an equivalent and useful characterization of absolute continuity of a *finite* measure ν with respect to a measure μ. This is often called the 'ϵ-δ criterion' of absolute continuity.

PROPOSITION 6.1.1. *Let μ be a measure and ν a finite measure on a σ-field $\mathcal{A}$. Then, ν is absolutely continuous with respect to μ if and only if $\lim_{\mu(A)\downarrow 0} \nu(A) = 0$, in the sense that, given any $\epsilon > 0$, there is a $\delta > 0$ such that $A \in \mathcal{A}$ and $\mu(A) < \delta$ imply that $\nu(A) < \epsilon$.*

Proof. The 'if' part is immediate from the definition of absolute continuity. We prove the 'only if' part by contradiction. Assume $\nu \ll \mu$. Suppose, if possible, there exists $\epsilon > 0$, such that, for all $n \geq 1$, there exists $A_n \in \mathcal{A}$ with $\mu(A_n) < 2^{-n}$ but $\nu(A_n) \geq \epsilon$. With $A = \limsup_n A_n$, we have $\mu(A) = 0$ by Borel-Cantelli Lemma. However, $\bigcup_{k\geq n} A_k \downarrow \limsup_n A_n$ and so, using finiteness of ν and the fact that $\nu\left(\bigcup_{k\geq n} A_k\right) \geq \nu(A_n) \geq \epsilon$, for all n, we get $\nu(A) = \lim_n \nu\left(\bigcup_{k\geq n} A_k\right) \geq \epsilon$. This contradicts the hypothesis that $\nu \ll \mu$, thus completing the proof. □

Returning now to our main aim of this section, as stated above, we start with a result, which will yield two very important theorems. One is Radon-Nikodym Theorem and the other is what is called the Lebesgue Decomposition Theorem.

THEOREM 6.1.2. *Let $(\Omega, \mathcal{A}, \mu)$ be a σ-finite measure space. Given any σ-finite measure ν on $\mathcal{A}$, there is a non-negative measurable function f and a μ-null set $N \in \mathcal{A}$, such that,*

$$\nu(A) = \int_A f d\mu + \nu(A \cap N), \quad \text{for all } A \in \mathcal{A}. \tag{6.1.1}$$

We postpone the proof of Theorem 6.1.2 for a while. Instead, we first show how Radon-Nikodym Theorem and the other important result, namely, the Lebesgue Decomposition Theorem, follow from this.

THEOREM **6.1.3.** (Radon-Nikodym Theorem) Let μ and ν be σ-finite measures on a measurable space $(\Omega, \mathcal{A})$. Then, ν is absolutely continuous with respect to μ if and only if there is a non-negative measurable function f on Ω such that

$$\nu(A) = \int_A f d\mu, \quad \text{for all } A \in \mathcal{A}. \tag{6.1.2}$$

REMARK **6.1.4.** A careful reader must have noted that our statement of Radon-Nikodym Theorem here assumes both μ and ν to be σ-finite, while in the statement given earlier in Chapter 3 (after Definition 3.3.19), only μ was assumed to be σ-finite. Let us assure the quite justifiedly concerned reader that we will return to this issue later and iron it out.

Proof. For the 'if' part, there is nothing to prove, since it follows from a property of integrals stated in Proposition 3.2.14.

To prove the 'only if' part, assume $\nu \ll \mu$. Let f and N be as in Theorem 6.1.2. Observe that since N is a μ-null set, the hypothesis $\nu \ll \mu$ implies that N is ν-null also. But this will give $\nu(A \cap N) = 0$, for all $A \in \mathcal{A}$. Thus, equation (6.1.1) reduces to (6.1.2) and that completes the proof. □

With f and N as in Theorem 6.1.2, let us define non-negative set functions ν_c and ν_s on $\mathcal{A}$ by

$$\nu_c(A) = \int_A f d\mu \text{ and } \nu_s(A) = \nu(A \cap N), \text{ for } A \in \mathcal{A}. \tag{6.1.3}$$

It is clear that ν_c and ν_s are both σ-finite (why?) measures on $\mathcal{A}$ and, according to Theorem 6.1.2, $\nu(A) = \nu_c(A) + \nu_s(A)$, for all $A \in \mathcal{A}$. Further, the measure ν_s is completely supported on a μ-null set N (that is, $\nu_s(N^c) = 0$). This property of ν_s, which, in some sense, is diametrically opposite to being absolutely continuous with respect to μ, leads to the following definition.

DEFINITION **6.1.5.** Given measures μ and ν on a σ-field $\mathcal{A}$, we say that μ and ν are *mutually singular*, and we write $\nu \perp \mu$, if there is a μ-null set $N \in \mathcal{A}$ such that N^c is ν-null.

Essentially, mutual singularity of two measures μ and ν means that μ and ν are 'supported' on two disjoint $\mathcal{A}$-sets. It is clear that, unlike absolute continuity, mutual singularity is a symmetric relation. Also, it is easy to see that $\nu \ll \mu$ and $\nu \perp \mu$ both hold if and only if $\nu \equiv 0$.

Going back to the decomposition $\nu = \nu_c + \nu_s$ now, what we see is that ν is being expressed as a sum of two σ-finite measures ν_c and ν_s on $\mathcal{A}$, such that, $\nu_c \ll \mu$ and $\nu_s \perp \mu$. This is exactly what is captured in the following theorem.

THEOREM **6.1.6.** (Lebesgue Decomposition Theorem) If $(\Omega, \mathcal{A}, \mu)$ is a σ-finite measure space, then any σ-finite measure ν on $\mathcal{A}$ has a unique decomposition as $\nu = \nu_c + \nu_s$ with $\nu_c \ll \mu$ and $\nu_s \perp \mu$.

6.1. Radon-Nikodym Theorem, Lebesgue Decomposition

Proof. We have just seen above that Theorem 6.1.2 gives us such a decomposition. Thus, only uniqueness of the decomposition remains to be proved. Let $\nu = \bar{\nu}_c + \bar{\nu}_s$, with $\bar{\nu}_c \ll \mu$ and $\bar{\nu}_s \perp \mu$, be any decomposition. With ν_c and ν_s as above, we have $\nu_s \perp \mu$ and $\bar{\nu}_s \perp \mu$. So there are μ-null sets N and $\overline{N}$ with $\nu_s(N^c) = 0 = \bar{\nu}_s(\overline{N}^c)$. Since $M = N \cup \overline{N}$ is μ-null, we have $\nu_c(M) = 0 = \bar{\nu}_c(M)$, implying that

$$\nu(A \cap M) = \nu_c(A \cap M) + \nu_s(A \cap M) = \nu_s(A \cap M) = \nu_s(A), \text{ for any } A \in \mathcal{A},$$

where the last equality follows from the fact that $M^c \subset N^c$ and hence $\nu_s(M^c) = 0$. One can similarly get $\nu(A \cap M) = \bar{\nu}_s(A)$, for all $A \in \mathcal{A}$, thus proving $\nu_s \equiv \bar{\nu}_s$. On the other hand, for any $A \in \mathcal{A}$, $\nu(A \cap M^c) = \nu_c(A \cap M^c) = \nu_c(A)$, because M is μ null and hence ν_c-null also. A similar argument also gives that $\nu(A \cap M^c) = \bar{\nu}_c(A)$, for any $A \in \mathcal{A}$. This proves that $\nu_c \equiv \bar{\nu}_c$ and that completes the proof. □

REMARK **6.1.7.** A reader may wonder why was the argument not over once we showed $\nu_s = \bar{\nu}_s$, because we already know that $\nu_s + \nu_c = \nu = \bar{\nu}_s + \bar{\nu}_c$? Why did we have to still prove $\nu_c = \bar{\nu}_c$? Keep in mind that we are dealing with possible infinities here on both sides and so the usual rules of algebra (in particular, the law of cancellation) do not apply automatically.

Having thus derived two major results as simple consequences of Theorem 6.1.2, we now proceed towards the proof of that theorem. The argument leading to the proof is built up of several intermediate steps. In what follows, we go through these intermediate steps, that will finally lead to the proof of Theorem 6.1.2. One of the key steps involve introducing a new concept called "essential supremum", the existence of which turns out to play a vital role. Let us first get the motivation behind this concept.

To begin with, let us assume that we are given two finite measures μ and ν on $(\Omega, \mathcal{A})$. We will justify later that it is enough to prove Theorem 6.1.2 for this special case. We first observe that if there is a representation (6.1.1), as stipulated by Theorem 6.1.2, then the function f appearing in that representation must satisfy the following two properties:

(1) $\int_A f \, d\mu \leq \nu(A)$, for all $A \in \mathcal{A}$, and,

(2) If h is any non-negative measurable function such that $\int_A h \, d\mu \leq \nu(A)$, for all $A \in \mathcal{A}$, then $h \leq f$ a.e.$[\mu]$.

Property (1) is obvious from the representation (6.1.1). To prove the second property, let h satisfy the hypothesis in (2). Then, with N as in (6.1.1), one has

$$\int_A h \, d\mu = \int_{A \cap N^c} h \, d\mu \leq \nu(A \cap N^c) = \int_{A \cap N^c} f \, d\mu = \int_A f \, d\mu, \text{ for any } A \in \mathcal{A}.$$

By Proposition 3.2.19, this implies that $h \leq f$ a.e.$[\mu]$ and property (2) is proved.

We now claim that the converse of the above observation is also true, as stated in the following Lemma.

LEMMA **6.1.8.** *Let μ and ν be two finite measures on $(\Omega, \mathcal{A})$. If f is a non-negative measurable function satisfying properties (1) and (2) above, then there exists a μ-null set N such that representation (6.1.1) holds.*

Proof of Lemma 6.1.8. The key idea which plays an important role in getting the required μ-null set N is that, given a non-negative measurable function f as in the hypothesis of the Lemma, one can construct, for any $\epsilon > 0$, a set $N_\epsilon \in \mathcal{A}$ satisfying

(a) $\mu(N_\epsilon) = 0$, and,

(b) $\nu(N_\epsilon^c) - \int_{N_\epsilon^c} f d\mu \leq \epsilon \mu(\Omega)$

To carry on with the the proof, let us asssume for the time being that this is indeed possible and use it to prove the main assertion of the Lemma. Taking $N = \bigcup_{k \geq 1} N_{1/k}$, it is clear from property (a), that $N \in \mathcal{A}$ is μ-null. We claim now that representation (6.1.1) holds with the given f and the μ-null set N. It should be clear that to prove this claim, we only need to prove that $\nu(A \cap N^c) = \int_A f d\mu$, for every $A \in \mathcal{A}$. Towards this, we first note that $\nu(A) - \int_A f d\mu \leq \nu(B) - \int_B f d\mu$, for any $\mathcal{A}$-sets A, B with $A \subset B$. This is a trivial consequence of property (1) of f applied to the set $B \cap A^c$. In particular, since $N^c \subset N_{1/k}^c$ for all $k \geq 1$, we get $\nu(N^c) - \int_{N^c} f d\mu \leq \nu(N_{1/k}^c) - \int_{N_{1/k}^c} f d\mu \leq \mu(\Omega)/k$, for all $k \geq 1$, where the last inequality is a consequence of property (b) for $N_{1/k}$. As $\mu(\Omega) < \infty$, we get $\nu(N^c) - \int_{N^c} f d\mu = 0$. But this, in turn, implies that $\nu(A \cap N^c) - \int_{A \cap N^c} f d\mu = 0$, for any $A \in \mathcal{A}$ (since $A \cap N^c \subset N^c$), that is, $\nu(A \cap N^c) = \int_{A \cap N^c} f d\mu = \int_A f d\mu$, for every $A \in \mathcal{A}$, as was to be shown.

To complete the proof of Lemma 6.1.8, we now have to show that, for any given $\epsilon > 0$, we can get $N_\epsilon \in \mathcal{A}$, satisfying (a) and (b). For this, we first note that, since ν is a finite measure, property (1) of f implies that $\int_\Omega f d\mu < \infty$, and therefore, $f(\omega) < \infty$ for $[\mu]$-a.e. ω.

We now make a very important observation, which will be repeatedly used in the rest of the proof to produce the required set N_ϵ. Let $A \in \mathcal{A}$ with $\mu(A) > 0$. It is then clear that, for any $\epsilon > 0$, the measurable function $h = (f + \epsilon)I_A$ satisfies $\mu(h > f) = \mu(A) > 0$. Therefore, by property (2) of f, there must exist $B \in \mathcal{A}$, such that, $\int_B h d\mu > \nu(B)$. Noting that, on A^c, $h = 0 \leq f$, we get

$$\int_{B \cap A} (f + \epsilon) d\mu + \int_{B \cap A^c} f d\mu \geq \int_D h d\mu > \nu(B) = \nu(B \cap A) + \nu(B \cap A^c).$$

But property (1) of f implies that $\int_{B \cap A^c} f d\mu \leq \nu(B \cap A^c) < \infty$. So we must have $\int_{B \cap A} (f + \epsilon) d\mu > \nu(B \cap A)$.

Thus, what we have just shown is that, given any A with $\mu(A) > 0$ and any $\epsilon > 0$, there exists $\widetilde{A} \subset A$ such that

$$\int_{\widetilde{A}} (f + \epsilon) d\mu > \nu(\widetilde{A}). \tag{6.1.4}$$

We are now going to use this fact repeatedly. Fix $\epsilon > 0$. Our aim is to extract a disjoint sequence $\{A_n \in \mathcal{A}\}$, finite or infinite, such that $N_\epsilon = \left(\bigcup_n A_n \right)^c$ will be a μ-null set satisfying $\nu(N_\epsilon^c) - \int_{N_\epsilon^c} f d\mu \leq \epsilon \mu(\Omega)$. Towards this, we apply the above observation, first with $A = \Omega$, to get

$$\beta_0 = \sup \left\{ \int_A (f + \epsilon) d\mu - \nu(A) : A \in \mathcal{A} \right\} > 0.$$

Choose a set $A_1 \in \mathcal{A}$ with

$$\int_{A_1} (f + \epsilon) d\mu - \nu(A_1) \geq \tfrac{1}{2} \beta_0.$$

6.1. Radon-Nikodym Theorem, Lebesgue Decomposition

If $\mu(A_1^c) = 0$, we stop. If, on the other hand, $\mu(A_1^c) > 0$, then we apply the above observation, now with $A = A_1^c$, to get

$$\beta_1 = \sup\left\{\int_A (f+\epsilon)d\mu - \nu(A) : A \in \mathcal{A}, A \subset A_1^c\right\} > 0.$$

We choose $A_2 \in \mathcal{A}$ with $A_2 \subset A_1^c$, such that

$$\int_{A_2} (f+\epsilon)d\mu - \nu(A_2) \geq \tfrac{1}{2}\beta_1.$$

If $\mu(A_1^c \cap A_2^c) = 0$, then we stop. Otherwise, we use the same argument again, with $A = A_1^c \cap A_2^c$, to get $A_3 \in \mathcal{A}$ with $A_3 \subset A_1^c \cap A_2^c$ such that

$$\int_{A_3} (f+\epsilon)d\mu - \nu(A_3) \geq \tfrac{1}{2}\beta_2,$$

where

$$\beta_2 = \sup\left\{\int_A (f+\epsilon)d\mu - \nu(A) : A \in \mathcal{A}, A \subset A_1^c \cap A_2^c\right\} > 0.$$

Proceeding this way, we can produce a finite or infinite sequence $\{A_n\}$ of disjoint $\mathcal{A}$-sets such that, for each n,

(i) $\beta_{n-1} = \sup\{\int_A (f+\epsilon)d\mu - \nu(A) : A \in \mathcal{A}, A \subset A_1^c \cap \cdots \cap A_{n-1}^c\} > 0$,

(ii) $A_n \in \mathcal{A}$ with $A_n \subset A_1^c \cap \cdots \cap A_{n-1}^c$,

(iii) $\int_{A_n} (f+\epsilon)d\mu - \nu(A_n) \geq \tfrac{1}{2}\beta_{n-1}$.

Taking $N_\epsilon = \left(\bigcup_n A_n\right)^c$, we now prove the set $N_\epsilon \in \mathcal{A}$ satisfies the required properties (a) and (b).

As for (b), note first that (iii) above implies $\int_{A_n} f d\mu + \epsilon\mu(A_n) - \nu(A_n) > 0$, for each n. Since all three terms are finite, this gives $\nu(A_n) - \int_{A_n} f d\mu < \epsilon\mu(A_n)$, for each n. Adding over n and using the fact that the sets A_n are disjoint, one easily gets $\nu(N_\epsilon^c) - \int_{N_\epsilon^c} f d\mu \leq \epsilon\mu(N_\epsilon^c) \leq \epsilon\mu(\Omega)$.

As for (a), first recall that our procedure yields a finite sequence $\{A_n\}$, say, $A_1, \cdots A_k$, only if $\mu(A_1^c \cap \cdots \cap A_k^c) = 0$. Thus, (a) holds trivially in this case. Let us therefore consider the case when our procedure yielded an infinite sequence $\{A_n\}$, which of course means that we also have an infinite sequence $\{\beta_n\}$ of positive real numbers as in (i). We claim that $\beta_n \downarrow 0$. For this, it suffices to show that $\beta_n \leq \tfrac{1}{2}\beta_{n-1}$, for all $n \geq 1$. Suppose, if possible, this is not true, that is, $\beta_n > \tfrac{1}{2}\beta_{n-1}$, for some n. What this would mean is that, for some $n \geq 1$, there is a set $A \in \mathcal{A}$ with $A \subset A_1^c \cap \cdots \cap A_n^c$, such that, $\int_A (f+\epsilon)d\mu - \nu(A) > \tfrac{1}{2}\beta_{n-1}$. Taking $B = A \cup A_n \in \mathcal{A}$, and recalling that $\int_{A_n} (f+\epsilon)d\mu - \nu(A_n) > \tfrac{1}{2}\beta_{n-1}$, by our construction of A_n, we will get

$$\int_B (f+\epsilon)d\mu - \nu(B) = \int_A (f+\epsilon)d\mu \quad \nu(A) + \int_{A_n} (f+\epsilon)d\mu - \nu(A_n) > \beta_{n-1}.$$

Since $B \subset A_1^c \cap \cdots \cap A_{n-1}^c$, this contradicts the definition of β_{n-1}.

Having thus established that $\beta_n \downarrow 0$, in case $\{A_n\}$ is an infinite sequence, we now complete the proof of (a), as follows. Suppose, if possible, $\mu(N_\epsilon) > 0$. But then, by the observation made earlier (see (6.1.4)), there is a set $\widetilde{A} \in \mathcal{A}$ with $\widetilde{A} \subset N_\epsilon$ such that

$$\int_{\widetilde{A}} (f+\epsilon)d\mu - \nu(\widetilde{A}) > 0.$$

Since $\beta_n \downarrow 0$, this would mean that
$$\int_{\widetilde{A}} (f+\epsilon)d\mu - \nu(\widetilde{A}) > \beta_{n-1}, \text{ for some } n \geq 1.$$

But then, with $B = \widetilde{A} \cup A_n \in \mathcal{A}$, we clearly have $B \subset A_1^c \cap \cdots \cap A_{n-1}^c$, and, argunig as above, we will get
$$\int_B (f+\epsilon)d\mu - \nu(B) > \frac{3}{2}\beta_{n-1},$$

contradicting the definition of β_{n-1}. Thus, we must have $\mu(N_\epsilon) = 0$. This now completes the proof of (a) and, with that, the proof of Lemma 6.1.8. □

Let us now take a stock of where we stand as far as proving Theorem 6.1.2 is concerned. What Lemma 6.1.8 shows is that, for proving Theorem 6.1.2, at least for finite measures μ and ν, it is enough to get hold of a non-negative measurable function f satisfying the properties (1) and (2). Thus, the question that we have to address now is, given finite measures μ and ν on $(\Omega, \mathcal{A})$, whether we can always get such a non-negative measurable f. To address that question, it is important to understand what the properties (1) and (2) really demand. The property (1), in itself, is not an issue because the function $f \equiv 0$ satisfies it trivially. It is the property (2) which is more crucial. It roughly says that f is, in some sense, the 'largest' among all non-negative measurable functions satisfying (1). This may tempt the reader to momentarily think that there is a trivial solution, namely, taking f to be the pointwise supremum of the class of all non-negative measurable functions on $(\Omega, \mathcal{A})$ satisfying (1) (which is a non-empty class, as pointed out already). However, the reader will also quickly realize the problem with this simple-minded solution. Firstly, this pointwise supremum may not automatically give us a measurable function and secondly, property (1) may not hold for the pointwise supremum. Indeed, a quick reflection would tell us that going for the pointwise supremum is too strong a prescription for what is actually being asked for. To be precise, the pointwise supremum will dominate any function satisfying property (1) *at all points* ω. However, what is really required in property (2) is that f must dominate any h satisfying property (1), only for $[\mu]$-almost every ω. Further, the μ-null set of ω's, where domination may fail, is allowed to vary with h. All these simply point to the fact that the requred f doesn't have to be the pointwise supremum of all non-negative measurable functions satisfying (1). And, that brings us back to the question: what is the required function f, if not the pointwise supremum? The concept that we now define identifies exactly what we are looking for.

DEFINITION **6.1.9**. Let $(\Omega, \mathcal{A}, \mu)$ be a measure space and $\mathcal{H}$ be a non-empty class of $\mathcal{A}$-measurable functions. A function f, measurable w.r.t. $\mathcal{A}$, is called an *essential supremum* of the class $\mathcal{H}$, denoted ess sup $\mathcal{H}$, if it satisfies
(a) $h \leq f$ a.e.$[\mu]$ for all $h \in \mathcal{H}$, and,
(b) $f \leq g$ a.e.$[\mu]$, for any $\mathcal{A}$-measurable g, such that, $h \leq g$ a.e.$[\mu]$ for all $h \in \mathcal{H}$.

6.1. Radon-Nikodym Theorem, Lebesgue Decomposition

REMARK 6.1.10. It is clear from the definition that ess sup $\mathcal{H}$, if one exists, is unique upto μ-null sets. Further, if f is an essential supremum for a non-empty class $\mathcal{H}$, then any measurable f' with $f' = f$ a.e.$[\mu]$, is also an essential supremum for $\mathcal{H}$.

In proving Theorem 6.1.2 for finite measures μ and ν, by exhibiting a non-negative measurable f satisfying properties (1) and (2), it should now be clear that it is the essential supremum of all non-negative measurable functions satisfying (1), that will do the job, provided, of course, such an essential supremum exists. In what follows, we first resolve this existence issue by proving a general result which asserts that for any non-empty family of measurable functions on a σ-finite measure space, an essential supremum always exists. We follow it up by observing some simple properties of essential supremum and proving some related results. After all of these, we will finally return to the proof of Theorem 6.1.2 and complete it.

THEOREM 6.1.11. For any non-empty class $\mathcal{H}$ of measurable functions on a σ-finite measure space $(\Omega, \mathcal{A}, \mu)$, an essential supremum of $\mathcal{H}$ exists.

Proof. Firstly, by standard arguments that we have repeatedly seen before, it is enough to prove the result when μ is a finite measure. Secondly, when $\mu \equiv 0$, any measurable function will act as an ess sup $\mathcal{H}$. Therefore, we need to consider only a non-zero finite measure μ. But since any such measure is a positive scalar multiple of a probability measure, we may as well assume that the given measure μ is a probability measure.

Next, by virtue of the order preserving homeomorphism $x \mapsto \frac{1}{2} + \frac{1}{\pi}\tan^{-1} x$ on $[-\infty, \infty]$ onto $[0, 1]$, there is no loss of generality in assuming that $0 \leq h \leq 1$, for all $h \in \mathcal{H}$.

For each countable $\mathcal{T} \subset \mathcal{H}$, let $f_\mathcal{T}$ denote the pointwise supremum of $\{h : h \in \mathcal{T}\}$. Note that $f_\mathcal{T}$ is measurable and $0 \leq f_\mathcal{T} \leq 1$, for each countable $\mathcal{T} \subset \mathcal{H}$. Let

$$\gamma = \sup\left\{\int f_\mathcal{T} \, d\mu : \mathcal{T} \subset \mathcal{H}, \mathcal{T} \text{ countable}\right\}.$$

Since μ is a probability measure, we have $0 \leq \int f_\mathcal{T} \, d\mu \leq 1$, for each countable $\mathcal{T} \subset \mathcal{H}$, and therefore $0 \leq \gamma \leq 1$. Let $\{\mathcal{T}_n\}$ be a sequence of countable subsets of $\mathcal{H}$ with $\mathcal{T}_n \subset \mathcal{T}_{n+1}$, such that, $\int f_{\mathcal{T}_n} \, d\mu \uparrow \gamma$. Clearly, $\{f_{\mathcal{T}_n}\}$ is a non-decreasing sequence of non-negative measurable functions, and so we can use Monotone Convergence Theorem to see that $\gamma = \int f \, d\mu$, where $f = \lim_n f_{\mathcal{T}_n}$.

We now show that $f = \text{ess sup}\,\mathcal{H}$. Clearly, $f = f_\mathcal{S}$, where $\mathcal{S}$ is the countable set $\bigcup_n \mathcal{T}_n$. Now, for any $h \in \mathcal{H}$, we have $f \leq f_{\mathcal{S} \cup \{h\}}$, while $\int f_{\mathcal{S} \cup \{h\}} \, d\mu \leq \gamma = \int f \, d\mu$, by definition of γ. It follows that $f_{\mathcal{S} \cup \{h\}} \leq f$ $[\mu]$-a.e., implying $h \leq f$ $[\mu]$-a.e. On the other hand, if g is any measurable function such that, $h \leq g$ a.e.$[\mu]$ for all $h \in \mathcal{H}$, then $f = \sup\{h : h \in \mathcal{S}\} \leq g$ a.e.$[\mu]$. This completes the proof. □

While Theorem 6.1.11 guarantees the existence of an essential supremum for any non-empty class of measurable functions on a σ-finite measure space, it does not guarantee that such an essential supremum would belong to the given class of functions. However, if one goes through the proof carefully and examines how a candidate for ess sup $\mathcal{H}$ was constructed, one would be able to easily identify sufficient conditions on $\mathcal{H}$ that ensures that an ess sup $\mathcal{H}$ can be chosen from the class $\mathcal{H}$. This is precisely stated in the following corollary, whose proof is an easy consequence of the proof of Theorem 6.1.11.

COROLLARY **6.1.12.** If $\mathcal{H}$ is a non-empty class of measurable functions on a σ-finite measure space satisfying the two properties (a) $h_1, h_2 \in \mathcal{H} \implies \max\{h_1, h_2\} \in \mathcal{H}$ and (b) $h_n \in \mathcal{H}$, for $n \geq 1$, and $h_n \uparrow h \implies h \in \mathcal{H}$, then there is an $h \in \mathcal{H}$ such that $h = \text{ess sup } \mathcal{H}$.

As a nice application of the idea of essential supremum, let us recall that the union of an arbitrary collection of sets in a σ-field $\mathcal{A}$ need not belong to $\mathcal{A}$. However, given a σ-finite measure μ on $\mathcal{A}$, it is always possible to find a set in $\mathcal{A}$, which will work 'μ-almost' like a union of the sets in the given collection. Such a set in $\mathcal{A}$ is called the 'essential union' of the given collection of sets. Further, under certain conditions on the collection, one may even choose an essential union from the collection itself. All these follow as easy consequences of Theorem 6.1.11 and Corollary 6.1.12 and are given in the next exercise.

EXERCISE **6.1.13.** (a) Let $(\Omega, \mathcal{A}, \mu)$ be a σ-finite measure space. Show that, given any non-empty collection $\mathcal{C}$ of sets in $\mathcal{A}$, there is a set $A \in \mathcal{A}$, such that, (i) $C \subset A$ a.e.$[\mu]$, for all $C \in \mathcal{C}$ and (ii) $B \in \mathcal{A}$ and $C \subset B$ a.e.$[\mu]$, for all $C \in \mathcal{C}$ imply that $A \subset B$ a.e.$[\mu]$.
(Here the statement $C \subset D$ a.e.$[\mu]$ means that $\mu(C \cap D^c) = 0$. See this in contrast with $C \subset D$, which is equivalent to $C \cap D^c = \emptyset$.)
(b) Show that, if the collection $\mathcal{C}$ above is closed under countable unions, then it is possible to get an $A \in \mathcal{C}$ that satisfies (i) and (ii).

EXERCISE **6.1.14.** Define the notions of essential infimum and essential intersection and prove their existence in a σ-finite measure space.

EXERCISE **6.1.15.** Let $\mathcal{A}$ denote the countable-cocountable σ-field $\mathcal{A}$ on $\mathbf{R}$ and let μ denote the counting measure on $\mathcal{A}$. Show that, for the class of measurable functions given by $\mathcal{H} = \{I_{\{x\}} : 0 \leq x \leq 1\}$, there does not exist an ess sup $\mathcal{H}$. Why is this not a violation of Theorem 6.1.11?

We have now completed all the necessary intermediate steps for getting back to the long awaited proof of Theorem 6.1.2.

Proof of Theorem 6.1.2. The assertion is trivially true if μ is the zero measure. Taking $f \equiv 0$ and $N = \Omega$ would do the job in that case.
Next, let us observe that there is no loss of generality in assuming that both μ and ν are finite measures. Indeed, we can otherwise get a partition of Ω by a sequence $\{\Omega_n, n \geq 1\}$ of $\mathcal{A}$-sets, such that, $\mu(\Omega_n) < \infty$ and $\nu(\Omega_n) < \infty$, for each n. Then, for each n, the measures μ_n and ν_n, defined by $\mu_n(A) = \mu(A \cap \Omega_n)$, $A \in \mathcal{A}$, and

6.1. Radon-Nikodym Theorem, Lebesgue Decomposition

$\nu_n(A) = \nu(A \cap \Omega_n)$, $A \in \mathcal{A}$, are both finite measures, supported on Ω_n. Suppose, for each n, we can get a representation

$$\nu_n(A) = \int_A f_n d\mu_n + \nu_n(A \cap N_n), \text{ for all } A \in \mathcal{A},$$

where, for each n, f_n is a non-negative measurable function and N_n is a μ_n-null set, Since ν_n is supported on Ω_n, we may assume $N_n \subset \Omega_n$ (replacing N_n by $N_n \cap \Omega_n$, if necessary). It is then easy to see that the non-negative measurable function $f = \sum_n f_n I_{\Omega_n}$ and the μ-null set $N = \bigcup_n N_n$ will prove the assertion (6.1.1) for the original σ-finite measures μ and ν.

In view of the above, we now assume that μ and ν are both finite measures, with $\mu \not\equiv 0$, and proceed to prove the existence of a non-negative measurable f and a μ-null set N such that (6.1.1) holds. By virtue of Lemma 6.1.8, all we need to do is to prove the existence of a non-negative measurable f satisfying properties (1) and (2). We achieve this by following the track mentioned in the paragraph preceding Theorem 6.1.11, that is, we consider the class

$$\mathcal{H} = \{h : h \text{ non-negative } \mathcal{A}\text{-measurable}, \int_A h d\mu \leq \nu(A) \text{ for all } A \in \mathcal{A}\}.$$

$\mathcal{H}$ is non-empty, because $h \equiv 0$ belongs to $\mathcal{H}$. Next, if $h_1, h_2 \in \mathcal{H}$ and $h = \max\{h_1, h_2\}$, then $h = h_1 I_B + h_2 I_{B^c}$, where $B = \{h_1 > h_2\}$. Now, for any $A \in \mathcal{A}$, one has $\int_A h d\mu = \int_{A \cap B} h_1 d\mu + \int_{A \cap B^c} h_2 d\mu \leq \nu(A \cap B) + \nu(A \cap B^c) = \nu(A)$, showing that $h = \max\{h_1, h_2\} \in \mathcal{H}$. Finally, it is an easy consequence of the Monotone Convergence Theorem that if $h_n \in \mathcal{H}$ for all n and $h_n \uparrow h$, then $h \in \mathcal{A}$. It now follows from Corollary 6.1.12, that there is $f \in \mathcal{H}$ such that $f = \text{ess sup } \mathcal{H}$. One can easily see that this non-negative $\mathcal{A}$-measurable function f satisfies properties (1) and (2) and that completes the proof. □

We return to Radon-Nikodym Theorem and discuss an important concept that arises in the context of that theorem. As we recall, Radon-Nikodym Theorem says that if μ and ν are σ-finite measures on a σ-field $\mathcal{A}$, then $\nu \ll \mu$ iff there is a non-negative $\mathcal{A}$-measurable f such that $\nu(A) = \int_A f d\mu$, for each $A \in \mathcal{A}$. It was already noted that the function f is unique upto μ-null sets, that is, if any other non-negative $\mathcal{A}$-measurable function g satisfies $\nu(A) = \int_A g d\mu$, for each $A \in \mathcal{A}$, then $g = f$ a.e.$[\mu]$ (and consequently, a.e.$[\nu]$ as well).

DEFINITION **6.1.16.** For σ-finite measures μ and ν on $\mathcal{A}$ with $\nu \ll \mu$, a function f satisfying $\nu(A) = \int_A f d\mu$, for each $A \in \mathcal{A}$, is called a *Radon-Nikodym derivative* (*RN derivative*, in short) of ν with respect to μ and is denoted as $\frac{d\nu}{d\mu}$. Sometimes, instead of writing $f = \frac{d\nu}{d\mu}$, one also writes $d\nu = f d\mu$ to mean the same thing.

The next exercise is an easy consequence of Definition 6.1.16. It proves some simple properties of RN derivatives, that show that RN derivatives, when they exist, behave in a fashion very similar to the derivatives of functions that one learns in calculus.

EXERCISE 6.1.17. (a) If $\nu \ll \mu$ and $\alpha \in (0, \infty)$, then show that $\alpha\nu \ll \mu$ and that, in case μ and ν are also σ-finite, then $\dfrac{d(\alpha\nu)}{d\mu} = \alpha \dfrac{d\nu}{d\mu}$.
(b) Show that $\nu_1 \ll \mu$ and $\nu_2 \ll \mu$ imply that $(\nu_1 + \nu_2) \ll \mu$. Show also that if, in addition, μ, ν_1 and ν_2 are all σ-finite also, then $\dfrac{d(\nu_1 + \nu_2)}{d\mu} = \dfrac{d\nu_1}{d\mu} + \dfrac{d\nu_2}{d\mu}$.

The reader should realize that, since the RN derivatives are unique only upto μ-null sets, any statement involving such derivatives should be regarded as an a.e.$[\mu]$ statement, that is, equalities in the above exercise and also in statements below, are only a.e.$[\mu]$ equalities.

PROPOSITION 6.1.18. Let μ, ν, η be σ-finite measures on a measurable space $(\Omega, \mathcal{A})$.
(a) If $\nu \ll \mu$ and $f = \dfrac{d\nu}{d\mu}$, then $\int g d\nu = \int gf d\mu$, for any measurable function g on $(\Omega, \mathcal{A})$, where the equality of the two integrals means that if one of them exists, then so does the other and, in that case, they are equal.
(b) If $\eta \ll \nu$ and $\nu \ll \mu$. then $\eta \ll \mu$ and further, $\dfrac{d\eta}{d\mu} = \dfrac{d\eta}{d\nu} \cdot \dfrac{d\nu}{d\mu}$.
(c) If $\nu \ll \mu$ and $\dfrac{d\nu}{d\mu} > 0$ a.e.$[\mu]$, then $\mu \ll \nu$ and further, $\dfrac{d\mu}{d\nu} = \left(\dfrac{d\nu}{d\mu}\right)^{-1}$.

Proof. (a) By the definition of the RN derivative $\dfrac{d\nu}{d\mu}$, the assertion holds when $g = I_A$, for any $A \in \mathcal{A}$. Proof for general measurable g can now be completed by going through the standard steps.
(b) That the relation $\ll$ is a transitive relation is immediate from definition, while the other part follows from (a), by taking $g = I_A \dfrac{d\eta}{d\nu}$, for any $A \in \mathcal{A}$.
(c) Taking $\Omega_0 = \{\omega : \dfrac{d\nu}{d\mu}(\omega) > 0\}$, one can easily show, using (a), that the function $h = \left(\dfrac{d\nu}{d\mu}\right)^{-1} I_{\Omega_0}$ satisfies $\mu(A) = \int_A h d\nu$, for all $A \in \mathcal{A}$. □

DEFINITION 6.1.19. Measures μ and ν on a σ-field $\mathcal{A}$ are said to be *equivalent* (or, *mutually absolutely continuous*) and we write $\mu \sim \nu$ if both $\nu \ll \mu$ and $\mu \ll \nu$ hold.

EXERCISE 6.1.20. Show that, for σ-finite measures μ and ν, the condition $\mu \sim \nu$ is equivalent to each of the following two conditions:
(a) $\nu \ll \mu$ and $\dfrac{d\nu}{d\mu} > 0$ a.e.$[\mu]$ (b) $\mu \ll \nu$ and $\dfrac{d\mu}{d\nu} > 0$ a.e.$[\nu]$.

EXERCISE 6.1.21. For σ-finite measures $\nu \ll \mu$, show that $\dfrac{d\nu}{d\mu} > 0$ a.e.$[\nu]$.

EXERCISE 6.1.22. (a) Show that, given any σ-finite measure μ on a σ-field $\mathcal{A}$, there is a probability measure P on $\mathcal{A}$ such that $\mu \sim P$.
(b) Show that, given any sequence $\{\mu_n\}$ of σ-finite measures on a σ-field $\mathcal{A}$, there is a probability measure P on $\mathcal{A}$ such that, $\mu_n \ll P$ for all n.

EXERCISE 6.1.23. Let $\Omega = \mathbb{N}$ and let $\mathcal{A}$ be the σ-field on Ω, consisting of all subsets. For any two measures ν and μ on $\mathcal{A}$, show that $\nu \ll \mu$ iff $\nu\{n\} = 0$ whenever $\mu\{n\} = 0$. Show also that, if $\nu \ll \mu$ are σ-finite measures on $\mathcal{A}$, then

$$\frac{d\nu}{d\mu}(\cdot) = \frac{\nu(\{\cdot\})}{\mu(\{\cdot\})} I_{\{n : \mu(\{n\}) \neq 0\}}(\cdot)$$

6.1. Radon-Nikodym Theorem, Lebesgue Decomposition

In particular, denoting μ to be the counting measure on $\mathcal{A}$, show that $\nu \ll \mu$ for any measure ν on $\mathcal{A}$ and that, for any σ-finite measure ν on $\mathcal{A}$, the RN derivative $\frac{d\nu}{d\mu}$ is given by $\frac{d\nu}{d\mu}(n) = \nu(\{n\})$, $n \in \mathbb{N}$.

EXERCISE **6.1.24.** (a) On $(\mathbb{R}, \mathcal{B})$, consider the Radon measure ν determined by the distribution function G on $\mathbb{R}$, defined as $G(x) = x^2$, for $x \geq 0$, and $G(x) = -x^2$, for $x < 0$. Denoting the Lebesgue measure on $\mathcal{B}$ by λ, show that $\nu \ll \lambda$ and that $\frac{d\nu}{d\lambda}(x) = 2|x|$. Examine if $\lambda \ll \nu$.
(b) More generally, suppose that ν is the Radon measure on $(\mathbb{R}, \mathcal{B})$, determined by a distribution function G on $\mathbb{R}$, which is continuously differentiable everywhere. With λ as in (a), show that $\nu \ll \lambda$ and find the RN derivative. Determine conditions under which $\lambda \ll \nu$.

EXERCISE **6.1.25.** Let $(\Omega, \mathcal{A})$ denote the product space $\left(\underset{j=1}{\overset{k}{\times}} \Omega_j, \underset{j=1}{\overset{k}{\otimes}} \mathcal{A}_j \right)$.

(a) Consider the product measures $\mu = \underset{j=1}{\overset{k}{\otimes}} \mu_j$ and $\nu = \underset{j=1}{\overset{k}{\otimes}} \nu_j$ on $(\Omega, \mathcal{A})$, where μ_j and ν_j, for each $1 \leq j \leq k$, are σ-finite measures on $(\Omega_j, \mathcal{A}_j)$ with $\nu_j \ll \mu_j$. Show that $\nu \ll \mu$ and express $\frac{d\nu}{d\mu}$ in terms of $\frac{d\nu_j}{d\mu_j}$, $1 \leq j \leq k$.
(b) Fix any $l < k$ and let $\varphi_l : \omega = (\omega_1, \cdots, \omega_k) \mapsto (\omega_1, \cdots, \omega_l)$ denote the natural projection on Ω onto $\Omega_1 \times \cdots \times \Omega_l$. Show that, if ν and μ are any two σ-finite measures on $\mathcal{A}$ with $\nu \ll \mu$, then $\nu\varphi_l^{-1} \ll \mu\varphi_l^{-1}$. In the special case, when μ and ν are finite measures on $\mathcal{A}$ with $\nu \ll \mu$, find $\frac{d(\nu\varphi_l^{-1})}{d(\mu\varphi_l^{-1})}$.

The Radon-Nikodym theorem, that we have proved earlier, concerns two σ-finite measures, μ and ν with $\nu \ll \mu$. We are now going to extend it. We are going to show that a similar result holds when μ is a σ-finite measure, but ν is what is called a finite 'signed' measure, as defined below.

DEFINITION **6.1.26.** Let $(\Omega, \mathcal{A})$ be a measurable space. A *finite signed measure* ν on $\mathcal{A}$ is a set function $\nu : \mathcal{A} \to \mathbb{R}$ satisfying: (i) $\nu(\emptyset) = 0$ and (ii) countable additivity, that is, $\nu(\underset{n}{\bigcup} A_n) = \underset{n}{\sum} \nu(A_n)$, for any sequence $\{A_n\}$ of disjoint $\mathcal{A}$-sets.

A finite signed measure is like a finite measure except that it is allowed to take negative values. It is easy to see that if ν_1 and ν_2 are two finite measures on $\mathcal{A}$, then $\nu(A) = \nu_1(A) - \nu_2(A)$, for $A \in \mathcal{A}$, defines a finite signed measure on $\mathcal{A}$. It turns out that every finite signed measure arises this way (see Theorem 6.1.28 below). But before that, we make an interesting observation about finite signed measures. Any finite measure μ on $(\Omega, \mathcal{A})$ is automatically bounded; indeed, $\mu(A) \leq \mu(\Omega)$, for all $A \in \mathcal{A}$. Interestingly, the same is true of any finite signed measure as well.

LEMMA **6.1.27.** *Let $(\Omega, \mathcal{A})$ be a measurable space. Then, any finite signed measure ν on $\mathcal{A}$ is bounded.*

Proof. Since $-\nu$ is also a finite signed measure, it is enough to show that ν is bounded above. We prove this by showing that, if this were false, then we can get a sequence A_n, $n \geq 1$, of disjoint $\mathcal{A}$-sets, with $|\nu(A_n)| > 1$ for all n. This would be a contradiction to countable additivity, since it would prevent the series $\sum_n \nu(A_n)$ to converge to the finite limit $\nu(\bigcup_n A_n)$. Here is how we get such a sequence $\{A_n\}$. Firstly, if ν is unbounded above, then there is a set $B \in \mathcal{A}$, such that, $\nu(B) > |\nu(\Omega)| + 1 \geq \nu(\Omega) + 1$. This would, of course, mean that $\nu(B) > 1$ and also $\nu(B^c) = \nu(\Omega) - \nu(B) < -1$. Now note that, since ν is assumed to be unbounded above, at least one of the sets $\{\nu(A) : A \subset B, A \in \mathcal{A}\}$ and $\{\nu(A) : A \subset B^c, A \in \mathcal{A}\}$ must be unbounded above. In case the first set is not bounded above, we take $A_1 = B^c$, while in the other case, we take $A_1 = B$. One can easily see that, in either case, we get a set $A_1 \in \mathcal{A}$, such that, $|\nu(A_1)| > 1$ and $\{\nu(A) : A \subset A_1^c, A \in \mathcal{A}\}$ is not bounded above. Using the same argument as above, we can now get an $\mathcal{A}$-set $A_2 \subset A_1^c$, such that $|\nu(A_2)| > 1$ and also, the set $\{\nu(A) : A \subset A_1^c \cap A_2^c, A \in \mathcal{A}\}$ is unbounded above. Proceeding in the same manner, we can then get an $\mathcal{A}$-set $A_3 \subset A_1^c \cap A_2^c$ such that $|\nu(A_3)| > 1$ and also, the set $\{\nu(A) : A \subset A_1^c \cap A_2^c \cap A_3^c, A \in \mathcal{A}\}$ unbounded above. By repeating this argument, we will get the sequence $\{A_n\}$, as proposed. $\square$

We now proceed to prove the important result, as mentioned above, that every finite signed measure arises as the difference of two finite measures. This is known as the *Jordan Decomposition Theorem*.

THEOREM **6.1.28.** (Jordan Decomposition Theorem) Let $(\Omega, \mathcal{A})$ be a measurable space and ν be a finite signed measure on $\mathcal{A}$. Then there are finite measures ν^+ and ν^- on $\mathcal{A}$ with $\nu^+ \geq \nu$ and $\nu^- \geq -\nu$ such that $\nu(A) = \nu^+(A) - \nu^-(A)$ for all $A \in \mathcal{A}$.

REMARK **6.1.29.** Note that the decomposition as asserted by Theorem 6.1.28 is not unique because if ν^+ and ν^- are as in the theorem, then we can always write ν also as $\nu = (\nu^+ + \nu_0) - (\nu^- + \nu_0)$, where ν_0 is any non-zero finite measure. However, the decomposition that we get in our proof is optimal in some sense (see Exercise 6.1.31 below.)

Proof. Define ν^+ on $\mathcal{A}$ by $\nu^+(A) = \sup\{\nu(B) : B \subset A, B \in \mathcal{A}\}$. ν^+ is clearly a non-negative set function, which is also finite by virtue of Lemma 6.1.27. It is clear that $\nu^+(\emptyset) = 0$. We now proceed to show that ν^+ is countably additive. Towards this, let $\{A_n, n \geq 1\}$ be a sequence of disjoint $\mathcal{A}$-sets and let $A = \bigcup_n A_n$. For any $\mathcal{A}$-set $B \subset A$, the sets $B_n = A_n \cap B$, $n \geq 1$, are disjoint $\mathcal{A}$-sets and $\nu^+(A_n) \geq \nu(B_n)$ so that $\nu(B) = \sum_n \nu(B_n) \leq \sum_n \nu^+(A_n)$. Since this is true for any $\mathcal{A}$-set $B \subset A$, we get $\nu^+(A) \leq \sum_n \nu^+(A_n)$. On the other hand, fixing any $\epsilon > 0$ we can get, for every $n \geq 1$, a $\mathcal{A}$-set $B_n \subset A_n$ with $\nu^+(A_n) < \nu(B_n) + 2^{-n}\epsilon$. This will give $\sum_n \nu^+(A_n) \leq \nu(\bigcup_n B_n) + \epsilon \leq \nu^+(A) + \epsilon$, the last inequality being a consequence of $\bigcup_n B_n \subset A$. As $\epsilon > 0$ was arbitrary, we get $\sum_n \nu^+(A_n) \leq \nu^+(A)$. This, coupled with the already proved reverse inequality, establishes countable additivity of ν^+, proving thereby that ν^+ is a finite measure on $\mathcal{A}$.

6.1. Radon-Nikodym Theorem, Lebesgue Decomposition

Next note that, by definition, $\nu^+(A) \geq \nu(A)$ for any $A \in \mathcal{A}$. This would mean that $\nu^-(A) = \nu^+(A) - \nu(A)$, $A \in \mathcal{A}$, would also define a finite measure and would satisfy $\nu^-(A) \geq -\nu(A)$, for every $A \in \mathcal{A}$. The proof is complete. □

EXERCISE 6.1.30. *Show that ν^- as obtained in the proof of Theorem 6.1.28 is also given by $\nu^-(A) = -\inf\{\nu(B) : B \subset A, B \in \mathcal{A}\}$, $A \in \mathcal{A}$.*

EXERCISE 6.1.31. *Let ν be as in Theorem 6.1.28 and ν^+, ν^- be as in its proof. Show that if ν_1 is any measure on $\mathcal{A}$ with $\nu_1(A) \geq \nu(A)$ for all $A \in \mathcal{A}$, then $\nu_1 \geq \nu^+$. In other words, ν^+ is the smallest measure on $\mathcal{A}$, which dominates the finite signed measure ν.*
Show analogously that ν^- is the smallest measure on $\mathcal{A}$, which dominates the finite signed measure $-\nu$.

DEFINITION 6.1.32. Let $(\Omega, \mathcal{A}, \mu)$ be a σ-finite measure space. A finite signed measure ν on $\mathcal{A}$ is said to be *absolutely continuous* with respect to μ, and we write $\nu \ll \mu$ as before, if $A \in \mathcal{A}$ and $\mu(A) = 0$ imply that $\nu(A) = 0$.

EXERCISE 6.1.33. *Show that a finite signed measure ν is absolutely continuous with respect to a σ-finite measure μ iff both ν^+ and ν^- are absolutely continuous with respect to μ, where ν^+ and ν^- are as in the proof of the Jordan Decomposition Theorem.*

THEOREM 6.1.34. (Radon-Nikodym Theorem for finite signed measures) *Let $(\Omega, \mathcal{A})$ be a measurable space and μ a σ-finite measure on $\mathcal{A}$. A finite signed measure ν on $\mathcal{A}$ is absolutely continuous with respect to μ if and only if there exists a μ-integrable function f, unique upto μ-null sets, such that $\nu(A) = \int_A f\,d\mu$, for all $A \in \mathcal{A}$.*

Proof. The 'if' part is immediate from properties of integrals (Proposition 3.2.14). Also, Corollary 3.2.21 implies that such a function f, if one exists, is unique upto μ-null sets. Thus, we only have to prove that if $\nu \ll \mu$, then such an integrable function f exists. By Exercise 6.1.33, $\nu \ll \mu$ implies that $\nu^+ \ll \mu$ and $\nu^- \ll \mu$. Using Radon-Nikodym Theorem for the finite measures ν^+ and ν^-, we get non-negative integrable functions f_1 and f_2, such that $\nu^+(A) = \int_A f_1\,d\mu$ and $\nu^-(A) = \int_A f_2\,d\mu$, for all $A \in \mathcal{A}$. Note that both f_1 and f_2 are finite a.e.[μ], so that the function $f = f_1 - f_2$ is well defined a.e.[μ]. It is clear that f is an integrable function and does the required job. □

DEFINITION 6.1.35. The function f as in Theorem 6.1.34 is called the *RN derivative* of the finite signed measure ν with respect to μ and is denoted $\frac{d\nu}{d\mu}$.

EXERCISE 6.1.36. *Examine which properties of RN derivatives of (non-negative) σ-finite measures go through for RN derivatives of finite signed measures.*

We now present another important result which gives some more insight into the Jordan Decomposition Theorem. In particular, it shows that the finite measures ν^+ and ν^-, as constructed in the proof of Jordan Decomposition Theorem, are supported on disjoint subsets of Ω. Jordan Decomposition Theorem is often presented together with the next theorem and is referred to as *Hahn-Jordan Decomposition Theorem*. But since we are presenting it here as a separate theorem, we are going to assume what was proved in the Jordan Decomposition Theorem.

THEOREM **6.1.37.** (Hahn Decomposition Theorem) Let ν be a finite signed measure on a measurable space $(\Omega, \mathcal{A})$ and let ν^+ and ν^- be the finite measures on $\mathcal{A}$, defined as $\nu^+(A) = \sup\{\nu(B) : B \subset A, B \in \mathcal{A}\}$, $\nu^-(A) = -\inf\{\nu(B) : B \subset A, B \in \mathcal{A}\}$, for $A \in \mathcal{A}$. Then, there exists a partition of Ω into disjoint $\mathcal{A}$-sets Ω^+ and Ω^-, such that, $\nu^+(A) = \nu(A \cap \Omega^+)$ and $\nu^-(A) = -\nu(A \cap \Omega^-)$, for all $A \in \mathcal{A}$. Further, $\nu(A) \geq 0$ for all $A \in \mathcal{A}$ with $A \subset \Omega^+$ and $\nu(A) \leq 0$ for all $A \in \mathcal{A}$ with $A \subset \Omega^-$.

Proof. Consider the finite measure $\mu = \nu^+ + \nu^-$. It is clear that $\nu \ll \mu$. Use Theorem 6.1.34 and consider the μ-integrable function $f = \frac{d\nu}{d\mu}$. Then, the set functions ν_1 and ν_2 defined on $\mathcal{A}$ by $\nu_1(A) = \int_A f^+ d\mu$ and $\nu_2(A) = \int_A f^- d\mu$ are both finite measures on $\mathcal{A}$. It is clear from the definition that $\nu = \nu_1 - \nu_2$ and also that ν_1 dominates ν and ν_2 dominates $-\nu$. These imply that $\nu^+(A) \leq \nu_1(A)$ and $\nu^-(A) \leq \nu_2(A)$, for all $A \in \mathcal{A}$ (see Exercise 6.1.31). On the other hand, denoting $h = \frac{d\nu^+}{d\mu}$ and $g = \frac{d\nu^-}{d\mu}$, one has two non-negative μ-integrable functions h and g, so that, the function $h - g$ is well-defined $[\mu]$-a.e. and gives a μ-integrable function. Moreover, $\int_A (h - g) d\mu = \nu(A) = \int_A (f^+ - f^-) d\mu$, for all $A \in \mathcal{A}$. This implies that $h - g = f^+ - f^-$ a.e.$[\mu]$. Using the fact that f^+ and f^- are non-negative functions with disjoint support, one can conclude from the above that $h \geq f^+$, $g \geq f^-$ a.e.$[\mu]$. But this gives $\nu^+(A) \geq \nu_1(A)$ and $\nu^-(A) \geq \nu_2(A)$, for all $A \in \mathcal{A}$. These, in conjunction with the reverse inequalities proved earlier, give us $\nu^+ \equiv \nu_1$ and $\nu^- \equiv \nu_2$. One can now easily verify that the disjoint sets $\Omega^+ = \{f \geq 0\}$ and $\Omega^- = \{f < 0\}$ satisfy the required properties. □

We finally return to a question that was raised earlier in Remark 6.1.4, since we promised there to come back to it. In the Radon-Nikodym Theorem, as stated earlier (Theorem 6.1.3), it was assumed that the measures μ and ν are both σ-finite. The question that was raised in Remark 6.1.4 is whether σ-finiteness is needed. First, let us give an example to show that σ-finiteness of μ is necessary. Take μ to be the counting measure on $(\mathbb{R}, \mathcal{B})$, that is, $\mu(B) = |B|$, if B is finite, and $\mu(B) = \infty$, otherwise. It should be clear that any measure on $\mathcal{B}$ is absolutely continuous with respect to μ, because the only μ-null set is $\emptyset$. In particular, $\lambda \ll \mu$. However, if f is any non-negative measurable function such that $\lambda(B) = \int_B f d\mu$, then taking $B = \{x\}$ will give us $f(x) = 0$, for any x. This means that the function $f \equiv 0$ can be the only possible candidate for $\frac{d\lambda}{d\mu}$, which is clearly a contradiction. This shows that R-N Theorem may fail if μ is not σ-finite. However, the measure ν need not be σ-finite. In other words, if μ is σ-finite and $\nu \ll \mu$, then RN theorem holds even if ν is not assumed to be σ-finite. The argument is sketched in the next exercise.

EXERCISE **6.1.38.** *Suppose μ and ν are measures on a σ-field $\mathcal{A}$. Assume that μ is σ-finite measure and that $\nu \ll \mu$. For a set $A \in \mathcal{A}$, say that "ν is σ-finite on A" if A can be written as a countable union of $\mathcal{A}$-sets, each of which has a finite ν-measure. Consider the class $\mathcal{C} = \{A \in \mathcal{A} : \nu \text{ is } \sigma\text{-finite on } A\}$. Observe that $\mathcal{C}$ is non-empty and let Ω_0 be the essential union of $\mathcal{C}$ with respect to μ. Show that ν is σ-finite on Ω_0. Also, for any $\mathcal{A}$-set A with $A \cap \Omega_0 = \emptyset$, show that $\nu(A)$ is either zero or infinity. Use all of these to show that a R-N derivative $\frac{d\nu}{d\mu}$ exists.*

We end this section with a passing remark that may interest some readers. A reader, who has learnt about differentiability and derivatives of real functions of a real variable in Calculus, would be well within her rights to ask: why again use the word 'derivative' in the context of absolute continuity between measures? Is there a connection between derivatives as in Calculus and R-N derivatives? It turns out that there is indeed a connection. Without going into the details, we briefly mention it here. From real analysis, we know that any bounded right continuous function G on $\mathbb{R}$, which is also of bounded variation is the difference of two bounded distribution functions. It should be clear from this that any such function G determines a unique finite signed measure μ on $\mathcal{B}$. It turns out that the finite signed measure μ on $\mathcal{B}$ is absolutely continuous with respect to the Lebesgue measure λ iff the function G on $\mathbb{R}$ is an "absolutely continuous function". We are assuming here that the reader is familiar with the notion of absolute continuity of a real-valued function on $\mathbb{R}$. Further, in this case, G is differentiable [λ]-a.e. on $\mathbb{R}$ and the R-N derivative $\frac{d\mu}{d\lambda}$ equals the calculus derivative G' of the function G. A very deep and well-known result in analysis, known as *Lebesgue Differentiation Theorem*, is important in this connection. It asserts, among other things, that every absolutely continuous function is differentiable a.e.[λ]. An interested reader may explore this further on her own.

6.2 L_p-Spaces

In this section, we are going to study an important family of classes of measurable functions on a measure space $(\Omega, \mathcal{A}, \mu)$. Recall that for any extended real valued measurable function f and any $0 < p < \infty$, we had defined in (3.4.1)

$$\|f\|_p = \left(\int |f|^p d\mu\right)^{1/p}$$

and in (3.4.2), we had defined, for $p = \infty$,

$$\|f\|_\infty = \inf\{c \cdot \mu(|f| > c) - 0\}.$$

It is easy to see that $\|\alpha f\|_p = |\alpha| \|f\|_p$, for $0 < p \leq \infty$ and for any real α. Also, for $1 \leq p < \infty$, we have $\|f + g\|_p \leq \|f\|_p + \|g\|_p$, by Minkowski's inequality (Theorem 3.4.6). The same inequality for $p = \infty$ follows easily from the definition.

For $1 \leq p \leq \infty$, the class of functions $\{f : \|f\|_p < \infty\}$ on a measure space $(\Omega, \mathcal{A}, \mu)$ is called the L_p-*space* (also sometimes, L_p *space*) on $(\Omega, \mathcal{A}, \mu)$, written $L_p(\Omega, \mathcal{A}, \mu)$. When the underlying measure space is clear from the context, we may denote this class simply by L_p, as we are going to do in the sequel. The previous paragraph shows that the class L_p, for $1 \le p \le \infty$, is a real vector space. The notation $\|\cdot\|_p$ would seem to suggest that it is a norm on L_p. The results noted in the previous paragraph show that it does indeed have alomst all the properties required of a norm, except only that $\|f\|_p = 0$ does not imply $f \equiv 0$; instead, it implies $f = 0$ only a.e.$[\mu]$. We rectify this little problem, by just deciding to identify any two functions as the same element of our "vector space" L_p, as long as they agree outside of a μ-null set. This is a fairly standard idea in vector space theory and is usually called "*quotienting*". This will now allow us to identify any function f with the identically zero function as long as $f = 0$ a.e.$[\mu]$ and will make $(L_p, \|\cdot\|_p)$ a normed linear space over the field of reals. Just to make it absolutely clear, once and for all, the normed linear space L_p is, strictly speaking, not a space of functions, but a space consisting of equivalence classes of functions where two functions are equivalent iff they equal a.e.$[\mu]$. Having said that though, we will, in practice, always treat an element of L_p as a function, which could be any one function from the equivalence class that it represents.

As is always the case with a normed linear space, the first question that we are going to ask is whether, for $1 \le p \le \infty$, the normed linear space $(L_p, \|\cdot\|_p)$ is complete, that is, whether every Cauchy sequence in $(L_p, \|\cdot\|_p)$ is convergent in the $\|\cdot\|_p$-norm. As it turns out, the answer is 'yes' and that is going to be our next theorem. This will show that, the space $(L_p, \|\cdot\|_p)$, for each $p \in [1, \infty]$, is a complete normed linear space, or a "*Banach Space*", in the usual terminology.

THEOREM **6.2.1**. *Let $(\Omega, \mathcal{A}, \mu)$ be any measure space. Then, for every $p \in [1, \infty]$, the L_p-space of $(\Omega, \mathcal{A}, \mu)$ is complete in the norm $\|\cdot\|_p$.*

Proof. We give the proof only for $p \in [1, \infty)$. The proof for $p = \infty$ is much simpler and is given below as exercise (Exercise 6.2.3).

Let $\{f_n\}$ be a Cauchy sequence in $(L_p, \|\cdot\|_p)$. We have to show that there exists $f \in L_p$ such that $\|f_n - f\|_p \to 0$ as $n \to \infty$. Using the Cauchy property of the sequence $\{f_n\}$, one easily gets a sequence of integers $1 \le n_1 < n_2 < n_3 < \cdots$, such that, for each $k \ge 1$,
$$m, n \ge n_k \implies \|f_n - f_m\|_p < 4^{-k}. \tag{6.2.1}$$
Denoting $A_k = \{\omega : |f_{n_{k+1}}(\omega) - f_{n_k}(\omega)| > 2^{-k}\}$, $k \ge 1$, and using Chebycheff's inequality (Theorem 3.4.2), we get
$$\mu(A_k) \le (\|f_{n_{k+1}} - f_{n_k}\|_p^p)/2^{-kp} < (2^p)^{-k}, \text{ for all } k \ge 1,$$
implying $\sum_k \mu(A_k) < \infty$. Borel-Cantelli Lemma now gives $\mu(\limsup_k A_k) = 0$. It is easy to see that, for any $\omega \notin \limsup_k A_k$, the sequence $\{f_{n_k}(\omega)\}$ converges as $k \to \infty$. Defining $f = \limsup_k f_{n_k}$, we get a measurable function f, such that, $f_{n_k} \to f$ a.e.$[\mu]$, as $k \to \infty$. Fatou's Lemma (Lemma 3.3.2) now gives $\int |f|^p d\mu \le \liminf_k \int |f_{n_k}|^p d\mu \le \sup_n \int |f_n|^p d\mu < \infty$, showing that $f \in L_p$. We have used here the simple fact that the sequence $\{f_n\}$ being Cauchy in L_p forces $\{f_n\}$ to be bounded in L_p. We finally want to show that, the original sequence $\{f_n\}$ converges in L_p to f. Here again, since $\{f_n\}$ is Cauchy in L_p, it is enough to show that the subsequence $\{f_{n_k}\}$ converges to f in L_p, as $k \to \infty$.

6.2. L_p-Spaces

Fixing $k \geq 1$, we have $\|f_{n_j} - f_{n_k}\|_p^p < 4^{-kp}$, for all $j > k$, from (6.2.1). Letting now $j \to \infty$, we get, by Fatou's Lemma, $\|f - f_{n_k}\|_p^p \leq \liminf_j \|f_{n_j} - f_{n_k}\|_p^p \leq 4^{-kp}$. This inequality being true for all $k \geq 1$, we can let $k \to \infty$ there to conclude that $\|f - f_{n_k}\|_p^p \to 0$, as $k \to \infty$. This completes the proof for $p \in [1, \infty)$. □

A careful examination of the proof of Theorem 6.2.1 given above, yields the following result as an important and useful by-product.

PROPOSITION 6.2.2. *Given any Cauchy sequence $\{f_n\}$ in $(L_p, \|\cdot\|_p)$, there is a subsequence $\{f_{n_k}\}$ and a function $f \in L_p$ such that f_{n_k} converges to f a.e.$[\mu]$, as $k \to \infty$.*

EXERCISE 6.2.3. *Completeness of $(L_\infty, \|\cdot\|_\infty)$*
(a) Show that, for a measurable function g on $(\Omega, \mathcal{A}, \mu)$, one has $\|g\|_\infty \leq C$ if and only if $|g| \leq C$ a.e.$[\mu]$.
(b) Show that if a sequence $\{f_n\}$ in $L_\infty(\Omega, \mathcal{A}, \mu)$ is Cauchy in the $\|\cdot\|_\infty$-norm, then there is a set $\Omega_0 \in \mathcal{A}$ with $\mu(\Omega_0^c) = 0$, such that, the sequence $\{f_n(\omega)\}$ is **uniformly Cauchy** on Ω_0.
(c) For f_n, $n \geq 1$ and f in $L_\infty(\Omega, \mathcal{A}, \mu)$, show that $\|f_n - f\|_\infty \to 0$ as $n \to \infty$ if and only if there is a set $\Omega_0 \in \mathcal{A}$ with $\mu(\Omega_0^c) = 0$, such that, $f_n(\omega) \to f(\omega)$, as $n \to \infty$, **uniformly** on Ω_0.
(d) Prove that $(L_\infty, \|\cdot\|_\infty)$ is complete.

EXERCISE 6.2.4. *Let $\Omega = \{1, 2, \cdots, k\}$ be equipped with the σ-field $\mathcal{A} = \mathcal{P}(\Omega)$ and a measure μ on $\mathcal{A}$ given by $\mu\{i\} = \mu_i > 0$, for each i. For each $p \in [1, \infty]$, describe the appropriate norm $\|\cdot\|$ on the vector space $\mathbb{R}^k$ that identifies $(\mathbb{R}^k, \|\cdot\|)$ with the normed linear space $L_p(\Omega, \mathcal{A}, \mu)$.*

EXERCISE 6.2.5. *Let $\Omega = \mathbb{N}$ be equipped with the σ-field $\mathcal{A} = \mathcal{P}(\mathbb{N})$ and a finite measure μ on $\mathcal{A}$ given by $\mu\{i\} = \mu_i > 0$, for each i. Show that the L_p spaces are strictly decreasing with p.*

REMARK 6.2.6. It was observed in Chapter 3 (Theorem 3.4.9) that, in a finite measure space $(\Omega, \mathcal{A}, \mu)$, if $\|f\|_s < \infty$ for some $s > 0$, then $\|f\|_r < \infty$ for all $0 < r < s$. What this means is that for finite measure spaces, the L_p-classes decrease as p increases. As was mentioned there, things behave very differently when the measure is infinite. The next two exercises illustrate that, for infinite measure spaces, the L_p classes do not generally behave in any specific way as p increases. One does encounter very divergent possible scenarios.

EXERCISE 6.2.7. *Let $(\Omega, \mathcal{A})$ be as in Exercise 6.2.5 and let μ be the counting measure on $\mathcal{A}$.*
(a) For the function $f(n) = \frac{1}{n}$, show that $\|f\|_p < \infty$ for all $p > 1$, but $\|f\|_1 = \infty$.
(b) For $g(n) \equiv 1$, show that $\|g\|_\infty < \infty$, but $\|g\|_p = \infty$, for all $0 < p < \infty$.
(c) Show that $\|f\|_s < \infty$ implies $\|f\|_r < \infty$, for all $r > s$. [Note that, this is diametrically opposite to what is asserted in Theorem 3.4.9.]
(d) Conclude that the L_p spaces on $(\Omega, \mathcal{A}, \mu)$ are strictly increasing with p.

EXERCISE **6.2.8.** *Consider the measure space $(\mathbb{R}, \mathcal{B}, \lambda)$.*
(a) For the function $f(x) = \frac{1}{x} I_{(0<x<1)}$, identify the set of all $p > 0$, for which $\|f\|_p < \infty$.
(b) Do the same, as in (a), for the function $g(x) = xI_{(|x|>1)}$.
(c) Contrast the findings of (a) and (b) with the assertion of Theorem 3.4.9 and also the findings of Exercise 6.2.7.
(d) Use the findings of (a) and (b) to conclude that the L_p-classes on $(\mathbb{R}, \mathcal{B}, \lambda)$ are neither decreasing nor increasing with p.

We now turn to another interesting question concerning L_p spaces. Now that we know that the L_p space on any measure space is, for each $p \in [1, \infty]$, a Banach space, our objective here is to identify, when we can, what are called the "*dual spaces* " of the L_p spaces. The dual space of a normed linear space is an important concept in basic functional analysis. Let us, however, assure the reader that to understand what we are going to discuss, it is not essential to have prior knowledge of functional analysis. We will define all the relevant terminology from basic functional analysis, that we will need. We start with the notion of a "linear functional" on a real vector space.

DEFINITION **6.2.9.** *Let V be a real vector space. A linear functional T on V is a map $T : V \to \mathbb{R}$ satisfying*

$$T(\alpha v_1 + \beta v_2) = \alpha T(v_1) + \beta T(v_2), \text{ for all } v_1, v_2 \in V \text{ and } \alpha, \beta \in \mathbb{R}.$$

If V is a normed linear space, then a linear functional T on V is said to be a bounded linear functional if there is a finite constant C such that $|T(v)| \leq C\|v\|$ for all $v \in V$.

It is easy to see that boundedness of a linear functional T on a normed linear space V is equivalent to

$$\sup\{|T(v)| : \|v\| \leq 1\} < \infty.$$

The space V^* of all bounded linear functionals on a normed linear space V can be easily seen to be a real vector space, with natural definitions of addition and scalar multiplication on V^*. Further $\|T\| = \sup\{|T(v)| : \|v\| \leq 1\}$ makes V^* a normed linear space. A reader, who is not already familiar with these, can easily verify the claims as easy exercises. This normed linear space V^* is what is called the "*dual space of V*". Also, for $T \in V^*$, the norm $\|T\|$, defined above, is often called the "*dual norm*" of T.

An interesting fact about the dual space V^*, that makes the study of dual spaces important, is that V^* turns out to be the same as the set of all continuous linear functionals on V, that is, the set of all linear functionals on V, that are continuous as maps on the normed linear space V into $\mathbb{R}$. This is easy to see. Indeed if $T \in V^*$, and $v_n \to v$ in V, then $|T(v_n) - T(v)| \leq \|T\|\|v_n - v\| \to 0$ and hence T is continuous. Conversely, if T is a linear functional on V which is not bounded, then for every $n \geq 1$, there is a $v_n \in V$ with $\|v_n\| \leq 1$ such that $|T(v_n)| > n$. Clearly, $\frac{1}{n} v_n \to 0 \in V$, while $|T(\frac{1}{n} v_n)| > 1$, for all n, showing that T is not continuous.

6.2. L_p-Spaces

With these basic definitions, let us now focus on our main aim here. What we want is to try and identify, whenever we can, are the dual spaces of the normed linear spaces $V = L_p(\Omega, \mathcal{A}, \mu)$. Indeed, assuming that μ is a σ-finite measure, we are now going to show that, for $1 \leq p < \infty$, the dual space of $L_p(\Omega, \mathcal{A}, \mu)$ can be identified as $L_q(\Omega, \mathcal{A}, \mu)$, where q is the 'conjugate index' of p, that is $\frac{1}{p} + \frac{1}{q} = 1$. It is important to clearly state what we mean here by the phrase "can be identified as". What we mean is that, for $1 \leq p < \infty$, the dual space L_p^*, equipped with its dual norm $\|\cdot\|$, as defined above, is "isometrically isomorphic" to the normed real vector space $(L_q, \|\cdot\|_q)$, in case the underlying measure μ is σ-finite. We will use the notation $L_p^* \cong L_q$ to express this. Note that $1 < q = p/(p-1) < \infty$ for $1 < p < \infty$, while $q = \infty$ for $p = 1$.

REMARK 6.2.10. As stated above, we are going to prove here that $L_p^* \cong L_q$, for $1 \leq p < \infty$, assuming that the underlying measure μ is σ-finite. However, in many contexts, it becomes important to know if one can drop the σ-finiteness assumption and still have the same isometry. It turns out that the isometric isomorphism $L_p^* \cong L_q$ remains valid for $1 < p < \infty$, even when μ is not σ-finite. This is what the reader is asked to prove in Exercise 6.5.13. Unfortunately, the situation is not so nice for $p = 1$, when the measure space is not σ-finite. Exercise 6.5.14 provides an example of a non-σ-finite measure space, where $L_\infty \cong L_1^*$ holds, while Exercise 6.5.15 gives an example where such an isomorphism does not hold. An enthusiastic reader should find Exercise 6.5.17 exciting. This is a follow-up of the previous two exercises and it seeks to completely identify an appropriate complete normed real vector space, in place of L_∞, which will always be isometrically isomorphic to L_1^*, for any measure space. Of course, for σ-finite measure spaces (and also for a special class of non-σ-finite measure spaces) this space will just be the classical $(L_\infty, \|\cdot\|_\infty)$-space.

To get on with our proof of $L_p^* \cong L_q$ for a σ-finite measure space $(\Omega, \mathcal{A}, \mu)$, we first consider $1 < p < \infty$, so that its conjugate index is given by $q = \frac{p}{p-1}$. For any $g \in L^q$, one can easily see, by Hölder's inequality, that the map T_g on L_p, given by $T_g : f \mapsto \int fg d\mu$, defines a bounded linear functional on $(L_p, \|\cdot\|_p)$ and that $\|T_g\| \leq \|g\|_q$. Here, $\|T_g\|$ denotes the dual norm of T_g as a linear functional on the normed vector space L_p. Further, with $g \in L_q$ as above, if we consider the function $f = \operatorname{sgn}(g)|g|^{q/p}$, then it is easy to verify that $f \in L_p$; in fact, $\|f\|_p = \|g\|_q^{q/p}$. On the other hand, one can easily see that $T_g(f) = \|g\|_q^q$, implying that $|T_g(f)| = \|g\|_q \|f\|_p$. This shows that the inequality $\|T_g\| \leq \|g\|_q$, obtained above through Hölder's inequality, is actually an equality.

If $p = 1$, then $q = \infty$ and, for any $g \in L_\infty$, the map T_g defined on L_1 by $T_g(f) = \int fg d\mu$ is, again by Hölder's inequality, a bounded linear functional on L_1 with $|T_g(f)| \leq \|g\|_\infty \|f\|_1$. To show that $\|T_g\| = \|g\|_\infty$, we need to only consider g with $\|g\|_\infty > 0$ and show that $\|T_g\| \geq \|g\|_\infty$. Take any α with $0 < \alpha < \|g\|_\infty$. Since, μ is σ-finite, we can get a set $A \in \mathcal{A}$ with $0 < \mu(A) < \infty$, such that, $|g(\omega)| > \alpha$ on A. Taking $f = \operatorname{sgn}(g) I_A/\mu(A)$ it is clear that $\|f\|_1 = 1$ and $T_g(f) > \alpha$, showing that $\|T_g\| > \alpha$. Since this is true for all $\alpha < \|g\|_\infty$, it follows that $\|T_g\| \geq \|g\|_\infty$ and and hence $\|T_g\| = \|g\|_\infty$.

Let us summarize what we have proved so far. If $(\Omega, \mathcal{A}, \mu)$ is a σ-finite measure space, then, for any $p \in [1, \infty)$ and with q the conjugate of p, we have proved that, for every $g \in L_q(\Omega, \mathcal{A}, \mu)$, the map $T_g(f) = \int f g d\mu$, $f \in L_p$, defines a bounded linear functional on $L_p(\Omega, \mathcal{A}, \mu)$. Further, $\|T_g\| = \|g\|_q$, where $\|T_g\|$ denotes the dual norm of T_g. In other words, what we have essentially proved is that, in case the measure space is σ-finite, then $L_q \subsetsim L_p^*$, for all $1 \leq p < \infty$, where the notation $\subsetsim$ is used to mean "isometrically isomorphic to a subspace".

REMARK 6.2.11. The reader should note here that, in the argument given above for $L_q \subsetsim L_p^*$, we needed to use σ-finiteness of μ only in the case $p = 1$. Thus, what we have proved above actually says that, for any measure space, the 'isometric inclusion' $L_q \subsetsim L_p^*$ remains valid for $1 < p < \infty$. In case the inquisitive reader is wondering, what is special about the case $p > 1$ as opposed to $p = 1$, here is the insight. For $p > 1$, the conjugate index q is finite, which means that, whatever be the measure space, for any $g \in L_q$, there is a set $A \in \mathcal{A}$ of σ-finite measure, such that, $g \equiv 0$ on A^c (see Exercise 3.6.3). This means that $T_g(f) = \int f g d\mu = \int_A f g d\mu$, for all $f \in L_p$. Thus, for any $g \in L_q$, the entire game will be played out on the σ-finite measure space, obtained by restricting both the σ-field and the measure to A. On the other hand, in case $p = 1$, that is, $q = \infty$, a function $g \in L_\infty$ clearly need not be supported on a set of σ-finite measure. What happens as a result of this is the following. While, for every $g \in L_\infty$, Hölder's inequality implies that the map $f \mapsto T_g(f)$, as defined above, does give a bounded linear functional on L_1 with $\|T_g\| \leq \|g\|_\infty$, we may not have $\|T_g\| = \|g\|_\infty$. Indeed, it is precisely to prove this equality that we brought in the assumption of σ-finiteness of μ in our argument above. As mentioned in Remark 6.2.10, this issue will be taken up in Exercise 6.5.17 and a solution will be obtained for non-σ-finite measure spaces.

Returning to our proof, we now prove the converse of what has been proved, that is, we prove that, for $1 \leq p < \infty$, any $T \in L_p^*$ indeed equals T_g for some $g \in L_q$, where q is the conjugate of p. We continue assuming that μ is σ-finite.

As before, we first consider the case $1 < p < \infty$. From σ-finiteness of μ, we can get $\mathcal{A}$-sets Ω_n, $n \geq 1$, with $\mu(\Omega_n) < \infty$, for each n, such that $\Omega_n \uparrow \Omega$. Clearly, $I_{A \cap \Omega_n} \in L_p(\mu)$, for any $A \in \mathcal{A}$ and for each n. Let $T \in L_p^*$. For each n, define a real valued set function ν_n on $\mathcal{A}$ by $\nu_n(A) = T(I_{A \cap \Omega_n})$. Using linearity and continuity of T, one can easily check that the set function ν_n is countably additive and hence defines a finite signed measure on $\mathcal{A}$. Absolute continuity of ν_n with respect to μ is obvious, since $\mu(A) = 0$ implies $I_{A \cap \Omega_n}$ is identified with the zero element of L_p. We appeal to Radon-Nikodym theorem (Theorem 6.1.34) and denote $g_n = \frac{d\nu_n}{d\mu}$. Since $\nu_n(\Omega_n^c) = 0$ we have $g_n I_{\Omega_n^c} = 0$ a.e.$[\mu]$. Further, $\nu_{n+1}(A \cap \Omega_n) = \nu_n(A)$, for any A, implying that $g_{n+1} = g_n$ on Ω_n a.e.$[\mu]$. What all these mean is that $g = \lim g_n$ exists and equals g_n on Ω_n. From the definition of g_n, we have $T(I_{A \cap \Omega_n}) = \int I_A g_n d\mu$, for all $A \in \mathcal{A}$. Using linearity of integrals and then the Dominated Convergence Theorem, one can easily deduce that $T(f I_{\Omega_n}) = \int f g_n d\mu$ for all bounded measurable f. Recall

6.2. L_p-Spaces

that, any measurable function f is the pointwise limit of a sequence $\{f_n\}$ of real-valued simple measurable functions that satisfy $|f_n| \leq |f|$, for all n, and, that for any bounded measurable function f, the function fI_{Ω_n} is μ-integrable. Since $g_n = 0$ on Ω_n^c it follows that, for all $k \geq 1$, the bounded measurable function $h_{nk} = \text{sgn}(g_n)|g_n|^{q/p}I_{\{|g_n| \leq k\}} \in L_p(\mu)$ with $\|h_{nk}\|_p = \|g_n I_{\{|g_n| \leq k\}}\|_q^{q/p}$ and therefore

$$\int |g_n|^q I_{\{|g_n| \leq k\}} d\mu = T(h_{nk}) \leq \|T\| \|h_{nk}\|_p = \|T\| \|g_n I_{\{|g_n| \leq k\}}\|_q^{q/p},$$

which gives $\|g_n I_{\{|g_n| \leq k\}}\|_q \leq \|T\|$. Letting $k \to \infty$, we get, by the Monotone Convergence Theorem, that $\|g_n\|_q \leq \|T\|$. Since $|g_n| \uparrow |g|$, one uses the Monotone Convergence Theorem again to conclude that $g \in L_q$ and $\|g\|_q \leq \|T\|$. Next, given $f \in L_p$, the sequence $f_n = fI_{\{|f| \leq n\}}I_{\Omega_n}$ of bounded measurable functions converge to f in L_p. Using continuity of T and the Dominated Convergence Theorem, one gets $T(f) = \lim_n T(f_n) = \lim_n \int fI_{\{|f| \leq n\}} g_n d\mu = \int fg d\mu$. Thus we have proved that $T(f) = T_g(f)$, for all $f \in L_p$. But we already know that $\|T_g\| = \|g\|_q$. It follows that $\|T\| = \|g\|_q$.

For the case $p = 1$, let Ω_n, g_n and g be exactly as in the case $p > 1$, so that $T(fI_{\Omega_n}) = \int fg_n d\mu$ for all bounded measurable f. We show straight away that $g \in L_\infty$. There is nothing to prove if $g \equiv 0$. So let $\alpha > 0$ be such that $\mu(|g| > \alpha) > 0$. Using σ-finiteness of μ, we can get a set A with $0 < \mu(A) < \infty$, such that $|g(\omega)| > \alpha$ on A. We can also ensure that $A \subset \Omega_n$, for all large n. The function $f = \text{sgn}(g)I_A/\mu(A)$ is bounded and has $\|f\|_1 = 1$. One has, for all large n, $T(fI_{\Omega_n}) = \int fg_n d\mu = \int fgI_{\Omega_n} d\mu = \int |g|I_A/\mu(A)d\mu > \alpha$. This implies $\alpha \leq \|T\|$, because $\|fI_{\Omega_n}\|_1 = \|f\|_1 = 1$, for all large n. Since this holds for all α with $\mu(|g| > \alpha) > 0$, we conclude that $\|g\|_\infty \leq \|T\|$. That $T = T_g$ is now proved in exactly the same way as case $p > 1$. Once we get that, $\|T\| = \|g\|_\infty$ follows.

Having thus proved that, for $1 \leq p < \infty$, every $T \in L_p^*$ equals T_g for some $g \in L_q$ and $\|T\| = \|g\|_q$, it is easy to see that the g is unique upto μ-null sets. We have thus completed the proof of a very important result, stated at the beginning of this section, identifying the dual space of the Banach space $L_p(\Omega, \mathcal{A}, \mu)$, for $1 \leq p < \infty$, at least when μ is σ-finite. Here is the formal statement of this result, known as *Riesz Representation Theorem*.

THEOREM 6.2.12. (Riesz Representation Theorem) Let $(\Omega, \mathcal{A}, \mu)$ be a σ-finite measure space. Then, $L_p^*(\Omega, \mathcal{A}, \mu) \cong L_q(\Omega, \mathcal{A}, \mu)$, for every $1 \leq p < \infty$, where q denotes the conjugate index of p. The identification of a linear functional $T \in L_p^*$ with an element $y \in L_q$ is given by $T \equiv T_g$, where $T_g(f) = \int fg d\mu$, for $f \in L_p$.

REMARK 6.2.13. The reader must be wondering about the case $p = \infty$, which is excluded from the Theorem 6.2.12. The fact is that the analogue of the theorem does not hold for $p = \infty$. More specifically, $L_\infty^* \neq L_1$, except in some very special cases (see Exercise 6.2.14). The general description of L_∞^* will be too much of a digression from our main theme and is, therefore, omitted. However, an interested reader is encouraged to look at Exercise 6.5.20 to gain a bit of insight.

EXERCISE **6.2.14.** *Show that if μ is a finite measure on the power set of a finite set Ω, then $L_\infty^* = L_1$.*

6.3 Approximation of L_p functions

As the title of this section suggests, our focus here is going to be discussing approximations of functions belonging to the class L_p by various special and simpler classes of functions. In almost all of this section, we will restrict ourselves to only $(\mathbb{R}, \mathcal{B}, \lambda)$ as our underlying measure space and will present a number of interesting and useful approximations of L_p functions on this space. However, we would like to point out that the results we prove here can be easily extended to L_p functions on any $(\mathbb{R}^k, \mathcal{B}^k, \mu)$, where μ is any Radon measure and not just the Lebesgue measure. We leave that for the enthusiastic readers to verify.

We start with a very simple yet useful result on approximation of functions L_p spaces, that is valid on a somewhat general class of measure space.

THEOREM **6.3.1.** *Let $(\Omega, \mathcal{A}, \mu)$ be a measure space. Suppose that μ is σ-finite on some field $\mathfrak{F}$ that generates $\mathcal{A}$ and let*
$$\mathcal{H} = \Big\{h : h = \sum_{i=1}^{n} c_i I_{F_i}, \text{ where } c_i \in \mathbb{R} \text{ and } F_i \in \mathfrak{F} \text{ with } \mu(F_i) < \infty, \text{ for each } i\Big\}.$$
Then, for all $p \in [1, \infty)$, the class $\mathcal{H}$ is contained in L_p and dense in L_p.

Proof. That $\mathcal{H} \subset L_p$ is obvious. We only need to show that $\mathcal{H}$ is dense in L_p. Recall that, given any measurable function f, we can get a sequence $\{f_n\}$ of real-valued simple functions, with $|f_n| \leq |f|$, such that $f_n \to f$ pointwise. Now, if $f \in L_p$, then each $f_n \in L_p$ also and, further, $f_n \to f$ in L_p, by the Dominated Convergence Theorem. It, therfore, suffices to show that any real-valued simple function g, with $g \in L_p$, can be approximated in L_p-norm by functions in $\mathcal{H}$. Let us take such a g, say, $g = \sum_{i=1}^{n} c_i I_{A_i}$, where each $c_i \in \mathbb{R}$ and A_i, $1 \leq i \leq n$, is a partition of Ω by $\mathcal{A}$-sets. Fixing any $\epsilon > 0$, we are going to get a function $h \in \mathcal{H}$ such that $\|g - h\|_p < \epsilon$. Note first that, $g \in L_p$ implies $\mu(A_i) < \infty$, for each i with $c_i \neq 0$. We use part (b) of Theorem 2.5.2 to get, for each i with $c_i \neq 0$, a set $F_i \in \mathfrak{F}$, such that, $\mu(A_i \Delta F_i) < \big[\epsilon/(n|c_i|)\big]^p$. For any i with $c_i = 0$, we take $F_i = \emptyset$. One easily sees that $\|c_i I_{A_i} - c_i I_{F_i}\|_p < \epsilon/n$, for each i. If we now consider the function $h = \sum_{i=1}^{n} c_i I_{F_i}$, then clearly $h \in \mathcal{H}$ and using Minkowski's inequality (Theorem 3.4.6), one gets $\|g - h\|_p < \epsilon$. This completes the proof. □

For the rest of the section, we specialize only to the L_p spaces on $(\mathbb{R}, \mathcal{B}, \lambda)$ as stated earlier and get some very special approximation results. In classical mathematics literature, these results are often clubbed together and referred to as the *"Three Principles of J E Littlewood"*.

We start with a very useful result, known as *Egoroff's theorem*. We state and prove it here for $(\mathbb{R}, \mathcal{B}, \lambda)$ only, although the result holds in more abstract settings, as we will see in Chapter 7. (See Theorems 7.1.4 and 7.2.2).

6.3. Approximation of L_p functions

THEOREM **6.3.2.** [Egoroff's Theorem] Let f_n, $n \geq 1$, and f be real-valued borel measurable functions defined on a set $B \in \mathcal{B}$ with $\lambda(B) < \infty$, such that, $f_n \to f$ a.e.$[\lambda]$ on B. Then, for any $\epsilon > 0$, there is a Borel set $C \subset B$ with $\lambda(C) < \epsilon$, such that, $f_n \to f$ uniformly on $B \setminus C$.

Proof. The hypothesis implies that $\lambda(\bigcap_n \bigcup_{k \geq n} \{\omega \in B : |f_k(\omega) - f(\omega)| \geq \frac{1}{j}\}) = 0$, for each $j \geq 1$. Since $\lambda(B) < \infty$, continuity from above implies that, given $\epsilon > 0$, we can get n_j, for each j, such that, $\lambda(\bigcup_{k \geq n_j} \{\omega \in B : |f_k(\omega) - f(\omega)| \geq \frac{1}{j}\}) < \epsilon/2^j$. Then $C = \bigcup_j \bigcup_{k \geq n_j} \{\omega \in B : |f_k(\omega) - f(\omega)| \geq \frac{1}{j}\}$ is a Borel subset of B with $\lambda(C) < \epsilon$. From the definition of C, it is easy to see that $f_n \to f$ uniformly on $B \setminus C$. □

The next result which is extremely interesting (in fact, somewhat striking) in its own right, is a fundamental step towards the subsequent main results of this section. Its proof uses Egoroff's Theorem at a crucial stage.

THEOREM **6.3.3.** *Given any real-valued borel measurable function f on a Borel set $B \subset \mathbb{R}$, and any $\epsilon > 0$, there is a closed set $F \subset B$ with $\lambda(B \setminus F) < \epsilon$, such that, the function f restricted to F is continuous.*

Proof. Let us first prove the result assuming that $\lambda(B) < \infty$. Suppose first that f is a real-valued simple function on B, say, $f = \sum_{i=1}^{n} c_i I_{A_i}$ where each $c_i \in \mathbb{R}$ and the sets A_i, $1 \leq i \leq n$, form a partition of B by Borel subsets of B. Since $\lambda(A_i) \leq \lambda(B) < \infty$, for each i, we use regularity of λ (see Theorem 2.10.1) to get, for each i, a closed set $F_i \subset A_i$ such that, $\lambda(A_i \setminus F_i) < \frac{\epsilon}{n}$. Clearly, the finite union $F = \bigcup_i F_i$ is a closed subset of B with $\lambda(B \setminus F) < \epsilon$. The reader can easily verify that f restricted to F is continuous. Next, for a general real borel measurable function f on B, we can get a sequence $\{f_n\}$ of real-valued simple functions on B, such that, $f_n \to f$, pointwise on B. By Theorem 6.3.2, we can get a Borel set $C \subset B$ with $\lambda(C) < \frac{\epsilon}{2}$, such that, $f_n \to f$ uniformly on $\widetilde{B} = B \setminus C$. Since $\lambda(\widetilde{B}) < \infty$, we can apply what has been proved above to each simple function f_n on $\widetilde{B}$, to get, for each n, a closed set $F_n \subset \widetilde{B}$ with $\lambda(\widetilde{B} \setminus F_n) < \epsilon/2^{n+1}$, such that, f_n restricted to F_n is continuous. Taking $F = \bigcap_n F_n$, we get a closed set $F \subset \widetilde{B} \subset B$ with $\lambda(B \setminus F) \leq \lambda(C) + \sum_n \lambda(\widetilde{B} \setminus F_n) < \epsilon$, such that each f_n restricted to F is continuous. Since $f_n \to f$ uniformly on $\widetilde{B}$ and hence also on F, it follows that f restricted to F is continuous.

Having proved the result assuming $\lambda(B) < \infty$, the general case can now be easily handled as follows. For each $k \in \mathbb{Z}$, denote $B_k = B \cap (k, k+1]$, and use the earlier case to get a closed set $F_k \subset B_k$ with $\lambda(B_k \setminus F_k) < \epsilon/2^{|k|+2}$, such that, f restricted to F_k is continuous. The reader may easily verify that the set $F = \bigcup_{k \in \mathbb{Z}} F_k \subset B$ is a closed set with $\lambda(B \setminus F) < \epsilon$ and that f restricted to F is continuous. □

The above leads to a very important and well-known result, known as *Lusin's Theorem*, which asserts that every real borel mesaurable function on $\mathbb{R}$ can be approximated by a continuous function, as closely as we want, in a sense made precise in the statement below.

THEOREM **6.3.4.** (Lusin's Theorem) Given any real-valued borel measurable function f on $(\mathbb{R}, \mathcal{B}, \lambda)$ and any $\epsilon > 0$, there exists a real-valued continuous function g on $\mathbb{R}$, such that, $\lambda(f \neq g) < \epsilon$. Further, g can be so chosen as to satisfy $\sup_x |g(x)| \leq \sup_x |f(x)|$. Moreover, if f is given to be vanishing outside an open interval I, then the function g also can be so chosen as to vanish outside I.

Proof. By Theorem 6.3.3, there is a closed set $F \subset \mathbb{R}$ with $\lambda(F^c) < \epsilon$, such that, f restricted to F is continuous. Using a well-known fact about open subsets of $\mathbb{R}$, we can write the open set $V = F^c$ as a countable union $V = \bigcup_i (a_i, b_i)$ where (a_i, b_i), $i \geq 1$, are disjoint open intervals. Note that $\lambda(V) = \lambda(F^c) < \epsilon$ forces that all these open intervals are bounded. We now define a function g on $\mathbb{R}$ as follows. For $x \in F$, we put $g(x) = f(x)$. For $x \in F^c = V = \bigcup_i (a_i, b_i)$, we pick the unique i, such that, $x \in (a_i, b_i)$ and define $g(x) = \frac{b_i - x}{b_i - a_i} f(a_i) + \frac{x - a_i}{b_i - a_i} f(b_i)$. The reader may easily verify that g defines a real-valued continuous function on $\mathbb{R}$. Since $g = f$ on F, we have $\lambda(f \neq g) < \epsilon$.

The assertion that $\sup_x |g(x)| \leq \sup_x |f(x)|$ follows easily from the fact that, by the way g has been defined, the value of g at any point x either equals the value of f at x or is a convex combination of values of f at two points.

To complete the proof, suppose next that f is given to vanish outside an open interval I, bounded or unbounded. We may assume that $I \neq \mathbb{R}$, since otherwise, f is the identically zero function and there is nothing to do. We choose an open interval $I' \subsetneq I$ with $\lambda(I \setminus I') < \frac{\epsilon}{2}$ in the following way. In case $I = (-\infty, b)$, $b \in \mathbb{R}$, we take $I' = (-\infty, b')$, where $b - \frac{\epsilon}{2} < b' < b$, while in case $I = (a, \infty)$, $a \in \mathbb{R}$, we take $I' = (a', \infty)$, where $a < a' < a + \frac{\epsilon}{2}$. Finally, in case I is a bounded open interval (a, b), we take $I' = (a', b')$, where $a < a' < b' < b$ with $(a' - a) + (b - b') < \frac{\epsilon}{2}$. If we now define φ on $\mathbb{R}$ by putting $\varphi(x) = x$ for $x \in I'$, $\varphi(x) = 0$ for $x \notin I$ and defining it on $I \setminus I'$ by linear interpolation, we will get a continuous function φ with values in $[0, 1]$. Now, by what has already been proved, we can obtain a continuous function $\widetilde{g}$ on $\mathbb{R}$ such that $\lambda(f \neq \widetilde{g}) < \frac{\epsilon}{2}$ and $\sup |\widetilde{g}(x)| \leq \sup |f(x)|$. If we now consider the function $g = \varphi \circ \widetilde{g}$, we get a real-valued continuous function g on $\mathbb{R}$, that satisfies $g = f = 0$ outside I and $g = \widetilde{g}$ on I', so that $\lambda(f \neq g) \leq \lambda(I \setminus I') + \lambda(f \neq \widetilde{g}) < \epsilon$. Also, the condition $\sup_x |g(x)| \leq \sup_x |f(x)|$ clearly holds. □

REMARK **6.3.5.** Although we have called Theorem 6.3.4 as "Lusin's Theorem," it is often considered as just one version of Lusin's Theorem. Indeed, there are a number of similar results, one derived out of another, that are together regarded as various versions of Lusin's Theorem. Some of the results that follow (Theorem 6.3.6 and the assertion in Exercise 6.3.8) fall under that category.

In what follows, let C_c denote the class of all real-valued continuous functions on $\mathbb{R}$ with compact support. Recall that a continuous function is said to have *compact support* if it vanishes outside a compact set. Using Theorem 6.3.4, we are now going to show that C_c is dense in L_p for any $p \in [1, \infty)$. As already pointed out at the start of this section, L_p here stands for $L_p(\mathbb{R}, \mathcal{B}, \lambda)$. Clearly, any function in C_c belongs to L_p, for all $p \in [1, \infty]$.

THEOREM **6.3.6.** Let $1 \leq p < \infty$. Then, for any $f \in L_p$ and any $\epsilon > 0$, there exists $g \in C_c$ with $\sup_x |g(x)| \leq \sup_x |f(x)|$, such that, $\|f - g\|_p < \epsilon$.

Proof. For each n, let h_n be the function defined as $h_n(x) = f(x) I_{\{|x| < n, |f(x)| < n\}}$. Since $f \in L_p$, the Dominated Convergence Theorem implies that $\|f - h_n\|_p \to 0$ and, therefore, we can get n_0, such that, $\|f - h_{n_0}\|_p < \epsilon/2$. Denoting h_{n_0} as h, we have a bounded measurable function h on $\mathbb{R}$ that vanishes outside a bounded open interval. Using Theorem 6.3.4, we can get $g \in C_c$ with $\sup_x |g(x)| \leq \sup_x |h(x)|$, such that, $\lambda(g \neq h) < [\epsilon/(4n_0)]^p$. Since $|h(x) - g(x)| \leq 2 \sup_x |h(x)| \leq 2n_0$, we get $\|h - g\|_p^p \leq (2n_0)^p \lambda(h \neq g) \leq (\epsilon/2)^p$. Minkowski's Inequality (Theorem 3.4.6) now gives $\|f - g\|_p \leq \|f - h\|_p + \|h - g\|_p < \epsilon$, completing the proof. □

REMARK **6.3.7.** Considering the L_∞ function $f \equiv 1$, one easily sees that the assertion of Theorem 6.3.6 cannot be extended to $p = \infty$. In other words, the closure of C_c, in $\| \cdot \|_\infty$-norm, is not L_∞. In fact, with a little reflection, one can see that the closure of C_c, in $\| \cdot \|_\infty$-norm, is C_0, that is, the class of all continuous functions f on $\mathbb{R}$ that vanish at infinity, meaning $\lim_{x \to \pm \infty} f(x) = 0$.

EXERCISE **6.3.8.** Show that given any bounded real-valued borel measurable function f on $\mathbb{R}$ with $\lambda(f \neq 0) < \infty$, there is a sequence $\{g_n\}$ of functions in C_c with $\sup_x |g_n(x)| \leq \sup_x |f(x)|$, for all n, such that, $g_n \to f$ a.e.$[\lambda]$.

6.4 Complex Measurable Functions and L_p-Spaces

In the integration theory introduced in Chapter 3 and developed and studied so far, we have always considered integrating only real-valued measurable functions. However, there are a number of contexts where it is important and useful to have a theory of integration for complex-valued functions as well. In fact, in some areas of analysis, dealing with integration of complex-valued functions becomes essential. In quantum theory, for example, one has to deal with complex-valued wave functions. The theory of Fourier transforms is, of course, another such example. Coming to probability theory also, expectation of complex-valued random variables are sometimes not only unavoidable, but also pays rich dividends. The reader may recall (see Definition 4.5.12) that, for a real random variable X, we introduced the notion of its moment generating function $m_X(\theta) = E(e^{\theta X})$, defined for all θ, for which the expectation is finite. However, what we also noted is that the set of θ where $m_X(\theta)$ is finite, may depend on X and may sometimes be only $\theta = 0$ (see the exercise following Definition 4.5.12). However, if, instead of $E(e^{\theta X})$, we consider $E(e^{i\theta X})$, called the 'characteristic function' (see Definition 8.2.6), it not only remedies the problem of non-existence, but also provides us with an extremely important tool in understanding properties of the underlying probability distributions. This is studied in detail in Section 8.2. It must also be noted that a big incentive of roping in complex-valued function in the theory of integration is that it allows one to bring in some complex analysis also into play.

As with integration theory of real functions, we will have to first define the notion of measurability for complex-valued functions on a measure space. The definition, as given below, is quite simple and very natural. Any complex-valued function on a set Ω can be viewed either as a $\mathbb{R}^2$-valued function or as a pair of real-valued functions (representing the 'real' and 'imaginary' parts). One can take anyone of these two viewpoints to define measurability of a complex-valued function and, as we know already, they would be equivalent.

DEFINITION 6.4.1. *A complex-valued function f on a measurable space $(\Omega, \mathcal{A})$ is said to be a* complex measurable function *if $f = u + iv$, where u and v are two real valued measurable functions. As usual, the real functions u and v are called the* real *and* imaginary *parts of f respectively and one writes $u = \mathsf{Re}(f)$ and $v = \mathsf{Im}(f)$.*

For a complex-valued function f on Ω, we will use the standard notation $|f|$ to denote the complex modulus of f, namely, $|f(\omega)| = \sqrt{u^2(\omega) + v^2(\omega)}$, $\omega \in \Omega$, where $u = \mathsf{Re}(f)$ and $v = \mathsf{Im}(f)$. The inequalities $\max\{|u|, |v|\} \leq |f| \leq |u| + |v|$ are going to be frequently used in the sequel. We will also use $\bar{f}$ for the complex conjugate of f, that is, $\bar{f}(\omega) = u(\omega) - iv(\omega)$, $\omega \in \Omega$, with u and v as above.

It is clear that any real-valued measurable function can be viewed as a complex-valued measurable function whose imaginary part is the identicallly zero function. Here is a simple exercise that can be derived as an immediate consequence of the analogous result for real measurable functions. We put it here just for the records., but my not need to use it much in what follows.

EXERCISE 6.4.2. *Let f be a complex measurable function on a measurable space $(\Omega, \mathcal{A})$. Show that there is a sequence $\{f_n\}$ of complex-valued simple functions, converging pointwise to f. Show also that if f is bounded (in 'complex modulus'), then the convergence $f_n \to f$ can be made uniform. [A complex-valued measurable function is called 'simple' if it takes only finitely many values.]*

In the next proposition, we note some simple facts about complex measurable functions, all of which, except one, are easy consequences of analogous results for real merasurable functions.

PROPOSITION 6.4.3. *(a) If f is a complex measurable function on $(\Omega, \mathcal{A})$, then $\bar{f}$ is a complex measurable function, $|f|$ is a non-negative real valued measurable function and αf is a complex measurable function, for any complex number α.*
(b) If f and g any two complex measurable functions on $(\Omega, \mathcal{A})$, then both $f + g$ and fg are complex measurable functions.
(c) For any complex measurable function f on $(\Omega, \mathcal{A})$, there is a complex measurable function h with $|h| \equiv 1$, such that $f = h|f|$.
(d) If $\{f_n\}$ is a sequence of complex measurable functions on $(\Omega, \mathcal{A})$, converging pointwise to a function f, then f is complex measurable.

6.4. Complex Measurable Functions and L_p-Spaces

Proof. Parts (a), (b) and (d) can be easily derived from the corresponding results for real measurable functions and hence their proofs are omitted. The only exception is part (c), but its proof is also very simple. Taking $A = \{\omega : f(\omega) \neq 0\}$, it is clear that $A \in \mathcal{A}$ and the function h defined as $h = \dfrac{f}{|f|} I_A + I_{A^c}$ is a complex measurable function, which does the required job. □

We now go straight to integration of complex measurable functions on a measure space. It essentially uses the definition of integrals of real measurable functions. However, there is one important difference. The reader will recall that, for a real measurale function f on a measure space $(\Omega, \mathcal{A}, \mu)$, its integral $\int f d\mu$ is defined without requiring that the integral be finite (that is, a real number). However, for a complex measurable function, its integral will be defined only when we get a complex number as the value of the integral. Using the terminology from integration of real measurable functions, what this says is that we will define integrals for only "integrable" complex measurable functions.

DEFINITION 6.4.4. A complex measurable function f on a measure space $(\Omega, \mathcal{A}, \mu)$ is said to be μ-*integrable* (or, simply *integrable*) if both $u = \text{Re}(f)$ and $v = \text{Im}(f)$ are μ-integrable and, in that case, the integral of f is defined as $\int f d\mu = \int u d\mu + i \int v d\mu$.

It is clear from Definition 6.4.4 that, for an integrable complex measurable function f, its integral $\int f d\mu$ is a complex number. Further, using Proposition 3.2.13, one can easily see that if f is an integrable complex measurable function, then for every $A \in \mathcal{A}$, the complex measurable function $f I_A$ is also integrable, and with u, v as in Definition 6.4.4, $\int_A f d\mu$ is the complex number given by

$$\int_A f d\mu = \int_A u d\mu + i \int_A v d\mu. \tag{6.4.1}$$

The next proposition can be easily derived from Definition 6.4.4, using the inequalities $\max\{|u|, |v|\} \leq |f| \leq |u| + |v|$, where $u = \text{Re}(f)$ and $v = \text{Im}(f)$.

PROPOSITION 6.4.5. A complex measurable function f on a measure space $(\Omega, \mathcal{A}, \mu)$ is integrable if and only if the non-negative measurable function $|f|$ is integrable.

Integrals of complex measurable functions enjoy many of the properties that integrals of their real counterparts do and are stated in Propositions 6.4.6 and 6.4.7. Parts (a), (b) and (c) of Proposition 6.4.6 follow easily from the corresponding properties of integrals of real functions, while no proof is needed for part (d) and much of part (e). It is only the last step of part (e), namely, deducing $\int f_n d\mu \to \int f d\mu$ from $\int |f_n - f| d\mu \to 0$, where the inequality of Proposition 6.4.7 needs to be used. This inequality, though an exact replica of an inequality for integrals of real functions, doesn't directly follow from its real counterpart.

PROPOSITION 6.4.6. All functions in the following statements are complex measurable functions on a measure space $(\Omega, \mathcal{A}, \mu)$.
(a) (Linearity) If f, g are μ-integrable, then so is $\alpha f + \beta g$, for any pair of complex numbers α, β and $\int (\alpha f + \beta g) d\mu = \alpha \int f d\mu + \beta \int g d\mu$.

(b) If $f = g$, a.e.$[\mu]$, and f is μ-integrable, then so is g and $\int f d\mu = \int g d\mu$.

(c) If μ is a σ-finite measure and if f and g are μ-integrable functions satisfying $\int_A f d\mu = \int_A g d\mu$, for all $A \in \mathcal{A}$ with $\mu(A) < \infty$, then $f = g$, a.e.$[\mu]$.

(d) (Fatou's Lemma) For any sequence $\{f_n\}$ of complex measurable functions, $\int \liminf_n |f_n| \, d\mu \leq \liminf_n \int |f_n| \, d\mu$.

(e) (DCT) If f_n, $n \geq 1$, are complex measurable functions with $f_n \to f$, a.e.$[\mu]$ and if there is a non-negative μ-integrable function g such that $|f_n| \leq g$, a.e.$[\mu]$, for all $n \geq 1$, then the functions f_n, $n \geq 1$, are all μ-integrable and $\int |f_n - f| d\mu \to 0$. In particular, $\int f_n d\mu \to \int f d\mu$, as $n \to \infty$.

Now we come to the useful inequality mentioned above, when we pointed out that this is needed at the last step of the proof of part (e) of Proposition 6.4.6.

PROPOSITION **6.4.7.** $\left|\int f d\mu\right| \leq \int |f| d\mu$, for any integrable complex measurable function f on $(\Omega, \mathcal{A}, \mu)$.

Proof. There is nothing to prove if $\int f d\mu = 0$. In case $\int f d\mu \neq 0$, we denote $\alpha = (\int f d\mu)/(|\int f d\mu|)$ to have $|\alpha| = 1$ and $|\int f d\mu| = \overline{\alpha} \int f d\mu = \int \overline{\alpha} f d\mu$, by Proposition 6.4.6(a). Since $|\int f d\mu|$ is a non-negative real number, the above equality reduces to $|\int f d\mu| = |\int \operatorname{Re}(\overline{\alpha} f) d\mu|$. The required inequality now follows from $|\int \operatorname{Re}(\overline{\alpha} f) d\mu| \leq \int |\operatorname{Re}(\overline{\alpha} f)| d\mu \leq \int |\overline{\alpha} f| d\mu = \int |f| d\mu$. □

We are now ready to define the L_p-spaces on $(\Omega, \mathcal{A}, \mu)$ in the same way as we did earlier, except that they will now be classes of complex measurable functions. Since real measurable functions are special cases of complex measurable functions, these L_p-spaces are going to be an enlargement of the L_p-spaces defined so far. Since we are not going to use these L_p-spaces anywhere else later, we are not going to introduce any new notations for them. However, to avoid any possible confusion, we will simply refer to them as "complex L_p-spaces". We start with some notations that are just extensions of the same notations introduced earlier for real measurable functions.

For complex measurable functions f on a measure space $(\Omega, \mathcal{A}, \mu)$, we denote
$$\|f\|_p = \left(\int |f|^p d\mu\right)^{1/p}, \ 1 \leq p < \infty$$
$$\|f\|_\infty = \inf\{a \geq 0 : |f| \leq a \text{ a.e.}[\mu]\} \tag{6.4.2}$$

Note that the definitions of $\|f\|_p$, $1 \leq p \leq \infty$, given here are consistent with the earlier defintions, in case f is real valued. Further, note that $\|f\|_p$ for a complex measurable function f is just the $\|\cdot\|_p$-norm of the non-negative real valued measurable function $|f|$. Accordingly, it is clear that both Hölder's inequality (Theorem 3.4.3) and Minkowski's inequality (Theorem 3.4.6) will hold for complex measurable functions as well. We just state these here.

THEOREM **6.4.8.** Let f and g be any two complex measurable functions on a measure space $(\Omega, \mathcal{A}, \mu)$. Then, the following inequalities hold.

(a) (Hölder's Inequality) $\int |fg| d\mu \leq \|f\|_p \|g\|_q$, where $p, q \in [1, \infty]$ satisfy $\frac{1}{p} + \frac{1}{q} = 1$.

(b) (Minkowski's Inequality) $\|f + g\|_p \leq \|f\|_p + \|g\|_p$, for all $p \in [1, \infty]$.

6.4. Complex Measurable Functions and L_p-Spaces

DEFINITION **6.4.9.** For $1 \leq p \leq \infty$, the *Complex L_p-spaces* on a measure space $(\Omega, \mathcal{A}, \mu)$, are classes of complex measurable functions defined as

$$L_p(\Omega, \mathcal{A}, \mu) = \{f \text{ complex measurable} : \|f\|_p < \infty\}, \text{ for } 1 \leq p \leq \infty. \quad (6.4.3)$$

We will assume that the underlying measure space is fixed and simply write L_p instead of $L_p(\Omega, \mathcal{A}, \mu)$, for the classes defined by (6.4.3). The next exercise is a simple observation, which is sometimes useful.

EXERCISE **6.4.10.** *Let f be a complex measurable function. For any $p \in [1, \infty]$, show that $f \in L_p$ if and only if both $\mathsf{Re}(f) \in L_p$, $\mathsf{Im}(f) \in L_p$.*

One can easily see, using Minkowski's inequality (Theorem 6.4.8(b)), that L_p, for each $p \in [1, \infty]$, is a complex vector space. Further, agreeing to identify any two complex measurable functions that are μ-a.e. equal, one sees that $\|\cdot\|_p$ defines a norm on L_p. And, interestingly, though not unexpectedly, almost the same argument, used to prove Theorem 6.2.1, shows that the complex normed linear space $(L_p, \|\cdot\|_p)$ is complete, for each $p \in [1, \infty]$. We state the result here and only outline the proof.

THEOREM **6.4.11.** *Let $(\Omega, \mathcal{A}, \mu)$ be any measure space. Then, for every $p \in [1, \infty]$, the complex L_p-space of $(\Omega, \mathcal{A}, \mu)$ is complete in the norm $\|\cdot\|_p$.*

Proof (an outline). To deal with $p \in [1, \infty)$ first, we start with a Cauchy sequence $\{f_n\}$ in complex L_p-space and get a strictly increasing sequence $\{n_k\}$ of natural numbers satisfying the same condition as in (6.2.1). Noting that Chebycheff's inequality 3.4.2 holds verbatim for complex measurable functions and using Borel-Cantelli Lemma, in exactly the same way as in the proof of Theorem 6.2.1, one gets a complex measurable function f such that $f_{n_k} \to f$, a.s.-$[\mu]$. Fatou's Lemma (Proposition 6.4.6(d)) can be used to see that $f \in L_p$. Finally, another application of Fatou's Lemma, exactly as in the proof of Theorem 6.2.1, gives $\|f_{n_k} - f\|_p \to 0$. One simply uses the Cauchy property now to get $f_n \to f$ in L_p. The proof for $p = \infty$ is again verbatim the same as in Exercise 6.2.3. □

The next question we are going to address is the very important question of identifying the "dual spaces" of the complex L_p-spaces. The concept of dual spaces for real vector spaces was discussed in Section 6.2. The basic ideas are the same, except that now we are dealing with a complex vector space and that calls for a slight (and expected) modification in our definition of "linear functionals".

DEFINITION **6.4.12.** Let V be a complex vector space. A *linear functional* T on V is a map $T : V \to \mathbb{C}$ satisfying

$$T(\alpha v_1 + \beta v_2) = \alpha T(v_1) + \beta T(v_2), \text{ for all } v_1, v_2 \in V \text{ and } \alpha, \beta \in \mathbb{C}$$

If $(V, \|\cdot\|)$ is a normed linear space over complex field, then a linear functional T on V is said to be a *bounded linear functional* if there is a real constant $0 \leq C < \infty$ such that $|T(v)| \leq C\|v\|$, for all $v \in V$.

The fact that boundedness of a linear functional $T : V \to \mathbb{C}$ is equivalent to having $\sup\{|T(v)| : \|v\| \leq 1\} < \infty$ remains valid for a complex vector space V as well. The space V^* of all bounded linear functionals on a complex normed linear space V, equipped with natural addition and multiplication by complex scalars, is now a complex vector space. Further, $\|T\| = \sup\{|T(v)| : \|v\| \leq 1\}$, as before, defines a norm on V^*. This complex normed linear space $(V^*, \|\cdot\|)$ is called the "dual space" of the complex normed linear space V, with the norm on V^* being often called the "dual norm". Once again, one can use the same argument, as in Section 6.2, to show that V^* equals the set of all linear functionals $T : V \to \mathbb{C}$, that are continuous in the given norm on the complex vector space V.

We now claim that the Riesz Representation Theorem (Theorem 6.2.12) that was proved in Section 6.2 remains valid for complex L_p-spaces as well. More importantly, the arguments remain more or less the same only with some minor modifications here and there. Assuming, as in Theorem 6.2.12, that the underlying measure μ is σ-finite, we present a brisk outline of the argument here, highlighting mainly the minor modifications necessary for the complex L_p-spaces.

For any $p \in [1, \infty)$, with conjugate index q, one uses Hölder's inequality (Theorem 6.4.8(a)) to easily see that, for any $g \in L_q$, the map $T_g : f \mapsto \int f g \, d\mu$ is a bounded linear functional on L_p with the dual norm $\|T_g\|$ satisfying $\|T_g\| \leq \|g\|_q$. To show that $\|T_g\| = \|g\|_q$, we split the argument (as in Section 6.2) to two cases.

For $1 < p < \infty$ and with $g \in L_q$ as above, we use Proposition 6.4.3(c) to get a complex measurable function h with $|h| \equiv 1$, such that $g = h|g|$. If we now take $f = \overline{h}|g|^{q/p}$, it is then easy to see that $f \in L_p$ with $\|f\|_p = (\|g\|_q)^{q/p}$ and $T_g(f) = \|g\|_q^q$. This shows that $\|T_g\| = \|g\|_q$.

In case $p = 1$, we have $q = \infty$. With $g \in L_\infty$ with $\|g\|_\infty > 0$ and T_g as above, we show $\|T_g\| \geq \|g\|_\infty$ (and hence $\|T_g\| = \|g\|_\infty$), by showing, as in Section 6.2, that $\|T_g\| > \alpha$ for any $0 < \alpha < \|g\|_\infty$. Fixing such an α, we use σ-finiteness of μ to get $A \in \mathcal{A}$, with $0 < \mu(A) < \infty$, such that $|g(\omega)| > \alpha$, for $\omega \in A$. If we now take $f = \overline{h} I_A / \mu(A)$, where h is as chosen in the previous paragraph, then it is easy to see that $\|f\|_1 = 1$ and $T_g(f) > \alpha$.

We have thus shown that on a σ-finite measure space, $L_q \subseteq L_p^*$, for every $p \in [1, \infty)$, with q being the conjugate of p. The proof of the converse is, once again, almost the same as in given Section 6.2 for real function spaces, except for one minor modification. Here is a brisk outline.

We first consider the case $1 < p < \infty$. Let $T \in L_p^*$ and let us denote $\mathrm{Re}(T)$ and $\mathrm{Im}(T)$ by T^r and T^i respectively. With Ω_n, $n \geq 1$, as in proof of Theorem 6.2.12, the set functions $\nu_n^r(A) = T^r(I_{A \cap \Omega_n})$ and $\nu_n^i(A) = T^i(I_{A \cap \Omega_n})$, for each n, will define finite signed measures on $\mathcal{A}$, both absolutely continuous with respect to μ. Radon-Nikodym theorem (Theorem 6.1.34) will give, for each n, real measurable functions u_n and v_n, such that $T^r(I_{A \cap \Omega_n}) = \int I_{A \cap \Omega_n} u_n \, d\mu$ and $T^i(I_{A \cap \Omega_n}) = \int I_{A \cap \Omega_n} v_n \, d\mu$, for all $A \in \mathcal{A}$. As in the proof of Theorem 6.2.12, we get that the limits $u = \lim u_n$ and $v = \lim v_n$ exist and also, $u = u_n$ and $v = v_n$ on Ω_n, for each n. Further, the same argument as used there gives us that, for every bounded real measurable function h, we have $T^r(hI_{\Omega_n}) = \int hu_n d\mu$ and $T^i(hI_{\Omega_n}) = \int hv_n d\mu$. Repeating the argument used there verbatim and using the easy fact that $\max\{|T^r(f)|, |T^i(f)|\} \leq \|T\|\|f\|_p$, for $f \in L_p$, one can deduce that $u \in L_q$ and $v \in L_q$. Next, by arguments used there, we can show that $T^r(\mathrm{Re}(f)) = \int \mathrm{Re}(f) u \, d\mu$ and $T^r(\mathrm{Im}(f)) = \int \mathrm{Im}(f) u \, d\mu$, for any $f \in L_p$. One can similarly deduce that $T^i(\mathrm{Re}(f)) = \int \mathrm{Re}(f) v \, d\mu$ and $T^i(\mathrm{Im}(f)) = \int \mathrm{Im}(f) v \, d\mu$, for any $f \in L_p$. With $g = u + iv$, we clearly have $g \in L_q$ and, combining the above, one gets $T(f) = T^r(f) + iT^i(f) = \int fg d\mu$, for any $f \in L_p$. Thus we have shown that given any $T \in L_p^*$, there is $g \in L_q$ such that $T = T_g$. That $\|T\| = \|g\|_q$ now follows from what was done earlier.

Finally, we consider the case $p = 1$. For $T \in L_1^*$, let T^r and T^i be as before. Also, let $\Omega_n, u_n, v_n, n \geq 1$ and u, v be as above. From what was done above, we know $T^r(hI_{\Omega_n}) = \int hu_n \, d\mu$ and $T^i(hI_{\Omega_n}) = \int hv_n \, d\mu$, for each n, and every bounded real measurable function h. Arguing exactly as was done in the proof of Theorem 6.2.12, one can directly show that both $u \in L_\infty$ and $v \in L_\infty$. Taking now $g = u + iv$, we get $g \in L_\infty$ and, procceding exactly as above, one can show that $T(f) = \int fg d\mu$, for all complex $f \in L_1$, thus proving that $T = T_g$, whence it also follows that $\|T\| = \|g\|_\infty$.

We have thus proved that Riesz Representation Theorem holds for complex L_p-spaces as well, assuming, of course, that the underlying measure μ is σ-finite. We state it here, because while the statement is essentially the same as that of Theorem 6.2.12, it now includes complex L_p-spaces.

THEOREM 6.4.13. (Riesz Representation Theorem for Complex L_p-spaces)
Let L_p, $1 \leq p \leq \infty$, denote the complex L_p-spaces on a σ-finite measure space $(\Omega, \mathcal{A}, \mu)$. Then, for every $1 \leq p < \infty$, the dual space L_p^* can be identified as L_q, where q is the conjugate index of p. Further, the identification of a functional $T \in L_p^*$ with an element $g \in L_q$ is given by $T \equiv T_g$, where $T_g(f) = \int fg d\mu$, for $f \in L_p$.

We end this section with a brief discussion on an issue that could not have escaped attention of a careful reader. Going back to the above proof of the "converse part" of Theorem 6.4.13, we started with a $T \in L_p^*$, for some $p \in [1, \infty)$ and considered $T^r = \mathrm{Re}(T)$ and $T^i = \mathrm{Im}(T)$. This allowed us to define, for each n, the signed measures ν_n^r and ν_n^i, both absolutely continuous with respect to μ, so that we could apply the Radon-Nikodym Theorem. This was crucial because our final candidate for g that gave us the representation $T = T_g$, was manufactured out of the R-N-derivatives $\frac{d\nu_n^r}{d\mu}$ and $\frac{d\nu_n^i}{d\mu}$. What if, instead of going to the decomposition of T into its real and imaginary parts, we had used T itself to define the set function $\nu_n(A) = T(I_{A \cap \Omega_n})$, $A \in \mathcal{A}$? Clearly, the immediate problem is that this would give us a complex-valued set function. But, if we ignore that temporarily, linearity and continuity of T would certainly imply that this complex-valued set function is countably additive and, of course, $\nu_n(\emptyset) = 0$. Further, ν_n would also be "*absolutely continuous*" with respect to μ. So, if we had a "Radon-Nikodym Theorem" for a complex-valued countably additive set function, absolutely continuous with respect to a σ-finite measure, then perhaps we could have obtained our candidate for g in a more straight-forward way. All of these lead naturally to the following definition.

DEFINITION **6.4.14.** Let $(\Omega, \mathcal{A})$ be a measurable space. A complex-valued set function ν on the σ-field $\mathcal{A}$ is called a *complex measure* if it satisfies (i) $\nu(\emptyset) = 0$ and (ii) $\nu(\bigcup_n A_n) = \sum_n \nu(A_n)$, for any sequence $\{A_n\}$ of disjoint sets in $\mathcal{A}$.

Here is a simple exercise, that show that a complex measure enjoys many of the properties that a finite signed measure has.

EXERCISE **6.4.15.** *Let ν be a complex measure on a σ-field $\mathcal{A}$. Show that*
(a) $\nu(A_1 \cup \cdots \cup A_n) = \nu(A_1) + \cdots + \nu(A_n)$, for any collection of disjoint sets $A_1, \ldots, A_n$ in $\mathcal{A}$.
(b) If $\{A_n\}$ is any sequence of $\mathcal{A}$-sets with either $A_n \uparrow A$ or $A_n \downarrow A$, then $\nu(A_n) \to \nu(A)$.

It is clear from Definition 6.4.14, that a set function ν defined on a σ-field given by
$$\nu(A) = \nu^r(A) + i\nu^i(A), \ A \in \mathcal{A}, \tag{6.4.4}$$
where ν^r and ν^i are two finite signed measures on $\mathcal{A}$, defines a complex measure on $\mathcal{A}$. In fact, it is easy to see that every complex measure arises this way and the representation (6.4.4) is unique. In particular, every finite signed measure can be thought of as a complex measure.

An important class of complex measures on $(\Omega, \mathcal{A})$ arise as follows. Suppose μ is a measure on $\mathcal{A}$. Then, for any μ-integrable complex function f on $(\Omega, \mathcal{A}, \mu)$, consider the complex-valued set function $\nu(A) = \int_A f d\mu$, $A \in \mathcal{A}$. Using properties of integrals of complex functions (parts (a) and (e) of Proposition 6.4.6), one can easily see that ν defines a complex measure on $\mathcal{A}$. Further, by part (b) of Proposition 6.4.6, one can easily see that $A \in \mathcal{A}$ and $\mu(A) = 0$ implies $\nu(A) = 0$, that is, ν is "*absolutely continuous*" with respect to μ. A natural question is : do all complex measures on $\mathcal{A}$, that are absolutely continuous with respect to the measure μ arise this way? In other words, what we are asking is whether there is a Radon-Nikodym type characterization of μ-absolutely continuous complex measures. The answer is 'yes', provided, of course, that μ is a σ-finite measure. A careful reader already sees that the proof is easy. Given any complex measure ν which is absolutely continuous with respect to μ, one easily sees that the finite signed measures ν^r and ν^i obtained from the representation (6.4.4) of ν, are both absolutely continuous with respect to μ. Assuming that μ is σ-finite, Theorem 6.1.34 implies that there are μ-integrable real measurable functions u and v, unique upto μ-null sets, such that, $\nu^r(A) = \int_A u d\mu$ and $\nu^i(A) = \int_A v d\mu$, for all $A \in \mathcal{A}$. The function $f = u + iv$ then gives us a μ-integrable complex measurable function, such that, $\nu(A) = \int_A f d\mu$, $A \in \mathcal{A}$, and also, such an f is unique upto μ-null sets. We thus proved the following strengthening of Theorem 6.1.34.

THEOREM 6.4.16. (Radon-Nikodym Theorem for Complex Measures)
Let $(\Omega, \mathcal{A}, \mu)$ be a σ-finite measure space. Then, a complex measure ν on $\mathcal{A}$ is absolutely continuous with respect to μ, that is, $\nu(A) = 0$ for every $A \in \mathcal{A}$ with $\mu(A) = 0$ if and only if there is a μ-integrable complex measurable function f, unique upto μ-null sets and denoted $\frac{d\nu}{d\mu}$, such that $\nu(A) = \int_A f d\mu$, for all $A \in \mathcal{A}$.

The interested reader should now make use of the above theorem to try and see if she can simplify the proof of the "converse" part of Theorem 6.4.13.

Having come this far, a curious reader may now be interested in developing a theory of integration of complex measurable functions with respect to complex measures and study the properties. Another important issue, which is more immediate and worth pursuing, is whether one can prove Lusin's Theorem type results for complex L_p functions. We believe that a foundation has been created and therefore, we leave it for an interested reader to pursue further.

6.5 Additional Exercises

EXERCISE 6.5.1. *Show that for any real random variable X, its distribution P_X can be written as $\alpha_1 Q_1 + \alpha_2 Q_2 + \alpha_3 Q_3$ where $\alpha_1, \alpha_2, \alpha_3$ are non-negative with $\alpha_1 + \alpha_2 + \alpha_3 = 1$ and Q_1, Q_2, Q_3 are probability measures on $\mathcal{B}$ which are discrete, continuous singular and absolutely continuous respectively. Here, 'continuous singular' means singular with respect to λ, but having no point mass.*

EXERCISE 6.5.2. *Let $(\Omega, \mathcal{A})$ be a measurable space and μ, ν be two σ-finite measures on $\mathcal{A}$. Derive a version of the R-N derivative $\frac{d\mu}{d(\mu+\nu)}$. [Use an appropriate Lebesgue decomposition.]*

EXERCISE 6.5.3. *Let $D = \{(x_1, x_2) \in \mathbb{R}^2 : 0 \leq x_2 \leq x_1 \leq 1\}$ and let μ be the probability measure on $\mathcal{B}^2$ defined as $\mu(A) = 2\lambda^2(A \cap D)$, for $A \in \mathcal{B}^2$, where λ^2 is the Lebesgue measure on $\mathcal{B}^2$. Consider the first coordinate projection map $h : (x_1, x_2) \mapsto x_1$ on $\mathbb{R}^2$ and let $\nu = \mu h^{-1}$ denote the induced measure on $([0,1], \mathcal{B}([0,1]))$. Denoting λ_1 to be the restriction of the Lebesgue measure λ to $([0,1], \mathcal{B}([0,1]))$, show that $\lambda_1 \ll \nu$ and find the RN derivative $\frac{d\lambda_1}{d\nu}$.*

EXERCISE 6.5.4. *Let $(\Omega, \mathcal{A})$ be a measurable space and let μ and ν be finite measures on $\mathcal{A}$ with $\nu \ll \mu$ and $f = \frac{d\nu}{d\mu}$. Show that, for every $\alpha > 0$, the Hahn decomposition of the finite signed measure $\nu_\alpha = \nu - \alpha \mu$ gives disjoint sets Ω^+ and Ω^- such that $f \geq \alpha$ a.e.$[\mu]$ on Ω^+ and $f < \alpha$ a.e. $[\mu]$ on Ω^-.*

EXERCISE 6.5.5. *Let $(\Omega, \mathcal{A})$ be a measurable space and let μ be a σ-finite measure on $\mathcal{A}$. Show that a finite signed measure ν on $\mathcal{A}$ is absolutely continuous with respect to μ if and only if $\lim_{\mu(A) \downarrow 0} |\nu(A)| = 0$, that is, given any $\epsilon > 0$, there is a $\delta > 0$, such that $A \in \mathcal{A}$ and $\mu(A) < \delta$ imply $|\nu(A)| < \epsilon$.*

EXERCISE **6.5.6.** Let $(\Omega, \mathcal{A}, \mu)$ be a σ-finite measure space and let ν_1 and ν_2 be σ-finite measures on $\mathcal{A}$, both absolutely continuous with respect to μ, with $\frac{d\nu_1}{d\mu} = f_1$ and $\frac{d\nu_2}{d\mu} = f_2$. Show that ν_1 and ν_2 are mutually absolutely continuous if $\mu(\{f_1 = 0\} \triangle \{f_2 = 0\}) = 0$. Examine if the converse is true.

EXERCISE **6.5.7.** Let $(\Omega, \mathcal{A}, \mu)$ be a σ-finite measure space. A family $\{\nu_\alpha\}$ of finite signed measures on $\mathcal{A}$ is said to be **uniformly absolutely continuous** with respect to μ, if, for any $\epsilon > 0$, there is a δ, depending **only** on ϵ, such that, $A \in \mathcal{A}$ and $\mu(A) < \delta$ imply $|\nu_\alpha(A)| < \epsilon$, for all α. Show that, if $\nu_\alpha \ll \mu$, for each α and if the family $\{\nu_\alpha\}$ is **equicontinuous** from above at $\emptyset$, (that is, given any $\epsilon > 0$ and any sequence $\{A_n\}$ of $\mathcal{A}$-sets with $A_n \downarrow \emptyset$ there is an n_0, depending only on ϵ, such that $n \geq n_0$ implies $|\nu_\alpha(A_n)| < \epsilon$, for all α), then $\{\nu_\alpha\}$ is uniformly absolutely continuous with respect to μ.

EXERCISE **6.5.8.** Let $(\Omega, \mathcal{A}, \nu)$ be a σ-finite measure space and μ_i, for each $i \geq 1$, be finite measures on $\mathcal{A}$ with $\mu_i \ll \nu$. Suppose, for each $A \in \mathcal{A}$, the limit $\lim_{i \to \infty} \mu_i(A) = \mu(A)$ exists and is finite. Show that $\{\mu_i, i \geq 1\}$ is uniformly absolutely continuous with respect to ν and hence $\mu \ll \nu$. [Exercise 2.13.29 showed that μ is a measure on $\mathcal{A}$.]
[Hint: If uniform absolute continuity does not hold, then there exists $\epsilon > 0$, such that, for each $n \geq 1$, you have $B_n \in \mathcal{A}$ with $\nu(B_n) < 1/n^2$ and $\sup_i \mu_i(B_n) > \epsilon$. Denoting $A_n = \bigcup_{k \geq n} B_k$, $n \geq 1$, consider the non-increasing sequence $\{A_n\}$ and argue as in Exercise 2.13.29.]

EXERCISE **6.5.9.** Let $1 \leq p < \infty$ and let $\{f_n\}$ be a sequence in $L_p(\Omega, \mathcal{A}, \mu)$ such that the series $\sum_n \|f_{n+1} - f_n\|_p$ converges. Let $g_n = \sum_{k=1}^{n} |f_{k+1} - f_k|$, $k = 1, 2, \ldots$.
(a) Show that $\{g_n\}$ converges, a.e.$[\mu]$, to a function $g \in L_p$.
(b) Show that $\{g_n\}$ converges to g in L_p.
(c) Show that $\{\omega : \{g_n(\omega)\} \text{ converges}\} \subset \{\omega : \{f_n(\omega)\} \text{ converges}\}$
(d) Conclude that $\{f_n\}$ converges, a.e.$[\mu]$, to a function f.
(e) Use DCT to conclude finally that $f_n \to f$ in L_p.
(f) Use the above to give an alternative proof of completeness of $L_p(\Omega, \mathcal{A}, \mu)$.

EXERCISE **6.5.10.** For any $f : \mathbb{R} \to \mathbb{R}$ and any $x \in \mathbb{R}$, let $T_x f : \mathbb{R} \to \mathbb{R}$ be defined as $T_x f(y) = f(x + y)$. The map $f \mapsto T_x f$ is known as the 'shift' or 'translation' operator and clearly takes measurable functions to measurable functions. Let us denote $L_p(\mathbb{R}, \mathcal{B}, \lambda)$ by simply L_p, for $1 \leq p \leq \infty$.
(a) For any p, $1 \leq p \leq \infty$, show that $f \in L_p$ implies $T_x f \in L_p$ and further, $\|T_x f\|_p = \|f\|_p$, that is, $T_x : L_p \to L_p$ is a norm-preserving linear operator.
(b) Fix an $f \in L_p$ where $1 \leq p < \infty$. Show that the map $x \mapsto T_x f$ on $\mathbb{R}$ into L_p is a continuous map, that is, $x_n \to x$ in $\mathbb{R}$ implies $\|T_{x_n} f - T_x f\|_p \to 0$. [Hint: It is enough to prove it for $x = 0$ and for continuous f with compact support.]

EXERCISE **6.5.11.** Show that if $\mathcal{A}$ is a countably generated σ-field on a non-empty set Ω and μ is a σ-finite measure on $\mathcal{A}$, then $L_p(\Omega, \mathcal{A}, \mu)$, for each p, $1 \leq p \leq \infty$,

is separable (that is, L_p has a countable dense subset). Conclude, in particular, that $L_p(\mathbb{R}, \mathcal{B}, \lambda)$ (in fact, $L_p(\mathbb{R}^k, \mathcal{B}^k, \lambda^k)$, for each $k \geq 1$) is separable.
What if $\mathcal{B}$ on $\mathbb{R}$ is replaced by the larger σ-field consisting of all Lebesgue measurable subsets of $\mathbb{R}$?
(Kakutani and Kodaira showed that it is possible to extend Lebesgue measure λ to a σ-field $\mathcal{A}$ on $\mathbb{R}$, that contains all the Lebesgue measurable sets, such that, any dense set in $L_2(\mathbb{R}, \mathcal{A}, \lambda)$ has to be of the same cardinality as $\mathcal{P}(\mathbb{R})$.)

EXERCISE **6.5.12.** [*This exercise aims to prove the same thing that Exercise 4.6.4 did, except that the steps are different.*]
(a) Show that the function
$$f(x) = \begin{cases} 0 & \text{for } x \leq 0 \\ e^{-1/x^2} & \text{for } x \geq 0 \end{cases}$$
is a C^∞-function, that is, it has derivatives of all orders.
(b) Let $a < b$ be real numbers. Show that there is a C^∞-function which is strictly positive on (a, b) and zero outside.
(c) Let $a < b$ be real numbers. Show that there is a C^∞-function f such that $f(x) = 0$ for $x < a$, $f(x) = 1$ for $x > b$ and increases from 0 to 1 on $[a, b]$. Show also that there is a C^∞ function g such that $g = 1$ for $x \leq a$, $g = 0$ for $x \geq b$ and decreases from 1 to 0 on $[a, b]$.
(d) Show that, given real numbers $a < b < c < d$, there is a C^∞ function f such that $f(x) = 1$ for $x \in [b, c]$, $f(x) = 0$ for $x \notin [a, d]$ and $f(x)$ increases from 0 to 1 on $[a, b]$ and decreases from 1 to 0 on $[c, d]$.
(e) Given any closed bounded interval J, show that there is a uniformly bounded sequence of C^∞-functions with compact support, which converge to I_J.
(f) Show that, for each $1 \leq p < \infty$, the class of C^∞ functions with compact support is dense in L_p.

EXERCISE **6.5.13.** (*This Exercise complements Theorem 6.2.12, for $1 < p < \infty$, in case μ is not σ-finite.*)
Let $(\Omega, \mathcal{A}, \mu)$ be a measure space, not necessarily σ-finite, and let $L_p = L_p(\Omega, \mathcal{A}, \mu)$, for $1 < p < \infty$. Denoting q to be the conjugate index of p, it has already been shown that $L_q \subsetneq L_p^*$ (see Remark 6.2.11). To prove the reverse inclusion, you have to show that for any $T \in L_p^*$, there is a $g \in L_q$ such that $T(f) = \int fg d\mu$ for all $f \in L_p$. Show this by completing the following steps.
A set $A \in \mathcal{A}$ will be called a 'σ-finite set', if $A = \bigcup_n A_n$, where $A_n \in \mathcal{A}$ with $\mu(A_n) < \infty$, for each n. Let T be a continuous linear functional on L_p.
(a) Show that for any σ-finite set $A \in \mathcal{A}$, there is a $g_A \in L_q$, unique a.e.$[\mu]$, such that, (i) $\|g_A\|_q \leq \|T\|$, (ii) $g_A = 0$ on A^c and (iii) $T(f) = \int g_A f d\mu$ for all $f \in L_p$ with $f \equiv 0$ on A^c. [Restrict T to an appropriate subspace and apply the result for σ-finite measure spaces.]
(b) Show that, if $A, B \in \mathcal{A}$ are σ-finite sets and g_A, g_B are as in (a), then $g_A = g_B$ on $A \cap B$, a.e.$[\mu]$. Conclude, in particular, that the function $g_A I_A + g_B I_{B \setminus A}$ is a candidate for $g_{A \cup B}$.

Denote $c = \sup\{\|g_A\|_q : A \in \mathcal{A} \text{ is a } \sigma\text{-finite set}\}$. Clearly, $c \leq \|T\| < \infty$. Choose a sequence $\{A_n\}$ of σ-finite sets, such that, $\|g_{A_n}\|_q \uparrow c$ and let $A = \bigcup_n A_n$.
(c) Show that A is a σ-finite set and that $\|g_A\|_q = c$. [Use part (b).]
(d) Show that, if $B \in \mathcal{A}$ is any σ-finite set with $B \cap A = \emptyset$, then $g_B = 0$ a.e. $[\mu]$.
(e) Now complete the proof by showing that $T(f) = \int g_A f d\mu$ for all $f \in L_p$. [Use Exercise 3.6.3 and part (d).]

The next three Exercises illustate that as far as Riesz Representation Theorem for non-σ-finite measure spaces is concerned, the situation is not as neat for $p = 1$ as it is for $1 < p < \infty$.

EXERCISE **6.5.14.** Consider the measure space $(\Omega, \mathcal{A}, \mu)$, where $\Omega = [0,1]$, $\mathcal{A} = \mathcal{P}([0,1])$ and $\mu(A) = |A|$, the cardinality of A. Note that, the class L_∞ on this measure space consists precisely of all bounded real-valued functions on $[0,1]$. Also, for any $g \in L_\infty$, Hölder's inequality implies that the map defined by $T_g(f) = \int fgd\mu$, $f \in L_1$, gives a bounded linear functional on L_1 with $\|T_g\| \leq \|g\|_\infty$.
(a) Using the L_1-functions $f = I_{\{x\}}$, $x \in [0,1]$, show that $\|T_g\| = \|g\|_\infty$ and conclude that $L_\infty \subsetneq L_1^*$.
(b) Show that $f \in L_1$ if and only if there is a countable set $D = \{x_1, x_2, \ldots\}$ and a real sequence $\{c_n\}$ with $\sum_n |c_n| < \infty$, such that $f = \sum_n c_n I_{\{x_n\}}$, and, in that case, $\|f\|_1 = \sum_n |c_n|$.
(b) Let T be a bounded linear functional on L_1. Let g be the real valued function on $[0,1]$ defined by $g(x) = T(I_{\{x\}})$, $x \in [0,1]$. Show that, $g \in L_\infty$ and that $T(f) = \int fgd\mu$, for all $f \in L_1$.
(c) Having thus estabished that every $T \in L_1^*$ equals T_g, for some $g \in L_\infty$, use part (a) now to conclude that $L_1^* \subseteq L_\infty$ and hence $L_1^* \cong L_\infty$.

EXERCISE **6.5.15.** Consider the measure space $(\Omega, \mathcal{A}, \mu)$, where Ω and μ are as in Exercise 6.5.14, but the σ-field $\mathcal{A}$ is now the countable-cocountable σ-field.
(a) Show that a real-valued function on $[0,1]$ belongs to L_∞ if and only if there is a countable $D \subset [0,1]$ and $c \in \mathbb{R}$ such that $g \equiv c$ on D^c and $\|g\|_D - \sup_D |g(x)| < \infty$ and in that case $\|g\|_\infty = \max\{\|g\|_D, |c|\}$. [Use Exercise 3.6.18.]
(b) Now use argument similar to that used in Exercise 6.5.14, to show that, for every $g \in L_\infty$, the map $T_g : f \mapsto \int fgd\mu$, $f \in L_1$, defines a bounded linear functional on L_1 and that $\|T_g\| = \|g\|_\infty$.
(c) Show that the function $h(x) = x$ is not measurable, but, for every $f \in L_1$, the function $h(x)f(x) = xf(x)$ is measurable. [Use Exercise 3.6.18.]
(d) Show that, for every $f \in L_1$, the function $xf(x)$ is integrable and the map $T : f \mapsto \int xf(x)d\mu(x)$ defines a bounded linear functional on L_1.
(e) Show that, given any real $\mathcal{A}$-measurable function g on Ω, there exists an $f \in L_1$, such that, gf is μ-integrable, but $\int g(x)f(x)d\mu(x) \neq T(f)$, where T is as defined in (d).
(f) Conclude that $T \neq T_g$, for any $g \in L_\infty$ and thus, for this non-σ-finite measure space, $L_\infty \subsetneq L_1^*$ and the inclusion is strict.

6.5. Additional Exercises

EXERCISE 6.5.16. *Consider the measure space* $(\Omega, \mathcal{A}, \mu)$, *where* Ω *is a two-point set* $\Omega = \{\omega_1, \omega_2\}$, $\mathcal{A} = \mathcal{P}(\Omega)$ *and* μ *is given by* $\mu(\{\omega_1\}) = 1, \mu(\{\omega_2\}) = \infty$.
(a) Show that, if g is a real-valued function on Ω with $|g(\omega_1)| < |g(\omega_2)|$, then the map $T_g : f \mapsto \int fg d\mu$, $f \in L_1$, is a bounded linear functional on L_1, for which $\|T_g\| < \|g\|_\infty$.
(b) Conclude that here, unlike in the previous two exercises, even the inclusion $L_\infty \subsetneq L_1^*$ does not hold (meaning that $g \mapsto T_g$ is not an isometry on L_∞ into L_1^*).

The previous three exercises show that, for non-σ-finite measure spaces, the dual of L_1 cannot, in general, be identified with L_∞ as an isometrically isomorphic copy. The next exercise suggests an interesting (and perhaps very natural) alternative.

EXERCISE 6.5.17. *Let $(\Omega, \mathcal{A}, \mu)$ be a measure space and let $\mathcal{S}$ be the class of all σ-finite sets in $\mathcal{A}$ ('σ-finite sets' are defined in Exercise 6.5.13). For any $S \in \mathcal{S}$, denote $\mathcal{A}_S$ to be the restriction of $\mathcal{A}$ to S and μ_S, the restriction of μ to $\mathcal{A}_S$. Clearly, $(S, \mathcal{A}_S, \mu_S)$ is a σ-finite measure space for each $S \in \mathcal{S}$.
A collection $G = \{g_S : S \in \mathcal{S}\}$ where $g_S \in L_\infty(S, \mathcal{A}_S, \mu_S)$, for each $S \in \mathcal{S}$, is called a "germ" if $g_{S_1} = g_{S_2}$, $[\mu]$-a.e., on $S_1 \cap S_2$ and $\|G\| = \sup_{S \in \mathcal{S}} \|g_S\|_\infty < \infty$. Identify two germs $G = \{g_S\}$ and $H = \{h_S\}$ and write $G \sim H$, if $g_S = h_S$, a.e.$[\mu]$, for all $S \in \mathcal{S}$. Strictly speaking, we should write "a.e.$[\mu_S]$" instead of "a.e.$[\mu]$", but since that is clear from the context, we will take this notational liberty.
Let $L_{\infty\infty}$ denote the set of all germs (modulo the above identification). For $G = \{g_S\} \in L_{\infty\infty}$, $H = \{h_S\} \in L_{\infty\infty}$ and $c \in \mathbb{R}$, define $G + H = \{g_S + h_S\}$ and $c \cdot G = \{cg_S\}$.
(a) Show that the above operations '+' and '·' are well-defined and make $L_{\infty\infty}$ a vector space. Show also that $\|\cdot\|$ defines a norm on $L_{\infty\infty}$. Further, show that for any $G = \{g_S\}$, there is an $S \in \mathcal{S}$, such that $\|G\| = \|g_S\|_\infty$.
(b) Show that the normed linear space $(L_{\infty\infty}, \|\cdot\|)$ is complete.
(c) Let $G = \{g_S\} \in L_{\infty\infty}$. For any $f \in L_1(\Omega, \mathcal{A}, \mu)$, recall Exercise 3.6.3 and define $T_G(f) = \int fI_S g_S d\mu_S$, where S is any σ-finite set such that $\{f \neq 0\} \subset S$. Show that T_G is well-defined and defines a linear functional on L_1.
(d) Denoting $\|T_G\| = \sup\{|T_G(f)| : \|f\|_1 \leq 1\}$, for $G \in L_{\infty\infty}$, show first that $\|G\| \leq \|T_G\|$ and then use part (a) to show that $\|T_G\| \leq \|G\|$.
(e) Conclude that T_G, for each $G \in L_{\infty\infty}$, defines a bounded linear functional on $L_1(\Omega, \mathcal{A}, \mu)$ with $\|T_G\| = \|G\|$, implying that $L_{\infty\infty} \subsetneq L_1^*(\Omega, \mathcal{A}, \mu)$.
(f) Conversely, given any bounded linear functional T on $L_1(\Omega, \mathcal{A}, \mu)$, show that, for every $S \in \mathcal{S}$, there is a $g_S \in L_\infty(S, \mathcal{A}_S, \mu_S)$, such that, $T(fI_S) = T_{g_S}(f_S)$, for $f \in L_1(\Omega, \mathcal{A}, \mu)$, where f_S denotes the restriction of f to S.
(g) Show that $G = \{g_S : S \in \mathcal{S}\}$ is a germ with $\|G\| \leq \|T\|$ and that $T = T_G$. Show also that this G obtained above is unique modulo the identification '$\sim$'.
(h) Use part (e) now to deduce that $\|T\| = \|G\|$ and combine this with the observation of part (e) to finally conclude that $L_1^*(\Omega, \mathcal{A}, \mu) \cong L_{\infty\infty}$.*

A natural question that Exercise 6.5.17 raises is whether each germ $G = \{g_S\}$

can be identified with one bounded real-valued function g on Ω, in the sense that $G \sim \{g|_S : S \in \mathcal{S}\}$, where $g|_S$ denotes the restriction of g to S. We may then hope to identify L_1^* as a vector space of bounded real-valued functions on Ω, whose restrictions to $S \in \mathcal{S}$ are all measurable. While Exercise 6.5.19 shows that this is not true in general, Exercise 6.5.18 presents a simple sufficient condition for this to hold.

EXERCISE **6.5.18.** Let $(\Omega, \mathcal{A}, \mu)$ be a measure space. Suppose $\Omega = \bigcup_\alpha S_\alpha$ for a (possibly uncountable) family $\{S_\alpha\}$ of disjoint σ-finite sets, such that, $\mu(A) < \infty$ implies $A \cap S_\alpha \ne \emptyset$ for countably many α, and, $\mu(A \cap S_\alpha) = 0$ for all α implies $\mu(A) = 0$.
(a) Show that every σ-finite set S is contained in a countable union of the S_α.
(b) Denoting the restrictions of $\mathcal{A}$ and μ to S_α, for each α, by $\mathcal{A}_\alpha$ and μ_α respectively, show that $T \in L_1^*(\Omega, \mathcal{A}, \mu)$ if and only if there is $g_\alpha \in L_\infty(S_\alpha, \mathcal{A}_\alpha, \mu_\alpha)$, for each α, such that, $T(fI_{S_\alpha}) = \int fI_{S_\alpha} g_\alpha d\mu$. Show also that $\|T\| = \sup_\alpha \|g_\alpha\|_\infty$.
(c) Defining g on Ω by setting $g = g_\alpha$ on Ω_α and $\|g\| = \sup_\alpha \|g_\alpha\|_\infty$, show that $G = \{g|_S : S \in \mathcal{S}\}$ is a germ and $\|G\| = \|g\|$.
(d) Conversely, show that, given any germ G, there is a bounded real-valued g on Ω, unique upto μ-null sets, such that $G \sim \{g|_S : S \in \mathcal{S}\}$.
(f) Denoting $\widetilde{L}_\infty$ to be the space of all bounded real-valued functions g on Ω, such that $g_\alpha = g|_{S_\alpha}$, for each α, is $\mathcal{A}_\alpha$-measurable and, defining $\|g\| = \sup_\alpha \|g_\alpha\|_\infty$, for $g \in \widetilde{L}_\infty$, show that $(\widetilde{L}_\infty, \|\cdot\|)$ is a normed linear space and that $L_{\infty\infty} \cong \widetilde{L}_\infty$ and hence conclude that $L_1^* \cong \widetilde{L}_\infty$.

EXERCISE **6.5.19.** Let $\Omega = [0,1] \times [0,1]$ and let $\mathcal{A}$ be the Borel σ-field on Ω. Denoting λ and ν to be respectively the Lebesgue measure and the counting measure on $\mathcal{B}([0,1])$, consider the measure $\mu = \mu_1 + \mu_2$ on $\mathcal{A}$, where $\mu_1 = \lambda \otimes \nu$ and $\mu_2 = \nu \otimes \lambda$ (see Exercise 5.4.9). Observe that μ is not σ-finite. For any p, write just L_p for $L_p(\Omega, \mathcal{A}, \mu)$.
(a) Show that, for any $f \in L_1$, the functions $xf(x,y)$ and $yf(x,y)$ are integrable with respect to the measures μ_1 and μ_2 respectively.
(b) Consider the map $f \mapsto Tf = \int xf(x,y)d\mu_1(x,y) + \int yf(x,y)d\mu_2(x,y)$, $f \in L_1$. Show that T defines a bounded linear functional on L_1 and that $\|T\| = 1$.
(c) Let $S \in \mathcal{A}$ be a σ-finite set in $(\Omega, \mathcal{A}, \mu)$. Use Exercise 5.4.9 to get countable subsets D_1, D_2 of $[0,1]$, such that, $\lambda(S_y^*) = 0$ for all $y \notin D_1$ and $\lambda(S_x) = 0$ for all $x \notin D_2$. Consider the function g_S defined on S by
$$g_S(x,y) = \begin{cases} x & \text{if } y \in D_1, x \notin D_2 \\ y & \text{if } x \in D_2, y \notin D_1 \\ 0 & \text{for all other } (x,y) \in S \end{cases}$$
Show that $g \in L_\infty(S, \mathcal{A}_S, \mu_S)$.
(d) Denoting $\mathcal{S}$ to be the class of all σ-finite sets in $(\Omega, \mathcal{A}, \mu)$, show that the collection $G = \{g_S : S \in \mathcal{S}\}$ defines a germ and that T defined in (b) equals T_G, in the sense described in Exercise 6.5.17.
(e) Show that there does not exist any function g on $[0,1] \times [0,1]$ such that $Tf = \int fg d\mu$, for all $f \in L_1$.

6.5. Additional Exercises

The next exercise is in continuation of Remark 6.2.13 stating why L_∞^* was kept aside in our pursuit of identifying the dual spaces L_p^* culminating in the Riesz Representation Theorem. It seeks to throw some light on why things behave a bit differently when it comes to the dual of L_∞.

EXERCISE 6.5.20. *(a) Consider a two-point set $\Omega = \{\omega_1, \omega_2\}$, equipped with the σ-field $\mathcal{A} = \mathcal{P}(\Omega)$. Show that, for any choice of $0 < \alpha \leq \infty, 0 < \beta \leq \infty$, if we consider the measure μ on $\mathcal{A}$ given by $\mu(\{\omega_1\}) = \alpha, \mu(\{\omega_2\}) = \beta$, then $L_\infty = \mathbb{R}^2 = L_\infty^*$.*
(This shows that neither L_∞ nor its dual, in this case, depends on the measure μ as long as it gives positive mass to both points. Note that, in case α, β are both finite, $L_1 = \mathbb{R}^2$, but when $\alpha = \beta = \infty$, the only element of L_1 is the zero function.]
(b) Consider any measure space $(\Omega, \mathcal{A}, \mu)$ and the corresponding L_∞. Show that, for any finite measure ν with $\nu \ll \mu$, the map $T : f \mapsto Tf = \int f d\nu, f \in L_\infty$, defines a bounded linear functional on L_∞. On the contrary, if ν is a finite measure on $\mathcal{A}$, such that, $\nu \not\ll \mu$, then show that $\int f d\nu$ is not even well-defined for any $f \in L_\infty$!
(c) Consider $\Omega = \mathbb{N}$, equipped with the σ-field $\mathcal{A} = \mathcal{P}(\mathbb{N})$ and the measure μ given by $\mu(\{k\}) = 2^{-k}, k \in \mathbb{N}$. See that the L_∞-space is just the space ℓ_∞ of all bounded real sequences $\mathbf{c} = \{c_k\}$ with norm $\|c\|_\infty = \sup_k |c_k|$. Show that the space $\ell_{cl} = \{\mathbf{c} \in \ell_\infty : \lim_n \frac{1}{n}(c_1 + \cdots + c_n) \text{ exists}\}$ is a subspace of ℓ_∞ and that $T(\mathbf{c}) = \lim_n \frac{1}{n}(c_1 + \cdots + c_n), \mathbf{c} \in \ell_{cl}$ defines a bounded linear functional on ℓ_{cl} with $\|T\| \leq \|\mathbf{c}\|_\infty$. A basic result of Functional Analysis, called Hahn-Banach Theorem guarantees that one can extend T to a bounded linear functional on $\ell_\infty = L_\infty(\Omega, \mathcal{A}, \mu)$. Show, however, that there is no measure ν on $(\Omega, \mathcal{A})$ such that T is given by $T(\mathbf{c}) = \int \mathbf{c}\, d\nu = \sum_k c_k \nu(\{k\})$, for $\mathbf{c} \in \ell_\infty = L_\infty(\Omega, \mathcal{A}, \mu)$.

EXERCISE 6.5.21. *Let μ be a finite signed measure on a measurable space $(\Omega, \mathcal{A})$. Show that, for any sequence $\{A_n\}$ of disjoint sets in $\mathcal{A}$, the series $\sum_n \mu(A_n)$ is absolutely convergent (that is, $\sum_n |\mu(A_n)| < \infty$).*

EXERCISE 6.5.22. *[Due to Halmos and Savage] Let $\mathcal{M}$ be a collection of probability measures on a measurable space $(\Omega, \mathcal{A})$, such that, there is a σ-finite measure on $\mathcal{A}$ with $P \ll \mu$ for all $P \in \mathcal{M}$. Show then that there exists a sequence $\{P_n, n \geq 1\}$ in $\mathcal{M}$, such that, if P_0 is the probability on $\mathcal{A}$ defined as $P_0 = \sum_n 2^{-n} P_n$, then the class of P_0-null sets equals $\{A \in \mathcal{A} : A \text{ is } P\text{-null for all } P \in \mathcal{M}\}$. [Note that the probability P_0 need not be in the collection $\mathcal{M}$].*

Chapter 7

Convergence and Laws of Large Numbers

In this chapter, we start by introducing various notions of convergence for measurable functions on a measure space $(\Omega, \mathcal{A}, \mu)$. Pointwise convergence is a familiar notion of convergence for sequences of functions. However, this does not take into account the underlying measure space structure. We introduce and study notions of convergence which are more specific to and relevant in the context of sequences of measurable functions on a measure space. After a quick introduction to these notions of convergence on a general measure space, we specialize to the extremely important topic of various notions of convergence for random variables. We study this in great detail and as a natural culmination of that, we finally delve into two of the classical limit theorems in probability, popularly known as the *Laws of Large Numbers*. At the end, we are also going to have a brief discussion on notions of convergence for random vectors.

7.1 Convergence Concepts

One notion of convergence for a sequence of measurable functions, that we are already familiar with, is "almost everywhere convergence". Let us recall that, for measurable functions f_n, $n \geq 1$, and f on a measure space $(\Omega, \mathcal{A}, \mu)$, the statement that *the sequence $\{f_n\}$ converges to f μ-almost everywhere*, written $f_n \to f$ a.e.$[\mu]$, means that $\mu(\{\omega : f_n(\omega) \not\to f(\omega)\}) = 0$. Thus, a.e.$[\mu]$ convergence is really pointwise convergence outside a μ-null set. Just to keep things simple, we will restrict ourselves here to real-valued measurable functions only. Using (as and when necessary) the fact that a countable union of μ-null sets is μ-null, one can easily see that a.e.$[\mu]$ convergence satisfies the following standard properties:

(i) $f_n \to f$ a.e.$[\mu]$ and $f_n \to g$ a.e.$[\mu]$ imply that that $f = g$ a.e.$[\mu]$, that is, the limit in a.e.$[\mu]$ convergence is unique upto μ-null sets.

(ii) $f_n \to f$ a.e.$[\mu]$ implies that $\alpha f_n \to \alpha f$ a.e.$[\mu]$, for every $\alpha \in \mathbb{R}$.

(iii) $f_n \to f$ a.e.$[\mu]$ and $g_n \to g$ a.e.$[\mu]$ imply that $f_n + g_n \to f + g$ a.e.$[\mu]$ and $f_n g_n \to f g$ a.e.$[\mu]$.

(iv) $f_n \to f$ a.e.$[\mu]$ implies that $\varphi \circ f_n \to \varphi \circ f$ a.e.$[\mu]$, for every continuous function $\varphi : \mathbb{R} \to \mathbb{R}$.

7.1. Convergence Concepts

Another notion of convergence for measurable functions that we came across is that of convergence in L_p. For any $p \geq 1$, and for measurable functions f_n, $n \geq 1$, and f, all belonging to the L_p space of a measure space $(\Omega, \mathcal{A}, \mu)$, we say that the sequence $\{f_n\}$ *converges in* L_p *to* f, written $f_n \xrightarrow{L_p} f$, if $\|f_n - f\|_p \to 0$ as $n \to \infty$. The following show that convergence in L_p also satisfies many, though not all, of the standard properties:

(i) $f_n \xrightarrow{L_p} f$ and $f_n \xrightarrow{L_p} g$ imply that $f = g$ a.e.$[\mu]$, that is, limit in L_p convergence is unique upto μ-null sets and, therefore, unique in L_p.

(ii) $f_n \xrightarrow{L_p} f$ implies that $\alpha f_n \xrightarrow{L_p} \alpha f$, for any $\alpha \in \mathbb{R}$.

(iii) By virtue of Minkowski's inequality, $f_n \xrightarrow{L_p} f$ and $g_n \xrightarrow{L_p} g$ imply that $f_n + g_n \xrightarrow{L_p} f + g$.

(iv) By Hölder's inequality, $f_n \xrightarrow{L_p} f$ and $g_n \xrightarrow{L_q} g$, where $1 \leq p, q \leq \infty$ are conjugate indices, imply that $f_n g_n \xrightarrow{L_1} fg$.

We now introduce another useful notion of convergence for measurable functions, that we have not come across so far.

DEFINITION 7.1.1. A sequence $\{f_n\}$ of measurable functions on a measure space $(\Omega, \mathcal{A}, \mu)$ is said to *converge in measure* to a measurable function f, and we write $f_n \xrightarrow{\mu} f$, if for every $\epsilon > 0$, $\mu(|f_n - f| > \epsilon) \to 0$ as $n \to \infty$.

One can easily verify that convergence in measure satisfies properties analogous to properties (i) (ii) and (iii) for the other two notions of convergence.

It would be of some interest to know how the three notions of convergence are related to one another, in terms of direct implications. As an easy consequence of Chebycheff's inequality, one can see that L_p-convergence implies convergence in measure. Unfortunately, in a general measure space, this is the only direct implication that holds. However, there are some other relationships among these notions of convergence, as discussed in the next two paragraphs.

Firstly, if $f_n \xrightarrow{L_p} f$, then the sequence $\{f_n\}$ is Cauchy in L_p and as seen earlier (see Proposition 6.2.2 and the proof of Theorem 6.2.1), this implies that there is a subsequence $\{f_{n_k}\}$ that converges a.e.$[\mu]$ to f, the L_p limit of $\{f_n\}$. A by-product of this is that $f_n \xrightarrow{L_p} f$ and $f_n \to g$ a.e.$[\mu]$ imply that $f = g$ a.e.$[\mu]$.

Next, suppose $f_n \xrightarrow{\mu} f$. Using Definition 7.1.1, we can get a sequence of integers $1 \leq n_1 < n_2 < \cdots$ satisfying $\mu(|f_{n_k} - f| > 2^{-k}) < 2^{-k}$, for every k. But then, $\sum_k \mu(|f_{n_k} - f| > 2^{-k}) < \infty$ and so, by the Borel-Cantelli Lemma (Theorem 2.7.4), we have $\mu\left(\limsup_k\{|f_{n_k} - f| > 2^{-k}\}\right) = 0$. As in the proof of Theorem 6.2.1, it follows that $f_{n_k} \to f$ a.e.$[\mu]$. What we just saw is that convergence in μ-measure implies a.e.$[\mu]$ convergence of a subsequence. Once again, as a by-product, we have that $f_n \xrightarrow{\mu} f$ and $f_n \to g$ a.e.$[\mu]$ imply that $f = g$ a.e.$[\mu]$.

EXERCISE 7.1.2. Show that if $f_n \xrightarrow{L_r} f$ and $f_n \xrightarrow{L_s} g$, $r \neq s$, then $f = g$ a.e.$[\mu]$.

REMARK 7.1.3. We have discussed here L_p-convergence only for $p \geq 1$. That is the standard practice for reasons mentioned in Remark 3.4.3. However, the definition of L_p-convergence, as given above, would make perfect sense for $0 < p < 1$ as well. Further, using Exercise 3.4.8 (a), one can easily see that properties (i), (ii) and (iii) of L_p-convergence, stated earlier for only $p \geq 1$, hold also for $0 < p < 1$.

We end this section with a result, of which, Theorem 6.3.2 was only a special case. If one examines the proof of that theorem carefully, one would easily see that, for the main result, the fact that we were working with functions defined on a Borel subset B of $\mathbb{R}$, was of no real relevance at all. The key issues were a.e. convergence on B and $\lambda(B) < \infty$. Thus, an exact transplant of the same argument will give us the following abstract version of *Egoroff's Theorem*.

THEOREM 7.1.4. [Egoroff's Theorem for a Finite Measure Space] If f_n, $n \geq 1$, and f are real-valued measurable functions on a finite measure space $(\Omega, \mathcal{A}, \mu)$, with $f_n \to f$ a.e. $[\mu]$, then, for every $\epsilon > 0$, there is a set $A \in \mathcal{A}$ with $\mu(A) < \epsilon$, such that, $f_n \to f$ uniformly on A^c.

7.2 Convergence Concepts for Random Variables

The various concepts of convergence for sequences of measurable functions on a general measure space, as discussed in the previous section, naturally apply equally well for sequences of real random variables on a probability space. However, when placed in this special context, these concepts of convergence assume an enormous significance. Indeed, as is going to unfold, much more can be obtained in this special context, than we could get in the context of general measure spaces. We first revisit these various convergence concepts in the context of random variables and introduce some alternative terminologies that are commonly used in probability.

In what follows, X_n, $n \geq 1$, and X will denote real random variables, all defined on the same probability space $(\Omega, \mathcal{A}, P)$.

DEFINITION 7.2.1. The sequence $\{X_n\}$ is said to converge to X 'P-almost surely', written $X_n \xrightarrow{a.s.} X$, if $P(\{\omega : \lim X_n(\omega) = X(\omega)\}) = 1$. This is sometimes also expressed by saying that $\{X_n\}$ converges to X *with probability one*, written, $X_n \xrightarrow{wp\,1} X$.

Clearly, 'almost sure' convergence is nothing but 'almost everywhere' convergence in the context of a probability space. It is important here to point out that, in general, for a probability measure P, the usual practice for asserting that something holds outside of a P-null is to use the phrase '*almost surely*' (or the abbreviation 'a.s.') in place of '*almost everywhere*' (or the abbreviation 'a.e.').

Before proceeding to the other modes of convergence, let us put on record here a very important and useful result. It is called *Egoroff's Theorem for Random Variables* and is nothing but Theorem 7.1.4 stated in a probability space.

7.2. Convergence Concepts for Random Variables

THEOREM **7.2.2.** [Egoroff's Theorem for Random Variables] If X_n, $n \geq 1$, and X are real random variables on a probability space $(\Omega, \mathcal{A}, P)$, with $X_n \xrightarrow{a.s.} X$, then, for every $\epsilon > 0$, there is a set $A \in \mathcal{A}$ with $P(A) < \epsilon$, such that, $X_n(\omega) \to X(\omega)$ uniformly in $\omega \in A^c$.

We now proceed to define and study the other convergence cencepts for sequences of random variables on a probability space. As before, X_n, $n \geq 1$, and X are real random variables, all defined on the same probability space $(\Omega, \mathcal{A}, P)$

DEFINITION **7.2.3.** The sequence $\{X_n\}$ is said to converge to X 'in probability', written $X_n \xrightarrow{P} X$, if for every $\epsilon > 0$, $P(\{\omega : |X_n(\omega) - X(\omega)| > \epsilon\}) \to 0$ as $n \to \infty$. This is same as convergence in P-measure as was introduced in Definition 7.1.1.

DEFINITION **7.2.4.** For $0 < p < \infty$, the sequence $\{X_n\}$ is said to converge to X in p-th moment, if $X_n \in L_p$, for each n, and $E|X_n - X|^p \to 0$ as $n \to \infty$. In the special case $p = 1$, we also call it *convergence in mean*. This is clearly the same as L_p-convergence and is, therefore, written as $X_n \xrightarrow{L_p} X$.

Of course, convergence in L_∞ also makes sense for random variables, but in the sequel, we will only consider $0 < p < \infty$, unless otherwise mentioned. Here is an important fact about convergence in p-th moment for real random variables. It is a simple consequence of the fact that, in a probability space, the $\|\cdot\|_p$-norm increases with p. Clearly, this will have no analogue for L_p convergence on infinite measure spaces.

PROPOSITION **7.2.5.** If $\{X_n\}$ converges to X is s-th moment for some $s > 0$, then $\{X_n\}$ converges to X also in r-th moment, for every $0 < r < s$.

The next two propositions are just restating, in the context of random variables, results that were stated in the previous section (including Remark 7.1.3).

PROPOSITION **7.2.6.** (a) $X_n \xrightarrow{a.s.} X$ and $X_n \xrightarrow{a.s.} Y$ imply that $X = Y$ a.s.
(b) $X_n \xrightarrow{a.s.} X$ and $\alpha \in \mathbb{R}$ imply that $\alpha X_n \xrightarrow{a.s.} \alpha X$.
(c) $X_n \xrightarrow{a.s.} X$ and $Y_n \xrightarrow{a.s.} Y$ imply that $X_n + Y_n \xrightarrow{a.s.} X + Y$.
(d) $X_n \xrightarrow{a.s.} X$ and $Y_n \xrightarrow{a.s.} Y$ imply that $X_n Y_n \xrightarrow{a.s.} XY$.
(e) $X_n \xrightarrow{a.s.} X$ implies that $\varphi(X_n) \xrightarrow{a.s.} \varphi(X)$ for any continuous function $\varphi : \mathbb{R} \to \mathbb{R}$.
Analogues of (a), (b) and (c) above hold if a.s. convergence is replaced throughout either by convergence in probability or by convergence in p-th moment.

EXERCISE **7.2.7.** Show that if $X_n \xrightarrow{a.s.} X$, where $X_n, n \geq 1$ and X are random variables with none of them taking the value zero, then $1/X_n \xrightarrow{a.s.} 1/X$.

PROPOSITION **7.2.8.** (a) $X_n \xrightarrow{L_p} X$ for some $p > 0$ implies $X_n \xrightarrow{P} X$.
(b) $X_n \xrightarrow{L_p} X$ for some $p > 0$ implies that $X_{n_k} \xrightarrow{a.s.} X$ for some subsequence $\{n_k\}$.
(c) $X_n \xrightarrow{P} X$ implies that $X_{n_k} \xrightarrow{a.s.} X$ for some subsequence $\{n_k\}$.

REMARK **7.2.9.** As observed in the previous section, one consequence of Proposition 7.2.5 is that if a sequence $\{X_n\}$ converges in two different modes of convergence, then the two limits must be a.s. equal. The reader may also note that parts (a) and (c) of Proposition 7.2.8 together imply part (b).

The next Exercise is just a restatement of Exercise 7.1.2 in the context of random variables. However, Proposition 7.2.5 makes the proof here almost trivial.

EXERCISE **7.2.10.** *If a sequence of random variables converges in r-th mean as well as s-th mean, where $r \neq s$, then the two limits must be a.s. equal.*

We are now going to primarily focus on results which are true exclusively for convergence of random variables. We have already seen one such result in Proposition 7.2.5. Once again, $X_n, n \geq 1$, and X, in what follows, represent real random variables on some probability space.

We already know that convergence in probability of a sequence of random variables always implies a.s. convergence of a subsequence. We will see later that convergence in probability does not, in general, guarantee a.s. convergence of the whole sequence. But, here is an important result which gives a characterization of a.s. convergence through convergence in probability of an appropriate sequence. One of the important fallouts of this will be that a.s. convergence of a sequence of random variables always implies convergence in probability of the sequence.

THEOREM **7.2.11.** $X_n \xrightarrow{a.s.} X$ if and only if $\sup_{k \geq n} |X_k - X| \xrightarrow{P} 0$.

Proof. From the definition of almost sure convergence, one should be able to see that $X_n \xrightarrow{a.s.} X$ is equivalent to $P\left(\bigcup_{j \geq 1} \bigcap_{n \geq 1} \bigcup_{k \geq n} \{|X_k - X| > \frac{1}{j}\}\right) = 0$. To see this, the reader has to only convince herself that $X_n(\omega) \not\to X(\omega)$ if and only if ω belongs to the set $\bigcup_{j \geq 1} \bigcap_{n \geq 1} \bigcup_{k \geq n} \{|X_k - X| > \frac{1}{j}\}$. Of course, this last set has probability zero if and only if $P\left(\bigcap_{n \geq 1} \bigcup_{k \geq n} \{|X_k - X| > \frac{1}{j}\}\right) = 0$, for each $j \geq 1$, or equivalently, $P\left(\bigcap_{n \geq 1} \bigcup_{k \geq n} \{|X_k - X| > \epsilon\}\right) = 0$, for all $\epsilon > 0$. Noting that $\bigcup_{k \geq n} \{|X_k - X| > \epsilon\} \downarrow \bigcap_{n \geq 1} \bigcup_{k \geq n} \{|X_k - X| > \epsilon\}$ and using continuity from above, we conclude that $X_n \xrightarrow{a.s.} X$ if and only if $P\left(\bigcup_{k \geq n} \{|X_k - X| > \epsilon\}\right) \to 0$, as $n \to \infty$, for every $\epsilon > 0$. Since $\{\sup_{k \geq n} |X_k - X| > \epsilon\} = \bigcup_{k \geq n} \{|X_k - X| > \epsilon\}$, we conclude that $X_n \xrightarrow{a.s.} X$ if and only if $P\left(\sup_{k \geq n} |X_k - X| > \epsilon\right) \to 0$, as $n \to \infty$, for every $\epsilon > 0$, which proves the result. $\square$

As pointed out earlier, an important consequence of the above theorem is that a.s. convergence always implies convergence in probability. The proof is trivial since all one needs to use is the above theorem and the simple inequality that $P(|X_n - X| > \epsilon) \leq P\left(\sup_{k \geq n} |X_k - X| > \epsilon\right)$, for each n and any $\epsilon > 0$.

COROLLARY **7.2.12.** $X_n \xrightarrow{a.s.} X$ implies $X_n \xrightarrow{P} X$

EXERCISE **7.2.13.** *Show that if $\sum_n P(|X_n - X| > \epsilon) < \infty$, for every $\epsilon > 0$, then $X_n \xrightarrow{a.s.} X$.*

EXERCISE **7.2.14.** *If $\{X_n\}$ is a sequence of random variables which is almost surely monotone, then show that $X_n \xrightarrow{P} X$ implies $X_n \xrightarrow{a.s.} X$.*

EXAMPLE **7.2.15.** On $\Omega = (0, 1]$, let $\mathcal{A}$ be the Borel σ-field on $(0, 1]$ and P be the Lebesgue measure. Consider the sequence $\{X_n\}$ of random variables defined on the

7.2. Convergence Concepts for Random Variables

probability space $(\Omega, \mathcal{A}, P)$ as follows: $X_1 = I_{(0,\ 1/2]}$, $X_2 = I_{(1/2,\ 1]}$, $X_3 = I_{(0,\ 1/4]}$, $X_4 = I_{(1/4,\ 1/2]}$, $X_5 = I_{(1/2,\ 3/4]}$, $X_6 = I_{(3/4,\ 1]}$, and so on. The reader should be able to visualize how X_n's are defined, but, for the record, the general formula is

$$X_n = I_{(j/2^k, (j+1)/2^k]}, \quad \text{for } n = 2(2^{k-1} - 1) + j,\ 1 \leq j \leq 2^k,\ k = 1, 2, \cdots.$$

Each X_n takes values 0 and 1 only. Further, it should be clear that $P(|X_n| > 0) = P(X_n = 1) \to 0$ as $n \to \infty$, which, of course, means that $X_n \xrightarrow{P} 0$. However, the reader should be able to convince herself that, given any $\omega \in \Omega$, there will be an infinite number of n, for which $X_n(\omega) = 1$. For this, one needs to just observe that, given any ω, we will have, for each $k \geq 1$, some $1 \leq j \leq 2^k$, such that $\omega \in (j/2^k, (j+1)/2^k]$. What this would mean is that $X_n(\omega) \not\to 0$ for any $\omega \in \Omega$. In particular, $\{X_n\}$ cannot converge to zero a.s. As a matter of fact, the reader may convince herself that $\{X_n(\omega)\}$ does not converge at all, for any ω (why?).

Example 7.2.15 is an illustration that convergence in probability does not imply a.s. convergence, in general. In other words, the implication asserted in Corollary 7.2.12 is, in general, a strict implication. However, there is a special situation when the reverse implication also holds. To state that we need the following definition.

DEFINITION **7.2.16.** A probability space $(\Omega, \mathcal{A}, P)$ is called a *discrete probability space* if Ω is a countable set and $\mathcal{A}$ is the σ-field consisitng of all subsets.

PROPOSITION **7.2.17.** On a discrete probability space $(\Omega, \mathcal{A}, P)$, convergence in probability implies almost sure convergence and therefore the two are equivalent.

Proof. We only have to show that, on a discrete probability space, convergence in probability implies a.s. convergence. Suppose $X_n \xrightarrow{P} X$ on a discrete probability space $(\Omega, \mathcal{A}, P)$. Now, if $\{X_n\}$ does not converge a.s. to X, then there must exist an $\omega_0 \in \Omega$ with $P(\{\omega_0\}) = \alpha > 0$, such that $X_n(\omega_0) \not\to X(\omega_0)$, that is, for some $\epsilon > 0$ and some subsequence $\{n_k\}$, we have $|X_{n_k}(\omega_0) - X(\omega_0)| > \epsilon$, for all k. But, that would mean $P(\{|X_{n_k} - X| > \epsilon\}) \geq \alpha$, for all k, which contradicts the hypothesis that $X_n \xrightarrow{P} X$. □

We already know (Proposition 7.2.8(a)) that convergence in pth moment, for any p, implies convergence in probability. Here is an example which illustrates that this implication is also a strict implication, in general.

EXAMPLE **7.2.18.** On the probability space $(\Omega, \mathcal{A}, P)$ as in Example 7.2.15, consider the random variables $X_n = n^{1/p} I_{(0,\ 1/n]}$, $n \geq 1$. For each n, the random variable X_n takes the values 0 and $n^{1/p}$ only and $P(X_n \neq 0) = \frac{1}{n}$. So, for any $\epsilon > 0$, we have $P(|X_n| > \epsilon) \leq P(|X_n| > 0) = P(X_n \neq 0) = \frac{1}{n} \to 0$, implying that $X_n \xrightarrow{P} 0$. However $E(|X_n|^p) = 1$, for all n, and so $\{X_n\}$ does not converge to zero in p-th moment. We could have made things even worse by fixing an $\alpha > 1$ and considering the sequence $Y_n = n^{\alpha/p} I_{(0,\ 1/n]}$, $n \geq 1$. Once again, one has $P(|Y_n| > 0) = \frac{1}{n} \to 0$ and so $Y_n \xrightarrow{P} 0$. But, $E|Y_n|^p = n^{\alpha-1}$, which diverges, as $n \to \infty$.

Having thus seen that convergence in probability does not, in general, imply L_p-convergence, a natural question to ask is: what additional conditions, if any, can one put in to make this implication go through? We will address this later, but before that, let us turn our attention to another important question.

We know that if $X_n \xrightarrow{a.s.} X$, then $\varphi(X_n) \xrightarrow{a.s.} \varphi(X)$ for any continuous function $\varphi : \mathbb{R} \to \mathbb{R}$ (see Proposition 7.2.6(e)). However, no such result has been established for the other two modes of convergence. We are now going to show that a similar result holds for convergence in probability also. The proof of this is fairly straightforward under a slightly stronger assumption, namely, that φ is uniformly continuous. We handle that special case first because the main idea of the proof for this case will be useful when we take up proving the result for continuous (but not necessarily uniformly continuous) functions φ.

PROPOSITION **7.2.19.** $X_n \xrightarrow{P} X$ implies that $\varphi(X_n) \xrightarrow{P} \varphi(X)$ for any uniformly continuous $\varphi : \mathbb{R} \to \mathbb{R}$.

Proof. Let $\epsilon > 0$ be given. Uniform continuity of φ asserts existence of a $\delta > 0$ such that $|\varphi(x) - \varphi(y)| \leq \epsilon$, for any $x, y \in \mathbb{R}$ with $|x - y| \leq \delta$. Equivalently, $|\varphi(x) - \varphi(y)| > \epsilon$, for $x, y \in \mathbb{R}$, will imply $|x - y| > \delta$. Consequently, we get $P(|\varphi(X_n) - \varphi(X)| > \epsilon) \leq P(|X_n - X| > \delta)$ and, by the hypothesis, the last probability goes to zero, as $n \to \infty$ □

We now move on to proving the general result, where we assume the function $\varphi : \mathbb{R} \to \mathbb{R}$ to be just continuous, not necessarily uniformly. Before going to the proof, let us try to understand what is the main challenge here. This will help us understand how the challenge is dealt with in the actual proof. Suppppose we start with an $\epsilon > 0$, as in the previous proof. Since φ is only assumed to be continuous and not uniformly continuous, we will only be able to get, for each $x \in \mathbb{R}$, some $\delta(x) > 0$, such that, $|\varphi(y) - \varphi(x)| \leq \epsilon$, for all $y \in \mathbb{R}$ with $|y - x| \leq \delta(x)$. Therefore the argument of the above proof cannot be employed here, since we are now unable here to get a single $\delta > 0$, with which we can claim that $\{\omega : |\varphi(X_n(\omega)) - \varphi(X(\omega))| > \epsilon\} \subset \{\omega : |X_n(\omega) - X(\omega)| > \delta\}$, as was done in the case of uniformly continuous φ. Of course, there is a possible rescue scenario. Suppose, we were given an additional hypothesis that all the random variables X_n, $n \geq 1$, and X take values in a closed bounded interval, say, $[-M, M]$. This would mean that we need to work only with the restriction of φ to $[-M, M]$ and therefore, can make use of the fact that any continuous function on a closed bounded interval is necessarily uniformly continuous. This would have allowed us to employ the same argument as in the last proof. Unfortunately, no such additional restrictions on the random variables are given. Here is how we are going to deal with the situation. We are going to use the hyothesis $X_n \xrightarrow{P} X$ to show that, by choosing an appropriately large M, we can make the probabilities $P(\{|X_n| > M\} \bigcup \{|X| > M\})$ arbitrarily small, simultaneously *for all n*. Once this is achieved, we can then make use of uniform continuity of φ on $[-M, M]$.

7.2. Convergence Concepts for Random Variables

THEOREM 7.2.20. $X_n \xrightarrow{P} X$ implies that $\varphi(X_n) \xrightarrow{P} \varphi(X)$ for any continuous function $\varphi : \mathbb{R} \to \mathbb{R}$.

Proof. Fix any $\epsilon > 0$. We have to show that, for any given $\eta > 0$, we can get n_0, such that, $P(|\varphi(X_n) - \varphi(X)| > \epsilon) < \eta$, for all $n \geq n_0$. First, since X is a real random variable we can find an $M_0 > 0$ such that $P(|X| > M_0) < \frac{\eta}{4}$. Using the hypothesis $X_n \xrightarrow{P} X$, we choose n_1, such that, $P(|X_n - X| > 1) < \frac{\eta}{4}$, for all $n > n_1$. By the choice of M_0 and n_1, we get that

$$P(|X_n| > M_0 + 1) \leq P(|X| > M_0) + P(|X_n - X| > 1) < \frac{\eta}{2}, \text{ for all } n > n_1.$$

Next, since X_i, for each $1 \leq i \leq n_1$, is a real random variable, we can get $M_i > 0$, such that, $P(|X_i| > M_i) < \frac{\eta}{2}$. Taking now $M = \max\{M_1, \ldots, M_{n_1}, M_0 + 1\}$, one easily sees that $(|X| > M) < \frac{\eta}{4}$ and $P(|X_n| > M) < \frac{\eta}{2}$, for all n. Using uniform continuity of φ on $[-M, M]$, one can get a $\delta > 0$, such that $|\varphi(x) - \varphi(y)| \leq \epsilon$, for all $x, y \in [-M, M]$ with $|x - y| \leq \delta$. This imples that

$$P(|\varphi(X_n) - \varphi(X)| > \epsilon, |X_n| \leq M, |X| \leq M) \leq P|X_n - X| > \delta), \text{ for all } n,$$

and therefore, from the hypothesis $X_n \xrightarrow{P} X$, we can get an n_0 such that, $P(|\varphi(X_n) - \varphi(X)| > \epsilon, |X_n| \leq M, |X| \leq M) < \frac{\eta}{4}$, for all $n \geq n_0$. Using the facts that $(|X| > M) < \frac{\eta}{4}$ and $P(|X_n| > M) < \frac{\eta}{2}$, for all n, one can now easily get that $P(|\varphi(X_n) - \varphi(X)| > \epsilon) < \eta$, for all $n \geq n_0$, which completes the proof. □

REMARK 7.2.21. The reader must have realized that a crucial part of the argument in the proof of Theorem 7.2.20 was to show that if $\{X_n\}$ is a sequence of real random variables converging in probability, then, for every $\delta > 0$, one can get an $M > 0$, depending only on δ, such that, $P(|X_n| > M) < \delta$, for all n. What is important here is that a single M works for all the X_n, a property that is expressed by saying that the sequence $\{X_n\}$ is a '*tight*' sequence. This concept of '*tightness*' is a very important idea and we will return to it later (see Definition 8.5.1).

EXERCISE 7.2.22. *(a) As an immediate application of Theoerm 7.2.20, you get that if $X_n \xrightarrow{P} X$, then $X_n^2 \xrightarrow{P} X^2$. Using this or otherwise, show that if $X_n \xrightarrow{P} X$ and $Y_n \xrightarrow{P} Y$, then $X_n Y_n \xrightarrow{P} XY$.*
(b) Show that if none of the random variables $X_n, n \geq 1$, and X take the value zero, then $X_n \xrightarrow{P} X$ implies $1/X_n \xrightarrow{P} 1/X$.

EXERCISE 7.2.23. *If $\{a_n\}$ is a sequence of real numbers converging to some real number a and if $\{X_n\}$ converges to X in probability (resp. a.s., resp. in L^p), then show that $\{a_n X_n\}$ converges to aX in probability (resp. a.s., resp. in L^p).*

What we have seen so far is that a.s. convergence as well as convergence in L_p (for any p) both imply convergence in probability. Further, we have illustrated through examples that both implications are, in general, strict. We have also seen that both convergence in probability and convergence in L_p (for any p) imply a.s. convergence of a subsequence. What has not been examined yet is whether there is any direct relationship between almost sure convergence and convergence in L_p. If one examines Example 7.2.15, one can easily see that the sequence $\{X_n\}$

there converges to 0 in L_p, for all p; however, as we already saw a.s. convergence does not hold there. Similarly, in Example 7.2.18, $\{X_n\}$ (and also, $\{Y_n\}$) can be easily seen to converge to 0, almost surely, (in fact, everywhere), but, as we saw $\{X_n\}$ (or, $\{Y_n\}$) does not converge to 0 in L_p.

What these illustrate is that almost sure convergence and convergence in L_p are, by themselves, incompatible, that is, none imply the other, in general. Therefore implication in any direction would require some aditional condition. Dominated Convergence Theoerm, for example, provides an additional condition under which, almost sure convergence implies L_1-convergence. A very minor modification covers convergence in L_p, for $p \in [1, \infty)$, which is stated below. One needs only to appply DCT and the trivial inequality $|a - b|^p \le 2^{p-1}\big||a|^p - |b|^p\big|$, for any $a, b \in \mathbb{R}$ and for $p \in [1, \infty)$.

THEOREM 7.2.24. Suppose that $X_n \xrightarrow{a.s.} X$. If, for some $p \in [1, \infty)$, there is an integrable random variable Y such that $|X_n|^p \le Y$ for all n, then, the random variables X_n, $n \ge 1$, and X all belong to L_p and $X_n \xrightarrow{L_p} X$.

Interestingly enough, it turns out that, the hypothesis $X_n \xrightarrow{a.s.} X$ in Theorem 7.2.24 can be weakened to $X_n \xrightarrow{P} X$ and the same assertion still holds. However, rather than proving just this, we proceed to prove an even stronger result that, first of all, relaxes the hypothesis that there is an integrable random variable Y, such that, $|X_n|^p \le Y$, for all $n \ge 1$. But, more importantly, it gives an "if and only if" result.

But in order to get there, we need to introduce a new and very important concept. To understand it, let us recall that a random variable X on a probability space $(\Omega, \mathcal{A}, P)$ is integrable if and only if $\int_{\{|X|>\lambda\}} |X| dP \to 0$ as $\lambda \to \infty$. We want to use this criterion of integrability of a random variable to introduce the notion of 'uniform integrability' for a family of random variables.

DEFINITION 7.2.25. A family $\{X_\alpha\}$ of random variables on a probability space $(\Omega, \mathcal{A}, P)$ is said to be *uniformly integrable*, abbreviated u.i., if

$$\int_{\{|X_\alpha|>\lambda\}} |X_\alpha| dP \to 0, \text{ as } \lambda \to \infty, \text{ uniformly in } \alpha, \tag{7.2.1}$$

or equivalently, $\sup_\alpha \int_{\{|X_\alpha|>\lambda\}} |X_\alpha| dP \to 0$ as $\lambda \to \infty$.

Thus, uniform integrability of a family $\{X_\alpha\}$ of random variables requires that the tail integrals of all the $|X_\alpha|$ should go to zero, *uniformly* in α. This implies not only that each random variable in a uniformly integrable family is integrable, but also that the family is actually bounded in L_1, as the next proposition shows.

PROPOSITION 7.2.26. Any uniformly integrable family is bounded in L_1.

Proof. Uniform integrability of a family $\{X_\alpha\}$ will, by definition, imply that there exists a $\lambda_0 > 0$, such that, $\sup_\alpha \int_{\{|X_\alpha|>\lambda_0\}} |X_\alpha| dP < 1$. But then, one easily sees that $\sup_\alpha \|X_\alpha\|_1 < \lambda_0 + 1$, proving the asserted L_1-boundedness. □

7.2. Convergence Concepts for Random Variables

The following exercise is aimed at showing that for uniform integrabilithy of a family, L_1-boundedness is not a sufficient condition, that is, converse of Proposition 7.2.26 does not hold.

EXERCISE 7.2.27. On the probability space $(\Omega, \mathcal{A}, P) = ((0,1], \mathcal{B}((0,1]), \lambda)$, consider the random variables $X_n = nI_{(0,1/n)}$, $n \geq 1$. Show that the sequence $\{X_n\}$ is bounded in L_1, but not uniformly integrable.

The following proposition gives two simple, yet very useful, sufficient conditions for uniform integrability of a family.

PROPOSITION 7.2.28. Let $\{X_\alpha\}$ be a family of random variables on a probability space $(\Omega, \mathcal{A}, P)$.
(a) If there is an integrable random variable Y such that $|X_\alpha| \leq Y$ for all α, then $\{X_\alpha\}$ is uniformly integrable.
(b) If $\{X_\alpha\}$ is bounded in L_p for some $p \in (1, \infty]$, then $\{X_\alpha\}$ is uniformly integrable.

Proof. (a) The hypothesis implies $\int_{\{|X_\alpha|>\lambda\}} |X_\alpha|\, dP \leq \int_{\{|Y|>\lambda\}} |Y|\, dP$, for all α. The result follows from this and the integrability of Y.
(b) Denote $\sup_\alpha \|X_\alpha\|_p = K < \infty$. Hölder's inequality followed by Chebycheff's inequality gives $\int_{\{|X_\alpha|>\lambda\}} |X_\alpha|\, dP \leq K\bigl(P(|X_\alpha|>\lambda)\bigr)^{1/q} \leq K(K^p/\lambda^p)^{1/q}$, for all α. Since the last quantity goes to 0, as $\lambda \to \infty$, uniform integrability follows. □

The next two exercises are aimed at illustrating that the implications in both parts (a) and (b) of Proposition 7.2.28 are, in general, strict. In both exercises, the underlying probability space is $(\Omega, \mathcal{A}, P) = ((0,1], \mathcal{B}((0,1]), \lambda)$.

EXERCISE 7.2.29. On the probability space $(\Omega, \mathcal{A}, P)$ as described above, consider the random variables $X_n = nI_{(1/(n+1), 1/n)}$, $n \geq 1$. Show that $\{X_n, n \geq 1\}$ is uniformly integrable, but there is no integrable random variable Y such that $|X_n| \leq Y$ for all n. [Show that $\sup_n |X_n|$ is not integrable.]

EXERCISE 7.2.30. On the same probability space as in Exercise 7.2.29, consider the random variables $X_n = n I_{\left(0,\, (n\log(n+1))^{-1}\right)}$, $n \geq 1$. Show that $\{X_n, n \geq 1\}$ is uniformly integrable, but not bounded in L_p, for any $p > 1$.

Proposition 7.2.26 and Exercise 7.2.27 together show that L_1-boundedness is necessary but not sufficient for uniform integrability. What we are going to next show is that L_1-boundedness along with an additional condition, often described as "*uniform absolute continuity*" for a family of random variables turn out to be equivalent to uniform integrability of the family.

To get the motivation for the notion of uniform absolute continuity, let us recall that if a random variable X on a probability space $(\Omega, \mathcal{A}, P)$ is integrable, then the set function $\nu(A) = \int_A |X|\, dP$, $A \in \mathcal{A}$, defines a finite signed measure on $\mathcal{A}$, which is absolutely continuous with respect to P. But then we know that this absolute continuity $\nu \ll P$ is equivalent to the property that $\int_A |X|\, dP \to 0$, as $P(A) \to 0$ (see Proposition 6.1.1). It is, therefore, natural to call a family

$\{X_\alpha\}$ of random variables to be "*Uniformly Absolutely Continuous*" if it satisfies the property that $\int_A |X_\alpha| dP \to 0$, as $P(A) \downarrow 0$, *uniformly* in α (equivalently, $\sup_\alpha \int_A |X_\alpha| dP \to 0$, as $P(A) \to 0$). It is fairly easy to see that any uniformly integrable family satisfies this property. We now show that uniform absolute continuity coupled with L_1-boundedness is equivalent to uniform integrability.

THEOREM **7.2.31**. A family $\{X_\alpha\}$ of random variables on a probability space $(\Omega, \mathcal{A}, P)$ is uniformly integrable if and only if $\{X_\alpha\}$ is bounded in L_1 and satisfies the property that $\int_A |X_\alpha| dP \to 0$, as $P(A) \downarrow 0$, uniformly in α.

Proof. For the 'only if' part, we already know that uniform integrability implies L_1-boundedness. We now prove uniform absolute continuity property. Given $\epsilon > 0$, we first use uniform integrability to choose λ, so that, $\int_{\{|X_\alpha|>\lambda\}} |X_\alpha| dP < \epsilon/2$, for all α. Let us now take $\delta = \epsilon/(2\lambda)$. Then for any A with $P(A) < \delta$, we will have, for all α,

$$\int_A |X_\alpha| dP \leq \int_{A \cap \{|X_\alpha|>\lambda\}} |X_\alpha| dP + \int_{A \cap \{|X_\alpha| \leq \lambda\}} |X_\alpha| dP < \tfrac{\epsilon}{2} + \lambda P(A) = \epsilon,$$

which proves uniform absolute continuity, completing the proof of 'only if' part. To prove the 'if' part, given $\epsilon > 0$, we use uniform absolute continuity first to choose $\delta > 0$, so that, for any $A \in \mathcal{A}$ with $P(A) < \delta$, we have $\int_A |X_\alpha| dP < \epsilon$, for all α. If we now take any $\lambda > \sup_\alpha \|X_\alpha\|_1/\delta$, then, by Chebycheff's inequality, $P(\{|X_\alpha| > \lambda\}) < \delta$, for all α, and therefore, we will get $\int_{\{|X_\alpha|>\lambda\}} |X_\alpha| dP < \epsilon$, for all α. This completes the proof. $\square$

We finally come to the result that was talked about in the paragraph following Theorem 7.2.24. The result essentially shows that, for a sequence of random variables on a probability space, convergence in probability and uniform integrability together give a necessary and sufficient conditiion for L_1-convergence.

THEOREM **7.2.32**. Let $X_n, n \geq 1$, and X be random variables on a probability space.
(a) If $\{X_n\}$ is uniformly integrable and $X_n \xrightarrow{P} X$, then $X \in L_1$ and $X_n \xrightarrow{L_1} X$.
(h) Conversely, if $X_n \xrightarrow{L_1} X$, then $\{X_n\}$ is uniformly integrable and $X_n \xrightarrow{P} X$.

Proof. (a) We are given that $\{X_n\}$ is uniformly integrable and that $X_n \xrightarrow{P} X$. By Proposition 7.2.26, $\sup_n \|X_n\|_1 < \infty$. Next, from Proposition 7.2.8(c), we know that $X_{n_k} \xrightarrow{a.s.} X$, for some subsequence $\{n_k\}$. Applying Fatou's Lemma, we now get, $\|X\|_1 \leq \liminf_k \|X_{n_k}\|_1 \leq \sup_n \|X_n\|_1 < \infty$, which shows $X \in L_1$. To show L_1-convergence now, let $\epsilon > 0$ be given. From Theorem 7.2.31, we know that $\{X_n\}$ satisfies the uniform absolute continuity property. Using this and integrability of X, we can get a $\delta > 0$, such that, for any $A \in \mathcal{A}$ with $P(A) < \delta$, we have $\int_A |X_n| dP < \epsilon/4$, for every $n \geq 1$, and also $\int_A |X| dP < \epsilon/4$. From $X_n \xrightarrow{P} X$, we now get an n_0, such that $P(\{|X_n - X| > \epsilon/2\}) < \delta$, for all $n \geq n_0$. This will imply that, for all $n \geq n_0$, we have

$\int_{\{|X_n-X|>\epsilon/2\}} |X_n - X| dP \leq \int_{\{|X_n-X|>\epsilon/2\}} |X_n| dP + \int_{\{|X_n-X|>\epsilon/2\}} |X| dP < \epsilon/2.$

Since $\int_{\{|X_n-X| \leq \epsilon/2\}} |X_n - X| dP \leq \epsilon/2$ holds for all n, one easily obtains that $\|X_n - X\|_1 < \epsilon$, for all $n \geq n_0$, proving the convergence $X_n \xrightarrow{L_1} X$.

(b) We are given that $X_n \xrightarrow{L_1} X$. From Proposition 7.2.8(a), we get $X_n \xrightarrow{P} X$ immediately. We only have to prove now that $\{X_n\}$ is uniformly integrable. Let $\epsilon > 0$ be given. We first use integrability of X to get $\delta > 0$, such that, $\int_A |X| dP < \epsilon/2$ for any $A \in \mathcal{A}$ with $P(A) < \delta$. Next, the hypothesis implies that $\sup_n \|X_n\|_1 < \infty$. If we take a $\lambda_0 > \sup_n \|X_n\|_1/\delta$, then Chebycheff's inequality will give $P(|X_n| > \lambda) < \delta$, for all n. Using L_1-convergence, we can get an n_0, such that, $\|X_n - X\|_1 < \epsilon/2$ for all $n > n_0$. It now follows that, for all $n > n_0$, we have $\int_{\{|X_n| > \lambda_0\}} |X_n| dP \leq \|X_n - X\|_1 + \int_{\{|X_n| > \lambda_0\}} |X| dP < \epsilon$. Finally, for each $i = 1, \cdots, n_0$, we use integrability of X_i to get $\lambda_i > 0$, such that $\int_{\{|X_i| > \lambda_i\}} |X_i| dP < \epsilon$. It is clear now that, if we take $\lambda = \max\{\lambda_0, \lambda_1, \ldots, \lambda_{n_0}\}$, then we will have $\int_{\{|X_n| > \lambda\}} |X_n| dP < \epsilon$, for all n. This completes the proof. □

The next result is an easy generalization of the above theorem and we leave its proof as an exercise for the reader.

THEOREM 7.2.33. *Let $X_n, n \geq 1$, and X be random variables on a probability space and let $1 \leq p < \infty$.*
(a) *If $\{|X_n|^p\}$ is uniformly integrable and $X_n \xrightarrow{P} X$, then $X \in L_p$ and $X_n \xrightarrow{L_p} X$.*
(b) *Conversely, if $X_n \xrightarrow{L_p} X$, then $\{|X_n|^p\}$ is uniformly integrable and $X_n \xrightarrow{P} X$.*

Having extensively studied three very important convergence concepts for sequences of random variables and the relationships among them, we are now ready to get into one of the most significant and fundamental type of results in probability theory. These are bracketed as what are called *Laws of Large Numbers* and fall under classical limits theorems in probability. We need to point out here that there is another important concept of convergence for random variables, that has been left out in this chapter and is going to be taken up in Chapter 8.

7.3 Laws of Large Numbers

In this section we would discuss a special type of fundamental limit theorems in probability theory, broadly classified as the "*laws of large numbers*". We will limit ourselves here only to what is called the "classical set-up". The interested reader should be aware that there have been enormous developments in the area over a long period of time and many extensions of these types of results beyond the classical set-up are now fairly well-known.

Let us start by roughly explaining what the laws of large numbers in the "classical set up" say. Here is a very important phrase that will be used widely in the sequel. We call a sequence $\{X_n, n \geq 1\}$ of random variables on a probability space to be "*independent and identically distributed*", to be abbreviated as "*i.i.d.*" to mean that the random variables $X_n, n \geq 1$, are mutually independent and they all have the same distribution. A good (and practical) way of interpreting such a sequence is as if we are making independent observations from a common distribution repeatedly and X_n, for each n, represents the nth observation. In such a scenario, the simple average of the first n observations, namely,

$$\overline{X}_n = \tfrac{1}{n}(X_1 + \cdots + X_n)$$

is called the *"sample mean"*, based on the *"sample"* of first n observations.

What the laws of large numbers assert is that if the common distribution of the $\{X_n,\, n \geq 1\}$ has a finite mean μ, then $\overline{X}_n \to \mu$, as $n \to \infty$. Keeping in mind that while μ is a constant, the sequence $\{\overline{X}_n,\, n \geq 1\}$ is a sequence of random variables. So, the above statement is a statement on convergence of a sequence of random variables and, would, therefore, make sense only if we specify the mode of convergence. This gives rise to the two standard laws of large numbers. One, where the convergence is meant to be in probability, is called the 'Weak Law of Large Numbers', abbreviated as WLLN, while, the other, where the convergence is meant to be almost sure, is known as the 'Strong Law of Large Numbers', abbreviated as SLLN. We want to assure the reader that convergence in L_1 is not being ignored (see Exercise 7.6.5 (c)). The statistician's interpretation of these laws is that the sample mean based on n independent observations from a *"population"* converges to the common population mean μ, if finite, as the sample size gets larger and larger. The importance of this for statisticians lies in that this allows one to use $\overline{X}_n$ as a 'good' (*'consistent'* in the language of statistics) estimate for μ, in the sense that, for a large enough sample size n, the sample mean $\overline{X}_n$ would be close to the finite population mean μ.

Physicists have a different way of looking at the laws of large numbers. They consider an evolving dynamical system and think of $X_n,\, n \geq 1$, as the evaluations of a real function at the states at successive time points. Thus, $\overline{X}_n$ represents the time average of these values upto time n, while μ is the "space average" of the same real function over all possible states. So, the laws of large numbers, for a physicist, says that the time average approaches space average as time goes by.

In the next two subsections, we prove the Weak Law of Large Numbers (WLLN) and the Strong Law of Large Numbers (SLLN) for i.i.d. sequences. But one point that a careful reader must be wondering about must be addressed first. Since a.s. convergence always implies convergence in probability, the reader must have realized that SLLN would imply WLLN. It is, therefore, natural to ask why do we need to prove both? While this is indeed correct, it is, however, instructive to see the proofs separately because they contain ideas that may be valuable if one wants to go beyond the classical set-up. To be more specific, the reader will see that the arguments and techniques used here for getting the Strong Law are way more sophisticated than those used for the Weak Law. Accordingly, it seems more likely to be able to push through the ideas used for WLLN in the i.i.d. case to get analogous results under significantly relaxed conditions (see, for example, Remark 7.3.2). In contrast, one would be confronted with steeper hurdles in attempting similar analogues of classical SLLN. Simply put, results are of course important in mathematics, but, more often than not, steps leading to those results are equally important!

7.3.1 Weak Law of Large Numbers

As stated before, $\{X_n, n \geq 1\}$ will represent, unless otherwise specified, an i.i.d. sequence of random variables and it is assumed throughout that the common distribution has a finite mean μ. Then WLLN says $\overline{X}_n \xrightarrow{P} \mu$. Denoting $S_n = \sum_1^n X_i = n\overline{X}_n$, the WLLN simply says that, for every $\epsilon > 0$, the probabilities $P(\{|\frac{S_n}{n} - \mu| > \epsilon\})$ go to 0, as $n \to \infty$. Of course, one does not hope to prove this by actually computing the above probabilities and showing that they go to zero. Firstly, the WLLN is a general result where no specific assumptions are made about the common distribution of the X_n, except that their common mean μ is finite. Secondly, even if the common distribution of the X_n were specified, that would not have been of much help, except in a few very special cases. For most cases, it is almost impossible, to find the distribution of S_n in closed form, from the specified common distribution of the X_n. And these include some very simple forms of the common distribution! For example, if the $\{X_n\}$ represent the faces that show up in successive rolls of a fair die, so that each X_n takes values in $\{1, \cdots, 6\}$, with equal probability for each, the reader may try and see how easy it is to compute the exact distribution of S_{20} or even S_7!

Thus a possible alternative would be to try and see if we can get bounds for these probabilities which hold in general and may be used in proving that they go to 0. Of course, the simplest bound that comes to mind is through Chebycheff's inequality which would say

$$P(\{|\tfrac{S_n}{n} - \mu| > \epsilon\}) \leq \frac{E|(S_n/n) - \mu|}{\epsilon} = \frac{E|X_1 - \mu|}{\epsilon}.$$

However, this bound does not help us in getting the probability on the left-hand-side to go to zero. To salvage the situation, let us make, for the time being, a slightly stronger assumption about the common distribution, namely, that it has a finite second moment. This will give us, again using Chebycheff's inequality,

$$P(\{|\tfrac{S_n}{n} - \mu| > \epsilon\}) = P\{(S_n - n\mu)^2 \geq n^2\epsilon^2\} \leq \tfrac{1}{n^2\epsilon^2}V(S_n) = \tfrac{1}{n\epsilon^2}V(X).$$

The last expression clearly goes to zero, as $n \to \infty$. Thus, under the stronger assumption that the common distribution has a finite second moment, we have proved the WLLN.

Of course, this is not what was originally stated. To come back and prove the result under the original assumption of only finiteness of mean, we need to employ a a new and interesting technique, known as *truncation*. The technique of truncation has proved to be a very powerful general tool in proving limit theorems in probability. The main idea of this technique is to replace the original sequence $\{X_n\}$ by a 'truncated' sequence $\{Y_n\}$, prove an appropriate convergrence result for this truncated sequence and then try to derive the required result out of this. Of course, it is important that the truncation be done in a careful manner so as to meet both ends.

In the present context, the truncation $Y_n = X_n I_{\{|X_n| \leq n\}}$, $n \geq 1$, is one that seems to work. This means that, for each n, the random variable X_n is truncated to its values as long as they are inside the interval $[-n, n]$, while setting it to be 0, when it goes outside that interval. This truncated X_n, for each n, is what Y_n represents. It is clear that $\{Y_n, n \geq 1\}$ are independent. This follows from that fact that Y_n, for each n, is a borel measurable function of only X_n (see Definition 4.3.8 and Corollary 4.3.5). However, identicality of distribution may not be preserved by the truncation, because the intervals of truncation depend on n. Denoting $T_n = \sum_{i=1}^{n} Y_i$, we will show that

$$\frac{T_n - E(T_n)}{n} \xrightarrow{P} 0. \qquad (7.3.1)$$

But, before proving (7.3.1), let us assume it is true and first argue how this will yield the desired WLLN for $\{X_n\}$. Towards this, we first observe that

$$\sum_{n=1}^{\infty} P(X_n \neq Y_n) = \sum_{n=1}^{\infty} P(|X_n| > n) = \sum_{n=1}^{\infty} P(|X_1| > n) < \infty,$$

where the second equality comes from identicality of distributions, while the finiteness of the series comes from finiteness of $E(|X_1|)$. Applying Borel-Cantelli Lemma, we will get $P(\limsup_n \{X_n \neq Y_n\}) = 0$, or equivalently,

$$P(\{\omega : X_n(\omega) = Y_n(\omega) \text{ for all but finitely many } n\}) = 1. \qquad (7.3.2)$$

Note now that, for those ω for which, $X_n(\omega) = Y_n(\omega)$ for all but fnitely many n, say, for all $n \geq n_0(\omega)$, we will have $S_n(\omega) - T_n(\omega) = S_{n_0}(\omega) - T_{n_0}(\omega)$, for all $n \geq n_0(\omega)$, which will imply that $[S_n(\omega) - T_n(\omega)]/n \to 0$. Thus, (7.3.2) implies

$$\frac{S_n - T_n}{n} \xrightarrow{a.s.} 0. \qquad (7.3.3)$$

Next, $E(Y_n) = E(X_n I_{\{|X_n| \leq n\}}) = E(X_1 I_{\{|X_1| \leq n\}}) \to E(X_1) = \mu$, by DCT. We can now use the well-known fact that if a real sequence $\{a_n\}$ converges to a limit l, then $\frac{1}{n} \sum_{i=1}^{n} a_i \to l$, to get

$$\frac{E(T_n)}{n} = \frac{1}{n} \sum_{i=1}^{n} E(Y_i) \to \mu \qquad (7.3.4)$$

Assuming (7.3.1) and using (7.3.3) and (7.3.4), one can now easily deduce that

$$\frac{S_n}{n} \xrightarrow{P} \mu,$$

which proves WLLN for the i.i.d. sequence $\{X_n\}$.

What is now left for us is to prove what was assumed above, namely, (7.3.1). To prove this, let us try to use the same idea as was used when we assumed X_n to have finite second moment. Each Y_n, by definition, is a bounded random variable and therefore, has a finite second moment. Further,

$$V(Y_n) \leq E(Y_n^2) = E(X_n^2 I_{\{|X_n| \leq n\}}) = E(X_1^2 I_{\{|X_1| \leq n\}}),$$

where the last equality uses the fact that the $\{X_n\}$ are identically distributed. Using independence of the $\{Y_n\}$, we get $V(T_n) = \sum_{i=1}^{n} V(Y_i) \leq \sum_{i=1}^{n} E(X_1^2 I_{\{|X_1| \leq i\}})$. We now use Chebycheff's inequality to finally get

$$P(\{\tfrac{1}{n}|T_n - E(T_n)| > \epsilon\}) \leq \frac{1}{n^2 \epsilon^2} \sum_{i=1}^{n} E(X_1^2 I_{\{|X_1| \leq i\}}) \leq \frac{1}{n^2 \epsilon^2} \sum_{i=1}^{n} i\, E(|X_1| I_{\{|X_1| \leq i\}})$$

Given that the $\{X_n\}$ have a finite common mean, we can bound the last expression by $\frac{1}{n^2 \epsilon^2} E(|X_1|) \sum_{i=1}^{n} i$. But unfortunately, the sum $\sum_{i=1}^{n} i$ itself is of the order

7.3. Laws of Large Numbers

of n^2 and therefore the above inequality does not get us what we want, namely, that the probability on the left-hand-side goes to zero, as $n \to \infty$.

To tackle the problem, a new idea is brought into the calculations. Choose a sequence $\{a_n\}$ of positive reals such that $a_n \to \infty$ and $a_n/n \to 0$. For instance, the sequence $\{\sqrt{n}\}$ would do. We then decompose the sum $\sum_{i=1}^{n} E(X_1^2 I_{\{|X_1| \le i\}})$ as

$$\sum_{i \le a_n} E(X_1^2 I_{\{|X_1| \le i\}}) + \sum_{a_n < i \le n} E(X_1^2 I_{\{|X_1| \le a_n\}}) + \sum_{a_n < i \le n} E(X_1^2 I_{\{a_n < |X_1| \le i\}}) \quad (7.3.5)$$

The first two sums in (7.3.5) are together bounded above by

$$\sum_{i=1}^{n} E(X_1^2 I_{(|X_1| \le a_n)}) = n E(X_1^2 I_{(|X_1| \le a_n)}) \le n a_n E(|X_1|),$$

while the third sum in (7.3.5) is bounded above by

$$\sum_{a_n < i \le n} i E(|X_1| I_{(a_n < |X_1| \le i)}) \le E(|X_1| I_{(|X_1| > a_n)}) \sum_{i \le n} i \le n^2 E(|X_1| I_{(|X_1| > a_n)}).$$

Using all of these in the inequality $V(T_n) \le \sum_{i=1}^{n} E(X_1^2 I_{\{|X_1| \le i\}})$, one gets

$$V(T_n) \le n a_n E(|X_1|) + n^2 E(|X_1| I_{(|X_1| > a_n)}) \quad (7.3.6)$$

Now, using Chebycheff's inequality alongwith (7.3.6), we get

$$P\left(\left\{\left|\frac{T_n - E(T_n)}{n}\right| > \epsilon\right\}\right) \le \frac{a_n}{n} \frac{E(|X_1|)}{\epsilon^2} + \frac{1}{\epsilon^2} E(|X_1| I_{(|X_1| > a_n)}). \quad (7.3.7)$$

The first term on the right hand side of (7.3.7) goes to zero by the choice of $\{a_n\}$, while the second term goes to zero by the Dominated Convergence Theorem, since $a_n \to \infty$. This proves (7.3.1) and hence completes the proof of WLLN for any i.i.d. sequence $\{X_n\}$ with finite common mean. The reader should convince herself that our argument used only the finiteness of common mean of the $\{X_n\}$ (equivalently, integrability of X_1). We now state what we have just provred.

THEOREM **7.3.1.** (Weak Law of Large Numbers) If $\{X_n\}$ is a sequence of i.i.d. random variables with finite mean μ, then

$$\frac{1}{n} \sum_{i=1}^{n} X_i \xrightarrow{P} \mu.$$

REMARK **7.3.2.** A close scrutiny of our argument will convince the reader that the full force of mutual independence of the $\{X_n\}$ was really not necessary. All that was used is 'pairwise independence', that is, we only needed that X_i and X_j are independent, for every pair $i \ne j$.

7.3.2 Strong Law of Large Numbers

Once again, we consider a sequence $\{X_n\}$ of i.i.d. random variables with finite common mean μ and, as before, write $S_n = \sum_{i=1}^{n} X_i$. The aim now is to prove the almost sure convergence result that $\frac{S_n}{n} \xrightarrow{a.s.} \mu$. or equivalently, $P(\{\frac{S_n}{n} \to \mu\}) = 1$. It is worth noting here that there is an interesting result, known as *Kolmogorov's 0-1 Law* (see Theorem 7.4.5), according to which, the probability $P(\{\frac{S_n}{n} \to \mu\})$ has to be either 0 or 1. Our aim here is to show that it is 1.

For the WLLN, we only needed to show is that $P(|S_n/n - \mu| > \epsilon) \to 0$, for all $\epsilon > 0$, as needed for convergence in probability. For a.s. convergence though, we need to prove something stronger. Interestingly, it turns out that convergence of the series $\sum_n P(|S_n/n - \mu| > \epsilon)$, for all $\epsilon > 0$, will suffice (see Lemma 7.3.3 below). This still is much stronger than $P(|S_n/n - \mu| > \epsilon) \to 0$.

LEMMA **7.3.3.** Let $\{Z_n\}$ be a sequence of random variables. If, for every $\epsilon > 0$, the series $\sum_n P(|Z_n| > \epsilon)$ converges, then $Z_n \xrightarrow{\text{a.s.}} 0$.

Proof. Given the hypothesis, Borel-Cantelli Lemma implies that, for every $\epsilon > 0$, $P(\limsup_n \{|Z_n| > \epsilon\}) = 0$. But then, from the definition of $\limsup_n \{|Z_n| > \epsilon\}$ and using continuity of P from above, we get that, for every $\epsilon > 0$,

$$P(\{\sup_{k \geq n} |Z_k| > \epsilon\}) = P(\bigcup_{k \geq n} \{|Z_k| > \epsilon\}) \to P(\limsup_n \{|Z_n| > \epsilon\}) = 0.$$

This proves $\sup_{k \geq n} |Z_k| \xrightarrow{P} 0$, or equivalently (by Theorem 7.2.11), $Z_n \xrightarrow{\text{a.s.}} 0$. □

Lemma 7.3.3 says that, to prove SLLN, it will be enough for us to try and prove convergence of the series $\sum_n P(\{|S_n/n - \mu| > \epsilon\})$, for every $\epsilon > 0$. As pointed out in the context of WLLN, there is no question of trying to prove convergence of the above series by actually computing these probabilities. Just like we did there, we have to achieve this by getting appropriate bounds for the probabilities. In the context of WLLN, where we only needed to prove $P(\{|S_n/n - \mu| > \epsilon\}) \to 0$, we first tried to get it directly using Chebycheff's inequality and it worked there under the stronger assumption that the $\{X_n\}$ have a common finite second moment. The upper bound for $P(\{|S_n/n - \mu| > \epsilon\})$ that we got there, under the assumption of finite second moment, is $\frac{1}{n} V(X_1)/\epsilon^2$ and that was good enough for concluding $P(\{|S_n/n - \mu| > \epsilon\}) \to 0$. However, this bound clearly doesn't give convergence of the series $\sum_n P(\{|S_n/n - \mu| > \epsilon\})$.

This suggests that to get an appropriate upper bound for $P(\{|S_n/n - \mu| > \epsilon\})$ directly from Chebycheff's inequality, which will lead to convergence of the series, we perhaps need something stronger than the finite second moment condition. We are going to examine it assuming that the $\{X_n\}$ have a finite common fourth moment, that is, $E(X_n^4) = E(X_1^4) < \infty$. Chebycheff bound will now give

$$P\left(\left\{\left|\frac{S_n}{n} - \mu\right| > \epsilon\right\}\right) \leq \frac{1}{n^4 \epsilon^4} E(S_n - n\mu)^4. \tag{7.3.8}$$

Denoting $Y_n = X_n - \mu$, we get an i.i.d. sequence $\{Y_n\}$ with common mean 0 and finite fourth moment. Also, $S_n - n\mu = \sum_{i=1}^n Y_i$ and so $E(S_n - n\mu)^4 = E\left(\sum_{i=1}^n Y_i\right)^4$. Expanding $\left(\sum_{i=1}^n Y_i\right)^4$ and using additivity of expected value, one easily gets

$$E(S_n - n\mu)^4 = A_1 + A_2 + A_3 + A_4 + A_5,$$

where $A_1 = \sum_i E(Y_i^4)$, $A_2 = \sum_{i \neq j} 3E(Y_i^2 Y_j^2)$, $A_3 = \sum_{i \neq j} 4E(Y_i Y_j^3)$, $A_4 = \sum_{i \neq j \neq k} 6E(Y_i Y_j Y_k^2)$ and $A_5 = \sum_{i \neq j \neq k \neq l} E(Y_i Y_j Y_k Y_l)$. Using the fact that $\{Y_n\}$ are i.i.d. with zero mean and finite common moments upto the fourth order, one easily gets $A_1 = nE(Y_1^4)$, $A_2 = 3n(n-1)E(Y_1^2 Y_2^2)$ and $A_3 = A_4 = A_5 = 0$. Thus, (7.3.8) reduces to

$$P\left(\left\{\left|\frac{S_n}{n} - \mu\right| > \epsilon\right\}\right) \leq \frac{nE(Y_1^4) + 3n(n-1)E(Y_1^2 Y_2^2)}{n^4 \epsilon^4},$$

from which convergence of $\sum_n P(\{|S_n/n - \mu| > \epsilon\})$ follows and that gives us SLLN for the sequence $\{X_n\}$.

Having thus proved SLLN under the stronger assumption of finite common fourth moment, using just Chebycheff's inequality and Lemma 7.3.3, the question now is how do we get back to proving SLLN under the original condition of only finite common mean. For WLLN, we could cross this bridge using a truncation tehnique coupled with Chebycheff. For SLLN, however, just a truncation plus Chebycheff bounds does not work. We are going to need something new.

Towards this, let us first observe that the assertion of SLLN that $\frac{S_n}{n} \xrightarrow{a.s.} \mu$ is the same as $\frac{1}{n}\sum_{i=1}^{n}(X_i - \mu) \xrightarrow{a.s.} 0$, that is, a.s. convergence to zero of the 'Cesaro means' of the sequence $\{X_n - \mu\}$. This makes the following interesting result on real sequences very relevant in the context of SLLN. The result, known as "Kronecker's Lemma", gives a sufficient condition for the Cesaro means of a real sequence to converge to zero. It is a bit surprising that, even though it is purely a result on real sequences and has per se nothing to do with probability, one doesn't see many applications of this result in Real Analysis.

LEMMA 7.3.4. (Kronecker's Lemma) For any sequence $\{x_n\}$ of real numbers, the convergence of the series $\sum_n \frac{x_n}{n}$ implies that $\frac{1}{n}\sum_1^n x_i \to 0$, as $n \to \infty$.

Proof. Denoting $s_0 = 0$ and $s_n = \sum_{i=1}^{n} \frac{x_i}{i}$, $n \geq 1$, the hypothesis says that the limit $\lim_n s_n = s$ exists and is finite. We can now write the x_n as $x_n = n(s_n - s_{n-1})$, for all $n \geq 1$, which gives us

$$\sum_1^n x_i = \sum_1^n i s_i - \sum_1^n i s_{i-1} = \sum_1^n i s_i - \sum_1^n (i-1) s_{i-1} - \sum_1^n s_{i-1} = n s_n - \sum_{i=1}^n s_{i-1}.$$

Observing that $\lim_n \frac{1}{n}\sum_{i=1}^n s_{i-1} = \lim_n s_n = s$, the required result follows. □

In view of the observation made in the paragraph just before Kronecker's Lemma, this lemma now says that in order to prove SLLN, it is enough to prove that the "random series" $\sum_{n\geq 1} \frac{X_n - \mu}{n}$ converges almost surely. It is not that this makes the problem any easier, but what it does is to provide us with a somewhat different line of attack for the problem.

Historically, it was Kolmogorov who first looked at SLLN from this point of view and found an ingenious way of proving a.s. convergence of the above series. Basically, Kolmogorov looked at a random series whose terms are independent random variables and derived some sufficient conditions for the a.s. convergence of such a series. This was then applied to get the classical SLLN. We now proceed to describe Kolmogorov's results.

The following result is a first step. It is known as *Kolmogorov's maximal inequality* and is of fundamental importance in getting Kolmogorov's sufficient conditions for a.s. convergence of random series whose terms are independent random variables.

THEOREM **7.3.5.** (Kolmogorov's Maximal Inequality) Let $\xi_1, \xi_2, \cdots, \xi_n$ be independent random variables with $E(\xi_i) = 0$, for all $i = 1, \cdots, n$, and denote $T_k = \sum_{i=1}^{k} \xi_i$, for $1 \leq k \leq n$. Then, for any $\epsilon > 0$,

$$P(\{\max_{1 \leq k \leq n} |T_k| > \epsilon\}) \leq \frac{1}{\epsilon^2} E(T_n^2) = \frac{1}{\epsilon^2} \sum_{i=1}^{n} E(\xi_i^2). \tag{7.3.9}$$

REMARK **7.3.6.** A quick comparison of the inequality (7.3.9) with Chebycheff's inequality clearly shows that (7.3.9) is much stronger. While Chebycheff's inequality would only give $P(\{|T_n| > \epsilon\}) \leq \frac{1}{\epsilon^2} E(T_n^2)$, the inequality (7.3.9) provides the same upper bound for the probability of the much larger event $\{\max_{1 \leq k \leq n} |T_k| > \epsilon\}$. Of course, one should keep in mind that Chebycheff's inequality is very general and has much wider applicability, while Kolmogorov's maximal inequality works under very specific assumptions on the random variables T_k, namely, that they are partial sums of independent mean zero random variables.

Proof. Since $E(T_n) = 0$, one has $E(T_n^2) = V(T_n)$ which, by the independence of $\xi_1, \cdots, \xi_n$, equals $\sum_{1}^{n} V(\xi_i) = \sum_{1}^{n} E(\xi_i^2)$. So, we only need to prove the inequality part in (7.3.9) above.

Let us denote $A_1 = \{|T_1| > \epsilon\}$ and $A_k = \{|T_k| > \epsilon, |T_j| \leq \epsilon, 1 \leq j < k\}$, for $2 \leq k \leq n$. It is then clear that the events $A_1, \cdots, A_n$ are disjoint and that the event $A = \{\max_{1 \leq k \leq n} |T_k| > \epsilon\}$ equals $\bigcup_{k=1}^{n} A_k$, so that $P(A) = \sum_{k=1}^{n} P(A_k)$.

Now for each $k = 1, \cdots, n$, the definition of A_k gives us

$$P(A_k) \leq \frac{1}{\epsilon^2} E(T_k^2 I_{A_k}) = \frac{1}{\epsilon^2}\left[E(T_n^2 I_{A_k}) - E((T_n - T_k)^2 I_{A_k}) - 2E(T_k I_{A_k}(T_n - T_k))\right].$$

Now note that, for $k < n$, the random variable $T_k I_{A_k}$ depends only on (that is, a borel function of) the random variables $\xi_1, \cdots, \xi_k$, while $T_n - T_k = \xi_{k+1} + \cdots + \xi_n$, implying that $T_k I_{A_k}$ and $T_n - T_k$ are independent. This, together with the fact that $E(T_n - T_k) = 0$ gives $E(T_k I_{A_k}(T_n - T_k)) = 0$ for each $k < n$. For $k = n$, it is trivially true. This gives us $P(A_k) \leq \frac{1}{\epsilon^2} E(T_n^2 I_{A_k})$, for each k, and therefore,

$$P(A) \leq \frac{1}{\epsilon^2} \sum_{k=1}^{n} E(T_n^2 I_{A_k}) = \frac{1}{\epsilon^2} E(T_n^2 I_A) \leq \frac{1}{\epsilon^2} E(T_n^2),$$

giving us the required inequality and thus completing the proof. □

REMARK **7.3.7.** Here is an interesting and useful way of thinking of the event $\{\max_{1 \leq k \leq n} |T_k| > \epsilon\}$. Suppose we had an infinite sequence $\{\xi_i\}$ of random variables and suppose T_n, $n \geq 1$, denote the successive partial sums. Then $\{\max_{1 \leq k \leq n} |T_k| > \epsilon\}$ represents the event that the sequence of successive partial sums crosses the threshold of $\pm \epsilon$, that is, goes out of the finite window $[-\epsilon, \epsilon]$, before or at time n.

7.3. Laws of Large Numbers

Let us now discuss how Kolmogorov's maximal inequality is used for deriving a sufficient condition for a.s. convergence of a series of independent random variables with zero means. Let $\{\xi_n\}$ be a sequence of independent random variables on a probability space $(\Omega, \mathcal{A}, P)$, each of which has zero mean. Denote $T_n = \sum_1^n \xi_i$, for $n \geq 1$. From the definition of convergence of a series of real numbers, we know that, for any $\omega \in \Omega$, convergence the infinite series $\sum_1^\infty \xi_n(\omega)$ just means convergence of the sequence $\{T_n(\omega)\}$, which, in turn, is equivalent to the sequence $\{T_n(\omega)\}$ being a Cauchy sequence. What all these mean is that

$$\{\omega : \sum_1^\infty \xi_n(\omega) \text{ does not converge}\} = \bigcup_j \{\omega : \sup_{k,l \geq 0} |T_{n+k}(\omega) - T_{n+l}(\omega)| > \tfrac{1}{j} \text{ for all } n\}.$$

Therefore, proving a.s. convergence of $\sum \xi_n$ is equivalent to showing that

$$P(\{\omega : \sup_{k,l \geq 0} |T_{n+k}(\omega) - T_{n+l}(\omega)| > \epsilon \text{ for all } n\}) = 0, \text{ for all } \epsilon > 0. \quad (7.3.10)$$

Since, for any $\epsilon > 0$, the sets $\{\sup_{k,l \geq 0} |T_{n+k}(\omega) - T_{n+l}(\omega)| > \epsilon\}$ decrease with n, continuity of P from above implies that (7.3.10) is equivalent to

$$\lim_{n \to \infty} P(\{\sup_{k,l \geq 0} |T_{n+k} - T_{n+l}| > \epsilon\}) = 0, \text{ for all } \epsilon > 0. \quad (7.3.11)$$

Using triangle inequality, one can easily see that, for any $n \geq 1$ and any $\epsilon > 0$,

$$\{\sup_{k \geq 1} |T_{n+k} - T_n| > \epsilon\} \subset \{\sup_{k,l \geq 0} |T_{n+k} - T_{n+l}| > \epsilon\} \subset \{\sup_{k \geq 1} |T_{n+k} - T_n| > \epsilon/2\},$$

from which it follows that (7.3.11) is equivalent to

$$\lim_{n \to \infty} P(\{\sup_{k \geq 1} |T_{n+k} - T_n| > \epsilon\}) = 0, \text{ for all } \epsilon > 0. \quad (7.3.12)$$

Observe that $P(\{\sup_{k \geq 1} |T_{n+k} - T_n| > \epsilon\}) = \lim_m P(\{\max_{1 \leq k \leq m} |T_{n+k} - T_n| > \epsilon\})$, for any $n \geq 1$ and any $\epsilon > 0$. Now, Kolmogorov's maximal inequality gives us

$$P(\{\max_{1 \leq k \leq m} |T_{n+k} - T_n| > \epsilon\}) \leq \frac{1}{\epsilon^2} \sum_{j=n+1}^{n+m} V(\xi_j), \text{ for any } m \geq 1, n \geq 1 \text{ and } \epsilon > 0.$$

Using this in the previous equality, we finally get that, for any $\epsilon > 0$,

$$P(\{\sup_{k \geq 1} |T_{n+k} - T_n| > \epsilon\}) \leq \frac{1}{\epsilon^2} \sum_{j > n} V(\xi_j), \text{ for all } n \geq 1. \quad (7.3.13)$$

If we had assumed that the series $\sum_n V(\xi_n)$ converges, then that would imply that $\lim_n \sum_{j>n} V(\xi_j) = 0$ and therefore, (7.3.13) would guarantee that (7.3.12) holds. But we already saw that the condition (7.3.12) is equivalent to a.s. convergence of the random series $\sum_n \xi_n$. This gives us the following theorem.

THEOREM 7.3.8. *Let $\{\xi_n\}$ be a sequence of independent random variables on some probability space, with $E(\xi_n) = 0$ for all n. If $\sum_n V(\xi_n) < \infty$, then the random series $\sum_n \xi_n$ converges almost surely.*

The following is an immediate consequence of the above theorem. Its proof is left for the reader as an easy exercise.

COROLLARY **7.3.9.** If $\{\xi_n\}$ is a sequence of independent random variables on some probability space, for which both the series $\sum_n E(\xi_n)$ and $\sum_n V(\xi_n)$ converge, then the random series $\sum_n \xi_n$ converges almost surely.

SLLN for an i.i.d. sequence $\{X_n\}$, under the stronger assumption of finite common second moment, now follows very easily from Theorem 7.3.8 and Kronecker's Lemma (Lemma 7.3.4). Indeed, denoting, as always, the common finite mean of the $\{X_n\}$ by μ and taking $\xi_n = (X_n - \mu)/n$, $n \geq 1$, we get a sequence $\{\xi_n\}$ of independent random variables, each with zero mean. Denoting now the common variance of $\{X_n\}$ by $\sigma^2 < \infty$, we get $V(\xi_n) = \sigma^2/n^2$ for each n, so that the series $\sum_n V(\xi_n)$ converges. Theorem 7.3.8, therefore, gives a.s. convergence of the random series $\sum_n \left(\frac{X_n - \mu}{n}\right)$. An application of Kronecker's Lemma 7.3.4 now implies that $\frac{1}{n}\sum_{i=1}^{n}(X_i - \mu) \xrightarrow{\text{a.s.}} 0$, thus proving SLLN.

Finally, let us get to proving the classical SLLN under the original assumption of only common finite mean. As for WLLN, here also we are going to use the method of truncation. In fact, we are going to use the same truncation that was used for WLLN. Thus, the truncated sequence is $Y_n = X_n I_{\{|X_n| \leq n\}}$, $n \geq 1$. Denoting $\xi_n = \frac{Y_n - E(Y_n)}{n}$, $n \geq 1$, we get a sequence $\{\xi_n\}$ of independent random variables, with each ξ_n having zero mean. Also, for each n, we have

$$V(\xi_n) = \frac{1}{n^2}V(Y_n) \leq \frac{1}{n^2}E(Y_n^2) = \frac{1}{n^2}E(X_1^2 I_{\{|X_1| \leq n\}}),$$

where in the last equality, we used identicality of distribution of the $\{X_n\}$. Thus,

$$\sum_{n=1}^{\infty} V(\xi_n) \leq \sum_{n=1}^{\infty} \frac{1}{n^2} \sum_{m=1}^{n} E(X_1^2 I_{\{m-1 < |X_1| \leq m\}}) = \sum_{m=1}^{\infty} E(X_1^2 I_{\{m-1 < |X_1| \leq m\}}) \sum_{n=m}^{\infty} \frac{1}{n^2},$$

where interchanging the order of summation is valid due to non-negativity of the terms.

Note now that $\sum_{n=m}^{\infty} \frac{1}{n^2} = \frac{1}{m^2} + \sum_{n=m+1}^{\infty} \frac{1}{n^2} \leq \frac{1}{m^2} + \int_m^{\infty} \frac{1}{x^2} dx \leq \frac{2}{m}$, for all $m \geq 1$. Using this and then the inequality $E(X_1^2 I_{\{m-1 < |X_1| \leq m\}}) \leq mE(|X_1| I_{\{m-1 < |X_1| \leq m\}})$ in the above inequality for $\sum V(\xi_n)$, one gets

$$\sum_{n=1}^{\infty} V(\xi_n) \leq 2 \sum_{m=1}^{\infty} E(|X_1| I_{\{m-1 < |X_1| \leq m\}}) = 2E(|X_1|) < \infty.$$

Using Theorem 7.3.8, we conclude that the random series $\sum_n \xi_n$ converges a.s. But, then an application of Kronecker's Lemma 7.3.4, gives

$$\frac{1}{n}\sum_{i=1}^{n}(Y_i - E(Y_i)) \xrightarrow{\text{a.s.}} 0. \tag{7.3.14}$$

The argument from now to SLLN is an exact replica of what was done to get to WLLN from (7.3.1). Since the $\{Y_n\}$ are exactly the same as there, (7.3.2) holds and hence, with T_n as defined there, (7.3.3) also holds. One also has (7.3.4). Using all of these, one can deduce from (7.3.14) that $\frac{1}{n}\sum_{1}^{n} X_i \xrightarrow{\text{a.s.}} \mu$, This proves the classical SLLN, called *Kolmogorov's SLLN*, as stated in the following theorem.

7.3. Laws of Large Numbers

THEOREM **7.3.10.** (Kolmogorov's Strong Law of Large Numbers) If $\{X_n\}$ is a sequence of i.i.d. random variables with finite common mean μ, then

$$\frac{1}{n}\sum_{1}^{n} X_i \xrightarrow{a.s.} \mu.$$

We now want to pursue the issue of SLLN a bit further. Having just proved it for an i.i.d. sequence with a finite common mean, we want to ask what, if anything, happens if we have a sequence, which is i.i.d. but does not have finite common mean? There are, of course, two possibilities now. Either the common mean exists, but is not finite, or mean does not exist.

We are going to show that, in the first case, we still have

$$\frac{1}{n}\sum_{i=1}^{n} X_i \xrightarrow{a.s.} E(X_1). \tag{7.3.15}$$

To see this, suppose, for example, $E(X_1) = +\infty$. We may then fix any constant $C > 0$ and consider $Y_n = X_n I_{\{X_n \leq C\}}$, for $n \geq 1$. Clearly then, $\{Y_n\}$ is an i.i.d. sequence with finite common mean and therefore, by the SLLN we have just proved $\frac{1}{n}\sum_{i=1}^{n} Y_i \xrightarrow{a.s.} E(Y_1) = E(X_1 I_{\{X_1 \leq C\}})$. But, by the definition of the random variables Y_n, we have $\liminf_n \frac{1}{n}\sum_{i=1}^{n} X_i \geq \lim_n \frac{1}{n}\sum Y_i \stackrel{a.s.}{=} E(X_1 I_{\{X_1 \leq C\}})$. Letting now $C \uparrow \infty$ and noting that $E(X_1 I_{\{X_1 \leq C\}})$ is non-decreasing in C and so $\lim_{C \uparrow \infty} E(X_1 I_{\{X_1 \leq C\}})$ exists, one will get $\liminf_n \frac{1}{n}\sum_{i=1}^{n} X_i \geq \lim_{C \uparrow \infty} E(X_1 I_{\{X_1 \leq C\}})$ a.s.. From this, it will follow that $\frac{1}{n}\sum_{1}^{n} X_i \xrightarrow{a.s.} +\infty$, as required, provided, of course, we can show that $\lim_{C \uparrow \infty} E(X_1 I_{\{X_1 \leq C\}}) = E(X_1) = +\infty$. Towards this, note that, for any $C > 0$, $X_1 I_{\{X \leq C\}} = X_1^+ I_{\{X_1^+ \leq C\}} - X_1^-$ and that $0 \leq X_1^+ I_{\{X_1^+ \leq C\}} \uparrow X_1^+$ as $C \uparrow \infty$. Noting that $E(X_1^-) < \infty$, one now uses Monotone Convergence Theorem to get $E(X_1 I_{\{X_1 \leq C\}}) = E(X_1^+ I_{\{X_1^+ \leq C\}}) - E(X_1^-) \to E(X_1^+) - E(X_1^-) = E(X_1)$, as $C \uparrow \infty$. This completes the proof of (7.3.15), in case $E(X_1) = +\infty$. When the common mean equals $-\infty$, one can use the previous case for the sequence $\{-X_n\}$ and derive (7.3.15) straight from there.

In the other case, that is, when the common mean does not exist, the situation is somewhat unsatisfactory, in the sense that nothing can be said in general.

We are now going to discuss a very important application of SLLN. Consider any i.i.d. sequence $\{X_n, n \geq 1\}$ of random variables on some probability space $(\Omega, \mathcal{A}, P)$. For each $n \geq 1$ and $x \in \mathbb{R}$, define a function $F_n(x, \cdot)$ on Ω by

$$F_n(x, \omega) = \frac{\#\{1 \leq i \leq n : X_i(\omega) \leq x\}}{n}. \tag{7.3.16}$$

Noting that the numerator in (7.3.16) is just $\sum_{i=1}^{n} I_{\{X_i(\omega) \leq x\}}$, it is clear that $F_n(x, \cdot)$, for each $n \geq 1$ and $x \in \mathbb{R}$, is a random variable. Also, one can easily see that for each $\omega \in \Omega$ and each n, the function $F_n(\cdot, \omega)$ is a (random) distribution function on $\mathbb{R}$. It is actually the distribution function of a (random) discrete distribution on $\mathbb{R}$. To understand it, fix an n and an ω. Then $X_1(\omega), \cdots, X_n(\omega)$ can be thought of as the sample observations in a sample of size n. If, for each $X_i(\omega)$, $1 \leq i \leq n$, we assign a mass $1/n$ to the real number $X_i(\omega)$, we end up getting a discrete probability distribution on $\mathbb{R}$. It is random because it depends on the sample observations. The reader can easily convince herself that $F_n(\cdot, \omega)$ is the distribution function of this discrete distribution. For this reason, it is commonly called the '*empirical distribution function based on n observations*'. Let us now denote F to be the common distribution function of the i.i.d. sequence $\{X_n\}$. Then, it is clear that, for each $x \in \mathbb{R}$, the sequence $\{I_{\{X_n \leq x\}}\}$ is an i.i.d. sequence of random variables with finite common mean $F(x)$. We can, therefore, apply the classical SLLN to get

$$F_n(x) \xrightarrow{\text{a.s.}} F(x), \quad \text{for each } x \in \mathbb{R}. \tag{7.3.17}$$

As things stand, the P-null set involved in the a.s.-convergence in (7.3.17) is allowed to depend on x and, therefore, we cannot straight away claim to have a single P-null set, such that, for ω outside of that, $F_n(x, \omega) \to F(x)$, for all $x \in \mathbb{R}$. However, with a little more argument, this can indeed be achieved. Using the a.s.-convergence $F_n(r) \xrightarrow{\text{a.s.}} F(r)$, for each rational r, we get a countable collection $\{N_r, r \in \mathbb{Q}\}$ of P-null sets. Then $N = \bigcup \{N_r : r \in \mathbb{Q}\}$ is a P-null set, such that, for $\omega \notin N$, we have $F_n(r, \omega) \to F(r)$, as $n \to \infty$, for all $r \in \mathbb{Q}$. We leave it as an easy exercise for the reader to show now that, for $\omega \notin N$, one actually has $F_n(x, \omega) \to F(x)$ as $n \to \infty$, for all $x \in \mathbb{R}$. This would give us

$$P(\{\omega : F_n(x, \omega) \to F(x) \text{ for all } x \in \mathbb{R}\}) = 1$$

However, what is much more striking is that, we can show that the convergence $F_n(x) \to F(x)$ is, with probability one, uniform in x. This is known as *Glivenko-Cantelli Theorem* and is our next result.

THEOREM **7.3.11.** (Glivenko-Cantelli Theorem) Let $\{X_n\}$ be an i.i.d. sequence of random variables on a probability space $(\Omega, \mathcal{A}, P)$, with common distribution function F. Let $\{F_n\}$ be the sequence of empirical distribution functions, as defined in (7.3.16). Then there is a P-null set N such that, for $\omega \notin N$, $F_n(x, \omega) \to F(x)$ uniformly in $x \in \mathbb{R}$.

REMARK **7.3.12.** A more widely used way in which the assertion of Glivenko-cantelli Lemma is stated is as follows. With F and $\{F_n\}$ as in the above theorem,

$$\sup_{x \in \mathbb{R}} |F_n(x, \cdot) - F(x)| \xrightarrow{\text{a.s.}} 0, \quad \text{as } n \to \infty$$

Note that, because of the right-continuity of $|F_n(x, \cdot) - F(x)|$ in x, the above supremum over $x \in \mathbb{R}$ equals the supremum over $r \in \mathbb{Q}$ and is, therefore, measurable.

Proof. We first prove it assuming that F is continuous. For each integer $k \geq 1$ and each $1 \leq j \leq k-1$, let $x_{j,k} = \inf\{x \in \mathbb{R} : F(x) \geq j/k\}$. Using SLLN, we can get a single P-null set N with the property that if $\omega \notin N$, then for each $k \geq 1$, there exists an integer $n_k = n_k(\omega) \geq 1$, such that,

$$|F_n(x_{j,k}, \omega) - F(x_{j,k})| < 1/k, \quad \text{for all } n \geq n_k \text{ and all } 1 \leq j \leq k-1. \tag{7.3.18}$$

Taking $x_{0,k} = -\infty$ and $x_{k,k} = \infty$ for all $k \geq 1$, the above holds trivially for $x_{0,k}$ and $x_{k,k}$ also. Note that, by the assumed continuity of F, we will have $F(x_{j,k}) = F(x_{j-1,k}) = 1/k$, for all $k \geq 1$ and all $1 \leq j \leq k$. Now take any $x \in \mathbb{R}$ with $x_{j-1,k} < x < x_{j,k}$, for $k \geq 1$ and $1 \leq j \leq k$. We then have for all $n \geq n_k$,

$$F_n(x,\omega) \leq F_n(x_{j,k},\omega) < F(x_{j,k}) + \tfrac{1}{k} = F(x_{j-1,k}) + \tfrac{2}{k} \leq F(x) + \tfrac{2}{k},$$

while

$$F_n(x,\omega) \geq F_n(x_{j-1,k},\omega) > F(x_{j-1,k}) - \tfrac{1}{k} = F(x_{j,k}) - \tfrac{2}{k} \geq F(x) - \tfrac{2}{k}.$$

Thus, if $\omega \notin N$, then, for any $k \geq 1$, there is an $n_k = n_k(\omega)$, such that $n \geq n_k$ implies $|F_n(x,\omega) - F(x)| < 2/k$ for all $x \in \mathbb{R}$. This completes the proof in case F is continuous.

For the general case, let us first note that for any $x \in \mathbb{R}$, one can apply SLLN to the i.i.d. sequence $\{I_{\{X_n < x\}}\}$, whose finite common mean is $F(x-)$, to get

$$F_n(x-,\cdot) \xrightarrow{\text{a.s.}} F(x-), \text{ for all } x \in \mathbb{R}.$$

With $x_{j,k}$, $k \geq 1, 1 \leq j \leq k-1$, as defined above, what we now do is to get a P-null set N with the property that if $\omega \notin N$, then for each $k \geq 1$, there exists $n_k = n_k(\omega)$ such that not only (7.3.18) holds, but also

$$|F_n(x_{j,k}-,\omega) - F(x_{j,k}-)| < 1/k \text{ for all } n \geq n_k \text{ and all } 1 \leq j \leq k-1.$$

Now take any x with $x_{j-1,k} < x < x_{j,k}$ for some $j = 1, \ldots, k$. Then, for all $n \geq n_k$,

$$F_n(x,\omega) \leq F_n(x_{j,k}-,\omega) < F(x_{j,k}-) + \tfrac{1}{k} \leq F(x_{j-1,k}) + \tfrac{2}{k} \leq F(x) + \tfrac{2}{k},$$

while

$$F_n(x,\omega) \geq F_n(x_{j-1,k}-,\omega) > F(x_{j-1,k}-) - \tfrac{1}{k} \geq F(x_{j,k}-) - \tfrac{2}{k} \geq F(x) - \tfrac{2}{k}.$$

Thus, in this case also, we have shown that if $\omega \notin N$, then, for any $k \geq 1$, there is an $n_k = n_k(\omega)$, such that $n \geq n_k$ implies $|F_n(x,\omega) - F(x)| < 2/k$ for all $x \in \mathbb{R}$. This completes the proof. □

We end this section by mentioning that SLLN has another simple yet very interesting application, which is widely known as 'Borel's Normal Number theorem'. This result and a natural generalization of it are given in Exercise 7.6.12.

7.4 Three Series Theorem and Zero-One Law

In our proof of SLLN in the previous Subsection 7.3.2, one of the crucial steps was to prove a.s. convergence of an appropriate random series and then apply Kronecker's Lemma. For this, we first needed to find some conditions that imply convergence of a random series $\sum \xi_n$, with independent $\{\xi_n\}$. This tool was provided by the Corollary 7.3.9 of Theorem 7.3.8, which says that, for almost sure convergence of a random series $\sum \xi_n$, where the $\{\xi_n\}$ are independent, a sufficient condition is that the two series $\sum E(\xi_n)$ and $\sum V(\xi_n)$ both converge. This sufficient condition was enough for our purposes there, which were to prove convergence of an appropriate random series and then to derive SLLN from that. However, it would be a very natural question to ask whether these conditions are also necessary for a.s. convergence of a random series $\sum \xi_n$ with independent $\{\xi_n\}$. It turns out that, under an additional assumption that the $\{\xi_n\}$ are uniformly bounded in absolute value by a constant, these are indeed necessary also.

As a first step towards proving this, let us start with a sequence $\{\xi_n\}$ of independent random variables, each having zero mean. Let us also assume that they are uniformly bounded in absolute value by some constant $C > 0$, that is, $|\xi_n| \leq C$ for all n. We show that if the random series $\sum \xi_n$ converges a.s., then the series $\sum V(\xi_n)$ must converge.

Denoting $T_n = \sum_{i=1}^{n} \xi_i$, $n \geq 1$, the a.s. convergence of the series $\sum \xi_n$ would imply that, almost surely, the sequence $\{T_n\}$ must be bounded. This, in turn, assures that we can get $M > 0$, such that $P(\{|T_n| \leq M, \text{ for all } n\}) = \alpha \text{ (say)} > 0$. Let us denote $A_n = \{|T_1| \leq M, \ldots, |T_n| \leq M\}$, for each n. Then, for each n, we have $E(T_{n+1}^2 I_{A_n}) = E((T_n + \xi_{n+1})^2 I_{A_n}) = E(T_n^2 I_{A_n}) + E(\xi_{n+1}^2) P(A_n)$. In the second equality, we used independence of $\{\xi_n\}$ and the zero mean assumption to get $E(\xi_{n+1}^2 I_{A_n}) = E(\xi_{n+1}^2) P(A_n)$ and $E(T_n \xi_{n+1} I_{A_n}) = E(T_n I_{A_n}) E(\xi_{n+1}) = 0$. Clearly, $P(A_n) \geq \alpha$ for each n and so we finally get the inequality

$$E(T_{n+1}^2 I_{A_n}) \geq E(T_n^2 I_{A_n}) + \alpha E(\xi_{n+1}^2) \quad (7.4.1)$$

Next, using the hypothesis that $|\xi_n| \leq C$ for all n, and the definition of A_n, we clearly have $|T_{n+1}| \leq |T_n| + |\xi_{n+1}| \leq M + C$, on the set A_n and hence also on $A_n \cap A_{n+1}^c$. First writing $E(T_{n+1}^2 I_{A_n}) = E(T_{n+1}^2 I_{A_{n+1}}) + E(T_{n+1}^2 I_{A_n \cap A_{n+1}^c})$ and then using the above observation, we get the inequality

$$E(T_{n+1}^2 I_{A_n}) \leq E(T_{n+1}^2 I_{A_{n+1}}) + (M + C)^2 P(A_n \cap A_{n+1}^c). \quad (7.4.2)$$

Combining the inequalities (7.4.1) and (7.4.2), we get

$$\alpha E(\xi_{n+1}^2) \leq E(T_{n+1}^2 I_{A_{n+1}}) - E(T_n^2 I_{A_n}) + (M + C)^2 P(A_n \cap A_{n+1}^c),$$

which give us

$$\alpha \sum_{1}^{n} E(\xi_{i+1}^2) \leq E(T_{n+1}^2 I_{A_{n+1}}) - E(T_1^2 I_{A_1}) + (M + C)^2 \sum_{1}^{n} P(A_i \cap A_{i+1}^c)$$
$$\leq M^2 + (M + C)^2 P(A_1)$$

This being true for all n, we get $\sum V(\xi_n) \leq \sum E(\xi_n^2) \leq [M^2 + (M + C)^2 P(A_1)] / \alpha$. We have thus shown that $\sum V(\xi_n) < \infty$.

Let us now remove the assumption that $E(\xi_n) = 0$ and show that, under the same hypothesis that $\{\xi_n\}$ is an independent sequence with $|\xi_n| \leq C$ for all n, if the random series $\sum \xi_n$ converges a.s., then the two series $\sum E(\xi_n)$ and $\sum V(\xi_n)$ must both converge. This needs an interestingly cute construction.

Denoting by $(\Omega, \mathcal{A}, P)$ the probability space on which the sequence $\{\xi_n\}$ is defined, we consider the product space $(\Omega \times \Omega, \mathcal{A} \otimes \mathcal{A}, P \otimes P)$ and call it $(\widetilde{\Omega}, \widetilde{\mathcal{A}}, \widetilde{P})$.

7.4. Three Series Theorem and Zero-One Law

On $(\widetilde{\Omega}, \widetilde{\mathcal{A}}, \widetilde{P})$ consider two sequences of random variables $\{X_n\}$ and $\{Y_n\}$ defined as follows:
$$X_n(\omega, \omega') = \xi_n(\omega), \qquad Y_n(\omega, \omega') = \xi_n(\omega').$$

It should be clear from the construction that both $\{X_n\}$ and $\{Y_n\}$ are sequences of independent random variables, with X_n as well as Y_n, for each n, having the same distribution as ξ_n. Further, the sequence $\{X_n\}$ is independent of the sequence $\{Y_n\}$. The easiest way to see this is by identifying the σ-fields generated the two sequences. Essentially, what this construction has done is to manufacture 'two independent copies' of the sequence $\{\xi_n\}$ on one probability space.

Now, definition of $\widetilde{P}$ and $\{X_n\}$ imply $\widetilde{P}(\{(\omega, \omega') : \sum X_n(\omega, \omega') \text{ converges}\}) = \widetilde{P}(\{(\omega, \omega') : \sum \xi_n(\omega) \text{ converges}\}) = P(\{\omega : \sum \xi_n(\omega) \text{ converges}\}) = 1$, by our hypothesis, Thus, the random series $\sum X_n$ converges a.s.$[\widetilde{P}]$. The same will hold for $\sum Y_n$ as well. But these will imply that the random series $\sum (X_n - Y_n)$ converges a.s. $[\widetilde{P}]$.

Noting now that $\{X_n - Y_n\}$ is a sequence of independent random variables, each with zero mean and satisfying $|X_n - Y_n| \leq 2C$ for all n, we can invoke our earlier result and conclude that the series $\sum V(X_n - Y_n)$ must converge. Since $V(X_n - Y_n) = V(X_n) + V(Y_n) = 2V(\xi_n)$, we conclude that the series $\sum V(\xi_n)$ must converge. But this, in turn, implies, by Theorem 7.3.8, that the random series $\sum [\xi_n - E(\xi_n)]$ converges a.s. Using now our hypothesis that the random series $\sum \xi_n$ converges a.s., we can conclude that the series $\sum E(\xi_n)$ must converge. We have just proved the following theorem, which is a partial converse of Corollary 7.3.9.

THEOREM 7.4.1. *Let $\{\xi_n : n \geq 1\}$ be a sequence of independent random variables that are uniformly bounded in absolute value by a constant. If the random series $\sum \xi_n$ converges a.s., then the series $\sum E(\xi_n)$ and $\sum V(\xi_n)$ must both converge.*

We have already seen that the converse of the above theorem is true – in fact, the converse holds without even needing the assumption of uniform boundedness. All the above analysis leads to a very well-known result due to Kolmogorov. It gives a necessary and sufficient condition for a.s. convergence of a random series whose terms are independent random variables.

THEOREM 7.4.2. *(Kolmogorov's 3-Series Theorem) Let $\{X_n\}$ be a sequence of independent random variables. Then the following three are equivalent.*
(a) *The series $\sum X_n$ converges a s*
(b) *Each of the following three series converges for all $\alpha > 0$:*
  (i) $\sum P(|X_n| > \alpha)$, (ii) $\sum E[X_n I_{(|X_n| \leq \alpha)}]$, (iii) $\sum V[X_n I_{(|X_n| \leq \alpha)}]$
(c) *Each of the three series in (b) converge for some $\alpha > 0$.*

Proof. It is clear that we only need to prove (a)$\Rightarrow$(b) and (c) $\Rightarrow$ (a).

(a)$\Rightarrow$(b): Fix any $\alpha > 0$ and let $Y_n = I_{(|X_n| > \alpha)}$, $n \geq 1$. Clearly, $\{Y_n\}$ is a sequence of independent random variables which are uniformly bounded. The a.s. convergence of the series $\sum X_n$ implies that $P(|X_n| \leq \alpha$ for all large $n) = 1$

and therefore $P(Y_n = 0$ for all large $n) = 1$. This, of course, means that the series $\sum Y_n$ converges a.s. Convergence of the series $\sum E(Y_n) = \sum P(|X_n| > \alpha)$ is now guranteed by Theorem 7.4.1, thus proving convergence of series (i) in the statement. But this, by the Borel-Cantelli Lemma, gives $P(\limsup_n \{|X_n| > \alpha\}) = 0$, or equivalently, $P(|X_n| \leq \alpha$ for all but finitely many $n) = 1$. Putting $Z_n = X_n I_{\{|X_n| \leq \alpha\}}$, $n \geq 1$, we get a sequence $\{Z_n\}$ of independent and uniformly bounded random variables with $P(Z_n = X_n$ for all but finitely many $n) = 1$. Since we are given that $\sum X_n$ converges a.s., it follows that $\sum Z_n$ also converges a.s. But then, Theorem 7.4.1 applies to the sequence $\{Z_n\}$ and gives convergence of both the series (ii) and (iii) in the statement.

(c) $\Rightarrow$ (a): Suppose that the series (i), (ii) and (iii) converge for some $\alpha > 0$. Defining $Z_n = X_n I_{\{|X_n| \leq \alpha\}}$, $n \geq 1$ and using Corollary 7.3.9, a.s. convergence of the random series $\sum Z_n$ follows from convergence of the series (ii) and (iii). But then, by the Borel-Cantelli Lemma, convergence of the series (i) implies that $P(X_n = Z_n$ for all but finitely many $n) = 1$. It follows therefore that the random series $\sum X_n$ must converge a.s. $\square$

We end this section with one more fundamental result due to Kolmogorov that has been conceptually rather impactful in a wide range of contexts. To motivate the result, let us recall the proof of Theorem 7.4.1. To prove it, we first proved that if $\{\xi_n\}$ is a sequence of independent and uniformly bounded random variables, each with zero mean, then a.s. convergence of the random series $\sum \xi_n$ implies convergence of the series $\sum V(\xi_n)$. However, if one goes through the proof carefully, one would realize that full strength of the hypothesis that $\sum \xi_n$ converges a.s. was not really used. All that was used in the proof is that $P(\sum \xi_n$ converges$) > 0$. In other words, the same result would hold even under this weaker hypothsis instead of that of a.s. convergence of $\sum \xi_n$.

There is no doubt that, assuming $P(\sum \xi_n$ converges$) > 0$, at first sight, seems to be a much weaker hypothesis than that of a.s. convergence. However, the reader would perhaps be surprised if she is told that, in reality, it is actually no weaker at all than the hypothesis of a.s. convergence. Indeed, this rather striking fact is a consequence of a beautiful result of Kolmogorov, that we are now going to discuss. To be specific, what this result asserts is that under an appropriate set-up, a large class of events, that are often of interest, can have their probabilities take only the values either zero or one – no values strictly between 0 and 1 can be taken by the probabilities of these events. We are now going to describe in detail the setting and this special class of events, known as "*tail events*", before stating and proving Kolmogorov's result.

Let $\{X_n\}$ be a sequence of random variables on some probability space $(\Omega, \mathcal{A}, P)$. Consider, for each n, the σ-field $\mathcal{T}_n$ defined as $\mathcal{T}_n = \sigma(\{X_k : k \geq n\})$. Clearly, each $\mathcal{T}_n$ is a sub-σ-field of $\mathcal{A}$ and also, $\mathcal{T}_n$ decreases, as n increases.

DEFINITION 7.4.3. *The sub-σ-field of $\mathcal{A}$ given by $\mathcal{T} = \bigcap_n \mathcal{T}_n$ is called the* tail σ-field. *Sets belonging to $\mathcal{T}$ are called* tail events *and all those random variables on*

7.4. Three Series Theorem and Zero-One Law

$(\Omega, \mathcal{A}, P)$, that are $\mathcal{T}$-measurable, are called *tail random variables*.

Before going to the main theorem, here are some examples to understand which events are tail events and which are not. Roughly speaking, an event A is a 'tail event' iff given an $\omega \in \Omega$, any "tail" of the sequence $\{X_n(\omega)\}$ completely determines whether $\omega \in A$ or $\omega \notin A$. In other words, a tail event is an event whose occurence/non-occurence is determined completely by *any tail* of $\{X_n\}$.

EXAMPLE 7.4.4. The events $\{X_n > X_{n+1}$ for infinitely many $n\}$, $\{\limsup_n X_n > 5\}$ and $\{\sum X_n$ converges$\}$ all are examples of tail events. On the contrary, $\{X_n \geq X_1$ for infinitely many $n\}$, $\{|X_1 + \cdots + X_n| > 7$ for all large $n\}$ and $\{X_{2n} > X_n$ for some $n\}$ are some examples of events that are not tail events.

THEOREM 7.4.5. (Kolmogorov's Zero-One Law) If $\{X_n\}$ is a sequence of independent random variables, then $P(A)$ is either zero or one, for any tail event A.

Proof. For each n, let $\mathcal{A}_n = \sigma\{X_1, \ldots, X_n\}$ and $\mathcal{A}_\infty = \sigma\{X_n, n \geq 1\}$. Recall that $\mathcal{T} = \bigcap_n \mathcal{T}_n$, where $\mathcal{T}_n = \sigma(\{X_k : k \geq n\})$, $n \geq 1$. Clearly, $\mathcal{T}_n, n \geq 1$ and $\mathcal{T}$ are all sub-σ-fields of $\mathcal{A}_\infty$. From the independence of the X_n's and the definitions of $\mathcal{A}_n, \mathcal{T}_n, n \geq 1$, it follows that $\mathcal{A}_n$, for each n, is independent of $\mathcal{T}_{n+1}$ and hence independent of $\mathcal{T}$. As a consequence, $\bigcup_n \mathcal{A}_n$ is independent of $\mathcal{T}$. But $\bigcup_n \mathcal{A}_n$ is clearly a field that generates $\mathcal{A}_\infty$ and therefore, by Theorem 2.12.8, we get that $\mathcal{A}_\infty$ is independent of $\mathcal{T}$. Since $\mathcal{T} \subset \mathcal{A}_\infty$, we conclude that $\mathcal{T}$ is independent of $\mathcal{T}$. But that will mean that, for any $A \in \mathcal{T}$, we have $P(A) = P(A \cap A) = (P(A))^2$, implying that $P(A)$ must be either 0 or 1. □

Here is an immediate Corollary of Kolmorov's Zero-One Law, the proof of which is left as an exercise.

COROLLARY 7.4.6. If $\{X_n\}$ is a sequence of independent random variables, then any tail random variable must be degenerate, that is, almost surely equal to a constant.

The following result, which stands out as a beautiful application of Kolmogorov's zero-one law, is extremely interesting in its own right and, in fact, somewhat striking too.

THEOREM 7.4.7. (Jessen-Wintner Theorem) Let $\{X_n\}$ be a sequence of independent random variables, each of which is discrete. If the series $\sum_n X_n$ converges almost surely, then the distribution of $\sum_n X_n$ is of '*pure type*', that is, it is either discrete or singular continuous or absolutely continuous.

Proof. For simplicity, let us write $X = \sum_n X_n$. For each n, let $D_n \subset \mathbb{R}$ denote the countable set defined as $D_n = \{x \in \mathbb{R} : P(X_n = x) > 0\}$. This, of course, will mean that $P(X_n \notin D_n) = 0$. Denote $D = \bigcup_n D_n$ and let G be the additive subgroup of $\mathbb{R}$, generated by the countable set D. The reader should be able to see quite easily that G is countable. What is not so immediate is the very crucial observation that, for every Borel set $B \subset \mathbb{R}$, the event $\{X \in B + G\}$ is a tail event. We leave it as a not-very-difficult exercise for the reader to figure out why that is the case. Once this observation is accepted, Kolmogorov's zero-one law implies that $P(X \in B + G)$ equals either 0 or 1, for every Borel set $B \subset \mathbb{R}$. There are now exactly three distinct possibilities.

Case 1: There is a countable set $B \subset \mathbb{R}$ such that $P(X \in B + G) = 1$. But this means that, with probability one, X takes values in the countable set $B + G$ implying that X must be discrete.

If Case 1 does not hold, that is, if $P(X \in B + G) = 0$, for every countable set $B \subset \mathbb{R}$, then clearly $P(X = x) = 0$, for every $x \in \mathbb{R}$ and therefore X has a continuous distribution. In this situation, we now have two possibilities.

Case 2: There is an uncountable Borel set $B \subset \mathbb{R}$ with $\lambda(B) = 0$, such that $P(X \in B + G) = 1$. Using translation invariance of λ and countability of G, one can easily see that $\lambda(B + G) = 0$. What this means is that $P_X \perp \lambda$, that is, X has a singular continuous distribution.

Case 3: $P(X \in B + G) = 0$ for every λ-null Borel set $B \subset \mathbb{R}$. Since $0 \in G$, we have $B \subset B + G$ and therefore, $P(X \in B) \leq P(X \in B + G) = 0$, for every λ-null Borel set $B \subset \mathbb{R}$. But this means that $P_X \ll \lambda$, that is, X has an absolutely continuous distribution. □

7.5 Convergence Concepts for Random Vectors

All the three notions of convergence that have been discussed in Section 7.2 for sequences of random variables, have their natural analogues for sequences of random vectors. This extension is based on simply using the standard Euclidean norm $\|\cdot\|$ on $\mathbb{R}^k$ in place of the absolute value $|\cdot|$ in $\mathbb{R}$. In this connection, note that if $\mathbf{X}$ is a random vector then $\|\mathbf{X}\|$ is a non-negative random variable.

In what follows, $\mathbf{X}_n$, $n \geq 1$, and $\mathbf{X}$ denote k-dimensional random vectors on some probability space $(\Omega, \mathcal{A}, P)$.

DEFINITION 7.5.1. The sequence $\{\mathbf{X}_n\}$ is said to converge *almost surely* $[P]$, written $\mathbf{X}_n \xrightarrow{a.s.} \mathbf{X}$ if $P(\{\omega : \lim \|\mathbf{X}_n(\omega) - \mathbf{X}(\omega)\| = 0\}) = 1$. This is sometimes also expressed by saying that $\{\mathbf{X}_n\}$ converges to $\mathbf{X}$ *with probability one*, written, $\mathbf{X}_n \xrightarrow{wp1} \mathbf{X}$.

DEFINITION 7.5.2. The sequence $\{\mathbf{X}_n\}$ is said to converge to $\mathbf{X}$ *in probability*, written $\mathbf{X}_n \xrightarrow{P} \mathbf{X}$, if for every $\epsilon > 0$, $P\{\omega : \|\mathbf{X}_n(\omega) - \mathbf{X}(\omega)\| > \epsilon\} \to 0$, as $n \to \infty$.

DEFINITION 7.5.3. For $1 \leq p < \infty$, the sequence $\{\mathbf{X}_n\}$ is said to converge to $\mathbf{X}$ in L_p, written $\mathbf{X}_n \xrightarrow{L_p} \mathbf{X}$, if $E\|\mathbf{X}_n - \mathbf{X}\|^p \to 0$, as $n \to \infty$.

The above definitions essentially boil down to saying that a sequence $\{\mathbf{X}_n\}$ of k-dimensional random vectors converges to a random vector $\mathbf{X}$ in any one of the three modes if and only if the sequence $\{\|\mathbf{X}_n - \mathbf{X}\|\}$ of non-negative random variables converges to the zero random variable in the same mode. Also, denoting $\mathbf{X}_n = (X_{n1}, \ldots, X_{nk})$ and $\mathbf{X} = (X_1, \ldots, X_k)$, it is an easy exercise for the reader to see that convergence of $\{\mathbf{X}_n\}$ to $\mathbf{X}$ in any one of the three modes is equivalent to convergence of the sequence $\{X_{ni}\}$ of random variables to X_i in the same mode, for each $i = 1, \cdots, k$. Using this, one can easily extend many of the results proved in Section 7.2 to the case of random vectors. One can also formulate and prove, for random vectors, analogues of the Laws of Large Numbers, studied in Section 7.3. An interested reader should take up all of these as nice challenges and complete these on her own.

7.6 Additional Exercises

EXERCISE 7.6.1. Show that for random variables X_n, $n \geq 1$, and X on a probability space $(\Omega, \mathcal{A}, P)$, the sequence $\{X_n\}$ converges in probability to X if and only if $E\bigl(|X_n - X|/[1 + |X_n - X|]\bigr) \to 0$ as $n \to \infty$.

EXERCISE 7.6.2. Let $(\Omega, \mathcal{A}, P)$ be a probability space and $L_0 = L_0(\Omega, \mathcal{A}, P)$ denote the class of all real random variables on $(\Omega, \mathcal{A}, P)$. For $X \in L_0$, denote $\|X\|_0 = E\bigl(|X|/(1 + |X|)\bigr)$.
(a) Show that $\|X + Y\|_0 \leq \|X\|_0 + \|Y\|_0$, for all $X, Y \in L_0$.
(b) Show that L_0 is a real vector space and that $d_0(X, Y) = \|X - Y\|_0$, for $X, Y \in L_0$, defines a translation-invariant pseudo-metric on L_0, which becomes a metric, provided (as was done for L_p spaces) we identify any two random variables that are equal a.s.$[P]$.
(c) With the identification as stated in (b), show that (L_0, d_0) turns out to be a complete metric space.

REMARK 7.6.3. Exercise 7.6.1 shows that convergence in probability is just convergence in the metric d_0 as defined above. Of course, convergence in p-th moment, for any $p \in [1, \infty]$, is convergence in the metric given by the L_p-norm. In Chapter 8, we will see another notion of convergence, called 'convergence in distribution', which will also be shown to be the same as convergence in a metric called the 'Lévy metric' (see Exercise 8.10.2). It turns out that a.s. convergence is an outlier in the sense that it is *not* given by a metric. The reader should try to convince herself of why that is the case. In that regard, it may be of help to recall the well-known fact that in a metric space (S, d), a sequence $\{x_n\}$ in S converges to $x \in S$ if and only if every subsequence of $\{x_n\}$ has a further subsequence converging to x.

EXERCISE 7.6.4. Let $1 < p < \infty$. If $X_n \xrightarrow{L^p} X$ and if $f : \mathbb{R} \to \mathbb{R}$ is continuous with the sequence $\{f(X_n)\}$ bounded in L^p, then show that $f(X_n) \xrightarrow{L^p} f(X)$.

EXERCISE 7.6.5. (a) Show that if $\{X_n\}$ and $\{Y_n\}$ are two uniformly integrable sequences of random variables, then so is the sequence $\{X_n + Y_n\}$.
(b) Show that if $\{X_n\}$ is a uniformly integrable sequence of random variables, then so is the sequence $\{\frac{S_n}{n}\}$, where $S_n = X_1 + \cdots + X_n$, for $n \geq 1$.

(c) Show that if $\{X_n\}$ is a sequence of i.i.d. random variables with a finite common mean μ and S_n, $n \geq 1$, are as in (b), then $\frac{S_n}{n} \xrightarrow{L_1} \mu$.

EXERCISE 7.6.6. (a) Consider the series $\sum_n n^{-p} \xi_n$, for $p > 0$, where ξ_n, $n \geq 1$, are i.i.d. random variables, taking values ± 1, each with probability $\frac{1}{2}$. Show that the series converges a.s. if and only if $p > 1/2$.
(b) More generally, show that for any real sequence $\{a_n\}$ and with ξ_n, $n \geq 1$ as in part (a), the random series $\sum_n \xi_n a_n$ converges a.s. if and only if the series $\sum_n a_n^2$ converges.

EXERCISE 7.6.7. Show that if $\{X_n\}$ is any sequence of independent random variables and $\{a_n\}$ is any sequence of positive real numbers with $a_n \uparrow \infty$, then $\limsup_n (S_n/a_n)$ and $\liminf_n (S_n/a_n)$ must both be degenerate random variables (possibly extended real-valued), where $S_n = X_1 + \cdots + X_n$, $n \geq 1$.

EXERCISE 7.6.8. Let $\{X_n\}$ be a sequence of i.i.d. non-degenerate random variables. Show that $P(\sum_n X_n \text{ converges}) = 0$. Assuming further that the common distribution is symmetric and denoting $S_n = X_1 + \cdots + X_n$, $n \geq 1$, show that $P(\limsup_n S_n = +\infty) = P(\liminf_n S_n = -\infty) = 1$.

EXERCISE 7.6.9. (a) Show that if $\{X_n\}$ is an i.i.d. sequence of $N(0, 1)$ random variables, then $\limsup_n (X_n/\sqrt{2 \log n}) = 1$, a.s.$[P]$.
(b) Show that if $\{X_n\}$ is an i.i.d. sequence of $Exp(\lambda)$ random variables, for some $\lambda > 0$, then $\limsup_n (X_n/\log n) = \lambda^{-1}$, a.s.$[P]$.

EXERCISE 7.6.10. Consider a sequence $\{X_n\}$ of independent random variables with X_n having density function $f_n(x) = \lambda_n e^{-\lambda_n x}$, $0 < x < \infty$, for some $\lambda_n > 0$. Show that the random series $\sum_n X_n$ converges a.s.$[P]$ if and only if the series $\sum_n 1/\lambda_n$ converges.

EXERCISE 7.6.11. Consider a sequence $\{X_n\}$ of i.i.d. random variables and denote $S_n = X_1 + \cdots + X_n$, $M_n = \max\{|X_1|, \ldots, |X_n|\}$, $n \geq 1$.
(a) Show that $M_n/n \xrightarrow{P} 0$ if and only if $nP(|X_1| > n) \to 0$, while $M_n/n \xrightarrow{a.s.} 0$ if and only if $E(|X_1|) < \infty$.
(b) Show that $(S_n - C_n)/n$ converges to 0 a.s.$[P]$, for some real sequence $\{C_n\}$ if and only if $E(|X_1|) < \infty$ and, in that case, $C_n/n \to E(X_1)$.
(c) Assume $E(X_1) = 1$ and let $\{a_n\}$ be a bounded real sequence. Show that $\left(\sum_{i=1}^n a_i X_i\right)/n \to 1$, a.s.$[P]$ if and only if $\left(\sum_{i=1}^n a_i\right)/n \to 1$.

EXERCISE 7.6.12. [Borel's Normal Number Theorem] Consider the probability space $(\Omega, \mathcal{A}, P)$ where $\Omega = (0, 1]$, $\mathcal{A}$ is the Borel σ-field on $(0, 1]$ and P is the Lebesgue measure on $(0, 1]$.
(a) For every $\omega \in \Omega$, consider the non-terminating decimal expansion of ω and let $X_n(\omega)$ denote n-th digit in that expansion. Show that $\{X_n, n \geq 1\}$ are i.i.d. random variables, each taking values $\{0, 1, \ldots, 9\}$ with equal probabilities. A number $\omega \in (0, 1]$ is called a 'normal number' if, for every $j \in \{0, 1, \ldots, 9\}$, the limiting

7.6. Additional Exercises

relative frequency of the digit j in the decimal expansion of ω equals $1/10$. Show that almost every ω is a normal number.

(b) Fix any integer $r > 1$. Show that every $\omega \in \Omega$ admits a non-terminating expansion $\omega = \sum_{n=1}^{\infty} X_n(\omega) r^{-n}$, where $X_n(\omega)$ takes values in $\{0, 1, \ldots, r-1\}$, for each n. Refer to this expansion as 'r-expansion' of ω and $X_n(\omega)$ as the 'digit at the nth place' in this r-expansion. Call a number ω to be a 'r-normal number' if, for every $j \in \{0, 1, \ldots, r-1\}$, the limiting relative frequency of the digit j in the r-expansion of ω equals $1/r$. Show that almost every ω is a r-normal number.

(c) Show that almost every ω is a r-normal number for every integer $r > 1$. [Note, however, that it is not easy to identify **even one 'non-trivial'** example of a number $\omega \in (0, 1]$ which is r-normal. Interestingly, one does not know if the fractional parts of the numbers π and e are r-normal for any r.]

EXERCISE 7.6.13. Let $\{X_n\}$ be a sequence of independent real random variables with F_n denoting the probability distribution function of X_n, for $n \geq 1$.
(a) Show that $\sup_n X_n < \infty$ a.s.[P] if and only if, for some $x \in \mathbb{R}$, the series $\sum_n (1 - F_n(x))$ converges.
(b) Assume $X = \sup_n X_n < \infty$ with probability 1. Show that the distribution function of X is given by $F(x) = \prod_n F_n(x)$, $x \in \mathbb{R}$.
(c) Assuming the hypothesis of (b), show that $E(X^+) < \infty$ if and only if, for some $a \in \mathbb{R}$, the series $\sum_n E(X_n I_{\{X_n > a\}})$ converges.

EXERCISE 7.6.14. Prove the following generalization of Kronecker's Lemma: Let $\{a_n\}$ be a sequence of positive real numbers with $a_n \uparrow \infty$. If $\{x_n\}$ is any real sequence suxh that the series $\sum_n \frac{x_n}{a_n}$ converges, then $\frac{1}{a_n} \sum_{k=1}^{n} x_k \to 0$, as $n \to \infty$. [Setting $b_n = \sum_{j > n} \frac{x_j}{a_j}$, for $n \geq 0$, one has $b_n \to 0$ and $x_k = a_k(b_{k-1} - b_k)$, $k \geq 1$.]

EXERCISE 7.6.15. Let $\{X_n\}$ be a sequence of independent zero mean random variables and $\{a_n\}$ be a sequence of positive real numbers with $a_n \uparrow \infty$. Suppose $\sum_n E(\phi(X_n))/\phi(a_n) < \infty$, for some function $\phi : \mathbb{R} \to (0, \infty)$ that satisfies $\phi(x)/|x| \leq \phi(y)/|y|$ and $\phi(x)/x^2 \geq \phi(y)/y^2$ for any pair x, y with $|x| < |y|$.
(a) Show that $P(|X_n| > a_n$ for infinitely many $n) = 0$.
(b) Setting $Y_n = X_n I_{\{|X_n| \leq a_n\}}$, show that $\sum_n V(Y_n/a_n) < \infty$ and hence conclude that $\sum_n (Y_n - E(Y_n))/a_n$ converges almost surely.
(c) Using the hypothesis that $E(X_n) = 0$ for all n, show that $\sum_n |E(Y_n)|/a_n < \infty$.
(d) Deduce now that the random series $\sum_n X_n/a_n$ converyes almsot surely and conclude that $\frac{1}{a_n} \sum_{k=1}^{n} X_k \to 0$, almost surely. [Use result of Exercise 7.6.14.]

EXERCISE 7.6.16. [*Marcinkiewicz-Zygmund SLLN*]
(a) Prove that, for every $\beta > 1$, the finite constant $C_\beta = \frac{\beta}{\beta - 1}$ satisfies the inequalities $\sum_{n=j}^{\infty} n^{-\beta} \leq C_\beta \cdot j^{-(\beta-1)}$ and $\sum_{j=1}^{n} n^{-1/\beta} \leq C_\beta \cdot j^{-(\beta-1)/\beta}$, for all $j \geq 1$.
(b) For any random variable X and any $0 < p < \alpha$, show that

$$\sum_n n^{-\alpha/p} E\bigl(|X|^\alpha I_{\{|X|\le n^{1/p}\}}\bigr) \le \frac{\alpha}{\alpha - p} \cdot E\bigl(|X|^p\bigr).$$

In what follows, let $\{X_n\}$ be a sequence of i.i.d. random variables such that $E\bigl(|X_1|^p\bigr) < \infty$, for some $p \in (0,2)$, and let $Y_n = X_n I_{\{|X_n|\le n^{1/p}\}}$, $n \ge 1$.

(c) Show that $\sum_n n^{-1/p}\bigl(Y_n - E(Y_n)\bigr)$ converges almost surely. (Use (b) with $\alpha = 2$.)

(d) Show that $P\bigl(X_n = Y_n \text{ for all but finitely many } n\bigr) = 1$ and hence conclude that $\sum_n n^{-1/p}\bigl(X_n - E(Y_n)\bigr)$ converges almost surely.

(e) In case $p < 1$, deduce from (d) that the series $\sum_n n^{-1/p} X_n$ converges almost surely. (Use (b) with $\alpha = 1$.)

(f) Assume now $p > 1$ and $E(X_1) = 0$, so that $E(Y_n) = E\bigl(X_n I_{\{|X_n|>n^{1/p}\}}\bigr)$. Show that $\sum_n n^{-1/p}\bigl|E(Y_n)\bigr| < \infty$ and hence conclude that the series $\sum_n n^{-1/p} X_n$ converges almost surely.

(g) In general, if $1 < p < 2$, apply the above result to the sequence $\{X_n - E(X_n)\}$, in place of $\{X_n\}$, to conclude that the series $\sum_n n^{-1/p}\bigl(X_n - E(X_n)\bigr)$ converges almost surely.

(h) Show that the conclusion of (g) holds in case $p = 1$ also. (Go back to (d).)

Now use an appropriate version of Kronecker's Lemma to deduce the following SLLN due to Marcinkiewicz and Zygmund. This implies Kolmogorov's SLLN.

(i) If $\{X_n\}$ is any sequence of i.i.d random variables with $E|X_1|^p < \infty$, for some $p \in (0,2)$, then $\dfrac{S_n - nc}{n^{1/p}} \to 0$ almost surely, where $S_n = X_1 + \cdots + X_n$, $n \ge 1$ and $c = E(X_1)$ in case $p \ge 1$, while c can be any real number in case $0 < p < 1$.

(j) Finally prove the following converse of (i):

If $\{X_n\}$ is a sequence of i.i.d. random variables such its partial sums $S_n = X_1 + \cdots + X_n$, $n \ge 1$ satisfy $\dfrac{S_n - nc}{n^{1/p}} \to 0$ almost surely, for some real c and some $p \in (0,2)$, then $E\bigl(|X_1|^p\bigr) < \infty$. (Show that $n^{-1/p} X_n \to 0$ almost surely and then apply Borel-Cantelli Lemma.)

EXERCISE **7.6.17**. Give an example of a sequence $\{X_n\}$ of independent random variables such that the series $\sum_n X_n$ converges almost surely, but (i) at least one of the series $\sum E(X_n)$ and $\sum_n V(X_n)$ does not converge; (ii) none of the series $\sum E(X_n)$ and $\sum_n V(X_n)$ converges.

EXERCISE **7.6.18**. (a) Show that if $\{\mathbf{X}_n = (X_{n1},\ldots,X_{nk}),\ n \ge 1\}$ converges to $\mathbf{X} = (X_1,\ldots,X_k)$ in any of the three modes of convergence, then for any vector $(a_1,\ldots,a_k)$, the sequence $\{Y_n = \sum_i a_i X_{ni},\ n \ge 1\}$ of random variables converges to $Y = \sum_i a_i X_i$ in the same mode.

(b) Show that if $\{\mathbf{X}_n = (X_{n1},\ldots,X_{nk}),\ n \ge 1\}$ converges to $\mathbf{X} = (X_1,\ldots,X_k)$ in probability and $\varphi : \mathbb{R}^k \to \mathbb{R}^m$ is a continuous map, then $\{\mathbf{Y}_n = \varphi(\mathbf{X}_n),\ n \ge 1\}$ converges to $\mathbf{Y} = \varphi(\mathbf{X})$ also in probability.

Chapter 8

Weak Convergence and Central Limit Theorem

In Section 7.2, we discussed three different concepts of convergence for sequences of random variables – almost sure convergence, convergence in L_p and convergence in probability – and studied their properties. These culminated in many important results that include the Laws of Large Numbers and the Three Series Theorem. In this chapter, we are going to introduce and study another important and natural notion of convergence for sequences of random variables. In the sequel, we are going to use the standard notation F_Y to denote the distribution function of a real random variable Y. The notion of convergence that we are going to introduce, is essentially based on pointwise convergence of the corresponding sequence of distribution functions, except that *it is not demanded at all points*. The next section starts with the precise definition.

8.1 Weak Convergence

DEFINITION 8.1.1. A sequence $\{X_n\}$ of real random variables is said to *converge in distribution* to X and we write $X_n \xrightarrow{d} X$ if $F_{X_n}(a) \to F_X(a)$, as $n \to \infty$, for all points a that are continuity points of F_X. The same thing is sometimes expressed by saying that $\{X_n\}$ *converges weakly* to X.

It should be clear from the definition that, unlike the other three modes of convergence, this notion of convergence makes sense without requiring that all the random variables X_n, $n \geq 1$, and X be defined on the same probability space. Also, as was said earlier, this is essentially pointwise convergence of the sequence of functions F_{X_n} to the function F_X at points of continuity of F_X. A natural question that arises, of course, is about why pointwise convergence is being demanded only at continuity points of F_X and not at all points of $\mathbb{R}$. Here is an example that will convince the reader that demanding pointwise convergence at all points would be too stringent and would rule out many natural cases where convergence in distribution actually seem to be happening. The example will also bear out that giving exemption to discontinuity points of F_X from the requirement of pointwise convergence is the right thing to do.

EXAMPLE 8.1.2. Let $X_n = (2Z-1)/n$, $n \geq 1$, where Z is a random variable with $U(0,1)$ distribution. One can easily see that X_n, for each n, is uniformly distributed over the interval $(-\frac{1}{n}, \frac{1}{n})$. This shows that, as $n \to \infty$, the interval of possible values of X_n keep on shrinking, finally reducing to the singleton set $\{0\}$. Therefore, any reasonable notion of convergence would dictate that the sequence $\{X_n\}$ can only converge to the degenerate random variable $X \equiv 0$, whose distribution function is given by $F_X(a) = I_{[0,\infty)}(a)$, $a \in \mathbb{R}$. However, the reader may compute the distribution function F_{X_n}, for each n, and verify that $F_{X_n}(a) \to F_X(a)$ at all $a \neq 0$, but $F_{X_n}(0) = \frac{1}{2}$, for all n, and so $F_{X_n}(0) \not\to F_X(0) = 1$. Note that $a = 0$ is the only point of discontinuity for F_X.

Before proceeding further, let us make note of two important contrasts between convergence in distribution vis-a-vis the other three concepts of convergence. For this and for what follow, it is convenient to introduce the notation $X \stackrel{d}{=} Y$ to mean that the random variables X and Y have the same distribution. With this notation, it is easy to see that if $X_n \stackrel{d}{\to} X$, then $X_n \stackrel{d}{\to} Y$ if and only if $X \stackrel{d}{=} Y$. Thus, in place of the a.s. uniqueness of limits that were noted earlier for the other three modes of convergence, limits in weak convergence are unique only in distribution, something that is clearly much weaker than a.s. uniqueness.

Another stark difference is that, unlike the other three modes of convergence, convergence in distribution is not preserved under addition. Here is an example to show this. Consider the random variable Z that takes the values ± 1 with probabilities $\frac{1}{2}$ each. Taking now, $X_n = Z$, $Y_n = -Z$, for $n \geq 1$, one can easily see that $X_n \stackrel{d}{\to} Z$ and $Y_n \stackrel{d}{\to} Z$. However, $X_n + Y_n \equiv 0$, for all n, and therefore we do not have $X_n + Y_n \stackrel{d}{\to} Z + Z$. The reader must have realized that, instead of the particular Z as above, we could have taken any non-degenerate random variable Z with a 'symmetric distribution', that is, satisfying $Z \stackrel{d}{=} -Z$. By 'non-degenerate', we mean a random variable which is not almost surely a constant. The reader is now urged to think of an example to illustrate that, unlike a.s. convergence and convergence in probability, convergence in distribution is also not preserved under multipliction.

The next question that we want to address is : where does this new notion of convergence stand vis-a-vis the other three concepts of convergence, in terms of implications? It turns out that each of the other three modes of convergence implies convergence in distribution. This could be one of the reasons why this is called weak convergence. Since we already know that both a.s. convergence and L_p-convergence imply convergence in probability, we only need to show that the latter implies convergence in distribution, which is what the next result asserts. But before we state it, here is an easy exercise, that provides a simple yet useful characterization of convergence in distribution.

EXERCISE 8.1.3. Show that $X_n \stackrel{d}{\to} X$ if and only if $\liminf_n F_{X_n}(a) \geq F_X(a-)$ and $\limsup_n F_{X_n}(a) \leq F_X(a)$, for all $a \in \mathbb{R}$..

8.1. Weak Convergence

PROPOSITION 8.1.4. If a sequence $\{X_n\}$ of random variables converges in probability to X, then $\{X_n\}$ also converges in distribution to X.

Proof. For brevity in notation, let us write here F_n for F_{X_n} and F for F_X. In view of Exercise 8.1.3, we only need to show that if $\{X_n\}$ converges in probability to X, then

$$\limsup_n F_n(a) \leq F(a) \text{ and } \liminf_n F_n(a) \geq F(a-), \text{ for all } a \in \mathbb{R}. \quad (8.1.1)$$

and the required result will follow.

To prove (8.1.1), we note that $F_n(a) = P(X_n \leq a, X > a+\epsilon) + P(X_n \leq a, X \leq a+\epsilon)$, for any $a \in \mathbb{R}$ and any $\epsilon > 0$, which leads to the inequality

$$F_n(a) \leq P(|X_n - X| > \epsilon) + F(a + \epsilon) \text{ for any } a \in \mathbb{R} \text{ and any } \epsilon > 0. \quad (8.1.2)$$

Similarly, $P(X \leq a - \epsilon) = P(X \leq a - \epsilon, X_n > a) + P(X \leq a - \epsilon, X_n \leq a)$ gives

$$F_n(a) \geq F(a - \epsilon) - P(|X_n - X| > \epsilon) \text{ for any } a \in \mathbb{R} \text{ and any } \epsilon > 0. \quad (8.1.3)$$

Now, in (8.1.2), if we let $n \to \infty$ first and $\epsilon \downarrow 0$ next, then the hypothesis $X_n \xrightarrow{P} X$ and the right-continuity of F will give us $\limsup_n F_n(a) \leq F(a)$. Similarly, the other inequality in (8.1.1) is obtained from (8.1.3), by first letting $n \to \infty$ and then letting $\epsilon \downarrow 0$. This completes the proof. $\square$

REMARK 8.1.5. By considering the sequence $\{X_n\}$ given by $X_n = (-1)^n Z$, $n \geq 1$, where Z is any non-degenerate random variable with a symmetric distribution, one sees that the converse of the implication in Proposition 8.1.4 does not hold in general. However, the next result shows that there is one special case when the converse holds.

PROPOSITION 8.1.6. Let $\{X_n\}$ and X be random variables, all defined on the same probability space $(\Omega, \mathcal{A}, P)$. If $X_n \xrightarrow{d} X$ and if X is a degenerate random variable, that is, $P(X = c) = 1$ for some constant c, then $X_n \xrightarrow{P} X$.

Proof. Since X is degenerate, we have $P(X = c) = 1$ for some $c \in \mathbb{R}$. This means that $F_X(a)$ equals 0, for all $a < c$ and equals 1, for all $a \geq c$. Therefore, $X_n \xrightarrow{d} X$ implies that $F_{X_n}(a) \to 0$, for all $a < c$ and $F_{X_n} \to 1$, for all $a > c$. Now, for any $\epsilon > 0$, using the fact that $P(X = c) = 1$, we get that
$P(|X_n - X| > \epsilon) = P(X_n < c - \epsilon) + P(X_n > c + \epsilon) \leq F_{X_n}(c - \epsilon) + (1 - F_{X_n}(c + \epsilon)) \to 0$,
as $n \to \infty$ and that completes the proof. $\square$

It was noted above that, in general, convergence in distribution is not preserved under either addition or multiplication. However, there is one exception to this phenomenon, namely when at least one of the limits is degenerate. These are only two in a class of a number of similar results, that are often clubbed together and referred to as 'Slutsky type result' and each one of them is called a 'Slutsky Theorem. These results often turn out to be extremely useful in Probability as well as in Statistics. The next result and the corollary following it give Slutsky Theorem for addition. Note that, in view of Proposition 8.1.6, the convergence '$\xrightarrow{P}$' in the hypotheses of both the results are, indeed, equivalent to '$\xrightarrow{d}$'.

PROPOSITION 8.1.7. If $\{X_n\}$ and $\{Y_n\}$ are sequences of random variables, all defined on the same probability space, such that, $X_n \xrightarrow{d} X$ and $Y_n \xrightarrow{P} 0$, then $X_n + Y_n \xrightarrow{d} X$.

Proof. Let us denote the distribution functions of X_n, $X_n + Y_n$ and X by F_n, G_n and F respectively. We first prove that for any pair of real numbers $x < y$, both of which are continuity points of F,

$$\text{(i) } \limsup_n G_n(x) \leq F(y) \quad \text{and} \quad \text{(ii) } \liminf_n G_n(y) \geq F(x). \tag{8.1.4}$$

From $G_n(x) = P(X_n + Y_n \leq x, X_n > y) + P(X_n + Y_n \leq x, X_n \leq y)$, we easily get the inequality $G_n(x) \leq P(|Y_n| > y - x) + F_n(y)$. Now letting $n \to \infty$, we get (i). Similarly, $G_n(y) = 1 - P(X_n + Y_n > y, X_n \leq x) - P(X_n + Y_n > y, X_n > x)$ gives the inequality $G_n(y) \geq 1 - P(|Y_n| > y - x) - P(X_n > x) = F_n(x) - P(|Y_n| > y - x)$, from which we get (ii), by letting $n \to \infty$. To prove the required result now, we take any point $a \in \mathbb{R}$, which is a continuity point of F and choose a sequence $\epsilon_k \downarrow 0$ such that $a \pm \epsilon_k$, $k \geq 1$, are all continuity points of F. We can do this, since we know that F is continuous at all but at most a countable number of points in $\mathbb{R}$. Using (i) with $x = a$ and $y = a + \epsilon_k$, we get $\limsup_n G_n(a) \leq F(a + \epsilon_k)$. This being true for all k, we now let $k \to \infty$ to get $\limsup_n G_n(a) \leq F(a)$. On the other hand, using (ii) now with $x = a - \epsilon_k$ and $y = a$, we get $\liminf_n G_n(a) \geq F(a - \epsilon_k)$. Once again, letting $k \to \infty$ gives $\liminf_n G_n(a) \geq F(a)$. We have thus proved that $G_n(a) \to F(a)$, for any continuity point a of F, as was required. □

A simple extension of the above result is given is the following corollary, the proof of which is left as an exercise for the reader.

COROLLARY 8.1.8. (Slutsky Theorem) If $\{X_n\}$ and $\{Y_n\}$ are sequences of random variables, all defined on the same probability space and if $X_n \xrightarrow{d} X$ and $Y_n \xrightarrow{P} c$, for some $c \in \mathbb{R}$, then $X_n + Y_n \xrightarrow{d} X + c$.

We next go to a very interesting and often extremely useful result related to convergence in distribution. In view of Proposition 8.1.4, we know that convergence in distribution is the weakest among all the different concepts of convergence, in the sense that each of the other modes of convergence implies convergence in distribution. However, a very interesting result, due to Skorokhod, shows that it is always possible to have a realization of a convergence in distribution phenomenon through an almost sure convergence, provided, of course, one is willing to have this realization on a different probability space. The precise statement follows.

THEOREM 8.1.9. (Skorokhod Representation Theorem) Let $X_n \xrightarrow{d} X$. Then there exists a probability space $(\Omega, \mathcal{A}, P)$ and random variables $\{Y_n, n \geq 1\}$ and Y defined on $(\Omega, \mathcal{A}, P)$, with $Y_n \stackrel{d}{=} X_n$, for all n, and $Y \stackrel{d}{=} X$, such that $Y_n \xrightarrow{a.s.} Y$.

Proof. We will need to recall ideas introduced in the proof of Theorem 4.1.6. Denoting the distribution functions of X_n, $n \geq 1$, and X as F_n, $n \geq 1$, and F respectively, we define G_n and G on $(0, 1)$, as was done in (4.1.1), by

$$G_n(u) = \sup\{y : F_n(y) < u\}, \; n \geq 1, \; G(u) = \sup\{y : F(y) < u\}, \; \text{for } 0 < u < 1.$$

8.1. Weak Convergence

With $(\Omega, \mathcal{A}, P)$ and U as in Theorem 4.1.6, we have $Y_n = G_n(U) \stackrel{d}{=} X_n$, $n \geq 1$, and $Y = G(U) \stackrel{d}{=} X$. We are now going to show that $Y_n \stackrel{a.s.}{\longrightarrow} Y$.

First, we show that $\liminf_n G_n(u) \geq G(u)$ for all $u \in (0,1)$. Actually, we are going to fix an $\epsilon > 0$ and show that $\liminf G_n(u) \geq G(u) - \epsilon$, for all $u \in (0,1)$. Given any $u \in (0,1)$, we choose $y \in (G(u) - \epsilon, G(u))$, which is a continuity point of F. Since $y < G(u)$, the definition of G will imply that $F(y) < u$. From the hypothesis that $X_n \stackrel{d}{\longrightarrow} X$, we know that $F_n(y) \to F(y)$, and therefore, for all large n, we must have $F_n(y) < u$. But this, by the definition of G_n, will imply that $G_n(u) \geq y > y - \epsilon$, for all large n, thus proving that $\liminf_n G_n(u) \geq G(u) - \epsilon$, as required.

Next, we show that $\limsup_n G_n(u) \leq G(u+)$, for any $u \in (0,1)$. Again, fixing an $\epsilon > 0$, we are going to show that $\limsup_n G_n(u) \leq G(v) + \epsilon$, for all $u \in (0,1)$ and any $v \in (0,1)$ with $v > u$. So, given $u \in (0,1)$, we choose and fix an $\epsilon > 0$ and take any $v \in (0,1)$ with $v > u$. Let $y \in (G(v), G(v) + \epsilon)$ be a continuity point of F. The definition of G again gives us $F(y) \geq v > u$, so that, for all large n, we must have $F_n(y) > u$ and, hence, $G_n(u) \leq y < G(v) + \epsilon$, proving that $\limsup_n G_n(u) \leq G(v) + \epsilon$, as required.

The two inequalities that we just proved easily imply that $G_n(u) \to G(u)$, for any $u \in (0,1)$, which is a continuity point of G. Since G is non-decreasing, the set D of discontinuity points of G must be countable. From the definition of the random variables Y_n, $n \geq 1$ and Y and from what we have just proved, it follows that $P(Y_n \to Y) = P(U \notin D) = 1$, thus completing the proof. $\square$

From the definition of $X_n \stackrel{d}{\longrightarrow} X$, we know that if the limiting distribution function F_X is continuous everywhere, then the above convergence is nothing but pointwise convergence of $\{F_{X_n}\}$ to F_X everywhere. What the next result, known as Pólya's Theorem, says is that, in case F_X is continuous everywhere, the convergence of $\{F_{X_n}\}$ to F_X is actually uniform over $\mathbb{R}$.

THEOREM 8.1.10. (Pólya's Theorem) If $X_n \stackrel{d}{\longrightarrow} X$ and F_X is continuous everywhere, then $\sup_{a \in \mathbb{R}} |F_{X_n}(a) - F_X(a)| \to 0$ as $n \to \infty$, that is, $F_{X_n}(a) \to F_X(a)$ as $n \to \infty$, uniformly over $a \in \mathbb{R}$.

Proof. For notational simplicity, let us denote F_{X_n} by F_n and F_X by simply F. Let $\epsilon > 0$ be given. Using the facts that $\lim_{x \to -\infty} F(x) = 0$ and $\lim_{x \to +\infty} F(x) = 1$, we can get $M > 0$ such that $F(-M) < \frac{\epsilon}{2}$ and $F(M) > 1 - \frac{\epsilon}{2}$. Using pointwise convergence of $\{F_n\}$ to F, one can get an integer $n_1 \geq 1$, such that, $F_n(-M) < \frac{\epsilon}{2}$ and $F_n(M) > 1 - \frac{\epsilon}{2}$, for all $n \geq n_1$. Since F is continuous, and hence uniformly continuous on $[-M, M]$, we can get a finite partition of $[-M, M]$ given by, say, $-M = x_0 < x_1 < \cdots < x_k = M$, such that for each $i = 1, \cdots, k$, $x, y \in [x_{i-1}, x_i]$ would imply $|F(x) - F(y)| < \frac{\epsilon}{2}$. Using pointwise convergence again, we choose an integer $n_2 \geq 1$, such that, for all $n \geq n_2$, we have $|F_n(x_i) - F(x_i)| < \frac{\epsilon}{2}$, for each $i = 0, \ldots, k$. Taking $n_0 = \max\{n_1, n_2\}$, the choice of M gives $\sup_{|x| \geq M} |F_n(x) - F(x)| < \epsilon$, for all $n \geq n_0$. On the other hand, for any $x \in [x_{i-1}, x_i]$, $1 \leq i \leq k$, we have,

$F_n(x) - F(x) \leq F_n(x_i) - F(x_i) + F(x_i) - F(x)$ and at the same time, we also have $F_n(x) - F(x) \geq F_n(x_{i-1}) - F(x_{i-1}) + F(x_{i-1}) - F(x)$. By the choice of the partition and that of n_0, the above two inequalities clearly imply that, for all $n \geq n_0$, we will have $\sup\{|F_n(x) - F(x)| : x \in [x_{i-1}, x_i]\} < \epsilon$, for each $i = 1, \ldots, k$. This gives $\sup_{|x| \leq M} |F_n(x) - F(x)| < \epsilon$, for all $n \geq n_0$. That completes the proof. □

The following result, which is widely known as the 'Local Limit Theorem', is a nice and important application of Scheffé's Lemma (Theorem 3.3.12). This often acts as a very useful tool for getting distributional limits for a sequence of random variables with densities or a sequence of discrete random variables, whose supports are sets of real numbers in arithmetic progression.

THEOREM **8.1.11**. (Local Limit Theorem)
(a) Let X_n, $n \geq 1$, and X be random variables with pdfs f_n, $n \geq 1$, and f respectievly. If $f_n \to f$ a.e.$[\lambda]$, then $X_n \xrightarrow{d} X$.
(b) Let X_n, $n \geq 1$, be discrete random variables with X_n having pmf p_n and taking values in $L_n = \{\alpha_n + j\delta_n : j \in \mathbb{Z}\}$, where $\{\alpha_n\}$ is a real sequence and $\{\delta_n\}$ is a sequence of positive real numbers with $\delta_n \downarrow 0$. Suppose there is a probability density function f on $\mathbb{R}$ satisfying the property that, for any $x \in \mathbb{R}$ and any sequence $\{x_n\}$ with $x_n \in L_n$, $n \geq 1$, such that, $x_n \to x$, one has $\frac{p_n(x_n)}{\delta_n} \to f(x)$, as $n \to \infty$. Then $X_n \xrightarrow{d} X$, where X is a random variable with pdf f.

Proof. (a) Since $\int f(x)dx = 1 = \int f_n(x)dx$, for all n, an application of Scheffé's Lemma (Theorem 3.3.12) will give us $\int |f_n(x) - f(x)|dx \to 0$. Noting now that $|F_{X_n}(a) - F_X(a)| \leq \int_\mathbb{R} |f_n(x) - f(x)|dx$, for every $a \in \mathbb{R}$, we get the required result.
(b) For each $n \geq 1$ and $j \in \mathbb{Z}$, let $A_{n,j}$ be the interval $(\alpha_n + (j-1)\delta_n, \alpha_n + j\delta_n]$. Consider a sequence $\{Y_n\}$ of random variables, where Y_n has pdf
$$f_n(x) = \sum_{j \in \mathbb{Z}} (p_n(\alpha_n + j\delta_n)/\delta_n) I_{A_{n,j}}(x), \ x \in \mathbb{R}.$$
It is not difficult to see that the hypothesis in the statement in (b) implies that $f_n \to f$ pointwise. Therefore, by part (a), we will have $Y_n \xrightarrow{d} X$. Consider now the sequence $\{Z_n\}$ defined as $Z_n = \sum_{j \in \mathbb{Z}} (\alpha_n + j\delta_n) I_{A_{n,j}}(Y_n)$. From the definition of Z_n, one has $|Z_n - Y_n| \leq \delta_n$ everywhere, for each n, so that $Z_n - Y_n \to 0$ everywhere, as $n \to \infty$. An application of Proposition 8.1.7 now gives $Z_n \xrightarrow{d} X$. It is easy to see, however, that, $Z_n \stackrel{d}{=} X_n$, for each n. The result now follows. □

REMARK **8.1.12**. The earliest version of what is widely known as the *Central Limit Theorem*, was proved by de Moivre and Laplace. The proof essentially used part (b) of Theorem 8.1.11 and involved very hard computations. This very early result, known as 'de Moivre-Laplace Limit Theorem' (see Exercise 8.10.15), seems to have been largely forgotten.

We now return to a very important and practical question. We have the definition of convergence in distribution, but, in practice, applying the definition

8.1. Weak Convergence

to examine whether a given sequence $\{X_n\}$ converges in distribution, does not seem to be very helpful. One of the major difficulties is that computing the exact distribution functions of the X_n can be extremely difficult, if not impossible, in many situations. Further, identifying the continuity points of the possible limiting distribution and then checking the required convergence at continuity points pose additional challenges. This leads one to ask whether there are other criteria that characterize weak convergence and are hopefully a bit more tractable than the definition of weak convergence. We are now going to see one such criterion. To understand how it comes about, let us go back to the definition of weak convergence and reformulate it in a slightly different way. For each $a \in \mathbb{R}$, let us denote by φ_a the function $I_{(-\infty,a]}$. Then the definition of $X_n \xrightarrow{d} X$ is the same as requiring that $E(\varphi_a(X_n)) \to E(\varphi_a(X))$, as $n \to \infty$, for all continuity points a of F_X. This may seem like making a simple thing complicated. However, just to see where we are heading, let us make a simple observation. Suppose we are given that we have random variables X_n, $n \geq 1$, and X, that satisfy the property that $E(f(X_n)) \to E(f(X))$, as $n \to \infty$, for every bounded continuous function $f : \mathbb{R} \to \mathbb{R}$. We are going to show that this will imply $X_n \xrightarrow{d} X$. Towards this, let us denote $f_{x,y}$, for every pair of reals $x < y$, to be the bounded continuous function on $\mathbb{R}$, which equals 1 on $(-\infty, x]$, equals 0 on $[y, \infty)$ and is defined on (x, y) by linear interpolation. Then, from our hypothesis, we will have $E(f_{a,a+\epsilon}(X_n)) \to E(f_{a,a+\epsilon}(X))$, for every $a \in \mathbb{R}$ and every $\epsilon > 0$. Using now the simple pointwise inequality $\varphi_a \leq f_{a,a+\epsilon} \leq \varphi_{a+\epsilon}$ in the above, we get $\limsup_n E(\varphi_a(X_n)) \leq E(\varphi_{a+\epsilon}(X))$, that is, $\limsup_n F_{X_n}(a) \leq F_X(a+\epsilon)$, for every $a \in \mathbb{R}$ and every $\epsilon > 0$. Letting $\epsilon \downarrow 0$ gives $\limsup_n F_{X_n}(a) \leq F_X(a)$, for every $a \in \mathbb{R}$. Using the functions $f_{a-\epsilon,a}$ and using similar arguments, one can get $F_X(a-) \leq \liminf_n F_{X_n}(a)$, for every $a \in \mathbb{R}$. The above two inequalities imply that $X_n \xrightarrow{d} X$ (see Exercise 8.1.3).

We have thus proved that the condition "$E(f(X_n)) \to E(f(X))$, for all bounded continuous $f : \mathbb{R} \to \mathbb{R}$" is a sufficient condition for $X_n \xrightarrow{d} X$. The next result completes the circle by asserting that it is both necessary and sufficient.

THEOREM **8.1.13**. *A sequence $\{X_n\}$ converges in distribution to X if and only if $E(f(X_n)) \to E(f(X))$, as $n \to \infty$, for every bounded continuous $f : \mathbb{R} \to \mathbb{R}$.*

Proof. Since the 'if' part is what we proved above, we need only to prove the 'only if' part. Let us write F_n, $n \geq 1$, and F respectively for F_{X_n}, $n \geq 1$, and F_X. We assume that $X_n \xrightarrow{d} X$, that is, $F_n(a) \to F(a)$, for all continuity points a of F. Suppose we are given a continuous function $f : \mathbb{R} \to \mathbb{R}$ satisfying $|f(x)| \leq C$, for all $x \in \mathbb{R}$, where $C > 0$. We need to show that $E(f(X_n)) \to E(f(X))$. Let $\epsilon > 0$ be given. Denoting $C' = 8C$, we first choose $M > 0$ such that $\pm M$ are both continuity points of F and also $F(-M) < \epsilon/C'$ and and $F(M) > 1 - \epsilon/C'$. The hypothesis then implies that there is an integer $n_1 \geq 1$ such that, for all $n \geq n_1$, we have $F_n(-M) < \epsilon/C'$ and $F_n(M) > 1 - \epsilon/C'$. Taking $J = (-M, M]$, we have $|E(f(X_n)I_{\{X_n \notin J\}}) - E(f(X)I_{\{X \notin J\}})| \leq C\{P(X_n \notin J) + P(X \notin J)\}$, using triangle inequality. Our choice of C', M and n_1 gives $P(X \notin J) < 2\epsilon/C' = \epsilon/4$ and also $P(X_n \notin J) < 2\epsilon/C' = \epsilon/4$, for all $n \geq n_1$. Combining all these, we get

$$\left|E\big(f(X_n)I_{\{X_n \notin J\}}\big) - E\big(f(X)I_{\{X \notin J\}}\big)\right| < \epsilon/2, \text{ for all } n \geq n_1. \tag{8.1.5}$$

Next, by virtue of uniform continuity of f on $[-M, M]$, there exists $\delta > 0$, such that, $|f(x) - f(y)| < \epsilon/8$, for all $x, y \in [-M, M]$, with $|x - y| < \delta$. We now partition the interval $[-M, M]$ into a finite number of disjoint subintervals, each having length $< \delta/2$. Let us now pick, for each of these disjoint subintervals, one continuity point of F lying strictly inside the interval and let us denote the chosen points as $x_1 < \cdots < x_k$. Denoting $x_0 = -M$ and $x_{k+1} = M$, we get a finite partition $-M = x_0 < x_1 < \cdots < x_k < x_{k+1} = M$ of $[-M, M]$, such that, each x_i is a continuity point of F and also $x_{i+1} - x_i < \delta$, for each $i = 0, \ldots, k$. Denote $f(x_i) = a_i$, $0 \leq i \leq k$, and define a function g on $J = (M, M]$ by

$$g(x) = \sum_{i=0}^{k} a_i I_{(x_i, x_{i+1}]}(x), \quad x \in J.$$

Observe that, by the way the points $-M = x_0 < x_1 < \cdots < x_k < x_{k+1} = M$ were constructed, we have $|f(x) - g(x)| < \epsilon/8$, for every $x \in J$. It follows that $\left|E\big(f(X_n)I_{\{X_n \in J\}}\big) - E\big(f(X)I_{\{X \in J\}}\big)\right| \leq \left|E\big(g(X_n)I_{\{X_n \in J\}}\big) - E\big(g(X)I_{\{X \in J\}}\big)\right| + \frac{\epsilon}{4}$.
But, by the definition of g, we have

$$E\big(g(X_n)I_{\{X_n \in J\}}\big) - E\big(g(X)I_{\{X \in J\}}\big) = \sum_{i=0}^{k} a_i \big\{P\big(X_n \in (x_i, x_{i+1}]\big) - P\big(X \in (x_i, x_{i+1}]\big)\big\}$$

and the last sum clearly equals $\sum_{i=0}^{k} a_i \big\{\big(F_n(x_{i+1}) - F(x_{i+1})\big) - \big(F_n(x_i) - F(x_i)\big)\big\}$. Invoking the hypothesis that $X_n \xrightarrow{d} X$ and the fact that $x_0, \ldots, x_{k+1}$ are all continuity points of F, one easily sees that this sum goes to 0, as $n \to \infty$. Thus, we can get an integer $n_2 \geq 1$, such that, $\left|E\big(g(X_n)I_{\{X_n \in J\}}\big) - E\big(g(X)I_{\{X \in J\}}\big)\right| < \epsilon/4$, for all $n \geq n_2$, implying finally that

$$\left|E\big(f(X_n)I_{\{X_n \in J\}}\big) - E\big(f(X)I_{\{X \in J\}}\big)\right| < \epsilon/2, \text{ for all } n \geq n_2. \tag{8.1.6}$$

Taking now $n_0 = \max\{n_1, n_2\}$, one can use (8.1.5) and (8.1.6) to conclude that $|E(f(X_n)) - E(f(X))| < \epsilon$ for all $n \geq n_0$ and that completes the proof. □

EXERCISE **8.1.14**. Show that the assertion of Theorem 8.1.13 remains valid if 'bounded continuous functions' is replaced either by 'bounded uniformly continuous functions' or by 'continuous functions with compact support'.

REMARK **8.1.15**. Exercise 8.1.14 above shows that the necessary and sufficient condition provided in Theorem 8.1.13 can be improved by requiring the condition $E\big(f(X_n)\big) \to E\big(f(X)\big)$ to hold for more restricted classes of functions than all bounded continuous functions. One of the most drastic improvements is given in Exercise 8.10.11, which claims that considering just the class of all C^∞ functions with compact support also provides a necessary and sufficient condition for $X_n \xrightarrow{d} X$.

A very important consequence of Theorem 8.1.13 is what is known as '*continuous mapping theorem*', which asserts that convergence in distribution is preserved under application of continuous maps. This would have been almost impossible

8.1. Weak Convergence

to prove directly from the definition of weak convergence. We just state it here, since the proof is just a one line application of Theorem 8.1.13.

THEOREM 8.1.16. (Continuous Mapping Theorem) If $\{X_n\}$ converges in distribution to X, then $\{h(X_n)\}$ converges in distribution to $h(X)$, for any real valued continuous function h on $\mathbb{R}$.

Another interesting and useful consequence of Theorem 8.1.13 is the following result, often referred to as the "*method of moments*" in weak convergence.

COROLLARY 8.1.17. (Method of Moments) Suppose X_n, $n \geq 1$, and X are random variables with all of them taking values inside a closed bounded interval I. Then, $X_n \xrightarrow{d} X$ if and only if $E(X_n^k) \to E(X^k)$, as $n \to \infty$, for every $k \geq 1$.

Note that the assumed boundedness of all the random variables X_n, $n \geq 1$, and X guarantees all of them have finite moments of all orders.

Proof. Since, for every $k \geq 1$, we can easily get a bounded continuous function f on $\mathbb{R}$ such that $f(x) = x^k$, for $x \in I$, the 'only if' part follows from Theorem 8.1.13 and the fact that all the X_n, $n \geq 1$, and X take values only in I.

For the 'if' part, note that the hypothesis implies $E(p(X_n)) \to E(p(X))$, as $n \to \infty$, for every polynomial p on $\mathbb{R}$. We now recall Weierstrass Approximation Theorem, which says that given any real continuous function f on $\mathbb{R}$, there is a sequence $\{p_m, m \geq 1\}$ of polynomials on $\mathbb{R}$ converging to f uniformly on the closed bounded interval I. Invoking this, one can easily deduce that $E(f(X_n)) \to E(f(X))$, as $n \to \infty$, for every bounded continuous $f : \mathbb{R} \to \mathbb{R}$. Invoking Theorem 8.1.13, we get $X_n \xrightarrow{d} X$, completing the proof. □

EXERCISE 8.1.18. Let X_n, $n \geq 1$ and X be non-negative random variables. Show that $X_n \xrightarrow{d} X$ if and only if $E(e^{-tX_n}) \to E(e^{-tX})$, as $n \to \infty$, for every $t \in [0, \infty)$

REMARK 8.1.19. Sometimes we use the notation $X_n \xrightarrow{d} F$ to mean that $X_n \xrightarrow{d} X$ where X is a random variable having distribution function F. This is perfectly unambiguous, because limits in weak convergence are unique only upto distribution. In fact, we can even go further, as noted in the next definition.

DEFINITION 8.1.20. For probability distribution functions F_n, $n \geq 1$, and F on $\mathbb{R}$, we say that F_n *converges weakly* to F, written $F_n \xrightarrow{d} F$, if $F_n(x) \to F(x)$, as $n \to \infty$, for every continuity point x of F. This is clearly equivalent to saying that $X_n \xrightarrow{d} X$ where X_n, $n \geq 1$, and X are random variables with distribution functions F_n, $n \geq 1$, and F respectively.

We return once again to the issue of finding a tractable criterion for convergence in distribution. Theorem 8.1.13 gives a necessary and sufficient condition, but unfortunately, it is not very useful because, according to it, we are required to check that $E(f(X_n)) \to E(f(X))$, for every bounded continuous f on $\mathbb{R}$ (or at least, for all continuous f on $\mathbb{R}$ with compact support.)

In the next section, we are going to introduce a new concept which will allow us to develop another very important characterization of convergence in distribution. However, the importance of this new concept lies not only in providing us with an extremely useful tool to examine weak convergence, but also in giving deeper understanding of the underlying probability distributions themeselves.

8.2 Characteristic Functions

We start with some preliminaries. So far, we have only talked about real random variables. But, for the concept that we want to introduce now, we need to talk about complex random variables. The definitions that follow are very natural. For more details, the reader may revisit the first paragraph of Section 6.4

DEFINITION 8.2.1. (a) A *complex random variable* on a probability space is a complex valued function $Z = X + iY$ on Ω, where both X and Y are real random variables. As usual we will write $X = \text{Re}(Z)$ and $Y = \text{Im}(Z)$. Of course a real random variable can always be viewed as a complex random variable with zero imaginary part.
(b) Let Z be a complex random variable with $X = \text{Re}(Z)$ and $Y = \text{Im}(Z)$. If both X and Y have finite expectations, then the *expected value* of Z is defined to be the complex number $E(Z) = E(X) + iE(Y)$.

It should be clear to the reader that a complex random variable is just a complex measurable function, as defined in Section 6.4, on a probability space, and the expectation of a compex random variable is just its integral as defined there. So, all the properties of complex measurable functions as well as of their integrals carry over verbatim to complex random variables and their expected values. In particular, if Z is a complex random variable, then $|Z| = \sqrt{X^2 + Y^2}$ is a non-negative real random variable.

As special cases of Propositions 6.4.6(a) and 6.4.7, we get the following result that asserts two elementary properties of expected values of complex random variables that are analogues of their counterparts for real random variables.

PROPOSITION 8.2.2. (a) (Linearity of Expected Value) If Z_1 and Z_2 are complex random variables on the same probability space, then $\alpha Z_1 + \beta Z_2$, for any two complex numbers α and β, is again a complex random variable. Further, if $E(Z_1)$ and $E(Z_2)$ are both defined, then $E(\alpha Z_1 + \beta Z_2)$ is also defined and equals $\alpha E(Z_1) + \beta E(Z_2)$.
(b) If Z is a complex random variable and $E(Z)$ is defined, then $|E(Z)| \leq E(|Z|)$.

REMARK 8.2.3. As was noted in the context of Propositions 6.4.6 and 6.4.7, while the linearity of expectation for complex random variables follows directly from the same property for real random variables, part (b) of the above proposition is not a direct consequence of the analogous inequality for real random variables. It needs a separate proof. In case a curious reader is wondering why that is the case, here is the reason. A careful examination of the definition of expected value of a complex random variable would show that the inequality in part (b) essentially says that

8.2. Characteristic Functions

$\sqrt{(E(X))^2 + (E(Y))^2} \leq E(\sqrt{X^2 + Y^2})$, for any two real random variables X and Y with finite expectations. Clearly, this inequality does not follow directly from the inequalities that we have for real random variables.

The reader should also note that, as special cases of Proposition 6.4.6 and Theorem 6.4.8, the natural analogues of Fatous' Lemma, DCT, Hölder's inequality and Minkowski's inequality hold for complex random variables as well. These will be freely used in the sequel, whenever needed. The only thing that the reader should be mindful of is that, in the present context, '$|\cdot|$' would often represent 'modulus' instead of 'absolute value'.

So far, we have mostly viewed complex random variables as complex measurable functions and their expected values as integrals, as defined in Section 6.4. The results so far are all borrowed from there as special cases. But now, we are going to bring in the notion of independence, which falls exclusively in the domain of random variables and is extremely important.

DEFINITION 8.2.4. Complex random variables $Z_1, \ldots, Z_n$ on a probability space are said to be *independent* if the random vectors $(\text{Re}(Z_1), \text{Im}(Z_1)), \ldots, (\text{Re}(Z_n), \text{Im}(Z_n))$ are independent.
A family $\{Z_\alpha\}$ of complex random variables on a probability space are said to be *independent* if every finite subfamily is.

From the definition of expected values of complex random variables, using linearity and applying Theorem 4.4.11, one easily gets the following analogue of that theorem for complex random variables.

PROPOSITION 8.2.5. if Z and W are two independent complex random variables on a probability space, such that $E(Z)$ and $E(W)$ are both defined, then $E(ZW)$ is also defined and $E(ZW) = E(Z)E(W)$.

With all these preminaries out of the way, we now introduce one of the most important concepts in probability, namely, that of '*characteristic function*'.

DEFINITION 8.2.6. The *characteristic function (c.f.)* of a real random variable X is defined to be the complex valued function on $\mathbb{R}$ given by

$$\varphi_X(t) = E(e^{itX}), \quad t \in \mathbb{R}.$$

If F is the distribution function of X, we sometimes refer to φ_X as the characteristic function of F and even write it as φ_F.

For a real random variable X and any $t \in \mathbb{R}$, we have $e^{itX} = \cos tX + i \sin tX$. Since both $\cos tX$ and $\sin tX$, for each $t \in \mathbb{R}$, are bounded real random variables (and hence have finite expectations), the function $\varphi_X(t)$ is well defined for every $t \in \mathbb{R}$ and equals the complex number $\varphi_X(t) = E(\cos tX) + iE(\sin tX)$. As pointed out, $\varphi_X(\cdot)$, for any real random variable X, defines a complex valued function on the real line. Here are some elementary properties.

THEOREM 8.2.7. (Elementary Properties of Characteristic Functions)
(i) $\varphi_X(0) = 1$.
(ii) $|\varphi_X(t)| \leq 1$.
(iii) If $Y = aX + b$ where $a, b \in \mathbb{R}$, then $\varphi_Y(t) = e^{itb}\varphi_X(at)$.
(iv) $\varphi_{-X}(t) = \varphi_X(-t) = \overline{\varphi_X(t)}$.
(v) If X has a symmetric distribution, then φ_X is a real valued function.
(vi) $\varphi_X(\cdot)$ is a continuous function; in fact, it is uniformly continuous.
(vii) If X and Y are independent, then $\varphi_{X+Y}(t) = \varphi_X(t)\varphi_Y(t)$.

Proof. (i) is obvious from definition and (ii) follows from Proposition 8.2.2(b) and the fact that $|e^{itX}| \equiv 1$. Both (iii) and (iv) follow from the linearity property in Proposition 8.2.2(a), while (v) is a consequence of (iv). (vii) follows from $e^{it(X+Y)} = e^{itX}e^{itY}$ and the fact that independence of the real random variables X and Y imply independence of the complex random variables e^{itX} and e^{itY}.

Finally, for (vi), observe that, for any $t, h \in \mathbb{R}$, we have, by part (a) and then part (b) of Proposition 8.2.2,

$$|\varphi_X(t+h) - \varphi_X(t)| = |E[e^{itX}(e^{ihX} - 1)]| \leq E|e^{ihX} - 1|.$$

Noting now that $|e^{ihX} - 1|$ is a non-negative random variable bounded by 2, for all h and that $|e^{ihX} - 1| \to 0$, as $h \to 0$, we apply DCT to get that $E|e^{ihX} - 1| \to 0$, as $h \to 0$. This proves uniform continuity of $\varphi_X(\cdot)$. □

REMARK 8.2.8. The converse of (v) is also true. In fact, if φ_X is a real-valued function, then using (iv), we will indeed get $\varphi_X \equiv \varphi_{-X}$. However, in order to now claim that $X \stackrel{d}{=} -X$, we need to know whether equality of characteristic functions imply equality of distributions. It turns out that this is indeed correct, that is, characteristic function of a random variable does indeed determine its distribution uniquely, a fact that will be proved later.

EXERCISE 8.2.9. Show that if $\varphi_1, \ldots, \varphi_n$ are characteristic functions (of some probability distributions), then any convex combination of $\varphi_1, \ldots, \varphi_n$ (that is, any function $\varphi(t) = \theta_1\varphi_1(t) + \cdots + \theta_n\varphi_n(t)$, where $\theta_1, \ldots, \theta_n$ are any choice of non-negative reals that add up to 1) is a characteristic function.

A reader familiar with the notion of 'Fourier Transforms' would realize that with the way we have defined the characteristic function φ_F of a probability distribution function F on $\mathbb{R}$, the function $\varphi_F(-t)$ is nothing but the value at t of what, in Fourier Analysis, is called the '*Fourier Transform*' of the probability measure P on $\mathcal{B}$, given by F. Next, from the definition of $\varphi_X(t)$ as $E(e^{itX})$, it is clear that if X is a discrete random variable, with values in the countable set D and probability mass function $p(x)$, $x \in D$, then $\varphi_X(t) = \sum_{x \in D} e^{itx}p(x)$. Similarly, for an absolutely continuous random variable X with probability density function f, one has $\varphi_X(t) = \int_{\mathbb{R}} e^{itx}f(x)dx$. Again, one can identify the function $\varphi_X(-t)$ as nothing but what, in Fourier Analysis, is called the Fourier Transform of the (non-negative) integrable function f.

8.2. Characteristic Functions

Here is another important point to be noted. While computing the characteristic function of an absolutely continnuous random variable X by the formula $\varphi_X(t) = \int e^{itx} f(x)dx$, one often uses standard integral formulas known for Riemann integrals of real functions, overlooking the fact that the integrand here is a complex valued function. Strictly speaking, one should break it up as $\int \cos(tx)f(x)dx + i\int \sin(tx)f(x)$ and use standard Riemann integral formulas, if available, for these two integrals with real integrands. However, the fortunate part is that the short-cut algorithm of using the standard integral formulas directly for the integral $\int e^{itx} f(x)dx$ usually yields the correct answer. We will illustrate this in Example 8.2.10(c) below. Of course, while computing the characteristic function of a random variable with a symmetric distribution, one can use Theorem 8.2.7 (v) to get $\varphi_X(t) = E(\cos tX)$.

EXAMPLE 8.2.10. For future reference, we compute the characteristic functions of some of the special distributions discussed in Chapter 4, Example 4.1.10.

(a) If $X \sim \text{Bin}(n, \theta)$, one can then directly see, using binomial theorem, that

$$\varphi_X(t) = \sum_{x=0}^{n} e^{itx} \binom{n}{x} \theta^x (1-\theta)^x = (1 - \theta + \theta e^{it})^n.$$

Another way to get this is to see that $X \stackrel{d}{=} Y_1 + \cdots + Y_n$, where $Y_1, \cdots, Y_n$ are i.i.d., with each Y_i taking values 0 and 1 with probabilities $1 - \theta$ and θ respectively. Since, for each i, one clearly has $\varphi_{Y_i}(t) = 1 - \theta + \theta e^{it}$, the above formula for φ_X is obtained simply by using Theorem 8.2.7 (vii).

(b) For $X \sim \text{Poi}(\lambda)$, using the formula $\sum_{x=0}^{\infty} \frac{z^x}{x!} = e^z$ for complex numbers z, we get

$$\varphi_X(t) = e^{\lambda(e^{it}-1)}.$$

(c) If $X \sim \text{Exp}(\lambda)$, then using the density of X, we get $\varphi_X(t) = \int_0^{\infty} e^{itx} \lambda e^{-\lambda x} dx = \lambda \int_0^{\infty} e^{-(\lambda - it)x} dx$. Now, one recalls the standard formula $\int_0^{\infty} e^{-ax} dx = 1/a$, for $a > 0$, and use that (mechanically!) to conclude that

$$\varphi_X(t) = \lambda/(\lambda - it).$$

The problem here is that $(\lambda - it)$ is not a positive real number and, therefore, cannot be used for a in the standard integral formula, without further justification. As was pointed out in the last paragraph, one should actually write $\varphi_X(t) = \int_0^{\infty} \cos(tx) \lambda e^{-\lambda x} dx + i \int_0^{\infty} \sin(tx) \lambda e^{-\lambda x} dx$ and try to compute these two real integrals. We leave it as an easy exercise for the reader to see that if these two real integrals are computed using integration by parts and then substituted in the above formula, one gets the exact same result as was obtained by the (wrong?) short-cut.

(d) If $X \sim \text{Lap}(\lambda)$, then X has a symmetric distribution and so we get

$$\varphi_X(t) = E(\cos tX) = \lambda \int_0^{\infty} \cos(tx) e^{-\lambda x} dx = \lambda/(\lambda^2 + t^2),$$

where the final formula is obained using integration by parts.

(e) Let $X \sim N(0, 1)$. Using symmetry of the distribution again, we get

$$\varphi_X(t) = E(\cos tX) = (\sqrt{2}/\pi) \int_0^{\infty} \cos(tx) e^{-x^2/2} dx.$$

Using Exercise 3.6.17, one can see that φ_X is continuously differentiable everywhere and $\varphi'_X(t) = -(\sqrt{2}/\pi) \int_0^\infty x \sin(tx) e^{-x^2/2} dx$. Applying integration by parts to compute the last integral, one can derive that the function φ_X satisfies the differential equation $\varphi'_X(t) = -t\varphi_X(t)$. Since $\varphi_X(0) = 1$, this differential equation has a unique solution given by
$$\varphi_X(t) = e^{-t^2/2}.$$

Using Theorem 8.2.7 (iii), one can now easily derive the characteristic function of the $N(\mu, \sigma^2)$ distribution from the above and see that it equals $e^{i\mu t} e^{-\sigma^2 t^2/2}$.

From Example 8.2.10, the reader would notice that when X has a Binomial or a Poisson distribution, its characteristic function $\varphi_X(t)$ takes the value 1 at all points t that are integer multiples of 2π, while at all other points t, one has $|\varphi_X(t)| < 1$. On the other hand, for the other three distributions considered in Example 8.2.10, one can see that $|\varphi_X(t)| < 1$ for all $t \neq 0$. Our next result (Theorem 8.2.12) asserts that this is not a coincidence. It says that, for the characteristic function φ_X of a random variable X, one has $|\varphi_X(t)| = 1$ for some $t \neq 0$ if and only if either X is a degenerate random variable or X has an 'Arithmetic Distribution', that is, a discrete distribution with set of values contained in an (infinite) arithmetic progression. We start with an exercise, which culminates in the theorem that follows.

EXERCISE 8.2.11. Let X be a random variable with characteristic function φ_X.
(a) Show that if $\varphi_X(t_0) = 1$, for some $t_0 \neq 0$, then $P(\cos(t_0 X) = 1) = 1$ and hence deduce that $P(X \in \{2\pi k/t_0 : k \in \mathbb{Z}\}) = 1$.
(b) Show that if $|\varphi_X(t_0)| = 1$, for some $t_0 \neq 0$, then there is a $\theta \in [0, 2\pi)$ such that the characteristic function of $Y = X - \frac{\theta}{t_0}$ equals 1 at t_0. Hence deduce that $P(X \in \{(\theta + 2\pi k)/t_0 : k \in \mathbb{Z}\}) = 1$.
(c) Show that if there is a sequence $t_n \downarrow 0$ such that $|\varphi_X(t_n)| = 1$ for all n, then X must be a degenerate random variable.
(d) Let X be a random variable taking the values $0, \pm\sqrt{2}$ and ± 1, each with probability $1/5$. Show that $|\varphi_X(t)| < 1$ for all $t \neq 0$. Observe that this does not contradict (b).

The following result summarizes the observations made in Exercise 8.2.11.

THEOREM 8.2.12. Let X be a real random variable and let φ_X denote its characteristic function. Then one and only one of the following is true.
(i) $|\varphi_X(t)| = 1$ for all $t \in \mathbb{R}$. This happens if and only if X is degenerate.
(ii) There exists $t_0 > 0$ such that $|\varphi_X(t_0)| = 1$ and $|\varphi_X(t)| < 1$ for $0 < t < t_0$. This happens if and only if X has an arithmetic distribution taking values in the set $D = \{a + kd : k \in \mathbb{Z}\}$, for some real number a and $d = 2\pi/t_0$.
(iii) $|\varphi_X(t)| < 1$ for all $t \neq 0$. This happens if and only if neither X is degenerate nor X has an arithmetic distribution.

REMARK 8.2.13. We put in record below a couple of important points that should be noted in connection with Theorem 8.2.12 are mentioned below.

8.2. Characteristic Functions

(a) As stated above, to say that a random variable X has an arithmetic distribution means that X is a discrete random variable with $P(X \in \{a + kd : k \in \mathbb{Z}\}) = 1$, for some real a and some $d > 0$. It should be clear that d is not unique, since we can replace d by d/m, for any $m > 1$, and the required condition will still hold. However, replacing d with a multiple of d may destroy the validity of the requirement. This leads one to consider the largest d, for which the condition $P(X \in \{a + kd : k \in \mathbb{Z}\}) = 1$ holds. This largest such d is often called the 'period' of the distribution of X. The reader may convince herself that in case (ii) of Theorem 8.2.12, the number $d = 2\pi/t_0$, where t_0 is as in the hypothesis, is indeed the period of X.

(b) The reader must be careful not to confuse between discrete distributions and arithmetic distributions. Every arithmetic distribution is discrete, but not all discrete distributions are arithmetic. In fact, X in Exercise 8.2.11 (d) has a discrete, but not arithmetic, distribution, which is why it falls in category (iii) of Theorem 8.2.12.

We next come to and settle in the affirmative an important question that was raised earlier. It is clear from the definition that the characteristic function of a random variable is really determined by its distribution, which is why we often talk of characteristic function of a distribution. The question that was raised earlier is whether the converse is true. In other words, do characteristic functions determine the distributions uniquely? We are going to show now that the answer is 'yes'. Indeed, in what follows, we are going to show that for random variables X and Y, the equality $\varphi_X \equiv \varphi_Y$ implies that $X \stackrel{d}{=} Y$.

Let φ be the characteristic function of a random variable X on a probability space $(\Omega, \mathcal{A}, P)$. Then for any $u, t \in \mathbb{R}$, $e^{-iut}\varphi(t) = \int_\Omega e^{it(X-u)} dP$ and so

$$\frac{1}{\sqrt{2\pi n}} \int_\mathbb{R} e^{-iut} \varphi(t) e^{-\frac{t^2}{2n}} dt = \frac{1}{\sqrt{2\pi n}} \int_\mathbb{R} \left(\int_\Omega e^{it(X-u)} dP \right) e^{-\frac{t^2}{2n}} dt.$$

Since the measurable map $(\omega, t) \mapsto e^{it(X(\omega)-u)} e^{-\frac{t^2}{2n}}$ on $\Omega \times \mathbb{R}$ is integrable with respect to $P \otimes \lambda$, we can apply Fubini's Theorem to the intgral on the right hand side to write it as $\int_\Omega \frac{1}{\sqrt{2\pi n}} \int_\mathbb{R} e^{it(X-u)} e^{-t^2/(2n)} dt\, dP$. Now, for any $s \in \mathbb{R}$, the integral $\frac{1}{\sqrt{2\pi n}} \int_\mathbb{R} e^{its} e^{-t^2/(2n)} dt$ can easily be identified as the value of the characteristic function of the $N(0, n)$ distribution at the point s and hence equals $e^{-ns^2/2}$ (see Example 8.2.10 (e)). Plugging this in the above integral, we get the equality

$$\frac{1}{\sqrt{2\pi n}} \int_\mathbb{R} e^{-iut} \varphi(t) e^{-\frac{t^2}{2n}} dt = \int_\Omega e^{-n(X-u)^2/2} dP.$$

Multiplying both sides by $\sqrt{n/2\pi}$ finally gives us

$$(2\pi)^{-1} \int_\mathbb{R} e^{-iut} \varphi(t) e^{t^2/(2n)} dt = \sqrt{n/2\pi} \int_\Omega e^{-n(X-u)^2/2} dP.$$

Using Exercise 5.4.15, one can easily deduce that the right hand side of the above equality is just the probability density function at u of the random variable $X + \frac{1}{\sqrt{n}} Z$ where Z_n is a $N(0, 1)$ random variable independent of X.

What all of these now mean is that if X are Y are two random variables with the same characteristic function φ and if we consider two N(0,1) random variables Z and W, with Z independent of X and W independent of Y, then, for each n, the random variables $X + \frac{1}{\sqrt{n}}Z$ and $Y + \frac{1}{\sqrt{n}}W$ will have the same density function; in particular, we will have
$$X + \tfrac{1}{\sqrt{n}}Z \stackrel{d}{=} Y + \tfrac{1}{\sqrt{n}}W, \quad \text{for each } n.$$
By letting $n \to \infty$ and noting that both $\frac{1}{\sqrt{n}}Z \xrightarrow{a.s.} 0$ and $\frac{1}{\sqrt{n}}W \xrightarrow{a.s.} 0$, we can invoke Slutsky Theorem (Proposition 8.1.7) to finally conclude that $X \stackrel{d}{=} Y$. We have thus proved the following very important uniqueness result.

THEOREM 8.2.14. (Uniqueness Theorem) For any two random variables X and Y,
$$\varphi_X \equiv \varphi_Y \text{ if and only if } X \stackrel{d}{=} Y.$$

As an immediate consequence, we now get the converse of Theorem 8.2.7 (v).

PROPOSITION 8.2.15. A random variable X has a symmetric distribution if and only if its characteristic function φ_X is real-valued.

Proof. $X \stackrel{d}{=} -X \iff \varphi_X = \varphi_{-X} \iff \varphi_X = \overline{\varphi}_X \iff \operatorname{Im}(\varphi_X) = 0$ □

The idea behind our proof of the uniqueness theorem above (Theorem 8.2.14) was a very clever and ingenious idea. The sole aim was to somehow show that the characteristic function of a random variable X determines the density function of $X + \frac{1}{\sqrt{n}}Z$, where Z is a N(0,1) random variable independent of X. In fact, we obtained an explicit formula for this density function in terms of the characteristic function of X. Once we identify this aim, the actual derivation is quite short and straight-forward. And it serves our purpose, namely, establishing the uniqueness result. However, one may wonder whether there is an explicit formula for the distribution of X itself, in terms of its characteristic function. Such a formula, if there is one, would also establish the uniqueness result. In the next section, we are going to show that there is, indeed, such a formula, known as the 'Inversion Formula'. However, the formula as well as the path to it is somewhat complicated, as expected. But the benefit is that apart from proving the uniqueness result, it leads to a number of other important cosequences.

8.3 Inversion Formula

Let X be a random variable on some probability space $(\Omega, \mathcal{A}, P)$ and let F_X denote its probability distribution function. In the last section, we have already seen (Theorem 8.2.14) that F_X is completely determined by the characteristic function φ_X of X. As stated in the concluding paragraph of the previous section, what we are now going to do is to get a formula expressing F_X in terms of φ_X.

To start with, we need the following result from calculus, a proof of which is outlined in Exercise 8.10.3.

$$\lim_{T \to \infty} \int_0^T \frac{\sin(\theta t)}{t}\,dt = \tfrac{\pi}{2} \text{ or } -\tfrac{\pi}{2}, \text{ according as } \theta > 0 \text{ or } \theta < 0. \tag{8.3.1}$$

8.3. Inversion Formula

Using the fact that, for each θ, the function $\frac{\sin(\theta t)}{t}$ is an even function of t, the formula (8.3.1) can be rewritten as

$$\frac{1}{\pi} \lim_{T \to \infty} \int_{-T}^{T} \frac{\sin(\theta t)}{t} \, dt = I_{(0,\infty)}(\theta) - I_{(-\infty,0)}(\theta).$$

From this, one easily gets that, for any real x and any pair of real numbers $a < b$,

$$\frac{1}{\pi} \lim_{T \to \infty} \int_{-T}^{T} \frac{\sin((x-a)t) - \sin((x-b)t)}{t} \, dt = 2I_{(a,b)}(x) + I_{\{a\}}(x) + I_{\{b\}}(x). \quad (8.3.2)$$

Next note that, for any $\alpha \neq \beta$, we have $\frac{\cos(\alpha t) - \cos(\beta t)}{t} \to 0$ as $t \to 0$, and therefore, $\frac{\cos(\alpha t) - \cos(\beta t)}{t}$ defines a continuous function in t on all of $\mathbb{R}$, provided we define it to be zero at $t = 0$. Further, since it is an odd function of t, we get that, for any real x and any pair of real numbers $a < b$,

$$\int_{-T}^{T} \frac{\cos((x-a)t) - \cos((x-b)t)}{t} \, dt = 0, \quad \text{for each } T > 0. \quad (8.3.3)$$

From (8.3.2) and (8.3.3), one gets, for any real x and any pair of reals $a < b$,

$$\lim_{T \to \infty} \frac{1}{\pi} \int_{-T}^{T} \frac{e^{it(x-a)} - e^{it(x-b)}}{it} \, dt = 2I_{(a,b)}(x) + I_{\{a\}}(x) + I_{\{b\}}(x). \quad (8.3.4)$$

In (8.3.4), we now put the random variable X in place of the real x and then take expectation to get that, for every pair of real numbers $a < b$,

$$E\left(\lim_{T \to \infty} \frac{1}{\pi} \int_{-T}^{T} \frac{e^{it(X-a)} - e^{it(X-b)}}{it} \, dt\right) = 2P(a < X < b) + P(X = a) + P(X = b)$$

(8.3.5)

We now want to interchange the expectation and the limit in the left hand side of (8.3.5) and for that, we need to apply the DCT.
Towards this, first observe that, for any pair of reals $a < b$,

$$\left| \int_{-T}^{T} \frac{e^{it(X-a)} - e^{it(X-b)}}{it} \, dt \right| \leq \left| \int_{-T}^{T} \frac{\sin((X-a)t)}{t} \, dt \right| + \left| \int_{-T}^{T} \frac{\sin((X-b)t)}{t} \, dt \right|$$

The right hand side of the above equals $\left| \int_{-T(X-a)}^{T(X-a)} \frac{\sin u}{u} \, du \right| + \left| \int_{-T(X-b)}^{T(X-b)} \frac{\sin u}{u} \, du \right|.$

We now use the fact that $C = \sup_{M > 0} \left| \int_{-M}^{M} \frac{\sin u}{u} \, du \right| < \infty$ (see Exercise 8.10.3 (a)) to conclude that

$$\left| \frac{1}{\pi} \int_{-T}^{T} \frac{e^{it(X-a)} - e^{it(X-b)}}{it} \, dt \right| \leq 2C/\pi, \quad \text{for all } T > 0 \text{ and all } a < b.$$

This allows us to use DCT in the left hand side of (8.3.5) and we get

$$\lim_{T \to \infty} \frac{1}{\pi} E\left(\int_{-T}^{T} \frac{e^{it(X-a)} - e^{it(X-b)}}{it} \, dt \right) = 2P(a < X < b) + P(X = a) + P(X = b).$$

(8.3.6)

We next want to use Fubini's Theorem to interchange the expectation and the integral over $[-T, T]$ in the left hand side of (8.3.6). For this, we use some simple algebra and the well-known facts that $|e^{i\theta}| = 1$ and $|e^{i\theta} - 1| \leq |\theta|$, for any real θ, to get

$$\left|\frac{e^{it(X-a)} - e^{it(X-b)}}{it}\right| = \left|e^{it(X-b)} \frac{e^{it(b-a)} - 1}{it}\right| = \left|\frac{e^{it(b-a)} - 1}{it}\right| \leq |b-a|,$$

Thus, the integrand in the left hand side of (8.3.6) is bounded in modulus and hence integrable on the finite measure space $(\Omega \times [-T, T], \mathcal{A} \otimes \mathcal{B}([-T, T]), P \otimes \lambda)$. This allows us to apply Fubini's Theorem to get

$$\lim_{T \to \infty} \frac{1}{\pi} \int_{-T}^{T} \frac{E(e^{it(X-a)}) - E(e^{it(X-b)})}{it} \, dt = 2P(a<X<b) + P(X=a) + P(X=b).$$

Noting that $\frac{E(e^{it(X-a)}) - E(e^{it(X-b)})}{it} = \frac{e^{-ita} - e^{-itb}}{it} \varphi_X(t)$, we get the following theorem.

THEOREM 8.3.1. (Inversion Formula) Let φ_X be the characteristic function of a random variable X. Then for any pair of real numbers $a < b$

$$\lim_{T \to \infty} \frac{1}{2\pi} \int_{-T}^{T} \frac{e^{-ita} - e^{-itb}}{it} \varphi_X(t) dt = P(a < X < b) + \tfrac{1}{2}P(X=a) + \tfrac{1}{2}P(X=b).$$

Denoting the distribution function of X by F_X, it is clear that when $a < b$ are both continuity points of F_X, then the right hand side of the above inversion formula equals $F_X(b) - F_X(a)$. Letting $a \downarrow -\infty$ through continuity points of F_X, this shows that $F_X(b)$, at each continuity point b of F_X, is determined by φ_X. One can use this and the facts that F_X is right-continuous everywhere and that the set of continuity points of F_X excludes at most countably many points of $\mathbb{R}$, one can easily argue that F_X is determined by φ_X.

The inversion formula assumes a very impotant and useful significance in the special case when the characteristic function φ_X happens to be integrable on $\mathbb{R}$ with respect to Lebesgue measure. Under this hypothesis, we can, using the inequality $|e^{-ita} - e^{-itb}| \leq |(b-a)t|$, easily see that $\frac{e^{-ita} - e^{-itb}}{it} \varphi_X(t)$ is integrable in t over $(-\infty, \infty)$. This allows us to use the DCT and write the left hand side of the inversion formula as $\frac{1}{2\pi} \int_{-\infty}^{\infty} \frac{e^{-ita} - e^{-itb}}{it} \varphi_X(t) dt$. From the inversion formula we will, therefore, be able to conclude that, if $\varphi_X(\cdot)$ is integrable on $\mathbb{R}$, then for any pair $a < b$ of continuity points of F_X, we have

$$F_X(b) - F_X(a) = \frac{1}{2\pi} \int_{-\infty}^{\infty} \frac{e^{-ita} - e^{-itb}}{it} \varphi_X(t) dt = \frac{1}{2\pi} \int_{-\infty}^{\infty} \left(\int_a^b e^{-itx} dx \right) \varphi_X(t) dt$$

Further, integrability of φ_X also allows us to apply Fubini's Theorem and interchange the order of integration in the extreme right hand side of the above equality to conclude that, for every pair $a < b$ of continuity points F_X,

$$F_X(b) - F_X(a) = \int_a^b f(x) dx, \qquad (8.3.7)$$

where

$$f(x) = \frac{1}{2\pi} \int_{-\infty}^{\infty} e^{-itx} \varphi_X(t) dt, \quad \text{for } x \in \mathbb{R}. \qquad (8.3.8)$$

8.3. Inversion Formula

Using integrability of φ_X and the DCT, one can easily verify that f, as defined in (8.3.8), is a continuous function on $\mathbb{R}$. Next, for all pair of real numbers $a < b$, the map $(a, b) \mapsto F_X(b) - F_X(a)$ is right continuous in each variable, while the map $(a, b) \mapsto \int_a^b f(x)dx$ is continuous in both variables. Since these two maps agree for all pairs $a < b$ that belong to the set of continuity points of F_X, which excludes at most countably many points in $\mathbb{R}$, it follows that these two maps must agree for all pairs $a < b$ of real numbers. We have thus proved that

$$F_X(b) - F_X(a) = \int_a^b f(x)dx, \quad \text{for all real } a, b \text{ with } a < b.$$

Since f is continuous, it follows immediately that f must be non-negative real valued everywhere. Further, by virtue of the continuity of f, the fundamental theorem of calculus implies that F_X is continuously differentiable everywhere and $F'_X = f$. All of what we just obtained, assuming that φ_X is integrable, is summed up in the following theorem.

THEOREM **8.3.2**. Suppose that the characteristic function φ_X of a random variable X is integrable on $\mathbb{R}$. Then, the distribution function F_X of X is continuously differentiable everywhere and X is absolutely continuous with a continuous density given by

$$f_X(x) = F'_X(x) = \frac{1}{2\pi} \int_{-\infty}^{\infty} e^{-itx} \varphi_X(t) dt, \quad \text{for } x \in \mathbb{R}.$$

In Fourier Analysis, the integral in the above formula for $f(x)$ is usually called the 'Inverse Fourier Transform' of φ_X and the result of Theorem 8.3.2 is expressed as follows. If the Fourier Transform φ of a probability measure P on $\mathcal{B}(\mathbb{R})$ is integrable on $\mathbb{R}$ (that is, belongs to $L_1(\mathbb{R}, \mathcal{B}, \lambda)$), then $P \ll \lambda$ with a continuous RN derivative $\frac{dP}{d\lambda}$, given by the 'Inverse Fourier Transform' of φ. The next example illustrates a beautiful application of this result.

EXAMPLE **8.3.3**. Recall from Example 8.2.10 (d) that $X \sim \text{Lap}(\lambda)$, where $\lambda > 0$, has characteristic function $\varphi_X(t) = \lambda/(\lambda^2 + t^2)$, which is clearly integrable on $\mathbb{R}$. Let us consider the special case when $\lambda = 1$, so that the density of X is $f_X(x) = \frac{1}{2} e^{-|x|}$, $x \in \mathbb{R}$. Applying Theorem 8.3.2 therefore, we will get

$$\frac{1}{\pi} \int_{-\infty}^{\infty} \frac{e^{-itx}}{1+t^2} dt = e^{-|x|}, \quad \text{or equivalently,} \quad \int_{-\infty}^{\infty} e^{itx} \frac{1}{\pi(1+x^2)} dx = e^{-|t|}.$$

The second equality is obtained from the first simply by bringing the $1/\pi$ inside the integral and by replacing x with $-t$ and t with x. The interesting thing now is that $g(x) = [\pi(1+x^2)]^{-1}$ can easily be seen to be a probability density function on $\mathbb{R}$. The distribution given by this density is known as the Cauchy distribution. What we have therefore obtained through the second equality is that the characteristic function of the Cauchy distribution is given by $\varphi(t) = e^{-|t|}$.

EXERCISE **8.3.4**. (a) A random variable X is said to have a 'triangular distribution' on $(-a, a)$, if it has density function given by $f_X(x) = \frac{1}{a}\left(1 - \frac{|x|}{a}\right)$, for $x \in (-a, a)$. (Graph of f_X explains why the name 'triangular'.) Show that the

characteristic function of X is given by $\varphi_X(t) = \frac{2(1-\cos at)}{a^2 t^2}$, $t \neq 0$ and $\varphi_X(0) = 1$.
(b) Deduce that, for any $a > 0$, the function $g(x) = \frac{1}{\pi}\frac{1-\cos ax}{ax^2}$, $-\infty < x < \infty$, (defined by continuity at $x = 0$) is a probability density function, whose characteristic function is $\varphi(t) = \left(1 - \frac{|t|}{a}\right)I_{[-a,a]}(t)$.

8.4 Moments and Characteristic Function

We have already seen that the characteristic function φ_X of any random variable X is always a continuous function — in fact, uniformly continuous — on $\mathbb{R}$. The question that we are going to address now is about possible differentiability properties of φ_X, if any. As we are going to see, this is intimately connected to finiteness of moments of the underlying random variable. In fact, we will show that if X has a finite kth moment, then φ_X is k-times continuously differentiable and we will also get explicit formula for all the derivatives upto order k. One of the byproducts of this will be getting a simple formula for those moments of X that are finite, in terms of the derivatives of φ_X of the corresponding order.

To start with, let us assume that the random variable X has finite first moment, that is, $E(|X|) < \infty$. We are going to prove that, in this case, φ_X is differentiable and the derivative φ'_X is uniformly continuous. Indeed, we will have a formula for φ'_X, which will also show that $E(X) = \frac{1}{i}\varphi'_X(0)$.

Towards this, let us observe that, for any two real numbers t and h, with $h \neq 0$, we have

$$\frac{\varphi_X(t+h) - \varphi_X(t)}{h} = E\left(\frac{e^{i(t+h)X} - e^{itX}}{h}\right) = E\left(\frac{e^{ihX} - 1}{h}e^{itX}\right).$$

We know that $(e^{ihx} - 1)/h \to ix$ as $h \to 0$, for every real x. Using the hypothesis $E(|X|) < \infty$ and the fact that $|e^{i\theta} - 1| \leq |\theta|$ for all real θ, we can apply DCT to conclude that $(\varphi_X(t+h) - \varphi_X(t))/h \to iE(Xe^{itX})$, as $h \to 0$. This shows that φ_X is differentiable everywhere with its derivative being given by $\varphi'_X(t) = iE(Xe^{itX})$, for $t \in \mathbb{R}$. Next, note that for any two reals t and h,

$$|\varphi'_X(t+h) - \varphi'_X(t)| = |E(Xe^{itX}(e^{ihX} - 1))| \leq E(|X||e^{ihX} - 1|).$$

Since, for all h with, say, $|h| \leq 1$, we have $|X||e^{ihX} - 1| \leq |X|$, which has finite expectation, and, $|e^{ihX} - 1| \to 0$ as $h \to 0$, we use DCT to conclude that φ'_X is uniformly continuous. Finally, putting $t = 0$ in the fomula for φ'_X, we get $\varphi'_X = iE(X)$, or equivalently, $E(X) = \frac{1}{i}\varphi'_X(0)$.

Having proved the result under the assumption of finite first moment, the next step is to use induction to prove the general result. Assume, by way of induction, that, if X has a finite kth moment, that is, $E|X|^k < \infty$, then φ_X has uniformly continuous derivatives of all orders upto k, given by

$$\varphi_X^{(j)}(t) = i^j E(X^j e^{itX}), \quad \text{for } t \in \mathbb{R} \text{ and for } j = 1, \ldots, k.$$

Suppose now that X has a finite $(k+1)$th moment, that is, $E|X|^{k+1} < \infty$. Since finiteness of $E|X|^{k+1}$ implies that of $E|X|^k$, we have, by the induction hypothesis,

8.4. Moments and Characteristic Function

$\varphi_X^{(k)}(t) = i^k E(X^k e^{itX})$, so that for any two real numbers t and h, with $h \neq 0$,

$$\frac{\varphi_X^{(k)}(t+h) - \varphi_X^{(k)}(t)}{h} = i^k E\left(X^k \frac{e^{i(t+h)X} - e^{itX}}{h}\right) = i^k E\left(X^k \frac{e^{ihX} - 1}{h} e^{itX}\right).$$

Since the quantity under brackets is bounded in absolute value by $|X|^{k+1}$ for all t, h and goes to $iX^{k+1}e^{itX}$, as $h \to 0$, we use the hypothesis that $E(|X^{k+1}|) < \infty$ and apply DCT to conclude that the $(k+1)$-th derivative of φ_X exists everywhere and is given by $\varphi_X^{(k+1)}(t) = i^{k+1} E(X^{k+1} e^{itX})$, for $t \in \mathbb{R}$. Uniform continuity of $\varphi_X^{(k+1)}$ is also proved using $E(|X^{k+1}|) < \infty$ and applying DCT in exactly the same way as it was done for φ_X'. Finally, $E(X^{k+1}) = \frac{1}{i^{k+1}} \varphi_X^{(k+1)}(0)$ follows from the formula for $\varphi_X^{(k+1)}(t)$. We have thus proved the following.

THEOREM 8.4.1. Let φ_X be the characteristic function of a random variable X. If, for some integer $k \geq 1$, the k-th moment of X is finite, then φ_X is k-times differentiable with all the derivatives upto the kth order uniformly continuous on $\mathbb{R}$ and given by

$$\varphi_X^{(j)}(t) = i^j E(X^j e^{itX}), \quad \text{for } t \in \mathbb{R} \text{ and } j = 1, \ldots, k.$$

In particular,

$$E(X^j) = \frac{1}{i^j} \varphi_X^{(j)}(0), \quad \text{for } j = 1, \ldots, k.$$

REMARK 8.4.2. One may wonder whether the converse of Theorem 8.4.1 is true, that is, whether differentiability of φ_X upto order k, for some integer $k \geq 1$, implies that X has finite moment of order k. Unfortunately, this is not true. However, what is rather intriguing is that it may fail *only* for odd integers $k \geq 1$. The converse, as stated here, remains valid for all even integers. However, we omit both the statement and proof of this result.

For a random variable X with finite kth moment for some $k \geq 1$, the above Theorem 8.4.1 leads to a nice and useful finite Taylor expansion of φ_X around zero. The reader would recall from calculus, that if h is a real valued function defined on an open interval around zero such that, $h^{(k)}(0)$ exists, for some $k \geq 1$, then h has the expansion

$$h(t) = \sum_{j=0}^{k} \frac{h^{(j)}(0)}{j!} t^j + o(t^k),$$

for t in some open neighbourhood of 0. Even when h is a complex valued function, one can get the same expansion for h by just combining the corresponding expansions for the real and imaginary parts.

REMARK 8.4.3. The symbol $o(t^k)$ used above is meant to be interpreted as follows. A statement like "$f(t) = g(t) + o(t^k)$ in a neighbourhood of 0" simply means that $(f(t) - g(t))/t^k$ goes to zero as $t \to 0$.

In view of the above and Theorem 8.4.1, we have the following result.

THEOREM 8.4.4. If a random variable X has a finite k-th moment for some integer $k \geq 1$, then its characteristic function φ_X has a finite Taylor expansion given by

$$\varphi_X(t) = \sum_{j=0}^{k} \frac{i^j E(X^j)}{j!} t^j + o(t^k), \qquad (8.4.1)$$

for all t in an open neighbourhood of 0.

8.5 Characteristic Functions and Weak Convergence

In this section we are going to examine the role of characteristic functions in the context of convergence in distribution. Suppose $X_n \xrightarrow{d} X$. Since for each $t \in \mathbb{R}$, the functions $x \mapsto \cos(tx)$ and $x \mapsto \sin(tx)$ are both bounded continuous functions on $\mathbb{R}$, one can use Theorem 8.1.13 to get that $\varphi_{X_n}(t) \to \varphi_X(t)$, for each $t \in \mathbb{R}$. The question we are going to address is whether the converse of this is true, that is, whether pointwise convergence of φ_{X_n} to φ_X implies $X_n \xrightarrow{d} X$. Clearly, an affirmative answer to this would provide us with a very useful criterion for weak convergence. In what follows, we are going to prove that the answer is indeed affirmative, but the passage to that involves several intermediate steps.

We first introduce an important concept. We know that, if P is any probability measure on $\mathcal{B}$, then by continuity from below, given any $\epsilon > 0$, we can get $M > 0$ such that $P([-M, M]) > 1 - \epsilon$. Of course, M here is allowed to depend on both ϵ and P. The question is, given a family of probabilities on $\mathcal{B}$, is it possible to get, for every $\epsilon > 0$, an $M > 0$, depending *only* on ϵ, such that $P([-M, M]) > 1 - \epsilon$, for every probability P in the given family? That this would be possible for any finite family of probabilities, is easy to see. However, for the family $\{P_n, n \geq 1\}$, where P_n, for each n, is the discrete probability assigning mass $1/2$ to each of the points $\pm n$, the answer to our question can easily be seen to be 'No'. This motivates our next definition.

DEFINITION 8.5.1. A family $\{P_\alpha\}$ of probability measures on $\mathcal{B}$ is called a Tight Family, if for every $\epsilon > 0$, there is an $M > 0$, depending only on ϵ, such that $P_\alpha([-M, M]) > 1 - \epsilon$, for all α.
A family $\{X_\alpha\}$ of real random variables is said to be a tight family if the family of corresponding probability distributions $\{P_{X_\alpha}\}$ is tight.

REMARK 8.5.2. We have encountered the idea of tightness before. A crucial step in the proof of Theorem 7.2.20 was to show that any sequence of random variables converging in probability must be tight.

EXERCISE 8.5.3. Show that any finite family of real random variables is tight.

EXERCISE 8.5.4. Show that a family $\{P_\alpha\}$ of probability measures on $\mathcal{B}$ is tight if and only if the associated family $\{F_\alpha\}$ of distribution functions has the property that $F_\alpha(x)$ converges to 0 as $x \to -\infty$ and to 1 as $x \to \infty$, uniformy in α.

EXERCISE 8.5.5. Show that any family of real random variables bounded in L_p, for some $p > 0$, is a tight family.

EXERCISE 8.5.6. Show that any sequence of real random variables, that converges in distribution, is a tight family.

We are now going to prove a partial converse of Exercise 8.5.6. The result, in a slightly more general form, is widely known as *Helly Selection Theorem*.

8.5. Characteristic Functions and Weak Convergence

THEOREM 8.5.7. If $\{X_n\}$ is a tight sequence of real random variables, then there is a subsequence that converges in distribution.

Proof. Denote, for each n, the distribution function of X_n by F_n. Let $\{r_j : j \geq 1\}$ be an enumeration of rational numbers. Then, $\{F_n(r_1)\}$ is a sequence of real numbers in $[0, 1]$ and, therefore, by the Bolzano-Weierstrass theorem, has a convergent subsequence. This means that there is a sequence $1 \leq n(1,1) < n(1,2) < \cdots$ of positive integers, such that, $\lim_k F_{n(1,k)}(r_1) = l_1$ exists. Consider now the sequence $\{F_{n(1,k)}(r_2)\}$. By Bolzano-Weierstrass theorem again, there is a further subsequence $n(2, 1) < n(2, 2) < \cdots$ of $\{n(1, k)\}$, such that, $\lim_k F_{n(2,k)}(r_2) = l_2$ exists. Since $\{n(2, k)\}$ is a subsequence of $\{n(1, k)\}$, we still have $\lim_k F_{n(2,k)}(r_1) = \lim_k F_{n(1,k)}(r_1) = l_1$. We now apply Bolzano-Weierstrass to $\{F_{n(2,k)}(r_3)\}$ and so on. Repeating this process will give us, for every j, a subsequence $\{n(j, k)\}$ of positive integers and a sequence $\{l_j\}$ of real numberes in $[0, 1]$ such that

(i) $\{n(j, k)\}$ is a subsequence of $\{n(j-1, k)\}$, for each $j \geq 2$, and,

(ii) $\lim_k F_{n(j,k)}(r_i) = l_i$, for all $j \geq 1$ and all $i = 1, \ldots, j$.

If we now take $n(k) = n(k, k)$, $k \geq 1$, then $\{F_{n(k)}\}$ is a subsequence of the original sequence $\{F_n\}$. Further, it is clear from the construction that $\{F_{n(k)}\}$ is, except possibly for a first few terms, a subsequence of $\{F_{n(j,k)}\}$ for every $j \geq 1$. This will imply that $\lim_k F_{n(k)}(r_j) = l_j$, for every j. Note also that, if $r_j < r_{j'}$, then $F_{n(k)}(r_j) \leq F_{n(k)}(r_{j'})$ for all k and so, $l_j = \lim_k F_{n(k)}(r_j) \leq \lim_k F_{n(k)}(r_{j'}) = l_{j'}$. Thus, if we put $H(r_j) = l_j$, then H defines a nondecreasing function on the set of rationals with values in $[0, 1]$. The function $G : \mathbb{R} \to [0, 1]$ defined by

$$G(x) = \inf\{H(r) : r > x, \ r \text{ rational}\} \tag{8.5.1}$$

can be easily seen to be nondecreasing and right continuous (see Exercise 2.8.22). We now use tightness of $\{X_n\}$ to show that G is a probability distribution function. All we need to show is that $\lim_{x \to \infty} G(x) = 1$ and $\lim_{x \to -\infty} G(x) = 0$. For the first one, we need to show that, given any $\epsilon > 0$, we can get a real number a such that $G(a) \geq 1 - \epsilon$. We Fix $\epsilon > 0$ and use the assumed tightness to get $M > 0$ such that $P(X_n \in [-M, M]) > 1 - \epsilon$, for all n. But then, for any rational $r > M$, we will have $F_n(r) \geq F_n(M) > 1 - \epsilon$, for all n, implying that $H(r) = \lim_k F_{n(k)}(r) \geq 1 - \epsilon$. This being true for all rational $r > M$, we will have $G(M) \geq 1 - \epsilon$, that is, $a = M$ does the required job.

For the other part, we need to show that, given any $\epsilon > 0$, we can get a real number b with $G(b) \leq \epsilon$. Taking $\epsilon > 0$ and choosing $M > 0$ as above, if we take and fix a rational $r > M$, then we will have $F_n(-r) \leq F_n(-M) < \epsilon$, for all n. This will imply $H(-r) = \lim_k F_{n(k)}(-r) \leq \epsilon$. If we now take any $b < -r$, it will follow from the definition of G that $G(b) \leq \epsilon$.

Having thus proved that G is a probability distribution function, our aim now is to show that $F_{n(k)} \xrightarrow{d} G$. We do this, by simply proving that the two inequalities $\limsup_k F_{n(k)}(x) \leq G(x)$ and $\liminf_k F_{n(k)}(x) \geq G(x-)$ both hold, for each $x \in \mathbb{R}$

(see Exercise 8.1.3).

Towards this, we fix an $x \in \mathbb{R}$. By the definition of $G(x)$, we can get, for any $\epsilon > 0$, a rational $r > x$ such that $H(r) < G(x) + \epsilon$. Since $F_{n(k)}(r) \to H(r)$, we must have $F_{n(k)}(r) < G(x) + \epsilon$, for all large k. Since $x < r$, we will have $F_{n(k)}(x) < G(x) + \epsilon$ also, for all large k, implying $\limsup_{k} F_{n(k)}(x) \leq G(x) + \epsilon$. This being true for any $\epsilon > 0$, we conclude $\limsup_{k} F_{n(k)}(x) \leq G(x)$. Next, the definition of the left limit $G(x-)$ implies that, for any $\epsilon > 0$, we can get $y < x$ such that $G(y) > G(x-) - \epsilon$. But then for all rational $r \in (y,x)$, we have $H(r) \geq G(y) > G(x-) - \epsilon$, implying that $F_{n(k)}(r) > G(x-) - \epsilon$, for all large k. Since $x > r$, we will have $F_{n(k)}(x) > G(x-) - \epsilon$ also, for all large k and therefore $\liminf_{k} F_{n(k)}(x) \geq G(x-) - \epsilon$. Once again, $\epsilon > 0$ being arbitrary, we conclude that $\liminf_{k} F_{n(k)}(x) \geq G(x-)$. We have thus exhibited a subsequence $\{X_{n(k)}\}$ of $\{X_n\}$, which converges in distribution. □

The next theorem gives a sufficient condition for a sequence of random variables to be tight in terms of some property of their characteristic functions. This and Theorem 8.5.7 are going to form the backbone for the proof of the main theorem of this section.

THEOREM 8.5.8. *Let $\{X_n\}$ be a sequence of random variables such that $\lim \varphi_{X_n}(t) = g(t)$ exists for all $t \in \mathbb{R}$. If g is continuous at $t = 0$, then $\{X_n\}$ is tight.*

The proof of this theorem requires a simple yet interesting inequality. We state it in the next lemma, prove it and then use it to prove the Theorem.

LEMMA 8.5.9. *If Y is a random variable, whose characteristic function is φ, then*

$$P(\{|Y| \leq M\}) \geq \left|\frac{M}{2}\int_{-2/M}^{2/M} \varphi(t)dt\right| - 1, \quad \text{for all } M > 0. \tag{8.5.2}$$

Proof of Lemma. Let $(\Omega, \mathcal{A}, P)$ be the probability space on which Y is defined. For any $a > 0$, we have $\frac{1}{a}\int_{-a}^{a} \varphi(t)dt = \frac{1}{a}\int_{-a}^{a} E(e^{itY})dt$. Since $|e^{itY}| = 1$, the function $(t,\omega) \mapsto e^{itY(\omega)}$ is integrable on $[-a,a] \times \Omega$ with respect to the finite measure $\lambda \otimes P$ and so we can use Fubini to interchange the integral and the expectation to get the above to equal $E(\frac{1}{a}\int_{-a}^{a} e^{itY}dt)$. Noting now that $\int_{-a}^{a} e^{itY}dt = 2\int_{0}^{a} \cos(tY)dt = (2\sin(aY))/Y$, one finally gets $\frac{1}{a}\int_{-a}^{a} \varphi(t)dt = \frac{2}{a}E((\sin(aY))/Y)$, for any $a > 0$. We need to clarify here that the value of the function $y \mapsto \sin(ay)/y$ is defined to be equal to a at $y = 0$, which makes it a continuous function on $\mathbb{R}$. Taking $a = 2/M$ for $M > 0$, we get

$$\left|\frac{M}{2}\int_{-2/M}^{2/M} \varphi(t)dt\right| = \left|2E(\frac{\sin(2Y/M)}{2Y/M})\right| \leq 2E\left|\frac{\sin(2Y/M)}{2Y/M}\right| \tag{8.5.3}$$

Splitting the expectation in the last expression of (8.5.3) into expectations over the sets $\{|Y| \leq M\}$ and $\{|Y| > M\}$ gives

$$\left|\frac{M}{2}\int_{-2/M}^{2/M} \varphi(t)dt\right| \leq 2E\left(\left|\frac{\sin(2Y/M)}{2Y/M}\right|I_{\{|Y|\leq M\}}\right) + 2E\left(\left|\frac{\sin(2Y/M)}{2Y/M}\right|I_{\{|Y|>M\}}\right) \tag{8.5.4}$$

8.5. Characteristic Functions and Weak Convergence

Using the bound $|\sin\theta/\theta| \leq 1$ in the first term and $|\sin\theta/\theta| \leq 1/|\theta|$ in the second term, one easily sees that the sum on the right hand side of (8.5.4) is bounded above by $2P(\{|Y| \leq M\}) + P(\{|Y| > M\}) = 1 + P(\{|Y| \leq M\})$. The asserted inequality in (8.5.2) nows follows immediately. □

Proof of Theorem 8.5.8. Let us first observe that, since $\varphi_n(0) = 1$ for each n, we have $g(0) = \lim \varphi_n(0) = 1$. Fixing any $\epsilon > 0$, we use continuity of g at zero to get $M_0 > 0$ such that $\left|\frac{M_0}{4} \int_{-2/M_0}^{2/M_0} g(t) dt\right| > 1 - \frac{\epsilon}{4}$. Since, $\varphi_n(t) \to g(t)$ for all t, we apply DCT to get an $n_0 \geq 1$ such that $\frac{M_0}{4} \int_{-2/M_0}^{2/M_0} |\varphi_n(t) - g(t)| dt < \frac{\epsilon}{4}$, for all $n > n_0$, and as a consequence, we get

$$\left|\frac{M_0}{4} \int_{-2/M_0}^{2/M_0} \varphi_n(t) dt\right| > 1 - \frac{\epsilon}{2}, \text{ for all } n > n_0.$$

Lemma 8.5.9 applied to X_n, $n > n_0$, implies

$$P(\{|X_n| \leq M_0\}) > 1 - \epsilon \text{ for all } n > n_0.$$

Using Exercise 8.5.3, we can get $M_1 > 0$, such that, $P(\{|X_n| \leq M_1\}) > 1 - \epsilon$, for all $n \leq n_0$. With $M = \max\{M_0, M_1\}$, we have $P(\{|X_n| \leq M\}) > 1 - \epsilon$, for all n. Since $\epsilon > 0$ was arbitrary, we have established tightness of $\{X_n\}$. □

We finally come to the main theorem of this section, known as *Lévy Continuity Theorem*. As already mentioned, this theorem is essentially a culmination of Theorems 8.5.7 and 8.5.8.

THEOREM **8.5.10.** (Lévy Continuity Theorem) *Let $\{X_n\}$ be a sequence of random variables with characteristic functions $\{\varphi_n\}$. If the limit $\varphi(t) = \lim_n \varphi_n(t)$ exists for all $t \in \mathbb{R}$ and φ is continuous at $t = 0$, then φ is a characteristic function and $X_n \xrightarrow{d} F$ where F is the probability distribution function whose characteristic function is φ.*

Proof. By Theorem 8.5.8, the hypothesis implies that the sequence $\{X_n\}$ is tight, and hence any subsequence of $\{X_n\}$ is also tight. Therefore, by theorem 8.5.7, every subsequence $\{X_{n(k)}\}$ of $\{X_n\}$ has a further subsequence $\{X_{n(k,j)}\}$ that converges in distribution. But then by what was noted at the beginning of the section, $\{\varphi_{n(k,j)}\}$ must converge pointwise to the characteristic function of the limiting distribution of the sequemce $\{X_{n(k,j)}\}$. But, from the hypothesis, the pointwise limit of the sequence $\{\varphi_n\}$, and hence also that of its subsequence $\{\varphi_{n(k,j)}\}$ is φ. This means that φ must be the characteristic function of the limiting distribution of the sequence $\{X_{n(k,j)}\}$. Letting F denote this unique probability distribution function whose characteristic function is φ, what we have shown is not only that φ is the characteristic function of a probability distribution function F, but also that every subsequence of $\{X_n\}$ has a further subsequence that converges in distribution to this F. We claim now that $X_n \xrightarrow{d} F$. We prove this by contradiction. Suppose, if possible, $x \in \mathbb{R}$ is a continuity point of F such that $F_{X_n}(x) \not\to F(x)$. This, of course means that, there is an $\epsilon > 0$ and a subsequence $\{n(k)\}$ such that $|F_{X_{n(k)}}(x) - F_X(x)| > \epsilon$, for all k. But, by what we have already proved, this subsequence $\{n(k)\}$ must have a further subsequence $\{n(k,j)\}$ such that $X_{n(k,j)} \xrightarrow{d} F$, implying, in particular, that $F_{X_{n(k,j)}}(x) \to F(x)$, as $j \to \infty$, which, of course, is impossible. This proves our claim and thereby completes the proof. □

An immediate consequence of Theorem 8.5.10 is the next very important result that was promised at the beginning of this section. It establishes pointwise convergence of characteristic functions as an equivalent criterion for convergence in distribution and has been used extensively in proving convergence in distribution in many situations. We will see a classical application in Section 8.7.

COROLLARY **8.5.11.** Let X_n, $n \geq 1$, and X be real random variables with characteristic functions φ_n, $n \geq 1$, and φ respectively. Then

$$X_n \xrightarrow{d} X \quad \text{if and only if} \quad \varphi_n(t) \to \varphi(t), \text{ for all } t \in \mathbb{R}.$$

We end this section with an important property of Gaussian distributions on $\mathbb{R}$, which comes as a nice application of results proved in this section. What the property says is that Gaussian distributions are "closed" under weak convergence. Of course, here we are going to think of degenerate distributions also as Gaussian distributions with zero variance. Here is the precise result.

THEOREM **8.5.12.** Let $\{X_n\}$ be a sequence of random variables with $X_n \sim \mathsf{N}(\mu_n, \sigma_n^2)$. Then $X_n \xrightarrow{d} X$ if and only if $\mu = \lim \mu_n$ and $\sigma = \lim \sigma_n$ exist and, in that case, $X \sim \mathsf{N}(\mu, \sigma^2)$ or X is degenerate at μ according as $\sigma > 0$ or $\sigma = 0$.

Proof. Noting that $\varphi_{X_n}(t) = e^{it\mu_n - t^2 \sigma_n^2 / 2}$, for each n, the 'if' part follows easily from Lévy conitnuity theorem, or more precisely, from Corollary 8.5.11.

To prove the 'only if' part, we recall that $X_n \xrightarrow{d} X$ implies that $\{X_n\}$ is a tight family (see Exercise 8.5.6). We claim that this implies that the sequences $\{\mu_n\}$ and $\{\sigma_n\}$ must both be bounded. Let us accept this for the time being and go ahead to prove the assertion. Using Bolzano-Weirestrass cleverly, we can get a subsequence $\{n(k)\}$ such that both $\{\mu_{n(k)}\}$ and $\{\sigma_{n(k)}\}$ converge to, say, μ and σ respectively. Since $X_n \xrightarrow{d} X$ implies $X_{n(k)} \xrightarrow{d} X$, we use the 'if' part (or directly Corollary 8.5.11) to get that $X \sim \mathsf{N}(\mu, \sigma^2)$ or X is degenerate at μ, according as $\sigma > 0$ or $\sigma = 0$. Instead of starting with the sequence $\{X_n\}$ as we did above, we could have started with any subsequence, say, $\{X_{n'}\}$ and applied the same argument. This will allow us to conclude that, for any subsequence $\{n'\}$, there is a further subsequence $\{n''\}$ such that $\lim \mu_{n''}$ and $\lim \sigma_{n''}$ both exist. But since the hypothesis will imply that $X_{n''} \xrightarrow{d} X$, we must have $\lim \mu_{n''} = \mu$ and $\lim \sigma_{n''} = \sigma$. This will imply that $\lim_n \mu_n = \mu$ and $\lim_n \sigma_n = \sigma$. For this, we used the well-known fact about real sequences that, given a real sequence $\{a_n\}$, if there is a real number a, such that, every subsequence of $\{a_n\}$ has a further subsequence converging to a, then the sequence $\{a_n\}$ itself converges to a.

8.6. Characteristic Functions: Some Characterizations

To complete the proof, we now have to show that tightness of $\{X_n\}$ implies boundedness of both $\{\mu_n\}$ and $\{\sigma_n\}$. That μ_n must be bounded is easy to see, because $P(X_n > \mu_n) = P(X_n < \mu_n) = 1/2$, for every n, and therefore, unboundedness of $\{\mu_n\}$ will contradict tightness. To see boundedness of $\{\sigma_n\}$, we observe that $X_n \sim N(\mu_n, \sigma_n)$ implies that $(X_n - \mu_n)/\sigma_n \sim N(0,1)$ and therefore we have

$$P(\mu_n - \sigma_n \leq X_n \leq \mu_n + \sigma_n) = \int_{-1}^{1} \frac{1}{\sqrt{2\pi}} e^{-u^2/2} \, du = \beta < 1, \text{ for every } n. \quad (8.5.5)$$

Since $\{\mu_n\}$ has already been shown to be bounded, unboundedness of σ_n will imply that the intervals $[\mu_n - \sigma_n, \mu_n + \sigma_n]$ must, along some subsequence, increase to $\mathbb{R}$. But, in view of (8.5.5), that will contradict the tightness of $\{X_n\}$. Thus, $\{\sigma_n\}$ must be bounded. □

REMARK 8.5.13. The main upshot of Theorem 8.5.12 is that if a sequence of normal random variables converge in distribution, then the limit distribution is either a normal distribution or a degenerate distribution. From this theorem, one can derive that we actually have convergence of all the moments, which is left as an easy exercise for the reader. In Section 8.9, we will see a multidimensional extension of Theorem 8.5.12.

8.6 Characteristic Functions: Some Characterizations

In this section, we conclude our analysis of characteristic functions by discussing two very important results, both of which aim at addressing the issue of identifying a function as a characteristic function.

The first one is an interesting and useful result, due to Pólya, that provides a very nice and simple sufficient condition for a real-valued function on $\mathbb{R}$ to be a characteristic function of a (necessarily symmetric) probability distribution. It must be noted that Pólya's criterion provides only a sufficient condition. However, it has proved to be immensely useful, primarily because of two reasons. Firstly, the criterion is very simple and, therefore, can be easily used to identify an important class of nice and interesting characteristc functions (even though it may not always be easy to explicitly describe the underlying probability distributions whose characteristic functions they are). Secondly, it can be used to provide examples of some intriguing pathologies related to characteristic functions, some of which will be given as exercises in Section 8.10.

It is essentially through an interesting application of the Inversion Theorem for integrable characteristic functions (Theorem 8.3.2) and Lévy Continuity Theorem (Theorem 8.5.10), that one gets Pólya's criterion. Here is how it goes.

We start with the function $\phi(t) = \big(1 - |t|\big) I_{[-1,1]}(t)$. An easy application of Theorem 8.3.2 tells us that ϕ is a characteristic function (see Exercise 8.3.4 (b)). Using Exercise 8.2.9 now, one easily sees that, for any choice of real numbers $0 < a_1 < \cdots < a_n$ and positive real numbers $\theta_1, \ldots, \theta_n$ with $\sum_k \theta_k = 1$, the function φ given by

$$\varphi_n(t) = \sum_k \theta_k \phi\left(\tfrac{t}{a_k}\right), \tag{8.6.1}$$

is a characteristic function. It is clear that φ_n is an even function (that is, symmetric around 0). Further, a closer scrutiny of the graph of φ_n on $[0, \infty)$ reveals that the graph is that of a convex polygon that starts at $\varphi_n(0) = 1$ and decreases to being identically 0 on $[a_n, \infty)$. It has linear segments on intervals $[0, a_1], [a_1, a_2], \ldots, [a_{n-1}, a_n], [a_n, \infty)$ with negative slopes, which increase from one segment to the next until finally becoming 0 on $[a_n, \infty)$.

The important point to note now is that if φ is any even, polygonal function on $\mathbb{R}$, which is convex on $[0, \infty)$ and satisfies (i) $\{t > 0 : \varphi(t) = 0\} = [\alpha, \infty)$, for some $\alpha > 0$, and (ii) $\varphi(0) = 1$, then φ must be given by (8.6.1). Indeed, given such a φ, denote the vertices of the convex polygonal graph of φ on the non-negative half line $[0, \infty)$ as $(0, 1), (a_1, \varphi(a_1)), \ldots, (a_n, 0)$, where $a_1 < \cdots < a_n = \alpha$ and the slopes on the successive linear segments $[0, a_1), \ldots, [a_{n-1}, a_n), [a_n, \infty)$ as $s_1, \ldots, s_n, s_{n+1}$ respectively. Clearly, $s_1 < \cdots < s_n < s_{n+1} = 0$. Consider now the positive real numbers $\theta_k = -(s_k - s_{k+1})a_k$, for $k = 1, \ldots n$. Denoting $a_0 = 0$, simple algebra shows that

$$\sum_{k=1}^n \theta_k = \sum_{k=1}^n s_k(a_{k-1} - a_k) = \sum_{k=1}^n \left(\varphi(a_{k-1}) - \varphi(a_k)\right) = \varphi(0) = 1.$$

One can now easily see that φ is exactly of the form as given in (8.6.1). We can, therefore, conclude that any even, polygonal function φ on $\mathbb{R}$, which is convex on $[0, \infty)$ and satisfies properties (i) and (ii) above, is a characteristic function.

Finally, given any even, real valued function φ on $\mathbb{R}$, which is convex on $[0, \infty)$ and satisfies (a) $\lim_{|t| \to \infty} \varphi(t) = 0$ and (b) $\lim_{t \to 0} \varphi(t) = \varphi(0) = 1$, it is easy to get a sequence $\{\varphi_n\}$ of even, polygonal functions on $\mathbb{R}$, that are convex on $[0, \infty)$ and satisfy (i) and (ii) above, such that, the given φ is the pointwise limit of the sequence $\{\varphi_n\}$. But, by what we have already noted above, each φ_n is a characteristic function. In view of the assumed continuity of φ at 0, Lévy Continuity Theorem 8.5.10 asserts that φ must be a charateristic function.

We have thus proved the result known as 'Pólya Criterion' and stated below, that identifies a large class of real valued functions as characteristic functions. As noted already, this criterion gives only a sufficient condition.

THEOREM **8.6.1.** (Pólya Criterion) Any symmetric, real valued function φ on $\mathbb{R}$, which is convex on $[0, \infty)$, and satisfies (a) $\lim_{|t| \to \infty} \varphi(t) = 0$ and (b) $\lim_{t \to 0} \varphi(t) = \varphi(0) = 1$, is a characteristic function.

REMARK **8.6.2.** Pólya used his criterion to assert that the function $\varphi(t) = e^{-|t|^\alpha}$ is a characteristic function, for each $0 < \alpha \leq 1$. The reason why this was important is that if X is a random variable with this characteristic function, then, one can easily see that, for each $n \geq 1$, the characteristic function $\varphi_n(t)$ of the random variable $n^{-\frac{1}{\alpha}} X$ will satisfy $\varphi(t) = (\varphi_n(t))^n$. In other words, for each $n \geq 1$, we have $X \stackrel{d}{=} X_1 + \cdots + X_n$, where $X_1, \ldots, X_n$ are i.i.d. with $X_i \stackrel{d}{=} n^{-\frac{1}{\alpha}} X$. Such a random variable X is said to have a 'stable distribution of index α' and these distributions form a very important class of distributions. It must be noted, however, the functions

8.6. Characteristic Functions: Some Characterizations

$\varphi(t) = e^{-|t|^\alpha}$, for $1 < \alpha \leq 2$, are also characteristic functions of stable distributions, but Pólya Criterion does not apply.

We now go to the next and final result of this section. This result, known as 'Bochner's Theorem', is arguably one of the most well-known results on characteristic functions and it gives a complete characterization of characteristic functions of probability distributions on $\mathbb{R}$. It should, however, be worth pointing out here that, while the theorem is an extremely deep and nice result, it has not proved to be very useful from the viewpoint of applicability, mainly because the most important sufficient (and necessary) condition asserted by the theorem, namely, the 'non-negative definiteness' of a function, is often not very easy to verify.

Before stating the theorem, let us recall what exactly is meant by 'non-negative definiteness' for a complex valued function of a real variable.

DEFINITION 8.6.3. A function $\varphi : \mathbb{R} \to \mathbb{C}$ is called 'non-negative definite' if

$$\sum_{j=1}^{n}\sum_{k=1}^{n} \varphi(t_j - t_k)\alpha_j\bar{\alpha}_k \geq 0, \quad \text{for all } t_1, \ldots, t_n \in \mathbb{R} \text{ and } \alpha_1, \ldots, \alpha_n \in \mathbb{C}. \quad (8.6.2)$$

Here is a simple result giving an alternative formulation for non-negative definiteness, which will later be used in a crucial way in our proof of Bochner's Theorem. Fairly routine arguments involving approximation by simple functions and use of Dominated Convergence Theorem lead to the result. Of course, one also needs to use boundedness of a non-negative definite $\varphi : \mathbb{R} \to \mathbb{C}$, which can be proved arguing along similar lines as in the proof of Theorem 8.6.5. With all these, we leave the proof of Lemma 8.6.4 as an exercise for the reader.

LEMMA 8.6.4. The condition (8.6.2) in the definition of non-negative definiteness is equivalent to

$$\int_{-\infty}^{\infty}\int_{-\infty}^{\infty} \varphi(t-s)h(t)\overline{h}(s)dsdt \geq 0, \quad (8.6.3)$$

for all continuous and integrable functions $h : \mathbb{R} \to \mathbb{C}$.

THEOREM 8.6.5. (Bochner's Theorem) A function $\varphi : \mathbb{R} \to \mathbb{C}$ is the characteristic function of a probability distribution on $\mathbb{R}$ if and only if φ is continuous, non-negative definite and $\varphi(0) = 1$.

Proof. We already know that the characteristic function φ of any probability distribution on $\mathbb{R}$ is continuous and satisfies $\varphi(0) = 1$. That any characteristic function φ satisfies (8.6.2) is easy to see from the definition of characteristic functions and is left for the reader to verify.

We are, therefore, only left to prove the 'if' part. So, suppose we are given a continuous, non-negative definite $\varphi : \mathbb{C} \to \mathbb{R}$, satisfying $\varphi(0) = 1$. We first make a couple of easy observations.

First, non-negative definiteness of φ and $\varphi(0) = 1$ imply $\varphi(-t) = \overline{\varphi(t)}$. To see this, we take $t_1 = 0, t_2 = t$ and use (8.6.2) to get

$$|\alpha_1|^2 + |\alpha_2|^2 + \alpha_1\varphi(-t)\bar{\alpha}_2 + \bar{\alpha}_1\varphi(t)\alpha_2 \geq 0, \quad \text{for any } \alpha_1, \alpha_2 \in \mathbb{C}.$$

Taking $\alpha_1 = \alpha_2 = 1$ in the above gives $\varphi(t) + \varphi(-t)$ to be real, while taking $\alpha_1 = 1, \alpha_2 = i$ gives $i(\varphi(t) - \varphi(-t))$ to be real. That proves $\varphi(-t) = \overline{\varphi(t)}$.

Next, non-negative definiteness and $\varphi(0) = 1$ imply $|\varphi(t)| \leq 1$. To see this, we use (8.6.2) with $t_1 = 0, t_2 = t$ and $\alpha_1 = -\varphi(t), \alpha_2 = 1$ and use the above observation $\varphi(-t) = \overline{\varphi(t)}$ to get $|\varphi(t)|^2 + 1 - |\varphi(t)|^2 - |\varphi(t)|^2 \geq 0$ implying $|\varphi(t)|^2 \leq 1$. Another simple fact that will also be used is that, for any $t \in \mathbb{R}$ and $n \in \mathbb{N}$,

$$\int_{-\infty}^{\infty} \exp\left(-itx - \frac{x^2}{2n^2}\right) dx = \sqrt{2\pi}\, n \exp\left(-\tfrac{1}{2}t^2 n^2\right), \tag{8.6.4}$$

which is just a consequence of the known formula for the charateristic function of a random variable with $N(0, n^2)$ distribution.

We now fix $x \in \mathbb{R}$ and $\sigma > 0$ and consider the bounded, continuous and integrable function $h(u) = \exp(-iux - u^2\sigma^2)$. By Lemma 8.6.4, non-negative definiteness of φ will imply

$$\int_{-\infty}^{\infty}\int_{-\infty}^{\infty} \varphi(u-v) \exp\left(-i(u-v)x - (u^2+v^2)\sigma^2\right) du\, dv \geq 0.$$

With the change of variable $u + v = s, u - v = t$, the integral on the left equals

$$\tfrac{1}{2}\int_{-\infty}^{\infty}\int_{-\infty}^{\infty} \varphi(t) \exp\left(-itx - \tfrac{1}{2}(s^2+t^2)\sigma^2\right) ds\, dt = \tfrac{\sqrt{2\pi}}{2\sigma} \int_{-\infty}^{\infty} \varphi(t) \exp\left(-itx - \tfrac{t^2\sigma^2}{2}\right) dt,$$

where we made use of the fact that $\int_{-\infty}^{\infty} \exp(-\tfrac{1}{2}s^2\sigma^2)\, ds = \tfrac{\sqrt{2\pi}}{\sigma}$.

Thus we get $\int_{-\infty}^{\infty} \varphi(t) \exp\left(-itx - \tfrac{t^2\sigma^2}{2}\right) dt \geq 0$, for all $x \in \mathbb{R}$ and all $\sigma > 0$. Consider now the non-negative real valued function f_σ, for each $\sigma > 0$, defined as

$$f_\sigma(x) = \tfrac{1}{2\pi} \int_{-\infty}^{\infty} \varphi(t) \exp\left(-itx - \tfrac{t^2\sigma^2}{2}\right) dt. \tag{8.6.5}$$

Noting that $0 \leq f_\sigma(x) \exp(-\tfrac{x^2}{2n^2}) \uparrow f_\sigma(x)$, as $n \to \infty$ and using MCT, one has

$$\int_{-\infty}^{\infty} f_\sigma(x) dx = \lim_n \int_{-\infty}^{\infty} f_\sigma(x) \exp(-\tfrac{x^2}{2n^2})\, dx. \tag{8.6.6}$$

Using the definition of $f_\sigma(x)$ in the integral on the right-hand-side of (8.6.6) gives

$$\int_{-\infty}^{\infty} f_\sigma(x) \exp(-\tfrac{x^2}{2n^2})\, dx = \tfrac{1}{2\pi} \int_{-\infty}^{\infty}\int_{-\infty}^{\infty} \varphi(t) \exp\left(-itx - \tfrac{t^2\sigma^2}{2} - \tfrac{x^2}{2n^2}\right) dt\, dx$$

$$= \tfrac{1}{2\pi} \int_{-\infty}^{\infty} \varphi(t) \exp(-\tfrac{t^2\sigma^2}{2})\Big\{\int_{-\infty}^{\infty} \exp\left(-itx - \tfrac{x^2}{2n^2}\right) dx\Big\} dt,$$

where integrabiity of the function $(t,x) \mapsto \varphi(t) \exp\left(-itx - \tfrac{t^2\sigma^2}{2} - \tfrac{x^2}{2n^2}\right)$ with reapect to Lebesgue measure on $\mathbb{R}^2$ is used to interchange order of integration. In the last double integral above, we first use (8.6.4) for the inner integral and then make the substitution $t \mapsto u = nt$, which gives

$$\int_{-\infty}^{\infty} f_\sigma(x) \exp(-\tfrac{x^2}{2n^2})\, dx = \tfrac{1}{\sqrt{2\pi}} \int_{-\infty}^{\infty} \varphi(\tfrac{u}{n}) \exp\left(-\tfrac{1}{2}u^2(1 + \tfrac{\sigma^2}{n^2})\right) du.$$

In view of (8.6.6) and using Dominated Convergence Theorem, we finally get

$$\int_{-\infty}^{\infty} f_\sigma(x) dx = \lim_n \tfrac{1}{\sqrt{2\pi}} \int_{-\infty}^{\infty} \varphi(\tfrac{u}{n}) \exp\left(-\tfrac{1}{2}u^2(1 + \tfrac{\sigma^2}{n^2})\right) du = \varphi(0) = 1,$$

showing that f_σ, for each $\sigma > 0$, is indeed a probability density function on $\mathbb{R}$. Denoting φ_σ to be the characteristic function of the density f_σ, we have

$$\varphi_\sigma(s) = \int_{-\infty}^{\infty} \exp(isx) f_\sigma(x)\, dx = \lim_n \int_{-\infty}^{\infty} \exp(isx - \tfrac{x^2}{2n^2}) f_\sigma(x)\, dx$$

$$= \lim_n \tfrac{1}{2\pi} \int_{-\infty}^{\infty}\int_{-\infty}^{\infty} \exp(isx - \tfrac{x^2}{2n^2}) \varphi(t) \exp(-itx - \tfrac{t^2\sigma^2}{2})\, dt\, dx,$$

where Dominated Convergence Theorem was used for the second equality. We now make the change of variable $t \mapsto u = n(t-s)$ and use Fubini to get

$$\varphi_\sigma(s) = \lim_n \frac{1}{2\pi} \int_{-\infty}^{\infty} \frac{1}{n} \varphi(s + \tfrac{u}{n}) \exp\left(-\tfrac{1}{2}(s + \tfrac{u}{n})^2 \sigma^2\right) \left\{ \int_{-\infty}^{\infty} \exp\left(-i\tfrac{u}{n}x - \tfrac{x^2}{2n^2}\right) dx \right\} du.$$

Using (8.6.4) once again for the inner integral and then Dominated Convergence Theorem, one finally gets

$$\varphi_\sigma(s) = \lim_n \frac{1}{2\pi} \int_{-\infty}^{\infty} \varphi(s + \tfrac{u}{n}) \exp\left(-\tfrac{u^2}{2} - \tfrac{1}{2}(s + \tfrac{u}{n})^2 \sigma^2\right) du = \varphi(s)\exp(-\tfrac{1}{2}s^2\sigma^2).$$

Now, φ_σ, for each $\sigma > 0$, is a characteristic function and the above shows that $\varphi_\sigma \to \varphi$ pointwise, as $\sigma \to 0$. By Lévy continuity theorem, it follows that φ is a characteristic function and that completes the proof. □

We end this section with one final and important remark that the hypothesis of continuity of φ in the 'if' part of Bochner's Theorem cannot be dispensed with. This is easily illustrated by the following exercise.

EXERCISE 8.6.6. *Show that the function $\varphi(t) = I_\mathbb{Q}(t)$, $t \in \mathbb{R}$, where Q is the set of rationals, is a real valued non-negative definite function on $\mathbb{R}$, that satisfies $\varphi(0) = 1$, but φ is not a characteristic function.*

8.7 Classical Central Limit Theorem

We are now ready to prove one of the most fundamental classical results of probability known as the *Central Limit Theorem* (abbreviated as *CLT*). This result is indeed very much central in much of the classical statistical theory and methodology. Much like the law of large numbers, the classical CLT also considers convergence of the sample mean $\overline{X}_n = \sum_{j=1}^n X_j/n$ associated to an i.i.d. sequence $\{X_n\}$, but the aim here is to study convergence in distribution.

The Strong Law of Large Numbers says that if an i.i.d. sequence $\{X_n\}$ has a finite common mean μ, then the sample mean $\overline{X}_n$ converges to μ almost surely, and therefore converges to the constant μ in distribution as well. Therefore, if we are interested in getting some sort of a non-degenerate limiting distribution, we must consider an appropriately scaled sample mean.

What the classical CLT asserts is that if the underlying i.i.d. sequence has a finite and positive common variance, then an appropriately normalized and scaled sample mean converges in distribution to a Standard Normal distribution. The reader should note that the assertion of CLT is that irrespective of what the underlying common distribution of the i.i.d sequence $\{X_n\}$ is, the limiting distribution of the appropriately scaled and normalized sample mean is *always* normal. In a way, this result establishes some kind of a *universality* of the Normal distribution and, thereby, underlines the huge importance and significance of Normal distribution in probability and statistics. Since we have already developed all the necessary tools and machinery for this, we proceed straight away to state and prove the classical Central Limit Theorem.

Chapter 8. Weak Convergence and Central Limit Theorem

THEOREM 8.7.1. (Classical CLT) If $\{X_n : n \geq 1\}$ is a sequence of i.i.d. random variables with a finite common variance $\sigma^2 > 0$, then
$$\sqrt{n}((\overline{X}_n - \mu)/\sigma) \xrightarrow{d} N(0,1) \qquad (8.7.1)$$
where $\overline{X}_n = \frac{1}{n}\sum_{j=1}^{n} X_j$, $n \geq 1$, and μ is the (necessarily finite) common mean.

REMARK 8.7.2. (a) The random variable $\sqrt{n}((\overline{X}_n - \mu)/\sigma)$ clearly has mean 0. One can easily verify also that $\sqrt{n}((\overline{X}_n - \mu)/\sigma)$ has variance 1. However, Theorem 8.7.1 does not make any specific assumption on the common distribution of the X_n except that it has a finite positive variance. This means that proving (8.7.1) by computing the distribution functions of $\sqrt{n}((\overline{X}_n - \mu)/\sigma)$ and then showing that they converge to that of $N(0,1)$ is completely out of question. As mentioned earlier, Corollary 8.5.11 will be our main tool. One may still wonder how we would be able to compute the characteristic functions, if we do not know the distributions. The point is that we do not need to compute the characteristic functions. There are some nice analytical properties that characteristic functions have which will allow us to get through without having to actually compute them. This is one of the main reasons why characteristic functions prove to be such a powerful tool in probability.
(b) With $S_n = \sum_{j=1}^{n} X_j$, we have $\sqrt{n}((\overline{X}_n - \mu)/\sigma) = (S_n - n\mu)/(\sqrt{n}\sigma)$, so that the assertion (8.7.1) is equivalent to $(S_n - n\mu)/(\sqrt{n}\sigma) \xrightarrow{d} N(0,1)$.

Proof. Denoting $Y_n = (X_n - \mu)/\sigma$, $n \geq 1$, we get an i.i.d. sequence $\{Y_n\}$ with common mean 0 and common variance 1. Further, $\sqrt{n}((\overline{X}_n - \mu)/\sigma) = T_n/\sqrt{n}$ where $T_n = \sum_{i=1}^{n} Y_i$. Since we know that the characteristic function of $N(0,1)$ is $\varphi(t) = e^{-t^2/2}$ (see Example 8.2.10 (e)), Corollary 8.5.11 says that all we need to do is to prove
$$\varphi_{T_n/\sqrt{n}}(t) \to e^{-t^2/2}, \quad \text{for all } t \in \mathbb{R}. \qquad (8.7.2)$$
There is nothing to prove for $t = 0$. So, we consider only $t \neq 0$ for the rest of the proof. If φ_Y denotes the common characteristic function of $\{Y_n\}$, then, by elementary properties of characteristic functions (Theorem 8.2.7), we have $\varphi_{T_n/\sqrt{n}}(t) = (\varphi_Y(t/\sqrt{n}))^n$. Since the Y_n's have common mean 0 and common variance 1, Theorem 8.4.4 asserts that $\varphi_Y(t) = 1 - \frac{t^2}{2} + o(t^2)$, in a neighbourhood of 0. This would mean that, for any fixed $t \neq 0$, we can write
$$\varphi_Y(t/\sqrt{n}) = 1 - \frac{t^2}{2n} + R_n, \text{ where } nR_n \to 0 \text{ as } n \to \infty.$$
But then for any fixed $t \neq 0$, we will have $\varphi_{T_n/\sqrt{n}}(t) = \left(1 - \frac{t^2}{2n} + R_n\right)^n$, where $nR_n \to 0$ as $n \to \infty$. In view of $\left|\left(1 - \frac{t^2}{2n} + R_n\right)^n - \left(1 - \frac{t^2}{2n}\right)^n\right| \leq |nR_n| \to 0$ (see Exercise 8.7.3 for the inequality), we finally get that, for any $t \neq 0$,
$$\lim_n \varphi_{T_n/\sqrt{n}}(t) = \lim_n \left(1 - \frac{t^2}{2n} + R_n\right)^n = \lim_n \left(1 - \frac{t^2}{2n}\right)^n = e^{-t^2/2},$$
completing the proof of (8.7.2). □

EXERCISE 8.7.3. Show, by induction, that if $z_1, \ldots, z_n$ and $w_1, \ldots, w_n$ are complex numbers with moduli ≤ 1, then $|z_1 \cdots z_n - w_1 \cdots w_n| \leq \sum_{j=1}^{n} |z_j - w_j|$.

8.8 Lindeberg-Feller-Lévy Central Limit Theorem

In this section, we are going to prove a Central Limit Theorem which is a little bit more general than the classical theorem that was proved in the last section. But the main purpose cis much more than just proving a more general theorem. The most important aim is to present a beautiful and novel technique that does not use characteristic functions and Corollary 8.5.11 of the Lévy Continuity Theorem. This approach has, of late, found application in a much wider variety of contexts where characteristic function does not prove to be a convenient tool. An interested reader may, for example, look at some recent work establishing CLT for non-linear functions of random variables.

Instead of considering an i.i.d. sequence and then studying convergence of the associated sample means, as in the classical set-up, we are going to consider a somewhat more general setup, which will, as a special case, capture the classical setting also. Before stating the main theorem, we describe this general setup.

DEFINITION 8.8.1. By a *Triangular Array* of random variables is meant a family of random variables $\{X_{nk} : n \geq 1, 1 \leq k \leq k_n\}$, where $\{k_n\}$ is a non-decreasing sequence of positive integers with $k_n \to \infty$.

The name 'triangular array' arises out of the interpretation that it represents an infinite number of rows of random variables, with each row having a finite number of random variables, but the number of the random variables in the successive rows keep on increasing to ∞. We are now ready to state our main theorem.

THEOREM 8.8.2. (Lindeberg-Feller-Lévy Central Limit Theorem)
Let $\{X_{nk} : n \geq 1, 1 \leq k \leq k_n\}$ be a triangular array of random variables with all having zero means and with finite variances $V(X_{nk}) = \sigma_{nk}^2$. Assume that

(i) for each n, the random variables X_{nk}, $1 \leq k \leq k_n$ are independent, and

(ii) $s_n^2 = \sum_k \sigma_{nk}^2 > 0$, for all $n \geq 1$, and satisfy

$$\lim_{n \to \infty} \frac{1}{s_n^2} \sum_{k=1}^{k_n} \int_{\{|X_{nk}| > \epsilon s_n\}} X_{nk}^2 dP = 0, \quad \text{for any } \epsilon > 0. \qquad (8.8.1)$$

Then

$$\frac{1}{s_n} \sum_{k=1}^{k_n} X_{nk} \xrightarrow{d} \mathsf{N}(0,1). \qquad (8.8.2)$$

REMARK 8.8.3. The condition satisfied by $\{s_n\}$ as stated in (8.8.1) is popularly known as the 'Lindeberg Condition'. Note that $\sum_k X_{nk}$, for each n, has mean zero and variance s_n^2.

That the classical CLT is a special case of the Lindeberg-Feller-Lévy CLT is immediate. We put it on record here as a corollary to Theorem 8.8.2.

COROLLARY 8.8.4. Let $(X_n : n \geq 1)$ be an i.i.d. sequence with common mean μ and common finite variance $\sigma^2 > 0$. Denote $S_n = X_1 + \cdots + X_n$, for each $n \geq 1$. Then

$$\frac{S_n - n\mu}{\sqrt{n}\sigma} \xrightarrow{d} \mathsf{N}(0,1).$$

Proof. Consider the triangular array $\{X_{nk}, n \geq 1, k \leq k_n\}$ with $k_n = n$ and $X_{nk} = (X_k - \mu)/\sqrt{n}$, for each $n \geq 1$ and $1 \leq k \leq n$. One can easily see that they have zero means, $s_n^2 = \sigma^2$, for all n, and the random variables $\{X_{nk}, 1 \leq k \leq n\}$ are independent, for each n. Further, for each n and each $k = 1, \ldots, n$, one has $\int_{\{|X_{nk}|>\epsilon s_n\}} X_{nk}^2 \, dP = \frac{1}{n} \int_{\{|X_k-\mu|>\epsilon\sigma\sqrt{n}\}} (X_k - \mu)^2 \, dP$. Using $X_k \stackrel{d}{=} X_1$, for all k, it now follows easily that

$$\frac{1}{s_n^2} \sum_{k=1}^n \int_{\{|X_{nk}|>\epsilon s_n\}} X_{nk}^2 \, dP = \frac{1}{\sigma^2} \int_{\{|X_1-\mu|>\epsilon\sigma\sqrt{n}\}} (X_1 - \mu)^2 \, dP,$$

which, for every $\epsilon > 0$, goes to 0, as $n \to \infty$, because $E((X_1 - \mu)^2) < \infty$. Thus, the Lindeberg condition holds, and, of course, $\frac{1}{s_n} \sum_{k=1}^n X_{nk} = \frac{S_n - n\mu}{\sqrt{n}\sigma}$. □

Next, consider a triangular array as in Theorem 8.8.2, but instead of assuming that the s_n^2, $n \geq 1$, satisfy Lindeberg condition (8.8.1), assume that they satisfy

$$\frac{1}{s_n^{2+\delta}} \sum_{k=1}^{k_n} E[|X_{nk}|^{2+\delta}] \to 0, \quad \text{as } n \to \infty, \quad \text{for some } \delta > 0. \tag{8.8.3}$$

This is commonly known as the 'Lyapunov's condition'. It follows that Lindeberg condition holds, because, with $\delta > 0$ as in (8.8.3), we have, for any $\epsilon > 0$,

$$\frac{1}{s_n^2} \sum_k \int_{\{|X_{nk}|>\epsilon s_n\}} X_{nk}^2 \, dP \leq \frac{1}{s_n^2} \sum_k \int_{\{|X_{nk}|>\epsilon s_n\}} X_{nk}^2 \left(\frac{|X_{nk}|}{\epsilon s_n}\right)^\delta dP \leq \frac{1}{\epsilon^\delta} \frac{1}{s_n^{2+\delta}} \sum_k E\{|X_{nk}|^{2+\delta}\},$$

which goes to zero, as $n \to \infty$, by the Lyapunov's condition (8.8.3). This gives us, as a corollary to Theorem 8.8.2, the following CLT, known as Lyapunov's CLT, which is often very useful in practice.

COROLLARY **8.8.5.** (Lyapunov's Central Limit Theorem) Consider a triangular array $\{X_{nk} : n \geq 1, 1 \leq k \leq k_n\}$ of random variables, all having zero means, and with $V(X_{nk}) = \sigma_{nk}^2$. Assume that, for each n, the random variables X_{nk}, $1 \leq k \leq k_n$, are independent, and, with $s_n^2 = \sum_k \sigma_{nk}^2 > 0$, $n \geq 1$, Lyapunov's condition (8.8.3) is satisfied. Then

$$\frac{1}{s_n} \sum_{k=1}^{k_n} X_{nk} \xrightarrow{d} N(0, 1).$$

We now proceed towards the proof of Theorem 8.8.2. As mentioned at the beginning, we are not going to prove it by using the classical technique of characteristic functions. Instead, we are going to take a different route. It must be pointed out though that, it is not the case that the technique of characteristic functions and Lévy Continuity Theorem does not work here. Indeed, most standard proofs of this CLT are actually based on that method. The reason why we choose to present a different proof is that it involves a very novel and interesting idea, which, as pointed out earlier, has become more and more popular because of its wide applicability in situations not amenable to use of characteristic functions. The plan here is to use the criterion stated in Exercise 8.10.11 for convergence in distribution. For the sake of clarity, we will divide the argument into a number of steps. We start by stating a few simple facts, that will be used in the sequel.

Let f be a C^∞ function on $\mathbb{R}$ with compact support and let g be the function on $\mathbb{R}$ defined as $g(h) = \sup_x \left| f(x+h) - f(x) - hf'(x) - \frac{1}{2}h^2 f''(x) \right|$. Note that g

8.8. Lindeberg-Feller-Lévy Central Limit Theorem

is measurable, since we could have taken supremum over just rationals.

Fact 1: There is a constant C such that $g(h) \leq C \min\{h^2, |h|^3\}$, for all h.

Fact 2: For all real numbers x, h_1 and h_2,
$$\left|f(x+h_1) - f(x+h_2) - (h_1-h_2)f'(x) - \tfrac{1}{2}(h_1^2 - h_2^2)f''(x)\right| \leq g(h_1) + g(h_2).$$

Fact 3: If U, V, W are random variables with finite second moments, then
$$\left|Ef(U+V) - Ef(U+W) - E\bigl((V-W)f'(U)\bigr) - \tfrac{1}{2}E\bigl((V^2 - W^2)f''(U)\bigr)\right|$$
$$\leq Eg(V) + Eg(W).$$

Fact 4: If U, V, W are independent random variables with finite second moments and if $E(V) = E(W)$ and $E(V^2) = E(W^2)$, then
$$|Ef(U+V) - Ef(U+W)| \leq E\bigl(g(V)\bigr) + E\bigl(g(W)\bigr).$$

Fact 1 can be proved using finite Taylor expansion and boundedness of f'' and f'''. Fact 2 is easily obtained from Fact 1 by using triangle inequality. To prove Fact 3, first apply Fact 2 with $x = U$, $h_1 = V$ and $h_2 = W$ and then take expectations. Finally, Fact 4 comes directly from Fact 3, using that $f'(U)$ and $f''(U)$ are both independent of the vector (V, W).

Returning now to the proof of Theorem 8.8.2, we only have to show (see Exercise 8.10.11) that, for any C^∞ function f on $\mathbb{R}$ with compact support,
$$\left|Ef\bigl((\textstyle\sum_k X_{nk})/s_n\bigr) - Ef(Z)\right| \to 0, \quad \text{where } Z \sim N(0,1). \tag{8.8.4}$$

At this point, we are going to assume that the triangular array $\{X_{nk}\}$ is defined on a probability space $(\Omega, \mathcal{A}, P)$, on which we also have an i.i.d. sequence $\{W_n\}$ of random variables with common distribution $N(0,1)$ such that the sequence $\{W_n\}$ is independent of the triangular array $\{X_{nk}\}$. To see that there is no loss of generality in making this assumption, the reader needs to make use of some of the theory developed in Chapter 10. To be precise, we develop a theory in Chapter 10 that guarantees that given a triangular array $\{X_{nk}\}$ on some probability space, it is possible to construct a new probability space and two families $\{Y_{nk}\}$ and $\{W_n\}$ of random variables on that probability space, such that, $\{Y_{nk}\}$ and $\{W_n\}$ independent of each other, $\{W_n\}$ is an i.i.d. sequence of $N(0,1)$ random variables and, for each n, the vector $(Y_{nk}, 1 \leq k \leq k_n)$ has the same joint distribution as the given $(X_{nk}, 1 \leq k \leq k_n)$. It is obvious that proving (8.8.4) with $\{Y_{nk}\}$ in place of $\{X_{nk}\}$ is good enough. Having thus justified the assumption, we are now going to go back to using $\{X_{nk}\}$, and not $\{Y_{nk}\}$, to denote our triangular array.

With this i.i.d. $N(0,1)$ sequence $\{W_n\}$, independent of the triangular array $\{X_{nk}\}$ available now, we define a new triangular array $\{Z_{nk}\}$ by setting
$$Z_{nk} = \sigma_{nk} W_k, \quad k = 1, \ldots, k_n, \quad \text{for each } n \geq 1.$$

Clearly $Z_{nk} \sim N(0, \sigma_{nk}^2)$ and hence $(\sum_k Z_{nk})/s_n \sim N(0,1)$. To prove (8.8.4), it therefore suffices to prove that
$$\left|Ef\bigl(\tfrac{1}{s_n}\textstyle\sum_k X_{nk}\bigr) - Ef\bigl(\tfrac{1}{s_n}\textstyle\sum_k Z_{nk}\bigr)\right| \to 0, \quad \text{as } n \to \infty. \tag{8.8.5}$$

286 Chapter 8. Weak Convergence and Central Limit Theorem

To achieve this, the main idea is to replace the X's in the first sum with Z's, one by one, and estimate the error committed. If the total error can be shown to be negligible, then we are done. For the moment, let us fix an n and define

$$T_k = \frac{1}{s_n} \sum_{j=k}^{k_n} X_{nj}, \text{ for } 1 \leq k \leq k_n, \text{ and } T_{k_n+1} = 0,$$

$$U_k = \frac{1}{s_n} \sum_{j=1}^{k-1} Z_{nj}, \text{ for } 2 \leq k \leq k_n + 1, \text{ and } U_1 = 0.$$

From these definitions, it is clear that proving (8.8.5) is the same as proving that

$$\left| E\big(f(T_1)\big) - E\big(f(U_{k_n+1})\big) \right| \to 0, \text{ as } n \to \infty. \tag{8.8.6}$$

Now, from $Ef(T_1) - Ef(U_{k_n+1}) = \sum_{j=1}^{k_n} [Ef(T_j + U_j) - Ef(T_{j+1} + U_{j+1})]$, one gets

$$\left| Ef(T_1) - Ef(U_{k_n+1}) \right| \leq \sum_{j=1}^{k_n} \left| Ef(T_j + U_j) - Ef(T_{j+1} + U_{j+1}) \right| \tag{8.8.7}$$

For each j, if we put $U = \frac{1}{s_n}\big(\sum_1^{j-1} Z_{nk} + \sum_{j+1}^n X_{nk}\big)$, $V = X_{nj}/s_n$ and $W = Z_{nj}/s_n$, then we clearly get $\left|Ef(T_j + U_j) - Ef(T_{j+1} + U_{j+1})\right| = \left|Ef(U+V) - Ef(U+W)\right|$. One can easily see that U, V and W are independent with all having zero means and $E(V^2) = \sigma_{nj}^2/s_n^2 = E(W^2)$. This means that we can now use Fact 4 to deduce that, for every $n \geq 1$ and every $j = 1, \ldots, k_n$,

$$\left| E\big(f(T_j + U_j)\big) - E\big(f(T_{j+1} + U_{j+1})\big) \right| \leq E\big(g(X_{nj}/s_n)\big) + E\big(g(Z_{nj}/s_n)\big),$$

and hence, from (8.8.7),

$$\left| E\big(f(T_1)\big) - E\big(f(U_{k_n+1})\big) \right| \leq \sum_{k=1}^{k_n} E\big(g(X_{nk}/s_n)\big) + \sum_{k=1}^{k_n} E\big(g(Z_{nk}/s_n)\big).$$

Thus, the proof of (8.8.6) will be complete, if we can show that

(i) $\lim_n \sum_{k=1}^{k_n} E\big(g(X_{nk}/s_n)\big) = 0$, and (ii) $\lim_n \sum_{k=1}^{k_n} E\big(g(Z_{nk}/s_n)\big) = 0$.

To prove (i), we observe that, fixing any $\epsilon > 0$, we get

$$E\big(g(X_{nk}/s_n)\big) \leq E\big(g(X_{nk}/s_n) I_{\{|X_{nk}| > \epsilon s_n\}}\big) + E\big(g(X_{nk}/s_n) I_{\{|X_{nk}| \leq \epsilon s_n\}}\big)$$
$$\leq CE\big(|X_{nk}/s_n|^2 I_{\{|X_{nk}| > \epsilon s_n\}}\big) + CE\big(|X_{nk}/s_n|^3 I_{\{|X_{nk}| \leq \epsilon s_n\}}\big)$$
$$\leq C\frac{1}{s_n^2} E\big(X_{nk}^2 I_{\{|X_{nk}| > \epsilon s_n\}}\big) + C\epsilon E\big(|X_{nk}/s_n|^2\big),$$

where we used Fact 1 to get the second inequality. The first term, added over $k = 1, \ldots, k_n$, goes to zero as $n \to \infty$, by the Lindeberg condition (8.8.1), while the second term, added over $k = 1, \ldots, k_n$, equals $C\epsilon$. Thus, for all large n, $\sum_{k=1}^{k_n} E\big(g(X_{nk}/s_n)\big)$ is at most $2C\epsilon$. Since ϵ was arbitrary, (i) is proved.

To prove (ii), we first let the reader verify, through standard computation, that if $Z \sim N(0, \sigma^2)$, then $E(|Z|^3) = 2\sqrt{\frac{2}{\pi}} \sigma^3$. Using this for each $n \geq 1$ and each $k = 1, \ldots, k_n$, we have $E(|Z_{nk}|^3) = 2\sqrt{\frac{2}{\pi}} \sigma_{nk}^3$, which gives us, using Fact 1 again, that

$$\sum_{k=1}^{k_n} E\big(g(Z_{nk}/s_n)\big) \leq \frac{C}{s_n^3} \sum_{k=1}^{k_n} E(|Z_{nk}|^3) = \frac{2C\sqrt{2/\pi}}{s_n^3} \sum_{k=1}^{k_n} \sigma_{nk}^3 \leq (2C\sqrt{2/\pi}) \frac{\max_k \sigma_{nk}}{s_n}.$$

To show finally that $\big(\max_k \sigma_{nk}\big)/s_n \to 0$ as $n \to \infty$, we fix any $\epsilon \in (0,1)$ and observe that $\sigma_{nk}^2/s_n^2 \leq \epsilon + E(X_{nk}^2 I_{\{|X_{nk}|>\epsilon s_n\}})/s_n^2$, for each $n \geq 1$ and each $1 \leq k \leq k_n$, giving us

$$\big(\max_k \sigma_{nk}^2\big)/s_n \leq \epsilon + \max_k E(X_{nk}^2 I_{\{|X_{nk}|>\epsilon s_n\}})/s_n^2.$$

Lindeberg condition (8.8.1) clearly implies that the second term goes to zero and that means that we will have $\big(\max_k \sigma_{nk}\big)/s_n < 2\epsilon$ for all large n. Since $\epsilon > 0$ was arbitary, this proves (ii) and thus completes the proof of Theorem 8.8.2.

REMARK 8.8.6. Towards the end of the proof, we made an interesting observation, namely, that the Lindeberg condition (8.8.1) implies $\big(\max_k \sigma_{nk}\big)/s_n \to 0$ as $n \to \infty$. This last condition can be interpreted as roughly saying that in any row of random variables in the triangular array $\{X_{nk}\}$, no one random variable in that row makes a "dominant contribution" to the variance of the row sum. The reason why this condition is important is that, there is a partial converse to the Lindeberg-Feller-Lévy Central Limit Theorem, due to Feller and Lévy, which says that, if $\{X_{nk}\}$ is a triangular array, satisfying all the conditions in Theorem 8.8.2, except the Lindeberg condition, then Lindeberg condition is necessary for the asymptotic normality (8.8.2) to hold, provided one assumes the additional condition that $\big(\max_k \sigma_{nk}\big)/s_n \to 0$ as $n \to \infty$. Thus, in some sense, Lindeberg condition is the minimal condition for CLT to hold for a triangular array, where no one random variable in any row stands out as the 'dominating contributor' towards the variance of the row sum. We give below the exact statement of this partial converse, though we will not prove it here.

THEOREM 8.8.7. (Lévy-Feller's Converse to Lindeberg-Feller-Lévy's CLT) Suppose $\{X_{nk}, n \geq 1, 1 \leq k \leq k_n\}$ is a triangular array of random variables with all having zero means and with finite variances $V(X_{nk}) = \sigma_{nk}^2$. Assume that, for each n, the random variables X_{nk}, $1 \leq k \leq k_n$, are independent and that $\big(\max_k \sigma_{nk}\big)/s_n \to 0$ as $n \to \infty$, where $s_n^2 = \sum_k \sigma_{nk}^2$. Then,

$$\frac{1}{s_n}\sum_k X_{nk} \xrightarrow{d} N(0,1)$$

implies that

$$\lim_{n\to\infty} \frac{1}{s_n^2} \sum_{k=1}^{k_n} \int_{\{|X_{nk}|>\epsilon s_n\}} X_{nk}^2 \, dP = 0, \quad \text{for every } \epsilon > 0.$$

8.9 Weak Convergence in Higher Dimensions

In this section, we extend the notion of weak convergence and all the other related concepts to multi-dimensional random vectors. The definition of weak convergence in higher dimensions is a very natural extension from the one dimensional case.

DEFINITION 8.9.1. If $\mathbf{X}_n$, $n \geq 1$, and $\mathbf{X}$ are k-dimensional random vectors with k-dimensional joint distribution functions F_n, $n \geq 1$, and F respectively, then $\mathbf{X}_n$ is said to *converge in distribution* (or, *converge weakly*) to $\mathbf{X}$, written $\mathbf{X}_n \xrightarrow{d} \mathbf{X}$, if $F_n(\mathbf{x}) \to F(\mathbf{x})$ for all continuity points $\mathbf{x}$ of F.

Many of the properties of convergence in distribution that we discussed in one dimension have their analogues in higher dimensions as well. We state and prove here only one. An enthusiastic reader should identify the analogues of other properties and verify them.

THEOREM 8.9.2. $\mathbf{X}_n \xrightarrow{d} \mathbf{X}$ if and only if $E(f(\mathbf{X_n})) \to E(f(\mathbf{X}))$ for all bounded continuous functions $f : \mathbb{R}^k \to \mathbb{R}$.

Proof. We write the proof here only for the case $k = 2$. The proof for general k is based on the same ideas and may be verified by the reader. For notational simplicity, the 2-dimensional random vectors $\mathbf{X}_n$, $n \geq 1$, and $\mathbf{X}$ are written as (Y_n, Z_n), $n \geq 1$, and (Y, Z), where Y_n, Z_n, $n \geq 1$, and Y, Z are random variables.

To prove the 'if' part, let (y_0, z_0) be a continuity point of the joint distribution function F of (Y, Z). Given $\epsilon > 0$, we can and do choose $\delta > 0$ such that $F(y_0 + \delta, z_0 + \delta) - \epsilon < F(y_0, z_0) < F(y_0 - \delta, z_0 - \delta) + \epsilon$. Consider now the two bounded continuous functions $f_1 : \mathbb{R}^2 \to \mathbb{R}$ and $f_2 : \mathbb{R}^2 \to \mathbb{R}$, defined as

$$f_1(y, z) = \left(\left[1 - (y - y_0)/\delta\right]^+ \wedge 1\right) \cdot \left(\left[1 - (z - z_0)/\delta\right]^+ \wedge 1\right),$$
$$f_2(y, z) = \left(\left[1 - (y - y_0 + \delta)/\delta\right]^+ \wedge 1\right) \cdot \left(\left[1 - (z - z_0 + \delta)/\delta\right]^+ \wedge 1\right).$$

From the definitions of f_1 and f_2, one easily gets

$$E(f_1(Y_n, Z_n)) \geq F_n(y_0, z_0) \geq E(f_2(Y_n, Z_n)). \tag{8.9.1}$$

By hypothesis, $E(f_1(Y_n, Z_n)) \to E(f_1(Y, Z))$ and $E(f_2(Y_n, Z_n)) \to E(f_2(Y, Z))$. Thus, letting $n \to \infty$ in (8.9.1), we will get $\limsup_n F_n(y_0, z_0) \leq E(f_1(Y, Z))$ and $\liminf_n F_n(y_0, z_0) \geq E(f_2(Y, Z))$. One can now complete the proof of the 'if' part by just observing that the definitions of f_1 and f_2 and the choice of δ imply that

$$E(f_1(Y, Z)) \leq F(y_0 + \delta, z_0 + \delta) \leq F(y_0, z_0) + \epsilon,$$
$$E(f_2(Y, Z)) \geq F(y_0 - \delta, z_0 - \delta) \geq F(y_0, z_0) - \epsilon.$$

For the 'only if' part, let $f : \mathbb{R}^2 \to \mathbb{R}$ be continuous with $|f(y, z)| \leq C$ for all $(y, z) \in \mathbb{R}^2$. Given $\epsilon > 0$, choose $M > 0$ such that $\pm M$ are continuity points of the marginal distribution functions of both Y and Z and $P(|Y| > M) < \frac{\epsilon}{8C}$ and $P(|Z| > M) < \frac{\epsilon}{8C}$. Using uniform continuity of f on the compact set $K = [-M, M] \times [-M, M]$, we can get $\delta > 0$ such that $|f(y, z) - f(y', z')| < \frac{\epsilon}{8}$ for all (y, z) and (y', z') in K with $|y - y'| < \delta$ and $|z - z'| < \delta$. Choose and fix a finite partition of $[-M, M]$ into subintervals of length $< \delta/2$. For each of the subintervals, pick a continuity point of the marginal distribution of Y lying strictly inside the subinterval. Denote the chosen points by, say, $y_1 < \ldots < y_k$. Similarly, for each of the subintervals, pick a continuity point of the marginal distribution of Z lying strictly inside the subinterval and denote the points so

8.9. Weak Convergence in Higher Dimensions

chosen as $z_1 < \ldots < z_k$. Set $y_0 = z_0 = -M$ and $y_{k+1} = z_{k+1} = M$. It is easy to verify that each of the points (y_i, z_j), $0 \leq i, j \leq k+1$, is a continuity point of the joint distribution F and they partition the set $(-M, M] \times (-M, M]$ into k^2 disjoint rectangles of the form $(a, b] \times (c, d]$. One can now consider the function g defined on $(-M, M] \times (-M, M]$, by setting its value on each of these rectangles to be equal to the value of f at its north-east vertex (that is, vertex (b, d) for the rectangle $(a, b] \times (c, d]$). One can now repeat verbatim the proof for one dimensional case to get the required result. □

REMARK 8.9.3. Exactly as in the case of one dimension, one can replace, in Theorem 8.9.2, "bounded continuous functions" by "bounded uniformly continuous functions" or "continuous functions (or, *just* C^∞ functions) with compact support" .

As an immediate consequence of the characterization given by Theorem 8.9.2, one gets the following useful result.

THEOREM 8.9.4. (Continuous Mapping Theorem) Suppose $\mathbf{X}_n$, $n \geq 1$, and $\mathbf{X}$ are k-dimensional random vectors, such that, $\mathbf{X}_n \xrightarrow{d} \mathbf{X}$, then $g(\mathbf{X}_n) \xrightarrow{d} g(\mathbf{X})$, for any continuous map $g : \mathbb{R}^k \to \mathbb{R}^m$.

REMARK 8.9.5. From the Continuous Mapping Theorem, it follows, in particular, that if $\mathbf{X}_n = (X_{n1}, \ldots, X_{nk}) \xrightarrow{d} \mathbf{X} = (X_1, \ldots, X_k)$, then $X_{ni} \xrightarrow{d} X_i$, for each $1 \leq i \leq k$. This will, of course, follow also from a very powerful tool, known as Cramer-Wold Device, which we are going to see later.

Just like real random variables, one can define characteristic functions for k-dimensional random vectors also (equivalently, for probability distributions on $\mathbb{R}^k$) and they play an equally important role. To define this, we need the following notation. For $\mathbf{t} = (t_1, \cdots, t_k) \in \mathbb{R}^k$ and $\mathbf{x} = (x_1, \cdots, x_k) \in \mathbb{R}^k$, we write $\langle \mathbf{t}, \mathbf{x} \rangle$ to denote the usual inner product of $\mathbf{t}$ and $\mathbf{x}$, that is, $\langle \mathbf{t}, \mathbf{x} \rangle = \sum_{j=1}^k t_j x_j$. Note that, if $\mathbf{X}$ is a k-dimensional random vector, then $\langle \mathbf{t}, \mathbf{X} \rangle$, for every $\mathbf{t} \in \mathbb{R}^k$, is a real random variable.

DEFINITION 8.9.6. The *characteristic function* of a k-dimensional random vector $\mathbf{X}$ is the function $\varphi_\mathbf{X} : R^k \to \mathbb{C}$ defined as $\varphi_\mathbf{X}(\mathbf{t}) = E\bigl(e^{i\langle \mathbf{t}, \mathbf{X} \rangle}\bigr)$, $\mathbf{t} \in \mathbb{R}^k$.

It is clear that $\varphi_\mathbf{X}$, for a random vector $\mathbf{X}$, depends only on the joint distribution of $\mathbf{X}$, which is why it is justifiedly also called the characteristic function of the joint distribution function F of $\mathbf{X}$ and is sometimes denoted by φ_F. The following result includes easy analogues of the elementary properties stated in Theorem 8.2.7 and some other simple properties, that are relevant only for the multi-dimensional case. The proofs are easy and hence left as exercise.

THEOREM 8.9.7. (Elementary Properties)
(i) $\varphi_\mathbf{X}(\mathbf{0}) = 1$.
(ii) $|\varphi_\mathbf{X}(\mathbf{t})| \leq 1$, for all $\mathbf{t} \in \mathbb{R}^k$.
(iii) If $\mathbf{Y} = a\mathbf{X} + \mathbf{b}$, for $a \in \mathbb{R}$ and $\mathbf{b} \in \mathbb{R}^k$, then $\varphi_\mathbf{Y}(\mathbf{t}) = e^{i\langle \mathbf{t}, \mathbf{b} \rangle} \varphi_\mathbf{X}(a\mathbf{t})$.

(iv) $\varphi_{-\mathbf{X}}(\mathbf{t}) = \varphi_{\mathbf{X}}(-\mathbf{t}) = \overline{\varphi}_{\mathbf{X}}(\mathbf{t})$.
(v) If $\mathbf{X}$ has symmetric distribution, that is, $-(\mathbf{X}) \stackrel{d}{=} \mathbf{X}$, then $\varphi_{\mathbf{X}}$ is a real-valued function.
(vi) $\varphi_{\mathbf{X}}$ is a uniformly continuous function on $\mathbb{R}^k$.
(vii) If $\mathbf{X}$ and $\mathbf{Y}$ are two k-dimensional random vectors that are independent, then $\varphi_{\mathbf{X}+\mathbf{Y}}(\mathbf{t}) = \varphi_{\mathbf{X}}(\mathbf{t})\varphi_{\mathbf{Y}}(\mathbf{t})$.
(viii) If $\mathbf{X} = (X_1, \cdots, X_k)$ is a random vector such that, for some $1 \leq j < k$, the two vectors $\mathbf{Y} = (X_1, \ldots, X_j)$ and $\mathbf{Z} = (X_{j+1}, \ldots, X_k)$ are independent then $\varphi_{\mathbf{X}}(t_1, \cdots, t_k) = \varphi_{\mathbf{Y}}(t_1, \cdots, t_j)\varphi_{\mathbf{Z}}(t_{j+1}, \cdots, t_k)$.
In particular, if $\mathbf{X} = (X_1, \ldots, X_k)$, where X_j, $1 \leq j \leq k$ are independent, then
$$\varphi_{\mathbf{X}}(t_1, \cdots, t_k) = \prod_j \varphi_{X_j}(t_j).$$
(ix) If $\mathbf{X} = (X_1, X_2, \cdots, X_k)$ and $\mathbf{Y} = (X_1, X_2, \cdots, X_j)$, for some $1 \leq j < k$, then $\varphi_{\mathbf{Y}}(t_1, \cdots, t_j) = \varphi_{\mathbf{X}}(t_1, \cdots, t_j, 0, \cdots, 0)$.

Just like in the one-dimensional case, $\varphi_{\mathbf{X}}$ of a k-dimensional random vector $\mathbf{X}$ uniquely determines the joint distribution of $\mathbf{X}$. In fact, there is a straightforward analogue of the inversion formula. We state it here but omit the proof, because it is a fairly routine extension of the proof given in the one-dimensional case.

THEOREM **8.9.8.** (Inversion formula) Let $\mathbf{X}$ be a k-dimensional random vector with characteristic function φ. Then for any $\mathbf{a}, \mathbf{b} \in \mathbb{R}^k$ with $\mathbf{a} < \mathbf{b}$,

$$\lim_{T \to \infty} \frac{1}{(2\pi)^k} \int_{-T}^{T} \cdots \int_{-T}^{T} \prod_{j=1}^{k} \frac{e^{-it_j a_j} - e^{-it_j b_j}}{it_j} \varphi(t_1, \cdots, t_k) \, dt_1 \cdots dt_k$$
$$= P(\mathbf{a} < \mathbf{X} < \mathbf{b}) + \tfrac{1}{2} P(X_j \in \{a_j, b_j\} \text{ for each } j).$$

Going through similar arguments as in the one-dimensional case, one gets the following uniqueness result for the k-dimensional case as a consequence of the above inversion formula.

PROPOSITION **8.9.9.** (Uniqueness) The joint distribution of a k-dimensional random vector $\mathbf{X}$ is uniquely determined by its characteristic function $\varphi_{\mathbf{X}}$.

The uniqueness result for the k-dimensional case has a very interesting and far-reaching consequence. To understand it, consider a k-dimensional random vector $\mathbf{X}$. We know that the joint distribution of $\mathbf{X}$ is not determined by its one-dimensional marginal distributions in general. One exception is when the coordinates of $\mathbf{X}$ are independent real random variables. The question we want to investigate is whether we can specify some family of real random variables associated to $\mathbf{X}$ whose distributions will always determine the joint distribution of $\mathbf{X}$ completely.

Theorem 4.4.8 tells us that if $\mathbf{X}$ is a k-dimensional random vector, then the distributions of the family of real random variables $f(\mathbf{X})$, where f varies over all bounded continuous functions $f : \mathbb{R}^k \to \mathbb{R}$, completely determine the joint distribution of $\mathbf{X}$. However, it is not a very satisfactory solution, because it involves considering too large a class of random variables. We are now going to describe how the uniqueness result Proposition 8.9.9 helps us make a huge leap in this regard. Indeed, Proposition 8.9.9 asserts that the joint distribution of any k-dimensional random vector $\mathbf{X} = (X_1, \ldots, X_k)$ is completely determined by the distributions of all the real random variables that are linear combinations of $X_1, \ldots, X_k$. To see this, note that any linear combination of $X_1, \ldots, X_k$ is nothing but $\langle \mathbf{t}, \mathbf{X} \rangle$, for some $\mathbf{t} \in \mathbb{R}^k$. So, if we know the distribution of $\langle \mathbf{t}, \mathbf{X} \rangle$ for every $\mathbf{t} \in \mathbb{R}^k$, then we know $\varphi_{\langle \mathbf{t}, \mathbf{X} \rangle}$, for all $\mathbf{t} \in \mathbb{R}^k$. But that determines $\varphi_{\mathbf{X}}$, because $\varphi_{\mathbf{X}}(\mathbf{t}) = \varphi_{\langle \mathbf{t}, \mathbf{X} \rangle}(1)$, for each $\mathbf{t} \in \mathbb{R}^k$. Thus, we get the following result, usually referred to as the Cramér-Wold Device.

THEOREM 8.9.10. (Cramér-Wold Device) The joint distribution of a k-dimensional random vector $\mathbf{X} = (X_1, \ldots, X_k)$ is completely determined by the distributions of all the possible linear combinations $\sum_{j=1}^{k} t_j X_j$, for $(t_1, \ldots, t_k) \in \mathbb{R}^k$.

There is a very interesting and slightly different way of interpreting what the Cramér-Wold Device says. Since the distribution of any real random variable is determined by the corresponding distribution function, Theorem 8.9.10 is equivalent to saying that the distribution functions $F_{\langle \mathbf{t}, \mathbf{X} \rangle}$, as $\mathbf{t}$ varies over $\mathbb{R}^k$, completely determine the joint distribution of $\mathbf{X}$. We now introduce a notation. For $\mathbf{t} \in \mathbb{R}^k$ and $a \in \mathbb{R}$, the 'half-space' determined by $\mathbf{t}$ and a is defined to be subspace of $\mathbb{R}^k$ given by $L_{\mathbf{t},a} = \{\mathbf{x} \in \mathbb{R}^k : \langle \mathbf{t}, \mathbf{x} \rangle \leq a\}$. All these half-spaces $L_{\mathbf{t},a}$, for $\mathbf{t} \in \mathbb{R}^k, a \in \mathbb{R}$, are clearly Borel subsets of $\mathbb{R}^k$. It is now easy to see that $F_{\langle \mathbf{t}, \mathbf{X} \rangle}(a)$, for $\mathbf{t} \in \mathbb{R}^k$ and $a \in \mathbb{R}$, is just the probability $P(\mathbf{X} \in L_{\mathbf{t},a})$. Thus, what the Cramer-wold Device essentially says is that any probability P on $\mathcal{B}^k$ is determined by the probabilities of all the 'half-spaces' $L_{\mathbf{t},a}$, $\mathbf{t} \in \mathbb{R}^k, a \in \mathbb{R}$.

Another specially important application of Cramér-Wold device is that it allows one to define k-dimensional Gaussian distribution with a dispersion matrix Σ, which is assumed to be symmetric and just non-negative definite.

DEFINITION 8.9.11. Given $\boldsymbol{\mu} \in \mathbb{R}^k$ and a real symmetric non-negative definite $k \times k$ matrix Σ, a k-dimensional random vector $\mathbf{X}$ is said to have a *Gaussian distribution with mean vector* $\boldsymbol{\mu}$ *and dispersion matrix* Σ, if for every (column) vector $\mathbf{t} \in \mathbb{R}^k$, the random variable $\langle \mathbf{t}, \mathbf{X} \rangle$ has a Normal distribution with mean $\langle \mathbf{t}, \boldsymbol{\mu} \rangle$ and variance $\mathbf{t}' \Sigma \mathbf{t}$. In case $\mathbf{t}' \Sigma \mathbf{t} = 0$, this is to mean that $\langle \mathbf{t}, \mathbf{X} \rangle$ is degenerate at $\langle \mathbf{t}, \boldsymbol{\mu} \rangle$.

REMARK 8.9.12. Just to convince the reader that the terms 'mean vector' for $\boldsymbol{\mu}$ and 'dispersion matrix' for Σ used here are not misplaced, one can take, for each $i = 1, \ldots k$, the vector $\mathbf{t} = (t_1, \ldots, t_k)'$ with $t_i = 1$ and $t_j = 0$ for $j \neq i$, to see that $E(X_i) = \langle \mathbf{t}, \boldsymbol{\mu} \rangle = \mu_i$ and $V(X_i) = \mathbf{t}' \Sigma \mathbf{t} = \sigma_{ii}$, for each i. Next, , for $i \neq j$, if one takes the vector $\mathbf{t} = (t_1, \ldots, t_k)'$ with $t_i = t_j = 1$ and $t_m = 0$ for $m \neq i, j$, one gets $V(X_i + X_j) = \mathbf{t}' \Sigma \mathbf{t} = \sigma_{ii} + 2\sigma_{ij} + \sigma_{jj}$. It now follows that $C(X_i, X_j) = \sigma_{ij}$. Thus, we get $M(\mathbf{X}) = \boldsymbol{\mu}$ and $D(\mathbf{X}) = \Sigma$, so that use of the terms 'mean vector' and 'dispersion matrix' here is consistent with Definition 4.5.5.

In the special case when Σ is positive definite, we have $\mathbf{t}' \Sigma \mathbf{t} > 0$, for all $\mathbf{t} \neq \mathbf{0}$. However, if Σ is assumed to be only non-negative definite, then we only have $\mathbf{t}'\Sigma\mathbf{t} \geq 0$, for all $\mathbf{t}$. As mentioned earlier and repeated once again in the above definition, a normal random variable with zero variance is interpreted as a degenerate random variable. As seen in Exercise 4.2.4, the k-dimensional Gaussian distribution has a joint density function if and only if Σ is positive definite. This case is often underlined by saying that $\mathbf{X}$ has a 'non-degenerate' k-dimensional Gaussian distribution. The discussions on Gaussian distributions in Chapter 4 were limited only to the non-degenerate case. In case Σ is non-negative definite, but not positive definite, then there is a vector $\mathbf{t} \neq \mathbf{0}$, such that, $\langle \mathbf{t}, \mathbf{X} \rangle$ is degenerate (see Exercise 4.5.6 (c)), that is, $\langle \mathbf{t}, \mathbf{X} \rangle = \alpha$ a.s.$[P]$, where $\alpha = \langle \mathbf{t}, \boldsymbol{\mu} \rangle$. One can easily show that, for any $\mathbf{t} \neq \mathbf{0}$ and any $\alpha \in \mathbb{R}$, the set $\{\mathbf{x} \in \mathbb{R}^k : \langle \mathbf{t}, \mathbf{x} \rangle = \alpha\}$ has k-dimensional Lebesgue measure 0. This shows that, in case Σ is non-negative definite, but not positive definite, the corresponding Gaussian distribution is singular with respect to Lebesgue measure on $\mathbb{R}^k$ and hence usually referred to as a *'singular Gaussian distribution'*.

The next proposition, which generalizes some of the earlier results for non-degenerate Gaussian distributions, is an easy consequence of the Definition 8.9.11.

PROPOSITION 8.9.13. Let $\boldsymbol{\mu} \in \mathbb{R}^k$ and Σ a real symmetric non-negative definite $k \times k$ matrix. If $\mathbf{X}$ is a k-dimensional random vector having a Gaussian distribution with mean vector $\boldsymbol{\mu}$ and dispersion matrix Σ, then, for every $m \geq 1$ and every real $m \times k$ matrix $\mathbf{A}$ and every vector $\mathbf{c} \in \mathbb{R}^m$, the m-dimensional random vector $\mathbf{Y} = \mathbf{A}\mathbf{x} + \mathbf{c}$ has a m-dimensional Gaussian distribution with mean vector $\mathbf{A}\boldsymbol{\mu} + \mathbf{c}$ and dispersion matrix $\mathbf{A}\Sigma\mathbf{A}'$.
In particular, for any $1 \leq m \leq k$ and any $1 \leq k_1 < \ldots < k_m \leq k$, the m-dimensional random vector $(X_{k_1}, \ldots, X_{k_m})'$ is Gaussian with mean vector $(\mu_{k_1}, \ldots, \mu_{k_m})'$ and dispersion matrix $((\sigma_{k_i,k_j}))$.

We now come to the multi-dimensional version of the Lévy Continuity Theorem. We just state it here without proof. However, a reader who has understood the proof of Lévy Continuity Theorem on $\mathbb{R}$, may want to try proving the multi-dimensional version herself.

THEOREM 8.9.14. (Lévy Continuity Theorem in Higher Dimensions) For $k > 1$, let $\mathbf{X}_n$, $n \geq 1$, and $\mathbf{X}$ be k-dimensional random vectors with characteristic functions φ_n, $n \geq 1$, and φ respectively. Then,

$$\mathbf{X}_n \xrightarrow{d} \mathbf{X} \quad \text{if and only if} \quad \varphi_n(\mathbf{t}) \to \varphi(\mathbf{t}), \text{ for all } \mathbf{t} \in \mathbb{R}^k.$$

This result in conjunction with with Cramér-Wold Device shows that proving weak convergence in $\mathbb{R}^k$ reduces to proving weak convergence in one dimension, as stated below.

THEOREM 8.9.15. $\mathbf{X}_n \xrightarrow{d} \mathbf{X}$ if and only if $\langle \mathbf{t}, \mathbf{X}_n \rangle \xrightarrow{d} \langle \mathbf{t}, \mathbf{X} \rangle$, for all $\mathbf{t} \in \mathbb{R}^k$.

Using Theorem 8.9.15, a multidimensional version of classical Central Limit Theorem now follows easily from the one-dimensional version (Theorem 8.7.1).

8.10. Additional Exercises

THEOREM 8.9.16. (Classical Central Limit Theorem) Let $\mathbf{X}_n$, $n \geq 1$, be a sequence of i.i.d. k-dimensional random vectors with a common mean vector $\boldsymbol{\mu}$ and a common positive definite dispersion matrix $\boldsymbol{\Sigma}$. If $\overline{\mathbf{X}}_n = (\mathbf{X}_1 + \cdots + \mathbf{X}_n)/n$, then

$$\sqrt{n}\,\boldsymbol{\Sigma}^{-1}\left(\overline{\mathbf{X}}_n - \boldsymbol{\mu}\right) \xrightarrow{d} \mathbf{Z}$$

where $\mathbf{Z}$ has a k-dimensional normal distribution with $\mathbf{0}$ as mean vector and the identity matrix $\mathbf{I}$ as the dispersion matrix.

We end this section with an extension of Theorem 8.5.12 to higher dimensions. The assertion of Theorem 8.5.12 was that Gaussian distributions in one dimension are closed under weak convergence. What the next theorem says is that the same holds in higher dimensions as well. The proof, which is left as an exercise, just uses the one-dimensional result and the Cramér-Wold Device.

THEOREM 8.9.17. Let $\{\mathbf{X}_n\}$ be a sequence of k-dimensional Gaussian random vectors with means $\{\boldsymbol{\mu}_n\}$ and dispersion matrices $\{\boldsymbol{\Sigma}_n\}$. Then $\mathbf{X}_n \xrightarrow{d} \mathbf{X}$ if and only if both $\boldsymbol{\mu} = \lim_n \boldsymbol{\mu}_n$ and $\boldsymbol{\Sigma} = \lim_n \boldsymbol{\Sigma}_n$ exist and in that case $\mathbf{X}$ is Gaussian with mean vector $\boldsymbol{\mu}$ and dispersion matrix $\boldsymbol{\Sigma}$.

Here, both the limits $\lim_n \boldsymbol{\mu}_n$ and $\lim_n \boldsymbol{\Sigma}_n$ mean componentwise limits. Note that, since each $\boldsymbol{\Sigma}_n$ is non-negative definite, the limit $\boldsymbol{\Sigma} = \lim_n \boldsymbol{\Sigma}_n$ will also be non-negative definite.

8.10 Additional Exercises

EXERCISE 8.10.1. Show that for real random variables X and X_n, $n \geq 1$, convergence of $\{X_n\}$ in distribution to X is equivalent to each of the following:
(i) there exists a dense subset $D \subset \mathbb{R}$, such that, $P(a < X_n \leq b) \to P(a < X \leq b)$ for all pairs $a, b \in D$ with $a < b$.
(ii) for every pair a, b of real numbers with $a < b$ and every $0 < \epsilon < (b-a)/2$, there exists $n_0 \geq 1$, such that, for all $n \geq n_0$,
$$P(a + \epsilon < X < b - \epsilon) - \epsilon \leq P(a < X_n < b) \leq P(a - \epsilon < X < b + \epsilon) + \epsilon.$$
(iii) for every $\epsilon > 0$ and every $\delta > 0$, there exists $n_0 \geq 1$, such that, for all $n \geq n_0$ and all pairs $a < b$ of reals,
$$P(a + \delta < X < b - \delta) - \epsilon \leq P(a < X_n < b) \leq P(a - \delta < X < b + \delta) + \epsilon.$$

EXERCISE 8.10.2. [Lévy Metric on the set of probability distributions on $\mathbb{R}$] For any two probability distribution functions F, G on $\mathbb{R}$, define
$$\rho(F, G) = \inf\{\epsilon > 0 : G(x - \epsilon) - \epsilon \leq F(x) \leq G(x + \epsilon) + \epsilon, \text{ for all } x \in \mathbb{R}\}.$$
(a) Show that $\rho(\cdot, \cdot)$ defines a metric on the set of all probability distribution functions on $\mathbb{R}$. [In view of the one-one correspondence between probabiity distributions on $\mathbb{R}$ and probability distribution functions on $\mathbb{R}$, one may also think of ρ as defining a metric on the set all probability distributions on $\mathbb{R}$.]
(b) Show that for probability distribution functions F_n, $n \geq 1$ and F on $\mathbb{R}$,
$$F_n \xrightarrow{d} F \text{ if and only if } \rho(F_n, F) \to 0.$$

EXERCISE 8.10.3. *Going through steps(a)–(f) given below, prove the formula*

$$\lim_{T\to\infty} \int_0^T \frac{\sin\theta x}{x} dx = \pm\frac{\pi}{2} \quad \text{according as} \quad \theta > 0 \text{ or } \theta < 0. \tag{8.10.1}$$

(a) *Show that* $\left|\int_0^T \frac{\sin x}{x} dx\right| \leq \int_0^\pi \frac{\sin x}{x} dx$, *for any* $T > 0$.

(b) *Show that* $\frac{1}{2} + \cos x + \cdots + \cos nx = \frac{\sin(n+\frac{1}{2})x}{2\sin\frac{1}{2}x}$ *and use this identity to deduce that* $\int_0^\pi \frac{\sin(n+\frac{1}{2})x}{2\sin\frac{1}{2}x} dx = \frac{\pi}{2}$.

(c) *If f and its derivative f' are continuous on $[a,b]$, use integration by parts to show that* $\lim_{\lambda\to\infty} \int_a^b f(x)\sin\lambda x\, dx = 0$.

(d) *Take* $f(x) = \frac{1}{x} - \frac{1}{2\sin\frac{x}{2}}$ *on $[0,\pi]$ and use results proved in (b) and (c) to deduce that* $\lim_{n\to\infty} \int_0^\pi \frac{\sin(n+\frac{1}{2})x}{x} dx = \frac{\pi}{2}$.

(e) *Denoting* $J(T) = \int_0^T \frac{\sin x}{x} dx$, *for $T > 0$, use integration by parts to show that* $|J(T_2) - J(T_1)| < 2/T_1$, *for $0 < T_1 < T_2$*.

(f) *Use all of the above steps to finally derive (8.10.1).*

EXERCISE 8.10.4. *Show that $X_n \xrightarrow{d} X$ iff there is a dense subset $D \subset \mathbb{R}$ such that $\lim F_{X_n}(x) = F_X(x)$ for all $x \in D$.*

EXERCISE 8.10.5. *Let ξ_n be i.i.d. uniform $(0,1)$ random variables. Define $X_n = \min_{1\leq i\leq n} \xi_i$. Show that $nX_n \xrightarrow{d} X$ where $F_X(x) = (1 - e^{-x})^+$.*

EXERCISE 8.10.6. *Let X_n, $n \geq 1$, and X be random variables taking non-negative integer values, and let p_n, $n \geq 1$ and p the respective pmfs. Show that $X_n \xrightarrow{d} X$ iff $p_n(k) \to p(k)$ for all $k = 0, 1, \ldots$.*

EXERCISE 8.10.7. *For each n, let I_{jn}, $1 \leq j \leq n$ be disjoint subintervals of $[0,1]$, each of length $1/n$ and for each j, let $x_{jn} \in I_{jn}$. Consider the random variable X_n taking the values $x_{1n}, \ldots, x_{nn}$, each with probability $1/n$. Show that $X_n \xrightarrow{d} \text{Uniform}[0,1]$.*

EXERCISE 8.10.8. *Show that, given any probabilty distribution function F on $\mathbb{R}$, there is a sequence $\{F_n\}$ of discrete probability distribution functions such that $F_n \xrightarrow{d} F$.*

EXERCISE 8.10.9. *Show that if $X_n \xrightarrow{d} X$ and if $\{f_n\}$ is a sequence of bounded real-valued continuous functions on $\mathbb{R}$ converging uniformly to a function f, then $E(f_n(X_n)) \to E(f(X))$.*

EXERCISE 8.10.10. *A family $\{f_\alpha\}$ of real-valued functions on $\mathbb{R}$ is said to be equicontinuous (everywhere), if for all $x \in \mathbb{R}$ and $\epsilon > 0$, there exists $\delta = \delta(x,\epsilon)$, such that $x - \delta < y < x + \delta$ implies $|f_\alpha(y) - f_\alpha(x)| < \epsilon$, for all α. Show that if $X_n \xrightarrow{d} X$, then, for any uniformly bounded equicontinuous family $\{f_\alpha\}$ of functions, $E(f_\alpha(X_n)) \to E(f_\alpha(X_n))$, uniformly in α.*

EXERCISE 8.10.11. *Show that $X_n \xrightarrow{d} X$ if and only if $E(f(X_n)) \to E(f(X))$ for every C^∞ function $f : \mathbb{R} \to \mathbb{R}$ with compact support. (Use Exercise 4.6.4(d) for the 'if' part.)*

8.10. Additional Exercises

EXERCISE 8.10.12. *Show that $X_n \xrightarrow{d} X$ if and only if $E(G(X_n)) \to E(G(X))$ for all continuous probability distribution functions G on $\mathbb{R}$.*

EXERCISE 8.10.13. *(a) Show that if φ is a characteristic function, then so is the function $\tilde{\varphi}(t) = \varphi(t/a)$ for any positive real number a.*
(b) Show that if $\varphi_1, \ldots, \varphi_n$ are characteristic functions, the any convex combination $\varphi = \sum_{k=1}^{n} p_k \varphi_k$, where $p_1, \ldots, p_n$ are non-negative real numbers with $\sum_k p_k = 1$, is also a characteristic function.

EXERCISE 8.10.14. *For a probability measure P on $(\mathbb{R}, \mathcal{B})$, with φ denoting its characteristic function, prove the following* **Bochner's Formula** *for mass at x:*

$$P(\{x\}) = \lim_{T \to \infty} \frac{1}{2T} \int_{-T}^{T} e^{-itx} \varphi(t)\, dt, \quad \text{for } x \in \mathbb{R}.$$

EXERCISE 8.10.15. *[DeMoivre-Laplace CLT]*
(a) (This involves a delicate use of Stirling's approximation formula.) Fix $0 < p < 1$ and let $q = 1 - p$. Show that if $x \in \mathbb{R}$ and $\{k_n\}$ is any sequence of non-negative integers such that $(k_n - np)/\sqrt{npq} \to x$, then $\sqrt{npq}\binom{n}{k_n} p^{k_n} q^{n-k_n} \to \frac{1}{\sqrt{2\pi}} e^{-x^2/2}$.
(b) Use Theorem 8.1.11 (b) now to deduce that if $\{X_n\}$ is a sequence of random variables with $X_n \sim \text{Bin}(n,p)$, for each n, then $(X_n - np)/\sqrt{npq}$ converges in distribution to $N(0,1)$.

EXERCISE 8.10.16. *We know that if $\mathbf{X}_n = (X_{n1}, \ldots, X_{nk}) \xrightarrow{d} \mathbf{X} = (X_1, \ldots, X_k)$, then $X_{ni} \xrightarrow{d} X_i$, for each $1 \leq i \leq k$ (see Remark 8.9.5). Give a counter-example to show that, unlike the other modes of convergence, the converse of the above does not hold.*

EXERCISE 8.10.17. *[Slutsky Type Results]*
(a) Extend Proposition 8.1.6 to random vectors.
(b) Let $\mathbf{X}_n, n \geq 1$, $\mathbf{Y}_n, n \geq 1$ and $\mathbf{X}$ be k-dimensional ($k > 1$) random vectors, such that, $\mathbf{X}_n \xrightarrow{d} \mathbf{X}$ and $\mathbf{Y}_n \xrightarrow{P} \mathbf{0}$. Show that $\mathbf{X}_n + \mathbf{Y}_n \xrightarrow{d} \mathbf{X}$.
(c) Show that if $\mathbf{X}_n, n \geq 1$, $\mathbf{Y}_n, n \geq 1$ and $\mathbf{X}$ are k-dimensional random vectors, such that, $\mathbf{X}_n \xrightarrow{d} \mathbf{X}$ and $\mathbf{Y}_n \xrightarrow{P} \mathbf{c}$, for some $\mathbf{c} \in \mathbb{R}^k$, then $(\mathbf{X}_n, \mathbf{Y}_n) \xrightarrow{d} (\mathbf{X}, \mathbf{c})$. [Show $(\mathbf{X}_n, \mathbf{c}) \xrightarrow{d} (\mathbf{X}, \mathbf{c})$, $(\mathbf{X}_n, \mathbf{Y}_n) - (\mathbf{X}_n, \mathbf{c}) \xrightarrow{P} \mathbf{0}$ and then apply part (b).]
(d) Conclude that under the hypothesis of (c), $h(\mathbf{X}_n, \mathbf{Y}_n) \xrightarrow{d} h(\mathbf{X}, \mathbf{c})$, for any continuous $h : \mathbb{R}^{2k} \to \mathbb{R}^m$. In particular, if $\{X_n\}$ and $\{Y_n\}$ are sequences of random variables, such that, $X_n \xrightarrow{d} X$ and $Y_n \xrightarrow{d} c$, for some $c \in \mathbb{R}$, then $X_n Y_n \xrightarrow{d} cX$ and, if $c \neq 0$, then $X_n/Y_n \xrightarrow{d} X/c$.

Chapter 9

Conditioning : The Right Approach

A reader initiated to probability theory would know that conditional probabilities and conditional expectations play a very important role in this theory. In a first course in probability, one starts with the notion of 'conditional probability of an event A, given an event B', denoted as $P(A \mid B)$ and defined by the formula $P(A \mid B) = P(A \cap B)/P(B)$. Of course, the formula is valid *only* when $P(B) > 0$, which is an issue that we will address later, but ignore for the time being. An important point we want to note here is that, if one fixes a B with $P(B) > 0$ and considers $P(A \mid B)$ for all events A, then the conditional probability $P(\cdot \mid B)$ is easily seen to have all the properties of a probability. The next concept that appears is that of 'conditional distribution of a random variable X given another random variable Y'. If $y \in \mathbb{R}$ is such that $P(Y = y) > 0$, then one can define conditional probabilities of X taking different values, given $Y = y$. This defines a distribution for the random variable X, usually called the conditional distribution of X, given $Y = y$. The expectation with respect to this distribution, if it exists, is called the conditional expectation of X, given $Y = y$ and is written as $E(X \mid Y = y)$. As long as the 'conditioning' random variable Y is a discrete random variable, all of these work absolutely fine with y being one of the possible values of Y. But the moment one moves to a random variable Y with a continuous distribution, meaning that $P(Y = y) = 0$, for every y, there seems to be no way of extending the above simple minded ideas. In the extremely special case, namely, when (X, Y) has a joint density, an ad hoc formula for what is called 'conditional density' of X, given $Y = y$, is introduced in elementary probability. This conditional density is taken to represent the conditional distribution of X, given $Y = y$, and one even proceeds to define $E(X \mid Y = y)$ as the expected value with respect to this conditional density. However, this has no obvious straight forward connection with the original concept of conditional probabilities. Of course, one possible way of rationalizing this, at least heuristically, is to first consider conditional distributions of X, given that Y takes values in arbitrarily small open neighbourhoods of y (assuming that the latter events have positive probability) and then to arrive at the "conditional density" formula by taking limit as the neighbourhoods shrink to the point y. But however nice it seems heuristically, making this idea mathematically rigorous is, in general, hugely

9.1. Conditional Expectation

challenging and sometimes even impossible. So, essentially it remains only an ad hoc formula. Next, even if this ad hoc idea of conditional density is accepted, the issue of how to define conditional distributions and conditional expectations remains wide open in all other cases, except when Y is discrete or when (X,Y) has a joint density.

The reader would surely agree that the situation described above is rather unsatisfactory. Ideally, one should have a single general definition of conditional distributions and conditional expectations and the specific formulas in the cases discussed above should automatically follow from that as special cases. This is exactly where the measure theoretic framework comes in handy. We have already seen how it helped us develop a unified theory of distribution and expectation for random variables and how the special formulas discussed in elementary probability for discrete and absolutely continuous random variables could be derived out of that as particular cases. In this section, we are going to do exactly that for the concepts of conditional distributions and conditional expectations. Making use of the measure theoretic frame work, we are going to put the concept of condtioning in a proper and unified theoretical framework. The special cases discussed above will follow from that, with due justification, as upshots.

Before we start, there is one word of caution. In the approach to conditioning that we are going to discuss, conditional expectations would come before conditional probabilities and conditional distributions. To an innocent reader, this may seem a little awkward and disconcerting, since in the usual thought process, probabilities and distributions come first and then come expectations as integrals with respect to distributions. However, a moment's reflection should convince the reader that if we somehow had prior access to the concept of expectations, then probabilities could be defined in terms of expectations. After all, the probability $P(A)$ of an event A is just the expectation $E(I_A)$.

9.1 Conditional Expectation

To motivate our definition of conditional expectations, let us start with a situation that is extensively discussed in elementary probability theory. Consider two discrete random variables X and Y defined on a probability space with D_1 and D_2 as the countable sets of possible values of X and Y respectively. Let us write p for the joint pmf $p_{(X,Y)}$, that is, $p(x,y) = P(X=x, Y=y)$, $x \in D_1, y \in D_2$. We will write p_1 and p_2 for the marginal pmf's of X and Y respectively, that is, $p_1(x) = P(X=x) = \sum_{y \in D_2} p(x,y)$, $x \in D_1$ and $p_2(y) = P(Y=y) = \sum_{x \in D_1} p(x,y)$, $y \in D_2$. We want to point out here that the countable sets D_1 and D_2 have been chosen so as to ensure $p_1(x) > 0$, for all $x \in D_1$ and $p_2(y) > 0$, for all $y \in D_2$.

For $y \in D_2$, applying the classical definition of conditional probability, one first gets $P(X=x \mid Y=y) = p(x,y)/p_2(y)$, $x \in D_1$ and then uses it to define the conditional expectation of X, given $Y=y$, as $E(X \mid Y=y) = \sum_{x \in D_1} xp(x,y)/p_2(y)$. Assuming for the moment that these conditional expectations are all finite,

$E(X \mid Y = y)$ gives a real number, for each $y \in D_2$. In other words, if we put $g(y) = \sum_{x \in D_1} xp(x,y)/p_2(y)$, $y \in D_2$, then g defines a real valued function on D_2. This function g has an interesting property which is usually overlooked, primarily because viewing and studying $E(X \mid Y = y)$ as a function of y are not very commonplace in elmentary probability. The property is that,

$$E(XI_{\{Y \in B\}}) = E(g(Y)I_{\{Y \in B\}}), \text{ for any } B \subset \mathbb{R}. \tag{9.1.1}$$

This is easy to see. Firstly, since Y takes values only in D_2, there is no loss of generality in assuming that $B \subset D_2$. But then, the left hand side of (9.1.1) equals $\sum_{y \in B} \sum_{x \in D_1} xp(x,y)$, while the right hand side is $\sum_{y \in B} g(y)p_2(y)$. That these two are equal is clear from the definition of g. This was, of course, fairly trivial. But what is more striking and crucial for what we are going to do, is that the function g, that we have defined above, is the only real valued fucntion on D_2 that satisfies (9.1.1). The good thing is that this is also easy to see. Indeed, let h be any real function on D_2 satisfying $E(XI_{\{Y \in B\}}) = E(h(Y)I_{\{Y \in B\}})$, for all $B \subset \mathbb{R}$. Then for any $y \in D_2$, taking $B = \{y\}$, one easily sees that the above equality reduces to $\sum_{x \in D_1} xp(x,y) = h(y)p_2(y)$, implying that $h(y) = \sum_{x \in D_1} xp(x,y)/p_2(y) = g(y)$. What all these mean is that, instead of defining the conditional expectations $E(X \mid Y = y)$, for various values $y \in D_2$, using the classical formula of conditional probability, we could have also defined it simply as a real valued function on D_2 satisfying (9.1.1) and we would still get the same thing. And the advantage of the second approach is that it does not require that there be an apriori notion of conditional distribution of X, given $Y = y$ and therefore avoids the need to use the classical formula of conditional probability, whose validity is restricted to conditioning only by events of positive probability. In other words, a definition of conditional expectation along the proposed lines can be made without any constraints on the conditioning random variable Y, that are usually required for the validity of classical formula for conditional probability. However this new approach does bring up some technical requirements. Firstly, we know by now that we cannot allow in our defining formula (9.1.1), $I_{\{Y \in B\}}$ for any $B \subset \mathbb{R}$. We have to restrict to $I_{\{Y \in B\}}$ where $B \subset \mathbb{R}$ is a Borel set, or equivalently, to I_A, where $A \in \sigma(Y)$. Secondly, to be able to talk about $E(g(Y)I_A)$, $A \in \sigma(Y)$, we can allow only measurable functions $g : \mathbb{R} \to \mathbb{R}$. Note that our g above was defined only for $y \in D_2$, but we can easily extend g to all of $\mathbb{R}$, by simply definng, say, $g(y) = 0$, for $y \notin D_2$. This would give a borel measurable $g : \mathbb{R} \to \mathbb{R}$. Thus, what we have established is that, given random variables X and Y defined on some probability space, if we can get a borel measurable $g : \mathbb{R} \to \mathbb{R}$ satisfying

$$E(XI_A) = E(g(Y)I_A), \text{ for all } A \in \sigma(Y), \tag{9.1.2}$$

then $g(y)$ gives us a possible candidate for $E(X \mid Y = y)$, that agrees with the classical definition in the special case when both X and Y are discrete.

Once we agree upto this point, a new window opens up. We do not necessarily have to talk about conditional expectation of a random variable X, given a random variable Y. Assuming that $(\Omega, \mathcal{A}, P)$ is the underlying probability space, we

9.1. Conditional Expectation

can now talk about conditional expectation of a random variable X, given a sub-σ-field $\mathcal{G} \subset \mathcal{A}$. The restrictions discussed in the previous paragraph would now have to be simply modified as follows. The condition $A \in \sigma(Y)$ in (9.1.2) should now be replaced by $A \in \mathcal{G}$. Secondly, recall that the random variables $g(Y)$, with $g: \mathbb{R} \to \mathbb{R}$ borel measurable, represent precisely all the $\sigma(Y)$-measurable random variables. Thus, in order to define conditional expectation, given $\mathcal{G}$, we should replace the random variable $g(Y)$ in (9.1.2) by a $\mathcal{G}$-measurable random variable.

Let us touch upon one last point before we go to the formal definition. At the start, when we discussed the case of two discrete random variables X and Y and recalled the classical formula $E(X \mid Y = y) = \sum_{x \in D_1} x p(x,y)/p_2(y)$ for conditional expectation, we did realize that the sum here may not be defined. But we ignored the issue then by simply assuming that these sums all exist and are finite. However, it is clear that we need to have some apriori condition, which will guarantee that they all exist and are finite. One simple sufficient condition is to assume that the series $\sum_{x \in D_1} x p_1(x)$ converges absolutely, that is, the random variable X is integrable. We are going to incorporate this assumption in our general definition also and define conditional expectations only for integrable random variables. Indeed, just as in the classical formulation, the integrabiltiy condition will guarantee existence of conditional expectations in the proposed formulation also. With all these, here is now our definition of conditional expectation.

DEFINITION 9.1.1. For an integrable random variable X on a probability space $(\Omega, \mathcal{A}, P)$ and a sub-σ-field $\mathcal{G} \subset \mathcal{A}$, the *conditional expectation of X given $\mathcal{G}$*, written $E(X \mid \mathcal{G})$, is defined to be a $\mathcal{G}$-measurable random variable Y such that

$$\int_G Y\, dP = \int_G X\, dP \quad \text{for all } G \in \mathcal{G}. \tag{9.1.3}$$

In the special case when $\mathcal{G}$ is the σ-field generated by a family $\{Y_\alpha,\ \alpha \in \Lambda\}$ of random variables, $E(X \mid \mathcal{G})$ is called the *conditional expectation of X, given $\{Y_\alpha, \alpha \in \Lambda\}$*, written $E(X \mid Y_\alpha, \alpha \in \Lambda)$.

Our first job is to show that $E(X \mid \mathcal{G})$, as defined above, exists. This is where integrability of X comes into play. Since X is integrable, $\mu(A) = \int_A X\, dP$, $A \in \mathcal{A}$, defines a real valued set function on $\mathcal{A}$, which is a finite signed measure and is absolutely continuous with respect to P. Restricting both μ and P to the sub-σ-field $\mathcal{G} \subset \mathcal{A}$ will give us a finite signed measure μ on $\mathcal{G}$, which is absolutely continuous with respect to P on $\mathcal{G}$. But then, by the Radon-Nikodym Theorem (Theorem 6.1.34), we will get a $\mathcal{G}$-measurable integrable random variable Y such that $\mu(G) = \int_G Y\, dP$ for all $G \in \mathcal{G}$. From the definition of μ, it is clear that this is exactly the condition (9.1.3) required above. Thus, Y meets the requirements in Definition 9.1.1 and is, therefore, a candidate for the conditional expectation $E(X \mid \mathcal{G})$. As we know, such a Y is unique upto P-null sets. To summarize, what we have just shown is that $E(X \mid \mathcal{G})$ exists, for any integrable random variable X on $(\Omega, \mathcal{A}, P)$ and any sub-σ-field $\mathcal{G} \subset \mathcal{A}$. Further, $E(X \mid \mathcal{G})$ is integrable and unique upto P-null sets. Also, by taking $G = \Omega$ in (9.1.3), one gets

$$E\big(E(X \mid \mathcal{G})\big) = E(X) \qquad (9.1.4)$$

REMARK 9.1.2. While we saw above that integrability of the random variable X plays a crucial role in asserting existence of its conditional expectation $E(X \mid \mathcal{G})$, we would like to just point out here that the same can also be done for any non-negative random variable. Indeed, given any non-negative random variable X, the non-negative set function $\mu(A) = \int_A X dP$, $A \in \mathcal{A}$, defines a measure on $\mathcal{A}$ (σ-finite, if X is non-negative real valued), which is absolutely continuous with respect to P. The restriction of μ to a sub-σ-field $\mathcal{G}$ will give a measure, which is absloutely continuous with respect to P on $\mathcal{G}$. So, as before, by the Radon-Nikodym Theorem (see Theorem 6.1.3 and Exercise 6.1.38), we will have a $\mathcal{G}$-measurable non-negative random variable Y such that $\int_G Y dP = \mu(G)$, for all $G \in \mathcal{G}$, equivalently,

$$\int_G Y dP = \int_G X dP, \text{ for all } G \in \mathcal{G}.$$

Thus Y satisfies all the properties required in the Definition 9.1.1 of $E(X \mid \mathcal{G})$. This means that, conditional expectation $E(X \mid \mathcal{G})$ exists for all non-negative random variables X also. However, it may not be integrable. In fact, even if X is non-negative real valued, $E(X \mid \mathcal{G})$ may equal ∞ (possibly on a set of positive probability). In any case, all of these are just for the sake of records. In the sequel, whenever we talk about conditional expectations, it will be *only* for integrable random variables.

The next proposition lists some elementary properties of conditional expectations. In the statement of the proposition, all the random variables, whose conditional expectations appear, are assumed to be integrable, even though that may not be explicitly mentioned there. Also, all the equalities/inequalities are to be interpreted as holding a.s.$[P]$.

PROPOSITION 9.1.3. (Elementary Properties)
(a) If X is $\mathcal{G}$ measurable, then $E(X \mid \mathcal{G}) = X$; in particular, $E(X \mid A) = X$
(b) $E(h(Y_1,\ldots,Y_k) \mid Y_1,\ldots,Y_k) = h(Y_1,\ldots,Y_k)$, for random variables $Y_1,\cdots,Y_k$ and borel measurable function $h : \mathbb{R}^k \to \mathbb{R}$ such that $h(Y_1,\cdots,Y_k)$ is integrable.
(c) (Idempotence) $E[\,E(X \mid \mathcal{G}) \mid \mathcal{G}\,] = E(X \mid \mathcal{G})$.
(d) If $\sigma(X)$ is independent of $\mathcal{G}$, then $E(X \mid \mathcal{G}) = E(X)$; in particular, if $\mathcal{G} = \{\Omega, \emptyset\}$, then $E(X \mid \mathcal{G}) = E(X)$.
(e) (Linearity) $E(\alpha X_1 + \beta X_2 \mid \mathcal{G}) = \alpha E(X_1 \mid \mathcal{G}) + \beta E(X_2 \mid \mathcal{G})$.
(f) (Monotonicity) If $X_1 \leq X_2$ then $E(X_1 \mid \mathcal{G}) \leq E(X_2 \mid \mathcal{G})$; in particular, if X is non-negative, then so is $E(X \mid \mathcal{G})$.
(g) $|E(X \mid \mathcal{G})| \leq E(|X| \mid \mathcal{G})$ and, as a consequence, $E\big(\big|E(X \mid \mathcal{G})\big|\big) \leq E(|X|)$.
(h) If Z is bounded and $\mathcal{G}$-measurable, then $E(XZ \mid \mathcal{G}) = E(X \mid \mathcal{G}) \cdot Z$; more generally, the same holds under the weaker asumption that Z is $\mathcal{G}$-measurable and both Z and XZ are integrable.
(i) (Smoothing Property) If $\mathcal{H}$ and $\mathcal{G}$ are two sub-σ-fields with $\mathcal{H} \subset \mathcal{G}$, then $E\big(\,E(X \mid \mathcal{G}) \mid \mathcal{H}\,\big) = E(X \mid \mathcal{H})$.

9.1. Conditional Expectation

Proof. (a) is trivial, because $\mathcal{G}$-measurability of X implies that X itself satisfies the conditions required of Y in Definition 9.1.1. (b) and (c) both follow from (a). For (d), observe that the constant random variable $Y = E(X)$ is $\mathcal{G}$-measurable and independence of $\sigma(X)$ and $\mathcal{G}$ implies that for any $G \in \mathcal{G}$, we have $\int_G X dP = E(XI_G) = E(X)P(G) = \int_G E(X) dP$. The second part of (d) follows now because $\sigma(X)$ is always independent of $\mathcal{G} = \{\Omega, \emptyset\}$. For (e), denoting $Y_1 = E(X_1 \mid \mathcal{G})$ and $Y_2 = E(X_2 \mid \mathcal{G})$, the random variable $Y = \alpha Y_1 + \beta Y_2$ is $\mathcal{G}$-measurable and clearly satisfies (9.1.3) with $X = \alpha X_1 + \beta X_2$. For (f), denoting $Y_1 = E(X_1 \mid \mathcal{G})$ and $Y_2 = E(X_2 \mid \mathcal{G})$, condition (9.1.3) gives $\int_G Y_1 dP = \int_G X_1 dP \leq \int_G X_2 dP = \int_G Y_2 dP$, for all $G \in \mathcal{G}$. Since Y_1 and Y_2 are both $\mathcal{G}$-measurable, Proposition 3.2.19 asserts that $Y_1 \leq Y_2$, a.s.$[P]$. Using $-|X| \leq X \leq |X|$ and using (e) and (f), one gets the first inequality of (g). The other inequality in (g) follows by taking expectation of both sides and using (9.1.4). For (h), we prove it under the weaker hypothesis that Z is $\mathcal{G}$-measurable and both Z and XZ are integrable. Denoting $Y = E(X \mid \mathcal{G})$, the random variable YZ is $\mathcal{G}$-measurable under our hyothesis. We need only show that $\int_G YZ dP = \int_G XZ dP$, for all $G \in \mathcal{G}$. In case $Z = I_{\overline{G}}$ for $\overline{G} \in \mathcal{G}$, the equally holds by using the set $G \cap \overline{G} \in \mathcal{G}$ in (9.1.3). By linearity of integrals, this then extends easily to the case when Z is a real-valued $\mathcal{G}$-measurable simple function. Finally, for general Z, we know that there is a sequence $\{Z_n\}$ of real-valued $\mathcal{G}$-measurable simple functions with $|Z_n| \leq |Z|$, such that $Z_n \to Z$. The proof can now be easily completed using DCT. To prove (i), denoting $Y = E(X \mid \mathcal{G})$, we have that $Z = E(Y \mid \mathcal{H})$ is an $\mathcal{H}$-measurable random variable that satisfies $\int_H Z dP = \int_H Y dP = \int_H X dP$, for all $H \in \mathcal{H}$, where the last equality used the fact any $H \in \mathcal{H}$ belongs also to $\mathcal{G}$. This proves that $Z = E(X \mid \mathcal{H})$, as asserted. □

In all of the following four exercises, X denotes an integrable random variable on a probability space $(\Omega, \mathcal{A}, P)$ and $\mathcal{G}$ denotes a sub-σ-field of $\mathcal{A}$.

EXERCISE 9.1.4. Show that $Y = E(X \mid \mathcal{G})$ if and only if Y is a $\mathcal{G}$-measurable random variable that satisfies $E(YZ) = E(XZ)$ for all bounded $\mathcal{G}$-measurable random variables Z.

EXERCISE 9.1.5. Show that if Y is a $\mathcal{G}$-measurable integrable random variable such that the equality $\int_C Y dP = \int_C X dP$ holds for all C belonging to either a semifield or a π-class generating $\mathcal{G}$, then $Y = E(X \mid \mathcal{G})$.

EXERCISE 9.1.6. [*Notation:* If $\mathcal{H}_1$ and $\mathcal{H}_2$ are two σ-fields on Ω, then $\mathcal{H}_1 \vee \mathcal{H}_2$ denotes the smallest σ-field containing both $\mathcal{H}_1$ and $\mathcal{H}_2$.] Show that if $\mathcal{H}$ is a sub-σ-field of $\mathcal{A}$, which is independent of $\sigma(X) \vee \mathcal{G}$, then $E(X \mid \mathcal{G} \vee \mathcal{H}) = E(X \mid \mathcal{G})$. [*Hint:* Sets $G \cap H$, $G \in \mathcal{G}$ and $H \in \mathcal{H}$ constitute a π-class generating $\mathcal{G} \vee \mathcal{H}$]

EXERCISE 9.1.7. Let $\{\mathcal{G}_\alpha, \alpha \in \Lambda\}$ be any family of sub-σ-fields of $\mathcal{A}$. Show that the family $\{X_\alpha = E(X \mid \mathcal{G}_\alpha), \alpha \in \Lambda\}$ of random variables is uniformly integrable.

The reader must have recognized that some of the results of Proposition 9.1.3 are essentially saying that many of the properties of expectation do carry over to conditional expectation as well. The next result is another one along similar lines. As mentioned above, all such results are to be interpreted as a.s. results.

PROPOSITION **9.1.8.** Let $(\Omega, \mathcal{A}, P)$ be a probability space and $\mathcal{G} \subset \mathcal{A}$ a sub-σ-field.
(a) (MCT for conditional expectation) If $\{X_n\}$ is a sequence of non-negative random variables increasing to an integrable random variable X, then each X_n is integrable and $0 \leq E(X_n \mid \mathcal{G}) \uparrow E(X \mid \mathcal{G})$ a.s.
(b) (Fatou's lemma for conditional expectation). For any sequence $\{X_n\}$ of non-negative integrable random variables, $E(\liminf X_n \mid \mathcal{G}) \leq \liminf E(X_n \mid \mathcal{G})$ a.s.
(c) (DCT for conditional expectation) Let $\{X_n\}$ be a sequence of random variables such that $X_n \to X$ a.s. If Y is an integrable random variable such that $|X_n| \leq Y$, for all n, then $E(|X_n - X| \mid \mathcal{G}) \to 0$ a.s. In particular, $E(X_n \mid \mathcal{G}) \to E(X \mid \mathcal{G})$ a.s.

REMARK **9.1.9.** Recall that, under the hypothesis of (c), DCT tells us that $\{X_n\}$ converges in L_1 to X, from which, using parts (d) and (f) of Proposition 9.1.3, one easily gets L_1-convergence of $\{E(X_n \mid \mathcal{G})\}$ to $E(X \mid \mathcal{G})$.

REMARK **9.1.10.** If, as noted in Remark 9.1.2, we had defined conditional expectations for non-negative (but not necessarily integrable) random variables as well, then part (a) of Proposition 9.1.8 would still hold (without requiring integrability of X).

Proof. (a) That each X_n is integrable is clear from $0 \leq X_n \leq X$ and assumed integrability of X. Denoting $Y_n = E(X_n \mid \mathcal{G})$, Proposition 9.1.3 (f) asserts that each Y_n is non-negative a.s. and $\{Y_n\}$ is a.s. non-decreasing, so that, if we put $Y = \limsup_n Y_n$, then $Y_n \uparrow Y$ a.s. By definition, Y is clearly $\mathcal{G}$-measurable. Moreover, using MCT for integrals, we will get

$$\int_G Y\,dP = \lim_n \int_G Y_n\,dP = \lim_n \int_G X_n\,dP = \int_G X\,dP, \quad \text{for all } G \in \mathcal{G},$$

yielding that $Y = E(X \mid \mathcal{G})$ and hence proving the required result.
(b) From (a), one easily gets $E(\liminf X_n \mid \mathcal{G}) = \lim_n E(\inf_{k \geq n} X_k \mid \mathcal{G})$ a.s. Since, for each n, we have $\inf_{k \geq n} X_k \leq X_n$ and hence $E(\inf_{k \geq n} X_k \mid \mathcal{G}) \leq E(X_n \mid \mathcal{G})$, by part (f) of Proposition 9.1.3, we get the required result by letting $n \to \infty$.
(c) The hypotheses imply that $2Y - |X_n - X|$, $n \geq 1$, are non-negative random variables bounded above by $2Y$. Applying (b) and the fact that $|X_n - X| \to 0$ a.s., we get $2E(Y \mid \mathcal{G}) \leq 2E(Y \mid \mathcal{G}) - \limsup E(|X_n - X| \mid \mathcal{G})$. Since Y is integrable, $E(Y \mid \mathcal{G})$ is also integrable and hence finite a.s. Using this in the last inequality, we get $\limsup E(|X_n - X| \mid \mathcal{G}) \leq 0$ a.s. From this, one can easily deduce that $E(|X_n - X| \mid \mathcal{G}) \to 0$ a.s. Using parts (e) and (f) of Proposition 9.1.3 now, one gets $E(X_n \mid \mathcal{G}) \to E(X \mid \mathcal{G})$ a.s. □

The next result extends Jensen's inequality 3.4.10 to conditional expectations. It is a very useful result and will be particularly instrumental in deriving certain properties of conditional expectation as a linear operator. In order to prove it,

9.1. Conditional Expectation

we need to use a new fact, which is essentially an improvement on Lemma 3.4.11. This is stated and argued in the following remark.

REMARK 9.1.11. Recall that, according to Lemma 3.4.11, any convex functiion $\varphi : \mathbb{R} \to \mathbb{R}$ is continuous and further, for each $x \in \mathbb{R}$, there exists an affine function $\psi_x(y) = \alpha_x + \beta_x y$, such that $\psi_x(y) \leq \varphi(y)$ for all $y \in \mathbb{R}$, with equality holding at $y = x$. It is clear then that φ equals the pointwise supremum of the family $\{\psi_x, x \in \mathbb{R}\}$ of affine functions. However, what is actually true, though not at all clear, is that φ actually equals the pointwise supremum of just the countable family $\{\psi_r, r \in \mathbb{Q}\}$ of affine functions. To see this, let us first note that by the properties of the functions $\{\psi_r, r \in \mathbb{Q}\}$, their pointwise supremum $\sup_{r \in \mathbb{Q}} \psi_r$ agrees with φ on the set of rationals. Further, one can easily verify that the function $\sup_{r \in \mathbb{Q}} \psi_r$, being supremum of affine functions, is convex on $\mathbb{R}$ and therefore continuous, by Lemma 3.4.11. We already know that φ is continuous. This implies that φ must agree with the function $\sup_{r \in \mathbb{Q}} \psi_r$ everywhere on $\mathbb{R}$, proving the claim. The significance of this observation lies in the fact that any real-valued convex function on $\mathbb{R}$ is shown to equal the pointwise supremum of a countable family of affine functions. In the context of Jensen's inequality for conditional expectations, the importance of this lies in the fact that, somewhere in the proof, we will have to deal with only a countable number of null sets.

THEOREM 9.1.12. (Jensen's Inequality for Conditional Expectation) Let X be an integrable random variable on $(\Omega, \mathcal{A}, P)$ and $\mathcal{G}$ be a sub-σ-field of $\mathcal{A}$. Then for any convex function $\varphi : \mathbb{R} \to \mathbb{R}$ such that $\varphi(X)$ is integrable,

$$\varphi\left(E(X \mid \mathcal{G})\right) \leq E\left(\varphi(X) \mid \mathcal{G}\right).$$

Proof. Let $\{\psi_r, r \in \mathbb{Q}\}$ be the family of affine functions discussed in Remark 9.1.11. Then, for every rational r, we have $\varphi \geq \psi_r$ pointwise and therefore

$$E\big(\varphi(X) \mid \mathcal{G}\big) \geq E\big(\psi_r(X) \mid \mathcal{G}\big) = \psi_r\big(E(X \mid \mathcal{G})\big), \quad (9.1.5)$$

where the last equality follows from linearity of conditional expectation and the fact that ψ_r is an affine function. Now note that, as always is the case with conditional expectations, (9.1.5) is only an a.s. inequality. So, for each $r \in \mathbb{Q}$, let N_r be the P-null set, outside of which (9.1.5) holds. But then, $N = \bigcup_{r \in \mathbb{Q}} N_r$ is also a P-null set such that, outside of N, the inequality (9.1.5) holds for all $r \in \mathbb{Q}$, and therefore, we will have, outside of N,

$$E\big(\varphi(X) \mid \mathcal{G}\big) \geq \sup_r \psi_r\big(E(X \mid \mathcal{G})\big) = \varphi\big(E(X \mid \mathcal{G})\big), \quad (9.1.6)$$

where the last equality is a consequence of what was estabilished in Remark 9.1.11. This completes the proof. □

We now apply above Jensen's inequality to get a very important property of conditional expectations. We know that, for each $p \in [1, \infty)$, the function $x \mapsto |x|^p$ is a convex function on $\mathbb{R}$. Applying Jensen's inequality 9.1.12 with this convex function, we will get the inequality $|E(X \mid \mathcal{G})|^p \leq E(|X|^p \mid \mathcal{G})$ a.s., for each $p \in [1, \infty)$, any $X \in L_p = L_p(\Omega, \mathcal{A}, P)$ and any sub-σ-field $\mathcal{G} \subset \mathcal{A}$. Taking expectation of both sides, we get $E(|E(X \mid \mathcal{G})|^p) \leq E(E(|X|^p \mid \mathcal{G})) = E(|X|^p)$, using (9.1.4) for the last equality. This, of course, gives $\|E(X \mid \mathcal{G})\|_p \leq \|X\|_p$. Analogue of this for $p = \infty$ is trivial. Indeed, if $X \in L_\infty$, then $|X| \leq \|X\|_\infty$ a.s., by the definition and this implies, by parts (f) and (g) of Proposition 9.1.3, that $|E(X \mid \mathcal{G})| \leq \|X\|_\infty$ a.s. and that finally gives us $\|E(X \mid \mathcal{G})\|_\infty \leq \|X\|_\infty$. The following important result nicely captures the above findings.

THEOREM **9.1.13.** (Contraction Property of Conditional Expectation in L_p): Let $(\Omega, \mathcal{A}, P)$ be a probability space and $\mathcal{G} \subset \mathcal{A}$ a sub-σ-field. Then, for all $p \in [1, \infty]$,

$$X \in L_p \implies E(X \mid \mathcal{G}) \in L_p \text{ and further, } \|E(X \mid \mathcal{G})\|_p \leq \|X\|_p. \quad (9.1.7)$$

REMARK **9.1.14.** Theorem 9.1.13 and some parts of Proposition 9.1.3 can be viewed in the following way. Given a probability space $(\Omega, \mathcal{A}, P)$ and a sub-σ-field $\mathcal{G} \subset \mathcal{A}$, one can think of $E(\cdot \mid \mathcal{G})$, the so-called '*conditional expectation operator*', as a map on the Banach space $L_1(\Omega, \mathcal{A}, P)$ into (indeed, onto) its subspace $L_1(\Omega, \mathcal{G}, P)$. Parts of Proposition 9.1.3 asserted that it is a linear, idempotent and positive operator. And now Theorem 9.1.13 says further that for all $p \in [1, \infty]$, it is a contraction from $L_p(\Omega, \mathcal{A}, P)$ to $L_p(\Omega, \mathcal{G}, P)$. For $p = 2$, this map has a further special feature which would be clear after the next result.

THEOREM **9.1.15.** Let $(\Omega, \mathcal{A}, P)$ be a probability space and let $X \in L_2(\Omega, \mathcal{A}, P)$. Then, for any sub-$\sigma$-field $\mathcal{G} \subset \mathcal{A}$ and any $\mathcal{G}$-measurable random variable Y,

$$\|X - E(X \mid \mathcal{G})\|_2 \leq \|X - Y\|_2. \quad (9.1.8)$$

Moreover, strict inequality holds in (9.1.8) unless $Y = E(X \mid \mathcal{G})$.

Proof. Without loss of generality we shall assume that $Y \in L_2$, since otherwise, the right hand side of (9.1.8) will equal $+\infty$ and the inequality will hold trivially. Denoting $E(X \mid \mathcal{G})$ by Z, observe that $Z \in L_2$ by Theorem 9.1.13 and therefore, both $(X - Z)$ and $(Z - Y)$ belong to L_2, implying, by Hölder's inequality, that $(X - Z)(Z - Y)$ is integrable. Using (9.1.4), part (h) of Proposition 9.1.3 and noting that $Z - Y$ is $\mathcal{G}$-measurable, we get

$$E\big((X - Z)(Z - Y)\big) = E\big((Z - Y)E((X - Z) \mid \mathcal{G})\big) = E\big((Z - Y)(Z - Z)\big) = 0.$$

Consequently, one gets $E\big((X-Y)^2\big) = E\big((X-Z)^2\big) + E\big((Z-Y)^2\big) \geq E\big((X-Z)^2\big)$. This proves the required inequality $\|X - Y\|_2 \geq \|X - Z\|_2$. Also, from the last step, it is clear that the inequality is strict unless $Y = Z$. □

REMARK **9.1.16.** The inequality of Theorem 9.1.15 has a couple of interesting interpretations. Firstly, we already know that, for all $p \in [1, \infty]$, the L_p spaces on any measure space $(\Omega, \mathcal{A}, \mu)$ are Banach spaces, that is, complete normed linear spaces. It is perhaps worth mentioning that we are discussing *real* L_p spaces here. The case $p = 2$ is a bit more special in the sense that $\langle f, g \rangle = \int f g \, d\mu$, $f, g \in L_2$ defines an inner product on L_2 such that $\|f\|_2 = \sqrt{\langle f, f \rangle}$. This makes L_2 a complete

9.1. Conditional Expectation

inner product space, or, what is commonly called a 'Hilbert space'. In particular, the L_2 space of a probability space $(\Omega, \mathcal{A}, P)$ is a Hilbert space with the inner product $\langle X, Y \rangle = E(XY)$. Further, if $\mathcal{G} \subset \mathcal{A}$ is any sub-σ-field, then it is easy to see that $\mathcal{H}' = L_2(\Omega, \mathcal{G}, P)$ is a (closed) subspace of the Hilbert space $\mathcal{H} = L_2(\Omega, \mathcal{A}, P)$. Now, what Theorem 9.1.15 says is that, given any $X \in \mathcal{H}$, the unique $Z \in \mathcal{H}'$ that minimizes $\|X - Y\|_2$ over all $Y \in \mathcal{H}'$ is given by $Z = E(X \mid \mathcal{G})$. Further, using similar argument as was used in the proof, one can see that this minimizer $Z \in \mathcal{H}'$ also satisfies $\langle X - Z, Y \rangle = E((X - Z)Y) = 0$, for all $Y \in \mathcal{H}'$, that is, $X - Z$ is 'orthogonal' to $\mathcal{H}'$. In the language of Hilbert Space theory, one would describe this by saying that, for any $X \in \mathcal{H}$, the 'orthogonal projection' of X onto the subspace $\mathcal{H}'$ is given by $Z = E(X \mid \mathcal{G})$.

Another interpretation of Theorem 9.1.15, which is closely related to the above, is in the context of prediction in statistical theory. From a statistical standpoint, X can be viewed as an "unknown" quantity arising out of the radom experiment $(\Omega, \mathcal{A}, P)$. To say that X is $\mathcal{A}$-measurable means the σ-field $\mathcal{A}$ carries all the information on X, in the sense that, if one has enough information to know, for every $A \in \mathcal{A}$, whether A occurs or not, then one has complete information on X. However, if one does not have complete information (which is often the case), then the only option would be to "predict" X on the basis of available information. The sub-σ-field $\mathcal{G}$ represents the available information in the sense, as above, that the information available is just enough to be able to know, for every $G \in \mathcal{G}$, whether G occurs or not. Clearly any predictor based on information captured by $\mathcal{G}$ has to be a $\mathcal{G}$-measurable random variable and what Theorem 9.1.15 says is that among all these admissible predictors, $E(X \mid \mathcal{G})$ is the "best" in the sense that it comes closest to X in terms of the 'mean square error', that is, L_2-distance. In the two extreme special cases, we have $E(X \mid \mathcal{A}) = X$ and $E(X \mid \{\Omega, \emptyset\}) = E(X)$, that is, when one has complete information, the best predictor of X is X itself, whereas, if one has nil information (captured by trivial σ-field), the best predictor of X is the constant $E(X)$.

Having given a very general definition of conditional expectation, that takes full advantage of the measure theoretic set-up and, in particular, of Radon-Nikodym Theorem and having studied its properties rather extensively, let us get back to where we started from and try to reconnect. We started with the simple definition of conditional expectations that one encounters in a first course in probability. We considered two discrete random variables X and Y and used the classical definition of conditional probability to first define the conditional p.m.f. of X, given $Y = y$, and then defined $E(X \mid Y = y)$ as expectation with respect to the distribution given by this conditional p.m.f. A careful examination of what is being done there would convince the reader that it is not necessary for X to be discrete for the classical approach to work. As long as Y is discrete, we can define, for any y with $P(Y = y) > 0$, the conditional probabilities $P(X \in B \mid Y = y)$, for $B \in \mathcal{B}$, using classical formula. This would give us a probability on $\mathcal{B}$, that can be legitimately regarded as the conditional distribution of X, given $Y = y$. It will therefore be prefctly alright to accept the expectation with respect to this distribution as $E(X \mid Y = y)$. Doing this for each y with $P(Y = y) > 0$, will give us a real valued function on the countable set D_2 of possible values of Y. We can extend it to a measurable function on $\mathbb{R}$, if we want, by setting it to be 0 outside of D_2. Thus, this approach thows up a real-valued measurable function g on $\mathbb{R}$ such that $g(y)$ agrees with the classical candidate for $E(X \mid Y = y)$, for all y with $P(Y = y) > 0$.

In contrast, our definition of conditional expectation does not give a real measurable function on $\mathbb{R}$; instead, what we have is a random variable denoted $E(X \mid Y)$, which is a real $\sigma(Y)$-measurable function on Ω satisfying a certain property. But, there is no $E(X \mid Y = y)$ visible in our definition. This raises a big question now. How to reconcile between these two apparently very different notions of conditional expectaions, namely, the classical notion of conditional expectation as a real function of different real values of Y vis-a-vis the notion of conditional expectation as a random variable?

Let us first start with the classical notion, where we can define $g(y) = E(X \mid Y = y)$ for $y \in \mathbb{R}$ and we can ensure that $g : \mathbb{R} \to \mathbb{R}$ is measurable. Clearly then, $g(Y)$ will be a $\sigma(Y)$-measurable random variable and we have checked that, at least when X is also discrete, this $g(Y)$ indeed turns out to be $E(X \mid Y)$, as per Definition 9.1.1. On the other hand, $E(X \mid Y)$, as per Definition 9.1.1, always exists, as long as X is integrable. Now to define $E(X \mid Y = y)$, what we do is the following. Since by definition, $E(X \mid Y)$, for any integrable X, is a $\sigma(Y)$-measurable real random variable, Theorem 3.5.13 asserts that there has to be a borel measurable $g : \mathbb{R} \to \mathbb{R}$, such that $E(X \mid Y) = g(Y)$. We simply define $E(X \mid Y = y)$, for $y \in \mathbb{R}$, by setting $E(X \mid Y = y) = g(y)$. This presents us with the concept of $E(X \mid Y = y)$ as a measurable function of y. We had already seen that, in case X and Y are both discrete, then any measurable function $g : \mathbb{R} \to \mathbb{R}$, such that $g(y)$ agrees with the classical definition of $E(X \mid Y = y)$, for all y with $P(Y = y) > 0$, is a candidate for the g talked about above, that is, $g(Y)$ is a candidate for $E(X \mid Y)$ as in Definition 9.1.1.

To summarize, in the classical approach, one defines conditional expectations "locally" at each possible value of Y. But this is possible only when Y is discrete. For general Y, the measure theoretic setting and Radon-Nikodym Theorem are used to give a "global" definition of conditional expectation as a $\sigma(Y)$-measurable random variable and then get a local definition, if needed, out of the global, using Theorem 3.5.13. The next definition formally states this where conditioning is done, given a finite number of random variables, instead of just one. Of course, instead of Theorem 3.5.13, we need to use its extension as in Exercise 3.5.15 (b).

DEFINITION **9.1.17.** Given random varables $Y_1, \ldots, Y_k$ and an integrable random variable X, all on a probability space $(\Omega, \mathcal{A}, P)$, the integrable random variable $E(X \mid Y_1, \ldots, Y_k)$ is, by definition, $\sigma(\{Y_1, \ldots, Y_k\})$-measurable, and consequently (Exercise 3.5.15, part (b)), there exists a borel measurable function $g : \mathbb{R}^k \to \mathbb{R}$, such that, $E(X \mid Y_1, \ldots, Y_k) = g(Y_1, \ldots, Y_k)$. With any g as above, we define
$$E(X \mid Y_1 = y_1, \ldots, Y_k = y_k) = g(y_1, \ldots, y_k), \text{ for } (y_1, \ldots, y_k) \in \mathbb{R}^k. \quad (9.1.9)$$

9.2 Conditional Probabilities and Regular Conditional Probabilities

In the usual sequence of things, probability comes first and then expectation is defined as an integral with respect to probability. So, it would be natural for the reader to wonder why we defined conditional expectation first rather than talking first about conditional probability and then define conditional expectation as an integral with respect to that. The problem with that lies in the fact that, conditional probability here is not going to be limited to the classical notion, which is severely restricted by the requirement that the conditioning event should have positive probability. We have already seen that this constraint makes its appplicability limited to very special cases. To have wider applicability, we must have a notion of conditional probability that works with no such restriction. We are going to do that here and the reader will soon see that having obtained a general definition of conditional expectations without having to define conditional probabilities first will actually work to our advantage. The question of whether eventually conditional expectation can be realized as an integral with respect to conditional probability, is somewhat involved and will be taken up in due course. Without mincing words any more, let us go straight to the definition of conditional probability in this wider framework. The reader may rest assured that the classical definition of conditional probability will be captured as a special case. As pointed out at the start of previous section, the main idea here is that, for any event A, its probability can be seen as an expectation, namely, $P(A) = E(I_A)$.

DEFINITION 9.2.1. Let $(\Omega, \mathcal{A}, P)$ be a probability space and $\mathcal{G} \subset \mathcal{A}$ a sub-σ-field. For any $A \in \mathcal{A}$ the *conditional probability of A, given* $\mathcal{G}$, written $P(A \mid \mathcal{G})$, is defined as $P(A \mid \mathcal{G}) = E(I_A \mid \mathcal{G})$.

According to the definition, $P(A \mid \mathcal{G})$ is, therefore, just an integrable $\mathcal{G}$-measurable random variable Y that satisfies

$$P(A \cap G) = \int_G Y \, dP, \quad \text{for all } G \in \mathcal{G}. \tag{9.2.1}$$

Moreover, $P(A \mid \mathcal{G})$ is unique upto a P-null set. The next proposition gives a set of elementary facts about conditional probabilities, that are easy consequences of properties of conditional expectations and so we omit the proof. Since, $P(A \mid \mathcal{G})$, for each $A \in \mathcal{A}$, is only unique upto a P-null set, the equalities/inequalities in the Proposition are to be interpreted as holding only $[P]$-a.s.

PROPOSITION 9.2.2. (Elementary Properties of Conditional Probability)
(a) $0 \leq P(A \mid \mathcal{G}) \leq 1$, for all $A \in \mathcal{A}$, and $P(\Omega \mid \mathcal{G}) = 1$, $P(\emptyset \mid \mathcal{G}) = 0$.
(b) $A_1 \subset A_2$ implies that $P(A_1 \mid \mathcal{G}) \leq P(A_2 \mid \mathcal{G})$.
(c) If $\{A_n\}$ is a sequence of disjoint sets in $\mathcal{A}$, then $P(\bigcup_n A_n \mid \mathcal{G}) = \sum_n P(A_n \mid \mathcal{G})$.
In particular, $P(A \mid \mathcal{G}) + P(A^c \mid \mathcal{G}) = 1$, for all $A \in \mathcal{A}$.
(d) If $A \in \mathcal{A}$ is independent of $\mathcal{G}$, then $P(A \mid \mathcal{G}) = P(A)$.

Proposition 9.2.2, specially parts (a) and (c), seems to suggest that conditional probability, as a set function $P(\cdot \mid \mathcal{G})$ on $\mathcal{A}$, satisfies all the properties of a probability. Notwithstanding the fact that $P(A \mid \mathcal{G})$, for each $A \in \mathcal{A}$, is a random variable, it is natural to ask the following question. Do we have a random set function $P(\cdot \mid \mathcal{G})$ which, for each ω, is a probability on $\mathcal{A}$, such that, for fixed $A \in \mathcal{A}$, it gives is precisely the random variable $P(A \mid \mathcal{G})$? To get a deeper understanding of what is at stake here, let us recall, first of all, that $P(A \mid \mathcal{G})$, for each $A \in \mathcal{A}$, is only unique upto a P-null set, that is, for each $A \in \mathcal{A}$, we possibly have several choices for $P(A \mid \mathcal{G})$, with any two of them allowed to differ only on a P-null set. This means that, for every $A \in \mathcal{A}$, we can choose a 'version' of $P(A \mid \mathcal{G})$ as we like, by which we mean that we choose one from among all those $\mathcal{G}$-measurable random variables, each of which is a possible candidate for $P(A \mid \mathcal{G})$. In the sequel whenever we use the phrase "choose a version", it should be understood to mean exactly this. Having chosen a version for each $A \in \mathcal{A}$, suppose we now define a function P^* on $\Omega \times \mathcal{A}$ where $P^*(\cdot, A)$, for each $A \in \mathcal{A}$, represents our chosen version of the $\mathcal{G}$-measurable random variable $P(A \mid \mathcal{G})$. The important issue now is whether it is clear that we can choose our versions, for each $A \in \mathcal{A}$, judiciously enough so that, for each ω, the set function $P^*(\omega, \cdot)$ is a probability on $\mathcal{A}$. To put it in simple terms, can we choose our conditional probabilities $P(A \mid \mathcal{G})$, for various $A \in \mathcal{A}$, so that it is a probability in A? In the classical definition of $P(A \mid B)$, subject to $P(B) > 0$, we know that the set function $P(\cdot \mid B)$ is indeed a probability. However, the situation here is not going to be that simple at all. In fact, we are soon going to see the difficulties involved. That is, in fact, the reason why one does not attempt to define conditional probabilities first and then define conditional expectations as integrals, because conditional probabilities may not, in general, be probabilities after all. This is precisely the question we asked above. The proposition above may mislead the reader to believing something that may not actually be true, unless of course the reader is very careful.

To get to the real difficulties now, while it is true that properties (a) and (c) of Proposition 9.2.2 makes one tempted to believe that conditional probabilities behave like probabilities, one has to be very careful about the interpretation of those properties. One should not forget the warning that all the equalities/inequalities appearing there hold only a.s., even though it is not explicitly mentioned. In particular, what property (c) actually asserts is that given a disjoint sequence $\{A_n\}$ of $\mathcal{A}$-sets, the stated countable additivity holds outside of a null set. One may think and rightly so, that by suitably modifying the chosen versions for $P(A_n \mid \mathcal{G})$, $n \geq 1$, and for $P(\bigcup A_n \mid \mathcal{G})$ on that null set, we can ensure countable additivity for every ω. But the important thing is that the null set may possibly depend on the sequence $\{A_n\}$ and therefore, modifications on the chosen versions have to be done not on just one null set but on the union of all possible null sets that may arise from all possible disjoint sequences of $\mathcal{A}$-sets. Now perhaps, the reader can see the magnitude of the problem at hand. It is quite likely

9.2. Conditional Probabilities and Regular Conditional Probabilities

(and is indeed the case in most situations of interest) that one may encounter an uncountable number of null sets on which modifications of our chosen versions would have to be made. But, that won't be permissible, because an uncountable union of null sets may not belong to $\mathcal{A}$ in the first place and secondly may not be a null set, even if it belongs to $\mathcal{A}$. This should make it clear that it is by no means immediate that we have an affirmative answer to the question we asked at the beginning. As a matter of fact, in general, the answer is not affirmative. This leads to the following definition.

DEFINITION 9.2.3. Let $(\Omega, \mathcal{A}, P)$ be a probability space and let $\mathcal{G} \subset \mathcal{A}$ and $\mathcal{H} \subset \mathcal{A}$ be two sub-σ-fields of $\mathcal{A}$. A kernel $P^* : \Omega \times \mathcal{H} \to [0,1]$ is called a *Regular Conditional Probability*, abbreviated as *RCP*, on $\mathcal{H}$, given $\mathcal{G}$, if
(i) for every $\omega \in \Omega$, $P^*(\omega, \cdot)$ is a probability on $\mathcal{H}$, and,
(ii) for each $H \in \mathcal{H}$, $P^*(\cdot, H)$ is a version of $P(H \mid \mathcal{G})$.

Thus, in the language of Definition 9.2.3, the question that we asked above is about existence of RCP on $\mathcal{A}$, given $\mathcal{G}$. As was already pointed out, RCP may not exist in general. However, we are not going to give here an example that illustrates non-existence of a RCP. Instead, we will focus our attention to describing some special cases when a RCP does exist. But, before we get to that, let us first show that, if a Regular Conditional Probability exists, then conditional expectations are indeed integrals with respect to this Regular Conditional Probability. This is something that was promised in the beginning of the section.

THEOREM 9.2.4. Let $(\Omega, \mathcal{A}, P)$ be a probability space and let $\mathcal{G}$ and $\mathcal{H}$ be sub-σ-fields of $\mathcal{A}$. If P^* on $\Omega \times \mathcal{H}$ is a RCP on $\mathcal{H}$, given $\mathcal{G}$, then for any integrable $\mathcal{H}$-measurable random variable X, the integral $\int X(\omega')P^*(\cdot, d\omega')$ is a version of the conditional expectation $E(X \mid \mathcal{G})$.

Proof. By the definition of RCP, the result is true when $X = I_H$, for $H \in \mathcal{H}$. From this, one can easily get the result for real-valued $\mathcal{H}$-measurable simple functions X. The passage from here to integrable non-negative $\mathcal{H}$-measurable X is through approximating X by an increasing sequence of non-negative real-valued $\mathcal{H}$-measurable simple functions and applying MCT for integrals on one hand and for Conditional expectations on the other. Finally, a general integrable $\mathcal{H}$ random variable X is written as $X = X^+ - X^-$ and then linearity of integrals and also of conditional expectations are used to complete the proof. □

Now we proceed towards showing existence of RCP in some special settings. The next two lemmas are going to be crucial tools in our argument towards that. Dynkin's π-λ Theorem 2.3.10 is going to be used in proving both the lemmas.

LEMMA 9.2.5. Let $(\Omega, \mathcal{A}, P)$ be a probability space and let $\mathcal{G}$ and $\mathcal{H}$ be sub-σ-fields of $\mathcal{A}$. Suppose that $\mathcal{P}$ is π-system generating $\mathcal{H}$ and that $P^* : \Omega \times \mathcal{H} \to [0,1]$ satisfies (i) $P^*(\omega, \cdot)$ is a probability on $\mathcal{H}$, for every $\omega \in \Omega$, and (ii) $P^*(\cdot, S)$ is a version of $P(S \mid \mathcal{G})$, for every $S \in \mathcal{P}$. Then P^* is an RCP on $\mathcal{H}$ given $\mathcal{G}$.

Proof. As pointed out, we are going to use Dynkin's π-λ Theorem 2.3.10. Let $\mathcal{L} = \{H \in \mathcal{H} : P^*(\cdot, H) \text{ is a version of } P(H \mid \mathcal{G})\}$. By the hypothesis, we have $\mathcal{P} \subset \mathcal{L}$. We now show that $\mathcal{L}$ is a λ-system, which, by virtue of Theorem 2.3.10, will complete the proof.

That $\emptyset \in \mathcal{L}$ is trivial. Next, observe that $1 - P(H \mid \mathcal{G})$, for any $H \in \mathcal{H}$, is always a version of $P(H^c \mid \mathcal{G})$. Therefore, if for some $H \in \mathcal{H}$, $P^*(\cdot, H)$ is a version of $P(H \mid \mathcal{G})$, then $1 - P^*(\cdot, H)$ will be a version of $P(H^c \mid \mathcal{G})$. But by property of P^*, we have $1 - P^*(\cdot, H) = P^*(\cdot, H^c)$, implying that $P^*(\cdot, H^c)$ is a version of $P(H^c \mid \mathcal{G})$. This shows that $\mathcal{L}$ is closed under complementation. Finally, for any disjoint sequence $\{H_n\}$ of $\mathcal{H}$-sets, $\sum P(H_n \mid \mathcal{G})$ is always a version of $P(\bigcup_n H_n \mid \mathcal{G})$. Now, if $P^*(\cdot, H_n)$ is a version of $P(H_n \mid \mathcal{G})$, for each n, then using property (i) again, $P^*(\cdot, \bigcup_n H_n)$ will be a version of $P(\bigcup_n H_n \mid \mathcal{G})$. This shows that $\mathcal{L}$ is closed under countable disjoint unions and completes the proof. $\square$

Before going into the statement of the next Lemma, the reader may want to recall Definition 2.9.23 of probability distribution functions on $\mathbb{Q}^k$ and also Theorem 2.9.24, which asserts one-one correspondence between these and probability measures on $\mathcal{B}^k$. We will also use 'probability kernels' (as in Definition 5.3.7).

LEMMA 9.2.6. Let $(\Omega, \mathcal{G})$ be a measurable space. Suppose $F : \Omega \times \mathbb{Q}^k \to [0, 1]$ is a function that satisfies
(i) $F(\omega, \cdot)$ is a probability distribution function on $\mathbb{Q}^k$, for every $\omega \in \Omega$, and,
(ii) $F(\cdot, \mathbf{r})$ is $\mathcal{G}$-measurable, for every $\mathbf{r} \in \mathbb{Q}^k$.
For each $\omega \in \Omega$, let $P^*(\omega, \cdot)$ be the unique probability on $\mathcal{B}(\mathbb{R}^k)$ determined by $F(\omega, \cdot)$. Then P^* is a probability kernel on $\Omega \times \mathcal{B}(\mathbb{R}^k)$.

Proof. We only have to show that $P^*(\cdot, B)$ is $\mathcal{G}$-measurable, for each $B \in \mathcal{B}(\mathbb{R}^k)$. Let $\mathcal{L} = \{B \in \mathcal{B}(\mathbb{R}^k) : P^*(\cdot, B) \text{ is } \mathcal{G}\text{-measurable}\}$. That $\mathcal{L}$ is a λ-system can easily be seen using the fact that for each $\omega \in \Omega$, $P^*(\omega, \cdot)$ is a probability. Recalling the notation $\mathbf{I_{sr}}$ from Section 2.9 and Theorem 2.9.24, one can easily see, using property (ii), that $P^*(\cdot, I_{\mathbf{sr}})$ is $\mathcal{G}$-measurable for all pairs $\mathbf{s} \leq \mathbf{r}$ in $\mathbb{Q}^k$. Thus, $\mathcal{L}$ contains the class of all sets $I_{\mathbf{sr}}$ with $\mathbf{s} \leq \mathbf{r}$ in $\mathbb{Q}^k$, which is a π-system generating $\mathcal{B}(\mathbb{R}^k)$. Dynkin's π-λ Theorem 2.3.10 now completes the proof. $\square$

We are ready to prove results illustrating special cases when RCP exists. Our first result in this direction asserts existence of RCP when Ω is a borel subset of $\mathbb{R}^k$, equipped with the Borel σ-field $\mathcal{A}$ on Ω, and $\mathcal{G}$ is any sub-σ-field of $\mathcal{A}$. After this, we will extend these results a little more.

THEOREM 9.2.7. (Existence of RCP)
(a) Let $\Omega = \mathbb{R}^k$ with $\mathcal{A} = \mathcal{B}(\mathbb{R}^k)$. Given any probability P on $\mathcal{A}$ and any sub-σ-field $\mathcal{G}$, a Regular Conditional Probability on $\mathcal{A}$, given $\mathcal{G}$, exists.
(b) Let Ω be any Borel subset of $\mathbb{R}^k$ and let $\mathcal{A}$ be the Borel σ-field on Ω. Given any probability P on $\mathcal{A}$ and any sub-σ-field $\mathcal{G}$, a Regular Conditional Probability on $\mathcal{A}$, given $\mathcal{G}$, exists.

9.2. Conditional Probabilities and Regular Conditional Probabilities 311

Proof. (a) For every $\mathbf{r} \in \mathbb{Q}^k$, let $J_\mathbf{r} = \{\mathbf{x} \in R^k : \mathbf{x} \le \mathbf{r}\}$ and let $H(\cdot, \mathbf{r})$ be a version of $P(J_\mathbf{r} \mid \mathcal{G})\}$. For each pair $\mathbf{s}, \mathbf{r} \in \mathbb{Q}^k$, with $\mathbf{s} \le \mathbf{r}$, let

$$N_{\mathbf{s},\mathbf{r}} = \{\omega \in \Omega : \Delta_{\mathbf{s},\mathbf{r}} H(\omega, \cdot) < 0\},$$

where $\Delta_{\mathbf{s},\mathbf{r}} H(\omega, \cdot)$, for each fixed ω, is as in Definition 2.9.17. Also, for each $\mathbf{r} \in \mathbb{Q}^k$, denote $\mathbf{r}(n) = \mathbf{r} + (\frac{1}{n}, \ldots, \frac{1}{n})$, $n \ge 1$ and let

$$N_\mathbf{r} = \{\omega \in \Omega : H(\cdot, \mathbf{r}(n)) \not\to H(\cdot, \mathbf{r})\}.$$

Next, denoting $\mathbf{n}$, for each $n \ge 1$, to be the point with all k-coordinates equal to the natural number n, let

$$N_\infty = \{\omega \in \Omega : H(\omega, \mathbf{n}) \not\to 1 \text{ as } n \to \infty\}.$$

Finally, for each $r \in \mathbb{Q}$ and $1 \le j \le k$, let

$$N_{r,j} = \{\omega \in \Omega : H(\omega, \mathbf{r}_{n,j}) \not\to 0 \text{ as } n \to \infty\},$$

where $\mathbf{r}_{n,j}$ denotes the point whose j-th coordinate equals $-n$ and the remaining $(k-1)$ coordinates are all equal to r. From the definition of H, it should be clear that, the set N_∞ as well as all the sets in the collections $\{N_{\mathbf{r},\mathbf{s}} : \mathbf{s}, \mathbf{r} \in \mathbb{Q}^k\}$, $\{N_\mathbf{r} : \mathbf{r} \in \mathbb{Q}^k\}$ and $\{N_{r,j} : r \in \mathbb{Q}, 1 \le j \le k\}$ belong to $\mathcal{G}$ and are P-null sets. Since each of these collections contain only a countable number of sets, if we denote N to be the union of all these sets, then N will be a P-null set belonging to $\mathcal{G}$. Choose and fix any probability distribution function F_0 on R^k and define F on $\Omega \times \mathbb{Q}^k$ by

$$F(\omega, \cdot) = \begin{cases} H(\omega, \cdot), & \text{if } \omega \notin N \\ F_0(\cdot), & \text{if } \omega \in N \end{cases}$$

The function F on $\Omega \times \mathbb{Q}^k$ clearly satisfies properties (i) and (ii) of Lemma 9.2.6. Denoting P^* to be the probability kernel determined by F, as in Lemma 9.2.6, application of Lemma 9.2.5 implies that P^* is the required RCP on $\mathcal{A}$, given $\mathcal{G}$.

(b) We will use (a) to first construct a RCP on $(R^k, \mathcal{B}(R^k))$ and then modify that appropriately to get the required RCP on $(\Omega, \mathcal{A})$. But to do this, we need to "embed" $(\Omega, \mathcal{A}, P)$ and the sub-σ-field $\mathcal{G}$ into R^k. To this end, first note that, since $\mathcal{A}$ is just the restriction of $\mathcal{B}(R^k)$ to Ω, the given probability P on $\mathcal{A}$ can easily be "extended" to a probability $\widetilde{P}$ on $\mathcal{B}(R^k)$ with $\widetilde{P}(\Omega) = 1$, by simply setting $\widetilde{P}(B) = P(\Omega \cap B)$, $B \in \mathcal{B}(R^k)$. Also, given the sub-σ-field $\mathcal{G} \subset \mathcal{A}$, one can easily see that $\widetilde{\mathcal{G}} = \{B \in \mathcal{B}(R^k) : \Omega \cap B \in \mathcal{G}\}$ is a sub-σ-field of $\mathcal{B}(R^k)$, whose restriction to Ω is precisely $\mathcal{G}$. Now, that the "embedding" is done, we can use part (a) to get a RCP $\widetilde{P}^*$ on $\mathcal{B}(R^k)$, given $\widetilde{\mathcal{G}}$. Now, notice that, since $\widetilde{P}(\Omega) = 1$, we must have $\int_{R^k} \widetilde{P}^*(\omega, \Omega) \widetilde{P}(d\omega) = \widetilde{P}(\Omega) = 1$, which implies $\widetilde{P}^*(\omega, \Omega) = 1$, for all ω outside of a $\widetilde{P}$-null set $\widetilde{N}$ in $\widetilde{\mathcal{G}}$. Note that $\widetilde{N} \in \widetilde{\mathcal{G}}$ means that $N = \Omega \cap \widetilde{N} \in \mathcal{G}$ and $P(N) = \widetilde{P}(\widetilde{N}) = 0$. Now, if we choose and fix any probability P_0 on $\mathcal{A}$ and define P^* on $\Omega \times \mathcal{A}$ as

$$P^*(\omega, A) = \begin{cases} \widetilde{P}^*(\omega, A) & \text{if } \omega \in \Omega \setminus N \\ P_0(A) & \text{if } \omega \in N \end{cases}$$

then one can easily see that P^* defines a RCP on $\mathcal{A}$, given $\mathcal{G}$. □

Our next goal is to show that the above result can be extended so as to ensure the existence of a RCP on even some abstract probability spaces, as long as they are somewhat special. Recall that for any borel set $\Omega_0 \subset \mathbb{R}$, the Borel σ-field on Ω_0 is simply the class of all those borel sets in $\mathbb{R}$, that are subsets of Ω_0.

DEFINITION **9.2.8.** A measurable space $(\Omega, \mathcal{A})$ is said to be a *Standard Borel Space* if $(\Omega, \mathcal{A})$ is *borel isomorphic* to some borel subset Ω_0 of $\mathbb{R}$, equipped with its Borel σ-field, meaning that there is a bijection $f : \Omega \to \Omega_0$ such that the functions f and f^{-1} are both measurable. Such a map f is often called a *bi-measurable bijection*.

REMARK **9.2.9.** From the definition it is clear that if $(\Omega, \mathcal{A})$ is borel isomorphic to a borel set $\Omega_0 \subset \mathbb{R}$ and $f : \Omega \to \Omega_0$ is a borel isomorphism, then every $A \in \mathcal{A}$ is the inverse image, under f, of a unique borel subset B of Ω_0 and conversely, every borel subset of B of Ω_0 is the image, under f, of a unique $A \in \mathcal{A}$. In other words, $(\Omega, \mathcal{A})$ can be completely identified, as a measurable space, with $(\Omega_0, \mathcal{B}(\Omega_0))$, where $\mathcal{B}(\Omega_0)$ denotes the Borel σ-field on Ω_0. In particular, any sub-σ-field of $\mathcal{A}$ can be completely identified with a unique sub-σ-field of $\mathcal{B}(\Omega_0)$. Further, any probability P on $\mathcal{A}$ induces a probability P_0 on $\mathcal{B}(\Omega_0)$, which, in turn, induces P on $\mathcal{A}$, via the measurable map $f^{-1} : \Omega_0 \to \Omega$. Thus, even the probability space $(\Omega, \mathcal{A}, P)$ can be identified with the probability space $(\Omega_0, \mathcal{B}(\Omega_0), P_0)$.

EXERCISE **9.2.10.** If $(\Omega, \mathcal{A})$ is a standard Borel space and if $\Omega' \in \mathcal{A}$, then, denoting $\mathcal{A}'$ to be the restriction of $\mathcal{A}$ on Ω', show that $(\Omega', \mathcal{A}')$ is also a standard Borel space.

We are now going to show that results like Theorem 9.2.7, that guarantee existence of RCP, can be extended easily to abstract measurable spaces, as long they are standard Borel spaces.

THEOREM **9.2.11.** (Existence of RCP on Standard Borel Spaces)
(a) Let $(\Omega, \mathcal{A})$ be a standard Borel space. Then for any sub-σ-field $\mathcal{G} \subset \mathcal{A}$ and any probability on $\mathcal{A}$, a Regular Conditional Probability on $\mathcal{A}$, given $\mathcal{G}$, exists.
(b) Let $(\Omega, \mathcal{A}, P)$ be any probability space and $\mathcal{G}$ be any sub-σ-field of $\mathcal{A}$. Suppose $(E, \mathcal{E})$ is a standard Borel space and $f : (\Omega, \mathcal{A}) \to (E, \mathcal{E})$ is a measurable function. Then a Regular Conditional Probability on $\sigma(f)$ given $\mathcal{G}$ exists.

Proof. (a) In view of what was noted in Remark 9.2.9, the hypothesis implies that if Ω_0 is a borel subset of $\mathbb{R}$ such that $(\Omega, \mathcal{A})$ is borel isomorphic to $(\Omega_0, \mathcal{B}(\Omega_0))$ via the bi-measurable bijection $f : \Omega \to \Omega_0$, then $(\Omega, \mathcal{A}, P)$, as a probability space, can be completely identified with $(\Omega_0, \mathcal{A}_0, P_0)$, where $\mathcal{A}_0$ denotes $\mathcal{B}(\Omega_0)$, the Borel σ-field on Ω_0 and $P_0 = Pf^{-1}$ is the induced probability on $\mathcal{A}_0$. Further, $\mathcal{G}$ can be completely identified with a unique sub-σ-field $\mathcal{G}_0$ of $\mathcal{A}_0$; indeed, $\mathcal{G}_0$ is just the σ-field $\{f(G) : G \in \mathcal{G}\}$. By Theorem 9.2.7 (b), for the probability P_o on $\mathcal{A}_0$, there exists a RCP P_0^* on $\mathcal{A}_0$, given $\mathcal{G}_0$. Let us now define P^* on $\Omega \times \mathcal{A}$ by setting $P^*(\omega, A) = P_0^*(f(\omega), f(A))$. Using the fact that $f : (\Omega, \mathcal{A}) \to (\Omega_0, \mathcal{A}_0)$ is a bi-measurable bijection and using the relation between P and P_0, it is easy to

see that that P^* defines a RCP on $\mathcal{A}$, given $\mathcal{G}$, under the probability P.
(b) The hypothesis that $(E, \mathcal{E})$ is a standard Borel space means that there is a borel set $E_0 \subset \mathbb{R}$ such that, as a measurable space, $(E, \mathcal{E})$ can be completely identified with $(E_0, \mathcal{B}(E_0))$, through a bi-measurable isomorphism $g : E \to E_0$. Replacing f by $g \circ h$, if necessary, we may, therefore, assume that E itself is a borel subset of $\mathbb{R}$ and $\mathcal{E}$ is the Borel σ-field on E. Arguing as in the proof of part (a) of Theorem 9.2.7, one gets a function H on $\Omega \times \mathbb{Q}$, such that, $H(\omega, \cdot)$, for each $\omega \in \Omega$, is a probability distribution function on $\mathbb{Q}$, and $H(\cdot, r)$, for each $r \in \mathbb{Q} \cup \{\pm\infty\}$, is a version of $P\big(f^{-1}(I_{-\infty,r})\,\big|\,\mathcal{G}\big)$. Now, note that the class $\mathcal{S} = \{f^{-1}(I_{s,r}) : s \leq r,\, s, r \in \mathbb{Q} \cup \{\pm\infty\}\}$ is a semifield that generates $\sigma(f)$. Defining P^* on $\Omega \times \mathcal{S}$ by $P^*\big(\omega, f^{-1}(I_{s,r})\big) = H(\omega, r) - H(\omega, s)$, we have, on one hand, $P^*(\omega, \cdot)$, for each ω, is a probability on $\mathcal{S}$ and on the other, $P^*(\cdot, S)$, for each $S \in \mathcal{S}$, is a version of $P\big(S\,\big|\,\mathcal{G}\big)$. For each ω, we have a unique extension of the probability $P^*(\omega, \cdot)$ on the semifield $\mathcal{S}$ to a probability on $\sigma(\mathcal{S}) = \sigma(f)$, to be called $P^*(\omega, \cdot)$ again. This now gives us a P^* defined on $\Omega \times \sigma(f)$, such that, $P^*(\omega, \cdot)$ is a probability on $\sigma(f)$, for each ω. An easy application of Dynkin's π-λ Theorem will show that $P^*(\cdot, A)$ is a version of $P\big(A\,\big|\,\mathcal{G}\big)$, for each $A \in \sigma(f)$. Thus, P^* gives the required RCP on $\sigma(f)$, given $\mathcal{G}$. □

Having defined and explored the notion of RCP, let us examine where does the classical concept of conditional probability fit inside this framework. To go back to the classical notion of conditional probability, let us consider a probability space $(\Omega, \mathcal{A}, P)$ and take an event $B \in \mathcal{A}$ with $P(B) > 0$. Then, for each $A \in \mathcal{A}$, the classical definition of conditional probability is $P(A \mid B) = P(A \cap B)/P(B)$. It is, of course, easy to see that $P(\cdot \mid B)$ is a probability on $\mathcal{A}$. However, it is not a random probability, unlike what we come across when we discuss RCP. To see this classical concept in light of RCP, let us assume also that $P(B) < 1$. In that case, we can define $P(A \mid B^c)$, for all $A \in \mathcal{A}$, and once again $P(\cdot \mid B^c)$ defines a probability, still non-random, on $\mathcal{A}$. Now, let us put these two together in the form of the probability kernel

$$P^*(\omega, A) = P(A \mid B) I_B(\omega) + P(A \mid B^c) I_{B^c}(\omega).$$

It is an easy exercise to verify that this P^* defines a RCP on $\mathcal{A}$, given the σ-field $\mathcal{G} = \{\Omega, B, B^c, \emptyset\}$. The next exercise is a continuation of this.

EXERCISE **9.2.12**. *(Connections with elementary conditional probability)*
(a) With $B \in \mathcal{A}$ and $\mathcal{G} = \{\Omega, B, B^c, \emptyset\}$, how would you define RCP on $\mathcal{A}$, given $\mathcal{G}$, if we drop the assumption that $0 < P(B) < 1$?
(b) Find RCP on $\mathcal{A}$, given $\mathcal{G}$, when $\mathcal{G}$ is the σ-field generated by a finite partition of Ω by $\mathcal{A}$-sets.
(c) Do the same when $\mathcal{G}$ is the σ-field generated by a countably infinite partition of Ω by $\mathcal{A}$-sets.

9.3 Regular Conditional Distributions

We know that for a random variable X on a probability space $(\Omega, \mathcal{A}, P)$, its expectation $E(X)$ is defined to be the integral $\int_\Omega X dP$, provided of course the integral exists. But we have also seen that $E(X)$ can be obtained as an integral on $\mathbb{R}$ as well, through the change of variable formula $E(X) = \int_R x P_X(dx)$, where P_X is the probability on $\mathbb{R}$, induced by X and called the distribution of X. More generally, if $\mathbf{X} = (X_1, \ldots, X_k)$ is a k-dimensional random vector on a probability space $(\Omega, \mathcal{A}, P)$, then, for any real-valued measurable function h on $\mathbb{R}^k$, we have $E(h(\mathbf{X})) = \int_{\mathbb{R}^k} h(\mathbf{x}) P_\mathbf{X}(d\mathbf{x})$, where $P_\mathbf{X}$ is the joint distribution of $\mathbf{X}$.

One may wonder whether something similar is possible for conditional expectations as well. In this section, we are going to show that this is indeed possible, but now, the distribution of X (or the joint distribution of $\mathbf{X}$) has to be replaced by 'conditional distribution' of X (or 'conditional joint distribution' of $\mathbf{X}$). We are now going to formally define what we mean by 'conditional distribution'. For the sake of generality, we are going to define it not just for random variables or random vectors, but for general measurable maps on a probability space into a measurable space $(E, \mathcal{E})$, which are to be henceforth referred to as 'E-valued random variables'.

DEFINITION 9.3.1. Let $(E, \mathcal{E})$ be a measurable space and X be an E-valued random variable on a probability space $(\Omega, \mathcal{A}, P)$. If $\mathcal{G}$ is any sub-σ-field of $\mathcal{A}$, then by a *Regular Conditional Distribution* (abbreviated as RCD) of X, given $\mathcal{G}$, is meant a probability kernel Q on $\Omega \times \mathcal{E}$ such that $Q(\cdot, B)$, for each $B \in \mathcal{E}$, is a version of $P(X^{-1}(B) \mid \mathcal{G})$.

It is a matter of fact that RCD, as defined in 9.3.1, does not always exist in this generality. However, as we did with RCP, instead of illustrating examples of non-existence, we will explore and identify some special situations, when RCD does exist. But before we get to that, let us outright show first that RCD, when it exists, does what we expect it to do, namely, that conditional expectations can indeed be realized as integrals with respect to the regular conditional distribution. This will place conditional expectations on the same footing as (unconditional) expectations. When this happens, most of the properties of conditional expectation that we have proved in Section 9.1, using the definition of conditional expectation, follow as immediate consequences of properties of expectations. As a matter of fact, this actually allows us to get some more properties of conditional expectations. (see Exercise 9.3.6).

THEOREM 9.3.2. Let $(E, \mathcal{E})$ be a measurable space and let X an E-valued random variable on a probability space $(\Omega, \mathcal{A}, P)$. Suppose $\mathcal{G} \subset \mathcal{A}$ is a sub-σ-field such that a RCD Q of X, given $\mathcal{G}$, exists. Then, for any real measurable function h on E with $E(|h(X)|) < \infty$, the integral $\int_E h(x) Q(\cdot, dx)$ exists and is a version of $E(h(X) \mid \mathcal{G})$.

9.3. Regular Conditional Distributions

Proof. For $h = I_B$, with any $B \in \mathcal{E}$, the result is immediate from the definition of RCD. From this, passage to real-valued simple functions is an easy consequence of linearity of integrals and of conditional expectations. Using MCT for integrals as well as for conditional expectations, one now gets the result easily for any non-negative real-valued h. Finally, for a general real-valued h, the hypothesis $E(|h(X)|) < \infty$ implies that both $E(h^+(X) \mid \mathcal{G})$ and $E(h^-(X) \mid \mathcal{G})$ are finite a.s. and therefore, so also are both $\int_E h^+(x)\, Q(\cdot, dx)$ and $\int_E h^-(x)\, Q(\cdot, dx)$, by what has already been proved. This shows that the integral $\int_E h(x) Q(\cdot, dx)$ is well-defined and real-valued a.s. and equals $E(h(X) \mid \mathcal{G})$ by linearity of conditional expectation. $\square$

We are now going to show first that when $(E, \mathcal{E}) = (\mathbb{R}^k, \mathcal{B}^k)$, $k \geq 1$, a Regular Conditional Distribution, as defined in 9.3.1, exists. Note that the case $k = 1$ corresponds to X being a real random variable. Incidentally, the reader may look at Exercise 9.4.16 to see that RCD exists for general E-valued random variables, under some extra conditions.

THEOREM 9.3.3. (Existence of RCD for real Random Variables and Vectors) For any real random variable X or any k-dimensional random vector $\mathbf{X}$ on a probability space $(\Omega, \mathcal{A}, P)$ and any sub-σ-field $\mathcal{G} \subset \mathcal{A}$, a Regular Conditional Distribution, given $\mathcal{G}$, exists.

We prove the result only for a k-dimensional random vector $\mathbf{X}$, where $k > 1$. The proof for real random variables is no different, in fact, even simpler. The argument is essentially the same as for the proof of Theorem 9.2.7(a), except that we use the following slightly modified version of Lemma 9.2.5 at the end.

LEMMA 9.3.4. Let $\mathbf{X}$ be a k-dimensional random vector on a probability space $(\Omega, \mathcal{A}, P)$ and let $\mathcal{G} \subset \mathcal{A}$ be a sub-σ-field. Suppose $\mathcal{P}$ is π-system generating $\mathcal{B}^k$ and Q^* is a probability kernel on $\Omega \times \mathcal{B}^k$ such that, for every $B \in \mathcal{P}$, is a version of $P(\mathbf{X}^{-1}(B) \mid \mathcal{G})$. Then Q^* is a Regular Conditional Distribution of $\mathbf{X}$ given $\mathcal{G}$.

Since the proof of Lemma 9.3.4 is a verbatim copy of the proof of Lemma 9.2.5, we omit the proof and go straight to the proof of Theorem 9.3.3.

Proof of Theorem 9.3.3. For every $\mathbf{r} \in \mathbb{Q}^k$, let $J_\mathbf{r} = \{\mathbf{x} \in \mathbb{R}^k : \mathbf{x} \leq \mathbf{r}\}$ and let $H(\cdot, \mathbf{r})$ be a version of $P(X^{-1}(J_\mathbf{r}) \mid \mathcal{G})$. We can now proceed, exactly as in the proof of Theorem 9.2.7, to get a function F on $\Omega \times \mathbb{Q}^k$ that satisfies properties (i) and (ii) of Lemma 9.2.6. Denote Q^* to be the probability kernel determined by F as in Lemma 9.2.6. Now, we can just use Lemma 9.3.4 to conclude that Q^* is a required RCD of $\mathbf{X}$, given $\mathcal{G}$. That completes the prroof. $\square$

We state a slight generalization of Theorem 9.3.3. Any reader who has closely followed the proof of Theorem 9.2.11 (b), should see that the proof here is even easier. Simply the probability $Q(\omega, \cdot)$ on $\mathcal{B}$, given by the distribution function $H(\omega, \cdot)$ on $\mathbb{Q}$, appearing in that proof, gives the required RCD of Theorem 9.3.5.

THEOREM **9.3.5.** Let $(E, \mathcal{E})$ be a standard Borel space and X an E-valued random variable on a probability space $(\Omega, \mathcal{A}, P)$. Then for any sub-σ-field $\mathcal{G} \subset \mathcal{A}$, a Regular Conditional Distribution of X, given $\mathcal{G}$, exists.

Now that we have proved, among other things, existence of RCD for any real random variable, given any sub-σ-field, Theorem 9.3.2 guarantees that the conditional expectation of any integrable random variable, given any sub-σ-field can be written as an integral over $\mathbb{R}$ with respect to the RCD. This, as was pointed out earlier, makes it possible for us to get all the properties of conditional expectations as easy consequences of properties of expectations. One may perhaps get even new properties that were not proved in Section 9.1. We urge the reader to take up this exercise seriously.

EXERCISE **9.3.6.** *Undertake the task charted out above. See if you can deduce some additional properties; for example, conditional versions of Chebycheff's inquality, Hölder's inequality, Minkowski's inequality, etc., would be interesting!*

For the rest of this section we shall try to use the concepts of RCD to properly understand and give sense to some of the concepts that are customarily and sometimes loosely talked about in an elementary course in probability. With two random variables X and Y defined on a probability space $(\Omega, \mathcal{A}, P)$, one often talks about the 'conditional distribution of X given $Y = y$'. Of course, this makes perfect sense in case Y is a discrete random variable. This came up briefly towards the end of Section 9.1. The question we want to ask now is whether it is possible to make sense of it without any such restrictions on Y? Indeed, what we are going to show now is that, given any two real random variables, one can always get a function Q on $\mathbb{R} \times \mathcal{B}$ that satisfies the following properties:

(i) $Q(\cdot, B)$ is Borel measurable for each $B \in \mathcal{B}$,

(ii) $Q(y, \cdot)$ is a probability on $\mathcal{B}$ for every $y \in \mathbb{R}$, and,

(iii) the function $Q(Y(\cdot), \cdot)$ on $\Omega \times \mathcal{B}$ is a RCD of X given $\sigma(Y)$.

Once we do this, $Q(y, \cdot)$ can then, quite legitimately, be thought of as capturing the conditional distribution of X given $Y = y$ and, of course, expectation with respect to this distribution can then be interpreted as $E(X \mid Y = y)$.

But getting such a Q is quite simple. We know, by Theorem 9.3.3, that a RCD Q^* of X, given $\sigma(Y)$, exists. Since, for each $B \in \mathcal{B}$ fixed, $Q^*(\cdot, B)$ is $\sigma(Y)$ measurable, there is, by Theorem 3.5.13, a real borel measurable function $Q(\cdot, B)$ on $\mathbb{R}$ such that $Q^*(\omega, B) = Q(Y(\omega), B)$. This means that this Q on $\mathbb{R} \times \mathcal{B}$ is a candidate for what we were looking for. This, as explained earlier, gives a proper meaning to the concept of conditional distribution of X given $Y = y$, where X and Y are any two random variables. The reader would surely realize that the same argument goes through if we replaced random variables by random vectors. The next theorem states the result we have proved.

9.3. Regular Conditional Distributions

THEOREM **9.3.7**. Let $\mathbf{X}$ and $\mathbf{Y}$ be two random vectors of dimensions $k \geq 1$ and $l \geq 1$ respectively, defined on some probability space $(\Omega, \mathcal{A}, P)$. Then there exists a function Q on $\mathbb{R}^l \times \mathcal{B}^k$, satisfying the following properties:

(i) $Q(\cdot, B)$ is Borel measurable for each $B \in \mathcal{B}^k$,

(ii) $Q(\mathbf{y}, \cdot)$ is a probability on $\mathcal{B}^k$ for every $\mathbf{y} \in \mathbb{R}^l$, and

(iii) the function $Q(\mathbf{Y}(\cdot), \cdot)$ on $\Omega \times \mathcal{B}^k$ is a RCD of $\mathbf{X}$ given $\sigma(\mathbf{Y})$.

We are now going to apply what we have discussed above in two specific cases, that any reader having taken a first course in probability would be familiar with.

EXAMPLE **9.3.8**. Let X be any real random variable and let Y be a real valued discrete random variable, both defined on the same probability space $(\Omega, \mathcal{A}, P)$. Let D denote the countable set given by $D = \{y : P(Y = y) > 0\}$. Since Y is discrete, we know that $P(Y \notin D) = 0$. For each $y \in D$ and $A \in \mathcal{B}$, let $P(X \in A \mid Y = y)$ be the conditional probability given by the usual formula $P(X \in A, Y = y)/P(Y = y)$. Choose and fix any probability Q_0 on $\mathcal{B}$. If we now define a function Q on $\mathbb{R} \times \mathcal{B}$ by

$$Q(y, A) = \begin{cases} P(X \in A \mid Y = y) & \text{if } y \in D \\ Q_0(A) & \text{if } y \notin D \end{cases},$$

then it is easy to see that Q satisfies the properties (i)-(iii) as in Theorem 9.3.7.

REMARK **9.3.9**. (a) The reader can convince herself that the X in Example 9.3.8 does not have to be a real random variable. Even if X is an E-valued random variable where $(E, \mathcal{E})$ is any measurable space, one can follow the same algorithm to get a function Q on $\mathbb{R} \times \mathcal{E}$ that satisfies properties (i)-(iii) as in Exercise 9.4.17. In particular, this would imply that a RCD of any E-valued random variable, given $\sigma(Y)$, exists, as long as Y is a real discrete random variable.

(b) The reader should also be able to convince herself that replacing the real discrete random variable Y by a k-dimensional discrete random vector $\mathbf{Y}$ would not affect the final upshots.

EXAMPLE **9.3.10**. Let X, Y be two random variables on a probability space such that $P_{(X,Y)}$ is absolutely continuous with respect to the Lebesgue measure on $\mathcal{B}(\mathbb{R}^2)$. Let f be the joint density function of (X, Y). Denote $g(y) = \int f(x, y) dx$, for $y \in \mathbb{R}$. The function g gives the 'marginal' density function of Y. By Fubini's Theorem, the function g is borel measurable and so, $D = \{y : g(y) > 0\}$ is a borel set. Choose and fix any probability density function h_0 on $\mathbb{R}$ and define a function h on $\mathbb{R}^2$ by

$$h(x, y) = \begin{cases} f(x, y)/g(y) & \text{if } y \in D \\ h_0(x) & \text{if } y \notin D \end{cases}$$

One can easily see that, for each $y \in \mathbb{R}$, the function $h(\cdot, y)$ is a probability density function on $\mathbb{R}$. Denoting $Q(y, \cdot)$, for each $y \in \mathbb{R}$, to be the probability on $\mathcal{B}$ given by the density $h(\cdot, y)$, one gets a function Q on $\mathbb{R} \times \mathcal{B}$. It is left to the reader to verify that this Q satisfies properties (i)-(iii) of Theorem 9.3.7. Those familiar with basic probability would recall that the function $f(x, y)/g(y)$ above is extensively used and is called the "*conditional density of X given $Y = y$*", though it is usually presented as an ad hoc prescription, often with no clear justification or interpretation.

In Examples 9.3.8 and 9.3.10, we talked about objects that are routinely discussed and extensively used in elementary probability, but mostly without sound interpretations of what they stand for. What we have attempted here is to place these ideas on proper theoretical foundation, so that the reader now clearly knows what they actually mean.

EXERCISE **9.3.11.** *Extend what was discussed in Example 9.3.10 to the case when at least one or both of* $\mathbf{X}$ *and* $\mathbf{Y}$ *are random vectors, possibly of different dimensions, with an absolutely continuous joint distribution.*

We end this section with some observations about connections between the notion of RCPs that we discussed and the construction of probabilities on finite product spaces seen in Chapter 5. Consider the product space $\left(\underset{j=1}{\overset{n}{\times}} \Omega_j, \underset{j=1}{\overset{n}{\otimes}} \mathcal{A}_j \right)$ and the measure μ as constructed in Theorem 5.3.10. Of course, here we are going to specialize to the case when μ_1 is a probability and each μ_j, $2 \leq j \leq n$, is a probability kernel so that μ is a probability. Denote π_j, $1 \leq j \leq n$, to be the usual coordinate projections. From the definition of μ it is easy to see that for each $j = 2, \ldots, n$, the probability kernel μ_j on $\left(\underset{i=1}{\overset{j-1}{\times}} \Omega_i \right) \times \mathcal{A}_j$ is nothing but a RCD of π_j given $(\pi_1, \ldots, \pi_{j-1})$. Thus, in the special case when μ_1 is a probability measure and $\mu_j, 2 \leq j \leq n$, are probability kernels, Theorem 5.3.10 can be interpreted as asserting the following extremely useful result.

THEOREM **9.3.12.** *Let* $(\Omega_j, \mathcal{A}_j), 1 \leq j \leq n$, *be measurable spaces. Given any probability measure* μ_1 *on* $\mathcal{A}_1$ *and any probability kernels* μ_j *on* $\left(\underset{i=1}{\overset{j-1}{\times}} \Omega_i \right) \times \mathcal{A}_j$, *for* $2 \leq j \leq n$, *there exists a probability space* $(\Omega, \mathcal{A}, \mu)$ *and* Ω_j-*valued random variables* π_j, *for* $1 \leq j \leq n$, *on* $(\Omega, \mathcal{A}, \mu)$, *such that,* μ_1 *is the distribution of* π_1 *and* μ_j, *for* $j = 2, \ldots, n$, *is the Regular Conditional Distribution of* π_j, *given* $(\pi_1, \ldots, \pi_{j-1})$.

REMARK **9.3.13.** The assertion made in Theorem 9.3.12 is of special significance in case when we have one measurable space $(E, \mathcal{E})$ and $(\Omega_j, \mathcal{A}_j) = (E, \mathcal{E})$, for all j.

In continuation of what we just discussed, a very pertinent question that could have been asked in Chapter 5, but was not, is whether all probabilities μ on the product space arise in this manner. This is often referred to as the *"disintegration problem"*. The answer to this question in general, is 'no'. However, what we have learnt in this chapter tells us that the answer to the above question is, indeed, 'yes' at least when $(\Omega_j, \mathcal{A}_j) = (\mathbb{R}, \mathcal{B})$, for all j. This is not entirely trivial, which is why we have the next exercise, asking the reader to prove it.

EXERCISE **9.3.14.** *(a) Use the result on existence of RCD to show that every probability* μ *on* $(\mathbb{R}^2, \mathcal{B}^2)$ *arises through a probability on* $(\mathbb{R}, \mathcal{B})$ *and a probability kernel on* $\mathbb{R} \times \mathcal{B}$, *as in Theorem 5.3.10.*
(b) Prove the analogous result for any probability μ *on* $(\mathbb{R}^k, \mathcal{B}^k)$, *for any* $k \geq 2$.

9.4 Additional Exercises

EXERCISE 9.4.1. Let X be a non-negative integrable random variable on a probability space $(\Omega, \mathcal{A}, P)$ and let $\mathcal{G} \subset \mathcal{A}$ be a sub-σ-field. Directly from the definition of 'essential supremum', show that
$$E(X \mid \mathcal{G}) = \operatorname{ess\,sup}\{Z : Z \text{ non-negative, } \mathcal{G}\text{-measurable, } \int_G Z dP \leq \int_G X dP \ \forall\, G \in \mathcal{G}\}.$$

EXERCISE 9.4.2. Let $\mathbf{X}$ and $\mathbf{Y}$ be random vectors, of dimensions k and m respectively, on a probability space. Suppose $h : \mathbb{R}^{k+m} \to \mathbb{R}$ is a Borel measurable function, such that, $h(\mathbf{X}, \mathbf{Y})$ is integrable. Show that
(a) $h(\mathbf{x}, \mathbf{Y})$ is an integrable random variable for $P_\mathbf{X}$-a.e. $\mathbf{x} \in \mathbb{R}^k$,
(b) $g(\mathbf{x}) = E\big(h(\mathbf{x}, \mathbf{Y})\big)$ is a real-valued Borel measurable function (set $g(\mathbf{x}) = 0$, if $E\big(h(\mathbf{x}, \mathbf{Y})\big)$ does not exist),
(c) $g(\mathbf{X})$ is a version of $E\big(h(\mathbf{X}, \mathbf{Y}) \mid \mathbf{X}\big)$, if $\mathbf{X}$ and $\mathbf{Y}$ are independent,
(d) the assertion in (c) may not hold without the assumption of independence.

EXERCISE 9.4.3. Let X_n, $n \geq 1$ and Y be integrable random variables on a probability space $(\Omega, \mathcal{A}, P)$, such that, $X_n \geq Y$, for all $n \geq 1$, and let $\mathcal{G} \subset \mathcal{A}$ be a sub-σ-field.
(a) Show that if $X_n \uparrow X$ and X is integrable, then $E(X_n \mid \mathcal{G}) \uparrow E(X \mid \mathcal{G})$.
(b) Show that if $\liminf_n X_n$ is integrable, then $E\big(\liminf_n X_n \mid \mathcal{G}\big) \leq \liminf_n E(X_n \mid \mathcal{G})$.

EXERCISE 9.4.4. Let $\{X_n\}$ be a uniformly integrable sequence of random variables on a probability space $(\Omega, \mathcal{A}, P)$ and let $\mathcal{G} \subset \mathcal{A}$ be a sub-σ-field. Show that if $X_n \to X$, $[P]$-a.s., then X is integrable and $E(X_n \mid \mathcal{G}) \to E(X \mid \mathcal{G})$ $[P]$-a.s.

EXERCISE 9.4.5. For a random variable X on a probability space $(\Omega, \mathcal{A}, P)$ with finite second moment and for a sub-σ-field $\mathcal{G} \subset \mathcal{A}$, the "conditional variance" of X, given $\mathcal{G}$, is defined as $V(X \mid \mathcal{G}) = E\big((X - E(X \mid \mathcal{G}))^2 \mid \mathcal{G}\big)$.
(a) Show that $V(X \mid \mathcal{G}) \leq E\big((X - Y)^2 \mid \mathcal{G}\big)$ $[P]$-a.s., for any $\mathcal{G}$-measurable random variable Y.
(b) Show that $V(X \mid \mathcal{G}) \geq V(\min\{X, \alpha\} \mid \mathcal{G})$, for any real α.
(c) Show that $V(X) = E\big(V(X \mid \mathcal{G})\big) + V\big(E(X \mid \mathcal{G})\big)$. [This implies, in particular, that $E(X \mid \mathcal{G})$ has smaller variance than X – any intuition?]

EXERCISE 9.4.6. Let $\{X_n\}$ be a sequence of i.i.d. random variables.
(a) Let $S_k = X_1 + \cdots + X_k$, $k \geq 1$ and denote $\mathcal{G}_n = \sigma(\{S_k, k \geq n\})$. Assuming that the common mean of the $\{X_n\}$ is finite, show that $E(X_i \mid \mathcal{G}_n) = S_n/n$, for each $n \geq 1$ and for each $i = 1, \ldots, n$.
(b) Let $\phi : \mathbb{R}^m \to \mathbb{R}$ be a Borel measurable function, which is invariant under permutations. For each $k \geq m$, consider the random variable U_k^ϕ defined through the formula $\binom{k}{m} U_k^\phi = \sum^{(k)} \phi(X_{i_1}, \ldots, X_{i_m})$, where $\sum^{(k)}$ denotes the sum over all possible choices of $1 \leq i_1 < \cdots < i_m \leq k$. U_k^ϕ is called the "U-statistic" given by the "kernel" ϕ based on $X_1, \ldots, X_k$. Assuming that $\phi(X_1, \ldots, X_m)$ is integrable and denoting $\mathcal{G}_n = \sigma(\{U_k^\phi, k \geq n\})$, show that $E\big(\phi(X_{i_1}, \ldots, X_{i_m}) \mid \mathcal{G}_n\big) = U_n^\phi$, for each $n \geq m$ and every choice of $1 \leq i_1 < \cdots < i_m \leq n$.

[Note that (a) is a special case of (b) with $m = 1$ and $\phi(x) = x$. Try to see what U_k^ϕ is when $m = 2$ and $\phi(x_1, x_2) = (x_1 - x_2)^2$.]

REMARK 9.4.7. Using results from a very important area of probability known as "Martingale Theory", one can easily deduce a.s. convergence of the sequence $\{U_n^\phi\}$ of U-statistics, from part (b) of Exercise 9.4.6.

Another important point worth mentioning here is that the results of Exercise 9.4.6 remain valid even if the i.i.d. hypothesis on the sequence $\{X_n\}$ is replaced by a weaker hypothesis, namely, that $\{X_n\}$ is an "exchangeable" sequence, meaning that, for any $n \geq 1$ and any permutation π of $\{1, 2, \ldots, n\}$, the vector $(X_{\pi(1)}, \ldots, X_{\pi(n)})$ has the same joint distribution as $(X_1, \ldots, X_n)$. An interested reader may try proving the results under this weaker hypothsis.

EXERCISE 9.4.8. Let $(\Omega, \mathcal{A}, P)$ be a probability space and $\mathcal{G}, \mathcal{G}_1$ and $\mathcal{G}_2$ be sub-σ-fields of $\mathcal{A}$. We say that $\mathcal{G}_1$ and $\mathcal{G}_2$ are "conditionally independent", given $\mathcal{G}$, if $P(G_1 \cap G_2 \mid \mathcal{G}) = P(G_1 \mid \mathcal{G}) \cdot P(G_2 \mid \mathcal{G})$, for all $G_1 \in \mathcal{G}_1$ and $G_2 \in \mathcal{G}_2$. Show that this is equivalent to any one of the following:
(a) If $\mathcal{P}_i$, for $i = 1, 2$, is any π-system generating $\mathcal{G}_i$, then $P(G_1 \cap G_2 \mid \mathcal{G}) = P(G_1 \mid \mathcal{G}) \cdot P(G_2 \mid \mathcal{G})$, for all $G_1 \in \mathcal{P}_1$ and $G_2 \in \mathcal{P}_2$.
(b) $P(G_1 \mid \mathcal{G} \vee \mathcal{G}_2) = P(G_1 \mid \mathcal{G})$ for all $G_1 \in \mathcal{G}_1$. [See notation in Exercise 2.13.8.]
(c) If $\mathcal{P}_1$ is any π-system generating $\mathcal{G}_1$, then $P(G_1 \mid \mathcal{G} \vee \mathcal{G}_2) = P(G_1 \mid \mathcal{G})$ for all $G_1 \in \mathcal{P}_1$.
(d) $E(X \mid \mathcal{G} \vee \mathcal{G}_2) = E(X \mid \mathcal{G})$, for all integrable $\mathcal{G} \vee \mathcal{G}_1$-measurable random variables X.

EXERCISE 9.4.9. Let X, Y and Z be random variables on a probability space, such that, $X, Y - X$ and $Z - Y$ are mutually independent. Show that X and Z are conditionally independent, given Y.

EXERCISE 9.4.10. Let $(\Omega, \mathcal{A}, P)$ be a probability space and $\mathcal{G} \subset \mathcal{A}$ a sub-σ-field. Take a set $A \in \mathcal{A}$ and put $B = \{\omega : P(A \mid \mathcal{G})(\omega) > 0\}$. Show that B is a $\mathcal{G}$-set satisfying (i) $A \subset B$ a.s. $[P]$ and (ii) $D \in \mathcal{G}$ and $A \subset D$ a.s. $[P]$ implies that $B \subset D$ a.s. $[P]$.

EXERCISE 9.4.11. Let $(\Omega, \mathcal{A}, P)$ be a probability space and $\mathcal{G} \subset \mathcal{A}$ a sub-σ-field. Suppose that $h : (\Omega, \mathcal{A}) \to (\Omega, \mathcal{A})$ be a measurable map, such that, $Ph^{-1} = P$. Show then that $E(X \circ h \mid h^{-1}\mathcal{G}) = E(X \mid \mathcal{G}) \circ h$, $[P]$-a.s., for any integrable X.

EXERCISE 9.4.12. Let X be a real random variable on a probability space. Show that, if X has a symmetric distribution (that is, $X \stackrel{d}{=} -X$), then $Q(\cdot, B) = \frac{1}{2}I_B(X(\cdot)) + \frac{1}{2}I_B(-X(\cdot))$, $B \in \mathcal{B}$, defines a RCD of X, given $|X|$.

EXERCISE 9.4.13. Let $(\Omega, \mathcal{A}, P)$ be a probability space and let $T_g : \Omega \to \Omega$, $g \in G$, be a finite "group" of measurable maps.
(a) Show that this implies that T_g, for each $g \in G$, is a bijection and that T_g and T_g^{-1} are both measurable (that is, each T_g is a bi-measurable bijection).
Let $\mathcal{G} = \{A \in \mathcal{A} : T_g^{-1}(A) = A \text{ for all } g \in G\}$.

9.4. Additional Exercises

(b) Show that $\mathcal{G}$ is a σ-field (called the "G-invariant" sub-σ-field of $\mathcal{A}$). Assume that P is G-invariant, that is, $PT_g^{-1} = P$, for all $g \in G$.
(c) Show then that $Q(\omega, A) = \frac{1}{|G|} \sum_{g \in G} I_A \circ T_g(\omega)$, $\omega \in \Omega$, $A \in \mathcal{A}$, defines a RCP on $\mathcal{A}$, given $\mathcal{G}$. (Here, $|G|$ denotes the cardinality of G.)
[Can you see that the previous exercise is a speciual case of this?]

EXERCISE 9.4.14. Show that the following are standard Borel spaces: (i) $(\mathbb{R}^k, \mathcal{B}^k)$, (ii) $(\mathbb{R}^\infty, \mathcal{B}^\infty)$ (see Chapter 10 for the definition), (iii) $(S, \mathcal{B}(S))$, where S is a complete separable metric space and $\mathcal{B}(S)$ is the Borel σ-field on S, that is, the σ-field on S generated by all open subsets of S. [Hint: Show that a complete separable metric space is homeomorphic to a Borel (in fact, G_δ) subset of $[0,1]^\infty$]

EXERCISE 9.4.15. Show that if P^* denotes the RCP as in Theorem 9.2.11 (b), then for any integrable $\sigma(f)$-measurable random variable X, a version of $E(X \mid \mathcal{G})$ can be realized as an integral of X with respect to P^*.

EXERCISE 9.4.16. Assume the general set-up as in Definition 9.3.1
(a) Show that an RCD of X given $\mathcal{G}$ exists (i) if $\sigma(X)$ and $\mathcal{G}$ are independent, (ii) if X is $\mathcal{G}$-measurable.
(b) Show that an RCD of X given $\mathcal{G}$ exists if $\mathcal{G}$ is generated by a countable partition of Ω by $\mathcal{A}$-sets. In particular if Y is a discrete real random variable, then an RCD of X given $\mathcal{G} = \sigma(Y)$ exists.

EXERCISE 9.4.17. Let $(\Omega, \mathcal{A}, P)$ be a probability space and $(S, \mathcal{S})$ and $(E, \mathcal{E})$ are measurable spaces. Suppose X and Y are random variables on the probability space, taking values in S and E respectively, such that an RCD of X given $\sigma(Y)$ exists. Then there is a function Q on $E \times \mathcal{S}$ satisfying (i) $Q(\cdot, B)$ is $\mathcal{E}$-measurable for each $B \in \mathcal{S}$, (ii) $Q(e, \cdot)$ is a probability on $\mathcal{S}$ for every $e \in E$ and (iii) the function $Q(Y(\cdot), \cdot)$ on $\Omega \times \mathcal{S}$ is an RCD of X given $\sigma(Y)$.

EXERCISE 9.4.18. Let X be any real random variable with pdf f and let $Y = |X|$. Denote $D = \{y : f(y) + f(-y) > 0\}$ and let P_0 be any probability on $\mathbb{R}$. Show that $Q(y, \cdot) = \{(\delta_y(\cdot)f(y) + \delta_{-y}(\cdot)f(-y))/[f(y) + f(-y)]\}I_D(y) + P_0(\cdot)I_{\{D^c\}}(y)$ is a conditional distribution of X given $Y = y$.

EXERCISE 9.4.19. Let U, V be independent random variables, each with $U(0,1)$ distribution. Put $X = U$, $Y = U \vee V$. Let P_0 be any probability on $\mathcal{B}$. Show that the function Q on $\mathbb{R} \times \mathcal{B}$ defined by

$$Q(y, B) = \left[\int_B \tfrac{1}{2y} I_{(0,y)}(u) du + \tfrac{1}{2} \delta_{\{y\}}(B) \right] I_{(0,1)}(y) + P_0(B) I_{\{(0,1)^c\}}(y)$$

gives a conditional distribution of X given $Y = y$.

Chapter 10

Infinite Products

As a motivation for what we are going to discuss in this Chapter, let us recall something that we noted towards the end of Section 9.3. Let $(E, \mathcal{E})$ be a measurable space and suppose we are given a probability μ_1 on $(E, \mathcal{E})$ and, for each $2 \leq j \leq n$, a probability kernel μ_j on $E^{j-1} \times \mathcal{E}$. Theorem 9.3.12 (and Remark 9.3.13) then guarantees that there exists an n-dimensional random vector $\mathbf{X} = (X_1, \cdots, X_n)$ on a probability space, where each X_j is an E-valued random variable and μ_1 is the distribution of X_1, while μ_j, for $2 \leq j \leq n$, is the Regular Conditional Distribution of X_j, given $(X_1, \cdots, X_{j-1})$.

This leads one to ask whether the above construction can be pushed through to construct an infinite sequence $\{X_n\}$ of E-valued random variables, such that X_1 has a specified marginal distribution and, X_j, for each $j \geq 2$, has a specified conditional distribution, given $(X_1, \ldots, X_{j-1})$. To put it in more exact terms, suppose $(E, \mathcal{E})$ is a measurable space and we are given a probability μ_1 on $(E, \mathcal{E})$ and, for each $j \geq 2$, a probability kernel μ_j on $E^{j-1} \times \mathcal{E}$. The question is whether we can construct a probability space $(\Omega, \mathcal{A}, P)$ and a sequence $\{X_n\}$ of E-valued random variables on it, such that μ_1 is the distribution of X_1, and, μ_j, for each $j \geq 2$, is the (regular) conditional distribution of X_j, given $(X_1, \cdots, X_{j-1})$.

It turns out that this is indeed possible and that is one of the things we are going to demonstrate in this chapter. To get to that, let us recall that the existence of a probability space and a random vector $(X_1, \ldots, X_n)$ on it, as noted in Theorem 9.3.12 (and Remark 9.3.13), is essentially based on the construction of finite products of measurable spaces, as was done in Chapter 5. This suggests that the first step towards constructing a probability space and a sequence of random variables on it with the above specifications, should be to bring in the idea of infinite products of measurable spaces. To be more specific, given a measurable space $(E, \mathcal{E})$, if we were to construct a probability space and then define an infinite sequence of E-valued random variables on it with any prescribed distributions and conditional distributions, the first natural strategy will be to try and define the (countably) infinite product of $(E, \mathcal{E})$ with itself. This is what we are going to take up first. However, just like we did for finite products, we will do something a bit more general, namely, we will actually define the product space of an infinite sequence of, possibly different, measurable spaces.

10.1 Product of a Sequence of Measurable Spaces

Let $(\Omega_n, \mathcal{A}_n)$, $n \geq 1$, be an infinite sequence of measurable spaces. We are going to define the product of these measurable spaces to get what will be called the infinite product space. The natural candidate for the underlying set for such an infinite product space has to be the usual infinite cartesian product $\Omega = \underset{n=1}{\overset{\infty}{\times}} \Omega_n$. Recall that the infnite cartesian product Ω consists precisely of all infiniite sequences $\omega = (\omega_1, \omega_2, \ldots)$, where ω_n, for each n, belongs to Ω_n.

To define now the product σ-field on Ω, we start with what are called the 'finite dimensional cylinders'. A very simple way to describe what we mean by a finite dimensional cylinder is the following. For each $n \geq 1$, let φ_n denote the natural projection map on Ω onto $(\underset{j=1}{\overset{n}{\times}} \Omega_j)$, that is,

$$\varphi_n : \omega = (\omega_1, \omega_2, \ldots) \mapsto (\omega_1, \ldots, \omega_n).$$

With $\underset{j=1}{\overset{n}{\otimes}} \mathcal{A}_j$ denoting the n-fold product σ-field on $\underset{j=1}{\overset{n}{\times}} \Omega_j$, as in Chapter 5, it is easy to see that, for each n, the class

$$\mathfrak{F}_n = \{\varphi_n^{-1}(B) : B \in \underset{j=1}{\overset{n}{\otimes}} \mathcal{A}_j\}$$

is a σ-field on Ω. In fact, it is the smallest σ-field on Ω that makes φ_n measurable. The σ-field $\mathfrak{F}_n$ on Ω is called the class of all (measurable) n-*dimensional cylinder sets* and a set $A = \varphi_n^{-1}(B) \in \mathfrak{F}_n$ with $B \in \underset{j=1}{\overset{n}{\otimes}} \mathcal{A}_j$ is called an n-*dimensional cylinder with base* B. Note that $\varphi_n^{-1}(B_1) = \varphi_n^{-1}(B_2)$ iff $B_1 = B_2$, that is, two n-dimensional cylinder sets are equal only if they have the same base. Also, for any $n \geq 1$ and $B \in \underset{j=1}{\overset{n}{\otimes}} \mathcal{A}_j$, the set $B' = B \times \Omega_{n+1} \in \underset{j=1}{\overset{n+1}{\otimes}} \mathcal{A}_j$ and $\varphi_{n+1}^{-1}(B') = \varphi_n^{-1}(B)$. Thus, every n-dimensional cylinder is also an $(n+1)$-dimensional cylinder, that is, $\mathfrak{F}_n \subset \mathfrak{F}_{n+1}$. Thus, $\{\mathfrak{F}_n\}$ gives us an increasing sequence of σ-fields on Ω. Also, observing that an $(n+1)$-dimensional cylinder with base B' can be an n-dimensional cylinder only if $B' = B \times \Omega_{n+1}$, for some $B \in \underset{j=1}{\overset{n}{\otimes}} \mathcal{A}_j$, it follows that $\mathfrak{F}_n \subsetneq \mathfrak{F}_{n+1}$, unless $\mathcal{A}_{n+1}$ is the trivial σ-field $\{\Omega_{n+1}, \emptyset\}$.

The class $\mathfrak{F} = \underset{n}{\bigcup} \mathfrak{F}_n$ is called the class of all (measurable) *finite-dimensional cylinders*. Being the union of an increasing sequence of σ-fields, the class $\mathfrak{F}$ is clearly a field on Ω. We are now ready to define the product σ-field on Ω.

DEFINITION **10.1.1.** The σ-field on Ω generated by the field $\mathfrak{F}$ of all finite-dimensional cylinders is called the *product σ-field* on Ω and is denoted $\underset{n=1}{\overset{\infty}{\otimes}} \mathcal{A}_n$.

In what follows, the infinite product measurable sapce $(\underset{n=1}{\overset{\infty}{\times}} \Omega_n, \underset{n=1}{\overset{\infty}{\otimes}} \mathcal{A}_n)$ defined above will be denoted by $(\Omega, \mathcal{A})$. In the special case when $(\Omega_n, \mathcal{A}_n) = (E, \mathcal{E})$, for all n, the infinite product space is usually denoted by $(E^\infty, \mathcal{E}^\infty)$.

It is clear from the definition that the product σ-field $\mathcal{A}$ on Ω is nothing but the smallest σ-field on Ω that makes the n-dimensional projection maps φ_n measurable, for all n. Here is a simple exercise that presents some other equivalent descriptions of the product σ-field, that will sometimes be useful in the sequel.

EXERCISE **10.1.2.** *(a) For each n, let $\pi_n : \Omega \to \Omega_n$ be the usual n-th coordinate projection, that is, $\pi_n(\omega) = \omega_n$, for $\omega = (\omega_1, \omega_2, \ldots)$. Show that $\mathcal{A}$ is the smallest σ-field on Ω that makes all the coordinate projection maps π_n, $n \geq 1$, measurable.*
(b) For any measurable space $(\Omega', \mathcal{A}')$, show that a function $f : (\Omega', \mathcal{A}') \to (\Omega, \mathcal{A})$ is measurable iff the compositions $\pi_n \circ f$ are measurable for all n.
(c) An n-dimensional (measurable) rectangle in Ω is a subset of Ω of the form $A_1 \times \cdots \times A_n \times (\underset{j>n}{\times} \Omega_j)$, where $A_j \in \mathcal{A}_j$, for $1 \leq j \leq n$. Show that the collection $\mathcal{S}$ of all finite dimensional rectangles forms a semi-field on Ω, that generates $\mathcal{A}$.
(d) Denoting $\mathcal{C}$ to be the class of all sets of the form $\overset{\infty}{\underset{n=1}{\times}} A_n$ with $A_n \in \mathcal{A}_n$, for each n, show that $\mathcal{A} = \sigma(\mathcal{C})$.

Having now defined the infinite product space $(\Omega, \mathcal{A}) = \left(\overset{\infty}{\underset{n=1}{\times}} \Omega_n, \overset{\infty}{\underset{n=1}{\otimes}} \mathcal{A}_n \right)$, our next aim is to see how to construct probabilities on this product space subject to some specifications. This is what we take up in the next section.

10.2 Constructing Probabilities on Countably Infinite Product Spaces

We follow the notations used in the previous section, that is, $(\Omega_n, \mathcal{A}_n)$, $n \geq 1$, is a sequence of measurable spaces and $(\Omega, \mathcal{A})$ denotes the corresponding product measurable space. The first construction of probability on this infinite product space that we are going to discuss is an extension to infinite products of what Theorem 5.3.10 asserted for finite products, except that we will only consider probability measures and probability kernels here. Suppose we are given a probability measure μ_1 on $(\Omega_1, \mathcal{A}_1)$ and a probability kernel μ_n, for each $n \geq 2$, on $\left(\overset{n-1}{\underset{j=1}{\times}} \Omega_j \right) \times \mathcal{A}_n$. From Theorem 5.3.10, we know that, for each $n \geq 1$, there is a unique probability P_n on $\left(\overset{n}{\underset{j=1}{\times}} \Omega_j, \overset{n}{\underset{j=1}{\otimes}} \mathcal{A}_j \right)$ constructed using the probability μ_1 and the probability kernels μ_j, $2 \leq j \leq n$. We are first going use these P_n's to define a set function P on the field $\mathfrak{F}$ of finite-dimensional cylinders.

Given $A \in \mathfrak{F}$ we know that there is an $n \geq 1$ and a unique $B \in \overset{n}{\underset{j=1}{\otimes}} \mathcal{A}_j$, such that, $A = \varphi_n^{-1}(B)$. This is what we use to define P on $\mathfrak{F}$. We put

$$P(A) = P_n(B), \text{ if } A \in \mathfrak{F} \text{ with } A = \varphi_n^{-1}(B) \text{ for } B \in \overset{n}{\underset{j=1}{\otimes}} \mathcal{A}_j. \tag{10.2.1}$$

This set function P thus defined on $\mathfrak{F}$ is clearly well-defined, because an n-dimensional cylinder set A with base B can be viewed as an $(n+1)$-dimensional cylinder only with base $B \times \Omega_{n+1}$ and, of course, $P_{n+1}(B \times \Omega_{n+1}) = P_n(B)$. Thus, (10.2.1) gives us a non-negative set function P defined on $\mathfrak{F}$, that clearly satisfies $P(\Omega) = 1$ and $P(\emptyset) = 0$. To show that P is a probability on $\mathfrak{F}$, we first show that P is finitely additive on $\mathfrak{F}$. Let A_1, A_2 be disjoint sets in $\mathfrak{F}$. We

10.2. Constructing Probabilities on Countably Infinite Product Spaces

already know that any m-dimensional cylinder can always be considered also as an $(m+1)$-dimensional cylinder and hence as an n-dimensional cylinder, for any $n > m$. We may, therefore, assume, without any loss of genearlity, that both A_1 and A_2 are n-dimensional cylinders for some n, with bases, say, B_1 and B_2 respectively. But then $A_1 \cup A_2$ will also be an n-dimensional cylinder with base $B_1 \cup B_2$. Further, disjointness of A_1 and A_2 will clearly imply disjointness of their bases B_1 and B_2. Since P_n is a probability on $\bigotimes_{j=1}^{n} \mathcal{A}_j$, we get, from (10.2.1), that $P(A_1 \cup A_2) = P_n(B_1 \cup B_2) = P_n(B_1) + P_n(B_2) = P(A_1) + P(A_2)$.

Now that we have proved finite additivity of the set function P on the field $\mathfrak{F}$, we just need to show continuity of P from above at the empty set, to conclude that P is countably additive and hence a probability measure on the field $\mathfrak{F}$. To this end, let $\{A_n\}$ be a sequence of sets in $\mathfrak{F}$ with $A_n \downarrow \emptyset$. We need to show that $P(A_n) \to 0$. Since $\{A_n\}$ is a decreasing sequence of sets in the field $\mathfrak{F}$, finite additivity of P on $\mathfrak{F}$ implies that $\{P(A_n)\}$ is a decreasing sequence of non-negative reals and therefore $\lim_n P(A_n)$ exists. Suppose, if possible, $P(A_n) \not\to 0$. Then we must have $\lim_n P(A_n) > 0$. We are now going to show that this implies that $\bigcap_n A_n \neq \emptyset$, which will contradict the hypothesis that $A_n \downarrow \emptyset$. Towards this, we are going to assume, without any loss of generality, that for each n, the set A_n is an n-dimensional cylinder. To avoid distraction, justification for this is provided in Remark 10.2.2. With this assumption, let $A_n = \varphi_n^{-1}(B_n)$, where $B_n \in \bigotimes_{j=1}^{n} \mathcal{A}_j$, for each n. Then, by the definition of P on $\mathfrak{F}$, as in (10.2.1) and recalling the formula for P_n from Theorem 5.3.10 (a), we have, for every n,

$$P(A_n) = P_n(B_n) = \int_{\Omega_1} f_n^{(1)}(\omega_1) \mu_1(d\omega_1), \qquad (10.2.2)$$

where

$$f_n^{(1)}(\omega_1) = \int_{\Omega_2} \cdots \int_{\Omega_n} I_{B_n}(\omega_1, \ldots, \omega_n) \mu_n(\omega_1, \ldots, \omega_{n-1}, d\omega_n) \cdots \mu_2(\omega_1, d\omega_2).$$

Now, $A_{n+1} \subset A_n$ implies $B_{n+1} \subset B_n \times \Omega_{n+1}$ and therefore, $f_{n+1}^{(1)}(\omega_1) \leq f_n^{(1)}(\omega_1)$. So, $f^{(1)}(\omega_1) = \lim_n f_n^{(1)}(\omega_1)$ exists for all ω_1 and, by DCT and (10.2.2), we get

$$\lim_n P(A_n) = \lim_n \int_{\Omega_1} f_n^{(1)}(\omega_1) \mu_1(d\omega_1) = \int_{\Omega_1} f^{(1)}(\omega_1) \mu_1(d\omega_1).$$

The assumption that $\lim_n P(A_n) > 0$ would now imply that there must exist $\omega_1^* \in \Omega_1$ with $f^{(1)}(\omega_1^*) = \lim_n f_n^{(1)}(\omega_1^*) > 0$. Clearly $\omega_1^* \in B_1$, since otherwise $f_1^{(1)}(\omega_1^*) = 0$ and hence $f_n^{(1)}(\omega_1^*) = 0$, for all n. Next, for each $n \geq 2$, we have

$$f_n^{(1)}(\omega_1^*) = \int_{\Omega_1} f_n^{(2)}(\omega_2) \mu_2(\omega_1^*, d\omega_2), \qquad (10.2.3)$$

where, recalling again the formula given in Theorem 5.3.10,

$$f_n^{(2)}(\omega_2) = \int_{\Omega_3} \cdots \int_{\Omega_n} I_{B_n}(\omega_1^*, \omega_2, \ldots, \omega_n) \mu_n(\omega_1^*, \ldots, \omega_{n-1}, d\omega_n) \cdots \mu_3(\omega_1^*, \omega_2, d\omega_3).$$

By the same argument as before, $f_n^{(2)}(\omega_2)$ decreases with n, so that, for each ω_2, the limit $f^{(2)}(\omega_2) = \lim_n f_n^{(2)}(\omega_2)$ exists. Once again, DCT and (10.2.3) gives

$$f^{(1)}(\omega_1^*) = \lim_n f_n^{(1)}(\omega_1^*) = \lim_n \int_{\Omega_2} f_n^{(2)}(\omega_2) \mu_2(\omega_1^*, d\omega_2) = \int_{\Omega_2} f^{(2)}(\omega_2) \mu_2(\omega_1^*, d\omega_2).$$

Since $f^{(1)}(\omega_1^*) > 0$, we can get $\omega_2^* \in \Omega_2$ such that $f^{(2)}(\omega_2^*) = \lim_n f_n^{(2)}(\omega_2^*) > 0$. This, in turn, implies that $(\omega_1^*, \omega_2^*) \in B_2$, since otherwise, $f_2^{(2)}(\omega_2^*) = 0$ and hence $f_n^{(2)}(\omega_2^*) = 0$, for all $n \geq 2$.

Proceeding in this manner, one can get, for each n, a point $\omega_n^* \in \Omega_n$, such that, $(\omega_1^*, \ldots, \omega_n^*) \in B_n$, for every $n \geq 1$. But, this clearly implies that the point $\omega^* = (\omega_1^*, \omega_2^*, \ldots)$ will belong to $\varphi_n^{-1}(B_n) = A_n$, for every $n \geq 1$, and that would mean $\bigcap_n A_n \neq \emptyset$, contradicting the hypothesis that $A_n \downarrow \emptyset$. Since the contradiction came from assuming $\lim_n P(A_n) > 0$, we must have $P(A_n) \to 0$. This proves that P is countably additive on the field $\mathfrak{F}$ and hence defines a probability on $\mathfrak{F}$. An appeal to Caratheodory Extension Theorem now assures a unique extension of P to a probability on $\sigma(\mathfrak{F}) = \mathcal{A}$, the product σ-field on Ω.

We have thus proved a very important extension of Theorem 5.3.10 to infinite product spaces, as stated in the following theorem.

THEOREM **10.2.1.** (Tulcea's Theorem) Let $(\Omega_n, \mathcal{A}_n)$, $n \geq 1$, be a sequence of measurable spaces and let $(\Omega, \mathcal{A})$ be the product measurable space. If μ_1 is a probability on $(\Omega_1, \mathcal{A}_1)$ and, for each $n \geq 2$, μ_n is a probability kernel on $\left(\underset{j=1}{\overset{n-1}{\times}} \Omega_j\right) \times \mathcal{A}_n$, then there is a unique probability P on $\mathcal{A}$ such that, for any n and any n-dimensional measurable cylinder $A = \varphi_n^{-1}(B)$,

$$P(A) = \int_\Omega I_B(\omega_1, \ldots, \omega_n) \mu_n(\omega_1, \ldots, \omega_{n-1}, d\omega_n) \cdots \mu_2(\omega_1, d\omega_2) \mu_1(d\omega_1). \quad (10.2.4)$$

REMARK **10.2.2.** In the proof of Theorem 10.2.1, we claimed that, given a decreasing sequence $\{A_n\}$ of sets in $\mathfrak{F}$, there is no loss of generality in assuming that, for each n, the nth set A_n is an n-dimensional cylinder. The reason we did not give the justification there, was to avoid distraction from the main chain of thought. We are now going to justify that assumption by describing how, given any decreasing sequence $\{A_n\}$ of finite dimensional cylinders sets, one can construct a new decreasing sequence $\{A'_n\}$ of finite-dimensional cylinder sets in such a way that each A_n equals A'_m, for some m, and each A'_n equals either Ω or A_m, for some m. These will ensure that $\bigcap A'_n = \bigcap A_n$ and also $\lim_n P(A'_n) = \lim_n P(A_n)$. This means that, nothing is lost in our proof if we replace the given sequence $\{A_n\}$ by the sequence $\{A'_n\}$. The advantage is that the way we construct the sequence $\{A'_n\}$, the set A'_n, for each n, can be thought of as an n-dimensional cylinder. Here is the construction. Given the original decreasing sequence $\{A_n\}$ of finite-dimensional cylinder sets, we can always get integers $1 \leq k_1 < k_2 < \cdots < k_n < \cdots$, such that A_n is a k_n-dimensional cylinder set. For this, if necessary, one may use the fact that any n-dimensional cylinder set is also an m-dimensional cylinder set for any $m > n$. We now define the sequence $\{A'_n\}$ as follows:

$$A'_n = \begin{cases} \Omega & \text{for } n < k_1 \\ A_1 & \text{for } k_1 \leq n < k_2 \\ \vdots & \\ A_m & \text{for } k_m \leq n < k_{m+1} \\ \vdots & \end{cases}$$

10.2. Constructing Probabilities on Countably Infinite Product Spaces

The reader should be able to convince herself that the sequence $\{A'_n\}$ thus constructed ticks all the boxes.

Let us consider the special case when $(E, \mathcal{E})$ is a measurable space and $(\Omega_n, \mathcal{A}_n) = (E, \mathcal{E})$, for all n. As already mentioned in the previous section, the product space in this special case is denoted as $(E^\infty, \mathcal{E}^\infty)$. The sequence of coordinate projections on E^∞ are going to be denoted by $\{X_n\}$. Clearly, each X_n is an E-valued measurable map on $(E^\infty, \mathcal{E}^\infty)$. Tulcea's Theorem 10.2.1 applied to this special case, yields the following important result.

THEOREM **10.2.3**. *Let μ_1 be a probability on $(E, \mathcal{E})$ and let μ_n, for each $n \geq 2$, be a probability kernel on $E^{n-1} \times \mathcal{E}$. Then there exists a unique probability P on $(E^\infty, \mathcal{E}^\infty)$ such that the sequence $\{X_n\}$ of E-valued coordinate random variables on the probability space $(E^\infty, \mathcal{E}^\infty, P)$ satisfies the following:*
(i) X_1 has distribution μ_1 and (ii) μ_n, for each $n \geq 2$, is the Regular Conditional Distribution of X_n, given $(X_1, X_2, \cdots, X_{n-1})$.

Another important special case of Thoerem 10.2.1 is when, for each $n \geq 2$, the probability kernel μ_n is actually just a probability on $(\Omega_n, \mathcal{A}_n)$. In this case, the associated unique probability P on the product space $(\Omega, \mathcal{A})$ is called the 'product probability', usually written as $\underset{n}{\otimes} \mu_n$. This special case gives an extension of finite products of measures, as discussed in Section 5.2, to infinite products, except that we only consider probability measures here. Here is the theorem in this special case.

THEOREM **10.2.4**. *Let $(\Omega_n, \mathcal{A}_n, \mu_n)$, $n \geq 1$, be a sequence of probability spaces. Denoting $(\Omega, \mathcal{A})$ to be the product measurable space $\left(\underset{n=1}{\overset{\infty}{\times}} \Omega_n, \underset{n=1}{\overset{\infty}{\otimes}} \mathcal{A}_n \right)$, there is a unique probability P on $(\Omega, \mathcal{A})$, such that for any n and any collection of sets $B_1 \in \mathcal{A}_1, \cdots, B_n \in \mathcal{A}_n$,*

$$P\left(\underset{j=1}{\overset{n}{\times}} B_j \times \underset{j=n+1}{\overset{\infty}{\times}} \Omega_j \right) = \prod_{j=1}^{n} \mu_j(B_j). \tag{10.2.5}$$

REMARK **10.2.5**. Considering the special case when $(\Omega_n, \mathcal{A}_n) = (E, \mathcal{E})$, for all n, and applying Theorem 10.2.4, what one gets is the following. Denoting $\{X_n\}$, as before, to be the sequence of coordinate projections on E^∞, one easily sees that (10.2.5) simply says that if P is as in Theorem 10.2.4, then $\{X_n\}$ is a sequence of E-valued random variables on the probability space $(E^\infty, \mathcal{E}^\infty, P)$, with $X_n \sim \mu_n$ for each n, that satisfy the property that, for all $n \geq 1$ and all $B_1, \ldots, B_n \subset \mathcal{E}$,

$$P(X_1^{-1}(B_1) \cap \cdots \cap X_n^{-1}(B_n)) = \prod_{j=1}^{n} P(X_j^{-1}(B_j)). \tag{10.2.6}$$

In particular, when $(E, \mathcal{E}) = (\mathbb{R}, \mathcal{B})$, the sequence $\{X_n\}$, as above, gives a sequence of real random variables and the property (10.2.6) simply says that the X_n, $n \geq 1$, are mutually independent. Thus, an important upshot of Theorem 10.2.4 is that it guarantees the existence of a probability space and a sequence of real random variables defined on it, which are independent with each having its own prescribed distribution. When the prescribed distributions are all equal, that is, $\mu_n = Q$, say, for all n, we get an i.i.d. sequence of random variables with a specified common distribution Q. In this special case, the product probability P on $(\mathbb{R}^\infty, \mathcal{B}^\infty)$ is usually denoted by Q^∞. For the record, we state the observation made here in the following theorem.

THEOREM 10.2.6. *Given any sequence $\{\mu_n\}$ of probability measures on $(\mathbb{R}, \mathcal{B})$, there is a probability space $(\Omega, \mathcal{A}, P)$ and a sequence $\{X_n\}$ of real random variables defined on it, such that, for each n, X_n has distribution μ_n and $\{X_n, n \geq 1\}$ are mutually independent.*

Of course, Remark 10.2.5 already says that an assertion like Theorem 10.2.6 is true for not just real random variables, but E-valued random variables also, where independence is to be understood as independence of the σ-fields $\sigma(X_n)$, $n \geq 1$.

REMARK 10.2.7. Having just talked about product probabilities on $(\mathbb{R}^\infty, \mathcal{B}^\infty)$, we want to point out an important connect between the theory of infinite products and some of the results we proved in Subsection 7.3.2 and Section 7.4. These surely would not have escaped the attention of a careful reader. Firstly, the classical SLLN (Theorem 7.3.10) is clearly equivalent to asserting that for any probability Q on $(\mathbb{R}, \mathcal{B})$ with a finite mean μ, the product probability Q^∞ on $(\mathbb{R}^\infty, \mathcal{B}^\infty)$ is fully supported on the set $A = \left\{\mathbf{x} = (x_1, x_2, \cdots) \in \mathbb{R}^\infty : \frac{1}{n}\sum_{i=1}^{n} x_i \to \mu \text{ as } n \to \infty\right\}$. More generally, if Q is any probability on $(\mathbb{R}, \mathcal{B})$ with a finite kth moment θ_k, then the set $\left\{\mathbf{x} = (x_1, x_2, \cdots) \in \mathbb{R}^\infty : \frac{1}{n}\sum_{i=1}^{n} x_i^k \to \theta_k \text{ as } n \to \infty\right\}$ has Q^∞-measure 1. Also, Kolmogorov's zero-one law (Theorem 7.4.5) is equivalent to the assertion that if P is a product probability $\otimes \mu_n$ on $(\mathbb{R}^\infty, \mathcal{B}^\infty)$, then every "tail set" has P-probability 0 or 1. Defining "tail σ-field" and "tail sets" in this context are left to the reader.

We now mention another important application of Tulcea's theorem. What the theorem does is that it guarantees existence of what are called '*Markov Chains*', with specified initial distribution and transition probabilities. We first discuss the case of Markov Chains with what is called a '*discrete state space*'.

Let S be a countable set and consider a function defined on $S \times S$ and taking values in the interval $[0, 1]$ that satisfies

$$\sum_{y \in S} p(x, y) = 1, \quad \text{for all } x \in S. \tag{10.2.7}$$

It is a common practice to think of the function p as an $S \times S$ matrix $\mathbf{p} = ((p_{xy}))$, where $p_{xy} = p(x, y)$. When written in this form, the property (10.2.7) translates to requiring that all the row sums of the matrix $\mathbf{p}$ equal 1. A $S \times S$ matrix $\mathbf{p}$ with non-negative entries and having the above row sum property is commonly called a '*stochastic matrix*' on S. If we now take $(E, \mathcal{E})$ as $(S, \mathfrak{S})$, where $\mathfrak{S} = \mathcal{P}(S)$, the power set of S, it is then clear that, for each $n \geq 2$, the function μ_n on $S^{n-1} \times \mathfrak{S}$ given by

$$\mu_n\big((x_1, \ldots, x_{n-1}), B\big) = \sum_{y \in B} p(x_{n-1}, y)$$

defines a probability kernel on $S^{n-1} \times \mathfrak{S}$.

10.2. Constructing Probabilities on Countably Infinite Product Spaces

Given any probability μ on $(S, \mathfrak{S})$, let P denote the unique probability on $(S^\infty, \mathfrak{S}^\infty)$ constructed out of $\mu_1 = \mu$ and μ_n, $n \geq 2$, as asserted in Tulcea's Thorem 10.2.1. Considering the probability space $(\Omega, \mathcal{A}, P) = (S^\infty, \mathfrak{S}^\infty, P)$ and the sequence $\{X_n\}$ of coordinate random variables on it, we get a sequence of random variables, with each taking values in the countable set S, such that $P(X_1 = x) = \mu(x)$ for all $x \in S$, and, for any $n \geq 2$ and any $x_1, \ldots, x_{n-1}, y \in S$,
$$P(X_n = y \mid X_1 = x_1, \cdots, X_{n-1} = x_{n-1}) = p(x_{n-1}, y)$$
$$= P(X_n = y \mid X_{n-1} = x_{n-1}).$$

A reader familiar with the theory of Markov chains on a countable state space S would recognize that it is the above property of S-valued random variables $\{X_n\}$ that defines a Markov Chain. She would also realize that what Tulcea's Theorem 10.2.1 is doing is that it guarantees that given any probability μ on $(S, \mathfrak{S})$ and any $S \times S$ stochastic matrix $\mathbf{p} = ((p_{xy}))$, there exists a probability space and a Markov Chain $\{X_n : n \geq 1\}$ with state space S, initial distribution μ and $\mathbf{p} = ((p_{xy}))$ as its transition probability matrix.

Application of Tulcea's Theorem 10.2.1, however, is not limited to constructing Markov Chains on countable state spaces only. Similar construction can be done to ensure existence of Markov Chains on any arbitrary state space. Let $(E, \mathcal{E})$ be any measurable space and let p be a probability kernel on $E \times \mathcal{E}$. As in the discrete case, one can use the given kernel p to define, for each $n \geq 2$, a probability kernel μ_n on $E^{n-1} \times \mathcal{E}$ by setting
$$\mu_n(x_1, \cdots, x_{n-1}, \cdot) = p(x_{n-1}, \cdot), \text{ for } x_1, \ldots, x_{n-1} \in E.$$
Given any probability μ on $\mathcal{E}$, one can apply Tulcea's Theorem 10.2.1 to get a unique probability P on $(E^\infty, \mathcal{E}^\infty)$, based on $\mu_1 = \mu$ and the μ_n, $n \geq 2$. If one now considers the probability space $(E^\infty, \mathcal{E}^\infty, P)$ and the sequence $\{X_n\}$ of coordinate (E-valued) random variables on it, one easily sees that $\{X_n\}$ is a Markov Chain with state space E, intial distribution μ and transition probabilities given by p, that is, for any $n \geq 2$, any $x_1, \ldots, x_{n-1} \in E$ and any $B \in \mathcal{E}$, one has
$$P(X_n \in B \mid X_1 = x_1, \ldots, X_{n-1} = x_{n-1}) = p(x_{n-1}, B) = P(X_n \in B \mid X_{n-1} = x_{n-1}).$$
In the above, the conditional probabilities are to be understood in the sense stated in Theorem 9.3.7, in the context of regular conditional distributions.

Here is an interesting and very familiar example of a Markov Chain on the state space $E = (0, 1)$, equipped with its Borel σ-field.

EXAMPLE 10.2.8. Take $(E, \mathcal{E})$ to be the interval $(0, 1)$, equipped with its Borel σ-field $\mathcal{B}((0, 1))$. Take the initial μ to be the Lebesgue measure on $(0, 1)$, which, in basic probability theory, commonly goes by the name '*uniform distribution on* $(0, 1)$'. The transition probabilities are given by the probability kernel p on $(0, 1) \times \mathcal{B}((0, 1))$, defined as follows. For $x \in (0, 1)$ and $B \in \mathcal{B}((0, 1))$, set
$$p(x, B) = \tfrac{1}{1-x} \lambda([x, 1) \cap B) I_{(0, \frac{1}{2})}(x) + \tfrac{1}{x} \lambda((0, x] \cap B) I_{[\frac{1}{2}, 1)}(x).$$
The evolution of the associated Markov Chain may be loosely described as follows. Initially a point x_1 is chosen at random from $(0, 1)$. If this chosen point x_1 happens

to lie in $(0, \frac{1}{2})$, then the next point x_2 is chosen at random from $[x, 1)$, while if x_1 lies in $[\frac{1}{2}, 1)$, then x_2 is chosen at random from $(0, x]$. The same mechanism is repeated to choose the next point x_3, depending only on the location of x_2. In other words, at any stage, depending on the location of the last chosen point, a subsequent point is chosen at random from the larger of the two intervals determined by the last chosen point, independently of the points chosen prior to it.

We now turn to a slightly different perspective for construction of probabilities on the infinite product $(E^\infty, \mathcal{E}^\infty)$ of a measurable space $(E, \mathcal{E})$. In Tulcea's construction, we started with a probability measure μ_1 on $(E, \mathcal{E})$ and a sequence $\{\mu_j, j \geq 2\}$ where μ_j, for each $j \geq 2$, is a probability kernel on $E^{j-1} \times \mathcal{E}$. If one goes back to the main idea of the construction of the probability P of Theorem 10.2.1, with focus only on the special case when $(\Omega_n, \mathcal{A}_n) = (E, \mathcal{E})$ for all n, the first step was to consider, for each n, the probability P_n on $(E^n, \mathcal{E}^n)$ constructed, as in Theorem 5.3.10, from the given probability μ_1 and the probability kernels μ_j, $2 \leq j \leq n$. These P_n's were used to define a set function P on the field $\mathfrak{F}$ of all finite-dimensional cylinder sets. We then showed that P is a probability on $\mathfrak{F}$ and used Caratheodory Extension Theorem to get a unique extension of P to the required probability on $\sigma(\mathfrak{F}) = \mathcal{E}^\infty$. What we now want to ask is this. What if, instead of constructing the P_n, $n \geq 1$, from given μ_1 and μ_j, $j \geq 1$, we were just given a sequence of probabilities $\{P_n\}$ to start with, where P_n, for each n, is a probability on $\mathcal{E}^n$? Could we then use these to define a set function P on the field $\mathfrak{F}$ of all finite dimensional cylinder sets? Of course, for such a set function to be well-defined and also finitely additive, one would need some obvious relation among the different P_n's, known as '*consistency*' condition, but let us set that aside for a while and assume that all of that is true. Can we now go ahead and show that this P is countably additive on $\mathfrak{F}$? At this point, recall that, in the argument leading to Tulcea's Theorem 10.2.1, the fact that the P_n's were constructed using μ_1 and the μ_j, $j \geq 2$, played a crucial role in proving that P is countably additive. This suggests that something has to make up for the absence of μ_1 and the μ_j, $j \geq 2$, from the picture. But before we come to that, let us try to reformulate our question in a slightly different way that perhaps captures a more compelling point of view. Given a probability P_n on $(E^n, \mathcal{E}^n)$, for each $n \geq 1$, we are really asking whether it is possible to construct a sequence $\{X_n\}$ of E-valued random variables on some probability space such that, for each n, the joint distribution of $(X_1, \ldots, X_n)$ is the given P_n. It should be clear to the reader that if we could construct a probability P on $(E^\infty, \mathcal{E}^\infty)$, such that, for any $n \geq 1$ and any n-dimensional cylinder A with base $B \in \mathcal{E}^n$, one has $P(A) = P_n(B)$, then the sequence $\{X_n\}$ of coordinate random variables on the probability space $(E^\infty, \mathcal{E}^\infty, P)$ would give us such a required sequence.

Now that we have formulated the problem in this manner, it would be clear that, in order for there to be a sequence $\{X_n\}$ having P_n as the joint distribution of $(X_1, \ldots, X_n)$, for each n, the specified probabilities $P_n, n \geq 1$, cannot be completely arbitrary. Indeed, if P_{n+1} is the joint distribution of $(X_1, \ldots, X_{n+1})$,

10.2. Constructing Probabilities on Countably Infinite Product Spaces

then that of $(X_1, \ldots, X_n)$ must be given by $sP((X_1, \ldots, X_n) \in B) = P_{n+1}(B \times E)$. So, the specified joint distribution P_n must satisfy $P_n(B) = P_{n+1}(B \times E)$ for all $B \in \mathcal{E}^n$. And this must be true for all n. There is another way of expressing this condition. Denoting the 'natural' projection map on E^{n+1} onto E^n by φ_n^{n+1}, namely, $\varphi_n^{n+1}(x_1, \ldots, x_n, x_{n+1}) = (x_1, \ldots, x_n)$, the above condition is the same as requiring $P_n = P_{n+1}(\varphi_n^{n+1})^{-1}$, for all n. This is called the '*consistency*' condition. For the sake of records, we put it in the next definition.

DEFINITION **10.2.9.** Given a probability P_n on $(E^n, \mathcal{E}^n)$, for each $n \geq 1$, the family $\{P_n\}$ is said to satisfy the *consistency condition* if, for each $n \geq 1$,

$$P_n = P_{n+1}(\varphi_n^{n+1})^{-1}, \text{ equivalently, } P_n(B) = P_{n+1}(B \times E) \text{ for all } B \in \mathcal{E}^n \quad (10.2.8)$$

We just saw that in order to have a sequence $\{X_n\}$ of E-valued random variables on some probability space such that P_n, for each n, is the joint distribution of $(X_1, \ldots, X_n)$, it is necessary that family $\{P_n\}$ satisfy the consistency condition. The question is whether it is sufficient.

The answer to this question is 'no' in general. To understand where the problem comes, let us note that if we use the given P_n, $n \geq 1$, to define P on the field $\mathfrak{F}$ of finite dimensional cylinders as was done earlier, it is fairly easy to see that the consistency condition would ensure that the set function P is well-defined and finitely additive. The real problem comes in proving countable additivity when $(E, \mathcal{E})$ is an arbitrary measure space. It is only for some special measure spaces, with some additional structure, that one can actually push it through. In what follows, we are going to do it for the special case when $(E, \mathcal{E}) = (\mathbb{R}, \mathcal{B})$.

Suppose we are given, for each $n \geq 1$, a probability P_n on $(\mathbb{R}^n, \mathcal{B}^n)$ and suppose that the family $\{P_n, n \geq 1\}$ satisfy the consistency condition (10.2.8). We are going to construct a probability P on $(\mathbb{R}^\infty, \mathcal{B}^\infty)$ such that the sequence $\{X_n\}$ of coordinate random variables on $(\mathbb{R}^\infty, \mathcal{B}^\infty, P)$ will satisfy the property that P_n, for each $n \geq 1$, is the joint distribution of $(X_1, \ldots, X_n)$. Clearly, we will start with defining a set function P on the field $\mathfrak{F}$ of all finite dimensional cylinders by setting $P(A) = P_n(B)$, if A is an n-dimensional cylinder A with base $B \in \mathcal{B}^n$, that is, $A = \varphi_n^{-1}(B)$, where $\varphi_n : \mathbb{R}^\infty \to \mathbb{R}^n$ is the projection to first n coordinates, as defined in Section 10.1. As already pointed out, by virtue of the assumed consistency condition, the set function P is well-defined and finitely additive on $\mathfrak{F}$. So, we only need to prove countable additivity of P on $\mathfrak{F}$, or equivalently, to prove that $P(A_n) \to 0$, for any sequence $\{A_n\}$ of $\mathfrak{F}$-sets with $A_n \downarrow \emptyset$. Suppose, on the contrary, this is not true, that is, there is a sequence $\{A_n\}$ of $\mathfrak{F}$-sets with $A_n \downarrow \emptyset$, such that $P(A_n) \not\to 0$. Since $P(A_n)$ is decreasing in n, we must then have an $\epsilon > 0$ such that $P(A_n) > \epsilon$, for all n. As argued in Remark 10.2.2, we can assume, without loss of generality, that A_n, for each n, is an n-dimensional cylinder, that is, $A_n = \varphi_n^{-1}(B_n)$, for some $B_n \in \mathcal{B}^n$. Now, by the definition of P on $\mathfrak{F}$, the condition $P(A_n) > \epsilon$ is the same as $P_n(B_n) > \epsilon$. It is here that we use the special structure of the space $(\mathbb{R}^n, \mathcal{B}^n)$, that says that every probability on $(\mathbb{R}^n, \mathcal{B}^n)$ is regular (see Theorem 2.10.1). Using this, we can get, for each n, a compact set $K_n \subset B_n$ with $P_n(B_n \setminus K_n) < \frac{\epsilon}{2^{n+1}}$. But then $C_n = \varphi_n^{-1}(K_n) \in \mathfrak{F}$ satisfies $P(A_n \setminus C_n) = P(B_n \setminus K_n) < \frac{\epsilon}{2^{n+1}}$. By setting $D_n = \bigcap_{j=1}^{n} C_j$, we get a decreasing sequence $\{D_n\}$ of $\mathfrak{F}$-sets, such that, $D_n \subset A_n$ and $P(A_n \setminus D_n) \leq \sum_{j=1}^{n} P(A_j \setminus C_j) < \frac{\epsilon}{2}$; in particular, $D_n \neq \emptyset$, for each n. We choose, for each n, any point $\mathbf{x}(n) = (x_1(n), x_2(n), \ldots) \in D_n$. Now, by the definition of the sets D_n, the sequence $\{x_1(n)\}$ belongs to the compact set K_1, and, therefore, has a subsequence $\{x_1(n_{1k}), k \geq 1\}$, where $1 < n_{11} < n_{12} < \ldots$, such that, $\{x_1(n)\}$ converges to a point, say, $x_1 \in K_1$. Next, all the terms of the sequence $\{(x_1(n_{1k}), x_2(n_{1k})), k \geq 2\}$, once again by the definition of the D_n, belong to the compact set $K_2 \subset R^2$ and, therefore, we can get a further subsequence $2 < n_{21} < n_{22} < \ldots$ of $\{n_{1k}\}$, such that, $\{(x_1(n_{2k}), x_2(n_{2k})), k \geq 1\}$ converges to a point, say, $(x_1, x_2) \in K_2$. Note that the first coordinate of this limit has to be the $x_1 \in K_1$ obtained earlier, because $\{x_1(n_{2k})\}$ is a subsequence of $\{x_1(n_{1k}), k \geq 1\}$. Proceeding this way, we can get, for every $j \geq 1$, a sequence $j < n_{j1} < n_{j2} < \ldots$ of natural numbers and a real number x_j, such that,

(i) $\{n_{j+1,k}, k \geq 1\}$ is a subsequence of $\{n_{jk}, k \geq 1\}$, for each j, and

(ii) $\{(x_1(n_{jk}), \ldots, x_j(n_{jk}))\}$ converges to $(x_1, \cdots, x_j) \in K_j$, for each j.

With $\mathbf{x} = (x_1, x_2, \ldots)$, where the x_j are as above, we have $(x_1, \cdots, x_n) \in K_n$, for each n, which implies that $\mathbf{x} \in C_n = \varphi_n^{-1}(K_n) \subset A_n$, for each n, and hence $\mathbf{x} \in \bigcap_n A_n$, contradicting the hypothesis that $A_n \downarrow \emptyset$. This contradiction proves that P defined on $\mathfrak{F}$ is continuous from above at the emptyset and hence defines a probability on $\mathfrak{F}$.

Using Caratheodory Extension Theorem, we can now get a unique extension of P to a probability on $\sigma(\mathfrak{F}) = \mathcal{B}^\infty$. Through all of these, we have now proved the following theorem, usually referred to as *Kolmogorov Consistency Theorem*. The consistency condition in the statement is as in Definition 10.2.9

THEOREM **10.2.10.** (Kolmogorov Consistency Theorem) For each $n \geq 1$, let P_n be a probability on $\mathcal{B}(\mathbb{R}^n)$. Then there is a unique probability P on $(\mathbb{R}^\infty, \mathcal{B}^\infty)$ such that $P(\varphi_n^{-1}(B)) = P_n(B)$, for every n and every $B \in \mathcal{B}(\mathbb{R}^n)$ if and only if the family $\{P_n, n \geq 1\}$ satisfies the consistency condition.

As pointed out before, an immediate consequence of Kolmogorov Consistency Theorem is that given any family of probabilities P_n on $(\mathbb{R}^n, \mathcal{B}^n)$, $n \geq 1$, we can construct a sequence $\{X_n\}$ of real random variables on some probability space, with $(X_1, \ldots, X_n)$ having joint distribution P_n, for each $n \geq 1$, provided that $\{P_n, n \geq 1\}$ satisfies the consistency condition.

We describe here one such interesting example. Let $\{\mu_n, n \geq 1\}$ be any real sequence and let C be a symmetric positive definite kernel on $\mathbb{N} \times \mathbb{N}$, meaning that, for each $n \geq 1$, the matrix $\Sigma_n = \left((C(i,j))\right)_{1 \leq i,j \leq n}$ is a symmetric positive definite matrix. Denoting P_n, for each n, to be the n-dimensional Gaussian distribution with mean vector $(\mu_1, \ldots, \mu_n)'$ and dispersion matrix Σ_n, one can easily see, from properties of Gaussian distributions, that the family $\{P_n, n \geq 1\}$ satisfies the consistency condition. Therefore, by Theorem 10.2.10, one can see that there is a unique probability P on $(\mathbb{R}^\infty, \mathcal{B}^\infty)$ such that the sequence $\{X_n\}$ of coordinate random variables on the probability space $(\mathbb{R}^\infty, \mathcal{B}^\infty, P)$ will satisfy the property that for every $n \geq 1$, the random vector $(X_1, \ldots, X_n)'$ has Gaussian distribution with mean vector $(\mu_1, \ldots, \mu_n)'$ and dispersion matrix Σ_n.

REMARK 10.2.11. Here is an important observation contrasting Tulcea's Theoerm with Kolmogorov Consistency Theoerm. First of all, we claim that Tulcea's Theorem for the special case when $(\Omega_n, \mathcal{A}_n) = (\mathbb{R}, \mathcal{B})$, $n \geq 1$, is an immediate consequence of Kolmogorov Consistency Theorem. Indeed, given a probability μ_1 on $(\mathbb{R}, \mathcal{B})$ and probability kernels μ_n on $\mathbb{R}^{n-1} \times \mathcal{B}$, for $n \geq 2$, the probabilities P_n constructed on $(\mathbb{R}^n, \mathcal{B}^n)$, for $n \geq 1$, as in Theorem 5.3.10, can be easily seen to satisfy the consistency condition. Therefore, we can get the required probability P on $(\mathbb{R}^\infty, \mathcal{B}^\infty)$,.. using Kolmogorov Consistency Theorem. This observation may lead one to wonder whether Tulcea's Theorem is weaker than Kolmogorov Consistency Theorem. However, a moment's reflection will convince the reader that this is not at all the case. One must not forget that Tulcea's Theorem holds for any sequence $(\Omega_n, \mathcal{A}_n)$, $n \geq 1$, of measurable spaces and not just when $(\Omega_n, \mathcal{A}_n) = (\mathbb{R}, \mathcal{B})$ for all n.

10.3 Probabilities on Uncountable Product Spaces

Having made the journey from finite products of measurable spaces to countably infinite products, we now want to go a step further and discuss arbitrary, possibly uncountable, products. We consider an arbitrary family $(\Omega_t, \mathcal{A}_t)$, $t \in T$, of measurable spaces, where the index set T is possibly uncountable. Our first aim would be to describe what is meant by the product of these measurable spaces. Later, we will take up the issue of constructing probabilities on these product spaces. The first step, of course, has to be to identify the underlying product set. As always and as is natural, the product set will be the cartesian product $\Omega = \underset{t \in T}{\times} \Omega_t$. In case the reader is not familiar with arbitrary cartesian products, this is just the set of all functions $\omega : T \to \bigcup_{t \in T} \Omega_t$, such that, $\omega(t) \in \Omega_t$, $t \in T$. For any $S \subset T$, let $\varphi_S : \underset{t \subset T}{\times} \Omega_t \to \underset{t \in S}{\times} \Omega_t$ denote the natural projection. What this means is that $\varphi_S(\omega)$, for any $\omega \in \Omega$, is nothing but the 'restriction' of the function ω on T to the subset S. In case $S \subset T$ is a finite set and if B is a set in the finite product σ-field $\underset{t \in S}{\otimes} \mathcal{A}_t$, then the set $\varphi_S^{-1}(B) \subset \Omega$ is called a finite dimensional cylinder with base B.

It is easy to check that the class $\mathfrak{F}$ consisting of all finite dimensional cylinders form a field on Ω. The σ-field $\sigma(\mathfrak{F})$ is called the product σ-field on Ω and is denoted by $\mathcal{A} = \underset{t \in T}{\otimes} \mathcal{A}_t$. The following are easy exercises, giving alternative, and often useful, descriptions of the product σ-field $\mathcal{A}$.

EXERCISE 10.3.1. For each $t \in T$, let e_t denote the 'evaluation map' on Ω onto Ω_t, namely, $e_t : \omega \mapsto \omega(t)$. Show that $\mathcal{A} = \sigma(e_t, t \in T)$. Show further that, if $(E, \mathcal{E})$ is any measurable space, then a function $f : (E, \mathcal{E}) \to (\Omega, \mathcal{A})$ is measurable iff $f \circ e_t$ is measurable, for every $t \in T$.

EXERCISE 10.3.2. For $S \subset T$, let $\mathcal{A}_S = \{\varphi_S^{-1}(A) : A \in \otimes_{t \in S} \mathcal{A}_t\}$. Show that $\mathcal{A} = \bigcup\{\mathcal{A}_S : S \subset T \text{ countable}\}$. (see Exercise 2.13.7.)

REMARK 10.3.3. Exercise 10.3.2 has an interesting interpretation, which gives an extremely useful description of sets in $\mathcal{A}$. Roughly, what it says is that every set in $\mathcal{A}$ is essentially a set in a countable product σ-field. In fact, what we are going to now see is that it plays a very crucial role in construction of probabilities on $\mathcal{A}$

In view of Exercise 10.3.2 (see also Remark 10.3.3), it turns out that, for constructing a probability on the product space $(\Omega, \mathcal{A})$, one just needs to construct probabilities on all countable products in a 'consistent' manner. This is exactly what is made precise in Lemma 10.3.4. The following notation will be used in the statement of the lemma. For $S \subset T$, we have already used φ_S to denote the 'restriction' map $\varphi_S : \omega \mapsto \omega|_S$, where $\omega|_S$ stands for 'restriction of ω to S. Now, for two subsets S_1 and S_2 of T with $S_1 \subset S_2$, we use the notation $\varphi_{S_1}^{S_2}$ to denote the 'restriction map from S_2 to S_1'. In this notation, one may denote φ_S, for $S \subset T$, also as φ_S^T, although we shall not do that.

LEMMA 10.3.4. Suppose for each countable $S \subset T$, there is a probability P_S on $\otimes_{t \in S} \mathcal{A}_t$. Further, suppose that, the family $\{P_S : S \subset T, S \text{ countable}\}$ is consistent in the sense that, for every pair S_1 and S_2 of countable subsets of T with $S_1 \subset S_2$, one has $P_{S_2}(\varphi_{S_1}^{S_2})^{-1} = P_{S_1}$. Then there is a unique probability P on $(\Omega, \mathcal{A})$ such that

$$P(\varphi_S^{-1}(A)) = P_S(A), \qquad (10.3.1)$$

for every countable $S \subset T$ and every $A \in \otimes_{t \in S} \mathcal{A}_t$.

Proof. Noting that every set $\widetilde{A} \in \mathcal{A}$ equals $\varphi_S^{-1}(A)$, for some countable $S \subset T$ and some $A \in \otimes_{t \in S} \mathcal{A}_t$ (see Exercise 10.3.2), the assumed consistency guarantees that (10.3.1) gives a well-defined non-negative set function P on $\mathcal{A}$, that clearly satisfies $P(\emptyset) = 0$, $P(\Omega) = 1$. To see that P is a probability on $\mathcal{A}$, we only need to prove countable additivity. If $\widetilde{A}_n$, $n \geq 1$, are disjoint $\mathcal{A}$-sets, then, by Exercise 10.3.2 again, there are countable sets $S_n \subset T$, $n \geq 1$, such that $\widetilde{A}_n \in \mathcal{A}_{S_n}$, for each n. But then $S = \bigcup_n S_n$ is countable set and clearly, $\widetilde{A}_n \in \mathcal{A}_S$, for all n, that is, for each n, there is $A_n \in \otimes_{t \in S} \mathcal{A}_t$ such that $\widetilde{A}_n = \varphi_S^{-1}(A_n)$. Disjointness of the $\widetilde{A}_n$, $n \geq 1$, implies that the A_n, $n \geq 1$, are disjoint. Since $\bigcup_n \widetilde{A}_n = \varphi_S^{-1}(\bigcup_n A_n)$, one completes the proof using 10.3.1 and countable additivity of P_S. □

The following two results are now immediate consequences of the above Lemma, using Theorem 10.2.4 for the first one and Kolmogorov Consistency

10.3. Probabilities on Uncountable Product Spaces

Theorem 10.2.10 for the second one. In the statements of the theorems, the notations φ_S and $\varphi_{S_1}^{S_2}$, for $S \subset T$ and $S_1 \subset S_2 \subset T$ respectively, are used to mean appropriate 'restriction' maps as defined earlier.

THEOREM 10.3.5. Let $(\Omega_t, \mathcal{A}_t, P_t)$, $t \in T$, be a family of probability spaces and let $\Omega = \underset{t \in T}{\times} \Omega_t$, $\mathcal{A} = \underset{t \in T}{\otimes} \mathcal{A}_t$. Then there is a unique probability P on $(\Omega, \mathcal{A})$, such that, for any $t_1, \ldots, t_n \in T$ and any choice of $A_1 \in \mathcal{A}_{t_1}, \ldots, A_n \in \mathcal{A}_{t_n}$,

$$P\big((\varphi_{\{t_1,\ldots,t_n\}})^{-1}(A_1 \times \cdots \times A_n)\big) = \prod_{i=1}^n P_{t_i}(A_i).$$

THEOREM 10.3.6. Let T be any arbitrary index set and, for every finite subset $S \subset T$, let P_S be a probability on $(\mathbb{R}^S, \mathcal{B}^S)$. Suppose that the family of probabilities $\{P_S : S \subset T, S \text{ finite}\}$ satisfies the consistency condition that $P_{S_2}(\varphi_{S_1}^{S_2})^{-1} = P_{S_1}$, for every pair S_1 and S_2 of finite subsets of T, with $S_1 \subset S_2$. Then there is a unique probability P on $(\mathbb{R}^T, \mathcal{B}^T)$ such that $P(\varphi_S^{-1}(A)) = P_S(A)$, for every finite set $S \subset T$ and every $A \in \mathcal{B}^S$.

Denoting X_t, $t \in T$, to be the usual coordinate projections or evaluation maps on Ω in Theorem 10.3.5, one can easily see that the theorem really asserts that, given arbitrary probablities P_t on $(\Omega_t, \mathcal{A}_t)$, for each $t \in T$, it is possible to construct a probability space and a family $\{X_t, t \in T\}$ of mutually independent random variables on it, such that, X_t, for each $t \in T$, takes values in Ω_t and has the specified P_t as its distribution.

Similarly, denoting X_t, $t \in T$, to be the coordinate projections or evaluation maps on $\mathbb{R}^T$, the upshot of Theorem 10.3.6 is that, given a consistent family of finite dimensional distributions, that is, a consistent family of probabilities P_S on $\mathcal{B}^S$, with S varying over all finite subsets of T, one can construct a probability space and a family of real random variables $\{X_t, t \in T\}$ on it, such that, they have the specified finite dimensional distributions, that is, for every finite $S \subset T$, the random vector $\{X_t, t \in S\}$ has joint distribution P_S.

As an important application of the second observation above, we consider a special case. Let $T \subset \mathbb{R}$ be any interval, μ any real function on T and let C be any real symmetric non-negative definite kernel on $T \times T$, meaning that, for any choice of a finite set of points $t_1 < \cdots < t_n$ in T, the matrix $\Sigma_{t_1,\ldots,t_n} = ((C(t_i,t_j)))$ is a real symmetric non-negative definite matrix. If, for each $t_1 < \cdots < t_n$ in T, one denotes $P_{\{t_1,\ldots,t_n\}}$ to be the n-dimensional Gaussian distribution with mean vector $(\mu(t_1), \ldots, \mu(t_n))'$ and dispersion matrix $\Sigma_{t_1,\ldots,t_n}$, then, from properties of Gaussian distributions (see Proposition 8.9.13), it follows that the family $\{P_{\{t_1,\ldots,t_n\}}\}$, as $t_1 < \cdots < t_n$ vary over all possible choices of finite sets of points from T, satisfies the consistency condition stated in Theorem 10.3.6. With the unique probability P on $(\mathbb{R}^T, \mathcal{B}^T)$ that one gets from Theorem 10.3.6, it will follow from the theorem that the family $\{X_t, t \in T\}$ of coordinate random variables on the probability space $(\mathbb{R}^T, \mathcal{B}^T, P)$ has the property that for any $t_1 < \cdots < t_n$ in T, the joint distribution of the random vector $(X_{t_1}, \ldots, X_{t_n})'$ is an n-diemnsional Gaussian distribution with mean vector $(\mu(t_1), \ldots, \mu(t_n))'$ and dispersion matrix $\Sigma_{t_1,\ldots,t_n}$. In the theory of stochastic processes, such a family $\{X_t, t \in T\}$ is called a '*Gaussian Process*' on index set T, with mean function μ and covariance kernel C. Thus, Kolmogorov Consistency Theorem 10.2.10 and Theorem 10.3.6 assert the existence of a Gaussian Process on any interval $T \subset \mathbb{R}$ as its index set, with any given real function on T as its mean function and any given real symmetric non-negative definite kernel on $T \times T$ as its covariance kernel. Note here that, by not requiring C to be strictly positive definite, we are allowing possible degeneracy in the Gaussian distributions. By far, the most important Gaussian process is what is known as the '*Standard Brownian Motion on* $[0, \infty)$', as defined below.

DEFINITION **10.3.7.** *Standard Brownian Motion* (abbreviated as *SBM*) on $[0, \infty)$ is defined to be a family $\{B_t, t \in [0, \infty)\}$ of real random variables on some probability space $(\Omega, \mathcal{A}, P)$, such that,

(i) $B_0 = 0$ $[P]$-a.s.,
(ii) $B_t - B_s \sim \mathsf{N}(0, t-s)$, for all $0 \le s < t$,
(iii) $B_{t_1}, B_{t_2} - B_{t_1}, \ldots, B_{t_n} - B_{t_{n-1}}$ are independent, for any $0 < t_1 < \cdots < t_n$, and,
(iv) for $[P]$-a.e. $\omega \in \Omega$, the map $t \mapsto B_t(\omega)$ is continuous on $[0, \infty)$.

The map $t \mapsto B_t(\omega)$ is often referred to as the ω-trajectory or ω-path and the property (iv) is paraphrased by saying that '*P*-almost all trajectories are continuous'.

Properties (i), (ii) and (iii) in the above definition are equivalent to saying that if $\{B_t, t \in [0, \infty)\}$ is a SBM, then for any choice of a finite set of points $t_1 < \cdots < t_n$ in $[0, \infty)$, the n-dimensional random vector $(B_{t_1}, B_{t_2} - B_{t_1}, \ldots, B_{t_n} - B_{t_{n-1}})'$ has a Gaussian distribution with mean vector $(0, \ldots, 0)'$ and the diagonal matrix with diagonal entries $t_1, t_2 - t_1, \ldots, t_n - t_{n-1}$ as its dispersion matrix. Further, the Gaussian distribution is non-degenerate, except when $t_1 = 0$. Since the map $(B_{t_1}, B_{t_2} - B_{t_1}, \ldots, B_{t_n} - B_{t_{n-1}}) \mapsto (B_{t_1}, B_{t_2}, \ldots, B_{t_n})$ is a non-singular linear transformation, one can use Proposition 8.9.13 to see that the above is equivalent to saying that, for any choice of $t_1 < \cdots < t_n$ in $[0, \infty)$, the n-dimensional random vector $(B_{t_1}, B_{t_2}, \ldots, B_{t_n})'$ has a Gaussian distribution with mean vector $(0, \ldots, 0)'$ and dispersion matrix $\Sigma_{t_1,\ldots,t_n} = ((t_i \wedge t_j))$, which is non-degenerate, by Propositions 4.2.6 and 4.5.7, unless $t_1 = 0$. Thus, properties (i), (ii) and (iii) of SBM together are essentially equivalent to saying that it is a Gaussian process with mean function $\mu(\cdot) \equiv 0$ and covariance kernel $C(s, t) = s \wedge t$. Although non-negative definiteness of the above kernel $C(\cdot, \cdot)$ is automatic from the above, the reader may verify independently that $C(s,t) = s \wedge t$ is a symmetric non-negative definite kernel on $[0, \infty) \times [0, \infty)$. One can now invoke Theorem 10.3.6 to conclude that there is a unique probability P on $(\Omega, \mathcal{A}) = (\mathbb{R}^{[0, \infty)}, \mathcal{B}^{[0, \infty)})$, such that, the family $\{X_t, t \in [0, \infty)\}$ of coordinate random variables on $(\Omega, \mathcal{A}, P)$ satisfies properties (i), (ii) and (iii) of a SBM. However, this still does not give us a SBM, because the very important property (iv), namely, the continuity of P-a.e. trajectory, is missing. If the set C of all real continuous functions on $[0, \infty)$ were a set in the σ-field $\mathcal{B}^{[0, \infty)}$ and if one could prove that $P(C) = 1$, then by restricting the σ-field $\mathcal{B}^{[0,\infty)}$ and the probability P to the set C, one would have obtained a probability space on which the same family of coordinate random variables would satisfy all the properties (i)–(iv), and thus we could have constructed the Standard Brownian Motion. However using Exercise 10.3.2, the reader would be easily able to see that the set C does not even belong to $\mathcal{B}^{[0,\infty)}$, let alone having P-probability one. Interestingly, one can show that (Exercise 10.5.9), if D denotes the complement of C in $\mathbb{R}^{[0,\infty)}$, then, $P^*(D) = 1$, for any probability P on $\mathcal{B}^{[0,\infty)}$, where P^* denotes the outer probabillity associated to P. All of these point to the fact that there is no simple way of constructing a SBM with all its properties, just from Kolmogorov Consistency Theorem (more specifically Theorem 10.3.6), that was used to get above the Gaussian process on $(\mathbb{R}^{[0,\infty)}, \mathcal{B}^{[0,\infty)})$. The breakthrough is provided by a very important result due to Kolmogorov and Chentsov, which says that we can have a "modification" of the Gaussian process of coordinate random variables, as constructed above, to get a SBM on $(\mathbb{R}^{[0,\infty)}, \mathcal{B}^{[0,\infty)}, P)$.

10.4 Kolmogorov-Chentsov Continuity Criterion

In this section, we are going to see a very important result. It deals with the following question. Suppose we have a family $\{X_t, t \in I\}$ of real random variables defined on some probability space, where the index set I is an interval in $\mathbb{R}$. We would like to have a 'realization' of this family that satisfies the additional condition that its paths are continuous, a.s. $[P]$. What is meant by a 'realization' here is to have a new family $\{Y_t, t \in I\}$, possibly on the same probability space, such that all the finite dimensional distributions of the family $\{Y_t, t \in I\}$ are the same as those of $\{X_t, t \in I\}$ and, in addition, the maps $t \mapsto Y_t(\omega)$ are continuous on I, for $[P]$-a.e. ω. The important question is whether we can have some nice sufficient conditions on the finite dimensional distributions of $\{X_t, t \in I\}$, which will guarantee that such a realization is possible. The theorem that we are going to see, gives a very simple sufficient condition. For the sake of simplicity, we will state and prove this when $I = [0, \infty)$, but the reader can easily verify that the entire argument goes through when I is any interval in $\mathbb{R}$, thus implying that the same criterion works there also.

We start with a Lemma that plays a crucial role in the proof of the main Theorem. The proof of the Lemma will repeatedly use a simple and basic fact, which is left as an Exercise below for the reader to verify.

EXERCISE 10.4.1. Let r be a dyadic rational in $[0, 1]$, that is, $r = \frac{j}{2^k}$ for some $k \geq 1$ and $0 \leq j \leq 2^k$. Show that, if $0 < r < 2^{-n}$, then there exist integers $n_k > \cdots > n_1 > n$, such that, $r = 2^{-n_1} + \cdots + 2^{-n_k}$.

LEMMA 10.4.2. Let D denote the set of dyadic rationals in $[0, 1]$ and let $x : D \to \mathbb{R}$ be a function satisfying the property that, there exists integer $n_0 \geq 1$, such that,

$$n \geq n_0 \text{ and } 1 \leq k \leq 2^n \text{ implies that } \left|x(\tfrac{k}{2^n}) - x(\tfrac{k-1}{2^n})\right| < 2^{-\gamma n}, \qquad (10.4.1)$$

where $\gamma > 0$ is a constant. Then there exists a constant $K = K_\gamma$, such that,

$$|x(t) - x(s)| \leq K(t-s)^\gamma, \text{ for all } s, t \in D \text{ with } 0 < t - s \leq 2^{-n_0}. \quad (10.4.2)$$

In particular, x is uniformly continuous on D. Further, x extends to a function y on $[0, 1]$ which satisfies

$$|y(t) - y(s)| \leq K(t-s)^\gamma, \text{ for all } s, t \in [0, 1] \text{ with } 0 < t - s \leq 2^{-n_0}. \quad (10.4.3)$$

Proof. Let $s, t \in D$ with $0 < t - s \leq 2^{-n_0}$. Let n be the largest positive integer, such that, $t - s \leq 2^{-n}$, that is, $n \geq n_0$ satisfies $2^{-(n+1)} < t - s \leq 2^{-n}$. Let $i_0 = [2^n s]$ and $j_0 = [2^n t]$, where $[x]$ denotes the 'integer part' of a real number x. This would mean that i_0 and j_0 are integers in $\{0, 1, \ldots, 2^n\}$, such that, $s_0 = \frac{i_0}{2^n} \leq s < \frac{i_0+1}{2^n}$ and $t_0 = \frac{j_0}{2^n} \leq t < \frac{j_0+1}{2^n}$. Clearly, $j_0 \geq i_0$ and hence $t_0 \geq s_0$.

Since $0 \leq s - s_0 < 2^{-n}$, we have, using the result of Exercise 10.4.1, that either $s = s_0$ or $s = s_0 + 2^{-i_1} + \cdots + 2^{-i_l}$, for some integers $i_l > \cdots > i_1 > n$. Similarly, since $0 \leq t - t_0 < 2^{-n}$, we have, either $t = t_0$ or $t = t_0 + 2^{-j_1} + \cdots + 2^{-j_k}$, for some integers $j_k > \cdots > j_1 > n$.

Note now that, $0 \leq s - s_0 < 2^{-n}$ and $0 \leq t - t_0 < 2^{-n}$ give $|(t-t_0) - (s-s_0)| < 2^{-n}$. But, $|(t-t_0) - (s-s_0)| = |(t-s) - (t_0-s_0)| \geq (t_0-s_0) - (t-s)$ and so we get that $t_0 - s_0 < 2^{-n} + (t-s) \leq 2^{-(n-1)}$, that is, $j_0 - i_0 < 2$. This means that either $j_0 = i_0$ or $j_0 = i_0 + 1$, that is, either $t_0 = s_0 = \frac{i_0}{2^n}$ or $s_0 = \frac{i_0}{2^n}$ and $t_0 = \frac{i_0+1}{2^n}$. Using the hypothesis (10.4.1) of the Lemma, we get $|x(t_0) - x(s_0)| < 2^{-\gamma n}$.

Next, denoting $s_1 = s_0 + 2^{-i_1}$, and $s_j = s_{j-1} + 2^{-i_j}$, $j = 2, \ldots, l$, we use (10.4.1) again to get $|x(s_1) - x(s_0)| < 2^{-\gamma i_1}, |x(s_2) - x(s_1)| < 2^{-\gamma i_2}, \ldots, |x(s_l) - x(s_{l-1})| < 2^{-\gamma i_l}$. Recall, as noted already, that we have either $s = s_0$ or $s = s_l$, so that, in any case, we will get $|x(s) - x(s_0)| \leq 2^{-\gamma i_1} + \cdots + 2^{-\gamma i_l} < \sum_{i > n} 2^{-\gamma i} = 2^{-\gamma(n+1)}(1 - 2^{-\gamma})^{-1}$.

Using similar argument, we can derive that $|x(t) - x(t_0)| < 2^{-\gamma(n+1)}(1 - 2^{-\gamma})^{-1}$.

Combining all these, we finally get

$$|x(t) - x(s)| \leq |x(t_0) - x(t)| + |x(s_0) - x(s)| + |x(t_0) - x(s_0)|$$
$$< 2 \cdot 2^{-\gamma(n+1)}(1 - 2^{-\gamma})^{-1} + 2^{-\gamma n} = K \cdot 2^{-\gamma(n+1)},$$

where $K = K_\gamma = 2^\gamma + 2(1 - 2^{-\gamma})^{-1}$. Recall that, by our choice of n, we have $2^{-(n+1)} < t - s$, so that $K \cdot 2^{-\gamma(n+1)} < K \cdot (t-s)^\gamma$. Thus we have proved (10.4.2).

That x is uniformly continuous on D is just an immediate consequence of what we just proved. To get the required extension y on $[0, 1]$, we take any $t \in [0, 1]$ and pick a sequence $\{t_n\}$ in D with $t_n \to t$. From uniform continuity of x on D, one can easily see that the sequence $\{x(t_n)\}$ is Cauchy. We define $y(t) = \lim_n x(t_n)$. The reader may easily verify that y is well defined (that is, the value $y(t)$ does not depend on any particular chosen sequnece $\{t_n\}$ in D converging to t). That y is an extension of x is immediate from continuity of x on D. Finally, given any $s, t \in [0, 1]$ with $0 < t - s \leq 2^{-n_0}$, we can pick sequences $\{s_n\}$ and $\{t_n\}$ in $D \cap (s, t)$ with $s_n \downarrow s$, $t_n \uparrow t$ and $s_n < t_n$ for every n. Using (10.4.2), we will immediately get $|x(t_n) - x(s_n)| \leq K(t_n - s_n)^\gamma$. Letting $n \to \infty$, it will follow that $|y(t) - y(s)| \leq K(t-s)^\gamma$, proving (10.4.3). That completes the proof of the Lemma. $\square$

10.4. Kolmogorov-Chentsov Continuity Criterion

THEOREM **10.4.3.** (Kolmogorov-Chentsov Continuity Criterion) Let $\{X_t, t \in [0, \infty)\}$ be a family of real random variables on a probability space $(\Omega, \mathcal{A}, P)$, that satisfies the property that there exist positive constants α, β and C, such that

$$E|X_t - X_s|^\alpha \le C|t-s|^{1+\beta}, \quad \text{for any } s, t \in [0, \infty). \tag{10.4.4}$$

Then, there exists, on the same probability space, a family $\{Y_t, t \in [0, \infty)\}$ of random variables with $Y_t = X_t$ a.s.$[P]$, for each $t \in [0, \infty)$, such that for all ω outside a P-null set N, the map $t \mapsto Y_t(\omega)$ is continuous.

Proof. For notational simplicity, we will write the random variables X_t as $X(t)$ and write $X_t(\omega)$ as $X(t, \omega)$, for $t \in [0, \infty)$ and $\omega \in \Omega$. The same notational convention will be followed for the family $\{Y_t\}$ also.

With the constants α and β as given, we choose and fix a $\gamma \in (0, \beta/\alpha)$, so that $\delta = \beta - \alpha\gamma > 0$. Now for any $n \ge 1$, and $k = 1, \ldots, 2^n$, one has, using Chebycheff's inequality and the hypothesis (10.4.4),

$$P\big(|X(\tfrac{k}{2^n}) - X(\tfrac{k-1}{2^n})| \ge 2^{-\gamma n}\big) \le 2^{\alpha\gamma n} \cdot E\big(|X(\tfrac{k}{2^n}) - X(\tfrac{k-1}{2^n})|^\alpha\big) \le C \cdot 2^{-(1+\delta)n}.$$

This gives us $P\big(\bigcup_{k=1}^{2^n} \{|X(\tfrac{k}{2^n}) - X(\tfrac{k-1}{2^n})| \ge 2^{-\gamma n}\}\big) \le 2^n C \cdot 2^{-(1+\delta)n} = C \cdot 2^{-\delta n}$, for every $n \ge 1$, implying that

$$\sum_{n=1}^\infty P\big(\bigcup_{k=1}^{2^n} \{|X(\tfrac{k}{2^n}) - X(\tfrac{k-1}{2^n})| \ge 2^{-\gamma n}\}\big) < \infty.$$

Therefore, by the Borel-Cantelli Lemma, there exists a P-null set N_0 such that for every $\omega \notin N_0$, there exists an integer $n_0 = n_0(\omega) \ge 1$, so that

$$|X(\tfrac{k}{2^n}, \omega) - X(\tfrac{k-1}{2^n}, \omega)| < 2^{-\gamma n}, \quad \text{for all } n \ge n_0 \text{ and } k = 1, \ldots, 2^n.$$

With D as in the statement of Lemma 10.4.2, what this says is that, for each $\omega \notin N_0$, the function $X(\cdot, \omega)$ on D satisfies the hypothesis (10.4.1) of the Lemma 10.4.2. We can therefore use the Lemma to get, for each $\omega \notin N_0$, a function $Y_0(\cdot, \omega)$ on $[0, 1]$, such that, $Y_0(t, \omega) = X(t, \omega)$, for $t \in D$, and

$$|Y_0(t, \omega) - Y_0(s, \omega)| \le K(t-s)^\gamma, \quad \text{for all } s, t \in [0, 1] \text{ with } 0 < t - s \le 2^{-n_0}.$$

Similarly, for every integer $j \ge 1$, we may apply the same argument to the family $\{X(j+t), t \in [0, 1]\}$ and get a P-null set N_j such that for $\omega \notin N_j$ there exists an integer $n_j = n_j(\omega) \ge 1$, and a function $Y_j(\cdot, \omega)$ on $[0, 1]\}$ with the properties that $Y_j(t, \omega) = X(j+t, \omega)$, for $t \in D$, and

$$|Y_j(t, \omega) - Y_j(s, \omega)| \le K(t-s)^\gamma, \quad \text{for all } s, t \in [0, 1] \text{ with } 0 < t - s \le 2^{-n_j}.$$

Consider the P-null set $N = \bigcup_{j \ge 0} N_j$ and define a family $\{Y(t), t \in [0, \infty)\}$ by

$$Y(t, \omega) = \sum_{j \ge 0} Y_j(t - j, \omega) I_{[j, j+1)}(t), \quad \text{for } \omega \notin N,$$

and, $\quad Y(\cdot, \omega) \equiv 0, \text{ for } \omega \in N.$

From the construction of the function $y(\cdot)$ in Lemma 10.4.2 and the above definition, it can be argued that $Y(t)$, for each $t \in [0, \infty)$, is a random variable. We leave this slightly non-trivial task as exercise for the reader. That granted, we have a family $\{Y(t), t \in [0, \infty)\}$ of real random variables on $(\Omega, \mathcal{A}, P)$.

We now show that for every ω, the map $t \mapsto Y(t, \omega)$ is continuous. In fact, we show something stronger, namely, that given any integer $M > 0$, there exists, for each ω, an $\epsilon = \epsilon(\omega) > 0$, such that

$$|Y(t,\omega) - Y(s,\omega)| \leq 2K_\gamma (t-s)^\gamma, \text{ for all } s, t \in [0, M] \text{ with } 0 < t - s < \epsilon. \quad (10.4.5)$$

There is nothing to prove if $\omega \in N$. For $\omega \notin N$, let $\epsilon = \min\{2^{-n_j(\omega)}, 0 \leq j < M\}$. It is easy to see that if $s, t \in [0, M]$ satisfy $0 < t - s < \epsilon$ then there are two possibilities. Either $j \leq s < t \leq j+1$ for $0 \leq j < M$ or $j-1 \leq s \leq j \leq t \leq j+1$ for $1 \leq j < M$. There is nothing to prove in the first case. By construction, $|Y(t,\omega) - Y(s,\omega)| \leq K_\gamma (t-s)^\gamma$ since $0 < t - s < \epsilon \leq 2^{-n_j}$. In the second case, we use $|Y(t) - Y(s)| \leq |Y(t) - Y(j)| + |Y(j) - Y(s)|$ to get the required result.

We are only left to show that $Y(t) = X(t)$ a.s.$[P]$ for each t. If t is a dyadic rational in $[0, \infty)$, then, by construction, $Y(t,\omega) = X(t,\omega)$ for $\omega \notin N$, that is, $Y_t = X_t$ a.s.$[P]$. If t is not a dyadic rational, we take any sequence $\{t_n\}$ of dyadic rationals with $t_n \to t$. The continuity property proved already shows that $Y(t_n, \omega) \to Y(t, \omega)$ for all ω and hence $X(t_n) \to Y(t)$ a.s.$[P]$. However, using the hypothesis (10.4.4) and Chebycheff's inequality one can easily see that $X(t_n) \xrightarrow{P} X(t)$. But, we already have $X(t_n) \xrightarrow{a.s.} Y(t)$. It follows that $Y(t) = X(t)$ a.s.$[P]$. The proof is now complete. □

REMARK 10.4.4. A function $f : [0, \infty) \to \mathbb{R}$ is said to be *locally Hölder continuous* with exponent $\gamma > 0$ if for every integer $M > 0$, there exists $\epsilon > 0$ such that $\sup\{|f(t) - f(s)|/|t - s|^\gamma : t, s \in [0, M], 0 < |t - s| < \epsilon\} < \infty$. The above proof (see (10.4.5)) shows that the family $\{Y(t)\}$ constructed there actually has the property that for every ω, the function $t \mapsto Y(t, \omega)$ is locally Hölder continuous with exponent γ, which is more than just continuity (or even uniform continuity).

Returning finally to how one now finishes the construction of a Standard Brownian Motion, recall that, towards the end of Section 10.3, we were able to construct, using Kolmogorov Consistency Theorem 10.2.10 and Theorem 10.3.6, a probability space $(\Omega, \mathcal{A}, P)$ and a family $\{X_t, t \in [0, \infty)\}$ of real random variables on it that satisfies properties (i)–(iii) of a Standard Brownian Motion. From property (ii) and using properties of Normal distribution, one easily gets

$$E(|X_t - X_s|^4) = 3(t-s)^2, \text{ for any } 0 < s < t.$$

This means that the family $\{X_t, t \in [0, \infty)\}$ satisfies the hypothesis (10.4.4) of Theorem 10.4.3, with $\alpha = 4$, $\beta = 1$ and $C = 3$. One can therefore, apply Theorem 10.4.3, to get a family $\{Y_t, t \in [0, \infty)\}$ of random variables on the same probability space, which has continuous trajectories a.s.$[P]$. The fact that $Y_t = X_t$ a.s.$[P]$, for each t, easily implies that $(Y_{t_1}, \ldots, Y_{t_n}) = (X_{t_1}, \ldots, X_{t_n})$ a.s.$[P]$, for each $0 \leq t_1 < t_2 < \cdots < t_n$ and this guarantees that the family $\{Y_t, t \in [0, \infty)\}$ will have the same finite dimensional distributions as those of the family $\{X_t, t \in [0, \infty)\}$. Thus, $\{Y_t, t \in [0, \infty)\}$ will satisfy properties (i)–(iii) of SBM. This means that, we have constructed a Standard Brownian Motion $\{Y_t, t \in [0, \infty)\}$ on the probability space $(\Omega, \mathcal{A}, P)$. It may be noted that, in view of the Remark 10.4.4, the $\{Y_t, t \in [0, \infty)\}$, thus constructed, has paths which are actually locally Hölder continuous with exponent γ, for any $\gamma < 1/4$. This can actually be strengthened to show that actually P-almost every path is locally Hölder continuous with exponent γ, for any $\gamma < 1/2$. We will not prove it here, but in the next Chapter, where we will undertake a brief journey with Standard Brownian Motion and explore some interesting and important properties of SBM.

10.5   Additional Exercises

EXERCISE **10.5.1.** Show that each of the following subsets of $\mathbb{R}^\infty$ belong to the product σ-field $\mathcal{B}^\infty$.
(i) $\ell_p = \{\mathbf{x} = (x_1, x_2, \ldots) \in \mathbb{R}^\infty : \sum_n |x_n|^p < \infty\}$, for $0 < p < \infty$.
(ii) $\ell_\infty = \{\mathbf{x} = (x_1, x_2, \ldots) \in \mathbb{R}^\infty : \sup_n |x_n| < \infty\}$.
(iii) $\mathbf{c}_0 = \{\mathbf{x} = (x_1, x_2, \ldots) \in \mathbb{R}^\infty : \lim_n x_n = 0\}$.
(iv) $\mathbf{c} = \{\mathbf{x} = (x_1, x_2, \ldots) \in \mathbb{R}^\infty : \{x_n\} \text{ is a Cauchy sequence}\}$.

EXERCISE **10.5.2.** Consider $T : \mathbb{R}^\infty \to \mathbb{R}^\infty$ given by $T : (x_1, x_2, \ldots) \mapsto (x_2, x_3, \ldots)$.
(a) Show that $T : (\mathbb{R}^\infty, \mathcal{B}^\infty) \to (\mathbb{R}^\infty, \mathcal{B}^\infty)$ is measurable and describe $\sigma(T)$.
(b) Denoting the n-fold composition $T \circ \cdots \circ T$ by T^n, describe $\sigma(T^n)$ and also the sigma-field $\bigcap_n \sigma(T^n)$.

EXERCISE **10.5.3.** Let $\mathbf{X} = \{X_n, n \geq 1\}$ be a sequence of real random variables on some probability space $(\Omega, \mathcal{A}, P)$.
(a) Show that there is a unique probability $\widetilde{P}$ on $(\mathbb{R}^\infty, \mathcal{B}^\infty)$ such that, for all $n \geq 1$ and all $B \in \mathcal{B}^n$, we have $P\big((X_1, \ldots, X_n) \in B\big) = \widetilde{P}\big(\varphi_n^{-1}(B)\big)$, where $\varphi_n : \mathbb{R}^\infty \to \mathbb{R}^n$ denotes the usual projection map $(x_1, x_2, \ldots) \mapsto (x_1, \ldots, x_n)$. The probability $\widetilde{P}$ is called the **distribution** (also called **law**) of the sequence $\mathbf{X}$ and will be denoted $P_\mathbf{X}$.
(b) Let $\mathbf{X}$ and $\mathbf{Y}$ be two sequences of random variables, defined on possibly different probability spaces $(\Omega, \mathcal{A}, P)$ and $(\Omega', \mathcal{A}', P')$. Show that, if, for each n, $P\big((X_1, \ldots, X_n) \in B\big) = P'\big((Y_1, \ldots, Y_n) \in B\big)$, for all $B \in \mathcal{B}^n$, then $P_\mathbf{X} = P_\mathbf{Y}$. This says that two sequences $\mathbf{X}$ and $\mathbf{Y}$ have the same law, if they have the same finite dimensional distributions. We write $\mathbf{X} \stackrel{d}{=} \mathbf{Y}$.
(c) Let $\mathbf{X}$ and $\mathbf{Y}$ be as in (b) with $\mathbf{X} \stackrel{d}{=} \mathbf{Y}$. Show then that, (i) $P(\mathbf{X} \in A) = P'(\mathbf{Y} \in A)$, for all $A \in \mathcal{B}^\infty$, and (ii) $h(\mathbf{X}) \stackrel{d}{=} h(\mathbf{Y})$, for any measurable function $h : (\mathbb{R}^\infty, \mathcal{B}^\infty) \to (\mathbb{R}^\infty, \mathcal{B}^\infty)$.
[Note that if $\mathbf{X}$ is a sequence of real random variables on a probability space $(\Omega, \mathcal{A}, P)$, then, by Exercise 10.1.2, $\{\mathbf{X} \in A\} \in \mathcal{A}$, for all $A \in \mathcal{B}^\infty$ and $h(\mathbf{X})$ defines a sequence of real random variaables on $(\Omega, \mathcal{A}, P)$, for any measurable $h : (\mathbb{R}^\infty, \mathcal{B}^\infty) \to (\mathbb{R}^\infty, \mathcal{B}^\infty)$.]
(e) Let $\{X_n, n \geq 1\}$ and $\{Y_n, n \geq 1\}$ be two sequences of real random variables defined on possibly different probability spaces as in (b), such that, for each n, $P\big((X_1, \ldots, X_n) \in B\big) = P'\big((Y_1, \ldots, Y_n) \in B\big)$, for all $B \in \mathcal{B}^n$. Show that if $X_n \to 0$ a.s.$[P]$, then $Y_n \to 0$ a.s.$[P']$, and more generally, if

$X_n \to X$ a.s.$[P]$, then there is a real random variable Y with $Y \stackrel{d}{=} X$ so that $Y_n \to Y$ a.s.$[P']$. [Hint: For the last part, observe that the hypothesis implies $\{X_n\}$ is Cauchy a.s.$[P]$.]

EXERCISE **10.5.4.** *[This is a slight generalization of the idea of countably infinite product spaces and Kolmogorov Consistency Theorem discussed in Section 10.2.] Throughout this exercise, D will denote a countably infinite set.*
(a) *Given a family $\{(\Omega_t, \mathcal{A}_t), t \in D\}$ of measurable spaces, define an appropriate class $\mathfrak{F}$ of subsets of $\underset{t \in D}{\times} \Omega_t$, consisting of all measurable finite-dimensional cylinder sets and show that it is a field on $\underset{t \in D}{\times} \Omega_t$.*
(b) *Denoting $\underset{t \in D}{\otimes} \mathcal{A}_t$ to be $\sigma(\mathfrak{F})$, show that it is the smallest σ-field on $\underset{t \in D}{\times} \Omega_t$ that makes all the usual finite-dimensional projection maps $\varphi_S : \underset{t \in D}{\times} \Omega_t \to \underset{t \in S}{\times} \Omega_t$, as S varies over all non-empty finite subsets of D, measurable.*
(c) *For any measurable space $(\Omega, \mathcal{A})$, show that $f : (\Omega, \mathcal{A}) \to \big(\underset{t \in D}{\times} \Omega_t, \underset{t \in D}{\otimes} \mathcal{A}_t\big)$ is measurable if and only if $e_t \circ f : (\Omega, \mathcal{A}) \to (\Omega_t, \mathcal{A}_t)$ is measurable for each $t \in D$, where e_t, for each $t \in D$, denotes the usual 'evaluation map'.*
For the rest of the exercise, take $(\Omega_t, \mathcal{A}_t) = (\mathbb{R}, \mathcal{B})$, for all $t \in D$, and use $(\mathbb{R}^D, \mathcal{B}^D)$ to denote the 'D-fold' product measurable space.
(d) *Let P_S, for each non-empty finite $S \subset D$, be a probability on $(\mathbb{R}^S, \mathcal{B}^S)$. Define an appropriate notion of 'consistency' for the family $\{P_S : \emptyset \neq S \subset D, S \text{ finite}\}$ and show that, given a consistent family, there is a unique probability P on $(\mathbb{R}^D, \mathcal{B}^D)$, such that $P(\varphi_S^{-1}(B)) = P_S(B)$, for all finite $S \subset D$ and $B \in \mathcal{B}^S$.*
(e) *Show that given any family $\mathbf{X} = \{X_t, t \in D\}$ of real random variables on a probability space $(\Omega, \mathcal{A}, P)$, there is a unique probability $P_\mathbf{X}$ on $(\mathbb{R}^D, \mathcal{B}^D)$, such that, for any finite $S \subset D$ and any $B \in \mathcal{B}^S$, $P\big((X_t, t \in S) \in B\big) = P_\mathbf{X}\big(\varphi_S^{-1}(B)\big)$.*
(f) *Show that if $\mathbf{X} = \{X_t, t \in D\}$ and $\mathbf{Y} = \{Y_t, t \in D\}$ are two families of real random variables on possibly different probability spaces $(\Omega, \mathcal{A}, P)$ and $(\Omega', \mathcal{A}', P')$, such that, $P\big((X_t, t \in S) \in B\big) = P'\big((Y_t, t \in S) \in B\big)$, for every finite $S \subset D$ and $B \in \mathcal{B}^S$, then $P_\mathbf{X} = P_\mathbf{Y}$ (we say that $\mathbf{X} \stackrel{d}{=} \mathbf{Y}$) and, in that case, $P(\mathbf{X} \in A) = P'(\mathbf{Y} \in A)$, for all $A \in \mathcal{B}^D$.*

EXERCISE **10.5.5.** *For $n \geq 1$, let S_n denote the permutation group on $\{1, \ldots, n\}$. For $\pi \in S_n$, consider $T_\pi : \mathbb{R}^\infty \to \mathbb{R}^\infty$ given by $T : (x_1, x_2 \ldots) \mapsto (y_1, y_2, \ldots)$, where $y_i = x_{\pi(i)}$, $1 \leq i \leq n$, and $y_i = x_i$ for $i > n$.*
(a) *Show that, for each $n \geq 1$ and $\pi \in S_n$, the map $T_\pi : (\mathbb{R}^\infty, \mathcal{B}^\infty) \to (\mathbb{R}^\infty, \mathcal{B}^\infty)$ is measurable.*
(b) *For a fixed $n \geq 1$, let $\mathcal{I}_n = \{A \in \mathcal{B}^\infty : T_\pi^{-1}(A) = A \text{ for all } \pi \in S_n\}$. Show that $\mathcal{I}_n$ is a σ-field on $\mathbb{R}^\infty$.*
(c) *The intersection σ-field $\mathcal{I} = \bigcap_n \mathcal{I}_n$ is called the **exchangeable** σ-field. Can you give a general description of the sets that belong to the σ-field $\mathcal{I}$? Also, give some examples of specific sets that in the σ-field $\mathcal{I}$.*

EXERCISE **10.5.6.** *Denoting P and Q to be respectively the $N(\mu, \sigma^2)$ and $N(\nu, \eta^2)$ probability distributions on $(\mathbb{R}, \mathcal{B})$, show that the probabilities P^∞ and Q^∞ on*

10.5. Additional Exercises

$(\mathbb{R}^\infty, \mathcal{B}^\infty)$ are mutually singular unless $\mu = \nu$ and $\sigma = \eta$. [Hint: Assume first that $\sigma = \eta = 1$ and show that P^∞ and Q^∞ are mutually singular unless $\mu = \nu$.]

EXERCISE 10.5.7. Show that both (i) $C(s,t) = e^{-(t-s)^2/2}$, for $t, s \in [0, \infty)$, and (ii) $C(s,t) = e^{-|t-s|}$, for $t, s \in [0, \infty)$, define symmetric non-negative definite kernels on $[0, \infty) \times [0, \infty)$. [Hint: Think of characteristic functions.]

EXERCISE 10.5.8. Show that if C_1 and C_2 are two symmetric non-negative definite kernels on $[0, \infty) \times [0, \infty)$, then so is $C(s,t) = C_1(s,t)C_2(s,t)$. [Hint: Think of independent Gaussian processes.]

EXERCISE 10.5.9. Consider the measurable space $\left(\mathbb{R}^{[0,\infty)}, \mathcal{B}^{[0,\infty)}\right)$.
(a) Denoting C to be the set of all real valued continuous functions on $[0, \infty)$, show that $C \notin \mathcal{B}^{[0,\infty)}$.
(b) Denoting D to be the complement of C in $\mathbb{R}^{[0,\infty)}$, show that $P^*(D) = 1$, for any probability P on $\mathcal{B}^{[0,\infty)}$, where P^* is the outer probability given by P.

EXERCISE 10.5.10. This is an example, due to Halmos, to exhibit where Consistency Theorem fails, thereby highlighting that the hypotheses of Kolmogrov's Consistency Theorem 10.2.10 cannot be done away with.
Call a set $M \subset [0,1]$ 'thick' (terminology due to Halmos) if M has outer Lebesgue measure 1 and inner Lebesgue measure 0.
(a) Construct a sequence M_n, $n \geq 1$ of thick sets such that $M_n \downarrow \emptyset$ and consider the infinite product $\Omega = \prod_n M_n$.
(b) Using the fact that each M_n is a thick set, show that the Lebesgue measure λ on $[0,1]$ can be lifted, for each n, to a probability measure P_n on $M_1 \times \cdots \times M_n$ onto its diagonal.
(c) Show that this family $\{P_n, n \geq 1\}$, viewed as a sequence of probabilities on the finite-dimensional cylinder sets in Ω, is a consistent family.
(d) Show that there is no probability P on Ω such that P_n, for each n, is the projection of P to the n-dimensional cylinders.

Chapter 11

Brownian Motion : A Brief Journey

The Standard Brownian Motion on $[0, \infty)$ was already introduced in the last chapter. Since one sometimes talks about Standard Brownian Motion on a closed bounded interval $[0, T]$ also, we start by redefining SBM, for the sake of records. In what follows I will denote either $[0, \infty)$ or the closed bounded interval $[0, T]$ for some $T \in (0, \infty)$. Also, a family $\{X_t, t \in I\}$ of real random variables on a probability space will be called a real "stochastic process", indexed by I. It often helps to think of the index set I as representing "time" and the stochastic process $\{X_t, t \in I\}$ as modelling the random motion of a particle in one-dimension with time. Also, following the standard and commonly used notation. we will usually denote a SBM by $\{B_t, t \in I\}$.

DEFINITION 11.0.1. A *Standard Brownian Motion* on I (abbreviated *SBM* on I), where $I = [0, \infty)$ or $[0, T]$, is a real stochastic process $\{B_t, t \in I\}$ defined on some probability space $(\Omega, \mathcal{A}, P)$ and satisfying the following properties.
  (i) $B_0 = 0$ a.s.$[P]$,
  (ii) $B_t - B_s \sim \mathsf{N}(0, t-s)$, for all $s < t$ in I,
  (iii) $B_{t_1}, B_{t_2} - B_{t_1}, \ldots, B_{t_n} - B_{t_{n-1}}$ are independent, for any $t_1 < \cdots < t_n$ in I, and,
  (iv) for $[P]$-a.e. $\omega \in \Omega$, the map $t \mapsto B_t(\omega)$ is continuous on I.

In the last chapter, we used Kolmogorov's Consistency Theorem, coupled with Kolmogorov-Chentsov Continuity Theorem, to establish existence of SBM on $[0, \infty)$. This clearly establishes existence of SBM on any $[0, T]$ as well.
In the process of proving existence of SBM on $[0, \infty)$, we showed that the properties (i)–(iii) for SBM on $[0, \infty)$ are equivalent to its being a Gaussian process with mean function $\mu(\cdot) \equiv 0$ and covariance kernel $C(s, t) = s \wedge t$. It is fairly obvious that the same would be true for SBM on $[0, T]$ as well. Since this is a very important observation, essentially providing us with an equivalent definition of SBM, we record it below as a Theorem. In the sequel, unless we specifically use $[0, \infty)$ or $[0, T]$ for the index set, we will use I to mean either of them. Also, we will often use the notation $X(t, \omega)$ instead of $X_t(\omega)$, in order to emphasize that it is a function of both t and ω.

THEOREM 11.0.2. A real stochastic process $\{X_t, t \in I\}$ defined on a probability space $(\Omega, \mathcal{A}, P)$ is a SBM on I if and only if (i) $\{X_t, t \in I\}$ is a Gaussian process on I with mean function $\mu(\cdot) \equiv 0$ and covariance kerenel $C(s,t) = s \wedge t$, and (ii) for P-a.e. ω, the trajectory $t \mapsto X(t, \omega)$ is continuous on I.

Theorem 11.0.2 often turns out to be very handy in showing that a given process $\{X_t, t \in I\}$ is a SBM on I. In fact, this will be well illustrated in the proof of the next result which shows how one can make certain transformations on a given SBM on $[0, \infty)$ to produce new SBMs.

PROPOSITION 11.0.3. Let $\{B_t, t \in [0, \infty)\}$ be a SBM. Then
(a) (*scaling*) for any real $a \neq 0$, the stochastic process $\{X_t, t \in [0, \infty)\}$, defined as $X_t = aB_{t/a^2}$, $t \geq 0$, is a SBM on $[0, \infty)$. In particular, $\{-B_t, t \in [0, \infty)\}$ is a SBM.
(b) (*time-reversal*) for any $T > 0$, the stochastic process $\{X_t, t \in [0, T]\}$, defined as $X_t = B_{T-t} - B_T$, $t \in [0, T]$, is a SBM on $[0, T]$.
(c) (*time-inversion*) the stochastic process $\{X_t, t \in [0, \infty)\}$, defined as $X_0 \equiv 0$ and $X_t = tB_{1/t}$, $t > 0$, is a SBM on $[0, \infty)$.
(d) (*markov property*) for any $s \geq 0$, the stochastic process $\{X_t, t \in [0, \infty)\}$, defined as $X_t = B_{s+t} - B_s$, $t \geq 0$, is a SBM on $[0, \infty)$, which is independent of $\{B_u, 0 \leq u \leq s\}$.

Proof. First of all, from the definition of the processes $\{X_t\}$ in (a)–(d) and properties of Multivariate Gaussian distribution, it is easy to see that, in each of the cases, the finite-dimensional distributions of $\{X_t\}$ are all Gaussian, implying that the processes are Gaussian processes. Also, from the mean function and covariance kernel of the SBM $\{B_t\}$ and using simple properties of expectation and covariance, one can easily see that each of the processes in (a)–(d) have mean function $\mu(\cdot) \equiv 0$ and covariance kernel $C(s,t) = s \wedge t$. Next, the continuity of P-a.e. trajectory of $\{B_t, t \in [0, \infty)\}$ clearly guarantees the same for the processes $\{X_t\}$ as defined in (a), (b) and (d). As for the process $\{X_t\}$ defined in (c) also, continuity on $(0, \infty)$ of P-a.e. trajectory is clear, thereby leaving only a.s. continuity at $t = 0$ in question. The easiest way to settle that is by appealing to what is called the Strong Law of Large Number for $\{B_t\}$ (Proposition 11.1.1), which asserts that $B_t/t \to 0$ a.s.$[P]$, as $t \to \infty$. From this, a simple change of variable gives $tB_{1/t} \to 0$ a.s.$[P]$, as $t \to 0$, proving the required continuity of $\{X_t\}$ at $t = 0$. The reader is encouraged to see Exercise 11.0.5 for a direct proof of $[P]$-a.e. continuity of $\{X_t\}$ in (c) at $t = 0$, that does not need Proposition 11.1.1. Having thus proved that each of the processes $\{X_t\}$, as in (a)–(d), is a SBM, we are only left with proving the independence, asserted in (d), between $\{X_t, , t \in [0, \infty)\}$ and $\{B_u, 0 \leq u \leq s\}$.

Fixing any $u_1 < \cdots < u_k < s$ and any $0 \leq t_1 < \cdots < t_n < \infty$, the independence of the σ-fields $\mathcal{G}_1 = \sigma(\{B_{u_1}, B_{u_2} - B_{u_1}, \ldots, B_{u_k} - B_{u_{k-1}}, B_s - B_{u_k}\})$ and $\mathcal{G}_2 = \sigma(\{B_{s+t_1} - B_s, B_{s+t_2} - B_{s+t_1}, \ldots, B_{s+t_n} - B_{s+t_{n-1}}\})$ is immediate from property (iii) in Definition 11.0.1. Since $(B_{u_1}, \ldots, B_{u_k}, B_s)$ is $\mathcal{G}_1$-measurable and $(X_{t_1}, \ldots, X_{t_n})$ is $\mathcal{G}_2$-measurable, it follows that $(X_{t_1}, \ldots, X_{t_n})$ is independent of $(B_{u_1}, \ldots, B_{u_k}, B_s)$. As it holds for any $u_1 < \cdots < u_k < s$ and $0 \leq t_1 < \cdots < t_n$, one concludes that the class $\mathfrak{F}_1 = \bigcup \sigma(B_{u_1}, \ldots, B_{u_k}, B_s)$, where the union is over all $u_1 < \cdots < u_k < s$, is independent of $\mathfrak{F}_2 = \bigcup \sigma(B_{s+t_1} - B_s, \ldots, B_{s+t_n} - B_s)$, where the union is over all $0 \leq t_1 < \cdots < t_n$. But $\mathfrak{F}_1$ and $\mathfrak{F}_2$ are clearly fields that generate the σ-fields $\sigma(\{B_u, 0 \leq u \leq s\})$ and $\sigma(\{X_t, t \in [0, \infty)\})$ respectively, from which the independence asserted in (d) follows by using Theorem 2.12.8. □

REMARK 11.0.4. In the above proof of Proposition 11.0.3, Strong Law of Large Numbers for SBM was used to get $tB_{1/t} \to 0$ a.s.$[P]$, as $t \to 0$, and thus establish continuity at $t = 0$ of $[P]$-a.e. trajectory of the process $\{X_t\}$ defined in (c). The following exercise gives a direct proof that simply uses the fact that $\{X_t\}$ and $\{B_t\}$ have the same finite-dimensional distributions on $(0, \infty)$. This may then be used to get an alternative proof of SLLN.

EXERCISE 11.0.5. [*You may use Exercise 10.5.4 for this.*]
Let $\{B_t\}$ be a SBM on $[0, \infty)$, defined on a probability space $(\Omega, \mathcal{A}, P)$. Denote by D, the countable set $\mathbb{Q} \cap (0, 1)$.
(a) Denoting $Y_r = \sup\{|B_s| : s \in D, s < r\}$, for each $r \in D$, show that
$$P\Big(\bigcap_{j \geq 1} \bigcup_{r \in D} \{Y_r \leq 1/j\}\Big) = 1.$$
(b) Denoting $X_t = tB_{1/t}$, for $t \in (0, \infty)$, show that $\{X_r, r \in D\} \stackrel{d}{=} \{B_r, r \in D\}$, in the sense described in Exercise 10.5.4.
(c) Denoting $Z_r = \sup\{|X_s| : s \in D, s < r\}$, for each $r \in D$, deduce that
$$P\Big(\bigcap_{j \geq 1} \bigcup_{r \in D} \{Z_r \leq 1/j\}\Big) = 1.$$
(d) Using the continuity of $[P]$-a.e. trajectory of $\{X_t\}$ on $(0, \infty)$, deduce that $P(Z_r = \sup\{|X_t| : 0 < t < r\}$, for all $r \in D) = 1$.
(e) From (c) and (d), conclude that $X_t \to 0$, as $t \to 0$, a.s.$[P]$.
(f) Use (e) to derive that $B_t/t \to 0$, as $t \to \infty$, a.s.$[P]$.

The simple properties stated in Proposition 11.0.3 have important ramifications. It will, therefore, be worthwhile to understand what these properties essentially say. Property (a) says that scaling both the time and space of a SBM in an "appropriate way" yields another SBM. Property (b) says that, from a time point $T > 0$, if one looks at the displacement of SBM in the "backward" direction of time, one again gets a SBM on $[0, T]$. Property (c) is somewhat interesting and one of its major takeaways is that any feature of a SBM "near" time 0 always leads to some statement on the behaviour of SBM near time "∞" and vice versa. Property (d) says that if one shifts the time and space axes so as to make the point (s, B_s) as the new origin, then the "future" motion from that point onwards would be once again that of a SBM, independently of the past leg of the motion upto and including time s. While this is, strictly speaking, not the Markov property, but it is labelled that way because this is at the heart of the Markov property for SBM, as we will see in Section 11.2. All these properties will prove to be very useful in deriving other important properties of SBM.

Towards the end of Chapter 10, it was pointed out that application of Kolmogorov-Chentsov continuity criteria in the context of SBM actually yields much more than continuity of trajectories. It actually shows that almost every trajectory of SBM is locally Hölder continuous everywhere with exponent γ, for any $\gamma < \frac{1}{4}$. It was also mentioned that this can actually be stretched to include all exponents $\gamma < \frac{1}{2}$. The proof of that is quite involved and we will not present it. Instead, what we want to now illustrate is that while Brownian trajectories exhibit nice behaviour in terms of Hölder continuity, they are otherwise very erratic on many other counts. In the next section, we prove results that show that a typical Brownian path behaves in a rather weird manner, so much so that the reader would soon find it difficult, if not impossible, to comprehend how the graph of a typical trajectory of SBM looks like.

11.1 Path Properties of SBM

In what follows, $\{B_t, t \in [0, \infty)\}$ will denote a Standard Brownian Motion defined on some probability space $(\Omega, \mathcal{A}, P)$. As mentioned earlier, we will often use the two notations $B_t(\cdot)$ and $B(t, \cdot)$ interchangeably. Recall that, for each $\omega \in \Omega$, the ω-trajectory or ω-path is the real valued function $B(\cdot, \omega)$ on $[0, \infty)$, which is, by property (iv) of Definition 11.0.1 continuous, for P-a.e. ω. By discarding a P-null set, if necessary, we assume that, for all ω, the trajectory $B(\cdot, \omega)$ is continuous.

We start with a simple yet useful result, which is usally referred to as the *Strong Law of Large Numbers* for SBM on $[0, \infty)$.

PROPOSITION **11.1.1.** (SLLN for SBM) $\dfrac{B_t}{t} \to 0$ as $t \to \infty$, P-almost surely.

A proof of Proposition 11.1.1 was already given in Exercise 11.0.5 (f), but we are going to see a direct proof, based on an application of Kolmogorov's maximal inequality (Theorem 7.3.5), as captured in following Lemma.

LEMMA **11.1.2.** *For any $0 \le a < b < \infty$ and any $\lambda > 0$,*

$$P\big(\max_{a \le t \le b} |B_t| > \lambda\big) \le 4b/\lambda^2 \tag{11.1.1}$$

Proof of Lemma. For each $n \ge 1$, let $t_{k,n} = a + (\frac{k}{2^n})(b - a)$, $k = 0, 1, \ldots, 2^n$ and let $\xi_{k,n} = B(t_{k,n}) - B(t_{k-1,n})$, $k = 1, \ldots, 2^n$. Since, for each $n \ge 1$, the random variables $\xi_{k,n}$, $1 \le k \le 2^n$ are i.i.d. with zero mean and finite variance, Kolmogorov's Maximal Inequality (Theorem 7.3.5) gives

$$P\big(\max_{1 \le k \le 2^n} |B(t_k, n) - B(a)| > \lambda/2\big) \le 4E\big((B(b) - B(a))^2\big)/\lambda^2 = 4(b-a)/\lambda^2.$$

By continuity of paths, $\max_{1 \le k \le 2^n} |B(t_k, n) - B(a)| \uparrow \max_{a \le t \le b} |B(t) - B(a)|$ as $n \to \infty$. Using continuity of probability, we get $P\big(\max_{a \le t \le b} |B(t) - B(a)| > \lambda/2\big) \le 4(b-a)/\lambda^2$. Noting that $P\big(\max_{a \le t \le b} |B_t| > \lambda\big) \le P\big(\max_{a \le t \le b} |B(t) - B(a)| > \lambda/2\big) + P\big(|B(a)| > \lambda/2\big)$, and that Chebycheff's inequality makes the second term on the right hand side of this inequality bounded above by $4a/\lambda^2$, one easily gets the iequality (11.1.1). □

Proof of SLLN. Take and fix $0 < \rho < 1$ and $\theta > 1$ with $\theta\rho^2 > 1$. For each $n \geq 1$, denote $Z_n = \max\{|B_t|/t : \theta^n \leq t \leq \theta^{n+1}\}$. We then get that, for every $n \geq 1$,

$$P(Z_n > \rho^n) \leq P\Big(\max_{\theta^n \leq t \leq \theta^{n+1}} |B_t| > (\theta\rho)^n\Big) \leq 4\theta/(\theta\rho^2)^n,$$

where Lemma 11.1.2 was used in the second inequality. Since $\theta\rho^2 > 1$, we get that the series $\sum_n P(Z_n > \rho^n)$ converges, whence, by using Borel-Cantelli Lemma, one concludes that, for P-a.e. ω, there exists $n_0 = n_0(\omega)$, such that $Z_n(\omega) \leq \rho^n$ for all $n \geq n_0$. Noting that $\rho^n \to 0$, the proof can now be easily completed. $\square$

We are now going to see a series of results, each bringing out a different facet of the very erratic nature of a typical Brownian path.

PROPOSITION 11.1.3. *For P-a.e. ω, the ω-trajectory of $\{B_t, t \in [0, \infty)\}$ is not monotone on any non-degenerate interval in $[0, \infty)$.*

Proof. First, since $\{-B_t, t \in [0, \infty)\}$ is also a SBM (by Proposition 11.0.2 (a)), it suffices to show that, for P-a.e. ω, the ω-trajectory of $\{B_t, t \in [0, \infty)\}$ is not non-decreasing on any non-degenerate interval in $[0, \infty)$. Next, every non-degenerate interval in $[0, \infty)$ contains an interval $[a, b]$ with rational $a < b$. Since there are only countably many intervals $[a, b]$ with rational $a < b$, it is enough for us to fix one such interval $[a, b]$ and prove that there is a P-null set $N_{a,b}$, such that, for $\omega \notin N_{a,b}$, the trajectory $B(\cdot, \omega)$ is not non-decreasing on $[a, b]$. Finally, we come to the most important part. From parts (a) and (d) of Proposition 11.0.2, one can easily see that the process $\{X_t = (b-a)^{-1/2}(B_{(b-a)t+a} - B_a), t \in [0, 1]\}$ is a SBM on $[0, 1]$ and $B(\cdot, \omega)$ is non-decreasing on $[a, b]$ if and only if $X(\cdot, \omega)$ is non-decreasing on $[0, 1]$. What all these mean is that, our task is reduced to simply showing that, for a SBM $\{X_t, t \in [0, 1]\}$ on $[0, 1]$, there is a P-null set N, such that, for all $\omega \notin N$, the trajectory $X(\cdot, \omega)$ is not non-decreasing on $[0, 1]$. Denoting $A_n = \{\omega : X(\frac{i}{n}, \omega) - X(\frac{i-1}{n}, \omega) \geq 0, \text{ for } i = 1, \ldots, n\}$ and $N = \bigcap_n A_n$, it is clear that, for $\omega \notin N$, the trajectory $t \mapsto X(t, \omega)$ is not non-decreasing on $[0, 1]$. Using the Definition 11.0.1 of SBM, one easily gets $P(A_n) = 2^{-n}$, for each $n \geq 1$, and that implies that $P(N) = 0$, completing the proof. $\square$

The last proposition was the first illustration of how badly behaved a typical trajectory of a SBM is. The reader should try to visualize, if she can, the graph of a continuous real-valued function on $[0, \infty)$, which is not monotone on any interval. However, that's only the beginning of the story. The next proposition brings out another feature of how erratic a typical Brownian path can be. To state the result, we introduce a few notations. For each $t > 0$, let us denote

$$m_t = \min\{B_u : 0 \leq u \leq t\} \text{ and } M_t = \max\{B_u : 0 \leq u \leq t\}.$$

Note that, by the continuity of all trajectories, $m_t = \inf\{B_r : 0 \leq r \leq t, r \text{ rational}\}$ and $M_t = \sup\{B_r : 0 \leq r \leq t, r \text{ rational}\}$, ensuring that both m_t and M_t are real random variables.

PROPOSITION 11.1.4. $P(\{\omega : m_t(\omega) < 0 < M_t(\omega) \text{ for all } t > 0\}) = 1$.

11.1. Path Properties of SBM

The reader may wonder whether the set $\{\omega : m_t(\omega) < 0 < M_t(\omega) \text{ for all } t > 0\}$ is an event (since it involves conditions on an uncountable number of random variables). While this concern is justified, the reader may rest assured that this issue will be resolved inside the proof.

Proof. In view of the fact $\{-B_t, t \in [0, \infty)\}$ is also a SBM (Proposition 11.0.2 (a)), it is enough to prove that $P(M_t > 0 \text{ for all } t > 0) = 1$. Fixing a sequence $\{t_n\}$ *strictly* decreasing to 0, one easily sees that $\{\omega : M_t(\omega) > 0 \text{ for all } t > 0\}$ equals $\{\omega : B(t_n, \omega) > 0 \text{ for infinitely many } n\}$, which shows, in particular, that the set $\{\omega : M_t(\omega) > 0 \text{ for all } t > 0\}$ is an event. We, therefore, have to just show that $P\bigl(\limsup_n A_n\bigr) = 1$, where $A_n = \{\omega : B(t_n, \omega) > 0\}$. This is going to be an interesting application of Kolmogorov's Zero-One Law (Theorem 7.4.5). First of all, we clearly have, for each n, $P(A_n) = \frac{1}{2}$, and, therefore, $P\bigl(\bigcup_{k \geq n} A_k\bigr) \geq \frac{1}{2}$. Since $P\bigl(\limsup_n A_n\bigr) = \lim_n P\bigl(\bigcup_{k \geq n} A_k\bigr)$, we get $P\bigl(\limsup_n A_n\bigr) \geq \frac{1}{2}$. Now that we have shown $P\bigl(\limsup_n A_n\bigr) \geq \frac{1}{2}$, we only need to show that the event $\{\limsup_n A_n\}$ is a tail event for some sequence of independent random variables, in order for us to conclude that $P\bigl(\limsup_n A_n\bigr)$ indeed equals 1. Denoting $X_n = B(t_n) - B(t_{n+1})$, we have a sequence $\{X_n\}$ of random variables, which are independent (by property (iii) in Definition 11.0.1). Note that $B(t_n) = \sum_{k \geq n} X_k$ and also that $\limsup_n A_n = \{B(t_n) = \sum_{k \geq n} X_k > 0 \text{ for infinitely many } n \geq m\}$, for every $m \geq 1$. This show that the event $\limsup_n A_n$ belongs to the tail σ-field of the sequence $\{X_n\}$. We conlcude that $P\bigl(\limsup_n A_n\bigr) = 1$ and that, as already pointed out, completes the proof. $\square$

What the last result says is that a typical Brownian path takes both positive and negative values in every neighbourhood of $t = 0$, no matter how small. Because of continuity of paths, this is the same as saying that a typical Brownian path has infinitely many crossings of level zero in every neighbourhood of $t = 0$. Of course, the reader may not find this behaviour to be too unfamiliar. She may perhaps have seen the graph of the continuous function $f(t) = t \sin(\frac{1}{t})$, $t > 0$ and $f(0) = 0$, which has a similar behaviour near $t = 0$. However, a typical Brownian path behaves in a far more bizzare way. For the function f, all other zeros of f, except $t = 0$, are isolated points in the "zero-set" of f. On the contrary, we will see later that none of the zeros of a typical Brownian path is an isolated point in its zero-set. So, if one now tries to visualize the graph of a typical Brownian path, one would realize that it is almost impossisble to comprehend what a typical path does everytime it leaves level zero.

The next result is an immediate consequence of Proposition 11.1.4 and is left as an exercise for the reader to prove. It is an illustration of something that was pointed out earlier, namely, that, by virtue of the "time-inversion" property (part (c) of Proposition 11.0.2), a property of Brownian trajectory "near time 0" often leads us automatically to an appropriate property "near time ∞". To state the result, we need the following notations. For each $t > 0$, denote $\widetilde{m}_t = \inf\{B_u : u \geq t\}$ and $\widetilde{M}_t = \sup\{B_u : u \geq t\}$. Once again, one can easily see that both $\widetilde{m}_t$ and $\widetilde{M}_t$ are random variables (possibly extended real-valued), for each t.

PROPOSITION 11.1.5. $P(\{\omega : \widetilde{m}_t(\omega) < 0 < \widetilde{M}_t(\omega) \text{ for all } t > 0\}) = 1.$

This last result implies that, as $t \to \infty$, a typical Brownian path crosses the level zero infinitely many times. A relevant question would be: what happens to the amplitude of fluctuation on the two sides of level zero, as $t \to \infty$? Since B_t has a Normal distribution with 0 mean and variance t, one would expect the amplitude of fluctuation to gradually blow up as $t \to \infty$. The next result confirms that intuition.

PROPOSITION 11.1.6. $P(\{\limsup_{n\to\infty} B_n = +\infty, \liminf_{n\to\infty} B_n = -\infty\}) = 1.$

For the proof of Proposition 11.1.6, we will need the following inequality which is often found to be useful in providing nice estimates for the tail probabilities of a N(0, 1) distribution. The proof of the inequality is a simple exercise in calculus and, therefore, left for the reader to verify.

LEMMA 11.1.7. $\dfrac{a}{1+a^2} e^{-a^2/2} \leq \int_a^\infty e^{-x^2/2} dx \leq \dfrac{1}{a} e^{-a^2/2}$, for any $a > 0$.

Proof of Proposition 11.1.6. Once again, in view of the fact that $\{-B_t\}$ is a SBM, it suffices to prove only that $P(\{\limsup_{n\to\infty} B_n = +\infty\}) = 1$. Fix an integer $m \geq 9$ and, for each $n \geq 1$, consider the events

$$A_n = \{B(m^n) < -3\sqrt{m^n \log n}\}, \quad C_n = \{B(m^{n+1}) - B(m^n) \geq \sqrt{2(m-1)m^n \log n}\}$$

Since $B(m^n) \sim N(0, m^n)$ and $B(m^{n+1}) - B(m^n) \sim N(0, (m-1)m^n)$, it follows that $P(A_n) = P(Z < -3\sqrt{\log n})$ and $P(C_n) = P(Z \geq \sqrt{2 \log n})$, where $Z \sim N(0,1)$. Using both inequalities given in Lemma 11.1.7, one can easily see that the series $\sum_n P(A_n)$ converges, while the series $\sum_n P(C_n)$ diverges. Noting that C_n, $n \geq 1$, are independent events and using both the Borel-Cantelli lemmas one gets

$$P\big(B(m^n) \geq -3\sqrt{m^n \log n} \text{ for all but finitely many } n\big) = 1, \text{ and}$$

$$P\big(B(m^{n+1}) - B(m^n) \geq \sqrt{2(m-1)m^n \log n} \text{ for infinitely many } n\big) = 1.$$

Taking the intersection of the above two sets of probability one, one gets

$$P\big(B(m^{n+1}) \geq [\sqrt{2(m-1)} - 3]\sqrt{m^n \log n} \text{ for infinitely many } n\big) = 1.$$

Recalling that we chose $m \geq 9$, one can easily deduce from the above that $P(\limsup_n B_n = +\infty) = 1$, thus completing the proof. □

Now that we have proved that the amplitude of fluctuation above and below level zero grows as $t \to \infty$, one would next wonder about the rate at which it grows. Since B_t has variance t, one may feel that the rate of growth is likely to

11.1. Path Properties of SBM

be $\sqrt{t}$. It turns out that the actual rate of growth is a little faster than $\sqrt{t}$. The precise rate of growth is given by a very well-known result called "Khintchine's Law of Iterated Logarithm". Of course, as we know already, any property of Brownian paths as $t \to \infty$ leads to a counterpart as $t \to 0$, due to the time-inversion property of SBM. Both results are stated together in the next theorem, without a proof.

THEOREM 11.1.8. (Khintchine's Law of Iterated Logarithm)

$$P\left(\limsup_{t \to \infty} \frac{B_t}{\sqrt{2t \log \log t}} = 1, \liminf_{t \to \infty} \frac{B_t}{\sqrt{2t \log \log t}} = -1\right) = 1, \text{ and,}$$

$$P\left(\limsup_{t \to 0} \frac{B_t}{\sqrt{2t \log \log(1/t)}} = 1, \liminf_{t \to 0} \frac{B_t}{\sqrt{2t \log \log(1/t)}} = -1\right) = 1.$$

We next focus on studying another aspect of a typical Brownian path, namely, its "total variation" on closed bounded subintervals of $[0, \infty)$. The results that we will see, would once again be testimony to how wild the oscillations of a typical Brownian path are. We quickly recall the definitions of some relevant concepts. Let f be a real-valued function on a closed bounded interval $[a, b]$. A "partition" π of the interval $[a, b]$ means a finite set of points $\pi = \{t_0 < t_1 < \cdots < t_{k-1} < t_k\}$, where $t_0 = a$ and $t_k = b$. The "norm" of the partition π, denoted $\|\pi\|$, is defined to be $\|\pi\| = \max_i(t_i - t_{i-1})$. Given such a partition π, the variation of f on $[a, b]$ with respect to the partition π is defined as

$$V(f, \pi, [a, b]) = \sum_{i=1}^{k} |f(t_i) - f(t_{i-1})|$$

and "total variation" of f over $[a, b]$ is defined as

$$V(f, [a, b]) = \sup_{\pi} V(f, \pi, [a, b]),$$

where the supremum is over all (finite) partitions π of $[a, b]$.

DEFINITION 11.1.9. A real-valued function f on a closed bounded interval $[a, b]$ is said to be *of bounded variation* on $[a, b]$ if $V(f, [a, b]) < \infty$. Otherwise, f is said to be of *unbounded variation* on $[a, b]$.

It is a standard exercise in Real Analysis that if $\{\pi_n\}$ is any sequence of (finite) partitions of $[a, b]$, such that, $\pi_n \subset \pi_{n+1}$, for all n, and $\|\pi_n\| \downarrow 0$, then for any continuous function f on $[a, b]$, one has $V(f, \pi_n, [a, b]) \uparrow V(f, [a, b])$. A reader who has not seen this before, should try and prove it. It is, of course, clear from the definition that if f is of bounded variation on $[a, b]$, then it is also of bounded variation on any subinterval $[c, d] \subset [a, b]$.

Returning to SBM, here is the result on the variation of Brownian paths on closed bounded subintervals of $[0, \infty)$.

PROPOSITION 11.1.10. *P-almost every Brownian path is of unbounded variation on every non-degenerate closed bounded subinterval of $[0, \infty)$.*

It is important to understand the statement of the above proposition clearly. It says that there is *one P-null set N* such that, for all $\omega \notin N$, the trajectory $B(\cdot, \omega)$ is of unbounded variation *on every* closed bounded subinterval of $[0, \infty)$.

Proof. Since any non-degenerate closed bounded subinterval of $[0, \infty)$ contains a subinterval $[a, b]$ with $a < b$ rational, it is enough to fix one interval $[a, b]$ with rational endpoints $a < b$ and show that

$$P(\omega : V(B(\cdot, \omega), [a, b]) = \infty) = 1.$$

To show this, we are going to use the standard fact mentioned in the paragraph following Definition 11.1.9. Choosing an appropriate sequence $\{\pi_n\}$ of partitions of $[a, b]$ with $\pi_n \subset \pi_{n+1}$, for all n, and $\|\pi_n\| \downarrow 0$, we will show that $V(B(\cdot, \omega), \pi_n, [a, b]) \uparrow +\infty$, for P-almost every ω. Denoting $\beta = (b-a)$, we choose $\pi_n = \{t_{i,n}, 0 \le i \le 2^n\}$, where $t_{i,n} = a + \frac{i\beta}{2^n}, 0 \le i \le 2^n$. It is clear that $\{\pi_n\}$ satisfies the required properties. Also, denoting $X_n(\omega) = V(B(\cdot, \omega), \pi_n, [a, b])$, it is clear that $X_n = \sum_{i=1}^{2^n} Y_{i,n}$, where $\{Y_{i,n}, 1 \le i \le 2^n\}$ are i.i.d. random variables with each $Y_{i,n} \stackrel{d}{=} \sqrt{\beta} 2^{-n/2} |Z|$, where $Z \sim N(0,1)$. This gives us $E(X_n) = 2^{n/2} C$, where $C = \sqrt{\beta} E(|Z|) > 0$, and so $E(X_n) \to \infty$, as $n \to \infty$. Thus, given any $K > 0$, there is n_0 such that $E(X_n) \ge K$ for all $n \ge n_0$, which can easily be seen to imply that $P(X_n \ge K) \ge P(|X_n - E(X_n)| \le E(X_n) - K)$, for all $n \ge n_0$. But then Chebycheff's inequality will imply that, for $n \ge n_0$, we have $P(X_n \ge K) \ge 1 - \frac{V(X_n)}{(E(X_n)-K)^2}$. Observing now that $V(X_n) = \beta V(|Z|) < \infty$ for all n, we finally have $P(X_n \ge K) \ge 1 - \frac{\beta V(|Z|)}{(E(X_n)-K)^2}$, for all $n \ge n_0$. Letting $n \to \infty$ now, we get $\lim_n P(X_n \ge K) = 1$. Noting that this holds for any $K > 0$ and using the fact that $\{X_n\}$ is a monotonically non-decreasing sequence of random variables, it is not difficult to deduce now that, P-almost surely, $X_n \uparrow +\infty$, as $n \to \infty$. We leave this last step as an exercise for the reader. □

The fact that almost every trajectory of SBM is of unbounded variation on all non-degenerate closed bounded intervals not only reinforces the extremely oscillatory nature of Brownian paths, but it also acted as a stumbling block for all efforts to develop a standard theory of integration with respect to Brownian path, in the spirit of Riemann-Stieltjes integral. The deadlock was finally removed by works of Norbert Wiener and later Kiyoshi Ito (also parallelly, Wolfgang Döeblin) who developed a completely different theory of integration, which has since been known as Stochastic Integral. One among the many major components around which this integration theory is built up, is what is known as the "Quadratic Variation". We will first define this concept for any real function and then prove an interesting result on quadratic variation of Brownian paths.

Once again, let f be a real-valued function on a closed bounded interval $[a, b]$. For a partition $\pi = \{a = t_0 < t_1 < \cdots < t_k = b\}$, the Quadratic Variation of f on $[a, b]$ with respect to the partition π, is defined as $Q(f, \pi, [a, b]) = \sum_{i=1}^{k} (f(t_i) - f(t_{i-1}))^2$. Given a nested sequence $\{\pi_n\}$ of partitions of $[a, b]$ with $\|\pi_n\| \downarrow 0$, if the limit

11.1. Path Properties of SBM

$\lim_n Q(f, \pi_n, [a, b])$ exists and is finite, then this is called the "Quadratic Variation of f on $[a, b]$ along the sequence $\{\pi_n\}$" and is denoted by $Q(f, \{\pi_n\}, [a, b])$. If $Q(f, \{\pi_n\}, [a, b])$ exists for all nested sequences $\{\pi_n\}$ with $\|\pi_n\| \downarrow 0$ and are all equal, then this common value is called the "Quadratic Variation of f on $[a, b]$", denoted $Q(f, [a, b])$. Here is a very simple exercise.

EXERCISE 11.1.11. *Show that a real continuous function of bounded variation on a closed bounded interval has quadratic variation 0 on that interval.*

Let us now consider a closed bounded interval $[a, b] \subset [0, \infty)$, with $a < b$. If $\pi = \{a = t_0 < t_1 < \cdots < t_k = b\}$ is a partition of $[a, b]$, then, for each ω, we have $Q(B(\cdot, \omega), \pi, [a, b]) = \sum_{i=1}^{k} Y_i^2(\omega)$, where $\{Y_i, 1 \le i \le k\}$ are independent random variables with $Y_i \sim N(0, t_i - t_{i-1})$. This immediately gives $E(Y_i^2) = t_i - t_{i-1}$, for each i. Also, using properties of normal distribution, it not difficult to verify that $E(Y_i^4) = 3(t_i - t_{i-1})^2$ (left as exercise) and hence $V(Y_i^2) = 2(t_i - t_{i-1})^2$. From all these, we get $E(Q(B(\cdot), \pi, [a, b])) = b - a$, $V(Q(B(\cdot), \pi, [a, b])) \le 2(b-a)\|\pi\|$. The following result is now an immediate consequence of this analysis.

PROPOSITION 11.1.12. *Let I be a closed bounded interval. Then, for any nested sequence $\{\pi_n\}$ of partitions of I with $\|\pi_n\| \downarrow 0$,*

$$Q(B(\cdot, \omega), \pi_n, I) \xrightarrow{L_2} \lambda(I), \text{ as } n \to \infty,$$

where λ denotes the Lebesgue measure.

Unfortunately, the convergence asserted in the above proposition is only convergence in L_2 and not almost sure, although we will have almost sure convergence along some subsequence of $\{\pi_n\}$. So, strictly speaking, we cannot say that the quadratic variation of a typical Brownian path on I equals $\lambda(I)$. A slightly stronger condition on $\|\pi_n\|$, than just $\|\pi_n\| \downarrow 0$, guarantees almost sure convergence along the sequence $\{\pi_n\}$ (see Exercise 11.3.5).

The result that we are going to see next is one of the most well-known results on properties of Brownian trajectories and is perhaps the most striking one. In short, what it says is that almost every trajectory of SBM is non-differentiable at every point of $[0, \infty)$. As a matter of fact, the result actually says something much stronger. It says that, for almost every trajectory of SBM, the slopes from both sides remain unbounded at every point. Before we can get to the statement, we need some definitions.

DEFINITION 11.1.13. *Let f be a real-valued function on a non-degenerate interval I. For a $t \in I$, which is an accumulation point of I from the right, the Right Upper and Right Lower Dini Derivatives at t are defined as*

$$D^* f(t+) = \limsup_{h \downarrow 0} \frac{f(t+h) - f(t)}{h} \text{ and } D_* f(t+) = \liminf_{h \downarrow 0} \frac{f(t+h) - f(t)}{h}.$$

Similarly, for a $t \in I$, which is an accumulation point of I from the left, the Left Upper and Left Lower Dini Derivatives at t are defined as

$$D^*f(t-) = \limsup_{h \uparrow 0} \frac{f(t+h) - f(t)}{h} \quad \text{and} \quad D_*f(t-) = \liminf_{h \uparrow 0} \frac{f(t+h) - f(t)}{h}.$$

It is clear from the definition that $-\infty \leq D_*f(t+) \leq D^*f(t+) \leq +\infty$ and $-\infty \leq D_*f(t-) \leq D^*f(t-) \leq +\infty$. Further, the right-hand derivative $Df(t+)$ (resp., left-hand derivative $Df(t-)$) exists if and only if $-\infty < D_*f(t+) = D^*f(t+) < +\infty$ (resp., $-\infty < D_*f(t-) = D^*f(t-) < +\infty$) and in that case, $Df(t+) = D_*f(t+) = D^*f(t+)$ (resp., $Df(t-) = D_*f(t-) = D^*f(t-)$). Finally, the derivative $Df(t)$ exists if and only if all four Dini derivatives are equal and finite and in that case, $Df(t)$ equals the common value.

We now state the result on SBM, which says that almost every trajectory is not only non-differentiable everywhere, but is actually much more badly behaved.

THEOREM 11.1.14. (Paley-Wiener-Zygmund) There is a P-null set N such that if $\omega \notin N$, then, for every $t \in [0, \infty)$, either $D^*B(t+,\omega) = +\infty$ or $D_*B(t+,\omega) = -\infty$ and also, for every $t \in (0, \infty)$, either $D^*B(t-,\omega) = +\infty$ or $D_*B(t-,\omega) = -\infty$.

Before getting to the proof, let us make some useful observations. We first claim that it suffices to prove that there is a P-null set N_0, such that, if $\omega \notin N_0$, then, for each $t \in [0, 1)$, either $D^*B(t+,\omega) = +\infty$ or $D_*B(t+,\omega) = -\infty$ and also, for every $t \in (0, 1]$, either $D^*B(t-,\omega) = +\infty$ or $D_*B(t-,\omega) = -\infty$. This is because, part (d) of Proposition 11.0.3 would allow us to apply the above result to the process $\{B^n(t) = B(n+t) - B(n), t \in [0,1]\}$, for each $n \geq 1$, thereby giving us only a countable number of P-null sets, such that, outside the union of these countably many P-null sets, the assertion of the theorem will hold. Next, in view of the time-reversal property of SBM (part (b) of Proposition 11.0.3), it is actually enough to just show that there is a P-null set N, such that, if $\omega \notin N$, then, for each $t \in [0, 1)$, either $D^*B(t+,\omega) = +\infty$ or $D_*B(t+,\omega) = -\infty$. We now proceed to prove exactly this. The proof presented here is attributed to Dvorëtzky, Erdös and Kakutani.

Proof. Let $A = \{\omega : -\infty < D_*B(t+,\omega) \leq D^*B(t+,\omega) < +\infty,$ for some $t \in [0,1)\}$. We cannot show that A itself is a P-null set, because A may not even belong to the underlying σ-field. What we will do instead is to get a set N with $A \subset N$, such that, N belongs to the underlying σ-field and $P(N) = 0$.

Towards this, note first that $\omega \in A$ implies existence of some $t \in [0, 1)$ such that the quotient $\frac{B(t+h,\omega) - B(t,\omega)}{h}$ remains bounded for all sufficiently small $h > 0$. Equivalently, $\omega \in A$ implies that there exists $t \in [0, 1)$ and positive integers j and k (all three depending possibly on ω), such that,

$$|B(t+h,\omega) - B(t,\omega)| \leq jh, \quad \text{for all } 0 \leq h \leq 1/k. \tag{11.1.2}$$

Without loss of generality, we assume that the integer k above also satisfies $t + \frac{1}{k} < 1$ (since this may involve only taking a larger k, if necessary). Now, for every integer $n > 4k$, we will have an i (again, depending possibly on ω), with $1 \leq i \leq n$, such that, $t \in \left[\frac{i-1}{n}, \frac{i}{n}\right)$. Using the inequalities $t + \frac{1}{k} < 1$ and $n > 4k$, one can easily see that $\frac{i+3}{n} = \frac{i-1}{n} + \frac{4}{n} < t + \frac{1}{k} < 1$, implying that $i \leq n - 4$.

11.1. Path Properties of SBM

What this means is that, the inequality (11.1.2) would hold with $t+h = \frac{i+m}{n}$, for $m = 0, 1, 2, 3$. In other words, for $\omega \in A$, we will have $t \in [0, 1)$, positive integers j and k with $t + \frac{1}{k} < 1$, such that, for every $n > 4k$, there is a positive integer $i \leq n - 4$ with $t \in \left[\frac{i-1}{n}, \frac{i}{n}\right)$ and

$$\left|B\left(\tfrac{i+m}{n}, \omega\right) - B(t, \omega)\right| \leq \left(\tfrac{i+m}{n} - t\right) j < \tfrac{m+1}{n} j, \text{ for } m = 0, 1, 2, 3.$$

Using triangle iequality, the above easily leads to

$$\left|B\left(\tfrac{i+m}{n}, \omega\right) - B\left(\tfrac{i+m-1}{n}, \omega\right)\right| \leq \tfrac{2m+1}{n} j, \text{ for } m = 1, 2, 3. \tag{11.1.3}$$

For positive integers i, j, n, let us denote

$$A_{i,j,n} = \left\{\omega : \left|B\left(\tfrac{i+m}{n}, \omega\right) - B\left(\tfrac{i+m-1}{n}, \omega\right)\right| \leq \tfrac{2m+1}{n} j, \text{ for } m = 1, 2, 3\right\}.$$

Clearly, $A_{i,j,n}$, for every i, j, n, belongs to the underlying σ-field and what we have shown is that, for every $\omega \in A$, there exist integers $j \geq 1$, $k \geq 1$, such that, for all $n > 4k$, there is an i, with $1 \leq i \leq n - 4$, for which $\omega \in A_{i,j,n}$. In other words, we have proved that

$$A \subset \bigcup_{j \geq 1} \bigcup_{k \geq 1} \bigcap_{n > 4k} \bigcup_{i=1}^{n-4} A_{i,j,n}.$$

The set on the right-hand side above clearly belongs to the underlying σ-field. To complete the proof, we only need to show that it is a P-null set. For this, it suffices to show that $P\left(\bigcap_{n > 4k} \bigcup_{i=1}^{n-4} A_{i,j,n}\right) = 0$, for each $j \geq 1$ and $k \geq 1$. For this, let us observe that the three increments appearing in the definition of $A_{i,j,n}$, namely, $B\left(\tfrac{i+1}{n}\right) - B\left(\tfrac{i}{n}\right)$, $B\left(\tfrac{i+2}{n}\right) - B\left(\tfrac{i+1}{n}\right)$ and $B\left(\tfrac{i+3}{n}\right) - B\left(\tfrac{i+2}{n}\right)$, are independent and each of these increments have a $N(0, \tfrac{1}{n})$ distribution. Using these, one easily gets $P(A_{i,j,n}) = P\left(|Z| \leq \tfrac{3j}{\sqrt{n}}\right) \times P\left(|Z| \leq \tfrac{5j}{\sqrt{n}}\right) \times P\left(|Z| \leq \tfrac{7j}{\sqrt{n}}\right)$, where $Z \sim N(0, 1)$. Using now the (rather crude) bound $P(|Z| \leq \alpha) = \tfrac{2}{\sqrt{2\pi}} \int_0^\alpha e^{-x^2/2} dx \leq \alpha$, for $\alpha > 0$, one gets $P(A_{i,j,n}) \leq \tfrac{105 j^3}{n^{3/2}}$, for all i, j and n, which gives

$$P\left(\bigcup_{i=1}^{n-4} A_{i,j,n}\right) \leq \frac{105 j^3}{\sqrt{n}}, \text{ for all } n > 4k \text{ and all } j \geq 1.$$

From this, we finally get

$$P\left(\bigcap_{n > 4k} \bigcup_{i=1}^{n-4} A_{i,j,n}\right) \leq \inf_{n > 4k} P\left(\bigcup_{i=1}^{n-4} A_{i,j,n}\right) \leq \inf_{n > 4k} \frac{105 j^3}{\sqrt{n}} = 0,$$

for all $j \geq 1$ and $k \geq 1$, as was to be proved. $\square$

REMARK 11.1.15. A careful look into the statement of Theorem 11.1.14 will reveal that what it asserts can also be seen in the following way. It says that almost all trajectories of SBM are NOT locally Hölder continuous with exponent 1 at any point $t \in [0, \infty)$. A slight modification of the argument used in the above proof allows one to prove a stronger result, namely, that almost every Brownian trajectory is not locally Hölder continuous anywhere on $[0, \infty)$ for any exponent $\gamma > 1/2$. This is the Exercise 11.3.9. Seen in conjunction with the remark made at the end of Chapter 10, it turns out that $\gamma = 1/2$ acts as a threshold, where there is a regime change in terms of local Hölder continuity. The behaviour for $\gamma = 1/2$ is a well-known result due to Lévy, but it is a little too complicated to present here.

We end this section by returning to a point that was earlier made about the "zero-set" of a Brownian trajectory. As stated at the outset of this section, we assume *all* trajectories to be continuous. Let us start by formally defining the zero-set and first observing some immediate properties. For each ω, the set

$$\mathbf{Z}(\omega) = \{t \in [0, \infty) : B_t(\omega) = 0\} \qquad (11.1.4)$$

defines the zero-set of the ω-trajectory. By continuity of Brownian trajectories, $\mathbf{Z}(\omega)$ is clearly a closed set, for each ω. Also, from Propositions 11.1.4 and 11.1.5, one easily sees that, for P-almost every ω, the set $\mathbf{Z}(\omega)$ is unbounded and has $t = 0$ as a limit point. Next, since $P(B_t = 0) = 0$ for each $t > 0$, we have $\int_0^\infty \int_\Omega I_{\{(t,\omega):B_t(\omega)=0\}} dP(\omega) dt = 0$. An application of Fubini's Theorem then implies that, the set $\mathbf{Z}(\omega)$ has zero Lebesgue measure, for $[P]$-a.e. ω. Let us finally turn to the claim that was made earlier. From Proposition 11.1.4, we already know that $t = 0$ is a limit point of $\mathbf{Z}(\omega)$, for $[P]$-a.e. ω. But, what we claimed earlier is that, for $[P]$-a.e. ω, every point of $\mathbf{Z}(\omega)$, and not just $t = 0$, is a limit point of $\mathbf{Z}(\omega)$. This will be proved if we can show that, for $[P]$-a.e. ω, any point of $\mathbf{Z}(\omega)$, which is isolated in $\mathbf{Z}(\omega)$ from the left, has to be a limit point of $\mathbf{Z}(\omega)$ from the right. The idea for proving this is to first observe that if $t_0 \in (0, \infty)$ is a point in $\mathbf{Z}(\omega)$, which is isolated in $\mathbf{Z}(\omega)$ from the left, then there must be a rational number $0 < r < t_0$, such that, $t_0 = \min\{t > r : B_t(\omega) = 0\}$. Suppose now that we consider, for each rational number $r \in (0, \infty)$, the random variable $\tau_r(\omega) = \inf\{t > r : B_t(\omega) = 0\}$, where we follow the convention of setting infimum of empty set as $+\infty$. Proposition 11.1.5 and continuity of Brownian trajectories guarantee that $\tau_r(\omega)$ is finite for $[P]$-a.e. ω and that the infimum is actually a minimum. So, by discarding a P-null set, if necessary, we assume, in what follows, that $\tau_r(\omega) < \infty$ and consequently, $\tau_r(\omega) \in \mathbf{Z}(\omega)$, for all ω. The reader surely realizes that we are considering τ_r, only for rational r, because this way we have to discard only countably many P-null sets. Recalling now what we observed above, any point of $\mathbf{Z}(\omega)$, which is isolated in $\mathbf{Z}(\omega)$ from the left, must equal $\tau_r(\omega)$, for some rational $r \in (0, \infty)$ (where r, of course, may depend on ω). Therefore, if we can prove that, for every rational $r \in (0, \infty)$, the point $\tau_r(\omega)$ is, for $[P]$-a.e. ω, a limit point of $\mathbf{Z}(\omega)$ from the right, then we would have proved our original claim. This is where we get to the most important part of the argument. From part (d) of the Prposition 11.0.3, we know that. for any $s \geq 0$, the process $\{B_{s+t} - B_s, t \in [0, \infty)\}$ is a SBM. Suppose, for the time being, that the same is true even when if we replace the fixed time point $s \geq 0$ by the finite random time τ_r taking values in $(0, \infty)$. That would mean that the process $\{X_t(\omega) = B_{\tau_r+t} - B_{\tau_r}, t \geq 0\}$ is a SBM, so that the zero set of $\{X_t, t \geq 0\}$ will, $[P]$-a.e., have $t = 0$ as a limit point. Noting that $B_{\tau_r} \equiv 0$, the zero set of $\{X_t, t \geq 0\}$ is easily seen to equal $\{t \geq 0 : \tau_r(\omega) + t \in \mathbf{Z}(\omega)\}$. Thus, we get that $\tau_r(\omega)$ is, for $[P]$-a.e. ω, a limit point of $\mathbf{Z}(\omega)$ from the right, as was to be proved.

Of course, the most crucial part of the above argument rests on the assumption that the assertion of part (d) of Proposition 11.0.3 remains true even when

11.2. Markov and Strong Markov Properties

we replace a fixed time $s \geq 0$ by the random time τ_r taking values in $(0, \infty)$. That this is a valid assumption will be proved in the next section, in the context of what is called the "strong markov property" of Brownian motion. What we will see is that, while such an assumption may not hold for any random time, it is valid for a certain class of random times and that the random time τ_r defined above is in that class. Assuming this for the time being, let us summarize all our above observations on the zero-set of a typical trajectory of a SBM.

PROPOSITION 11.1.16. *For P-almost every ω, the zero set $\mathbf{Z}(\omega)$, as defined in (11.1.4), is a closed, unbounded subset of $[0, \infty)$, with zero Lebesgue measure and it is dense in itself, meaning that, each point of $\mathbf{Z}(\omega)$ is a limit point of $\mathbf{Z}(\omega)$.*

11.2 Markov and Strong Markov Properties

Let us start by recalling part (d) of Proposition 11.0.3, which says that, for any $s \geq 0$, the process $\{X_t = B_{s+t} - B_s, t \in [0, \infty)\}$ is a SBM, which is independent of $\{B_u, 0 \leq u \leq s\}$. For any fixed $s \geq 0$, the family $\{B_u, 0 \leq u \leq s\}$ captures the evolution of the SBM upto time s, whereas $\{B_{s+t}, t \geq 0\}$ captures the evolution post time s. The fact that $\{X_t, t \geq 0\}$ is a SBM and is also independent of $\{B_u, 0 \leq u \leq s\}$ and, in particular, of B_s, will naturally lead us to conclude that, given $\{B_u, 0 \leq u \leq s\}$, the process $\{B_{s+t}, t \geq 0\} \equiv \{X_t + B_s, t \geq 0\}$ would "conditionally" evolve like a SBM "plus" B_s added to it, that is, like a process which evolves like a SBM in every other way, except that it "starts at" B_s, instead of at 0. To summarize, part (d) of Proposition 11.0.3 can be viewed as essentially saying that, given the history of evolution of the SBM upto and including time s, the evolution post time s is "conditionally distributed" like what may be called a Brownian motion starting at its "present position" B_s. This is indeed correct and it is this which is known as the Markov property of Brownian motion. However, to formulate this property in a proper language and subsequently to prove it, we need to make certain things a bit more precise. First, we need to define what we mean by a Brownian motion "starting at" points other than 0. While the informal idea must be clear from the above discussion, here is the formal definition.

DEFINITION 11.2.1. *For any $x \subset \mathbb{R}$, a Brownian motion starting at x (abbreviated as BM starting at x) is a process $\{B_t^x, t \subset [0, \infty)\}$, defined on some probability space $(\Omega, \mathcal{A}, P)$, and satisfying*

(i) $B_0^x \equiv x$,

(ii) $B_t^x - B_s^x \sim \mathsf{N}(0, t-s)$, for all $0 \leq s < t$,

(iii) $B_{t_1}^x, B_{t_2}^x - B_{t_1}^x, \ldots, B_{t_n}^x - B_{t_{n-1}}^x$ *are independent, for any $0 \leq t_1 < \cdots < t_n$, and,*

(iv) *the map $t \mapsto B_t^x(\omega)$ is continuous on $[0, \infty)$, for all $\omega \in \Omega$.*

One can see that $\{B_t^x, t \in [0, \infty)\}$ differs from a SBM only in property (i). In fact, using this notation, SBM is nothing but $\{B_t^0, t \in [0, \infty)\}$, a BM starting at 0. One can easily see that, as with SBM, an equivalent characterization of a BM starting at x, is that it is a Gaussian process on $[0,\infty)$, with continuous trajectories and having mean function $\mu(\cdot) \equiv x$ and covariance kernel $C(s,t) = s \wedge t$. This is left as an exercise. Using this (or otherwise), one can easily see that, if $\{B_t, t \in [0,\infty)\}$ is a SBM, then $\{x + B_t, t \in [0,\infty)\}$ is a BM starting at x and conversely, if $\{B_t^x, t \in [0,\infty)\}$ is a BM starting at x, then $\{B_t^x - x, t \in [0,\infty)\}$ is a SBM. Here is an easy exercise, which will be used later in proving the Markov property. It is worth pointing out here that even though at the start of the section, we were confined only to SBM and its markov property, we are actually going to prove Markov property of BM starting at x, for any $x \in \mathbb{R}$.

EXERCISE 11.2.2. With $\{B_t^x, t \in [0,\infty)\}$ as in Definition 11.2.1, show that, for any $s \geq 0$, the process $\{B_{s+t}^x - B_s^x, t \geq 0\}$ is a SBM, which is independent of $\{B_u^x, 0 \leq u \leq s\}$.

Coming now to Markov property, it should be clear from whatever we have discussed so far, that Markov property of SBM, or more generally, of BM starting at x, seeks to describe, for any $s \geq 0$, the conditional distribution of the evolution post time s, given the history of evolution upto and including time s. Now, the evolution post time s, is captured by the process $\{B_{s+t}, t \in [0,\infty)\}$ for SBM, and, more generally, by $\{B_{s+t}^x, t \in [0,\infty)\}$ for BM starting at x. In any case, to talk of these conditional distributions in a proper way (see Section 9.3), we need to first fix a measurable space $(E, \mathcal{E})$, such that, the 'post-time-s' evolutions can be regarded as E-valued random variables. In view of the continuity of all trajectories BM, a very natural choice for E happens to be the space C, the set of all real-valued continuous functions on $[0,\infty)$. We now proceed to define an appropriate σ-field on C and establish some preliminary facts that will eventually enable us to formulate and prove the Markov property, as envisaged at the start of this section, in the language of Section 9.3.

For the rest of this section, C will denote, as above, the set of all real-valued continuous functions on $[0,\infty)$ (one often uses the notation $C([0,\infty), \mathbb{R})$ for C). Note that, property (iv) of Definition 11.2.1 asserts that, for any $x \in \mathbb{R}$, the map $\omega \mapsto B^x(\cdot, \omega)$ is a map on Ω into C. We now define an appropriate σ-field on C, which will make this map a measurable map. For this, let $\{e_t, t \in [0,\infty)\}$ denote the usual evaluation maps on C, that is, $e_t(f) = f(t)$, for $f \in C, t \in [0,\infty)$. We now define $\mathcal{C}$ to be the smallest σ-field on C that makes the evaluation maps $e_t : C \to \mathbb{R}$ measurable, for all $t \in [0,\infty)$. Here is a simple, yet very useful exercise.

EXERCISE 11.2.3. Given any measurable space $(E, \mathcal{E})$ and a map $\varphi : E \to C$, show that φ is $(\mathcal{E}, \mathcal{C})$-measurable if and only if $e_t \circ \varphi : (E, \mathcal{E}) \to (\mathbb{R}, \mathcal{B})$ is measurable for all $t \in [0,\infty)$.

Now, if $\{B_t^x, t \in [0,\infty)\}$ is a BM starting at x defined on $(\Omega, \mathcal{A}, P)$, then by defintion, $e_t \circ B^x(\cdot, \omega) \mapsto B_t^x(\omega)$ is real random variable, for each t. From

11.2. Markov and Strong Markov Properties

this, it follows, as an immediate consequence of Exercise 11.2.3, that, for each $x \in \mathbb{R}$, the map $\omega \mapsto B^x(\cdot, \omega)$ is a measurable map on $(\Omega, \mathcal{A}, P)$ into $(C, \mathcal{C})$. In other words, $B^x(\cdot)$, for each $x \in \mathbb{R}$, can be thought of as a C-valued random variable on $(\Omega, \mathcal{A}, P)$. Denoting the induced probability on $(C, \mathcal{C})$ by P_x, we get a family of probabilities $\{P_x, x \in \mathbb{R}\}$ on $(C, \mathcal{C})$, where P_x can be thought of as the distribution or law of a Brownian motion starting at x. In particular, for each $x \in \mathbb{R}$, the usual coordinate process on $(C, \mathcal{C}, P_x)$ will be a BM starting at x.

REMARK 11.2.4. There is another way of viewing the space C and the σ-field $\mathcal{C}$, which is given in Exercise 11.3.6. One can define a metric on C which makes C a complete, separable metric space, with $\mathcal{C}$ as its Borel σ-field, that is, the σ-field on C, generated by all open sets. This guarantees the existence of an RCD for any $(C, \mathcal{C})$-valued random variable (see Theorem 9.3.5). However, we will not need this, because, while proving the stated Markov property, we will be able to explicitly describe a probability kernel and then directly show that it defines a version of the required regular conditional distribution.

Here again is another simple exercise, both parts of which can be easily proved by using the criterion given in Exercise 11.2.3. But before that, let us establish a notation that is going to be consistently used in the sequel. If x is a real number and f is a real valued function on $[0, \infty)$, we will use the notation $x+f$ throughout to denote the real valued function on $[0, \infty)$ given by $x + f(t)$, $t \in [0, \infty)$.

EXERCISE 11.2.5. (a) Show that, for each $x \in \mathbb{R}$, the map $\varphi_x : C \to C$ given by $\varphi_x : f \mapsto x + f$ is a measurable map on $(C, \mathcal{C})$ to $(C, \mathcal{C})$.
(b) Show that the map $\varphi : (\mathbb{R} \times C, \mathcal{B} \otimes \mathcal{C}) \to (C, \mathcal{C})$ given by $\varphi : (x, f) \mapsto x + f$ is a measurable map.

Here is a very important consequence of Exercise 11.2.5. We put it as a Lemma, which will be used a few times later.

LEMMA 11.2.6. Let Q be a probability on $(C, \mathcal{C})$. Then, for every $x \in \mathbb{R}$ and $E \in \mathcal{C}$, the set $\{f \in C : x + f \in E\} \in \mathcal{C}$ and $Q_x(E) = Q(\{f : x + f \in E\})$, $E \in \mathcal{C}$, defines a probability on $\mathcal{C}$. Further, for every $E \in \mathcal{C}$, the map $x \mapsto Q_x(E)$ is a measurable map on $(\mathbb{R}, \mathcal{B})$

Proof. For any $x \in \mathbb{R}$, we have $\{f : x + f \in E\} = \varphi_x^{-1}(E)$, where φ_x is as in Exercise 11.2.5, part (a). This implies that $\{f : x + f \in E\} \in \mathcal{C}$, for any $E \in \mathcal{C}$, and also that $Q_x(E) = Q(\varphi_x^{-1}(E))$, $E \in \mathcal{C}$, defines a probability on $\mathcal{C}$. For the other part, use the map φ in Exercise 11.2.5, part (b), to see that, for any $E \in \mathcal{C}$, the set $A = \{(x, f) \in \mathbb{R} \times C : x + f \in E\}$ belongs to the product σ-field $\mathcal{B} \otimes \mathcal{C}$. From what we saw in Chapter 5, specifically in Propositions 5.1.7 and 5.1.8, we have, for every $x \in \mathbb{R}$, the x-section of A, namely, $A(x) = \{f : x+f \in E\}$ belongs to $\mathcal{C}$ and further, the map $x \mapsto Q(A(x))$ is measurable on $(\mathbb{R}, \mathcal{B})$. By noting that $Q(A(x)) = Q_x(E)$, the proof is complete. □

Recall the family $\{P_x, x \in \mathbb{R}\}$ of probabilities on $(C, \mathcal{C})$ defined earlier. For each $x \in \mathbb{R}$, the coordinate process on $(C, \mathcal{C}, P_x)$ is a BM starting at x. In particular, on $(C, \mathcal{C}, P_0)$, the coordinate process $\{X_t, t \in [0, \infty)\}$ is a SBM and therefore, by what was noted earlier the process $\{x + X_t, t \in [0, \infty)\}$, for each $x \in \mathbb{R}$, will be a BM starting at x. What all these mean is that for any $x \in \mathbb{R}$, we have $P_x(E) = P_0(\{f : x + f \in E\})$, for every $E \in \mathcal{C}$. As an immediate application of Lemma 11.2.6, with $Q = P_0$, we get the following result.

PROPOSITION 11.2.7. For each $E \in \mathcal{C}$, the map $x \mapsto P_x(E)$ is a measurable map on $(\mathbb{R}, \mathcal{B})$.

We are now ready to state the Markov property of Brownian motion.

THEOREM 11.2.8. (Markov property of Brownian motion)
For $x \in \mathbb{R}$, let $\{B_t^x, t \in [0, \infty)\}$, defined on some probability space $(\Omega, \mathcal{A}, P)$, be a Brownian motion starting at x. Then, for any $s \geq 0$, the kernel

$$Q(\omega, E) = P_{B_s^x(\omega)}(E), \quad \omega \in \Omega, \ E \in \mathcal{C} \quad (11.2.1)$$

gives a version of the regular conditional distribution of $\{B_{s+t}^x, t \in [0, \infty)\}$, given $\{B_u^x, 0 \leq u \leq s\}$.

From whatever we have developed so far, we can prove the above theorem directly. But instead, what we are going to do is to prove a general lemma and get the theorem as a straightforward application of the lemma. The reason for doing this is that the lemma is not directly related to Brownian motion and, therefore, may be useful in some other contexts also. In fact, we will use the same lemma later for proving also the strong Markov property of Brownian motion.

LEMMA 11.2.9. Let $(\Omega, \mathcal{A}, P)$ be a probability space and $\mathcal{G} \subset \mathcal{A}$ be a sub-σ-field. Consider a C-valued random variable Y on $(\Omega, \mathcal{A}, P)$, which is independent of $\mathcal{G}$. For a $\mathcal{G}$-measurable real random variable X on Ω, denote Z to be the C-valued random variable defined as $Z = X + Y$ (which, according to the notational convention established just before Exercise 11.2.5, means that $Z(\omega)$, for each ω, is the function given by $Z(\cdot, \omega) = X(\omega) + Y(\cdot, \omega)$). If Q_0 is the probability on $(C, \mathcal{C})$ defined as $Q_0 = PY^{-1}$, then the kernel

$$Q(\omega, E) = Q_0(\{f \in C : X(\omega) + f \in E\}), \quad \omega \in \Omega, \ E \in \mathcal{C}, \quad (11.2.2)$$

gives a version of the regular conditional distribution of Z, given $\mathcal{G}$.

Proof of Lemma 11.2.9. Using the criterion given by Exercise 11.2.3, it is easy to see that Z is a C-valued random variable. Coming to the kernel Q defined in (11.2.2), it is clear that $Q(\omega, \cdot)$, for each $\omega \in \Omega$, is a probability on $\mathcal{C}$. Next, by Lemma 11.2.6, the map $x \mapsto Q_0(\{f : x + f \in E\})$, for each $E \in \mathcal{C}$, is a measurable map on $(\mathbb{R}, \mathcal{B})$, from which it follows that, for each $E \in \mathcal{C}$, the map $Q(\cdot, E)$ is $\mathcal{G}$-measurable. Thus, we have shown that Q defined by (11.2.2) is a probability kernel on $\Omega \times \mathcal{C}$, with $Q(\cdot, E)$ measurable with respect to $\mathcal{G}$, for each $E \in \mathcal{C}$. To complete the proof, we now have to show that, for any $G \in \mathcal{G}$ and any $E \in \mathcal{C}$,

11.2. Markov and Strong Markov Properties

$$P\big(G\cap\{\omega:Z(\omega,\cdot)\in E\}\big)=\int_G Q(\omega,E)dP(\omega). \qquad (11.2.3)$$

Towards this, let us first observe that the map $\varphi:(\Omega,\mathcal{A},P)\to(\Omega\times C,\mathcal{G}\otimes\mathcal{C})$ given by $\omega\mapsto(\omega,Y(\cdot,\omega))$ is a measurable map and $P\varphi^{-1}=P\otimes Q_0$. For this, one simply considers any measurable rectangle $G\times E\subset\Omega\times C$, with $G\in\mathcal{G}$, $E\in\mathcal{C}$ and sees that $\varphi^{-1}(G\times E)=G\cap Y^{-1}(E)\in\mathcal{A}$ and moreover, from the independence of Y and $\mathcal{G}$ and using $Q_0=PY^{-1}$, one gets $P\varphi^{-1}(G\times E)=P(G)\times Q_0(E)$. Coming now to the proof of (11.2.3), let us fix $G\in\mathcal{G}$, $E\in\mathcal{C}$ and denote $H=\{(\omega,f)\in\Omega\times C:\omega\in G, X(\omega)+f\in E\}$. We claim that $H\in\mathcal{G}\otimes\mathcal{C}$. Observing that $H=(G\times C)\cap\{(\omega,f):X(\omega)+f\in E\}$, it is clear that the claim will be proved if we can show that the map $(\omega,f)\mapsto X(\omega)+f$ is a measurable map on $(\Omega\times C,\mathcal{G}\otimes\mathcal{C})$ to $(C,\mathcal{C})$. But that can be easily seen by using once again the criterion given in Exercise 11.2.3. With H as defined above, one easily sees that $G\cap\{\omega:Z(\cdot,\omega)\in E\}=\{\omega:(\omega,Y(\cdot,\omega))\in H\}$ and now, recalling the map φ defined above, one finally gets $G\cap\{\omega:Z(\cdot,\omega)\in E\}=\{\omega:\varphi(\omega)\in H\}$. Thus, the left-hand side of (11.2.3) equals $P\varphi^{-1}(H)=P\otimes Q_0(H)=\int Q_0(H(\omega))dP(\omega)$, where we used the fact that $P\varphi^{-1}=P\otimes Q_0$ and denoted the ω-section of H by $H(\omega)$. From the defintion of H, one easily gets

$$Q_0\big(H(\omega)\big)=I_G(\omega)Q_0(\{f\in C:X(\omega)+f\in E\})=I_G(\omega)Q(\omega,E),$$

which proves the equality (11.2.3) and this completes the proof. □

Proof of Theorem 11.2.8. Fixing $s\ge 0$, denote $\mathcal{G}=\sigma(\{B^x_u, 0\le u\le s\})$ and $Y(\cdot,\omega)=B^x(s+\cdot)-B^x_s(\omega)$. By Exercise 11.2.2, we have that Y is a C-valued random variable on $(\Omega,\mathcal{A},P)$, which is independent of $\mathcal{G}$ and also $PY^{-1}=P_0$, the distribution of SBM. Now, $X=B^x_s$ is a $\mathcal{G}$-measurable random variable and $Z(\cdot,\omega)=X(\omega)+Y(\cdot,\omega)=B^x(s+\cdot,\omega)$. The result now follows immediately from Lemma 11.2.9, on recalling that $P_0(\{f\in C:y+f\in E\})=P_y(E)$, for any $y\in\mathbb{R}$ and any $E\in\mathcal{C}$, as was noted just before stating Proposition 11.2.7. □

We next proceed to prove a stronger version of Markov property, known as the "Strong Markov property", for Brownian motion. The Markov property that we just proved says that, for any fixed $s\ge 0$, the evolution of the Brownian motion after time s, given the history upto and including time s, is conditionally distributed like a Brownian motion starting at its state at the "present" time s. What we now want to ask is whether the same property holds, if one replaces the "fixed" time s by a "random" time τ. By a random time, we mean here a random variable taking values in $[0,\infty)$. Of course, one realizes that, with a random time τ, describing the history of evolution upto and including time τ is not as straightforward as that for the history upto and including a fixed time s, which is captured simply by the family of random variables $\{B^x_u, 0\le u\le s\}$. Notwithstanding that hurdle, let us come back to the question we asked. Firstly, the bad news is that the result is not true in this generality, that is, for all random times. However, the good news is that there is a special class of random times, for which the stated property indeed holds. This last result is what is called the strong Markov property of Brownian motion.

We start by first describing the class of random times for which the strong Markov property holds. Let us clearly spell out the underlying setting that we are going to work on. We have a probability space $(\Omega, \mathcal{A}, P)$, on which a SBM $\{B_t, t \in [0, \infty)\}$ is defined. We assume, as before, that $B_0 \equiv 0$ and that all trajectories of the SBM are continuous. We then know that, for each $x \in \mathbb{R}$, the process $\{B_t^x = x + B_t, t \in [0, \infty)\}$ defines a BM starting at x.

For each $t \in [0, \infty)$, we consider the σ-field $\mathcal{A}_t = \sigma(\{B_u, u \leq t\})$. Note that, each $\mathcal{A}_t$ is a sub-σ-field of $\mathcal{A}$ and also $\mathcal{A}_s \subset \mathcal{A}_t$, for any $s \leq t$. In the general theory of stochastic processes, such a non-decreasing family of σ-fields is referred to as a "filtration". The particular filtration $\{\mathcal{A}_t, t \geq 0\}$ that we have defined is known as the "natural filtration" of the underlying SBM. We also define $\mathcal{A}_\infty = \sigma(\{B_t, t \in [0, \infty)\})$. It is clear that $\mathcal{A}_\infty = \bigvee_t \mathcal{A}_t$, that is, the smallest σ-field containing all the $\mathcal{A}_t, t \geq 0$.

At this point, note that, for each $x \in \mathbb{R}$, we have $\mathcal{A}_t = \sigma(\{B_u^x, u \leq t\})$, $t \geq 0$, that is, $\{\mathcal{A}_t, t \geq 0\}$ is also the "natural filtration" of $\{B_t^x, t \in [0, \infty)\}$. We are now going to define a special class of random times, to be called "optional times" for which the strong Markov property of Brownian motion will be shown to hold. These are going to be random variables on $(\Omega, \mathcal{A}, P)$ taking values in $[0, \infty]$, that satisfy a special property. It may be noted that, we are allowing ∞ as a possible value for these random times, since sometimes that cannot be avoided. However, when it comes to proving strong Markov property, we will only consider finite-valued random times.

DEFINITION 11.2.10. *A random variable τ on $(\Omega, \mathcal{A}, P)$, taking values in $[0, \infty]$, is called an* optional time *with respect to the filtration $\{\mathcal{A}_t\}$ (also, sometimes called an $\{\mathcal{A}_t\}$-optional time), if $\{\tau < t\} \in \mathcal{A}_t$, for all $t > 0$.*

The condition $\{\tau < t\} \in \mathcal{A}_t$, for $t > 0$, simply says that the event $\{\tau < t\}$, for any $t > 0$, depends only on the information on the evolution of the BM upto and including time t. We define below another special class of random times that satisfy a slightly stronger condition than optional times.

DEFINITION 11.2.11. *A random variable τ on $(\Omega, \mathcal{A}, P)$, taking values in $[0, \infty]$, is called a* stopping time *with respect to the filtration $\{\mathcal{A}_t\}$ (also, sometimes called an $\{\mathcal{A}_t\}$-stopping time), if $\{\tau \leq t\} \in \mathcal{A}_t$, for all $t \geq 0$.*

The difference between optional times and stopping times should be clear from the above definitions. It should, for example, be clear that, for any $t \geq 0$, the event $\{\tau = t\}$ belongs to $\mathcal{A}_t$ if τ is a stopping time, but not necessarily so if τ is an optional time. To make the distinction clearer, let us define, for each $t \geq 0$, the σ-field $\mathcal{A}_{t+} = \bigcap_{s > t} \mathcal{A}_s$. It is clear from the definition that $\{\mathcal{A}_{t+}, t \geq 0\}$ is a filtration, that is, $\mathcal{A}_{s+} \subset \mathcal{A}_{t+}$, for any $s < t$. It is also clear that $\mathcal{A}_t \subset \mathcal{A}_{t+}$ and the inclusion is strict. The σ-field $\mathcal{A}_t$ only captures the evolution of the BM upto and including time t, while $\mathcal{A}_{t+}$ captures not only that, but also captures a "sneak peek" into the "immediate future" after time t. To give a concrete

11.2. Markov and Strong Markov Properties

example, the event $\limsup_n \{B_{t+\frac{1}{n}} > a\}$, for some $a \in \mathbb{R}$, will belong to $\mathcal{A}_{t+}$, but not to $\mathcal{A}_t$. The enthusiastic reader may try to think of other examples of events that are in $\mathcal{A}_{t+}$, but not in $\mathcal{A}_t$. Here are a couple of easy exercises.

EXERCISE 11.2.12. *(a) Any $\{\mathcal{A}_t\}$-stopping time is also an $\{\mathcal{A}_t\}$-optional time.*
(b) Any $\{\mathcal{A}_t\}$-optional time τ (and hence also, any $\{\mathcal{A}_t\}$-stopping time τ) is an $\mathcal{A}_\infty$-measurable random variable.
(c) A random variable τ taking values in $[0, \infty]$ is an $\{\mathcal{A}_t\}$-optional time if and only if it is a $\{\mathcal{A}_{t+}\}$-stopping time.
(d) Any constant random variable, say, $\tau \equiv t_0$ is a $\{\mathcal{A}_t\}$-stopping time.

In light of the clear distinction, as discussed above, between what the two σ-fields $\mathcal{A}_{t+}$ and $\mathcal{A}_t$, for any t, capture, part (c) of Exercise 11.2.12 underlines the main distinction between an optional time and a stopping time.

The most important and common examples of optional/stopping times are what are called "hitting times" or "times of first entrance". Among other things, these examples will also reveal why we have to allow ∞ as a possible value for such times. Fix a set $A \subset \mathbb{R}$ and define τ_A on Ω by $\tau_A(\omega) = \inf\{t \geq 0 : B_t(\omega) \in A\}$. This is called the "hitting time" or the "first entrance time" of the set A by the SBM. Of course, one could have used B_t^x also in place of SBM to define the first enrance time by BM starting at x. The first point to note here is that the infimum will not be defined for an ω for which the set $\{t \geq 0 : B_t(\omega) \in A\}$ is empty, and we cannot apriori rule out this possibility. So, to have τ_A to be defined for all ω, one follows the convention of setting the infimum of empty set as $+\infty$. Now, the important question is whether τ_A is an optional time, at least for borel sets $A \subset \mathbb{R}$. The answer is 'yes' but the proof is quite delicate and involves completing the σ-fields. We skip the proof for general borel sets. However, the proof is easy for closed or open subsets of $\mathbb{R}$ (see Exercise 11.3.10).

We are next going to define σ-fields associated to optional/stopping times τ, that are meant to capture "history of evolution upto and including time τ.

DEFINITION 11.2.13. For an $\{\mathcal{A}_t\}$-optional time τ, the class of sets
$$\mathcal{A}_{\tau+} = \{A \in \mathcal{A}_\infty : A \cap \{\tau < t\} \in \mathcal{A}_t, \text{ for all } t > 0\}$$
is called the *pre-$\tau+$ σ-field*.
For an $\{\mathcal{A}_t\}$-stopping time τ, the class of sets
$$\mathcal{A}_\tau = \{A \subset \mathcal{A}_\infty : A \cap \{\tau \leq t\} \in \mathcal{A}_t, \text{ for all } t \geq 0\}$$
is called the *pre-τ σ-field*.

It is clear from the definition that both classes $\mathcal{A}_{\tau+}$ and $\mathcal{A}_\tau$ are contained in the σ-field $\mathcal{A}_\infty$. Also, since any stopping time τ is also an optional time, we can define $\mathcal{A}_{\tau+}$ for a stopping time τ as well and it can easily be seen that, in that case, $\mathcal{A}_\tau \subset \mathcal{A}_{\tau+}$. But, the most important issue is whether the classes $\mathcal{A}_{\tau+}$ and $\mathcal{A}_\tau$, as in Definition 11.2.13 above, are indeed σ-fields, as they are claimed to be. This and some other facts are fairly easy to verify and are given in the following exercise.

EXERCISE **11.2.14.** (a) Show that the classes $\mathcal{A}_{\tau+}$ and $\mathcal{A}_\tau$, as in Definition 11.2.13, are σ-fields.
(b) Show that, if τ is a stopping time (and hence also an optional time), then $\mathcal{A}_\tau \subset \mathcal{A}_{\tau+}$.
(c) Show that, if τ is an optional time, then the σ-field $\mathcal{A}_{\tau+}$ is the same as the class $\{A \in \mathcal{A}_\infty : A \cap \{\tau \leq t\} \in \mathcal{A}_{t+}, \text{ for all } t \geq 0\}$.
(d) Show that, if τ and η are two optional times (resp., stopping times) with $\tau \leq \eta$ everywhere, then $\mathcal{A}_{\tau+} \subset \mathcal{A}_{\eta+}$ (resp., $\mathcal{A}_\tau \subset \mathcal{A}_\eta$).
(e) Show that, if τ is a constant stopping time, say, $\tau \equiv t_0$, then $\mathcal{A}_\tau = \mathcal{A}_{t_0}$ and $\mathcal{A}_{\tau+} = \mathcal{A}_{t_0+}$.
(f) Show that if τ is an optional time (resp., a stopping time), then τ is measurable with respect to the σ-field $\mathcal{A}_{\tau+}$ (resp., the σ-field $\mathcal{A}_\tau$).

We now describe a very important result related to finite optional times. It says that any finite optional time is the monotonically decreasing limit of a sequence of finite stopping times, each taking only countably many values. This fact will play a crucial role in many of our subsequent results, including our proof of strong Markov property for Brownian motion. The construction is very simple and is given below.

For each $n \geq 1$, consider the function φ_n defined on $[0, \infty)$ by

$$\varphi_n(t) = ([2^n t] + 1)/2^n, \text{ for } t \in [0, \infty).$$

In other words, $\varphi_n(t) = k/2^n$, for $(k-1)/2^n \leq t < k/2^n$, $k = 1, 2, \ldots$. It is clear that, for each $n \geq 1$, the function φ_n takes values in the set $\{k/2^n, k \geq 1\}$ and $\varphi_n(t) > t$, for all $t \in [0, \infty)$. Further, $\varphi_n(t) \downarrow t$, for all $t \in [0, \infty)$.

Now, given a finite optional time τ, that is, an optional time taking values in $[0, \infty)$, we define $\tau_n = \varphi_n(\tau)$. It is then clear from the above that τ_n, for each n, takes values only in $\{k/2^n, k \geq 1\}$. Also, for each ω, we have $\tau(\omega) < \tau_n(\omega)$, for all n and $\tau_n(\omega) \downarrow \tau(\omega)$, as $n \to \infty$. To see that each τ_n is a stopping time is easy. Note first that, for any $n \geq 1$, $k \geq 1$, we have $\{\tau_n = \frac{k}{2^n}\} = \{\frac{k-1}{2^n} \leq \tau < \frac{k}{2^n}\} \in \mathcal{A}_{\frac{k}{2^n}}$, since τ is an $\{\mathcal{A}_t\}$-optional time. It follows now that, for any n and any $t \geq 0$, $\{\tau_n \leq t\} = \bigcup_{k:\frac{k}{2^n} \leq t} \{\tau_n = \frac{k}{2^n}\} \in \mathcal{A}_t$, proving that τ_n is a stopping time, for each n.

PROPOSITION **11.2.15.** Let τ be a finite $\{\mathcal{A}_t\}$-optional time and let $\{\tau_n\}$ be the sequence of stopping times defined above. Then $\mathcal{A}_{\tau+} = \bigcap_n \mathcal{A}_{\tau_n}$.

Proof. Firstly, let $A \in \mathcal{A}_{\tau+}$. Then, $A \in \mathcal{A}_\infty$ and, for any $n \geq 1$ and any $t \geq 0$,

$$A \cap \{\tau_n \leq t\} = A \cap \Big(\bigcup_{\frac{k}{2^n} \leq t} \{\tau_n = \frac{k}{2^n}\} \Big) = \bigcup_{\frac{k}{2^n} \leq t} \Big(A \cap \{\frac{k-1}{2^n} \leq \tau < \frac{k}{2^n}\} \Big) \in \mathcal{A}_t,$$

since $A \cap \{\frac{k-1}{2^n} \leq \tau < \frac{k}{2^n}\} \in \mathcal{A}_{\frac{k}{2^n}} \subset \mathcal{A}_t$, for all k with $\frac{k}{2^n} \leq t$. This proves that $A \in \mathcal{A}_{\tau_n}$. Since this is true for each n, we have $\mathcal{A}_{\tau+} \subset \bigcap_n \mathcal{A}_{\tau_n}$.

To prove the other inclusion, take $A \in \bigcap_n \mathcal{A}_{\tau_n}$. Clearly, we have $A \in \mathcal{A}_\infty$. To prove that $A \in \mathcal{A}_{\tau+}$, it is enough to show now that $A \cap \{\tau \leq t\} \in \mathcal{A}_{t+}$, for each $t \geq 0$ (see part (c) of Exercise 11.2.14). Fix a $t \geq 0$ and let $t_n = \varphi_n(t)$, $n \geq 1$,

11.2. Markov and Strong Markov Properties

where φ_n, $n \geq 1$, are as defined earlier. Then we know that $t < t_n$, for each n, and $t_n \downarrow t$, from which, it follows that $A \cap \{\tau < t_n\} \downarrow A \cap \{\tau \leq t\}$. Now, from the definition of τ_n, one easily sees that $\{\tau < t_n\} = \{\tau_n \leq t_n\}$, for each n. Since $A \in \bigcap_n \mathcal{A}_{\tau_n}$, we have $A \cap \{\tau < t_n\} = A \cap \{\tau_n \leq t_n\} \in \mathcal{A}_{t_n}$, for each n. Now, to show that $A \cap \{\tau \leq t\} \in \mathcal{A}_{t+}$, we need to show that $A \cap \{\tau \leq t\} \in \mathcal{A}_s$, for any $s > t$. So, let us take an $s > t$. We know then that there is an n_0, such that, $t_n < s$ and hence $\mathcal{A}_{t_n} \subset \mathcal{A}_s$, for all $n \geq n_0$. This, coupled with our previous observation, will imply that $A \cap \{\tau < t_n\} \in \mathcal{A}_s$, for all $n \geq n_0$. It now follows that $A \cap \{\tau \leq t\} = \bigcap_{n \geq n_0} A \cap \{\tau < t_n\} \in \mathcal{A}_s$ and that completes the proof. □

The following result, which will play an important part in the proof of strong markov property, comes as an easy consequence of Proposition 11.2.15. Recall that, on our probability space $(\Omega, \mathcal{A}, P)$, we have a SBM $\{B_t, t \in [0, \infty)\}$ and $\{B_t^x = x + B_t, t \in [0, \infty)\}$, for $x \in \mathbb{R}$, represents a BM starting at x.

THEOREM **11.2.16.** For any finite optional time τ, the random variable B_τ^x, defined as $B_\tau^x(\omega) = B^x(\tau(\omega), \omega)$, $\omega \in \Omega$, is measurable with respect to the σ-field $\mathcal{A}_{\tau+}$.

Proof. Letting τ_n, $n \geq 1$, denote the same sequence of stopping times as was used in the proof of Proposition 11.2.15. We first show that $B_{\tau_n}^x$, for each n, is measurable with respect to $\mathcal{A}_{\tau_n}$, that is, fixing a Borel set $A \subset \mathbb{R}$, we show that $\{B_{\tau_n}^x \in A\} \in \mathcal{A}_{\tau_n}$, for each n. Clearly, $\{B_{\tau_n}^x \in A\} = \bigcup_k \{B_{\frac{k}{2^n}}^x \in A, \tau_n = \frac{k}{2^n}\} \in \mathcal{A}_\infty$, since, $\{B_{\frac{k}{2^n}}^x \in A, \tau_n = \frac{k}{2^n}\} \in \mathcal{A}_{\frac{k}{2^n}} \subset \mathcal{A}_\infty$, for each k. Next, for any $t \geq 0$ and any $k \geq 1$ with $\frac{k}{2^n} \leq t$, we have $\{B_{\frac{k}{2^n}}^x \in A, \tau_n = \frac{k}{2^n}\} \in \mathcal{A}_{\frac{k}{2^n}} \subset \mathcal{A}_t$, which gives $\{B_{\frac{k}{2^n}}^x \in A, \tau_n \leq t\} = \bigcup_{\frac{k}{2^n} \leq t} \{B_{\frac{k}{2^n}}^x \in A, \tau_n = \frac{k}{2^n}\} \in \mathcal{A}_t$. Thus, we have proved that $B_{\tau_n}^x$ is measurable with respect to $\mathcal{A}_{\tau_n}$, for each n. Now, $\tau_n \downarrow \tau$ implies, by the continuity of Brownian trajectories, that $B_\tau^x = \lim_n B_{\tau_n}^x$. Since $\mathcal{A}_{\tau_n}$ is a decreasing sequence of σ-fields (see part (d), Exercise 11.2.14), it follows that B_τ^x is measurable with respect to $\bigcap_n \mathcal{A}_{\tau_n}$, which equals $\mathcal{A}_{\tau+}$, by Proposition 11.2.15, thus completing the proof. □

We are now ready to state and prove the strong Markov property of BM. To get to it, let us recall that our proof of Markov property was essentially based on the fact that, for any $s \geq 0$, the process $\{B_{s+t}^x - B_s^x\}$ is a SBM, independent of $\{B_u^x, 0 \leq u \leq s\}$. For proving the strong Markov property, we will derive an analogue of this, with a finite optional time τ replacing the fixed time s. Once we have that, then Lemma 11.2.9 will take over to yield the strong Markov property.

THEOREM **11.2.17.** Let τ be any finite optional time. Then, for any $x \in \mathbb{R}$, the process $\{B_{\tau+t}^x - B_\tau^x, t \geq 0\}$ is a SBM, independent of $\mathcal{A}_{\tau+}$.

Proof. We need to show that, for any $A \in \mathcal{A}_{\tau+}$ and any $E \in \mathcal{C}$,

$$P(A \cap \{\omega : B^x(\tau(\omega) + \cdot, \omega) - B^x(\tau(\omega), \omega) \in E\}) = P(A)P_0(E), \quad (11.2.4)$$

where P_0 is the probability on $\mathcal{C}$, introduced earlier, that denotes the distribution of SBM. To prove this, we fix $A \in \mathcal{A}_{\tau+}$ and show that the equality (11.2.4) holds for all $E \in \mathcal{C}$. We are going to use Dynkin's $\pi - \lambda$ Theorem (Theorem 2.3.10) to do this. Firstly, it is easy to see that, for any fixed $A \in \mathcal{A}_{\tau+}$, the class of $E \in \mathcal{C}$ for which (11.2.4) holds, forms a λ-system. Now, we consider the class $\mathcal{P}$ of subsets of C that consists of the empty set and all sets E of the form $E = \{f \in C : f(t_j) \in (a_j, b_j),\ j = 1, \ldots, k\}$, where $\{t_1, \ldots, t_k\}$ vary over all possible choices of a finite set of points in $[0, \infty)$ and $(a_j, b_j),\ j = 1, \ldots, k$ are any choice of bounded open intervals. Clearly, the class $\mathcal{P}$ forms a π-system. Also, recalling the evaluation maps $e_t, t \in [0, \infty)$ on C, one can see that the class $\mathcal{P}$ is nothing but all possible finite intersections of sets the form $e_t^{-1}\{(a,b)\}$, and hence generates $\mathcal{C}$. We now show that, for any fixed $A \in \mathcal{A}_{\tau+}$, the equality (11.2.4) holds for all $E \in \mathcal{P}$, whence it will follow that (11.2.4) holds for all $E \in \mathcal{C}$, thus proving the theorem. Towards this, let us observe that (11.2.4) can equivalently be written as

$$\int_A I_E\big(B^x(\tau(\omega) + \cdot, \omega) - B^x(\tau(\omega), \omega)\big) dP(\omega) = P(A) \int_C I_E dP_0. \qquad (11.2.5)$$

Let us take any $\emptyset \neq E \in \mathcal{P}$, say, $E = \{f \in C : f(t_j) \in (a_j, b_j),\ j = 1, \ldots, k\}$, where $t_1, \ldots, t_k$ are points in $[0, \infty)$ and $(a_j, b_j),\ j = 1, \ldots, k$ are bounded open intervals. Then clearly, $I_E(f) = \prod_{j=1}^{k} I_{(a_j, b_j)}(f(t_j))$, for $f \in C$. Now, we know that given any bounded open interval (a, b) in $\mathbb{R}$, we can get a sequence $\{g_n\}$ of continuous functions on $\mathbb{R}$, with $0 \leq g_n \leq 1$, such that $g_n \uparrow I_{(a,b)}$. Thus, for $E \in \mathcal{P}$ as above, we can get, for each $j = 1, \ldots, k$, a sequence $\{g_{nj}\}$ of continuous functions on $\mathbb{R}$, with $0 \leq g_{nj} \leq 1$, such that

$$\gamma_n(f) = \prod_{j=1}^{k} g_{nj}(f(t_j)) \uparrow I_E(f), \quad \text{for } f \in C.$$

Now, we are going to prove a result below (Lemma 11.2.18), which will imply

$$\int_A \gamma_n\big(B^x(\tau(\omega) + \cdot, \omega) - B^x(\tau(\omega), \omega)\big) dP(\omega) = P(A) \int_C \gamma_n\, dP_0, \quad \text{for each } n \geq 1.$$

From this, one now easily gets (11.2.5) for $E \in \mathcal{P}$, by using MCT and that completes the proof. $\square$

LEMMA 11.2.18. *Let τ be a finite optional time and let $x \in \mathbb{R}$. Then for any $A \in \mathcal{A}_{\tau+}$ and any function $\gamma : C \to \mathbb{R}$ of the form $\gamma(f) = \prod_{j=1}^{k} g_j(f(t_j))$, for some $\{t_1, \ldots, t_k\} \subset [0, \infty)$ and bounded continuous functions $g_j : \mathbb{R} \to \mathbb{R},\ j = 1, \ldots, k$,*

$$\int_A \gamma\big(B^x(\tau(\omega) + \cdot, \omega) - B^x(\tau(\omega), \omega)\big) dP(\omega) = P(A) \int_C \gamma\, dP_0. \qquad (11.2.6)$$

Proof. For the finite optional time τ, let $\{\tau_n\}$ be the sequence of stopping times, as defined earlier (and used in the proof of Proposition 11.2.15). For $A \in \mathcal{A}_{\tau+}$, and, for each $n \geq 1$ and $k \geq 1$, denote

$$A_{n,k} = A \cap \left\{\tau_n = \tfrac{k}{2^n}\right\} = A \cap \left\{\tfrac{k-1}{2^n} \leq \tau < \tfrac{k}{2^n}\right\}.$$

For each $n \geq 1$, we then have

11.2. Markov and Strong Markov Properties

$$\int_A \gamma\big(B^x(\tau_n(\omega)+\cdot,\omega)-B^x(\tau_n(\omega),\omega)\big)\,dP(\omega)$$

$$=\sum_k \int_{A_{n,k}} \gamma\big(B^x(\tau_n(\omega)+\cdot,\omega)-B^x(\tau_n(\omega),\omega)\big)\,dP(\omega)$$

$$=\sum_k \int_{A_{n,k}} \gamma\big(B^x(\tfrac{k}{2^n}+\cdot,\omega)-B^x(\tfrac{k}{2^n},\omega)\big)\,dP(\omega)$$

Now, for any $n \geq 1$, we have $A_{n,k} \in \mathcal{A}_{\frac{k}{2^n}}$ and $B^x(\frac{k}{2^n}+\cdot) - B^x(\frac{k}{2^n})$ is a SBM, independent of $\mathcal{A}_{\frac{k}{2^n}}$, for each $k \geq 1$ (see Exercise 11.2.2). Using this, one easily sees that the last sum above reduces to $P(A)\int_C \gamma\,dP_0$, giving us

$$\int_A \gamma\big(B^x(\tau_n(\omega)+\cdot,\omega)-B^x(\tau_n(\omega),\omega)\big)\,dP(\omega) = P(A)\int_C \gamma\,dP_0, \text{ for every } n \geq 1.$$

Since $\tau_n \downarrow \tau$, we use the continuity of Brownian trajectories, the form of the function γ and DCT to see that the integral on the left-hand side of the above equality goes to $\int_A \gamma\big(B^x(\tau(\omega)+\cdot,\omega) - B^x(\tau(\omega),\omega)\big)dP(\omega)$, as $n \to \infty$, yielding the required result. $\square$

REMARK 11.2.19. Let $\{B_t, t \in [0,\infty)\}$ be a SBM and, for each rational r, let $\tau_r = \inf\{t > r : B_t = 0\}$. By part (d) of Exercise 11.3.10, τ_r is an optional time (in fact, a stopping time). Further, by discarding a P-null set, if necessary, we may and do assume that τ_r is finite everywhere. Observing that $B(\tau_r) \equiv 0$, Theorem 11.2.17 will imply that $\{B(\tau_r+t), t \in [0,\infty)\}$ is a SBM. This finishes the unfinished part of the argument in the proof of Proposition 11.1.16, while aiming to show that every point in the zero-set of a.e. trajectory of SBM is a limit point of the zero-set.

Having proved Theorem 11.2.17, we get the strong Markov property as an easy consequence of Lemma 11.2.9. Given any finite optional time τ, we apply Lemma 11.2.9 with $\mathcal{G} = \mathcal{A}_{\tau+}$, the C-valued random variable $Y(\cdot) = B^x_{\tau+\cdot} - B^x_\tau$, and the real random variable $X = B^x_\tau$, which is measurable with respect to $\mathcal{A}_{\tau+}$, by Theorem 11.2.16. Also, by Theorem 11.2.17, Y is independent of $\mathcal{A}_{\tau+}$ and has distribution P_0. Lemma 11.2.9 then gives us a version of the regular conditional distribution of the C-valued random variable $Z(\cdot) = X + Y(\cdot) = B^x_{\tau+\cdot}$, which is captured in the next result.

THEOREM 11.2.20. (Strong Markov Property of BM)
On $(\Omega, \mathcal{A}, P)$, let $\{B^x_t, t \in [0,\infty)\}$ be the BM starting at $x \in \mathbb{R}$. Then, for any finite optional time τ, the kernel

$$Q(\omega, E) = P_{B^x_\tau(\omega)}(E), \quad \omega \in \Omega, \ E \in \mathcal{C}$$

is a version of the regular conditional distribution of $\{B^x_{\tau+t}, t \in [0,\infty)\}$, given $\mathcal{A}_{\tau+}$.

The strong Markov property, even for constant stopping times, gives a slight improvement over the Markov property proved earlier (Theorem 11.2.8). To see this, if we apply the strong Markov property with the constant stopping time $\tau \equiv s$, for any $s \geq 0$, then noting that $\mathcal{A}_{\tau+} = \mathcal{A}_{s+}$, we get the following result.

THEOREM 11.2.21. For any $x \in \mathbb{R}$ and any $s \geq 0$, the kernel

$$Q(\omega, E) = P_{B^x_s(\omega)}(E), \quad \omega \in \Omega, \ E \in \mathcal{C}$$

is a version of the regular conditional distribution of $\{B^x_{s+t}, t \in [0,\infty)\}$, given $\mathcal{A}_{s+}$.

Here is an interesting consequence of the strong Markov property, known as "André's Reflection Principle". Let $\{B_t, t \in [0,\infty)\}$ be a SBM on some probability space $(\Omega, \mathcal{A}, P)$ and let $\mathcal{A}_t = \sigma(\{B_u, u \leq t\}), t \in [0,\infty)$, denote the natural filtration associated to the SBM. For a finite $\{\mathcal{A}_t\}$-optional time τ, consider the process $\{\widetilde{B}_t, t \in [0,\infty)\}$ defined as

$$\widetilde{B}_t(\omega) = B_t(\omega) I_{\{t \leq \tau(\omega)\}} + \bigl(2B_\tau(\omega) - B_t(\omega)\bigr) I_{\{t > \tau(\omega)\}}, \text{ for } \omega \in \Omega, t \geq 0. \quad (11.2.7)$$

A close examination of the trajectories of the process will show that for each ω, its ω-trajectory coincides with that of the SBM upto the point $\tau(\omega)$, while beyond $\tau(\omega)$, its ω-trajectory is just the reflection of the ω-trajectory of SBM around the horizontal line at the level $B_\tau(\omega)$. The interesting point is that the process $\{\widetilde{B}_t, t \in [0,\infty)\}$ can also be written as

$$\widetilde{B}_t(\omega) = B_{\tau \wedge t}(\omega) + \bigl(B_\tau(\omega) - B_t(\omega)\bigr) I_{\{t > \tau(\omega)\}}, \text{ for } \omega \in \Omega, t \geq 0,$$

while the SBM $\{B_t, t \in [0,\infty)\}$ can be written as

$$B_t(\omega) = B_{\tau \wedge t}(\omega) + \bigl(B_t(\omega) - B_\tau(\omega)\bigr) I_{\{t > \tau(\omega)\}}, \text{ for } \omega \in \Omega, t \geq 0.$$

One can show that $\{B_{\tau \wedge t}, t \in [0,\infty)\}$ is a C-valued random variable measurable with respect to $\mathcal{A}_{\tau+}$. From Theorem 11.2.17, we know that $\{B_{\tau+s} - B_\tau, s \in [0,\infty)\}$ is a SBM, independent of $\mathcal{A}_{\tau+}$ and so, by the symmetry of SBM (Proposition 11.0.3, part (a)), $\{B_\tau - B_{\tau+s}, s \in [0,\infty)\}$ is also a SBM, independent of $\mathcal{A}_{\tau+}$. Using all of these, it is not difficult to verify (see Exercise 11.3.13) that

$$\bigl(\{B_{\tau \wedge t}, t \geq 0\}, \{(B_\tau - B_t) I_{\{t > \tau\}}, t \geq 0\}\bigr) \stackrel{d}{=} \bigl(\{B_{\tau \wedge t}, t \geq 0\}, \{(B_t - B_\tau) I_{\{t > \tau\}}, t \geq 0\}\bigr),$$

whence it follows that $\{\widetilde{B}_t, t \in [0,\infty)\} \stackrel{d}{=} \{B_t, t \in [0,\infty)\}$.

PROPOSITION 11.2.22. (André's Reflection Principle) Let $\{B_t, t \in [0,\infty)\}$ be a SBM and $\{\mathcal{A}_t, t \in [0,\infty)\}$ its natural filtration. Then for any finite $\{\mathcal{A}_t\}$-optional time τ, the process $\{\widetilde{B}_t, t \in [0,\infty)\}$ defined by (11.2.7) is also a SBM.

Here is an important application of the Reflection Principle. Consider a SBM $\{B_t, t \in [0,\infty)\}$. Assume that all trajectories are continuous. Let $M_t = \max\{B_s : 0 \leq s \leq t\}$, for $t \geq 0$. We have already noted that M_t is a real random variable and clearly $M_t \geq 0$. Denoting $\tau_a = \inf\{t > 0 : B_t \geq a\}$, for $a > 0$, we clearly have $\{M_t \geq a\} = \{\tau_a \leq t\}$. Using properties of Brownian trajectories, the reader should be able to convince herself that $0 < \tau_a < \infty$, $[P]$-a.s. Also, note that τ_a is a stopping time with respect to the natural filtration of the SBM (Exercise 11.3.10(a)). Now, one easily has, for any $a > 0$,

$$P(M_t \geq a) = P(B_t \geq a) + P(M_t \geq a, B_t < a) = P(B_t \geq a) + P(\tau_a < t, B_t < a).$$

If we now consider the process $\{\widetilde{B}_t, t \in [0,\infty)\}$ as in (11.2.7) with $\tau = \tau_a$, then it should be clear that $\{\tau_a < t, B_t < a\} = \{\tau_a < t, \widetilde{B}_t > a\}$. Further, if we define $\widetilde{\tau}_a = \inf\{t > 0 : \widetilde{B}_t \geq a\}$, then one can easily see that $\widetilde{\tau}_a \equiv \tau_a$, so that, $\{\tau_a < t, \widetilde{B}_t > a\} = \{\widetilde{\tau}_a < t, \widetilde{B}_t > a\} = \{\widetilde{B}_t > a\}$. Using the Reflection Principle, the probabilty of the last event is the same as that of $\{B_t > a\}$. All of these finally give us $P(M_t \geq a) = 2P(B_t \geq a) = P(|B_t| \geq a)$. We have thus proved that

11.2. Markov and Strong Markov Properties

$$M_t \stackrel{\mathrm{d}}{=} |B_t|, \quad \text{for each } t > 0.$$

The above result has another interesting and useful consequence. We already noted that, $\{M_t \geq a\} = \{\tau_a \leq t\}$. Using the above, we now get the c.d.f. of the non-negative random variable τ_a to be given by

$$P(\tau_a \leq t) = P(|B_t| \geq a) = \frac{2}{\sqrt{2\pi t}} \int_a^\infty e^{-\frac{x^2}{2t}} \, dx, \quad \text{for each } t > 0. \tag{11.2.8}$$

From this, it is not difficult to show that τ_a has an absolutely continuous distribution and one can easily find its density function. This and some more interesting facts are included in the Exercise 11.3.14.

Everything must come to an end and so do we, with just a few words on the topic that we took up in this chapter – ideally, this should have appeared at the start of the chapter, but we decided against it to continue with the flow that started at the end of the previous chapter.

The reader must be wondering that SBM has $E(X_t) = 0$ which means that on the average it is not moving but we are still calling it motion. Yes, it is the motion of a particle on which no 'external physical forces' are acting!

Brownian Motion is a glorious symphony being played by diverse disciplines together. Its origins lay in the observation of the Scottish Botanist Robert Brown: pollen particles in water were moving in spite of taking all precautions to avoid influence of any external force. His credit lies not only in performing meticulous experiments and drawing wider attention for the phenomenon but also on insisting for an explanation.

Several years later, the French stock market analyst Louis Bachelier, while modelling the price fluctuation as a process (X_t), independently observed that to reach a fluctuation z at time $t + s$, there must have been a fluctuation x at time t and during the remaining time period s, there must have been further fluctuation of amount $z - x$. This is a version of Markov property and Bachelier 'deduced' BM.

Independently, a few years later Physicists Einstein and Smoluchowski, while trying to understand the 'possible molecular nature of matter', discovered the following. Even though each hit of a suspended particle by molecules of fluid creates 'almost unobservable' displacement, the continuous bombardment of the suspended particle by the surrounding fluid molecules does lead to an observable displacement. They quantified the displacement (on the basis of which, the Avagadro number was soon calculated) leading to BM.

The importance of this discovery was realized by Bertrand Russel, who posed the problem of a clear mathematical formulation of this to his post-doctoral student Norbert Wiener. Few years later Wiener laid a rigorous mathematical construction of the motion. Once measure theory, in spaces more general than Euclidean, was understood and Kolmogorov laid the measure theoretic foundation of probability, the topic acquired wings. As J L Doob says 'Wiener constructed this process rigorously more than a decade before probabilists made their subject respectable and he applied the process both inside and outside mathematics in many important problems'. Irving Segal says 'the novelty of Wiener's Brownian motion theory was such that it was not at all widely appreciated at that time and the few who did, such as H Cramér in Sweden and P Lévy in France were outside the United States.' Wiener discovered several properties of this process.

Wiener realized that X_t, being the cumulative disturbance to the particle during the interval $[0,t]$, has the flavour of 'distribution function' F, though at any time t, this cumulative disturbance function X_t, unlike a number $F(t)$, is a random variable. Further, this X_t, unfortunately, is not a well behaved function of t (see Proposition 11.1.10). Nonetheless, the 'ensemble' of paths behaves well and you can perform integration. Remembering the standard $\int I_{(a,b]} dF(t) = F(b) - F(a)$, one can associate $X_b - X_a$ as a dX_t integral of an indicator $I_{(a,b]}$ for a bounded interval $(a,b]$. Viewing $f = I_{(a,b]}$ as an element of $L^2([0,\infty))$ (with Lebesgue measure), its integral $I(f)$ is an element of $L^2(\Omega, P)$ (the space where BM is defined). It is a simple matter to see, using covariance kernel of BM, that this preserves the inner products (see Remark 9.1.16) and hence can be extended as a linear map to all of $L^2([0,\infty))$. Thus for $f \in L^2([0,\infty))$ we have $I(f) \in L^2(\Omega, P)$. This is the celebrated Wiener integral. Thus 'reasonable functions' of t can be integrated – a completely new theory of integration appears in the horizon now!

The next crucial step was taken by Kiyosi Ito who said 'reasonable functions depending on both t and ω' can also be integrated. In other words, certain stochastic processes can be integrated w.r.t. the BM. Indefinite integral can also be defined to get actually a process. This led to a rich theory of stochastic integral and stochastic differential equations. Solutions of these SDE led to models to various phenomena; be it physics or biology or ecology. Incidentally, this programme, from a somewhat different view point, was also carried out by Wolfgang Doeblin, which unfortunately remained unknown for more than five decades.

There are several other aspects to BM. For instance, the BM transition function is a fundamental solution of heat equation (See Exercise 11.3.4), the infinitesimal generator of the brownian semigroup is half times the Laplacian, and so on. All these lead to connections between differential equations, BM and physics.

With these comments we invite the interested reader to explore the exciting developments using SBM. Since internet provides a 'continuously updating source' for references, we are not providing any.

11.3 Additional Exercises

EXERCISE **11.3.1**. *Show that if $\{B_t, t \in [0,1]\}$ is a SBM on $[0,1]$, then the process $\{X_t, t \in [0,\infty)\}$ defined by $X_t = (1+t)B_{t/(1+t)} - tB_1, t \in [0,\infty)$, is a SBM on $[0,\infty)$.*

EXERCISE **11.3.2**. *Exercise 11.3.1 illustrates one way of constructing a SBM on $[0,\infty)$ using a SBM on $[0,1]$. This exercise gives a construction of SBM on*

11.3. Additional Exercises

$[0, \infty)$ by successively gluing i.i.d. copies of SBM on $[0,1]$.
(a) Argue that if you have a SBM on $[0,1]$ on some probability space, then you can construct a probability space with an i.i.d. sequence $\{X^n(t), t \in [0,1]\}$, $n \geq 1$, of SBMs on $[0,1]$ defined on it.
With $\{X^n(t), t \in [0,1]\}$, $n \geq 1$, as in (a), define $\{B_t, t \in [0,\infty)\}$ as follows: $B_t = X^1(t)$, $t \in [0,1]$ and $B_t = X^1(1) + \cdots + X^{[t]}(1) + X^{[t]+1}(t - [t])$, $t \in (1, \infty)$, where $[t]$ denotes the 'integer part' of t.
(b) Show that the process $\{B_t, t \in [0, \infty)\}$, thus constructed, is a SBM on $[0, \infty)$.

EXERCISE 11.3.3. Let $\{B_t, t \in [0,1]\}$ be a SBM on $[0,1]$. Show that the process $\{B_t^{(0)}, t \in [0,1]\}$ defined by $B_t^{(0)} = B_t - tB_1$, $t \in [0,1]$, is a Gaussian process with continuous trajectories and $B_0^{(0)} = B_1^{(0)} = 0$. Find its mean function and covariance kernel. [The process $\{B_t^{(0)}, t \in [0,1]\}$ is called the "Brownian Bridge" and is important in non-parametric statistical theory.]

EXERCISE 11.3.4. (a) Consider a SBM $\{B_t, t \in [0,\infty)\}$ and let $\phi(t, \cdot)$ denote the density function of B_t, for $t > 0$. Show that the function $\phi(t, x)$, $t > 0, x \in \mathbb{R}$ satisfies the partial differential equation $\frac{\partial \phi}{\partial t} = \frac{1}{2} \frac{\partial^2 \phi}{\partial x^2}$, called the 'Heat Equation'.
(b) Let f be an integrable function on $(\mathbb{R}, \mathcal{B}, \lambda)$. Define a real-valued function u on $[0, \infty) \times \mathbb{R}$ by $u(t, x) = E(f(B_t^x))$, where $\{B_t^x, t \in [0, \infty)\}$, for each $x \in \mathbb{R}$, is a BM starting at x. Show that u satisfies $\frac{\partial u}{\partial t} = \frac{1}{2} \frac{\partial^2 u}{\partial x^2}$ on $(0, \infty) \times \mathbb{R}$ and $u(0, x) = f(x)$, for all $x \in \mathbb{R}$. Show that, in case f is bounded and continuous, $u(t, x) \to f(x)$ as $t \to 0$. One says that u is a solution of the 'Heat Equation' for the given 'initial value' $f(\cdot)$.
[The above partial differential equation models 'heat transfer' with time on an insulated infinite rod. $f(\cdot)$ denotes heat at different positions on the rod at time $t = 0$, while $u(t, \cdot)$ represents heat distribution at time t. The 'transition density' $p(t, x, y)$ of BM, called the 'fundamental solution' of the heat equation, represents heat at location x at time t, given unit heat at location y at time 0.]

EXERCISE 11.3.5. Let $[a, b] \subset [0, \infty)$. Show that if $\{\pi_n\}$ is a nested sequence of partitions of $[a, b]$ such that $\sum_n \|\pi_n\| < \infty$, then $Q(B(\cdot, \omega), \pi_n, [a, b]) \xrightarrow{a.s.} (b - a)$, as $n \to \infty$.

EXERCISE 11.3.6. Let C and $\mathcal{C}$ be as defined in Section 11.2. For $f, g \in C$, define
$$\rho(f, g) = \sum_k 2^{-k} \frac{\rho_k(f,g)}{1+\rho_k(f,g)},$$
where, $\rho_k(f, g) = \sup\{|f(t) - g(t)| : 0 \leq t \leq k\}$, for each $k \geq 1$.
(a) Show that, ρ defines a metric on C such that $\rho(f_n, f) \to 0$ if and only if $\{f_n\}$ converges to f, uniformly on every compact subset of $[0, \infty)$.
(b) Show that C, equipped with ρ, is a complete, separable metric space.
(c) Show that $d(f, g) = \sum_k 2^{-k} \min\{\rho_k(f, g), 1\}$, $f, g \in C$ defines a metric that is equivalent to ρ.
(d) Show that $\mathcal{C}$ is the smallest σ-field on C, containig all open ρ-balls and hence conclude that $\mathcal{C}$ is the Borel σ-field on C.

EXERCISE 11.3.7. Let f be a real-valued continuous function on $[0,\infty)$. Given any non-degenerate interval $I \subset [0,\infty)$, we say that "f has a local maximum in I" if there is $x \in I$ and an $\epsilon > 0$ such that $f(x) \geq f(y)$ for all $y \in I \cap (x-\epsilon, x+\epsilon)$. A local maximum $x \in I$ is said to be "a strict local maximum" if there is $\epsilon > 0$ such that $f(x) > f(y)$ for all $y \in I$ with $0 < |y - x| < \epsilon$.
(a) Show that if f is not monotone on any interval, then f has a local maximum on every interval $[t_1, t_2] \subset [0,\infty)$.
(b) Show that if every local maxima of f is strict, then the set of local maxima of f is countable.
Let $\{B_t, t \in [0,\infty)\}$ be a SBM on some probability space.
(c) For $0 \leq s_1 < s_2 < t_1 < t_2 < \infty$, let $M_1 = \max\{B_t : t \in [s_1, s_2]\}$ and $M_2 = \max\{B_t : t \in [t_1, t_2]\}$. We know that both M_1 and M_2 are random variables. Show that the random variables $X = M_1 - B_{s_2}$, $Y = B_{t_1} - B_{s_2}$ and $Z = M_2 - B_{t_1}$ are mutually independent and hence deduce that $P(M_1 = M_2) = 0$.
(d) Conclude that, for almost every trajectory of $\{B_t\}$, all its local maxima are strict and that the set of its local maxima is a countable dense subset of $[0,\infty)$.

EXERCISE 11.3.8. Let $\{B_t, t \in [0,1]\}$ be a SBM on $[0,1]$.
(a) Show that, if f is any real-valued continuous function on $[0,1]$ with $f(0) = 0$, then for every $\epsilon > 0$, $P(\{|B(t,\omega) - f(t)| < \epsilon, \text{ for all } 0 \leq t \leq 1\}) > 0$.
(b) Use (a) to conclude that the 'closed support' of the 'law' of $\{B_t, t \in [0,1]\}$ is the set of real-valued continuous functions f on $[0,1]$ with $f(0) = 0$ (equipped with the metric $\rho(f,g) = \sup_t |f(t) - g(t)|$).

EXERCISE 11.3.9. Let $\{B_t, t \geq 0\}$ be a SBM on some probability space $(\Omega, \mathcal{A}, P)$. Fix an integer $m \geq 2$ and denote $\gamma = \gamma(m) = \frac{1}{2} + \frac{1}{m-1}$.
(a) For each $j \geq 1$ and $k \geq 1$, let $D_{j,k}$ denote the set of ω for which there exists $t \in [0, 1-\frac{1}{k}]$ such that $|B(t+h, \omega) - B(t, \omega)| \leq jh^\gamma$ for all $h \in (0, \frac{1}{k})$. Show that
$$D_{j,k} \subset \bigcap_{n > (m+1)k} \bigcup_{1 \leq i < n-m} N_{n,i,j,k}, \text{ where}$$
$$N_{n,i,j,k} = \bigcap_{l=1}^{m} \{\omega : |B(\tfrac{i+l}{n}, \omega) - B(\tfrac{i+l-1}{n}, \omega)| \leq j[(l+1)^\gamma + l^\gamma]/n^\gamma\}.$$
(b) Show that $P(N_{n,i,j,k}) \leq C_{j,m} \cdot n^{-m(\gamma - \frac{1}{2})}$, for some constant $C_{j,m}$, depending only on j and m, and hence conclude that $P(D_{j,k}) = 0$.
(c) Denote D_m to be the set of all those ω for which there exists a $t \in [0,1)$ and a $\delta > 0$ with $t + \delta < 1$, such that $\sup_{0 < h < \delta} |B(t+h, \omega) - B(t, \omega)|/h^\gamma < \infty$. Show that $D_m \subset \bigcup_{j \geq 1} \bigcup_{k \geq 1} D_{j,k}$ and hence conclude that $D_m \subset N_m$, for a P-null set N_m.
(d) Use the result proved in (c) to deduce that, there is a P-null set N such that for $\omega \notin N$, the Brownian trajectory $B(\cdot, \omega)$ is **nowhere** locally Hölder continuous on $[0,\infty)$, for **any** exponent $\gamma > \frac{1}{2}$.

EXERCISE 11.3.10. (a) Show that if $A \subset \mathbb{R}$ is a closed set, then τ_A defined as $\tau_A(\omega) = \inf\{t \geq 0 : B_t(\omega) \in A\}$ is an $\{\mathcal{A}_t\}$-stopping time.
(b) Show that for $A \subset \mathbb{R}$ open, τ_A defined as $\tau_A(\omega) = \inf\{t \geq 0 : B_t(\omega) \in A\}$ is an $\{\mathcal{A}_t\}$-optional time.

11.3. Additional Exercises

(c) Give an example to illustrate that τ_A, for an open $A \subset \mathbb{R}$, may not be an $\{\mathcal{A}_t\}$-stopping time.
(d) Show that τ defined as $\tau(\omega) = \inf\{t \geq t_0 : B_t(\omega) \in A\}$, for some fixed $t_0 > 0$, and a closed (resp., open) set $A \subset \mathbb{R}$, is a stopping time (resp., an optional time).

EXERCISE 11.3.11. *(a) Show that if τ and η are optional (resp., stopping) times, then $\tau \wedge \eta$, $\tau \vee \eta$ and $\tau + \eta$ are all optional (resp., stopping) times.
(b) Show that if τ and η are optional (resp., stopping times), then $\mathcal{A}_{(\tau \wedge \eta)+} = \mathcal{A}_{\tau+} \cap \mathcal{A}_{\eta+}$ (resp., $\mathcal{A}_{\tau \wedge \eta} = \mathcal{A}_\tau \cap \mathcal{A}_\eta$).*

EXERCISE 11.3.12. *Using Lemma 11.2.18 and an appropriate Monotone Class Theorem for functions, try and give a simpler proof of Therem 11.2.17.*

EXERCISE 11.3.13. *With $\{B_t, t \in [0, \infty)\}$ and τ as in Proposition 11.2.22, show that*

$$\left(\{B_{\tau \wedge t}, t \geq 0\}, \{(B_\tau - B_t)I_{\{t > \tau\}}, t \geq 0\}\right) \stackrel{d}{=} \left(\{B_{\tau \wedge t}, t \geq 0\}, \{(B_t - B_\tau)I_{\{t > \tau\}}, t \geq 0\}\right).$$

[Note that, on each side, you have a pair of C-valued random variables.]

EXERCISE 11.3.14. *(a) From the formula given in (11.2.8), show that τ_a, for $a > 0$, has an absolutely continuous distribution and find its density function.
(b) Find the distribution of τ_a, for $a < 0$.
(c) Show that, for $0 < a < b$, the random variable $\tau_b - \tau_a$ is a strictly positive random variable, independent of τ_a and has the same distribution as that of τ_{b-a}.
(d) Show that the process $\{\tau_a, a \geq 0\}$ is a process with non-negative strictly increasing trajectories and has independent and stationary increments.*

In Chapter 10, we saw a construction of SBM using Kolmogorov Consistency Theorem, coupled with Kolmogorov-Chentsov Continuity Theorem. However, there are several other constructions of SBM, each based on a different idea. The next two exercises describe two of these constructions, which are fairly simple. Exercise 11.3.1 (or 11.3.2) tells us that it is enough to construct a SBM on $[0, 1]$.

EXERCISE 11.3.15. *(Lévy's Construction by building 'skeletons' of SBM over finer and finer mesh and then taking limit.)*
(a) Show that, if $\{B_t, t \in [0,1]\}$ is a SBM, then, for any $s < t$ in $[0, 1]$, the conditional distribution of $B_{(s+t)/2}$, given $B_s = x$, $B_t = y$, is $N(\frac{x+y}{2}, \frac{t-s}{4})$.

The above forms the central idea of the construction. What it says is that, if, for some $s < t$ in $[0, 1]$, we have already constructed B_s and B_t of a SBM, then taking $B_{(s+t)/2} = \frac{1}{2}(B_s + B_t) + \frac{\sqrt{t-s}}{2}Z$, with $Z \sim N(0,1)$ independent of (B_s, B_t), ensures that $(B_s, B_{(s+t)/2}, B_t)$ are jointly distributed as if they are from a SBM. Let $D_n = \{\frac{k}{2^n} : k = 0, \ldots, 2^n\}$, for each $n \geq 0$, and let $D = \bigcup_n D_n$. Observe that $D_n \subsetneq D_{n+1}$, $D_n \uparrow D$ and D is dense in $[0, 1]$.
$(\Omega, \mathcal{A}, P)$ be a probability space and let $\{Z_t, t \in D\}$ be i.i.d. $N(0, 1)$ random variables defined on it. Define now a sequence $\{X_n(t), t \in [0,1]\}$, $n \geq 0$, of processes recursively as follows.

First, set $X_0(0) \equiv 0$, $X_0(1) = Z_1$, $X_0(t) = tX_0(1)$, $t \in (0,1)$. Having defined $\{X_n(t)\}$, define $\{X_{n+1}(t)\}$ by first setting $X_{n+1}(t) = X_n(t)$, for $t \in D_n$, while setting $X_{n+1}(t) = \frac{1}{2}\left(X_{n+1}(\frac{k-1}{2^n}) + X_{n+1}(\frac{k}{2^n})\right) + 2^{-(\frac{n}{2}+1)}Z_t$, for $t = \frac{2k-1}{2^{n+1}}$, and finally using linear interpolation for $t \notin D_{n+1}$. Observe that, for each n, the process $\{X_n(t), t \in [0,1]\}$ has continuous trajectories and $X_n(0) \equiv 0$.

(b) Show that, for each $n \geq 0$, the random vector $(X_n(t), t \in D_n)$ has a Gaussian distribution with mean vector $\mathbf{0}$ and dispersion matrix $((s \wedge t))_{s,t \in D_n}$.

(c) Show that, $\sup_{t \in [0,1]} |X_{n+1}(t) - X_n(t)| = 2^{-(\frac{n}{2}+1)} \max_{1 \leq k \leq 2^n} |Z_{\frac{2k-1}{2^n}}|$, for every $n \geq 0$.

(d) Use Lemma 11.1.7 to show that $\sum_n P\left(\sup_{t \in [0,1]} |X_{n+1}(t) - X_n(t)| > 2^{-n/4}\right) < \infty$
and hence conclude that $\{X_n(\cdot)\}$ converges uniformly on $[0,1]$, a.s. $[P]$.

(e) Use parts (b) and (d) now to get a SBM $\{X(t), t \in [0,1]\}$ on $(\Omega, \mathcal{A}, P)$.

Exercise 11.3.16. (*Paley-Wiener Construction through Fourier Expansion of Brownian paths.*)
A SBM on $[0, \pi]$ is constructed by considering the Fourier Series Expansion of its trajectories and showing a.s. uniform convergence of the series. Using scaling property, one can of course transform this into a SBM on $[0,1]$, if needed.

Start with a probability space $(\Omega, \mathcal{A}, P)$ with an i.i.d. sequence $\{Z_k, k \geq 0\}$ of $N(0,1)$ random variables defined on it. For each pair $1 \leq m < n$ of integers, define $\{R_{m,n}(t), t \in [0, \pi]\}$ and $r_{m,n}$ by

$$R_{m,n}(t) = \sum_{k=m}^{n-1} \frac{\sin kt}{k} Z_k, \quad \text{and} \quad r_{m,n} = \sup\{|R_{m,n}(t)| : 0 \leq t \leq \pi\}.$$

(a) Show that $r_{m,n}^2 \leq \sum_{k=m}^{n-1} \frac{Z_k^2}{k^2} + 2 \sum_{l=1}^{n-m-1} \left|\sum_{j=m}^{n-l-1} \frac{Z_j Z_{j+l}}{j(j+l)}\right|$, for all $1 \leq m < n$.

(b) Show that, for all $1 \leq m < n$,
$$E(r_{m,n}^2) \leq \sum_{k=m}^{n-1} \frac{1}{k^2} + 2 \sum_{l=1}^{n-m-1} \left(\sum_{j=m}^{n-l-1} \frac{1}{j^2(j+l)^2}\right)^{1/2} \leq \frac{n-m}{m^2} + 2(n-m)\sqrt{\frac{n-m}{m^4}}.$$

(c) From (b), deduce that, $E(r_{m,2m}) \leq \sqrt{3}m^{-1/4}$, for every integer $m \geq 1$, and hence conclude that $\sum_{n \geq 1} r_{2^{n-1},2^n} < \infty$, a-e.$[P]$.

Define $\{B_t, t \in [0, \pi]\}$ on $(\Omega, \mathcal{A}, P)$ by
$$B_t = \frac{t}{\sqrt{\pi}} Z_0 + \sum_{k \geq 1} \sqrt{\frac{2}{\pi}} \frac{\sin kt}{k} Z_k = \frac{t}{\sqrt{\pi}} Z_0 + \sum_{n \geq 1} \sum_{k=2^{n-1}}^{2^n-1} \sqrt{\frac{2}{\pi}} \frac{\sin kt}{k} Z_k, \quad \text{for } 0 \leq t \leq \pi.$$

(d) Deduce from (c) that, P-almost surely, the above series converges uniformly on $[0, \pi]$ and hence defines a real-valued process $\{B_t, t \in [0, \pi]\}$ on $(\Omega, \mathcal{A}, P)$, with continuous trajectories, P-almost surely.

(e) Show that the process $\{B_t\}$ must be Gaussian with identically zero mean function and covariance kernel $C(s,t) = \frac{ts}{\pi} + \frac{2}{\pi} \sum_{k \geq 1} \frac{\sin kt \, \sin ks}{k^2} = s \wedge t$, $s, t \in [0, \pi]$.

11.4 Supplementary Section : RKHS

Having covered some of the intriguing facets of Brownian motion, we felt tempted to throw light on another very important theory surrounding Brownian motion

11.4. Supplementary Section : RKHS

and that is the reason for this supplementary section. As noted in our concluding comments at the end of Section 11.2, there are several stories that emanate from Brownian motion, like diffusions, Gaussian free fields, etc, to name a few. We leave it to the interested reader to explore. In this section, we discuss a topic, that has connections to several areas of not only mathematics, but also of physics. The p-th (symmetric) Homogeneous subspace $\mathcal{H}_p$ that will appear later, has the interpretation of 'p particle Boson system'; the direct sum representation of Theorem 11.4.19 (b) is called the 'Fock space'. Since these are L_2 subspaces that are invariant under translation (of time axis), this has connection to the spectral theory of translation semigroup on the Brownian space.

This being a supplementary section, our intention here is just to give an exposure, which means that we just describe the steps leading to the final result, without proving any of the steps. This should not be a huge cause for concern, because most of the theorems to be stated are simple and belong to that part of mathematics where 'once a statement is made, it is not difficult to prove'. We do not mean that they are trivial, by any means. What we mean is that they are results of a profound, yet very natural, thought process. The backbone of the theory is Functional Analysis and the aim is to get a beautiful decomposition of the L_2 of the 'Wiener space' into orthogonal subspaces. By the Wiener space, we mean the space $C([0, 1])$ of real-valued continuous functions on $[0, 1]$, equipped with its Borel σ-field under the 'sup-norm' and the 'Wiener measure', that is, the law of SBM on $[0, 1]$.

We already came across the notion of 'non-negative definite kernels' in Section 10.3 as covariance kernels of stochastic processes, but let us start with a formal definition. In what follows, T denotes a non-empty set.

DEFINITION **11.4.1.** A real-valued function R defined on $T \times T$ is called a *non-negative definite kernel* on T if, for any choice of a finite set of points $t_1, \ldots, t_n \in T$, the matrix $((R(t_i, t_j))$ is an $n \times n$ symmetric, non-negative definite matrix.

The following result states what we have already seen before and explains why non-negative definite kernels are of importance in probability theory.

THEOREM **11.4.2.** (a) For any real-valued stochastic process $\{X_t, t \in T\}$, its covariance kernel defined by $R(s,t) = C(X_s, X_t)$ is a non-negative definite kernel.
(b) Given any non-negative definite kernel, it is the covariance kernel of a mean zero Gaussian process.

The next result, that connects any non-negative definite kernel to a unique Hilbert space which "reproduces" the given kernel, is of fundamental importance in all that is to follow.

THEOREM **11.4.3.** (a) Given a non-negative definite kernel R on T, there is a Hilbert space $H(R)$ such that
 (i) $f \in H(R)$ implies that f is a real function defined on T.

(ii) For each $t \in T$, the function $s \mapsto R(s,t)$ belongs to $H(R)$ and these functions $\{R(\cdot,t), t \in T\}$ span $H(R)$.

(iii) $f \in H(R)$ and $t \in T$ imply that $f(t) = \langle f, R(.,t) \rangle$.

(iv) If H' is another Hilbert space satisfying (i), (ii) and (iii), then $H' \equiv H(R)$.

(b) If $\{X_t, t \in T\}$ is a zero mean Gaussian process on $(\Omega, \mathcal{A}, P)$ with covariance kernel R and $L(X)$ is the closed linear span of $\{X_t, t \in T\}$ in $L^2(\Omega, \mathcal{A}, P)$, then there is a canonical isometry between $L(X)$ and $H(R)$ given by $X_t \mapsto R(\cdot, t)$.

One can take part (b) of Theorem 11.4.3 as a cue to prove part (a). Theorem 11.4.3 leads one to define an important concept, which was first systematically developed by Aronszajn and Bergman.

DEFINITION 11.4.4. Given a non-negative definite kernel R on a non-empty set T, the Hilbert space $H(R)$ (as in Theorem 11.4.3) is called the *Reproducing Kernel Hilbert Space* (or *RKHS*, in short) of R (or, of the process $\{X_t, t \in T\}$ with covariance kernel R), because of the reproducing property of its functions as in (a)(iii) of Theorem 11.4.3.

We would like to point out here that the theory of RKHS has been known to play a prominent role not only in probability theory, but also in statistics, machine learning and, more recently, in artificial intelligence.

The next result gives a complete description of the RKHS of the covariance kernel of SBM on $[0, 1]$.

THEOREM 11.4.5. For the non-negative definite kernel R on $T = [0, 1]$, given by $R(s, t) = s \wedge t$, its RKHS $H(R)$ consists precisely of all absolutely continuous functions f on $[0, 1]$, for which $f' \in L_2([0, 1])$, and the inner product on $H(R)$ is given by $\langle f, g \rangle = \int_0^1 f'(u) g'(u) du$, for $f, g \in H(R)$.

We now make two defintions, of which the first is a standard one that any reader exposed to basic Hilbert space theory will be familiar with. The second definition is that of 'tensor products' and will be fundamental to what we discuss next.

DEFINITION 11.4.6. A set S of elements of a Hilbert space H is said to be an *Orthonormal Set* if $\|x\| = 1$ for each $x \in S$ and $\langle x, y \rangle = 0$ for all $x \neq y$ in S. A *Complete Orthonormal Set* (or *CONS*, in short) in H is an orthonormal set S such that no orthonormal set in H contains S as a proper subset.

DEFINITION 11.4.7. Let H^i, for $1 \leq i \leq p$, be a Hilbert space with $\{e_n^i : n \geq 1\}$ as a CONS. The Hilbert space H with the symbols $\{e_{n_1}^1 \otimes \cdots \otimes e_{n_p}^p : n_i \geq 1\}$ as its CONS, is called the *Tensor product* of the Hilbert spaces H^i, $1 \leq i \leq p$ and is denoted $\overset{p}{\underset{1}{\otimes}} H^i$. If $u^i = \sum a_n^i e_n^i \in H^i$, $1 \leq i \leq p$, then $\overset{p}{\underset{1}{\otimes}} u^i$, called the *tensor product* of the elements u^i, $1 \leq i \leq p$, is the element of the tensor product $\overset{p}{\underset{1}{\otimes}} H^i$ given by $\sum a_{n_1}^1 a_{n_2}^2 \cdots a_{n_p}^p e_{n_1}^1 \otimes \cdots \otimes e_{n_p}^p$.

11.4. Supplementary Section : RKHS

The following result is a fairly easy consequence of the above definition of tensor products.

THEOREM 11.4.8. Let H^i, $1 \leq i \leq p$ and $\otimes H^i$ be as in the definition above.
(a) $\langle \otimes u^i, \otimes v^i \rangle = \prod \langle u^i, v^i \rangle$, for all $\otimes u^i, \otimes v^i \in \otimes H^i$. In particular, $\langle u^i, v^i \rangle = 0$, for any one index i, implies that $\langle \otimes u^i, \otimes v^i \rangle = 0$.
(b) The map $(u^1, \cdots, u^p) \mapsto \otimes u^i$ is a multilinear map on $H^1 \times \cdots \times H^p$ into $\otimes H^i$.
(c) The set of all finite linear combinations of $\otimes u^i$ is dense in $\otimes H^i$.
(d) If $\{f_n^i : n \geq 1\}$ is a CONS in H^i, for each index i, then $\{\otimes f_{n_i}^i : n_i \geq 1\}$ is a CONS in $\otimes H^i$.

REMARK 11.4.9. One can do all of these without apriori fixing a basis – either by completing the linear span of the abstract symbols $\otimes u^i$ with an appropriate inner-product (after making usual identifications), or by defining it in an axiomatic way as a Hilbert space satisfying certain conditions (but then one has to show the existence).

An interesting situation arises in the special case when all the Hilbert spaces H^i are the same space H.

DEFINITION 11.4.10. For each permutation π of $\{1, \cdots, p\}$, we define the map π on $\otimes^p H$ to itself by setting $\pi(x^1 \otimes \cdots \otimes x^p) = x^{\pi(1)} \otimes \cdots \otimes x^{\pi(p)}$.

Although we are using the same notation π for the permutation and also for the map on $\otimes^p H$ to itself, it should not cause any confusion. From the context, it should be clear whether we are talking about the permutation π or the Hilbert space map π.

THEOREM 11.4.11. (a) The Hilbert space map π is well defined for every permutation π. Also, if η is the inverse permutation of π, then the corresponding Hilbert space map η is the adjoint of π.
(b) The Hilbert space map $\sigma(x) = \frac{1}{p!} \sum_\pi \pi(x)$, where the sum is over all permutations π, is self-adjoint and idempotent and, therefore, an orthogonal projection.
(c) Denoting H^p to be the range of the orthogonal projection σ in (b), the subspace H^p is spanned by the vectors x^p as x ranges over H, where x^p denotes the p-fold tensor product of x with itself.
(d) Let $\{e_n : n \geq 1\}$ be a CONS in H. For each choice of integers $1 \leq \lambda_1 < \cdots < \lambda_r$ and of integers $n_1, \cdots, n_r$, with $\sum n_i = p$, denote

$$e_{\lambda_1, \cdots, \lambda_r}^{n_1, \cdots, n_r} = \sqrt{\frac{p!}{n_1! \cdots n_r!}} \sigma(e_{\lambda_1}^{n_1} \cdots e_{\lambda_r}^{n_r})$$

Then, the set of vectors $\{e_{\lambda_1, \cdots, \lambda_r}^{n_1, \cdots, n_r}\}$ form a CONS in H^p.

DEFINITION 11.4.12. H^p, as defined in part (c) of Theorem 11.4.11 is called the p-fold Symetric Tensor Product of H with itself.

We now take up the special case of tensor products of L_2 spaces. Here is the first basic result.

THEOREM 11.4.13. (a) Consider the Hilbert spaces $H_i = L_2(\Omega_i, \mathcal{A}_i, \mu_i)$, $1 \leq i \leq p$. There is a canonical isometry between the tensor product $\otimes_1^p L^2(\Omega_i, \mathcal{A}_i, \mu_i)$ and the Hilbert space $L^2(\Omega, \mathcal{A}, \mu)$, where $(\Omega, \mathcal{A}, \mu)$ denotes the usual product space $(\times_i \Omega_i, \otimes_i \mathcal{A}_i, \otimes_i \mu_i)$.
(b) If $H = L^2(\Omega, \mathcal{A}, \mu)$, then $\otimes^p H \cong L^2(\Omega^p, \mathcal{A}^p, \mu^p)$ and H^p is isometrically isomorphic to the subspace of $L^2(\Omega^p, \mathcal{A}^p, \mu^p)$, consisting of the symmetric (i.e., permutation invariant) functions.

The next step is to apply all that has been developed so far to the special case when the underlying Hilbert space H is the RKHS of a given non-negative definite kernel. Given a non-negative definite kernel R on a non-empty set T and any integer $p > 1$, we define a kernel K on T^p by the formula

$$K\big((s_1, \cdots, s_p), (t_1, \cdots t_p)\big) = \prod_i R(s_i, t_i).$$

THEOREM 11.4.14. (a) K is a non-negative definite kernel on T^p.
(b) If π is any permutation of $\{1, \ldots, p\}$, then for any $f \in H(K)$, the function f_π defined by $f_\pi(t_1, \cdots, t_p) = f(t_{\pi(1)}, \cdots, t_{\pi(p)})$, also belongs to $H(K)$. Further, the map $f \mapsto f_\pi$ is an isometry of $H(K)$ onto itself.
(c) If η is the inverse permutation to a permutation π, then the isometry $f \mapsto f_\eta$ is the adjoint of the isometry $f \mapsto f_\pi$.
(d) $S(K) = \{f \in H(K) : f_\pi = f \text{ for all permutations } \pi\}$ is a closed subspace of $H(K)$ and the map $f \mapsto Pf = \frac{1}{p!}\sum f_\pi$ is the projection map on $H(K)$ onto $S(K)$.
(e) The kernel $\tilde{K}$ on T^p defined as

$$\tilde{K}\big((s_1, \cdots, s_p), (t_1, \cdots, t_p)\big) = \frac{1}{p!}\sum_\pi K\big((s_{\pi(1)}, \cdots, s_{\pi(p)}), (t_1, \cdots, t_p)\big),$$

is a non-negative definite kernel on T^p and its RKHS is $S(K)$.
(f) There is a unique isometry Ψ from $\otimes^p H(R)$ to $H(K)$, such that

$$\Psi(R(., t_1) \otimes \cdots \otimes R(., t_p)) = K(.\,; t_1, \cdots, t_p).$$

Further, $P \circ \Psi = \Psi \circ \sigma$ on $\otimes^p H(R)$.

Recall the Hilbert space H^p, the p-fold symmetric tensor product of a Hilbert space H with itself, as given in Definition 11.4.12. We are now going to define what is called the 'symmetric Fock space' of H.

DEFINITION 11.4.15. The infinite direct sum $R \oplus H \oplus H^2 \oplus \cdots \cdots$ is called the Symmetric Fock Space of H and is denoted by $\exp(H)$. For $x \in H$, we denote by e^x the vector $1 \oplus x \oplus \frac{x^2}{\sqrt{2!}} \oplus \cdots \oplus \frac{x^p}{\sqrt{p!}} \oplus \cdots$.
[Recall that x^p, for $p > 1$, denotes the p-fold tensor product of x with itself.]

THEOREM 11.4.16. (a) $e^x \in \exp(H)$ and $\langle e^x, e^y \rangle = e^{\langle x, y \rangle}$, for all $x, y \in H$.
(b) The set $\{e^x\,;\, x \in H\}$ spans $\exp(H)$.
(c) If one regards H as RKHS with $T = H$ and the non-negative definite kernel $R(x, y) = \langle x, y \rangle_H$, then $K(x, y) = e^{\langle x, y \rangle}$ is also a non-negative definite kernel on H and its RKHS $H(K)$ is isometrically isomorphic to $\exp(H)$, in a canonical way.

11.4. Supplementary Section : RKHS

We are now about to reach our final aim, that is, to describe a decomposition of the L_2 of Wiener space into orthogonal subspaces, famously known as the 'homogeneous chaos decomposition'. Towards this, the family of functions defined below will play an important role. These functions are fairly well-known in the theory of differential equations as eigenfunctions of the differential opeator $L(u) = u'' - xu'$.

Consider the real-valued functions H_n, $n \geq 0$, defined on $\mathbb{R}$ by
$$H_n(x) = (-1)^n e^{x^2/2} D^n \left(e^{-x^2/2}\right), \ n \geq 0$$
It turns out that H_n, for each $n \geq 0$, is a polynomial of degree n (see Theorem 11.4.17). The polynomials $\{H_n, n \geq 0\}$ are known as the (un-normalized) *Hermite Polynomials*. Here are the first few of these polynomials.
$$H_0(x) \equiv 1, \quad H_1(x) = x, \quad H_2(x) = x^2 - 1, \quad H_3(x) = x^3 - 3x$$
$$H_4(x) = x^4 - 6x^2 + 3, \quad H_5(x) = x^5 - 10x^3 + 15x$$
The polynomials $\{h_n, n \geq 0\}$ defined as $h_n(x) = \frac{1}{\sqrt{n!}} H_n(x)$, $n \geq 0$, are called the (normalized) *Hermite Polynomials*.

THEOREM 11.4.17. (a) H_n, for each $n \geq 0$, is a polynomial of degree exactly n with 1 as the leading coefficient. Further, for each odd n, the polynomial H_n has only odd powered terms in x with non-zero coefficients, while for each even n, it has only even powered terms in x with non-zero coefficients.
(b) The polynomials H_n, $n \geq 0$, satisfy the recurrence relation
$$H_n(x) = xH_{n-1}(x) - (n-1)H_{n-2}(x)$$
(c) The family $\{h_n, n \geq 0\}$ forms a CONS for the L^2 space on $\mathbb{R}$, equipped with the $N(0,1)$-probability on its Borel σ-field.

Suppose now that $\{X_t, t \in T\}$ is a mean zero Gaussian process on a probability space $(\Omega, \mathcal{A}, P)$ with $\mathcal{A} = \sigma(\{X_t, t \in T\})$. Denote $L(X)$ to be the L^2 span of $\{X_t, t \in T\}$ and let $\{\psi_i, i \geq 1\}$ be a CONS in $L(X)$. Consider the subspaces $\{\hat{G}_p, p \geq 0\}$, $\{G_p, p \geq 0\}$ and $\{\mathcal{H}_p, p \geq 0\}$ of L^2, as defined below.

DEFINITION 11.4.18. Denote $\hat{G}_0$ to be the space of constant functions and $\hat{G}_p$, for $p \geq 1$, to be the space of all polynomials in $\{\psi_i\}$ of degree less than or equal to p. Set $G_0 = \hat{G}_0$ and G_p, for $p \geq 1$ to be the orthogonal complement of $\hat{G}_{p-1}$ in $\hat{G}_p$. Finally, let $\mathcal{H}_p$, for each $p \geq 0$, denote the closure of G_p.
G_p, for each $p \geq 0$, is called the *p-th polynomial chaos* and $\mathcal{H}_p$, for each $p \geq 0$, is called the *p-th Homogeneous Chaos*.

THEOREM 11.4.19. (a) For any choice of integers $1 \leq \lambda_1 < \lambda_2 < \cdots < \lambda_r$, the polynomial in $\{\psi_i\}$ given by
$$\gamma = \sum a_{n_1, \cdots, n_r} h_{n_1}(\psi_{\lambda_1}) \cdots h_{n_r}(\psi_{\lambda_r}), \tag{11.4.1}$$
where the sum is over all choices of r-tuples of non-negative integers $(n_1, \cdots, n_r)$ with $\sum n_i = p$ and real numbers $a_{n_1, \cdots, n_r}$, belongs to G_p. Conversely, every polynomial $\gamma \in G_p$ is of the form (11.4.1).
(b) $L^2(\Omega, \mathcal{F}, P) = \bigoplus_{p \geq 0} \mathcal{H}_p = \bigoplus_{p \geq 0} \left(\otimes L^p(X)\right) = \bigoplus_{p \geq 0} \left(\otimes H^p(\mathbb{R})\right)$, where $L^p(X)$ and $H^p(\mathbb{R})$ denote the p-fold symmetric tensor products of $L(X)$ and $H(\mathbb{R})$ respectively.

REMARK 11.4.20. We need to point out here that the proof of part (b) uses what is known as 'Martingale convergence theorem', a result that we have not discussed in the book.

We now state our final result that gives the promised decomposition of L_2 of the Wiener space as an infinite direct sum of Homogeneous Chaos. This comes as a direct consequence of part (b) of Theorem 11.4.19.

THEOREM 11.4.21. If P is Wiener measure on $C[0,1]$, then $L^2(P) = \exp L^2([0,1])$.

We like to end with a note that all the above analysis that was done for the SBM on $[0,1]$ can be done for SBM on $[0,\infty)$, with P being the probability on $C\big([0,\infty)\big)$ representing the distribution of SBM on $[0,\infty)$.

Index

L_p-space, $L_p(\Omega, \mathcal{A}, \mu)$, 193
    dual space of −, 197
    Completeness of −, 194, 207
    Complex −, 207
σ-field, 34
    − generated by, 39
    μ-complete −, 49
    μ-completion of −, 49
    countable-cocountable −, 21
    countably generated −, 57
    tail −, tail events, 244

absloutely continuous, 116
absolutely continuous
    mutually − (equivalent), 188
almost everywhere, a.e., 106

Bernstein Polynomials, 148
BM starting at x, $\{B_t^x, t \in [0, \infty)\}$, 357
    André's Reflection Principle, 368
    Markov proprerty of −, 360
    natural filtration of −, 362
    optional time, 362
    stopping time, 362
    strong Markov property of −, 367
Borel
    − σ-field on $\mathbb{R}$, 57
    − measurable function, 92
    − subset of $\mathbb{R}$, 57
    − σ-field on $\mathbb{R}^k$, 67
    − subset of $\mathbb{R}^k$, 67
    standard − space, 312

Cantor
    − Distribution, 133
    − distribution function, 133
    − set, 65
characteristic function, 261, 289
    differentiability of −, 271
    Inverse Fourier Transform, 269

Inversion Formula, 268, 290
Pólya Criterion, 278
Taylor expansion for −, 271
Uniqueness Theorem, 266
conditional distribution
    existence of regular −, 315, 316
    regular −, 314
conditional expectation
    L_p-contraction property of −, 304
    definition and existence of −, 299
    Jenesen's inequality for −, 303
    MCT, Fatou, DCT for −, 302
conditional probability, 307
    existence of regular −, 310, 312
    regular −, 309
convergence
    − P-almost surely, 220
    − μ-almost everywhere, 218
    − in L_p, 219
    − in pth moment, 221
    − in distribution, 251
    − in mean, 221
    − in measure, 219
    − in probability, 221
    weak −, 251
Cramér-Wold Device, 291

distribution function
    − on $\mathbb{Q}$, 66
    − on $\mathbb{R}$, 61
    − on $\mathbb{R}^2$, 68
    − on $\mathbb{R}^k$, 72

essential supremum, 184

field, 20
    − generated by, 21
    finite-cofinite field, 21
    semi−, 18

Gaussian process, 336
  covariance kernel, 336
  mean function, 336

Hermite Polynomials, 379
Hilbert Space
  CONS in a $-$, 376
  Reproducing Kernel $-$, 376
  Tensor product of $-$, 376
Homogeneous Chaos, 379

independence of
  $-$ classes of events, 81
  $-$ complex random variables, 261
  $-$ events, 80
  $-$ random variables, 143
inequality
  Cauchy-Schwarz $-$, 119
  Chebycheff's $-$, 117
  Hölder's $-$, 118, 206
  Jensen's $-$, 121
  Kolmogorov's Maximal $-$, 236
  Markov's $-$, 117
  Minkowski's $-$, 119, 206
integral
  $-$ of a complex function, 205
  $-$ of a real function, 102

Law
  Kolmogorov's Zero-One $-$, 245
Law of Large Numbers
  Kolmogorov's Strong $-$, 239
  Weak $-$, 233
Lebesgue
  $-$ σ-field, 58
  $-$ -Stieltje's measure on $\mathbb{R}$, 61
  $-$ measurable function, 97
  $-$ measurable sets, 58
  $-$ measure on $\mathbb{R}$, 58
  $-$ measure on $\mathbb{R}^k$, 73
Lemma
  Fatou's $-$, 109
  Borel-Cantelli $-$, 56
  Kronecker's $-$, 235
  Scheffé's $-$, 114
  Second Borel-Cantelli $-$, 84
locally Hölder continuous, 340

measurable
  $(\mathcal{A}, \mathcal{E})$-measurable map, 123

$-$ space, 92
complex $-$ function, 204
real $-$ function, 92
measure, 18
  $-$ kernel, σ-finite, 171
  σ-finite measure, 19
  complex measure, 210
  convolution of σ-finite $-$, 176
  finite measure, 19
  finite signed measure, 189
  induced measure, 127
  outer measure, 31
  probability measure, 19
measure space, 92
  probability space, 92
  finite/σ-finite measure space, 92
Method of Moments, 259
monotone class, 45
mutually singular measures, 180

non-negative definite
  $-$ function, 279
  $-$ kernel, 335, 375

probability distribution
  $-$ continuous, 79
  $-$ discrete, 79
  $-$ function on $\mathbb{Q}$, 66
  $-$ function on $\mathbb{Q}^k$, 74
  $-$ function on $\mathbb{R}$, 62
  $-$ function on $\mathbb{R}^k$, 73
  convolution of $-$, 177
product
  countable $-$ σ-field, 323
  countable $-$ space, 323
  finite $-$ σ-field, 162
  finite $-$ of σ-finite measures, 169
  finite $-$ of measurable spaces, 169
  uncountable $-$ σ-field, 333
  uncountable $-$ space, 333

Radon measure
  $-$ on $\mathcal{B}(\mathbb{R})$, 61
  $-$ on $\mathcal{B}(\mathbb{R}^k)$, 68
  regularity of $-$, 75
Radon-Nikodym derivative, 187, 191
Random Variable, 135
  k-th moment, variance of $-$, 153
  complex $-$, 260
  covariance between $-$, 153

INDEX

383

Distribution Function of −, 135
Distribution of −, 135
Expected Value of −, 146
Moment Generating Fn of −, 157
probability density fn of −, 137
probability mass function of −, 137
triangular array of −, 283
Random Vector, 139
dispersion matrix of −, 153
Gaussian −, degenerate, 291
Gaussian/Normal distribution, 140
Joint Distribution Fn. of −, 139
Joint Distribution of −, 139
joint prob density fn. of −, 140
joint prob mass fn. of −, 139
mean vector of −, 153

Standard Brownian Motion, SBM, 336
SLLN for −, 347
Khinchine's LIL for −, 351
non-differentiability of paths, 354
quadratic variation of paths, 353
unbounded variation of paths, 351
zero-sets of Brownian paths, 357
Symmetric Fock Space, 378

Theorem
Bochner's −, 279
Borel's Normal Number −, 65, 248
Bounded Convergence −, 111
Caratheodory Extension −, 42
Change of Variable −, 127
Classical Central Limit −, 282, 293
Continuous Mapping −, 259, 289
Dominated Convergence −, 110
Dynkin's π-λ −, 48

λ-system, 47
π-system, 47
Egoroff's −, 201, 220, 221
Fubini's −, 165
Glivenko-Cantelli −, 240
Hahn Decomposition −, 192
Helly Selection −, 273
Jessen-Wintner −, 245
Jordan Decomposition −, 190
Kolmogorov Consistency −, 332
Kolmogorov's 3-Series −, 243
Kolmogorov-Chentsov −, 339
Lévy Continuity −, 275, 292
Lebesgue Decomposition −, 180
Lindeberg-Feller-Lévy CLT, 283
Local Limit −, 256
Lusin's −, 202
Lyapunov's Central Limit −, 284
Monotone Class − for fns., 128
Monotone Class − for sets, 45
Monotone Convergence −, 109
Pólya's −, 255
Radon-Nikodym −, 180, 191, 211
Riesz Representation −, 199, 209
Skorokhod Representation −, 254
Slutsky −, 254, 295
Tulcea's −, 326
tight family of probabilities, 272

uniform absolute continuity, uniformly
absolutely continuous, 228
uniform integrability, uniformly
integrable, 226

Vitali set, 88

GPSR Compliance

The European Union's (EU) General Product Safety Regulation (GPSR) is a set of rules that requires consumer products to be safe and our obligations to ensure this.

If you have any concerns about our products, you can contact us on ProductSafety@springernature.com

In case Publisher is established outside the EU, the EU authorized representative is:

Springer Nature Customer Service Center GmbH
Europaplatz 3
69115 Heidelberg, Germany

Batch number: 08202983

Printed by Printforce, the Netherlands